Operator Theory: Advances and Applications
Vol. 163

Editor:
I. Gohberg

Editorial Office:
School of Mathematical
Sciences
Tel Aviv University
Ramat Aviv, Israel

Editorial Board:
D. Alpay (Beer-Sheva)
J. Arazy (Haifa)
A. Atzmon (Tel Aviv)
J. A. Ball (Blacksburg)
A. Ben-Artzi (Tel Aviv)
H. Bercovici (Bloomington)
A. Böttcher (Chemnitz)
K. Clancey (Athens, USA)
L. A. Coburn (Buffalo)
R. E. Curto (Iowa City)
K. R. Davidson (Waterloo, Ontario)
R. G. Douglas (College Station)
A. Dijksma (Groningen)
H. Dym (Rehovot)
P. A. Fuhrmann (Beer Sheva)
B. Gramsch (Mainz)
J. A. Helton (La Jolla)
M. A. Kaashoek (Amsterdam)
H. G. Kaper (Argonne)
S. T. Kuroda (Tokyo)
P. Lancaster (Calgary)
L. E. Lerer (Haifa)
B. Mityagin (Columbus)
V. Olshevsky (Storrs)
M. Putinar (Santa Barbara)
L. Rodman (Williamsburg)
J. Rovnyak (Charlottesville)
D. E. Sarason (Berkeley)
I. M. Spitkovsky (Williamsburg)
S. Treil (Providence)
H. Upmeier (Marburg)
S. M. Verduyn Lunel (Leiden)
D. Voiculescu (Berkeley)
D. Xia (Nashville)
D. Yafaev (Rennes)

Honorary and Advisory
Editorial Board:
C. Foias (Bloomington)
P. R. Halmos (Santa Clara)
T. Kailath (Stanford)
H. Langer (Vienna)
P. D. Lax (New York)
M. S. Livsic (Beer Sheva)
H. Widom (Santa Cruz)

Operator Theory and Indefinite Inner Product Spaces

Presented on the occasion of the retirement of Heinz Langer in the Colloquium on Operator Theory, Vienna, March 2004

Matthias Langer
Annemarie Luger
Harald Woracek
Editors

Birkhäuser Verlag
Basel · Boston · Berlin

Editors:

Matthias Langer
Department of Mathematics
University of Strathclyde
26 Richmond Street
Glasgow G1 1XH
UK
e-mail: ml@maths.strath.ac.uk

Annemarie Luger
Harald Woracek
Institut für Analysis und Scientific Computing
Technische Universität Wien
Wiedner Hauptstrasse 8–10 / 101
1040 Wien
Austria
e-mail: aluger@mail.zserv.tuwien.ac.at
harald.woracek@tuwien.ac.at

2000 Mathematics Subject Classification Primary 46C20, 47B50; Secondary 34L05, 47A57, 47A75

A CIP catalogue record for this book is available from the
Library of Congress, Washington D.C., USA

Bibliographic information published by Die Deutsche Bibliothek
Die Deutsche Bibliothek lists this publication in the Deutsche Nationalbibliografie; detailed bibliographic data is available in the Internet at <http://dnb.ddb.de>.

ISBN 3-7643-7515-9 Birkhäuser Verlag, Basel – Boston – Berlin

Part of Springer Science+Business Media
Printed on acid-free paper produced from chlorine-free pulp. TCF ∞
Cover design: Heinz Hiltbrunner, Basel
Printed in Germany
ISBN-10: 3-7643-7515-9
ISBN-13: 978-3-7643-7515-7
e-ISBN: 3-7643-7516-7

9 8 7 6 5 4 3 2 1
www.birkhauser.ch

Heinz Langer

Table of Contents

Introduction

Research articles

Survey Article

Preface

In this volume we present a collection of research papers which mainly follow lectures given at the "Colloquium on Operator Theory". This conference was held at the Vienna University of Technology in March 2004 on the occasion of the retirement of Heinz Langer. The present volume of the series "Operator Theory: Advances and Applications" is dedicated to him and his scientific work.

The book starts with an introductory part which provides some information about the colloquium itself. We have also included the laudation given by Aad Dijksma and a list of the recent publications of Heinz Langer, which updates his bibliography given in OT 106. The main part of the book consists of fifteen original research papers, which deal with various aspects of operator theory and indefinite inner product spaces. It concludes with a historical survey on the theory of positive definite functions.

It would not have been possible to bring together so many colleagues in Vienna without financial support provided by several organizations. We wish to thank the

Außeninstitut der TU Wien
Österreichische Forschungsgemeinschaft (ÖFG)
Research Training Network HPRN-CT-2000-00116 of the European Union
Vienna Convention Bureau
Österreichisch-Ukrainisches Kooperationsbüro
Österreichische Mathematische Gesellschaft (ÖMG)

and everybody who has contributed to make this conference and this book possible. Our special thanks also go to the referees who did a lot of work and in some cases made valuable and essential suggestions, which decisively improved the quality of the papers. Photos were provided by Dr. Mathias Beiglböck (conference photo) and Peter Geisenberger (photo of Heinz Langer).

As organizers of the "Colloquium on Operator Theory" it was a great pleasure for us that so many colleagues participated in the conference; as editors of the present volume we are happy that a large part of them decided to contribute to these proceedings. We greatly acknowledge the interesting experience.

Vienna, July 2005

Matthias Langer
Annemarie Luger
Harald Woracek
(the editors)

Laudation

By Aad Dijksma, Vienna, March 4, 2004

Dear Vice-Rector Kaiser, Dean Dorninger, and Professor Inge Troch, dear Heinz, dear colleagues, Ladies and Gentlemen.

Introduction

Heinz Langer is a world leading expert in spectral analysis and its applications, in particular in operator theory on spaces with an indefinite inner product. He has moved mathematical boundaries and opened new areas of research. He has the talent to focus on what is fundamental and to give directions to what is possible. By sharing his ideas he has stimulated many to explore unknown mathematical territory. Heinz is co-author of a book with M.G. Krein and I.S. Iokhvidov and he has published more than 170 papers with 45 co-authors from all over the world. He has directed the research of about 25 Ph.D. students. His results are still being applied and quoted in international journals in mathematics and theoretical physics. As one of his collaborators I can attest to the fact that Heinz is still very active.

When the organizers of this workshop invited me to give this laudation speech I immediately said yes. I deem it an honor to pay tribute to Heinz. Not only because I admire his work, but also because I have learned a lot from him and still do and because I consider him a close friend. I feel privileged and pleased to speak on this occasion. First I will outline Heinz's biography and highlight his successful years at the Vienna University of Technology. Then I will try to clarify what I meant when I said that Heinz has moved mathematical boundaries and opened new areas of research by discussing some of the main themes in his work. Finally, I will speak about Heinz's connection with The Netherlands and especially with Groningen.

Biography

Heinz Langer was born in Dresden on August 8, 1935. He lived in Dresden for almost 55 years. He attended the Gymnasium there, studied mathematics at the Technical University of Dresden, where he became an assistant and where he obtained his Ph.D. in 1960 and his Habilitation in 1965. He was appointed professor in mathematics at his Alma Mater in 1966, at the age of 31. At that time there was officially no research group in analysis, let alone functional analysis, so Heinz joined the group in stochastics headed by his Ph.D. advisor Prof. P.H. Müller. He did research in and lectured on semigroups, one-dimensional Markov processes, and the spectral theory of Krein-Feller differential operators, but with Ph.D. students and colleagues from abroad he could work in his favorite topic: operator theory. Heinz declined a position at the prestigious Mathematical Institute of the Academy of Sciences of the GDR, because he liked to teach and to work with Ph.D. and post-doctoral students.

At the beginning of his career Heinz spent two years abroad, first as a postdoc in Odessa in 1961/62 upon the invitation of M.G. Krein and later in 1966/67, shortly after his appointment as professor, on a fellowship of the National Research Council of Canada which was arranged by Professor I. Halperin in Toronto. The contact with Krein began with Heinz posting a handwritten manuscript in German, in the mailbox at the central station of Dresden on December 31, 1959. I will say more about the contents of the manuscript when I speak about the main themes in Heinz's work. The subsequent stay in Odessa and the many shorter visits after that had a tremendous influence on Heinz, his career in mathematics, and, I think, on his style of doing mathematics. When Heinz mentions Krein it is with great affection and respect, and I know that Krein thought of Heinz as one his most brilliant students and collaborators.

In between the two trips abroad Heinz had married Elke and in May 1967 their daughter Henriette was born.

In the 1970's and 1980's Heinz spent several extended periods away from home in, for example, Jyväskylä, Stockholm, Uppsala, Linköping, Antwerp, Groningen, Amsterdam, and Regensburg. In all these places Heinz left his mark. He established some form of research co-operation, set up Ph.D. projects and made it possible for students and colleagues to come to Dresden. My account on Heinz's connection with Groningen later on serves as an example of his stimulating influence on other mathematicians. Heinz remained in Dresden until 1989. In October of that year, shortly before the fall of the Berlin wall, he and his family gave up hearth and home, left the GDR and went to Regensburg. Turbulent times followed. Thanks to Albert Schneider Heinz obtained a professorship for one year at the University of Dortmund. After that year, with the support of Reinhard Mennicken, Heinz became professor at the University of Regensburg. Finally, in August 1991, Heinz moved to Vienna where he began a new and successful period in his life and in his mathematical career.

Heinz was offered the prestigious chair "Anwendungsorientierte Analysis" at the Vienna University of Technology, previously held by Professor Edmund Hlawka, but under a different name. The University could hardly have found a more worthy successor. For several years Heinz was chairman of the Institute. He was pragmatic and apparently did his job well because he was re-elected. In no time the Vienna University of Technology became an internationally renowned research center of operator theory with young people doing challenging research. In Vienna Heinz guided 7 Ph.D. and 3 post-doctoral students. The center attracted many visitors from all over the world for visits to do research and for workshops. During the past 12 years Heinz has organized 4 workshops, one in co-operation with the Schrödinger Institute and one in 2001 when the Vienna University of Technology awarded Professor Israel Gohberg from Israel an honorary doctorate. He obtained various long term research grants including funds for Ph.D. and post-doctoral projects. Heinz has clear and practical ideas and a keen sense on what to apply for and on how to formulate it. Two of these grants were from the Austrian "Fonds zur Förderung der Wissenschaftlichen Forschung (FWF)," the last of which was awarded very recently for a project on canonical systems. As to a third grant: Heinz is the leader of the Austrian-German team in a Research Training Network of the European Union, in which 10 universities from 9 different countries participate.

In recognition of his work Heinz was elected corresponding member of the Austrian Academy of Sciences.

Main themes in his work

Heinz's lifelong mathematical interest has been in the theory of operators, in particular operators on indefinite inner product spaces and its applications. This subject was suggested by Professor P.H. Müller, his thesis advisor. Heinz has initiated many new projects and made fundamental contributions to their development. His work draws the attention of several mathematicians and physicists. To illustrate this I elaborate on four of the main themes in his work:

1. From the Ph.D. period: the invariant subspace theorem.
2. From the Habilitation period: definitizable operators.
3. From the period 1975–1985: extension theory.
4. From the Vienna period: block operator matrices.

These main themes are interconnected and overlap in time. The division in periods is made only to facilitate the exposition.

The invariant subspace theorem

Now I come back to the handwritten German manuscript. For his Ph.D. thesis, so in the late fifties and early sixties, Heinz read the papers by L.S. Pontryagin, M.G. Krein, and I.S. Iokhvidov. Pontryagin in 1944 published his famous theorem that a self-adjoint operator A in a Π_κ space with a κ-dimensional negative subspace has a maximal invariant non-positive subspace and that the spectrum of A restricted to this subspace lies in the closed upper half-plane. In 1956 Krein gave a different proof of Pontryagin's theorem using a fixed point theorem but then for unitary operators. Iokhvidov used the Cayley transform to show that the theorems of Pontryagin and Krein are equivalent. Heinz generalized Pontryagin's theorem to self-adjoint operators on a Krein space. The sole assumption was a simple compact corner condition to have control over the operator when restricted to the possibly infinite-dimensional negative subspace of the Krein space. It was a truly remarkable achievement. Heinz wrote it up by hand and in German and sent the manuscript to Krein on New Year's eve in 1959. It would later be the main result in Heinz's Ph.D. thesis. Iokhvidov in Odessa had tried to prove the same theorem before but he had not succeeded. So naturally Krein was very interested in Heinz's proof and he must have been impressed because in one of his Crimean lecture notes he refers to the generalized invariant subspace theorem as the Pontryagin-Langer theorem. In any case, Krein reacted by inviting Heinz to come to Odessa. The visit marked the beginning of a fruitful co-operation that lasted for more than 25 years, until the death of Krein in 1989, and resulted in 13 joint papers. But the story goes on: One of the first things they discovered was a beautiful and unexpected application of the invariant subspace theorem. With the help of this theorem they proved that a self-adjoint quadratic operator pencil has a root with some specified spectral properties. The two papers "On some mathematical principles in the linear theory of damped oscillations", which deal with this factorization, led to new publications in spectral theory and applications to mechanics and physics. They are probably their most frequently cited joint papers.

Definitizable operators

In 1962 Krein and Heinz proved that a self-adjoint operator in a Pontryagin space has a generalized spectral function. Krein showed that an integral operator with a positive kernel gives rise to a positive operator on a Krein space and that this operator also has a generalized spectral function. Heinz discovered that these are two examples of a special class of self-adjoint operators on a Krein space, namely the class of definitizable operators. The concept and the name are due to Heinz. A self-adjoint operator A is definitizable if it has a nonempty resolvent set and for some polynomial p, $p(A)$ is nonnegative. Like the compact corner condition in the Pontryagin-Langer theorem, definitizability is a way to keep control over the nonpositivity of the inner product of the Krein space. In his Habilitation submitted in 1965 Heinz shows that a definitizable operator has a generalized spectral function and he applies his theory to quadratic operator pencils. It is no exaggeration when I say that this work is a genuine corner stone in the spectral theory of operators in spaces with an indefinite metric. Like its Hilbert space counterpart, the spectral function in a Krein space has a wide range of applications such as to quadratic pencils, just mentioned, Sturm-Liouville problems with indefinite weight, elliptic problems, and variational principles.

Extension theory

The research of Heinz and Krein in the extension theory of symmetric operators in Pontryagin and Krein spaces began with generalizing Krein's theory of generalized resolvents, resolvent matrices, and entire operators to an indefinite setting with applications involving new classes of meromorphic functions with finitely many poles and canonical systems. It culminated in the 4 seminal "Über einige Fortsetzungsprobleme ...." papers published jointly with Krein in the period 1977–85. These papers are still quoted and applied today. In fact, they could be called trendsetters, because there is a growing interest in generalizing, where possible, positive definite results to an indefinite setting. Of the many examples I only mention the indefinite version of the de Branges theory of entire functions and canonical systems. The research in this area is carried out by the group around Heinz here at the Vienna University of Technology. The results are important and of a high quality. Heinz can be proud of the research team he leaves behind.

Block operator matrices

This topic was taken up by Heinz jointly with Reinhard Mennicken in Regensburg and came to bloom in Vienna. 2×2 block operator matrices are common in the theory of operators on Krein spaces. But now the emphasis is different. The problem is to describe the spectral properties of an operator defined on a product of two

Hilbert spaces and given as a 2×2 block operator matrix in terms of the properties of the operator entries of the matrix. Typical examples come from mathematical physics or system theory, where the entries are differential operators of different order and hence unbounded. One of the first problems is to define the domain of definition of the block operator matrix. Many papers have appeared since 1991. They concern, for example, the location of the essential spectrum, the solution of a Riccati equation, and block diagonalization. A new concept initiated and further developed jointly with others in the last 5 years is that of the quadratic numerical range. It is a new tool for localizing the spectrum of a block operator matrix.

The Netherland connection

Let me begin with how I came to meet Heinz. I wrote a letter to him with some results on differential operators with eigenvalue depending boundary conditions on February 25, 1980. Heinz responded on March 19, 1980 with detailed answers to my questions. Some years before, Rien Kaashoek had visited Heinz in Dresden and organized a return visit for Heinz to Amsterdam. This took place in October 1981 and Heinz used the opportunity to come to Groningen as well. This visit was an immediate success, we were on first name basis within a few minutes, and his visit to my house and family went as smooth as pie. With Henk de Snoo we started to work on classes of meromorphic functions which arose in extension theory of symmetric operators in spaces with an indefinite metric and applied the theory to self-adjoint boundary eigenvalue problems with eigenvalue depending boundary conditions. At one time Heinz predicted that what we were about to start would lead to many publications. Before I realized it, I asked "Oh, how many?" (I dislike vague remarks.) Heinz actually paused to think and answered "About 10." I was impressed. How could he predict that many? But how right he was: Many papers have appeared since then. First jointly with Henk de Snoo, even more than 10, and later also with others. Since that first visit in 1981, we have met at numerous working visits, conferences, and workshops. When near the Mediterranean we would go out to swim. Out of all these contacts grew a friendship which I treasure very much. During his many visits to Groningen, Heinz generously shared many ideas with us and stimulated us to work on them by ourselves. They resulted in a master thesis and a Ph.D. thesis about extension theory and interpolation, some papers with postdoctoral students about extension theory and commutant lifting, and even a book about Schur functions, operator colligations, and reproducing kernel Pontryagin spaces which my co-authors D. Alpay, J. Rovnyak, and H.S.V. de Snoo and I dedicated to him in appreciation, admiration and amity.

Right now Heinz and I together with Daniel Alpay (from Israel), Tomas Azizov, and Yuri Shondin (both from Russia) and others are working on two projects. One is about an indefinite version of the Schur algorithm. The other concerns singular perturbations of self-adjoint operators with applications to quantum physics and to, for example, the Bessel and Laguerre differential operators. I very much enjoy working with Heinz. His enthusiasm for the problem at hand is stimulating and I am impressed by the ease with which Heinz finds the right wording for a paper. Our style of working together depends on where we are. When in Vienna, we sit on a couch in Heinz's office, we have scratch paper on our knees, and our hands are blue from writing, so to speak. When in Groningen, we stand before a big green blackboard and our hands are white from the chalk.

Upon the invitation of Rien Kaashoek, Heinz regularly visited the Free University in Amsterdam. Heinz took part in the reading and examination committee for some of Rien's Ph.D. students. He also gave a series of lectures at the Thomas Stieltjes Institute. Heinz was an important participant in the INTAS-projects which Rien arranged. As to mathematics, Heinz's visits to the Free University resulted in joint works about solutions of the Riccati equation with André Ran and his students. Heinz was co-promotor for one of them. Heinz's relation with the Netherlands is not restricted to the Free University and the State University of Groningen only. At present Heinz is a member of the international evaluation committee to evaluate the output over the period 1996–2001 of the Mathematics Departments of all Dutch Universities. It is yet another indication that Heinz is recognized as an authority in mathematics with an international reputation.

Concluding remark

I have said a lot and inadvertently I may have omitted things I should have said, but I should finish. Allow me to make one last remark: some years ago at some place Heinz and I had to fill in some forms and we discovered that at the space where we had to write down our profession we had entered different things. I had written "Professor" and Heinz had filled in "Mathematician." Heinz may have retired as a professor but I hope and wish that he will not retire as a mathematician for many years to come, for mathematics and for the sake of all of us, I am sure.

Speech of Heinz Langer

By Heinz Langer, Vienna, March 4, 2004

Magnifizenz, Spectabilis, Inge and Aad, Ladies and Gentlemen!

I feel very honoured by all these nice words, thank you very much indeed. And I am deeply touched by the fact that so many friends and colleagues have come to this conference, although at the beginning it was intended to be just a small meeting. My thanks go to the organizers, Institutsvorstand Inge Troch, Harald Woracek, Annemarie Luger, Matthias Langer, and Fritz Vogl, who have put a lot of efforts in bringing us together for these days.

Many things have been said about the 45 years of my life as a mathematician. All this may look very smooth from outside, but in fact I am not somebody who likes to make long-term plans for the future. Looking back, I see two special features which have determined my life.

The first one is that I often made decisions which at the time were rather unpopular and not understandable for others.

Let me give three examples. At the time of my diploma at the end of the 1950's, under the strong influence of Bourbaki, abstract linear topological spaces and, in particular, locally convex spaces were very much in fashion. However, I chose a more classical topic, namely operator theory in Hilbert spaces which at that time by some colleagues was considered a bit outdated, but from today's viewpoint it turned out to be a very good decision. And I am very grateful to my teacher P.H. Müller that he proposed to me to study indefinite inner product spaces.

Another example is my stay in Odessa at the beginning of the sixties. Krein had invited me to spend a year with him within an existing exchange programme between the GDR and the Soviet Union. Officially the German-Soviet friendship was a big issue, but many people were feeling differently, and not all of my friends could understand why I wanted to follow this invitation. But I insisted: even after my first application in 1960 had been rejected for political reasons, I tried again and succeeded one year later in 1961. In fact, this year in Odessa was one of the most important turning points of my life. I was deeply impressed by Krein's personality, and his school of Functional Analysis in Odessa did not only influence me strongly as a mathematician, but was also the origin of true friendships which last until today. And I am very happy that some of these friends are here.

The third example is my decision not to return to the GDR in October 1989 and to give up a secure position for a rather uncertain future. But, of course, the price for this security was very high, and this brings me to the second special feature, which I mentioned before: during a great part of my life I had to find ways to cope with the boundary conditions imposed by the political system of the GDR. So it was not always possible to do what you wanted and when you wanted it.

For example,

- I was not allowed to study physics (which was maybe not the worst thing in the end),
- I was allowed to go to Odessa only one year later,
- contacts to colleagues in the West were very restricted, and permissions to travel abroad were often denied or delayed, like my first trip to Canada: I got the invitation from Israel Halperin to Toronto for 1965. This was not so long after the Berlin wall had been built, and so this first invitation was not even considered seriously by the authorities. However, Israel Halperin was very insistent and renewed the invitation for the next year, then even by a letter to the minister. And he succeeded.

This was somehow a typical situation, and I am very grateful also to other colleagues who made it possible for me to visit them in the West by not giving up after a first or even a second failure: Ilppo Simo Louhivaara who at the time was rector in Jyväskylä, Rudi Hirschfeld in Antwerp who visited the GDR with an official delegation from Belgium and on this occasion established contacts with me, Göran Borg, who was the rector of the Royal Institute of Technology in Stockholm, and Björn Textorius, Åke Pleijel in Uppsala, Hrvoje Kraljevic and the late Branko Najman in Zagreb, Rien Kaashoek and Israel Gohberg in Amsterdam, and finally Reinhard Mennicken in Regensburg. All these contacts to colleagues from abroad were very important for me, also because they allowed me to continue to work in operator theory while in Dresden I held a professorship in probability theory.

And even in the most critical situation, when I arrived in West Germany in October 1989 and did not have a job, colleagues from there, in particular, Reinhard Mennicken and Albert Schneider from Dortmund helped me to keep my period of unemployment to 3 weeks.

Now you may wonder how it came to this happy end in Austria. Well, I don't know myself, and this is one of the big miracles in my life. Certainly, it had nothing to do with the fact that Austria had been present for me already since my childhood like a fairy tale. I am still keeping an Edelweiss in Meyers Ostalpen from my father which he found in 1926 at the Karlingerboden near Kaprun, and in our living room we had a painting called "Im Stubaital". After the Second world war it was basically impossible to visit Austria from East Germany, and my first visit in 1955 via West Germany was illegal from the point of view of GDR and kind of adventurous. Hitchhiking with a friend from Munich to Salzburg, but in separate cars, we had agreed to meet at a bridge over the Salzach mentioned in my Baedeker from the 1920's. I arrived first, but the bridge had disappeared .... Only 35 years later, in 1990, I could return to Austria, when I applied for the Chair previously held by Edmund Hlawka.

That I finally got this position was for me like a dream came through. In the past almost 13 years here at the University of Technology in Vienna I had the freedom to follow my own scientific interests, to build up a research group in operator theory, and to cooperate with colleagues from all over the world. I am very grateful to the University and the former Institute of Analysis and Technical Mathematics for giving me this opportunity and supporting me in many ways. It has always been a pleasure to work in the friendly atmosphere here. And I also enjoyed teaching several generations of students of electrical engineering and mathematics.

You may now have got the impression that I am just another example of a Saxonian as described by Franz Grillparzer in his "Loblied auf Österreich" where it reads

S'ist möglich, daß in Sachsen und beim Rhein
Es Leute gibt, die mehr in Büchern lasen

than people in Austria, I should add. However, I really enjoy living in Vienna with all its theaters and museums, opera houses and concert halls, cafés and vineyards, and nearby mountains and valleys. And, being attentive, you can even find a number of close cultural ties between Vienna and Dresden, like the Albertina or Gottfried Semper's architecture.

I was happy when I received the Austrian citizenship 13 years ago. Indeed, I do now share the passionate love of the Austrians for this beautiful country and its rich culture. And I have at least tried to adopt the Austrian mentality which in his "Loblied" Grillparzer describes as follows:

Allein, was not tut und was Gott gefällt,
Der klare Blick, der off'ne, richt'ge Sinn,
Da tritt der Österreicher hin vor jeden,
denkt sich sein Teil und läßt die andern reden!

Programme of the "Colloquium on Operator Theory"

Thursday, 4 March	
9'15	Opening: H. Kaiser (Vice-Rector of the Vienna University of Technology) D. Dorninger (Dean of the Faculty for Mathematics and Geoinformation) I. Troch (Head of the Institute for Analysis and Scientific Computing) Laudation: A. Dijksma Heinz Langer
10'30	I. Gohberg: Continuous analogue of orthogonal polynomials
	Coffee break
Chair: A. Luger	
11'30	C. Tretter: Spectral theory of block operator matrices and applications in mathematical physics
12'15	V. Adamyan: Matrix continuous analogues of orthogonal trigonometric polynomials
12'45	Organizational remarks, conference photo
	Lunch break
Chair: R. Mennicken	
14'30	P. Lancaster: An inverse quadratic eigenvalue problem
15'00	H. de Snoo: Singular Sturm–Liouville problems nonlinear in the eigenvalue parameter
15'30	M. Kaltenbäck: Indefinite canonical systems and the inverse spectral theorem
	Coffee break
Chair: A. Fleige	
16'30	F. Szafraniec: q-disease in operator theory: some cases
17'00	B. Textorius: Directing mappings in Krein spaces
17'30	H. Winkler: Isometric isomorphisms between strings and canonical systems

Friday, 5 March

Chair: B. Kirstein

9'00	A. Dijksma: The algorithm of Issai Schur in an indefinite setting
9'45	K.-H. Förster: On matrix and operator polynomials with nonnegative coefficients

Coffee break

Chair: O. Staffans

11'00	A. Ran: Some remarks on LQ-optimal control: asymptotics and the inverse problem
11'30	A. Gheondea: The indefinite Caratheodory problem
12'00	D. Alpay: Rational functions and backward shift operators in the hyperholomorphic case

Coffee break

Chair: B. Textorius

13'00	M. Brown: Inverse resonance problems for the Sturm–Liouville problem and for the Jacobi matrix
13'30	P. Binding: Oscillation of indefinite Sturm–Liouville eigenfunctions
14'00	M. Möller: The spectrum of the multiplication operator associated with a family of operators in a Banach space

Afternoon in Vienna

Conference dinner

Saturday, 6 March, morning		
	Chair: M. Langer	
9'00	P. KURASOV: Pontryagin type models for soliton potentials: inverse scattering method for operator extensions	
		Chair: H. de Snoo
9'30	D. VOLOK: De Branges–Rovnyak spaces and Schur functions: the hyperholomorphic case	S. HASSI: Boundary relations and Weyl families of symmetric operators
10'00	X. MARY: Subdualities and associated (reproducing) kernels	A. BATKAI: Polynomial stability of operator semigroups
10'30	Y. SHONDIN: Pontryagin space boundary value problems for a singular differential expression	V. STRAUSS: On J-symmetric operators with square similar to a bounded symmetric operator
	Coffee break	
	Chair: A. Ran	
11'30	B. ĆURGUS: Indefinite Sturm–Liouville problems with eigenparameter dependent boundary conditions	
		Chair: P. Binding
12'00	C. TRUNK: Spectral points of type π_+ for closed operators in Krein spaces	D. POPOVICI: Moment theorems for commuting multi-operators
12'30	J. BEHRNDT: Finite-dimensional perturbations of locally definitizable selfadjoint operators in Krein spaces	H. LUNDMARK: Direct and inverse spectral problem for a non-selfadjoint third order generalization of the discrete string equation
	Lunch break	

Saturday, 6 March, afternoon		
	Chair: C. Tretter	
14'30	V. Pivovarchik: Shifted Hermite–Biehler functions	
		Chair: C. Trunk
15'00	G. Wanjala: The Schur transform at the boundary point $z = 1$	A. Lasarow: Some basic facts on orthogonal rational matrix-valued functions on the unit circle
15'30	L. Kérchy: Canonical factorization of vectors with respect to an operator	L. Mikayelyan: Orthogonal polynomials on the unit circle with respect to a rational weight function
	Coffee break	
Chair: H. Woracek		
16'30	Z. Sasvári: The extension problem for positive definite functions. A historical survey	
17'15	M. Kaashoek: Metric constrained interpolation problems and control theory	
	Closing	

Participants of the "Colloquium on Operator Theory"

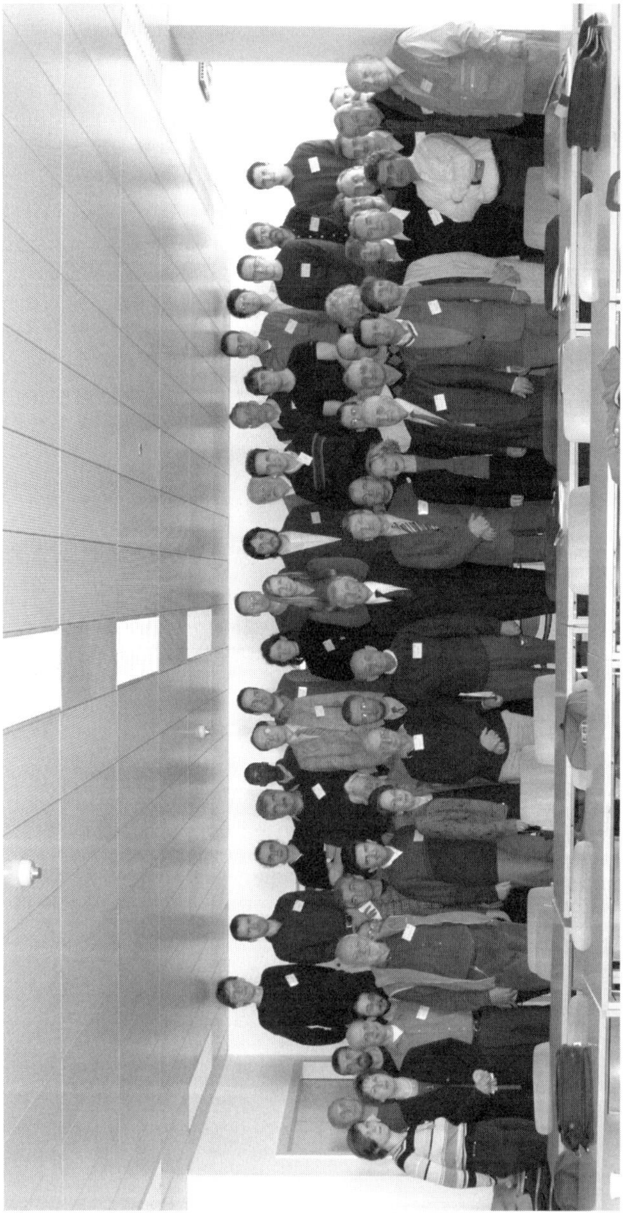

56
58
31
17
20
2
30
47
12
38
32
21
29
13
5
7
45
18
51
27
22
44
40
1
23
49
43
41
39
10
35
25
16
57
15
3
14
42
8
55
24
33
50
54
37
9
4
19
46
26
53
6
34
59
36
28

List of Participants

1. **Vadim Adamyan**, I.I. Mechnikov Odessa National University, Ukraine,
   vadamyan@paco.net, vma@dtp.odessa.ua
2. **Daniel Alpay**, Department of Mathematics, Ben-Gurion University of the Negev, Beer-Sheva, Israel,
   dany@math.bgu.ac.il
3. **Andras Batkai**, Department of Applied Analysis, ELTE TTK, Budapest, Hungary,
   batka@cs.elte.hu
4. **Jussi Behrndt**, Institut für Mathematik, TU-Berlin, Germany,
   behrndt@math.tu-berlin.de
5. **Christa Binder**, Institut für Analysis und Scientific Computing, TU Wien, Austria,
   chbinder@mail.zserv.tuwien.ac.at
6. **Paul Binding**, Department of Mathematics and Statistics, University of Calgary, Canada,
   binding@ucalgary.ca
7. **Bernhard Bodenstorfer**, Frankfurt/Main, Germany,
   bernhard.bodenstorfer@ecb.int
8. **Malcolm Brown**, Department of Computer Science, Cardiff University, United Kingdom,
   malcolm.brown@cs.cf.ac.uk
9. **Branko Ćurgus**, Department of Mathematics, Western Washington University, United States,
   curgus@cc.wwu.edu
10. **Aad Dijksma**, Department of Mathematics, University of Groningen, The Netherlands,
    dijksma@math.rug.nl
11. **David Eschwé**, Wien, Austria,
    deschwe@gosch.com
12. **Andreas Fleige**, Dortmund, Germany,
    andreas.fleige@continentale.de
13. **Karl-Heinz Förster**, Institut für Mathematik, TU-Berlin, Germany,
    foerster@math.tu-berlin.de
14. **Aurelian Gheondea**, Department of Mathematics, Bilkent University, Ankara, Turkey,
    aurelian@fen.bilkent.edu.tr
15. **Israel Gohberg**, School of Mathematical Sciences, Tel-Aviv University, Israel,
    gohberg@math.tau.ac.il

16. **Seppo Hassi**, Department of Mathematics and Statistics, University of Vaasa, Finland,
sha@uwasa.fi

17. **Rien Kaashoek**, Department of Mathematics, Vrije Universiteit, Amsterdam, The Netherlands,
kaash@cs.vu.nl

18. **Michael Kaltenbäck**, Institut für Analysis und Scientific Computing, TU Wien, Austria,
mbaeck@geometrie.tuwien.ac.at

19. **Victor Katsnelson**, Faculty of Mathematics and Computer Science, The Weizmann Institute of Science, Rehovot, Israel,
victor.katsnelson@weizmann.ac.il

20. **László Kérchy**, University of Szeged, Hungary,
kerchy@sol.math.u-szeged.hu

21. **Bernd Kirstein**, Mathematisches Institut, Universität Leipzig, Germany,
kirstein@mathematik.uni-leipzig.de

22. **Hrvoje Kraljevic**, Zagreb, Croatia,
hrk@math.hr

23. **Pavel Kurasov**, Department of Mathematics, Lund Institute of Technology, Sweden,
kurasov@maths.lth.se

24. **Peter Lancaster**, University of Calgary, Canada,
lancaste@ucalgary.ca

25. **Heinz Langer**, Institut für Analysis und Scientific Computing, TU Wien, Austria,
hlanger@mail.zserv.tuwien.ac.at

26. **Matthias Langer**, Institut für Analysis und Scientific Computing, TU Wien, Austria,
mlanger@mail.zserv.tuwien.ac.at

27. **Andreas Lasarow**, Departement Computerwetenschappen, K. U. Leuven, Heverlee, Belgium,
andreas.lasarow@cs.kuleuven.ac.be

28. **Annemarie Luger**, Institut für Analysis und Scientific Computing, TU Wien, Austria,
aluger@mail.zserv.tuwien.ac.at

29. **Hans Lundmark**, Department of Mathematics, Linköping University, Sweden,
halun@mai.liu.se

30. **Xavier Mary**, ENSAE – CREST LS, 3, avenue Pierre Larousse, 92245 Malakoff Cedex, Paris, France,
xavier.mary@ensae.fr

31. **Vladimir Matsaev**, School of Mathematics, Tel Aviv University, Israel, matsaev@post.tau.ac.il
32. **Reinhard Mennicken**, Naturwissenschaftliche Fakultät I, Universität Regensburg, Germany, reinhard.mennicken@mathematik.uni-regensburg.de
33. **Levon Mikaelyan**, Department of Informatics and Applied Mathematics, Yerevan State University, Armenia, mikaelyanl@ysu.am
34. **Manfred Möller**, School of Mathematics, University of the Witwatersrand, Johannesburg, South Africa, manfred@maths.wits.ac.za
35. **Maria Magdalena Nafalska**, Institut für Mathematik, TU-Berlin, Germany, m_nafalska@freenet.de
36. **Vjacheslav Pivovarchik**, Odessa State Academy of Structure and Architecture, Ukraine, v.pivovarchik@paco.net, vnp@dtp.odessa.ua
37. **Dan Popovici**, Department of Mathematics and Computer Science, University of the West Timişoara, Romania, popovici@math.uvt.ro
38. **Andre Ran**, Department of Mathematics/FEW, Vrije Universiteit, Amsterdam, The Netherlands, ran@cs.vu.nl
39. **Adrian Sandovici**, Department of Mathematics, University of Groningen, The Netherlands, adrian@math.rug.nl
40. **Zoltán Sasvári**, Institut für Mathematische Stochastik, TU Dresden, Germany, sasvari@math.tu-dresden.de
41. **Wilfried Schenk**, Institut für Mathematische Stochastik, TU Dresden, Germany, schenk@math.tu-dresden.de
42. **Yuri G. Shondin**, Faculty of Physics, Nizhny Novgorod State Pedagogical University, Russia, shondin@sinn.ru
43. **Henk de Snoo**, Department of Mathematics, University of Groningen, The Netherlands, desnoo@math.rug.nl
44. **Olof Staffans**, Abo Akademi, Finland, staffans@abo.fi

45. **Vladimir Strauss**, Depto de Matemáticas Puras y Aplicadas, Universidad Simón, Caracas, Venezuela,
str@usb.ve

46. **Franciszek Hugon Szafraniec**, Instytut Matematyki, Uniwersytet Jagielloński, Kraków, Poland,
fhszafra@im.uj.edu.pl, umszafra@cyf-kr.edu.pl

47. **Björn Textorius**, Department of Mathematics, Linköping University, Sweden,
bjtex@mai.liu.se

48. **Igor Tkachenko**, Polytechnic University of Valencia, Spain,
imtk@mat.upv.es

49. **Christiane Tretter**, FB 3 – Mathematik, Universität Bremen, Germany,
ctretter@math.uni-bremen.de

50. **Inge Troch**, Institut für Analysis und Scientific Computing, TU Wien, Austria,
inge.troch@tuwien.ac.at

51. **Carsten Trunk**, Institut für Mathematik, TU-Berlin, Germany,
trunk@math.tu-berlin.de

52. **Fritz Vogl**, Institut für Analysis und Scientific Computing, TU Wien, Austria,
fvogl@osiris.tuwien.ac.at

53. **Dan Volok**, Department of Mathematics, Ben-Gurion University of the Negev, Beer-Sheva, Israel,
volok@cs.bgu.ac.il

54. **Markus Wagenhofer**, FB 3 – Mathematik, Universität Bremen, Germany,
wagenhofer@math.uni-bremen.de

55. **Gerald Wanjala**, Department of Mathematics, University of Groningen, The Netherlands,
gerald@math.rug.nl

56. **Henrik Winkler**, Department of Mathematics, University of Groningen, The Netherlands,
winkler@math.rug.nl

57. **Monika Winklmeier**, FB 3 – Mathematik, Universität Bremen, Germany,
winklmr@math.uni-bremen.de

58. **Harald Woracek**, Institut für Analysis und Scientific Computing, TU Wien, Austria,
harald.woracek@tuwien.ac.at

59. **Rosalinde Pohl** (Secretary)

Bibliography of Heinz Langer

In the volume OT 106 of the series "Operator Theory: Advances and Applications" which appeared in 1998, a bibliography of Heinz Langer up to this year was compiled. The following list updates this bibliography; it collects his recent papers.

[131] Direct and inverse spectral problems for generalized strings, *Integral Equations Operator Theory* **30** (1998), 409–431 (with H. Winkler)

[132] A factorization result for generalized Nevanlinna functions, *Integral Equations Operator Theory* **36** (2000), 121–125 (with A. Dijksma, A. Luger, and Yu. Shondin)

[133] Classical Nevanlinna–Pick interpolation with real interpolation points, *Oper. Theory Adv. Appl.* **115** (2000), 1–50 (with D. Alpay and A. Dijksma)

[134] The spectral shift function for certain block operator matrices, *Math. Nachr.* **211** (2000), 5–24 (with V. Adamjan)

[135] A class of 2×2–matrix functions, *Glas. Matem. Ser. III* **35(55)** (2000), 149–160 (with A. Luger)

[136] Linearization and compact perturbation of self-adjoint analytic operator functions, *Oper. Theory Adv. Appl.* **118** (2000), 255–285 (with A. Markus and V. Matsaev)

[137] Self-adjoint differential operators with inner singularities and Pontryagin spaces, *Oper. Theory Adv. Appl.* **118** (2000), 105–175 (with A. Dijksma, Yu. Shondin, and C. Zeinstra)

[138] On singular critical points of positive operators in a Krein space, *Proc. Amer. Math. Soc.* **128** (2000), 2621–2626 (with B. Ćurgus and A. Gheondea)

[139] Variational principles for real eigenvalues of selfadjoint operator pencils, *Integral Equations Operator Theory* **38** (2000), 190–206 (with D. Eschwé and P. Binding)

[140] Dissipative eigenvalue problems for a Sturm–Liouville operator with a singular potential, *Proc. Roy. Soc. Edinburgh Sect. A* **130** (2000), 1237–1257 (with B. Bodenstorfer and A. Dijksma)

[141] Diagonalization of certain block operator matrices and applications to Dirac operators, *Oper. Theory Adv. Appl.* **122** (2001), 331–358 (with C. Tretter)

[142] Existence and uniqueness of contractive solutions of some Riccati equations, *J. Funct. Anal.* **179** (2001), 448–473 (with V. Adamyan and C. Tretter)

[143] A spectral theory for a λ-rational Sturm–Liouville problem, *J. Differential Equations* **171** (2001), 315–345 (with V. Adamyan and M. Langer)

[144] Compact perturbation of definite type spectra of self–adjoint quadratic operator pencils, *Integral Equations Operator Theory* **39** (2001), 127–152 (with V. Adamyan and M. Möller)

[145] A new concept for block operator matrices: The quadratic numerical range, *Linear Algebra Appl.* **330** (2001), 89–112 (with A. Markus, V. Matsaev, and C. Tretter)

[146] Corners of numerical ranges, *Oper. Theory Adv. Appl.* **124** (2001), 385–400 (with A. Markus and C. Tretter)

[147] Elliptic eigenvalue problems with eigenparameter dependent boundary conditions, *J. Differential Equations* **174** (2001), 30–54 (with P. Binding, R. Hryniv, and B. Najman)

[148] A relation for the spectral shift function of two self–adjoint extensions, *Oper. Theory Adv. Appl.* **127** (2001), 437–445 (with V. A. Yavrian and H. de Snoo)

[149] The Schur algorithm for generalized Schur functions I: coisometric realizations, *Oper. Theory Adv. Appl.* **129** (2001), 1–36 (with D. Alpay, T. Ya. Azizov, and A. Dijksma)

[150] Invariant Subspaces of infinite-dimensional Hamiltonians and solutions of the corresponding Riccati equations, *Oper. Theory Adv. Appl.* **130** (2001), 235–254 (with A. C. M. Ran and B. A. van de Rotten)

[151] On the Loewner problem in the class N_κ, *Proc. Amer. Math. Soc.* **130** (2002), 2057–2066 (with D. Alpay and A. Dijksma)

[152] Triple variational principles for eigenvalues of self-adjoint operators and operator functions, *SIAM J. Math. Anal.* **34** (2002), 228–238 (with D. Eschwé)

[153] Variational principles for eigenvalues of block operator matrices, *Indiana Univ. Math. J.* **51** (2002), 1427–1459 (with M. Langer and C. Tretter)

[154] The Schur algorithm for generalized Schur functions II: Jordan chains and transformations of characteristic functions, *Monatsh. Math.* **138** (2003), 1–29 (with D. Alpay, T. Ya. Azizov, and A. Dijksma)

[155] Self-adjoint block operator matrices with non-separated diagonal entries and their Schur complements, *J. Funct. Anal.* **199** (2003), 427–451 (with A. Markus, V. Matsaev, and C. Tretter)

[156] The Schur algorithm for generalized Schur functions III: J-unitary matrix polynomials on the circle, *Linear Algebra Appl.* **369** (2003), 113–144 (with D. Alpay, T. Ya. Azizov, and A. Dijksma)

[157] Continuous embeddings, completions and complementation in Krein spaces, *Rad. Mat.* **12** (2003), 37–79 (with B. Ćurgus)

[158] A basic interpolation problem for generalized Schur functions and coisometric realizations, *Oper. Theory Adv. Appl.* **143** (2003), 39–76 (with D. Alpay, T. Ya. Azizov, A. Dijksma, and G. Wanjala)

[159] Rank one perturbations at infinite coupling in Pontryagin spaces, *J. Funct. Anal.* **209** (2004), 206–246 (with A. Dijksma and Yu. Shondin)

[160] Continuations of Hermitian indefinite functions and canonical systems: an example, *Methods Funct. Anal. Topology* **10** (2004), 39–53 (with M. Langer and Z. Sasvári)

[161] Solution of a multiple Nevanlinna–Pick problem via orthogonal rational functions, *J. Math. Anal. Appl.* **293** (2004), 605–632 (with A. Lasarow)

[162] Oscillation results for Sturm–Liouville problems with an indefinite weight function, *J. Comput. Appl. Math.* **171** (2004), 93–101 (with P. Binding and M. Möller)

[163] Factorization of J-unitary matrix polynomials on the line and a Schur algorithm for generalized Nevanlinna functions, *Linear Algebra Appl.* **387** (2004), 313–342 (with D. Alpay and A. Dijksma)

[164] The Schur algorithm for generalized Schur functions IV: unitary realizations, *Oper. Theory Adv. Appl.* **149** (2004), 23–45 (with D. Alpay, T. Ya. Azizov A. Dijksma, and G. Wanjala)

[165] A Krein space approach to PT-symmetry, *Czechoslovak J. Phys* **54** (2004), 1113–1120 (with C. Tretter)

[166] Minimal realizations of a scalar generalized Nevanlinna function related to their basic factorization, *Oper. Theory Adv. Appl.* **154** (2004), 69–90 (with A. Dijksma, A. Luger, and Yu. Shondin)

[167] Partial non-stationary perturbation determinants, *Oper. Theory Adv. Appl.* **154** (2004), 1–18 (with V. Adamyan)

[168] Spectrum of definite type of self-adjoint operators in Krein spaces, *Linear and Multilinear Algebra* **53** (2005), 115–136 (with M. Langer, A. Markus, and C. Tretter)

[169] Spectral problems for operator matrices, *Math. Nachr.* **278** (2005), 1408–1429 (with A. Batkai, P. Binding, A. Dijksma, and R. Hryniv)

[170] Bounded normal operators in a Pontryagin space, *Oper. Theory Adv. Appl.* **162** (2006), 231–251 (with F. Szafraniec)

[171] Partial non-stationary perturbation determinants for a class of J-symmetric operators, *Oper. Theory Adv. Appl.* **162** (2006) 1–18 (with V. Adamyan)

Operator Theory:
Advances and Applications, Vol. 163, 1–22

General Solution of the Stieltjes Truncated Matrix Moment Problem

Vadim M. Adamyan and Igor M. Tkachenko

To Heinz Langer with admiration and gratitude for fruitful co-operation

Abstract. The description of all solutions of the truncated Stieltjes matrix moment problem consisting in finding all $s \times s$ matrix measures $d\boldsymbol{\sigma}(t)$ on $[0, \infty)$ with given first $2n+1$ power $s \times s$ matrix moments $(\mathbf{C}_j)_{j=0}^{n}$ is obtained in a general case, when the block Hankel matrix $\mathbf{\Gamma}_n := (\mathbf{C}_{j+k})_{j,k=0}^{n}$ may be non-invertible. Special attention is paid to the description of canonical solutions for which $d\boldsymbol{\sigma}(t)$ is a sum of at most $sn + s$ point matrix "masses" with the minimal sum of their ranks.

Mathematics Subject Classification (2000). Primary 30E05, 30E10; Secondary 82C70, 82D10.

Keywords. Stieltjes moments problem, matrix functions, Nevanlinna's formula.

1. Introduction

In this paper we consider the following problem:

Given a set of Hermitian $s \times s$ matrices

$$\{\mathbf{C}_0, \mathbf{C}_1, \mathbf{C}_2, \ldots, \mathbf{C}_{2n}\}, \quad n = 0, 1, 2, \ldots. \tag{1.1}$$

Find all non-negative matrix measures $d\boldsymbol{\sigma}(t)$ *such that*

$$\int_0^\infty t^k \, d\boldsymbol{\sigma}(t) = \mathbf{C}_k \ , \ k = 0, 1, 2, \ldots, 2n. \tag{1.2}$$

This is a matrix version of the well-known *Stieltjes truncated moment problem.* The Stieltjes problem was considered in different settings in many monographs and papers starting from the classical memoirs by Stieltjes himself [14, 15]. Classical

The authors appreciate the fair work and helpful suggestions and corrections of the referees of this paper.

results on the topic are contained in the books [3, 10, 13] and the papers [12, 11]; more recent developments on the truncated moment problems can be found in [5, 7, 6, 1, 8, 9]. The above matrix problem was considered recently in [2]. The main results of that paper can be summarized in the following two theorems.

Theorem 1.1. *A system of Hermitian matrices* $\{\mathbf{C}_0, \mathbf{C}_1, \mathbf{C}_2, \dots, \mathbf{C}_{2n}\}$, $n = 0, 1, 2, \dots$, *admits the representation*

$$\int_0^\infty t^k \, d\boldsymbol{\sigma}(t) = \mathbf{C}_k \ , \ \ k = 0, 1, 2, \dots, 2n, \tag{1.3}$$

if and only if

a) *the block Hankel matrix* $\mathbf{\Gamma}_n := (\mathbf{C}_{k+j})_{k,j=0}^n$ *is non-negative;*

b) *for any set* $\boldsymbol{\xi}_0, \dots, \boldsymbol{\xi}_r \in \mathbb{C}_s$, $0 \le r \le n-1$, *and with* $(\boldsymbol{\xi}, \boldsymbol{\eta})$ *being the standard scalar product in* $\mathbb{C}_s$, *the condition*

$$\sum\nolimits_{j,k=0}^{r} (\mathbf{C}_{j+k} \boldsymbol{\xi}_k, \boldsymbol{\xi}_j) = 0 \tag{1.4}$$

implies

$$\sum\nolimits_{j,k=0}^{r} (\mathbf{C}_{j+k+2} \boldsymbol{\xi}_k, \boldsymbol{\xi}_j) = 0; \tag{1.5}$$

c) *the block Hankel matrix* $\mathbf{\Gamma}_{n-1}^{(1)} := (\mathbf{C}_{k+j+1})_{k,j=0}^{n-1}$ *is non-negative and for any set* $\boldsymbol{\xi}_0, \dots, \boldsymbol{\xi}_r \in \mathbb{C}_s$, $0 \le r \le n-1$, *the condition*

$$\sum\nolimits_{j,k=0}^{r} (\mathbf{C}_{j+k+1} \boldsymbol{\xi}_k, \boldsymbol{\xi}_j) = 0 \tag{1.6}$$

implies (1.5).

In what follows we will denote by $\mathfrak{R}$ the Nevanlinna class of holomorphic dissipative $s \times s$ matrix functions on the upper half-plane, i.e., matrix functions with non-negative imaginary parts and by $\mathfrak{S}$ the subset of $\mathfrak{R}$ consisting of all Nevanlinna $s \times s$ matrix functions $\mathbf{t}(z)$, $\mathrm{Im} z > 0$, which admit the integral representation

$$\mathbf{t}(z) = \int_0^\infty \frac{1}{t - z} d\boldsymbol{\rho}(t)$$

with a non-decreasing $s \times s$ matrix function ($\mathbb{C}_s$ operator) $\boldsymbol{\rho}(t)$ such that the condition

$$\int_0^\infty d(\boldsymbol{\rho}(t)\mathbf{h}, \mathbf{h}) < \infty$$

holds for any $\mathbf{h} \in \mathbb{C}_s$.

Theorem 1.2. *Let the conditions of Theorem* 1.1 *hold and* $\det \mathbf{\Gamma}_n > 0$, *and let* $\mathbf{\Delta}_{\Xi}(z)$ *be the upper-left block of the matrix function*

$$\mathbf{\Gamma}_n \left(\mathbf{\Gamma}_{\Xi;n}^{(1)}(z) - z\mathbf{\Gamma}_n \right)^{-1} \mathbf{\Gamma}_n \ ,$$

where

$$\mathbf{\Gamma}_{\Xi;n}^{(1)}(z) := \begin{pmatrix} & & \mathbf{C}_{n+1} \\ & \mathbf{\Gamma}_{n-1}^{(1)} & \vdots \\ & & \mathbf{C}_{2n} \\ \mathbf{C}_{n+1} & \dots \quad \mathbf{C}_{2n} & \Xi(z)^{-1} + z \end{pmatrix}$$

and $\Xi \in \mathfrak{S}$. *Then the relation*

$$\int_0^\infty \frac{1}{t-z} d\boldsymbol{\sigma}_\Xi(t) = \mathbf{\Delta}_\Xi(z), \ \mathrm{Im} z > 0,$$

by means of the Stieltjes inversion formula

$$(\boldsymbol{\sigma}_\Xi(\beta)\mathbf{h},\mathbf{h}) - (\boldsymbol{\sigma}_\Xi(\alpha)\mathbf{h},\mathbf{h}) = -\frac{1}{\pi}\lim_{\eta\downarrow 0}\int_\alpha^\beta \mathrm{Im}\,(\mathbf{\Delta}_\Xi(t+i\eta)\mathbf{h},\mathbf{h})\,dt,$$
$$\boldsymbol{\sigma}_\Xi(\alpha-0) = \boldsymbol{\sigma}_\Xi(\alpha+0), \ \boldsymbol{\sigma}_\Xi(\beta-0) = \boldsymbol{\sigma}_\Xi(\beta+0),$$
$$\mathbf{h} \in \mathbb{C}_s, \ -\infty < \alpha < \beta < \infty,$$

establishes a one-to-one correspondence between a set of all solutions $\boldsymbol{\sigma}_\Xi(t)$ *of the truncated Stieltjes matrix moment problem with the given moments* $\{\mathbf{C}_0,\dots,\mathbf{C}_{2n}\}$ *and the subset of all Nevanlinna matrix functions* Ξ *from* $\mathfrak{S}$ *which satisfy the condition*

$$\Xi(\lambda)^{-1} - \sum\nolimits_{j,k=0}^{n-1} \mathbf{C}_{n+j+1}\mathbf{b}_{jk}(\lambda)\mathbf{C}_{n+k+1} > 0, \ \lambda < 0\ , \tag{1.7}$$

with $s \times s$ *matrices* $\mathbf{b}_{jk}(z)$ *defined by the equality*

$$\left(\mathbf{\Gamma}_{n-1}^{(1)} - z\right)^{-1} = (\mathbf{b}_{jk}(z))_{j,k=0}^{n-1}, \ z \notin [0,\infty).$$

The aim of the present work is to describe all solutions of the Stieltjes truncated matrix moment problem in the *degenerate case*: $\det \mathbf{\Gamma}_n = 0$.

In Section 2 following ideas of M.G. Krein [11] we show how the truncated matrix moment problems can be reduced to problems of the extension theory for symmetric operators.

In Section 3 we consider the so-called completely degenerate truncated moment problem, we find here its unique solution in an explicit form.

A special class of the canonical solutions of the truncated Stieltjes problem is described in Section 4. For this class of solutions the sought non-decreasing matrix function $\boldsymbol{\sigma}(t)$ has no more than $ns + s$ points of growth. An algorithm for the construction of such solutions is given here as well.

In the last section we give, omitting the detailed proof, the description of all solutions of the problem under consideration.

2. Reduction to the extension theory problem

From now on we will assume that the conditions of Theorem 1.1 hold. Notice that due to the conditions a) and c) of this theorem, the quadratic forms of the matrices $\mathbf{C}_j$ are non-negative, or briefly $\mathbf{C}_j \geq 0, \ j = 0,\dots,2n$. Without loss of generality we can assume that all matrix moments $\mathbf{C}_j$ are invertible, i.e., that $\mathbf{C}_j > 0,\ j = 0, \dots, 2n$. Indeed, let $\mathbb{N}_j \subset \mathbb{C}_s$ be the null-spaces of $\mathbf{C}_j$. Due to the condition a) of Theorem 1.1 for any vector $\boldsymbol{\eta} \in \mathbb{C}_s$ the equality $\mathbf{C}_0\boldsymbol{\eta} = 0$ implies $\mathbf{C}_j\boldsymbol{\eta} = 0,\ j = 1, \dots, 2n$. Besides, since for any vector $\boldsymbol{\eta} \in \mathbb{C}_s$ and any solution $\boldsymbol{\sigma}(t)$ of the moment problem all integrals

$$\int_0^\infty t^k d(\boldsymbol{\sigma}(t)\,\boldsymbol{\eta}, \boldsymbol{\eta}),\ \ 1 \le k \le 2n$$

vanish simultaneously, any equality $\mathbf{C}_j\boldsymbol{\eta} = 0$ implies $\mathbf{C}_k\boldsymbol{\eta} = 0,\ 1 \le k \le 2n$. Hence, $\mathbb{N}_0 \subset \mathbb{N}_1 = \mathbb{N}_k,\ k = 1, \dots, 2n$ and for a suitable basis in $\mathbb{C}_s$ the matrices $\mathbf{C}_j$ can be reduced to the form

$$\mathbf{C}_0 = \begin{pmatrix} \widetilde{\mathbf{C}}_0 & 0 \\ 0 & \widetilde{\widetilde{\mathbf{C}}}_0 \end{pmatrix};\ \mathbf{C}_j = \begin{pmatrix} \widetilde{\mathbf{C}}_j & 0 \\ 0 & 0 \end{pmatrix},\ j = 1, \dots, 2n, \tag{2.1}$$

where by construction $\det \widetilde{\mathbf{C}}_j > 0,\ j = 0, 1, \dots, 2n$.

If the truncated Stieltjes problem for invertible matrix moments $\widetilde{\mathbf{C}}_j$ is solvable, then the initial problem is solvable as well, and its general solution $\boldsymbol{\sigma}(t)$ can be presented as

$$\boldsymbol{\sigma}(t) = \mathbf{U}\begin{pmatrix} \widetilde{\boldsymbol{\sigma}}(t) & 0 \\ 0 & \widetilde{\widetilde{\mathbf{C}}}_0\vartheta(t) \end{pmatrix}\mathbf{U}^*\ ,$$

where $\mathbf{U}$ is a fixed $s \times s$-unitary matrix,

$$\vartheta(t) = \begin{cases} 0, & t \le 0, \\ 1, & t > 0, \end{cases}$$

and $\widetilde{\boldsymbol{\sigma}}(t)$ runs through the set of solutions of the truncated Stieltjes matrix moment problem for the moments $\widetilde{\mathbf{C}}_j$.

Let a non-decreasing matrix function $\boldsymbol{\sigma}(t)$, $0 \le t < \infty$, be a solution of the truncated Stieltjes matrix moment problem, i.e.,

$$\int_0^\infty t^k\ d\boldsymbol{\sigma}(t) = \mathbf{C}_k\ ,\ \ k = 0, 1, 2, \dots, 2n. \tag{2.2}$$

Consider the set of continuous vector functions $\mathbf{f}(t)$, $0 \le t < \infty$, with values in $\mathbb{C}_s$, which satisfy the condition

$$\int_0^\infty (d\boldsymbol{\sigma}(t)\,\mathbf{f}(t), \mathbf{f}(t)) < \infty. \tag{2.3}$$

Construct a pre-Hilbert space $\mathcal{L}$ of such vector functions taking the bilinear functional

$$\langle \mathbf{f}, \mathbf{g} \rangle = \int_0^\infty (d\boldsymbol{\sigma}(t)\,\mathbf{f}(t), \mathbf{g}(t)) \tag{2.4}$$

as the scalar product. Notice that by (1.2) the vector polynomials

$$\mathbf{f}(t) = \boldsymbol{\xi}_0 + t \cdot \boldsymbol{\xi}_1 + \cdots + t^n \cdot \boldsymbol{\xi}_n, \qquad \boldsymbol{\xi}_0, \dots, \boldsymbol{\xi}_n \in \mathbb{C}_s, \tag{2.5}$$

of degree n belong to $\mathcal{L}$. We will denote the linear subset of these polynomials by $\mathcal{P}_n$.

Let $\mathcal{L}_0$ be the subspace of $\mathcal{L}$ consisting of all vector functions $\mathbf{f}$ such that

$$\|\mathbf{f}\| := \sqrt{\langle \mathbf{f}, \mathbf{f} \rangle} = 0.$$

If $\mathbf{g} = \mathbf{f} + \mathbf{f}_0$, where $\mathbf{f} \in \mathcal{L}$, $\mathbf{f}_0 \in \mathcal{L}_0$, than, due to the Schwartz inequality $\langle \mathbf{f}, \mathbf{f}_0 \rangle = 0$ and, hence, $\|\mathbf{g}\| = \|\mathbf{f}\|$. Let us denote by $\mathcal{L}_1$ the factor space $\mathcal{L}/\mathcal{L}_0$. For the class of elements $\widehat{\mathbf{g}} = \mathbf{f} + \mathcal{L}_0$ of this factor space we set $\|\widehat{\mathbf{g}}\|_{\mathcal{L}_1} = \|\mathbf{f}\|$. Taking the completion of $\mathcal{L}_1$ with respect to this norm, we obtain the Hilbert space $\mathrm{L}^2_\sigma(\mathbb{C}_s)$. We keep the same symbol $\langle .,. \rangle$ for the scalar product in $\mathrm{L}^2_\sigma(\mathbb{C}_s)$. Let L_n be the subspace of $\mathrm{L}^2_\sigma(\mathbb{C}_s)$ generated by the subset of vector polynomials $\mathcal{P}_n$. By (2.2) and (2.4) for $\mathbf{f}, \mathbf{g} \in \mathcal{P}_n$,

$$\mathbf{f}(t) = \sum_{l=0}^{n} t^r \cdot \boldsymbol{\xi}_r, \quad \mathbf{g}(t) = \sum_{l=0}^{n} t^r \cdot \boldsymbol{\eta}_r, \quad \boldsymbol{\xi}_0, \dots, \boldsymbol{\eta}_n \in \mathbb{C}_s ,$$

we have:

$$\langle \mathbf{f}, \mathbf{g} \rangle = \sum_{j,k=0}^{n} (\mathbf{C}_{j+k} \boldsymbol{\xi}_k, \boldsymbol{\eta}_j). \tag{2.6}$$

Therefore the scalar products on L_n in the spaces $\mathrm{L}^2_\sigma(\mathbb{C}_s)$ for all non-decreasing matrix functions $\boldsymbol{\sigma}(t)$ which satisfy (1.2) must coincide. Take any $\tilde{\boldsymbol{\sigma}}(t)$ satisfying (1.2) and consider the non-negative symmetric operator $\widetilde{A}_0$ of multiplication by the independent variable t on the subset of all continuous compact vector functions in the related space $\mathrm{L}^2_{\tilde{\sigma}}(\mathbb{C}_s)$. The closure $\widetilde{A}$ of $\widetilde{A}_0$ is a self-adjoint operator [4]. Let us denote by L_{n-1} the subspace of L_n generated by vector polynomials of the degree $\le n-1$. The restriction A_0 of the operator $\widetilde{A}$ to L_{n-1} is a non-negative symmetric operator which by definition of $\widetilde{A}$ actually does not depend on the choice of a solution of the truncated Stieltjes moment problem. Therefore each solution $\tilde{\boldsymbol{\sigma}}(t)$ of this problem generates some non-negative self-adjoint extension $\widetilde{A}$ of A_0.

Let $\widetilde{\mathbf{E}}_t$, $0 \le t < \infty$, be now the spectral function of some simple non-negative extension $\widetilde{A}$ [1] of A_0. For the canonical orthonormal basis $\{\mathbf{e}_1, \dots, \mathbf{e}_s\}$ in $\mathbb{C}_s$ we introduce the set of classes $\{\widehat{\mathbf{e}}_{10}, \dots, \widehat{\mathbf{e}}_{s0}\} \subset \mathrm{L}_n$ which contain zero degree vector monomials

$$\mathbf{e}_{10}(t) \equiv \mathbf{e}_1, \dots, \mathbf{e}_{s0}(t) \equiv \mathbf{e}_s,$$

respectively. Let us consider the non-decreasing $s \times s$ matrix function $\tilde{\boldsymbol{\sigma}}(t) = (\widetilde{\sigma}_{\mu\nu}(t))^s_{\mu,\nu=1}$, $0 \le t < \infty$,

$$\widetilde{\sigma}_{\mu\nu}(t) := \left\langle \widetilde{\mathbf{E}}_t \widehat{\mathbf{e}}_{\nu 0}, \widehat{\mathbf{e}}_{\mu 0} \right\rangle_{\mathrm{L}_n}, \quad 1 \le \mu, \nu \le s. \tag{2.7}$$

By definition of A_0 and $\widetilde{A}$, for the classes $\{\widehat{\mathbf{e}}_{1k}, \dots, \widehat{\mathbf{e}}_{sk}\} \subset \mathrm{L}_n$ which contain the vector monomials

$$\widehat{\mathbf{e}}_{1k}(t) = t^k \mathbf{e}_1, \dots, \widehat{\mathbf{e}}_{sk}(t) = t^k \mathbf{e}_s, \ 0 \le k \le n,$$

[1] A self-adjoint extension A of A_0 acting in a Hilbert space $\mathfrak{H} \supseteq \mathrm{L}_n$ is called simple if A has no invariant subspaces in $\mathfrak{H} \ominus \mathrm{L}_n$.

it holds:

$$\widehat{\mathbf{e}}_{\mu k} = A_0^k \widehat{\mathbf{e}}_{\mu 0} = \widetilde{A}^k \widehat{\mathbf{e}}_{\mu 0}, \quad 1 \le \mu \le s, \quad 0 \le k \le n.$$

Hence, for $0 \le j, k \le n$ and $1 \le \mu, \nu \le s$ we have

$$\begin{aligned}(\mathbf{C}_{j+k})_{\mu,\nu} &= (\mathbf{C}_{j+k}\mathbf{e}_\nu, \mathbf{e}_\mu)_{\mathbb{C}_s} = \\ &= \langle \widehat{\mathbf{e}}_{\nu k}, \widehat{\mathbf{e}}_{\mu j} \rangle_{\mathrm{L}_n} = \left\langle \widetilde{A}^k \widehat{\mathbf{e}}_{\nu 0}, \widetilde{A}^j \widehat{\mathbf{e}}_{\mu 0} \right\rangle_{\mathrm{L}_n} \\ &= \int_0^\infty t^{j+k} \, d\left\langle \widetilde{\mathbf{E}}_t \widehat{\mathbf{e}}_{\nu 0}, \widehat{\mathbf{e}}_{\mu 0} \right\rangle_{\mathrm{L}_n} = \int_0^\infty t^{j+k} \, d\widetilde{\sigma}_{\mu\nu}(t).\end{aligned}$$

We proved the following

Proposition 2.1. *The formula*

$$\tilde{\boldsymbol{\sigma}}_{\mu\nu}(t) = \left\langle \widetilde{\mathbf{E}}_t \widehat{\mathbf{e}}_{\nu 0}, \widehat{\mathbf{e}}_{\mu 0} \right\rangle_{\mathrm{L}_n}, \quad 0 \le t < \infty, \tag{2.8}$$

establishes a one-to-one correspondence between the set of spectral functions $\widetilde{\mathbf{E}}_t$ *of non-negative simple extensions* $\widetilde{A}$ *of* A_0 *and the set of solutions* $\tilde{\boldsymbol{\sigma}}(t)$ *of the truncated Stieltjes matrix moment problem.*

3. Solution in the completely degenerate case

The non-decreasing matrix functions $\boldsymbol{\sigma}(t)$ which satisfy (2.2) and for which

$$\mathrm{L}^2_\sigma(\mathbb{C}_s) = \mathrm{L}_n$$

are called *canonical*. Equivalently, solutions of the truncated matrix Stieltjes problem given by the expression (2.7) are *canonical* if $\widetilde{\mathbf{E}}_t$ is the spectral function of some non-negative canonical self-adjoint extension $\widetilde{A}$ of A_0, i.e., a non-negative self-adjoint extension without leaving L_n. Due to (2.6), a canonical $\boldsymbol{\sigma}(t)$ is a non-decreasing matrix function which has only a finite number of points of growth and the sum of the ranks of all jumps of $\boldsymbol{\sigma}$ at such points is $\le ns$. It was proven in [2] that the set of canonical solutions of the truncated matrix Stieltjes moment problem is non-empty whenever this problem is solvable, i.e., whenever the conditions of Theorem 1.1 hold. Formula (2.8) which establishes a one-to-one correspondence between the set of non-negative self-adjoint extensions of A_0 without leaving L_n and the set of canonical solutions of the Stieltjes problem, makes it possible to find, under the conditions of Theorem 1.1, an explicit algebraic formula for the description of the sought canonical solutions. To this end we can use as a starting point (2.7) and the relation

$$\begin{aligned}\int_0^\infty \frac{1}{t-z} d\widetilde{\sigma}_{\mu\nu}(t) &= \int_0^\infty \frac{1}{t-z} d\left\langle \widetilde{\mathbf{E}}_t \widehat{\mathbf{e}}_{\nu 0}, \widehat{\mathbf{e}}_{\mu 0} \right\rangle_{\mathrm{L}_n} = \\ &= \left\langle (\widetilde{A} - z)^{-1} \widehat{\mathbf{e}}_{\nu 0}, \widehat{\mathbf{e}}_{\mu 0} \right\rangle_{\mathrm{L}_n}.\end{aligned} \tag{3.1}$$

From now on we will assume that $\det \mathbf{\Gamma}_n = 0$, i.e., we will consider the *degenerate* case of the above problems.

Let $\mathbb{L}_n(\mathbb{C}_s)$ denote the $(n+1)s$-dimensional linear space of column vectors

$$\boldsymbol{\xi} = \begin{pmatrix} \boldsymbol{\xi}_0 & \cdots & \boldsymbol{\xi}_n \end{pmatrix}^\top, \ \boldsymbol{\xi}_0, \dots, \boldsymbol{\xi}_n \in \mathbb{C}_s, \tag{3.2}$$

($\top$ stands for the transposition operation) with the scalar product

$$(\boldsymbol{\xi}, \boldsymbol{\eta}) = \sum_{j=0}^{n} (\boldsymbol{\xi}_j, \boldsymbol{\eta}_j)_{\mathbb{C}_s} .$$

We will denote as before by L_n the same linear vector space but with the scalar product

$$\langle \boldsymbol{\xi}, \boldsymbol{\eta} \rangle = (\boldsymbol{\Gamma}_n \boldsymbol{\xi}, \boldsymbol{\eta}) = \sum_{j,k=0}^{n} \left(\mathbf{C}_{j+k} \boldsymbol{\xi}_k, \boldsymbol{\eta}_j \right)_{\mathbb{C}_s} .$$

L_n was considered above as the space of vector polynomials.

Let $\mathbb{L}_{n-1}(\mathbb{C}_s)$ be the subspace of $\mathbb{L}_n(\mathbb{C}_s)$ which consists of vectors (3.2) but with $\boldsymbol{\xi}_n = \mathbf{0}$ and let

$$\mathcal{N}_n = \mathbb{L}_n(\mathbb{C}_s) \ominus \mathbb{L}_{n-1}(\mathbb{C}_s) .$$

We denote by P_n the orthogonal projector in $\mathbb{L}_n(\mathbb{C}_s)$ onto $\mathcal{N}_n$. In the natural basis of subspaces of $\mathbb{L}_n(\mathbb{C}_s)$ this projector is evidently given as the following $(n+1) \times (n+1)$ block operator matrix

$$P_n = \begin{pmatrix} 0 & \dots & 0 \\ \vdots & \ddots & \vdots \\ 0 & \dots & \mathbf{I} \end{pmatrix},$$

where $\mathbf{I}$ is the $s \times s$ unit matrix. Let us denote by Q_n the orthogonal projector in $\mathbb{L}_n(\mathbb{C}_s)$ onto the null-space $\mathcal{M}_n$ of the operator $\boldsymbol{\Gamma}_n$ and let $Q_n^\perp = I - Q_n$. We call the truncated matrix Stieltjes moment problem *completely degenerate* if $P_n \mathcal{M}_n = \mathcal{N}_n$. We will also use the same definition for non-negative Hankel matrices $\boldsymbol{\Gamma}_n$ having the property: $P_n \mathcal{M}_n = \mathcal{N}_n$. In the completely degenerate case for each vector

$$\mathbf{e}_{ni} = \begin{pmatrix} \mathbf{0} \\ \vdots \\ \mathbf{0} \\ \mathbf{e}_i \end{pmatrix}$$

of the canonical orthonormal basis $\{\mathbf{e}_{n1}, \dots, \mathbf{e}_{ns}\}$ in $\mathcal{N}_n$ there is a non-zero vector

$$\widehat{\boldsymbol{\xi}}_{ni} = \begin{pmatrix} \boldsymbol{\xi}_{ni,0} \\ \vdots \\ \boldsymbol{\xi}_{ni,n-1} \\ \mathbf{e}_i \end{pmatrix}$$

such that

$$\mathbf{\Gamma}_n\widehat{\boldsymbol{\xi}}_{ni} = \begin{pmatrix} & & & \mathbf{C}_n \\ & \mathbf{\Gamma}_{n-1} & & \vdots \\ & & & \mathbf{C}_{2n-1} \\ \mathbf{C}_n & \dots & \mathbf{C}_{2n-1} & \mathbf{C}_{2n} \end{pmatrix} \begin{pmatrix} \boldsymbol{\xi}_{ni,0} \\ \vdots \\ \boldsymbol{\xi}_{ni,n-1} \\ \mathbf{e}_i \end{pmatrix} = 0, \ i = 1, \dots, s. \tag{3.3}$$

Note that at least some of $\boldsymbol{\xi}_{ni,k} \neq \mathbf{0}$ in (3.3), $k = 0, \dots, n-1$. Indeed, otherwise we would have that

$$\mathbf{C}_n\mathbf{e}_i = \cdots = \mathbf{C}_{2n}\mathbf{e}_i = 0.$$

But by our assumption all matrices $\mathbf{C}_0, \dots, \mathbf{C}_{2n}$ are invertible, a contradiction. As follows, the class from L_n containing the monomial $t^n\mathbf{e}_i$ contains also the vector polynomial

$$\mathbf{d}_{ni}(t) = -\boldsymbol{\xi}_{ni,0} - \cdots - t^{n-1}\boldsymbol{\xi}_{ni,n-1}.$$

Hence, in the completely degenerate case $\mathrm{L}_n = \mathrm{L}_{n-1}$. This equality implies that the symmetric operator A_0 from L_{n-1} into L_n defined above by relations

$$A_0\widehat{\mathbf{e}}_k = \widehat{\mathbf{e}}_{k+1}, \ k = 0, \dots, n-1,$$

actually is self-adjoint. As follows, in the completely degenerate case the solution of the truncated Stieltjes problem is unique.

For vectors $\widehat{\boldsymbol{\xi}}_{n0}, \dots, \widehat{\boldsymbol{\xi}}_{ns}$ appearing in (3.3) let us introduce vector polynomials

$$\phi_i(z) := \boldsymbol{\xi}_{ni,0} + \cdots + z^{n-1}\boldsymbol{\xi}_{ni,n-1} + z^n\mathbf{e}_i, \ i = 1, \dots, s,$$

and the $s \times s$ matrix function $\mathbf{D}(z)$, the corresponding columns of which are $\boldsymbol{\phi}_i(z)$. Since

$$\mathbf{D}(z) \underset{z\to\infty}{=} z^n\mathbf{I} + o\left(|z|^n\right), \tag{3.4}$$

the matrix function is invertible everywhere except of at most sn points. Let $\sigma(\mathbf{D})$ be the spectrum of the matrix function $\mathbf{D}(z)$, i.e., the set of those $z \in \mathbb{C}$ for which there exists a non-zero $\mathbf{h} \in \mathbb{C}_s$ such that $\mathbf{D}(z)\mathbf{h} = 0$.

Proposition 3.1. *In the completely degenerate case the spectrum $\sigma(A_0)$ of the non-negative self-adjoint operator A_0 coincides with $\sigma(\mathbf{D})$ and for $z \notin \sigma(\mathbf{D})$ the resolvent $(A_0 - z)^{-1}$ of A_0 acts on a vector polynomial $\mathbf{g}(t)$ of degree $\leq n-1$ representing a class of polynomials from L_{n-1} by the formula*

$$\left((A_0 - z)^{-1}\mathbf{g}\right)(t) = \frac{1}{t-z}\left(\mathbf{g}(t) - \mathbf{D}(t)\mathbf{D}(z)^{-1}\mathbf{g}(z)\right). \tag{3.5}$$

Proof. In the completely degenerate case each vector of L_n can be represented as a vector polynomial of degree $n-1$. Correspondingly A_0 acts on such vector polynomials as the multiplication operator by t. Let us consider the equation

$$(t-z)\,\mathbf{f}(t)\Big(= ((A_0 - z)\mathbf{f})(t)\Big) = \mathbf{g}(t), \ z \notin \sigma(\mathbf{D}), \tag{3.6}$$

and assume that the degree of the representing polynomial $\mathbf{g}(t)$ is $\leq n$ and that the degree of the sought polynomial $\mathbf{f}(t)$ is $\leq n-1$. If $\mathbf{g}(z)=0$ then we put

$$\mathbf{f}(t)=(t-z)^{-1}\,\mathbf{g}(t).$$

Otherwise we can substitute on the right-hand side of the equation (3.6) instead of $\mathbf{g}(t)$ the equivalent polynomial

$$\widetilde{\mathbf{g}}(t)=\mathbf{g}(t)-\mathbf{D}(t)\mathbf{D}(z)^{-1}\mathbf{g}(z),$$

and get as a solution of (3.6) the vector polynomial

$$\mathbf{f}(t)=\frac{1}{(t-z)}\left(\mathbf{g}(t)-\mathbf{D}(t)\mathbf{D}(z)^{-1}\mathbf{g}(z)\right),$$

the degree of which is, evidently, $\leq n-1$. Hence for $z\notin\sigma(\mathbf{D})$ and any vector polynomial $\mathbf{g}(\cdot)\in \mathrm{L}_n$ the equation (3.6) has a solution in the form of a vector polynomial from L_{n-1}. Therefore $\mathbb{C}\setminus\sigma(\mathbf{D})$ belongs to the resolvent set of A_0 and $(A_0-z)^{-1}$ for $z\in\mathbb{C}\setminus\sigma(\mathbf{D})$ is given by (3.5).

On the other hand, if $z\in\sigma(\mathbf{D})$, then there is a non-zero vector $\mathbf{h}\in\mathbb{C}_s$ such that $\mathbf{D}(z)\mathbf{h}=0$. Note that by (3.4)

$$\mathbf{D}(t)\mathbf{h}\underset{t\to\infty}{=}t^n\mathbf{h}+o(t^n)$$

and thus $\mathbf{D}(t)\mathbf{h}\neq 0$ identically. If $(t-z)^{-l+1}\mathbf{D}(t)\mathbf{h},\ 1\leq l\leq n$, belongs to the zero class in L_n, but $\mathbf{f}_0(t):=(t-z)^{-l}\mathbf{D}(t)\mathbf{h}$ is a non-zero element from L_n, then $\mathbf{f}_0(t)$ is a non-trivial solution of the homogeneous equation $(A_0-z)\,\mathbf{f}=0$. □

Remark 3.2. Since A_0 is a non-negative operator, we have $\sigma(\mathbf{D})\subset[0,\infty)$.

Let $\boldsymbol{\sigma}_0(t)$ be the (unique) solution of the truncated Stieltjes matrix moment problem and let

$$\mathbf{E}(z)=\int_0^\infty\frac{1}{t-z}d\boldsymbol{\sigma}_0(t)\left(\mathbf{D}(t)-\mathbf{D}(z)\right). \tag{3.7}$$

Note that if

$$\mathbf{D}(z)=z^n\mathbf{I}+z^{n-1}A_{n-1}+\cdots+zA_1+A_0,$$

then

$$\mathbf{E}(z)=\sum_0^{n-1}z^k\mathbf{C}_{n-1-k}+\sum_0^{n-2}z^k\mathbf{C}_{n-2-k}A_{n-1}+\cdots+\mathbf{C}_0A_1. \tag{3.8}$$

It stems from formula (3.5) applied to the classes which contain the zero-degree polynomials and the expression (3.1) that the following assertion is true.

Proposition 3.3. *In the completely degenerate case the unique solution $\sigma_0(t)$ of the truncated Stieltjes matrix moment problem satisfies the relation*

$$\int_0^\infty\frac{1}{t-z}d\boldsymbol{\sigma}_0(t)=-\mathbf{E}(z)\mathbf{D}(z)^{-1}. \tag{3.9}$$

The point of Proposition 3.3 is that it allows one to calculate "explicitly" the unique solution $\boldsymbol{\sigma}_0(t)$ of the truncated Stieltjes moment problem in the completely degenerate case. To this end it is enough to find, following the above prescription, the matrix polynomial $\mathbf{D}(z)$, take $\mathbf{E}(z)$ from (3.8) and apply the Stieltjes inversion formula to (3.9). By this formula for each $\mathbf{h} \in \mathbb{C}_s$ and $\beta > \alpha$ such that $\det \mathbf{D}(\alpha) \neq 0$, $\det \mathbf{D}(\beta) \neq 0$, we have

$$(\boldsymbol{\sigma}_0(\beta)\mathbf{h}, \mathbf{h}) - (\boldsymbol{\sigma}_0(\alpha)\mathbf{h}, \mathbf{h}) = -\frac{1}{\pi} \lim_{\eta \downarrow 0} \int_{\alpha}^{\beta} \operatorname{Im} \left(\mathbf{E}(t+i\eta)\mathbf{D}(t+i\eta)^{-1}\mathbf{h}, \mathbf{h} \right) dt. \quad (3.10)$$

4. Canonical solutions

Now we will omit the assumption that the projection $P_n\mathcal{M}_n$ of the null-space $\mathcal{M}_n$ onto the subspace $\mathcal{N}_n$ necessarily covers all this subspace.

Theorem 4.1. *Let $\mathbf{C}_0, \mathbf{C}_1, \mathbf{C}_2, \dots, \mathbf{C}_{2n}$ be any set of Hermitian $s \times s$ matrices which satisfy the conditions of Theorem 1.1 and let $\boldsymbol{\sigma}(t)$ be any solution of the truncated Stieltjes matrix moment problem for this set. Then the extended set of non-negative Hermitian matrices*

$$\mathbf{C}_0, \dots, \mathbf{C}_{2n},\ \mathbf{C}'_{2n+1} = \int_0^{\infty} t^{2n+1}\, d\boldsymbol{\sigma}(t)\,,\ \mathbf{C}'_{2n+2} = \int_0^{\infty} t^{2n+2}\, d\boldsymbol{\sigma}(t) \quad (4.1)$$

satisfies all conditions of Theorem 1.1 with n replaced by $n+1$ and for the extended set (4.1) the truncated Stieltjes moment problem is completely degenerate if and only if $\boldsymbol{\sigma}(t)$ is a canonical solution.

Proof. Since the Stieltjes problem for the set (4.1) is evidently solvable, this set satisfies the conditions of Theorem 1.1. Suppose that $\boldsymbol{\sigma}(t)$ is a canonical solution of the Stieltjes problem for the set $\mathbf{C}_0, \dots, \mathbf{C}_{2n}$. Recall that the solvability of the Stieltjes problem for this set guarantees the existence of canonical $\boldsymbol{\sigma}(t)$ [2]. By the definition of canonical solutions for $\boldsymbol{\sigma}(t)$ we have $\mathrm{L}_{n+1}=\mathrm{L}_n=\mathrm{L}^2_{\boldsymbol{\sigma}}(\mathbb{C}_s)$. Therefore for any non-zero vector $\mathbf{h} \in \mathbb{C}_s$ there is a vector polynomial

$$\mathbf{P}_{\mathbf{h}}(t) = t^n \xi_{\mathbf{h}n} + \dots + \xi_{\mathbf{h}0}$$

such that the vector polynomials $t^{n+1}\mathbf{h}$ and $\mathbf{P}_{\mathbf{h}}(t)$ belong to the same class or, more precisely, the polynomial $t^{n+1}\mathbf{h} - \mathbf{P}_{\mathbf{h}}(t)$ belongs to the zero-class in $\mathrm{L}^2_{\boldsymbol{\sigma}}(\mathbb{C}_s)$. Since the block Hankel matrix $\boldsymbol{\Gamma}_{n+1} = (\mathbf{C}_{j+k})_{j,k=0}^{n+1}$ is non-negative definite and

$$(\mathbf{C}_{2n+2}\mathbf{h}, \mathbf{h}) - (\mathbf{C}_{2n+1}\mathbf{h}, \xi_{\mathbf{h}n}) - (\mathbf{C}_{2n+1}\xi_{\mathbf{h}n}, \mathbf{h}) - \dots + (\mathbf{C}_{2n+1}\xi_{\mathbf{h}0}, \xi_{\mathbf{h}0}) =$$
$$\int_0^{\infty} \left(d\boldsymbol{\sigma}(t) \left[t^{n+1}\mathbf{h} - \mathbf{P}_{\mathbf{h}}(t)\right], \left[t^{n+1}\mathbf{h} - \mathbf{P}_{\mathbf{h}}(t)\right]\right) = 0,$$

we see that the vector

$$\begin{pmatrix} -\xi_{\mathbf{h}0} \\ \vdots \\ -\xi_{\mathbf{h}n} \\ \mathbf{h} \end{pmatrix} \quad (4.2)$$

belongs to the null-space of $\mathbf{\Gamma}_{n+1}$. As this is true for any $\mathbf{h} \in \mathbb{C}_s$ we conclude that for the set (4.1) the Stieltjes problem is completely degenerate.

Let us assume now that the solution $\boldsymbol{\sigma}(t)$ is such that for the extended set (4.1) the Stieltjes problem is completely degenerate. Then for each vector $\mathbf{h} \in \mathbb{C}_s$ there is a vector of the form (4.2) from the null-space of $\mathbf{\Gamma}_{n+1}$ or, equivalently, there is a vector polynomial

$$t^{n+1}\mathbf{h} - t^n \xi_{\mathbf{h}n} - \cdots - \xi_{\mathbf{h}0}$$

from the zero-class in $\mathrm{L}^2_{\boldsymbol{\sigma}}(\mathbb{C}_s)$. Therefore, for the given solution $\boldsymbol{\sigma}(t)$ we have $\mathrm{L}_{n+1} = \mathrm{L}_n = \mathrm{L}^2_{\boldsymbol{\sigma}}(\mathbb{C}_s)$ This implies that $\boldsymbol{\sigma}(t)$ is a canonical solution. □

Theorem 4.1 gives us a hint for the description of all canonical solutions of the Stieltjes moment problem. To this end it is enough first to find all those extensions $\mathbf{C}_0, \mathbf{C}_1, \ldots, \mathbf{C}_{2n}, \mathbf{C}_{2n+1}, \mathbf{C}_{2n+2}$ of the given set of moments $\mathbf{C}_0, \mathbf{C}_1, \ldots, \mathbf{C}_{2n}$, for which corresponding Stieltjes problems turn out to be completely degenerate, and then to find unique solutions of the arising completely degenerate problems as it has been done in the previous section.

We start the description of the demanded extensions of a given moment set with the following general proposition.

Theorem 4.2. *Let A be a bounded non-negative self-adjoint operator in a Hilbert space $\mathcal{H}_1$ and B be a bounded operator from $\mathcal{H}_1$ into a Hilbert space $\mathcal{H}_2$. Put $Q(s) = B(A+s)^{-1}B^*$, $0 < s < \infty$. The following conditions are equivalent:*

- *there are bounded non-negative self-adjoint extensions of the operator $A+B : \mathcal{H}_1 \to \mathcal{H}_1 \oplus \mathcal{H}_2$ to the Hilbert space $\mathcal{H} = \mathcal{H}_1 \oplus \mathcal{H}_2$;*
- *the operator Q_0 in $\mathcal{H}_2$ defined in a way that*

$$Q_0 f = \lim_{s \to 0+} Q(s) f, \quad f \in \mathcal{H}_2,$$

is bounded.

If these conditions hold, then bounded non-negative self-adjoint extensions $\widetilde{A}$ of $(A+B)_{|\mathcal{H}_1}$ have the form of block operator matrices

$$\widetilde{A} = \begin{pmatrix} A & B^* \\ B & D \end{pmatrix}, \tag{4.3}$$

where D runs through the set of non-negative self-adjoint operators in $\mathcal{H}_2$ and satisfies the condition

$$D - Q_0 \geq 0. \tag{4.4}$$

Proof. Since the operators A and B are bounded, we have that each bounded self-adjoint extension of $(A+B)_{|\mathcal{H}_1}$ to $\mathcal{H}$ can be represented in the form (4.3) with a bounded self-adjoint D. Note further that an extension $\widetilde{A}$ is non-negative if and only if for every $\lambda < 0$ the operator $\widetilde{A} - \lambda$ is strictly positive. By the

Schur-Frobenius factorization formula we have

$$\begin{aligned}\widetilde{A}-\lambda &= \begin{pmatrix} I & 0 \\ B(A-\lambda)^{-1} & I \end{pmatrix} \\ &\times \begin{pmatrix} A-\lambda & 0 \\ 0 & D-\lambda-B\,(A-\lambda)^{-1}\,B^* \end{pmatrix} \begin{pmatrix} I & (A-\lambda)^{-1}B^* \\ 0 & I \end{pmatrix}, \ \lambda<0. \end{aligned} \tag{4.5}$$

Therefore the extension $\widetilde{A}$ is non-negative if and only if

$$W(\lambda) := D-\lambda-B\,(A-\lambda)^{-1}\,B^* > 0,\ \lambda<0.$$

Let the operator Q_0 exist and be bounded. Since the operator function $W(\lambda)$ is non-increasing on the half-axis $(-\infty,0)$, the operator $\widetilde{A}$ is non-negative if and only if the inequality (4.4) holds.

Suppose now that for some $f\in\mathcal{H}_2$ we have $\lim_{s\to 0+}(Q(s)f,f)=+\infty$. As D is a bounded operator we conclude that for s small enough and the same $f\in\mathcal{H}_2$ the inequality $(W(-s)f,f)<0$ holds. Therefore there are no non-negative operators among bounded self-adjoint extensions of $(A+B)_{|\mathcal{H}_1}$. □

Remark 4.3. Let the operators A and B be as in Theorem 4.2 and the space $\mathcal{H}_1$ be finite-dimensional. Then among bounded self-adjoint extensions of $(A+B)_{|\mathcal{H}_1}$ to the Hilbert space $\mathcal{H}=\mathcal{H}_1\oplus\mathcal{H}_2$ there are non-negative operators if and only if the null-space of A is contained in the null-space of B.

Remark 4.4. Let us assume that the operator A in Theorem 4.2 has a bounded inverse. Then the condition $D\geq BA^{-1}B^*$ for a bounded self-adjoint D in (4.3) is necessary and sufficient for the non-negativity of $\widetilde{A}$ given by the block operator matrix (4.3).

Proposition 4.5. *Let operators $A:\mathcal{H}_1\to\mathcal{H}_1$ and $B:\mathcal{H}_1\to\mathcal{H}_2$ be as in Theorem 4.2 and assume that A has a bounded inverse. Then the operator $\widetilde{A}_0$ in the Hilbert space $\mathcal{H}=\mathcal{H}_1\oplus\mathcal{H}_2$ given as the block operator matrix*

$$\widetilde{A}_0 = \begin{pmatrix} A & B^* \\ B & B^*A^{-1}B \end{pmatrix} \tag{4.6}$$

is the unique non-negative self-adjoint extension of $(A+B)_{|\mathcal{H}_1}$ to $\mathcal{H}$ which satisfies the condition:

$$P_2\mathcal{M}_{\widetilde{A}_0} = \{0\}\oplus\mathcal{H}_2, \tag{4.7}$$

where P_2 is the orthogonal projector in $\mathcal{H}$ from $\mathcal{H}$ to $\{0\}\oplus\mathcal{H}_2$ and $\mathcal{M}_{\widetilde{A}_0}$ is the null-space of $\widetilde{A}_0$.

Proof. By assumptions of the proposition and Remark 4.4 the extension (4.6) is bounded, self-adjoint and non-negative with $D=BA^{-1}B^*$. Let us take an arbitrary vector $h\in\mathcal{H}_2$ and consider the vector

$$\widetilde{h} = \begin{pmatrix} -A^{-1}B^*h \\ h \end{pmatrix} \in \mathcal{H}.$$

It is evident that $\widetilde{A}_0\widetilde{h}=0$. Hence, (4.7) holds for the extension (4.6).

Now, let $\widetilde{A}$ be any bounded non-negative self-adjoint extension which satisfies the condition (4.7). Then for any $h \in \mathcal{H}_2$ there is a vector $h_1 \in \mathcal{H}_1$ such that $\widetilde{h} = h_1 + h \in \mathcal{H}$ is a null-vector of $\widetilde{A}$, that is

$$Ah_1 + B^*h = 0, \ Bh_1 + Dh = 0. \tag{4.8}$$

As A is invertible we deduce from (4.8) that the equality $\left(-BA^{-1}B^* + D\right)h = 0$ is true for any $h \in \mathcal{H}_2$. Hence $D = BA^{-1}B^*$. □

Having disposed of these preliminary steps, we can now return to the description of all canonical extensions of the truncated Stieltjes moment problem. Let $\mathbf{C}_0, \mathbf{C}_1, \mathbf{C}_2, \dots, \mathbf{C}_{2n}$ be any sequence of Hermitian $s \times s$ matrices, which satisfy all the conditions of Theorem 1.1.

We begin with the non-degenerate case and assume that the non-negative block Hankel matrix $\mathbf{\Gamma}_n = (\mathbf{C}_{j+k})_{j,k=0}^{n}$ is invertible. This implies that the non-negative block Hankel matrix $\mathbf{\Gamma}_{n-1}^{(1)} = (\mathbf{C}_{j+k+1})_{j,k=0}^{n-1}$ is invertible as well. Indeed, if $\mathbf{\Gamma}_{n-1}^{(1)}$ is non-invertible, then by the conditions of Theorem 1.1 the matrix $\mathbf{\Gamma}_{n-1}^{(2)} = (\mathbf{C}_{j+k+2})_{j,k=0}^{n-1}$ is also non-invertible. But $\mathbf{\Gamma}_{n-1}^{(2)}$ is a diagonal block of the positive definite matrix $\mathbf{\Gamma}_n$, a contradiction. Put

$$\mathbf{S}_1 = \begin{pmatrix} \mathbf{C}_{n+1} \\ \vdots \\ \mathbf{C}_{2n} \end{pmatrix}.$$

By Theorem 4.2 and Remark 4.4 for any non-negative definite $s \times s$ matrix $\mathbf{W}$ the $n \times n$ block Hankel matrix

$$\mathbf{\Gamma}_n^{(1)} = \begin{pmatrix} \mathbf{\Gamma}_{n-1}^{(1)} & \mathbf{S}_1 \\ \mathbf{S}_1^* & \mathbf{W} + \mathbf{S}_1^* \left[\mathbf{\Gamma}_{n-1}^{(1)}\right]^{-1} \mathbf{S}_1 \end{pmatrix} \tag{4.9}$$

is a non-negative definite $n \times n$ block Hankel extension of the block matrix $\mathbf{\Gamma}_{n-1}^{(1)}$.

Let

$$\mathbf{S} = \begin{pmatrix} \mathbf{C}_{n+1} \\ \vdots \\ \mathbf{C}_{2n} \\ \mathbf{W} + \mathbf{S}_1^* \left[\mathbf{\Gamma}_{n-1}^{(1)}\right]^{-1} \mathbf{S}_1 \end{pmatrix}.$$

By taking $\mathbf{W} + \mathbf{S}_1^* \left[\mathbf{\Gamma}_{n-1}^{(1)}\right]^{-1} \mathbf{S}_1$ as $\mathbf{C}_{2n+1}$ and the matrix $\mathbf{S}^*\mathbf{\Gamma}_n^{-1}\mathbf{S}$ as $\mathbf{C}_{2n+2}$, we construct the $(n+2) \times (n+2)$ block Hankel matrix

$$\mathbf{\Gamma}_{n+1} = (\mathbf{C}_{j+k})_{j,k=0}^{n+1} = \begin{pmatrix} \mathbf{\Gamma}_n & \mathbf{S} \\ \mathbf{S}^* & \mathbf{S}^*\mathbf{\Gamma}_n^{-1}\mathbf{S} \end{pmatrix}. \tag{4.10}$$

From Proposition 4.5 we can conclude that the matrix (4.10) is the unique non-negative definite extension of $\mathbf{\Gamma}_n$ for which the condition

$$P_{n+1}\mathcal{M}_{n+1} = \mathcal{N}_{n+1}$$

holds; as before $\mathcal{M}_{n+1}$ is the null-space of $\mathbf{\Gamma}_{n+1}$ and P_{n+1} is the orthogonal projector in the space $\mathbb{L}_{n+1}$ of column vectors

$$\boldsymbol{\xi} = \begin{pmatrix} \boldsymbol{\xi}_0 \\ \vdots \\ \boldsymbol{\xi}_n \\ \boldsymbol{\xi}_{n+1} \end{pmatrix}, \ \boldsymbol{\xi}_0, \dots, \boldsymbol{\xi}_{n+1} \in \mathbb{C}_s,$$

onto the subspace $\mathcal{N}_{n+1} = \mathbb{L}_{n+1} \ominus \mathbb{L}_n$ of column vectors with $\boldsymbol{\xi}_0 = \cdots = \boldsymbol{\xi}_n = 0$.

Proposition 4.6. *For the set of Hermitian $s \times s$ matrices $\mathbf{C}_0, \mathbf{C}_1, \dots, \mathbf{C}_{2n}$ which satisfy the conditions of Theorem* 1.1 *and such that the block Hankel matrix $\mathbf{\Gamma}_n = (\mathbf{C}_{j+k})_{j,k=0}^n$ is invertible, all extensions $\mathbf{C}_{2n+1}$, $\mathbf{C}_{2n+2}$ that generate completely degenerate truncated Stieltjes matrix moment problems are described by formulas*

$$\mathbf{C}_{2n+1} = \mathbf{W} + (\mathbf{C}_{n+1}, \dots, \mathbf{C}_{2n}) \left[\mathbf{\Gamma}_{n-1}^{(1)}\right]^{-1} \begin{pmatrix} \mathbf{C}_{n+1} \\ \vdots \\ \mathbf{C}_{2n} \end{pmatrix}, \tag{4.11}$$

$$\mathbf{C}_{2n+2} = (\mathbf{C}_{n+1}, \dots, \mathbf{C}_{2n+1})\, \mathbf{\Gamma}_n^{-1} \begin{pmatrix} \mathbf{C}_{n+1} \\ \vdots \\ \mathbf{C}_{2n+1} \end{pmatrix}, \tag{4.12}$$

where the "parameter" $\mathbf{W}$ runs through the set of all non-negative definite $s \times s$ matrices.

Proof. It remains to prove that (4.11) with (4.12) provides all demanded extensions. Let now $\mathbf{C}_{2n+1}$, $\mathbf{C}_{2n+2}$ be an extension generating a completely degenerate Stieltjes problem. By Theorem 1.1 this means, in particular, that the extended Hankel matrix $\mathbf{\Gamma}_n^{(1)}$ should be non-negative definite. According to Remark 4.4 this implies the expression (4.11) for $\mathbf{C}_{2n+1}$ with some non-negative definite matrix parameter $\mathbf{W}$. By Proposition 4.5, $\mathbf{C}_{2n+2}$ for a certain appropriate $\mathbf{C}_{2n+1}$ cannot differ from that given by expression (4.11). □

From now on we assume that the block Hankel matrix $\mathbf{\Gamma}_n$ is not invertible. The problem of describing all extensions $\mathbf{C}_{2n+1}$, $\mathbf{C}_{2n+2}$ which transform the given Stieltjes $(2n+1)$ moment problem into completely degenerate $(2n+3)$ ones, can be handled in the following way. We can take any sequence of Hermitian $s \times s$ matrices $\mathbf{G}_0, \mathbf{G}_1, \mathbf{G}_2, \dots, \mathbf{G}_{2n}$ such that the block Hankel matrices $\mathbf{\Delta}_n := (\mathbf{G}_{k+j})_{k,j=0}^n$ and $\mathbf{\Delta}_{n-1}^{(1)} := (\mathbf{G}_{k+j+1})_{k,j=0}^{n-1}$ are invertible and positive definite. Then for any $\varepsilon > 0$ the perturbed Hankel matrices

$$\mathbf{\Gamma}_n(\varepsilon) := \mathbf{\Gamma}_n + \varepsilon \mathbf{\Delta}_n, \ \mathbf{\Gamma}_{n-1}^{(1)}(\varepsilon) := \mathbf{\Gamma}_{n-1}^{(1)} + \varepsilon \mathbf{\Delta}_{n-1}^{(1)} \tag{4.13}$$

are invertible and positive definite. Taking for the set of moments $\mathbf{C}_0(\varepsilon) := \mathbf{C}_0 + \varepsilon\mathbf{G}_0, \dots, \mathbf{C}_{2n}(\varepsilon) := \mathbf{C}_{2n} + \varepsilon\mathbf{G}_{2n}$ the solutions (4.11), (4.12) of the extension problem under consideration, we can further obtain the demanded solutions by passing to the limit $\varepsilon \downarrow 0$.

To this end let us first show that under the conditions of Theorem 1.1 there exists a finite limit of non-negative matrices

$$\mathbf{S}_1^*(\varepsilon)\left[\mathbf{\Gamma}_{n-1}^{(1)}(\varepsilon)\right]^{-1}\mathbf{S}_1(\varepsilon), \qquad \left(\mathbf{S}_1(\varepsilon) = \begin{pmatrix} \mathbf{C}_{n+1}(\varepsilon) \\ \vdots \\ \mathbf{C}_{2n}(\varepsilon) \end{pmatrix}\right)$$

as $\varepsilon \downarrow 0$.

We denote by $P_1(\mathbf{\Gamma}_{n-1}^{(1)})$ the orthogonal projector onto the null-space $\mathcal{M}_{\mathbf{\Gamma}_{n-1}^{(1)}} \subset \mathbb{L}_{n-1}(\mathbb{C}_s)$ of the non-negative block Hankel matrix $\mathbf{\Gamma}_{n-1}^{(1)} := (\mathbf{C}_{k+j+1})_{k,j=0}^{n-1}$ and set $P_2(\mathbf{\Gamma}_{n-1}^{(1)}) = I - P_1(\mathbf{\Gamma}_{n-1}^{(1)})$. With respect to the representation

$$\mathbb{L}_{n-1}(\mathbb{C}_s) = \mathcal{M}_{\mathbf{\Gamma}_{n-1}^{(1)}}^{\perp} \oplus \mathcal{M}_{\mathbf{\Gamma}_{n-1}^{(1)}}$$

and by the definition of $\mathcal{M}_{\mathbf{\Gamma}_{n-1}^{(1)}}$ and (4.13) we can write

$$\mathbf{\Gamma}_{n-1}^{(1)}(\varepsilon) = \begin{pmatrix} \varepsilon\mathbf{\Delta}_{n-1,11}^{(1)} & \varepsilon\mathbf{\Delta}_{n-1,12}^{(1)} \\ \varepsilon\mathbf{\Delta}_{n-1,21}^{(1)} & \mathbf{\Gamma}_{n-1,22}^{(1)} + \varepsilon\mathbf{\Delta}_{n-1,22}^{(1)} \end{pmatrix},$$

$$\mathbf{\Gamma}_{n-1,ij}^{(1)} + \varepsilon\mathbf{\Delta}_{n-1,ij}^{(1)} = P_i(\mathbf{\Gamma}_{n-1}^{(1)})\mathbf{\Gamma}_{n-1}^{(1)}(\varepsilon)\,|_{P_j(\mathbf{\Gamma}_{n-1}^{(1)})\mathbb{L}_{n-1}(\mathbb{C}_s)}, \quad i,j = 1,2.$$

Put

$$\mathbf{\Upsilon}_{n-1}^{(1)} = \lim_{\lambda\uparrow 0} P_2\left(\mathbf{\Gamma}_{n-1}^{(1)}\right)\left(\mathbf{\Gamma}_{n-1}^{(1)} - \lambda\right)^{-1}|_{\mathcal{M}_{\mathbf{\Gamma}_{n-1}^{(1)}}^{\perp}}.$$

Note that the matrices $\mathbf{\Upsilon}_{n-1}^{(1)}$ and $\mathbf{\Delta}_{n-1,11}^{(1)}$ are invertible. The Schur-Frobenius factorization formula gives:

$$\mathbf{\Gamma}_{n-1}^{(1)}(\varepsilon)^{-1} = \frac{1}{\varepsilon}\begin{pmatrix} \left(\mathbf{\Delta}_{n-1,11}^{(1)}\right)^{-1} & 0 \\ 0 & 0 \end{pmatrix} +$$
$$\begin{pmatrix} \left(\mathbf{\Delta}_{n-1,11}^{(1)}\right)^{-1}\mathbf{\Delta}_{n-1,12}^{(1)}\mathbf{\Upsilon}_{n-1}^{(1)}\mathbf{\Delta}_{n-1,21}^{(1)}\left(\mathbf{\Delta}_{n-1,11}^{(1)}\right)^{-1} & -\left(\mathbf{\Delta}_{n-1,11}^{(1)}\right)^{-1}\mathbf{\Delta}_{n-1,12}^{(1)}\mathbf{\Upsilon}_{n-1}^{(1)} \\ \mathbf{\Upsilon}_{n-1}^{(1)}\mathbf{\Delta}_{n-1,21}^{(1)}\left(\mathbf{\Delta}_{n-1,11}^{(1)}\right)^{-1} & \mathbf{\Upsilon}_{n-1}^{(1)} \end{pmatrix}$$
$$+O(\varepsilon). \tag{4.14}$$

Note further that for any $\mathbf{h} \in \mathbb{C}_s$ we have

$$\left(\mathbf{S}_1^*(\varepsilon)\left[\mathbf{\Gamma}_{n-1}^{(1)}(\varepsilon)\right]^{-1}\mathbf{S}_1(\varepsilon)\mathbf{h},\mathbf{h}\right) = \left(\left[\mathbf{\Gamma}_{n-1}^{(1)}(\varepsilon)\right]^{-1}\mathbf{k}(\varepsilon),\mathbf{k}(\varepsilon)\right),$$

$$\mathbf{k}(\varepsilon) = \mathbf{\Gamma}_{n-1}^{(2)}(\varepsilon)\begin{pmatrix}0\\ \vdots\\ 0\\ \mathbf{h}\end{pmatrix}, \quad \mathbf{\Gamma}_{n-1}^{(2)}(\varepsilon) = (\mathbf{C}_{k+j+2} + \varepsilon\mathbf{G}_{k+j+2})_{k,j=0}^{n-1}.$$

Recall that by the assumptions of Theorem 1.1, the null space of $\mathbf{\Gamma}_{n-1}^{(1)}$ belongs to the null-space of

$$\mathbf{\Gamma}_{n-1}^{(2)}\left(=\mathbf{\Gamma}_{n-1}^{(2)}(0)\right).$$

Therefore taking into account (4.14), we get

$$\lim_{\varepsilon\downarrow 0}\mathbf{S}_1^*(\varepsilon)\left[\mathbf{\Gamma}_{n-1}^{(1)}(\varepsilon)\right]^{-1}\mathbf{S}_1(\varepsilon) = \mathbf{S}_1^*(0)P_2\left(\mathbf{\Gamma}_{n-1}^{(1)}\right)\mathbf{\Upsilon}_{n-1}^{(1)}P_2\left(\mathbf{\Gamma}_{n-1}^{(1)}\right)\mathbf{S}_1(0).$$

Note that by Proposition 4.6 for $\varepsilon > 0$ a self-adjoint Hankel extension $\mathbf{\Gamma}_n^{(1)}(\varepsilon)$ of $\mathbf{\Gamma}_{n-1}^{(1)}(\varepsilon)$ is non-negative definite if and only if

$$\mathbf{C}_{2n+1}(\varepsilon) = \mathbf{W} + \mathbf{S}_1^*(\varepsilon)\left[\mathbf{\Gamma}_{n-1}^{(1)}(\varepsilon)\right]^{-1}\mathbf{S}_1(\varepsilon),$$

where $\mathbf{W}$ is any non-negative definite $s \times s$ matrix. As follows, for each fixed non-negative definite $\mathbf{W}$ the limit Hankel matrix $\mathbf{\Gamma}_n^{(1)(0)}$ with the limit

$$\mathbf{C}_{2n+1}(0) = \mathbf{W} + \mathbf{S}_1^*(0)P_2\left(\mathbf{\Gamma}_{n-1}^{(1)}\right)\mathbf{\Upsilon}_{n-1}^{(1)}P_2\left(\mathbf{\Gamma}_{n-1}^{(1)}\right)\mathbf{S}_1(0) \tag{4.15}$$

is non-negative definite. On the other hand, as

$$\mathbf{S}_1^*(0)P_2\left(\mathbf{\Gamma}_{n-1}^{(1)}\right)\mathbf{\Upsilon}_{n-1}^{(1)}P_2\left(\mathbf{\Gamma}_{n-1}^{(1)}\right)\mathbf{S}_1(0) = \lim_{\lambda\uparrow 0}\mathbf{S}_1^*(0)\left[\mathbf{\Gamma}_{n-1}^{(1)} - \lambda\right]^{-1}\mathbf{S}_1(0),$$

it stems from Proposition 4.6 that only extensions $\mathbf{C}_{2n+1}$ given by (4.15) with $\mathbf{W} \geq 0$ provide the non-negativity of $\mathbf{\Gamma}_n^{(1)}$.

To find now the set of all appropriate extensions $\mathbf{C}_{2n+2}$ it remains to select among the non-negative definite $s \times s$ matrices $\mathbf{W}$ those for which there exist finite limits of the matrices

$$\mathbf{S}^*(\varepsilon)\mathbf{\Gamma}_n(\varepsilon)^{-1}\mathbf{S}(\varepsilon)$$

as $\varepsilon \downarrow 0$. To this end note that taking any $\mathbf{h} \in \mathbb{C}_s$ and setting

$$\mathbf{k} = \begin{pmatrix}0\\ \vdots\\ 0\\ \mathbf{h}\end{pmatrix}, \quad \begin{pmatrix}\mathbf{C}_{n+1}(\varepsilon)\mathbf{h}\\ \vdots\\ \mathbf{C}_{2n}(\varepsilon)\mathbf{h}\\ \left(\mathbf{W} + \mathbf{S}_1^*(\varepsilon)\left[\mathbf{\Gamma}_{n-1}^{(1)}(\varepsilon)\right]^{-1}\mathbf{S}_1(\varepsilon)\right)\mathbf{h}\end{pmatrix} = \mathbf{\Gamma}_n^{(1)}(\varepsilon)\mathbf{k},$$

we can write

$$\left(\mathbf{S}^*(\varepsilon)\mathbf{\Gamma}_n(\varepsilon)^{-1}\mathbf{S}(\varepsilon)\mathbf{h},\mathbf{h}\right)=\left(\mathbf{\Gamma}_n(\varepsilon)^{-1}\mathbf{\Gamma}_n^{(1)}(\varepsilon)\mathbf{k},\mathbf{\Gamma}_n^{(1)}(\varepsilon)\mathbf{k}\right). \tag{4.16}$$

Let $P_1(\mathbf{\Gamma}_n)$ be the orthogonal projector onto the null-space $\mathcal{M}_{\mathbf{\Gamma}_n}\subset\mathbb{L}_n(\mathbb{C}_s)$ of the non-negative block Hankel matrix $\mathbf{\Gamma}_n:=(\mathbf{C}_{k+j})_{k,j=0}^n$, $P_2(\mathbf{\Gamma}_n)=I-P_1(\mathbf{\Gamma}_n)$ and let

$$\mathbf{\Upsilon}_n=\lim_{\lambda\uparrow 0}P_2(\mathbf{\Gamma}_n)\left(\mathbf{\Gamma}_n-\lambda\right)^{-1}|_{\mathcal{M}_{\mathbf{\Gamma}_n}^{\perp}}.$$

Note that for any $\boldsymbol{\xi}\in\mathcal{M}_{\mathbf{\Gamma}_n}$ we have

$$\left(\mathbf{\Gamma}_n^{(1)}(0)\boldsymbol{\xi}\right)_j=\sum_{k=0}^{n-1}\mathbf{C}_{j+1+k}\xi_k=\left(\mathbf{\Gamma}_n\boldsymbol{\xi}\right)_{j+1}=0,\ 0\le j\le n-1. \tag{4.17}$$

Let P_n' be the orthogonal projector onto the subspace $P_n\mathcal{M}_n$. Taking into account that the matrices $\mathbf{\Gamma}_n^{(1)}(0)$, $\mathbf{\Gamma}_n$ are Hermitian, we deduce from (4.17) that

$$\mathbf{\Gamma}_n^{(1)}(0)P_1(\mathbf{\Gamma}_n)\mathbf{\Gamma}_n^{(1)}(0)=\mathbf{\Gamma}_n^{(1)}(0)P_1(\mathbf{\Gamma}_n)P_n'P_1(\mathbf{\Gamma}_n)\,\mathbf{\Gamma}_n^{(1)}(0). \tag{4.18}$$

Applying the same arguments as above to (4.16) and by virtue of (4.18), we obtain that

$$\begin{aligned}\left(\mathbf{S}^*(\varepsilon)\mathbf{\Gamma}_n(\varepsilon)^{-1}\mathbf{S}(\varepsilon)h,h\right)&\underset{\varepsilon\downarrow 0}{=}\tfrac{1}{\varepsilon}\left(P_n'P_1(\mathbf{\Gamma}_n)\mathbf{\Gamma}_n^{(1)}(0)h,P_n'\mathbf{\Gamma}_n^{(1)}(0)h\right)\\&+\left(\mathbf{\Upsilon}_nP_2(\mathbf{\Gamma}_n)\mathbf{\Gamma}_n^{(1)}(0)\mathbf{h},P_2(\mathbf{\Gamma}_n)\mathbf{\Gamma}_n^{(1)}(0)\mathbf{h}\right)+O(\varepsilon).\end{aligned} \tag{4.19}$$

Hence, the condition

$$P_n'P_1(\mathbf{\Gamma}_n)\mathbf{\Gamma}_n^{(1)}(0)P_n=0, \tag{4.20}$$

or, equivalently,

$$P_n\mathbf{\Gamma}_n^{(1)}(0)P_1(\mathbf{\Gamma}_n)P_n'=0,$$

is necessary and sufficient for the existence of a finite limit of the matrix function $\mathbf{S}^*(\varepsilon)\mathbf{\Gamma}_n(\varepsilon)^{-1}\mathbf{S}(\varepsilon)$ as $\varepsilon\downarrow 0$.

Remark 4.7. The condition (4.20) as well as the condition

$$P_n'\mathbf{\Gamma}_n^{(2)}(0)P_1(\mathbf{\Gamma}_n)P_n=0\ \left(P_n'P_1(\mathbf{\Gamma}_n)\mathbf{\Gamma}_n^{(2)}(0)P_n'=0\right) \tag{4.21}$$

are necessary for an extended set of matrices

$$\mathbf{C}_0,\dots,\mathbf{C}_{2n},\ \mathbf{W}+\mathbf{S}_1^*(0)P_2\left(\mathbf{\Gamma}_{n-1}^{(1)}\right)\mathbf{\Upsilon}_{n-1}^{(1)}P_2\left(\mathbf{\Gamma}_{n-1}^{(1)}\right)\mathbf{S}_1(0),\ \mathbf{C}_{2n+2}$$

to be, in the degenerate case, the set of the first $2n+3$ moments of some non-decreasing measure on the half-axis $(0,\infty)$.

Indeed, let the condition of the remark hold. Then by Theorem 1.1, both Hankel matrices $\mathbf{\Gamma}_n$ and $\mathbf{\Gamma}_{n+1}$ are non-negative definite. Let $\boldsymbol{\xi}\in\mathbb{L}_n(\mathbb{C}_s)$ be a non-zero null-vector of $\mathbf{\Gamma}_n$. Then for the vector

$$\widehat{\boldsymbol{\xi}}=\begin{pmatrix}\boldsymbol{\xi}\\ \mathbf{0}\end{pmatrix}\in\mathbb{L}_{n+1}(\mathbb{C}_s)$$

we have:

$$\left(\mathbf{\Gamma}_{n+1}\widehat{\boldsymbol{\xi}},\widehat{\boldsymbol{\xi}}\right)=(\mathbf{\Gamma}_n\boldsymbol{\xi},\boldsymbol{\xi})=0.$$

Since $\mathbf{\Gamma}_{n+1}\geq\mathbf{0}$, this means that $\widehat{\boldsymbol{\xi}}$ is a null-vector of $\mathbf{\Gamma}_{n+1}$. Therefore, $\boldsymbol{\xi}$ is also a null-vector of the Hankel matrix $\mathbf{\Gamma}_n^{(1)}(0)$, which implies (4.20). The condition (4.21) can be verified in the same manner.

Let us introduce the linear subspace

$$\mathbb{N}_n:=\left\{\mathbf{h}\in\mathbb{C}_s:\begin{pmatrix}0\\ \vdots\\ 0\\ \mathbf{h}\end{pmatrix}\in P_n\mathcal{M}_n\right\}. \tag{4.22}$$

By definition (4.22) for any $\mathbf{h}\in\mathbb{N}_n$ there is

$$\boldsymbol{\xi}(\mathbf{h})\in\mathbb{L}_{n-1}\left(\mathbb{C}_s\right),\ \boldsymbol{\xi}(\mathbf{h})=\left(\xi_j(\mathbf{h})\right)_{j=0}^{n-1}$$

such that

$$\begin{pmatrix}\boldsymbol{\xi}(\mathbf{h})\\ \mathbf{h}\end{pmatrix}\in\mathcal{M}_n.$$

The application of (4.20) and (4.21) to such a null-vector yields

$$\left[\mathbf{W}+\mathbf{S}_1^*(0)P_2\left(\mathbf{\Gamma}_{n-1}^{(1)}\right)\mathbf{\Upsilon}_{n-1}^{(1)}P_2\left(\mathbf{\Gamma}_{n-1}^{(1)}\right)\mathbf{S}_1(0)\right]\mathbf{h}=-\mathbf{C}_{n+1}\mathbf{h}-\ldots-\mathbf{C}_{2n}\mathbf{h}, \tag{4.23}$$

$$\mathbf{C}_{2n+2}\mathbf{h}=-\mathbf{C}_{n+2}\mathbf{h}-\ldots-\left[\mathbf{W}+\mathbf{S}_1^*(0)P_2\left(\mathbf{\Gamma}_{n-1}^{(1)}\right)\mathbf{\Upsilon}_{n-1}^{(1)}P_2\left(\mathbf{\Gamma}_{n-1}^{(1)}\right)\mathbf{S}_1(0)\right]\mathbf{h}. \tag{4.24}$$

Thus all admissible extensions

$$\mathbf{C}_{2n+1}=\mathbf{W}+\mathbf{S}_1^*(0)P_2\left(\mathbf{\Gamma}_{n-1}^{(1)}\right)\mathbf{\Upsilon}_{n-1}^{(1)}P_2\left(\mathbf{\Gamma}_{n-1}^{(1)}\right)\mathbf{S}_1(0),\quad \mathbf{C}_{2n+2}$$

of the moment set must coincide on the subspace $\mathcal{M}_n$. Since admissible matrices $\mathbf{C}_{2n+1}$, $\mathbf{C}_{2n+2}$ are Hermitian, they can differ only on the subspace $\mathbb{N}_n^{\perp}=\mathbb{C}_s\ominus\mathbb{N}_n$. We proved that for any set of Hermitian $s\times s$ matrices $\mathbf{C}_0,\mathbf{C}_1,\mathbf{C}_2,\ldots,\mathbf{C}_{2n}$ which satisfy the conditions of Theorem 1.1, the set of extensions $\mathbf{C}_{2n+1}$, $\mathbf{C}_{2n+2}$ which generates a completely degenerate Stieltjes problem, is always non-empty and *the variety of such extensions is exhausted by the expressions*

$$\mathbf{C}_{2n+1}=\mathbf{S}_1^*P_2\left(\mathbf{\Gamma}_{n-1}^{(1)}\right)\mathbf{\Upsilon}_{n-1}^{(1)}P_2\left(\mathbf{\Gamma}_{n-1}^{(1)}\right)\mathbf{S}_1+\mathbf{W},\ \mathbf{S}_1=\begin{pmatrix}\mathbf{C}_{n+1}\\ \vdots\\ \mathbf{C}_{2n}\end{pmatrix}, \tag{4.25}$$

$$\mathbf{C}_{2n+2}=\left(\mathbf{\Gamma}_n^{(1)}P_2(\mathbf{\Gamma}_n)\mathbf{\Upsilon}_nP_2(\mathbf{\Gamma}_n)\mathbf{\Gamma}_n^{(1)}\right)_{nn},\ \mathbf{\Gamma}_n^{(1)}=\left(\mathbf{C}_{j+k+1}\right)_{j,k=0}^{n}, \tag{4.26}$$

where $\mathbf{W}$ runs through the set of all non-negative $s\times s$-matrices which satisfy the condition $\mathbf{W}\mathbb{N}_n=\{0\}$ and $(\ldots)_{nn}$ stands for the nnth block of the corresponding block-matrix.

Proceeding from the assertion of Theorem 4.1 and the results of this and previous sections, we arrive at the following algorithm for finding all canonical solutions of the truncated Stieltjes matrix moment problem:

- for the set of matrix moments $\mathbf{C}_0, \mathbf{C}_1, \mathbf{C}_2, \ldots, \mathbf{C}_{2n}$ which satisfy the conditions of Theorem 1.1, find the orthogonal projectors $P_1\left(\mathbf{\Gamma}^{(1)}_{n-1}\right)$ and $P_1(\mathbf{\Gamma}_n)$ onto the null-spaces of Hankel matrices $\mathbf{\Gamma}^{(1)}_{n-1} = (\mathbf{C}_{j+k+1})^{n-1}_{j,k=0}$ and $\mathbf{\Gamma}_n = (\mathbf{C}_{j+k})^{n}_{j,k=0}$, respectively;
- find Hermitian matrices $\mathbf{\Upsilon}^{(1)}_{n-1}, \mathbf{\Upsilon}_n$ which satisfy the equations

$$\begin{aligned}
&\mathbf{\Gamma}^{(1)}_{n-1}\mathbf{\Upsilon}^{(1)}_{n-1} = P_2\left(\mathbf{\Gamma}^{(1)}_{n-1}\right) = I - P_1\left(\mathbf{\Gamma}^{(1)}_{n-1}\right),\\
&\mathbf{\Gamma}_n\mathbf{\Upsilon}_n = P_2\left(\mathbf{\Gamma}_n\right) = I - P_1\left(\mathbf{\Gamma}_n\right)
\end{aligned}$$

  and the conditions

$$\mathbf{\Upsilon}^{(1)}_{n-1}P_1\left(\mathbf{\Gamma}^{(1)}_{n-1}\right) = \mathbf{\Upsilon}_n P_1\left(\mathbf{\Gamma}_n\right) = 0;$$

- taking any matrix $\mathbf{W}\geq 0$ which satisfies the condition

$$\mathbf{W}P_n P_1\left(\mathbf{\Gamma}_n\right) = 0,$$

  define the extension $\mathbf{C}_{2n+1}$ by (4.25) and then the extension $\mathbf{C}_{2n+2}$ by formula (4.26);
- for the Hankel matrix $\mathbf{\Gamma}_{n+1} = (\mathbf{C}_{j+k})^{n+1}_{j,k=0}$ determine the $(n+2)\times s$ matrix

$$\widehat{\mathbf{X}} = \begin{pmatrix} \mathbf{X} \\ \mathbf{I} \end{pmatrix}, \; \mathbf{X} = \begin{pmatrix} X_0 \\ \vdots \\ X_n \end{pmatrix}, \tag{4.27}$$

  with $s\times s$ matrices X_j, $\mathbf{I}$, which satisfies the equation $\mathbf{\Gamma}_{n+1}\widehat{\mathbf{X}} = 0$; then determine the $(n+2)\times s$ matrix

$$\mathbf{Y} = \begin{pmatrix} Y_0 \\ \vdots \\ Y_{n-1} \\ Y_n \end{pmatrix}, \; Y_j = \sum_{k=0}^{j}\mathbf{C}_k X_{n+1-j+k}, \; X_{n+1} = \mathbf{I};$$

- for the matrix polynomials

$$\mathbf{D}(z) = z^{n+1}\mathbf{I} + \sum_{k=0}^{n} z^k X_k, \; \mathbf{E}(z) = \sum_{k=0}^{n} z^k Y_k$$

find all poles $t_{\min} = t_1 < t_2 < \cdots < t_\nu = t_{\max}$, $\nu \le ns$, of the rational matrix function $\mathbf{E}(z)\mathbf{D}(z)^{-1}$ and its residues $M_1, \ldots, M_\nu$ at these poles, and put

$$\boldsymbol{\sigma}(t) = \begin{cases} 0, & t \le t_1, \\ M_1, & t_1 < t \le t_2, \\ M_1 + M_2, & t_2 < t \le t_3, \\ \ldots & \\ M_1 + \cdots + M_\nu, & t > t_\nu. \end{cases}$$

5. Non-canonical solutions

By (4.25), (4.26) the set of all canonical solutions of the truncated matrix Stieltjes moment problem is parametrized in the degenerate but not the completely degenerate case by the set of $s \times s$ matrices $\mathbf{W} \ge 0$ which satisfy the condition $\mathbf{W}\mathbb{N}_n = \{0\}$. To find the description of all canonical solutions $\boldsymbol{\sigma}_{\mathbf{W}}(t)$, notice that for the extended Hankel matrix $\boldsymbol{\Gamma}_{n+1}$ the $(n+1) \times s$ matrix $\mathbf{X}$ (4.27) is of the form

$$\begin{aligned} &\mathbf{X} = \mathbf{X}_0 + \mathbf{X}_{\mathbf{W}}, \\ &\mathbf{X}_0 = \begin{pmatrix} X_{00} \\ \vdots \\ X_{0n} \end{pmatrix} := \boldsymbol{\Upsilon}_n \begin{pmatrix} \mathbf{C}_{n+1} \\ \vdots \\ \mathbf{C}_{2n+1} \end{pmatrix}, \\ &\mathbf{C}_{2n+1} = \mathbf{S}_1^* P_2\left(\boldsymbol{\Gamma}_{n-1}^{(1)}\right) \boldsymbol{\Upsilon}_{n-1}^{(1)} P_2\left(\boldsymbol{\Gamma}_{n-1}^{(1)}\right) \mathbf{S}_1; \\ &\mathbf{X}_{\mathbf{W}} = \begin{pmatrix} X_{10}\mathbf{W} \\ \vdots \\ X_{1n}\mathbf{W} \end{pmatrix}, \ \mathbf{X}_1 = \begin{pmatrix} X_{10} \\ \vdots \\ X_{1n} \end{pmatrix} := \boldsymbol{\Upsilon}_n \begin{pmatrix} 0 \\ \vdots \\ 0 \\ I - P_n' \end{pmatrix}. \end{aligned} \tag{5.1}$$

Put

$$\begin{aligned} &\mathbf{D}_{0n}(z) = z^{n+1}\mathbf{I} + \sum_{k=0}^{n} z^k X_{0k}, \ \mathbf{D}_{1n}(z) = \sum_{k=0}^{n} z^k X_{1k}; \\ &\mathbf{E}_{0n}(z) = \sum_{k=0}^{n} z^k Y_{0k}, \ Y_{0j} = \sum_{k=0}^{j} \mathbf{C}_k X_{0n+1-j+k}, \ X_{0n+1} = \mathbf{I}, \\ &\mathbf{E}_{1n}(z) = \sum_{k=0}^{n-1} z^k Y_{1k}, \ Y_{1j} = \sum_{k=0}^{j} \mathbf{C}_k X_{0n-j+k}. \end{aligned} \tag{5.2}$$

It stems from the above results that *the Nevanlinna type formula*

$$\int_0^\infty \frac{1}{t-z} d\boldsymbol{\sigma}_{\mathbf{W}}(t) = -\big(\mathbf{E}_{0n}(z) + \mathbf{E}_{1n}(z)\mathbf{W}\big)\big(\mathbf{D}_{0n}(z) + \mathbf{D}_{1n}(z)\mathbf{W}\big)^{-1} \tag{5.3}$$

establishes a one-to-one correspondence between the set of all canonical solutions $\boldsymbol{\sigma}_{\mathbf{W}}$ *of the truncated matrix Stieltjes moment problem and the set of* $s \times s$ *matrix* $\mathbf{W} \ge 0$ *which satisfy the condition* $\mathbf{W}\mathbb{N}_n = \{0\}$.

Notice also that the matrix functions $\mathbf{D}_{0n}(z), \mathbf{D}_{1n}(z), \mathbf{E}_{0n}(z), \mathbf{E}_{1n}(z)$ are independent of the choice of $\mathbf{W} \geq 0$, $\mathbf{W}\mathbb{N}_n = \{0\}$. Let us denote by $\widetilde{P}'_n$ the orthoprojector onto the subspace $\mathbb{N}_n$ in $\mathbb{C}_s$. By definition of these matrix functions or, irrespectively of this, from (5.3) it follows that

$$\mathbf{D}_{0n}(\overline{z})^* \mathbf{E}_{0n}(z) = \mathbf{E}_{0n}(\overline{z})^* \mathbf{D}_{0n}(z), \ \mathbf{D}_{1n}(\overline{z})^* \mathbf{E}_{1n}(z) = \mathbf{E}_{1n}(\overline{z})^* \mathbf{D}_{1n}(z);$$

$$\mathbf{D}_{0n}(\overline{z})^* \mathbf{E}_{1n}(z) - \mathbf{E}_{0n}(\overline{z})^* \mathbf{D}_{1n}(z) = I - \widetilde{P}'_n, \quad \operatorname{Im} z \geq 0. \tag{5.4}$$

Besides, by (5.3), for any $\mathbf{W} \geq 0$, $\mathbf{W}\widetilde{P}'_n = 0$ the matrix function $\mathbf{D}_{0n}(z) + \mathbf{D}_{1n}(z)\mathbf{W}$ is invertible in the upper half-plane and on the half-axis $(-\infty, 0)$. By applying these facts and the arguments used in [1], [2] and not adduced here, we come to the following description of all solution of the truncated matrix Stieltjes moment problem in the degenerate case.

Theorem 5.1. *Let the conditions of Theorem* 1.1 *hold. Then the formula*

$$\begin{aligned} &\int_0^\infty \frac{1}{t-z} d\boldsymbol{\sigma}_\Xi(t) \\ &- \left(\mathbf{E}_0(z) + \mathbf{E}_1(z)(\Xi(z)^{-1} + z\mathbf{I})(I - \widetilde{P}'_n)\right) \\ &\times \left(\mathbf{D}_0(z) + \mathbf{D}_1(z)\left(\Xi(z)^{-1} + z\mathbf{I}\right)(I - \widetilde{P}'_n)\right), \end{aligned} \tag{5.5}$$

establishes a one-to-one correspondence between the set of all solutions $\boldsymbol{\sigma}_\Xi(t)$ of the truncated matrix Stieltjes problem and the subset $\mathfrak{S}$ of the Nevanlinna matrix functions $\Xi(z)$ with values in $\mathbb{C}_s \ominus \mathbb{N}_n$.

Remark 5.2. The subset of canonical solutions is generated by substitution into (5.5) of matrix functions $\Xi(z) = (\mathbf{W} - z\mathbf{I})^{-1}$ with $\mathbf{W} \geq 0$ such that $\mathbb{N}_n \subset \ker \mathbf{W}$.

Acknowledgement

The financial support of the Polytechnic University of Valencia is gratefully acknowledged. V. Adamyan was also supported by the USA Civil Research and Development Foundation and the Government of Ukraine (CRDF grant UM1-2567-OD-03).

References

[1] V.M. Adamyan, I.M. Tkachenko, *Solution of the truncated matrix Hamburger moment problem according to M.G. Krein,* Operator Theory: Advances and Applications, **118** (2000), 33–52 (Proceedings of the Mark Krein International Conference on Operator Theory and Applications, Vol.II, Operator Theory and Related Topics), Birkhäuser, Basel.

[2] V.M. Adamyan, I.M. Tkachenko, *Solution of the Stieltjes Truncated Matrix Moment Problem,* Opuscula Mathematica, **25** (2005), no.1, 5–24.

[3] N.I. Akhiezer, *The classical moment problem and some related questions in analysis,* Hafner, N.Y., 1965.

[4] N.I. Akhiezer, I.M. Glazman, *Theory of Linear Operators in Hilbert Spaces*, Frederick Ungar Publishing Co., N.Y., 1966.

[5] V. Bolotnikov, *Degenerate Stieltjes moment problem and J-inner polynomials,* Z. Anal. Anwendungen **14** (1995), no. 3, 441–468.

[6] R.E. Curto and L.A. Fialkow, *Recursiveness, positivity, and truncated moment problems,* Houston J. Math., **17** (1991), 603–635.

[7] H. Dym, *On Hermitian Block Hankel Matrices, Matrix Polynomials, the Hamburger Moment Problem, Interpolation and Maximum Entropy,* Integral Equations Operator Theory, **12** (1989), 757–812.

[8] G.-N. Chen, Y.-J. Hu, *A unified treatment for the matrix Stieltjes moment problem in both nondegenerate and degenerate cases,* J. Math. Anal. Appl. **254** (2001), no. 1, 23–34.

[9] Y.-J. Hu, G.-N. Chen, *A unified treatment for the matrix Stieltjes moment problem,* Linear Algebra Appl. **380** (2004), 227–239.

[10] S. Karlin, W.S. Studden, *Tschebyscheff systems: with applications in analysis and statistics*, Interscience, 1966.

[11] M.G. Krein, *The theory of extensions of semi-bounded Hermitian operators and its applications*, Mat. Sbornik I, **20** (1947), 431–495; II, **21** (1947), 365–404 (in Russian).

[12] M.G. Krein, M.A. Krasnoselskii, *Fundamental theorem on the extension of Hermitian operators and certain of their applications to the theory of orthogonal polynomials and the problem of moments*, Uspekhi. Matem. Nauk (N.S.) **2** (1947), no. 3(19), 60–106 (in Russian).

[13] M.G. Krein, A.A. Nudel'man, *The Markov moment problem and extremal problems,* "Nauka", Moscow, 1973 (in Russian). English translation: Translation of Mathematical Monographs AMS, 50, 1977.

[14] T.J. Stieltjes, *Recherches sur les fractions continues*, Annales de la Faculté des Sciencies de Toulouse, **8** (1894), 1–122; **9** (1895), 1–47.

[15] T.J. Stieltjes, *Collected Papers*, G. van Dijk (ed.), Springer, Berlin, 1993.

Vadim M. Adamyan
Department of Theoretical Physics
I.I. Mechnikov Odessa National University
65026 Odessa
Ukraine
e-mail: `vadamyan@paco.net`

Igor M. Tkachenko
Department of Applied Mathematics
Polytechnic University of Valencia
46022 Valencia
Spain
e-mail: `imtk@mat.upv.es`

Operator Theory:
Advances and Applications, Vol. 163, 23–54

Q-functions of Quasi-selfadjoint Contractions

Yury Arlinskiĭ, Seppo Hassi and Henk de Snoo

To Heinz Langer on the occasion of his retirement

Abstract. A bounded everywhere defined operator T in a Hilbert space $\mathfrak{H}$ is said to be a *quasi-selfadjoint contraction* or (for short) a *qsc-operator*, if T is a contraction and ker $(T - T^*) \neq \{0\}$. For a closed linear subspace $\mathfrak{N}$ of $\mathfrak{H}$ containing ran $(T - T^*)$ the operator-valued function $Q_T(z) = P_{\mathfrak{N}}(T - zI)^{-1}\restriction \mathfrak{N}$, $|z| > 1$, where $P_{\mathfrak{N}}$ is the orthogonal projector from $\mathfrak{H}$ onto $\mathfrak{N}$, is said to be a *Q-function* of T acting on the subspace $\mathfrak{N}$. The main properties of such Q-functions are studied, in particular the underlying operator-theoretical aspects are considered by using some block representations of the contraction T and analytical characterizations for such functions $Q_T(z)$ are established. Also a reproducing kernel space model for $Q_T(z)$ is constructed. In the special case where T is selfadjoint $Q_T(z)$ coincides with the Q-function of the symmetric operator $A := T\restriction (\mathfrak{H} \ominus \mathfrak{N})$ and its selfadjoint extension $T = T^*$ in the usual sense.

Mathematics Subject Classification (2000). Primary: 47A10, 47A56, 47A64; Secondary 47A05, 47A06, 47B15.

Keywords. Symmetric contraction, contractive extension, quasi-selfadjoint operator, Q-function, operator model, resolvent.

1. Introduction

The concept of a Q-function was introduced by M.G. Kreĭn for the case of a densely symmetric operator S in a Hilbert space $\mathfrak{H}$ with equal defect numbers by means of a selfadjoint extension A of S, cf. [26], [28], [33], and also [29], [30], [31]. Such a function belongs to the class $\mathbf{N}$ of Nevanlinna (or Herglotz-Nevanlinna) functions, i.e., $Q(z) \in \mathbf{N}$ if it is holomorphic in the open upper and lower half-planes and satisfies the conditions $Q(\bar{z}) = Q(z)^*$ and $(\operatorname{Im} z)\,(\operatorname{Im} Q(z)) \geq 0$, $z \in \mathbb{C}_+ \cup \mathbb{C}_-$. The

This work was supported by the Research Institute for Technology at the University of Vaasa. The first author was also supported by the Academy of Finland (projects 203227, 208057) and the Dutch Organization for Scientific Research NWO (B 61–553).

Q-function plays an essential role in Kreĭn's resolvent formula, which describes all (generalized resolvents of) selfadjoint extensions of S. In fact, all generalized resolvents (canonical as well as exit space) were first described independently by M.A. Naimark [41] and M.G. Kreĭn [26]; see also [29] for further historical remarks. A characteristic property of a Q-function $Q(z)$ in the class of Nevanlinna functions is that $\operatorname{Im} Q(z)$ is invertible (at some or equivalently at every point $z \in \mathbb{C}_+ \cup \mathbb{C}_-$): every Nevanlinna function with this property is a Q-function of some simple symmetric operator S and a selfadjoint extension A of S in a Hilbert space $\mathfrak{H}$. Moreover, the simple (completely non-selfadjoint) symmetric operator S and its selfadjoint extension A are essentially unique in the sense that the Q-function of S determines S and A uniquely up to unitary equivalence. Another approach for describing selfadjoint as well as non-selfadjoint intermediate extensions of a symmetric operator is via a boundary value space and the corresponding Weyl function, see [21], [19], [18], and the references therein.

Two special subclasses of Q-functions, consisting of the so-called Q_μ- and Q_M-functions, which belong to the class $\mathbf{N}$ of Nevanlinna functions were defined and investigated by M.G. Kreĭn and I.E. Ovcharenko in [32]. Here the underlying symmetric operator is a non-densely defined contraction. In a recent paper [7] by the authors some extensions of Q_μ- and Q_M-functions were introduced; in fact, this paper contains also some corrections to the result stated in [32]. Some other type of Q-function associated to a non-densely defined symmetric contraction has been considered in [47], including the resolvent formulas for the selfadjoint (canonical and exit space) extensions.

In this paper a class of operator-valued Q-functions associated with a non-densely defined symmetric contraction A and its, in general, *non-selfadjoint* contractive extensions T is introduced. By definition a bounded operator T in the Hilbert space $\mathfrak{H}$ is a *quasi-selfadjoint contraction* or, for short, a *qsc*-operator if $\operatorname{dom} T = \mathfrak{H}$, $\|T\| \leq 1$, and $\ker\,(T - T^*) \neq \{0\}$. Let T be a *qsc*-operator and let $\mathfrak{N}$ be a proper subspace of $\mathfrak{H}$ which contains $\operatorname{ran}(T - T^*)$. Define the operator-valued function $Q(z)$ as follows

$$Q(z) = P_{\mathfrak{N}}(T - zI)^{-1}{\restriction}\,\mathfrak{N}, \quad |z| < 1. \tag{1.1}$$

In what follows the function $Q(z)$ in (1.1) will be called a Q-function of T with respect to the subspace $\mathfrak{N} \subset \mathfrak{H}$. Observe, that if T is selfadjoint then the function Q defined by (1.1) is an ordinary Q-function associated with T and the symmetric restriction $A := T{\restriction}\,\mathfrak{H}_0$ of T, where $\mathfrak{H}_0 = \mathfrak{H} \ominus \mathfrak{N}$. However, if T is not selfadjoint this function in general is not a Nevanlinna function. A *qsc*-operator T may be considered as a contractive, in general, non-selfadjoint extension of the symmetric contraction $A = T{\restriction}\,\mathfrak{H}_0$ which is also called a quasi-selfadjoint contractive extension of A; here A is symmetric due to $\mathfrak{H}_0 \subset \ker\,(T - T^*)$. Such kind of extensions were parametrized and investigated in [10] and [12]. The special case of selfadjoint contractive extensions was investigated by M.G. Kreĭn [27] and by M.G. Kreĭn and I.E. Ovcharenko [32]. In particular, in [32] two special Q-functions of the Nevanlinna class for the symmetric contraction were defined and studied and the resolvent formulas for selfadjoint contractive extensions (*sc*-extensions) were established. These formulas were extended in [10] and [12] for *qsc*-extensions. A boundary value space approach for describing extensions of dual pairs of densely defined operators appears in [37] and for dual pairs of linear relations and their canonical and generalized resolvents in [39], [40], see also [34], [35]. In the present paper the approach can be seen as a non-selfadjoint counterpart of the Q-function approach developed and systematically used in the papers of M.G. Kreĭn and H. Langer, cf., e.g., [29]–[31].

The contents of this paper will be briefly described. In Section 2 some preliminary notions are introduced. The extension theory for closed symmetric contractions is developed in Section 3. This includes a discussion of minimality of the underlying symmetric operator A and its contractive extensions. The Q-functions for intermediate contractive extensions as in (1.1) are introduced in Section 4, where also a number of associated nonnegative kernels will appear. A resolvent formula for *qsc*-extensions of a symmetric contraction A is derived in Section 5. It involves a Q-function of the form (1.1) for a given *qsc*-extension T of A. In Section 6 a model for such Q-functions is constructed by means of a *qsc*-operator acting in a reproducing kernel Hilbert space and it is proved that two $\mathfrak{N}$-minimal *qsc*-operators whose Q-functions in (1.1) coincide are unitarily equivalent. This model is used to establish some characteristic properties of Q-functions of *qsc*-operators in Section 7. In Section 8 linear fractional transformations of Q-functions are considered.

The results in the present paper can be connected with and augmented by the study of a certain class of passive systems. In particular, the Q-functions of quasi-selfadjoint operators investigated in the present paper are in one-to-one correspondence with the transfer functions of so-called passive quasi-selfadjoint systems, which are introduced and investigated in [8].

The work in this paper is a continuation of some investigations done by several authors in the early eighties, including Heinz Langer and Björn Textorius, being influenced by ideas developed by M.G. Kreĭn and I.E. Ovcharenko. It is a pleasure to record the mathematical indebtedness of the authors of the present paper to Heinz Langer over a period of many years.

2. Preliminaries

2.1. Basic notations

The class of all continuous linear operators defined on a complex Hilbert space $\mathfrak{H}_1$ and taking values in a complex Hilbert space $\mathfrak{H}_2$ is denoted by $\mathbf{L}(\mathfrak{H}_1, \mathfrak{H}_2)$ and $\mathbf{L}(\mathfrak{H}) := \mathbf{L}(\mathfrak{H}, \mathfrak{H})$. The domain, the range, and the null-space of a linear operator T are denoted by $\operatorname{dom} T$, $\operatorname{ran} T$, and $\ker T$. For $T \in \mathbf{L}(\mathfrak{H})$ the operators $T_R = (T + T^*)/2$, $T_I = (T - T^*)/2i$ are said to be the *real* and the *imaginary part* of T. For a contraction $T \in \mathbf{L}(\mathfrak{H}_1, \mathfrak{H}_2)$ the *defect operator* D_T of T is defined by

$$D_T := (I - T^*T)^{1/2}. \tag{2.1}$$

It is a nonnegative contraction and satisfies the well-known commutation relation

$$TD_T = D_{T^*}T, \tag{2.2}$$

cf. [46]. The closure of the range $\operatorname{ran} D_T$ is denoted by $\mathfrak{D}_T$ and $\rho(T)$ stands for the set of all regular points of a closed operator T. If R_l and R_r are two nonnegative operators in $\mathbf{L}(\mathfrak{H})$ and $S_0 \in \mathbf{L}(\mathfrak{H})$, then the symbol $\mathbf{B}(S_0, R_l, R_r)$ denotes the *operator ball* in $\mathbf{L}(\mathfrak{H})$ with the center S_0 and the left and right radii R_l and R_r, respectively, i.e., the set of all operators in $\mathbf{L}(\mathfrak{H})$ of the form $T = S_0 + R_l^{1/2} X R_r^{1/2}$, where X is a contraction from $\overline{\operatorname{ran}}\, R_r$ into $\overline{\operatorname{ran}}\, R_l$. It is well known, see [43], [44], that a necessary and sufficient condition for $T \in \mathbf{L}(\mathfrak{H})$ to belong to $\mathbf{B}(S_0, R_l, R_r)$ is the following:

$$|((T - S_0)f, g)|^2 \leq (R_r f, f)\,(R_l g, g) \quad \text{for all } f, g \in \mathfrak{H}. \tag{2.3}$$

If $R_l = R_r = R$ the corresponding operator ball is denoted by $\mathbf{B}(S_0, R)$.

2.2. Quasi-selfadjoint contractions

Recall that $T \in \mathbf{L}(\mathfrak{H})$ is a quasi-selfadjoint contraction (a *qsc*-operator) if

$$\operatorname{dom} T = \mathfrak{H}, \quad \|T\| \leq 1, \text{ and } \ker\,(T - T^*) \neq \{0\}.$$

A *qsc*-operator T is said to be a *quasi-selfadjoint contractive extension* or *qsc-extension* of a closed symmetric contraction A if

$$\operatorname{dom} A \subset \ker\,(T - T^*) \quad \text{or equivalently} \quad \operatorname{ran}\,(T - T^*) \subset (\operatorname{dom} A)^\perp,$$

cf. [10], [12]. Clearly, an operator $T \in \mathbf{L}(\mathfrak{H})$ is a *qsc*-extension of A if and only if

$$A \subset T \text{ and } A \subset T^*,$$

or, equivalently, if T is an *intermediate extension* of A. A *qsc*-operator T has always symmetric restrictions A for which T is a *qsc*-extension. Namely, with a subspace $\mathfrak{N} \supset \operatorname{ran}\,(T - T^*)$ define

$$\operatorname{dom} A = \mathfrak{H} \ominus \mathfrak{N}, \quad A = T \restriction \operatorname{dom} A.$$

Then $\operatorname{dom} A \subset \ker\,(T - T^*)$. A *qsc*-operator T is called *completely non-selfadjoint* if there is no non-zero invariant subspace on which the restriction of T is selfadjoint.

Lemma 2.1. [15] *A qsc-operator T is completely non-selfadjoint if and only if*

$$\overline{\operatorname{span}}\,\{\operatorname{ran} T^n (T - T^*) : n = 0, 1, \dots\} = \mathfrak{H}.$$

2.3. The classes $C(\alpha)$

Let $\alpha \in [0, \pi/2)$ and denote by $\mathcal{S}(\alpha)$ the following sector of the complex plane:

$$\mathcal{S}(\alpha) = \{\, z \in \mathbb{C} : |\arg z| \leq \alpha \,\}.$$

A linear operator S, in general unbounded, in a Hilbert space $\mathfrak{H}$ is said to be *sectorial* with vertex at the origin and *semiangle* α, if its *numerical range*

$$W(S) = \{\, (Sf, f) : \|f\| = 1,\ f \in \operatorname{dom} S \,\}$$

is contained in the sector $\mathcal{S}(\alpha)$, cf. [25]. This condition is equivalent to

$$|\mathrm{Im}\,(Sf,f)| \leq (\tan\alpha)\,\mathrm{Re}\,(Sf,f) \quad \text{for all } f \in \mathrm{dom}\, S.$$

If the resolvent set of S is not empty then S is called maximal sectorial. A maximal sectorial operator S is densely defined and its adjoint S^* is also a maximal sectorial operator.

A bounded operator T on a Hilbert space $\mathfrak{H}$ is said to belong to the class $C(\alpha)$, $\alpha \in (0,\pi/2)$, if

$$\|T\sin\alpha \pm i\cos\alpha\, I\| \leq 1, \tag{2.4}$$

cf. [3]. Clearly, T belongs to $C(\alpha)$ if and only if T^* belongs to $C(\alpha)$. Moreover, it follows from (2.4) that the operators belonging to $C(\alpha)$ are contractive. The condition (2.4) is equivalent to each of the following two conditions:

$$|(T_I f,f)| \leq \frac{\tan\alpha}{2}\,\|D_T f\|^2 \quad \text{for all } f \in \mathfrak{H}; \tag{2.5}$$

or

$$\begin{array}{l}\text{the operator } (I-T^*)(I+T) \text{ is sectorial with}\\ \text{vertex at the origin and semiangle } \alpha,\end{array} \tag{2.6}$$

cf. [4]. Note that the linear fractional transformation $T=(I-S)(I+S)^{-1}$ of a maximal sectorial operator S with vertex at the origin and semiangle α is an operator of the class $C(\alpha)$. Let

$$\widetilde{C} = \bigcup \{C(\alpha) :\ \alpha \in [0,\pi/2)\}.$$

Some properties of the operators in the class $\widetilde{C}$ were studied in [3], [4]. In particular, in [3] it was proved that $T \in \widetilde{C}$ implies that

$$\mathrm{ran}\, D_{T^n} = \mathrm{ran}\, D_{T^{*n}} = \mathrm{ran}\, D_{T_R}, \qquad n = 1,2,\ldots,$$

where T_R is the real part of T. Furthermore it was proved in [3] that the subspace $\mathfrak{D}_T$ reduces the operator T, that the operator $T{\restriction}\ker D_T$ is selfadjoint and unitary, and that $T{\restriction}\mathfrak{D}_T$ is a completely non-unitary contraction of the class C_{00}, i.e.,

$$\lim_{n\to\infty} T^n f = \lim_{n\to\infty} T^{*n} f = 0 \quad \text{for all} \quad f \in \mathfrak{D}_T,$$

cf. [46].

2.4. The Schur-Frobenius formula for the resolvent of a block operator

Let the Hilbert space $\mathfrak{H}$ be decomposed as $\mathfrak{H} = \mathfrak{H}_1 \oplus \mathfrak{H}_2$ and decompose $T \in \mathbf{L}(\mathfrak{H})$ accordingly:

$$T = \begin{pmatrix} T_{11} & T_{12} \\ T_{21} & T_{22} \end{pmatrix}, \quad T_{ij} \in \mathbf{L}(\mathfrak{H}_i,\mathfrak{H}_j). \tag{2.7}$$

Define the operator-valued functions

$$V_T(z) = T_{21}(T_{11}-zI)^{-1}T_{12} - T_{22}, \quad W_T(z) = -zI - V_T(z), \quad z \in \rho(T_{11}). \tag{2.8}$$

By the Schur-Frobenius formula the resolvent $(T-z)^{-1}$ of T can be rewritten in the block form

$$\begin{pmatrix} (T_{11}-zI)^{-1}\left(I+T_{12}W_T(z)^{-1}T_{21}(T_{11}-zI)^{-1}\right) & -(T_{11}-zI)^{-1}T_{12}W_T^{-1}(z) \\ -W_T^{-1}(z)T_{21}(T_{11}-zI)^{-1} & W_T^{-1}(z) \end{pmatrix} \tag{2.9}$$

for $z\in\rho(T)\cap\rho(T_{11})$. In particular,

$$P_{\mathfrak{H}_2}(T-zI)^{-1}\upharpoonright\mathfrak{H}_2=-\left(V_T(z)+zI\right)^{-1},\quad z\in\rho(T)\cap\rho(T_{11}). \tag{2.10}$$

2.5. Nevanlinna functions

Let $\mathfrak{N}$ be a Hilbert space. An operator-valued function $V(z)$, $z\in\mathbb{C}\setminus\mathbb{R}$, with values in $\mathbf{L}(\mathfrak{N})$ is said to be a Nevanlinna function or an R-function, cf. [24], if $V(z)$ is holomorphic on $\mathbb{C}\setminus\mathbb{R}$, $V^*(z)=V(\bar{z})$, and $\operatorname{Im} z\operatorname{Im} V(z)\geq 0$ for all $z\in\mathbb{C}\setminus\mathbb{R}$. The subclass of Nevanlinna functions $V(z)$ which are holomorphic on the domain $\operatorname{Ext}[-1,1]:=\mathbb{C}\setminus[-1,1]$ is denoted by $\mathbf{N}_{\mathfrak{N}}[-1,1]$. By the general theory of Nevanlinna functions, cf. [24], [15], every function $V(z)$ in $\mathbf{N}_{\mathfrak{N}}[-1,1]$ has an integral representation of the form

$$V(z)=\Gamma+\int_{-1}^{1}\frac{dG(t)}{t-z},$$

where Γ is a bounded selfadjoint operator on $\mathfrak{N}$ and the $\mathbf{L}(\mathfrak{N})$-valued function $G(t)$ is nondecreasing, nonnegative, normalized by $G(-1-0)=0$, and has finite total variation concentrated on $[-1,1]$. Clearly, $V(\infty):=s-\lim\limits_{z\to\infty}V(z)=\Gamma$. The next result is also well known, cf. [15].

Theorem 2.2. *Let $\mathfrak{N}$ be a Hilbert space and let $V(z)\in\mathbf{N}_{\mathfrak{N}}[-1,1]$. Then there exist a Hilbert space $\mathfrak{H}$, a selfadjoint contraction B on $\mathfrak{H}$, and $F\in\mathbf{L}(\mathfrak{N},\mathfrak{H})$, such that*

$$V(z)=V(\infty)+F^*(B-zI)^{-1}F,\quad z\in\operatorname{Ext}[-1,1]. \tag{2.11}$$

In what follows the subclass of functions $V(z)$ in $\mathbf{N}_{\mathfrak{N}}[-1,1]$ which have the limit values $V(\pm1)$ in $\mathbf{L}(\mathfrak{N})$ plays a central role. In this case Theorem 2.2 can be completed as follows.

Theorem 2.3. *Let $\mathfrak{N}$ be a Hilbert space and let $V(z)\in\mathbf{N}_{\mathfrak{N}}[-1,1]$. If for all $f\in\mathfrak{N}$ the limit values*

$$\lim_{x\uparrow-1}\left(V(x)f,f\right),\quad \lim_{x\downarrow1}\left(V(x)f,f\right) \tag{2.12}$$

are finite, then there exist a Hilbert space $\mathfrak{H}$, a selfadjoint contraction B in $\mathfrak{H}$, and an operator $G\in\mathbf{L}(\mathfrak{N},\mathfrak{D}_B)$, such that

$$V(z)=V(\infty)+G^*D_B^2(B-zI)^{-1}G,\quad z\in\operatorname{Ext}[-1,1]. \tag{2.13}$$

Conversely, for every function $V(z)$ of the form (2.13) *the limit values* (2.12) *exist for all $f\in\mathfrak{N}$ and are finite.*

Proof. By Theorem 2.2 $V(z)$ has the representation (2.11), where B is a selfadjoint contraction in a Hilbert space $\mathfrak{H}$ and $F \in \mathbf{L}(\mathfrak{N}, \mathfrak{H})$. Since the limits in (2.12) exist for all $f \in \mathfrak{N}$, one concludes that

$$\operatorname{ran} F \subset \operatorname{ran} (I-B)^{1/2} \cap \operatorname{ran} (I+B)^{1/2}.$$

Consequently, $\operatorname{ran} F \subset \operatorname{ran} D_B$ and this implies that $F = D_B G$ for some operator $G \in \mathbf{L}(\mathfrak{N}, \mathfrak{D}_B)$, cf. [23].

Conversely, if $V(z)$ is of the form (2.13) then $\operatorname{ran} D_B \subset \operatorname{ran} (B \pm I)^{1/2}$ and this implies the existence of the limit values (2.12) for all $f \in \mathfrak{N}$, cf. [32]. □

It follows from Theorem 2.3 that

$$\begin{aligned} V(-1) &:= s-\lim_{x \uparrow -1} V(x) = V(\infty) + G^*(I-B)G \in \mathbf{L}(\mathfrak{N}), \\ V(1) &:= s-\lim_{x \downarrow 1} V(x) = V(\infty) - G^*(I+B)G \in \mathbf{L}(\mathfrak{N}), \end{aligned} \tag{2.14}$$

so that

$$V(-1)+V(1) = 2V(\infty) - 2G^*BG, \quad V(-1)-V(1) = 2G^*G. \tag{2.15}$$

2.6. Nonnegative kernels, reproducing kernel Hilbert spaces, and sectorial kernels

An operator-valued function $K(z,\xi) : \Omega \times \Omega \to \mathbf{L}(\mathfrak{N})$, $\Omega \subset \mathbb{C}$, is said to be a *nonnegative kernel* [1], [13], [42], if

$$\sum_{i,j=1}^{n} (K(w_j,w_i)f_i, f_j)_{\mathfrak{N}} \geq 0$$

for every choice of points $\{w_i\}_{i=1}^n \subset \Omega$ and vectors $\{f_i\}_{i=1}^n \subset \mathfrak{N}$. With the kernel $K(z,\xi)$ is associated a *reproducing kernel Hilbert space* $\mathcal{H}_K$. It is the completion of the linear space of vectors of the form

$$\sum_{i=1}^{n} K(\cdot, w_i) f_i, \quad \{w_i\}_{i=1}^n \subset \Omega, \quad \{f_i\}_{i=1}^n \subset \mathfrak{N}, \quad n \in \mathbb{N},$$

with respect to the inner product

$$\left(\sum_{i=1}^{n} K(\cdot, w_i) f_i, \sum_{j=1}^{m} K(\cdot, \mu_j) g_j \right)_{\mathcal{H}_K} = \sum_{i=1}^{n} \sum_{j=1}^{m} (K(\mu_j, w_i) f_i, g_j)_{\mathfrak{N}}.$$

Then the Hilbert space $\mathcal{H}_K$ consists of the $\mathfrak{N}$-valued functions $f(\cdot)$ such that for every $h \in \mathfrak{N}$ the reproducing property holds:

$$(f(\cdot), K(\cdot, w)h)_{\mathcal{H}_K} = (f(w), h)_{\mathfrak{N}}, \quad w \in \Omega.$$

Observe that an $\mathbf{L}(\mathfrak{N})$-valued function $V(z)$ belongs to the Nevanlinna class $\mathbf{N}(\mathfrak{N})$ if and only if the function

$$K(z,\xi) = \frac{V(z) - V(\xi)^*}{z - \bar{\xi}}, \quad z, \xi \in \mathbb{C} \setminus \mathbb{R},$$

is a nonnegative kernel. Also note that the kernel associated with generalized resolvents (of selfadjoint exit space extensions) in a Hilbert space is given by

$$K(z,\xi)=\frac{V(z)-V(\xi)^*}{z-\bar{\xi}}-V(z)V(\xi)^*,\quad z,\xi\in\mathbb{C}\setminus\mathbb{R}.$$

An operator-valued function $K(z,\xi):\Omega\times\Omega\to\mathbf{L}(\mathfrak{N})$, $\Omega\subset\mathbb{C}$, is said to be an *α-sectorial kernel*, if

$$\sum_{i,j=1}^{n}(K(w_j,w_i)f_i,f_j)_{\mathfrak{N}}\in\mathcal{S}(\alpha)$$

for every choice of points $\{w_i\}_{i=1}^n\subset\Omega$ and vectors $\{f_i\}_{i=1}^n\subset\mathfrak{H}$, i.e.,

$$\left|\operatorname{Im}\sum_{i,j=1}^{n}(K(w_j,w_i)f_i,f_j)_{\mathfrak{N}}\right|\le(\tan\alpha)\operatorname{Re}\left(\sum_{i,j=1}^{n}(K(w_j,w_i)f_i,f_j)_{\mathfrak{N}}\right),$$

cf. [5]. For $\alpha=0$ the corresponding kernel is nonnegative.

3. qsc-extensions of closed symmetric contractions

3.1. A decomposition of closed symmetric contractions

Let A be a non-densely defined closed symmetric contraction in the Hilbert space $\mathfrak{H}$ with the domain $\operatorname{dom}A=:\mathfrak{H}_0$ and let $\mathfrak{N}:=\mathfrak{H}\ominus\operatorname{dom}A$. Let P_0 and $P_{\mathfrak{N}}$ be the orthogonal projections in $\mathfrak{H}$ onto $\mathfrak{H}_0$ and $\mathfrak{N}$, respectively. Then the operator $A_0=P_0A$ is contractive and selfadjoint in the subspace $\mathfrak{H}_0$. Let $D_{A_0}=(I-A_0^2)^{1/2}$ be the defect operator determined by A_0. The operator $A_{21}=P_{\mathfrak{N}}A$ is also contractive. Moreover, it follows from $A^*A\le I$ that $A_{21}^*A_{21}\le D_{A_0}^2$. Therefore, the identity

$$K_0D_{A_0}f=P_{\mathfrak{N}}Af,\quad f\in\operatorname{dom}A,$$

defines a contractive operator K_0 from $\mathfrak{D}_{A_0}:=\overline{\operatorname{ran}}\,D_{A_0}$ into $\mathfrak{N}$, cf. [20], [23]. This gives the following decomposition for the symmetric contraction A

$$A=A_0+K_0D_{A_0}=\begin{pmatrix}A_0\\K_0D_{A_0}\end{pmatrix}.\tag{3.1}$$

3.2. A matrix representation for qsc-extensions

Let the closed symmetric contraction A be defined on the subspace $\mathfrak{H}_0=\operatorname{dom}A$ and decompose A according to $\mathfrak{H}=\mathfrak{H}_0\oplus\mathfrak{N}$ as in (3.1). Let T be a qsc-extension of A, so that $A\subset T$ and $A\subset T^*$, and decompose $T=(T_{ij})$ also with respect to $\mathfrak{H}=\mathfrak{H}_0\oplus\mathfrak{N}$, cf. (2.7). Then clearly $T_{11}=A_0$, $T_{12}^*=T_{21}=K_0D_{A_0}$. The next result gives a parametrization of all qsc-extensions of A and some of its subclasses by means of block formulas, cf. [14], [17], [45], and [10], [12]. For completeness a short, simple proof is presented.

Theorem 3.1. *Let A be a closed symmetric contraction A in $\mathfrak{H}=\mathfrak{H}_0\oplus\mathfrak{N}$ with* $\operatorname{dom}A=\mathfrak{H}_0$ *and decompose A as in* (3.1)*. Then:*

(i) *the formula*

$$T = \begin{pmatrix} A_0 & D_{A_0}K_0^* \\ K_0 D_{A_0} & -K_0A_0K_0^* + D_{K_0^*}XD_{K_0^*} \end{pmatrix} : \begin{pmatrix} \mathfrak{H}_0 \\ \mathfrak{N} \end{pmatrix} \to \begin{pmatrix} \mathfrak{H}_0 \\ \mathfrak{N} \end{pmatrix} \tag{3.2}$$

gives a one-to-one correspondence between all qsc-extensions T of the symmetric contraction $A = A_0 + K_0 D_{A_0}$ and all contractions X in the subspace $\mathfrak{D}_{K_0^*} := \overline{\operatorname{ran}}\, D_{K_0^*} \subset \mathfrak{N}$;

(ii) *T in (3.2) belongs to the class $C(\alpha)$ if and only if X belongs to the class $C(\alpha)$, $\alpha \in (0, \pi/2)$;*

(iii) *T is a selfadjoint contractive extension of A if and only if X in (3.2) is a selfadjoint contraction in $\mathfrak{D}_{K_0^*}$.*

Proof. (i) Every operator $T \in \mathbf{L}(\mathfrak{H})$ satisfying the conditions $A \subset T$ and $A \subset T^*$ admits the block-matrix representation of the form

$$T = \begin{pmatrix} A_0 & D_{A_0}K_0^* \\ K_0 D_{A_0} & D \end{pmatrix} : \begin{pmatrix} \mathfrak{H}_0 \\ \mathfrak{N} \end{pmatrix} \to \begin{pmatrix} \mathfrak{H}_0 \\ \mathfrak{N} \end{pmatrix},$$

where $D \in \mathbf{L}(\mathfrak{N})$. Then $I - T^*T$ is given in the block form

$$I - T^*T = \begin{pmatrix} D_{A_0}^2 - D_{A_0}K_0^*K_0D_{A_0} & -A_0D_{A_0}K_0^* - D_{A_0}K_0^*D \\ -K_0D_{A_0}A_0 - D^*K_0D_{A_0} & D_{K_0^*}^2 - K_0A_0^2K_0^* - D^*D \end{pmatrix}.$$

Contractivity of T means that

$$0 \le \|D_{A_0}f - A_0K_0^*h\|^2 + \|D_{K_0^*}h\|^2 - \|K_0D_{A_0}f + Dh\|^2, \tag{3.3}$$

for all $f \in \mathfrak{H}_0$ and $h \in \mathfrak{N}$. Since $\operatorname{ran} K_0^* \subset \mathfrak{D}_{A_0}$ and $A_0\mathfrak{D}_{A_0} \subset \mathfrak{D}_{A_0}$, there exists a sequence $\{f_n\}_{n=1}^\infty \subset \mathfrak{D}_{A_0}$ such that for a given $h \in \mathfrak{N}$ the equality

$$\lim_{n\to\infty} D_{A_0}f_n = A_0K_0^*h$$

holds. Hence, it follows from (3.3) that $E := K_0A_0K_0^* + D$ satisfies

$$\|Eh\|^2 \le \|D_{K_0^*}h\|^2, \quad \|E^*h\|^2 \le \|D_{K_0^*}h\|^2, \quad h \in \mathfrak{N}, \tag{3.4}$$

where the second inequality follows from the first one by taking into account that T^* is a contraction, too. By the second inequality in (3.4) there exists a contraction $Z \in \mathbf{L}(\mathfrak{N}, \mathfrak{D}_{K_0^*})$ such that $E = D_{K_0^*}Z$, i.e., $D = -K_0A_0K_0^* + D_{K_0^*}Z$. By substituting this into (3.3) one obtains

$$0 \le \|D_{K_0}(D_{A_0}f - A_0K_0^*h) - K_0^*Zh\|^2 + \|D_{K_0^*}h\|^2 - \|Zh\|^2, \quad f \in \mathfrak{H}_0, \quad h \in \mathfrak{N}, \tag{3.5}$$

since by means of (2.2) one has

$$\begin{aligned} -\|K_0(D_{A_0}f - A_0K_0^*h) + D_{K_0^*}Zh\|^2 &= -\|K_0(D_{A_0}f - A_0K_0^*h)\|^2 \\ &\quad - \|Zh\|^2 + \|K_0^*Zh\|^2 - 2\mathrm{Re}\,\left(D_{K_0}(D_{A_0}f - A_0K_0^*h), K_0^*Zh\right). \end{aligned}$$

Due to the inclusion $\operatorname{ran} Z \subset \mathfrak{D}_{K_0^*}$, one can choose a sequence $\{f_n\}_{n=1}^\infty \subset \mathfrak{D}_{A_0}$ such that for a given $h \in \mathfrak{N}$ the equality

$$\lim_{n\to\infty} D_{K_0}D_{A_0}f_n = D_{K_0}A_0K_0^*h + K_0^*Zh \tag{3.6}$$

holds. Now (3.5) shows that $\|Zh\|^2 \le \|D_{K_0^*}h\|^2$ for all $h \in \mathfrak{N}$, so that $Z = XD_{K_0^*}$ for some contraction X in $\mathfrak{D}_{K_0^*}$. Therefore, $E = D_{K_0^*}Z = D_{K_0^*}XD_{K_0^*}$ and

$$D = -K_0A_0K_0^* + D_{K_0^*}XD_{K_0^*}. \tag{3.7}$$

Conversely, let D be of the form (3.7), where X is a contraction in $\mathfrak{D}_{K_0^*}$. Then $D_X^2 \ge 0$ implies that T given by (3.2) satisfies

$$\begin{aligned}&\left((I - T^*T)\begin{pmatrix} f \\ h\end{pmatrix}, \begin{pmatrix} f \\ h\end{pmatrix}\right)\\ &\qquad = \|D_{K_0}(D_{A_0}f - A_0K_0^*h) - K_0^*XD_{K_0^*}h\|^2 + \|D_XD_{K_0^*}h\|^2 \ge 0,\end{aligned} \tag{3.8}$$

cf. (3.5). Thus, every contraction X in $\mathfrak{D}_{K_0^*}$ defines a *qsc*-extension T of A via (3.2).

(ii) It follows from (3.2) and (3.8) that T satisfies (2.5) if and only if

$$\begin{aligned}&|(X_ID_{K_0^*}h, D_{K_0^*}h)|\\ &\qquad \le \frac{\tan\alpha}{2}\left(\|D_{K_0}(D_{A_0}f - A_0K_0^*h) - K_0^*XD_{K_0^*}h\|^2 + \|D_XD_{K_0^*}h\|^2\right)\end{aligned} \tag{3.9}$$

holds for all $f \in \mathfrak{H}_0$, $h \in \mathfrak{N}$. In view of (3.6) the condition (3.9) is equivalent to

$$|(X_Ih, h)| \le \frac{\tan\alpha}{2}\,\|D_Xh\|^2 \tag{3.10}$$

for all $h \in \mathfrak{D}_{K_0^*}$.

(iii) The statement is clear since T in (3.2) is selfadjoint if and only if X is selfadjoint in $\mathfrak{D}_{K_0^*}$. □

The class of all selfadjoint contractive (*sc*-) extensions of A in part (iii) of Theorem 3.1 forms an operator interval $[A_\mu, A_M]$. Using the block representation (3.2) the endpoints of $[A_\mu, A_M]$ are given by

$$A_\mu = \begin{pmatrix} A_0 & D_{A_0}K_0^* \\ K_0D_{A_0} & -K_0A_0K_0^* - D_{K_0^*}^2\end{pmatrix}, \tag{3.11}$$

and

$$A_M = \begin{pmatrix} A_0 & D_{A_0}K_0^* \\ K_0D_{A_0} & -K_0A_0K_0^* + D_{K_0^*}^2\end{pmatrix}, \tag{3.12}$$

with $X = -I{\upharpoonright}\mathfrak{D}_{K_0^*}$ and $X = I{\upharpoonright}\mathfrak{D}_{K_0^*}$, respectively. From the formulas (3.11) and (3.12) it is seen that

$$\frac{A_\mu + A_M}{2} = \begin{pmatrix} A_0 & D_{A_0}K_0^* \\ K_0D_{A_0} & -K_0A_0K_0^*\end{pmatrix}, \quad \frac{A_M - A_\mu}{2} = \begin{pmatrix} 0 & 0 \\ 0 & D_{K_0^*}^2\end{pmatrix}.$$

This means that all *qsc*-extensions in (3.2) of the symmetric contraction A form an operator ball

$$\mathbf{B}\left(\frac{A_\mu + A_M}{2}, \frac{A_M - A_\mu}{2}\right)$$

with center

$$(A_\mu + A_M)/2$$

and equal left and right radii

$$R_l = R_r = (A_M - A_\mu)^{1/2}/\sqrt{2}.$$

The one-to-one correspondence between all qsc-extensions T of A and all contractions X in Theorem 3.1 can be reformulated also as follows

$$T = \frac{A_\mu + A_M}{2} + \left(\frac{A_M - A_\mu}{2}\right)^{1/2} X \left(\frac{A_M - A_\mu}{2}\right)^{1/2}, \tag{3.13}$$

where the parameters X are contractions in the subspace $\overline{\operatorname{ran}}\,(A_M - A_\mu)$, cf. [10], [11], [12]. It is easy to see from (3.2), (3.11), and (3.12), that if T is a qsc-extension of A such that $T_R = (T + T^*)/2 = A_\mu\,(A_M)$, then in fact $T = A_\mu\,(A_M)$. Namely, $X = X_R + iX_I$ satisfies

$$\begin{cases} 0 \leq X^*X = X_R^2 + i(X_R X_I - X_I X_R) + X_I^2 \leq I, \\ 0 \leq XX^* = X_R^2 - i(X_R X_I - X_I X_R) + X_I^2 \leq I, \end{cases} \tag{3.14}$$

so that $0 \leq X_R^2 + X_I^2 \leq I$ and here clearly $X_R^2 = I$ implies $X_I = 0$.

The description of all contractive selfadjoint extensions of a symmetric contraction A as the operator interval $[A_\mu, A_M]$ is due to M.G. Kreĭn [27]. In that paper the notion of shorted operators was also introduced and used for instance to establish the following characterization for A_μ and A_M:

$$\operatorname{ran}\,(I + A_\mu)^{1/2} \cap \mathfrak{N} = \{0\}, \quad \operatorname{ran}\,(I - A_M)^{1/2} \cap \mathfrak{N} = \{0\}, \tag{3.15}$$

cf. [7], [22]. Block formulas for describing all contractive extensions of a dual pair appear in [14], [17], [45], a description in Crandall's form [16] in [2]. The one-to-one correspondence between all qsc-extensions T of A of the class $C(\alpha)$ and all operators X in $\overline{\operatorname{ran}}\,(A_M - A_\mu)$ belonging to the class $C(\alpha)$ by means of (3.13) was proved in a different way in [12], another proof based on (3.2) was given in [38].

3.3. Simplicity of the symmetric operator

According to [32] a closed symmetric contraction A is said to be *simple* if there is no non-zero subspace in $\operatorname{dom} A$ which is invariant under A. Since A is symmetric, simplicity of A is equivalent to A being completely non-selfadjoint, i.e., to A having no selfadjoint parts.

Lemma 3.2. *Let the closed symmetric contraction $A = A_0 + K_0 D_{A_0}$ in $\mathfrak{H} = \mathfrak{H}_0 \oplus \mathfrak{N}$, $\mathfrak{H}_0 = \operatorname{dom} A$, be decomposed as in (3.1) with $K_0 : \mathfrak{D}_{A_0} \to \mathfrak{N}$. Then A is simple if and only if the subspace*

$$\begin{aligned} \mathfrak{H}_0^s &:= \overline{\operatorname{span}}\,\{\,(A_0 - zI)^{-1} K_0^* \mathfrak{N} : \, z \in \rho(A_0)\,\} \\ &= \overline{\operatorname{span}}\,\{\,A_0^n K_0^* \mathfrak{N} : \, n = 0, 1, \dots\} \end{aligned} \tag{3.16}$$

coincides with $\mathfrak{H}_0$. In this case, $\mathfrak{D}_{A_0} = \mathfrak{H}_0$, $K_0 : \mathfrak{H}_0 \to \mathfrak{N}$, and $\|A_0 f\| < \|f\|$ for all $f \in \mathfrak{H}_0 \backslash \{0\}$.

Proof. Suppose that A is simple. Then clearly $\ker D_{A_0} = \{0\}$ or equivalently $\|A_0 f\| < \|f\|$ for all $f \in \mathfrak{H}_0 \setminus \{0\}$, so that $\mathfrak{D}_{A_0} = \mathfrak{H}_0$ and $K_0 : \mathfrak{H}_0 \to \mathfrak{N}$. Observe, that the subspace $\mathfrak{H}_0^s$ in (3.16) and therefore also $\mathfrak{H}_0 \ominus \mathfrak{H}_0^s$ is invariant under $A_0 = A_0^*$. Then the subspace $\mathfrak{H}_0 \ominus \mathfrak{H}_0^s$ is also invariant under D_{A_0}. Moreover,

$$\mathfrak{H}_0 \ominus \mathfrak{H}_0^s = \{\, f \in \mathfrak{H}_0 : \; K_0 A_0^n f = 0, \; n = 0, 1, \dots \,\}. \tag{3.17}$$

It follows that $K_0 D_{A_0} f = 0$ for all $f \in \mathfrak{H}_0 \ominus \mathfrak{H}_0^s$. Hence, in view of (3.1) $Af = A_0 f$ for all $f \in \mathfrak{H}_0 \ominus \mathfrak{H}_0^s$. This means that the subspace $\mathfrak{H}_0 \ominus \mathfrak{H}_0^s$ is invariant under A. Since A is a simple, one concludes that $\mathfrak{H}_0^s = \mathfrak{H}_0$.

Conversely, assume that $\mathfrak{H}_0^s = \mathfrak{H}_0$. Since $\operatorname{ran} K_0^* \subset \mathfrak{D}_{A_0}$ and $\mathfrak{D}_{A_0}$ is invariant under A_0, the definition of $\mathfrak{H}_0^s$ in (3.16) shows that $\mathfrak{H}_0^s \subset \mathfrak{D}_{A_0}$. Hence, the assumption implies that $\mathfrak{H}_0 = \mathfrak{D}_{A_0} = \overline{\operatorname{ran}}\, D_{A_0}$, so that $\ker D_{A_0} = \{0\}$. Now, suppose that $\widetilde{\mathfrak{H}}_0 \subset \mathfrak{H}_0$ is a subspace which is invariant under A. Then for every $f \in \widetilde{\mathfrak{H}}_0$ one has $Af = A_0 f + K_0 D_{A_0} f \in \widetilde{\mathfrak{H}}_0$, so that $K_0 D_{A_0} f = 0$ for all $f \in \widetilde{\mathfrak{H}}_0$ and $A{\upharpoonright}\,\widetilde{\mathfrak{H}}_0 = A_0{\upharpoonright}\,\widetilde{\mathfrak{H}}_0$. Hence, $\widetilde{\mathfrak{H}}_0$ is invariant under A_0 and D_{A_0}. Moreover, since $\ker D_{A_0} = \{0\}$ the image $D_{A_0}\widetilde{\mathfrak{H}}_0$ is dense in $\widetilde{\mathfrak{H}}_0$. This implies that $K_0 \widetilde{\mathfrak{H}}_0 = \{0\}$ and since $A_0^n \widetilde{\mathfrak{H}}_0 \subset \widetilde{\mathfrak{H}}_0$ one has $K_0 A_0^n \widetilde{\mathfrak{H}}_0 = \{0\}$ for all $n = 0, 1, \dots$, i.e.,

$$\widetilde{\mathfrak{H}}_0 \subset \{\, f \in \mathfrak{H}_0 : \; K_0 A_0^n f = 0, \; n = 0, 1, \dots \} = \mathfrak{H}_0 \ominus \mathfrak{H}_0^s = \{0\},$$

cf. (3.17). Therefore, A is simple. □

Let T be a *qsc*-extension of A in the Hilbert space $\mathfrak{H} = \mathfrak{H}_0 \oplus \mathfrak{N}$ with $\mathfrak{H}_0 = \operatorname{dom} A$. It is evident that the subspace

$$\mathfrak{H}'_T := \overline{\operatorname{span}} \left\{ (T - zI)^{-1}\mathfrak{N} : \, |z| > 1 \right\} = \overline{\operatorname{span}} \left\{ T^n \mathfrak{N} : \, n = 0, 1, 2, \dots \right\}, \tag{3.18}$$

is invariant under T, and that the subspace

$$\mathfrak{H}''_T := \mathfrak{H} \ominus \mathfrak{H}'_T, \tag{3.19}$$

is invariant under T^*. Since $\mathfrak{N} \subset \mathfrak{H}'_T$, one obtains

$$\mathfrak{H}''_T \subset \mathfrak{N}^\perp = \operatorname{dom} A \subset \ker\, (T - T^*).$$

Therefore the restriction of T^* to $\mathfrak{H}''_T$ is a selfadjoint operator in $\mathfrak{H}''_T$. The restriction $T{\upharpoonright}\,\mathfrak{H}'_T$ $(= P_{\mathfrak{H}'_T} T{\upharpoonright}\,\mathfrak{H}'_T)$ is called the $\mathfrak{N}$*-minimal part of* T. Moreover, T is said to be $\mathfrak{N}$*-minimal* if the equality $\mathfrak{H} = \mathfrak{H}'_T$ holds. If T be a *qsc*-extension of A then its adjoint T^* is also a *qsc*-extension of A and one can associate with it the subspace $\mathfrak{H}'_{T^*}$ and the corresponding $\mathfrak{N}$-minimal part of T^*. The next result shows the $\mathfrak{N}$-minimal parts of T and T^* are *qsc*-extensions of the simple part $A{\upharpoonright}\,\mathfrak{H}_0^s$ of A in the same subspace $\mathfrak{H}'_T = \mathfrak{H}'_{T^*}$.

Proposition 3.3. *Let A be a symmetric contraction in $\mathfrak{H} = \mathfrak{H}_0 \oplus \mathfrak{N}$ with $\mathfrak{H}_0 = \operatorname{dom} A$, let T be a qsc-extension of A in $\mathfrak{H}$, and let T^* be its adjoint. Then the subspaces $\mathfrak{H}'_T$, $\mathfrak{H}'_{T^*}$, and $\mathfrak{H}_0^s$ of $\mathfrak{H} = \mathfrak{H}_0 \oplus \mathfrak{N}$ as defined in* (3.18) *and* (3.16) *are connected by*

$$(\mathfrak{H}' :=)\; \mathfrak{H}'_T = \mathfrak{H}'_{T^*} = \mathfrak{H}_0^s \oplus \mathfrak{N}. \tag{3.20}$$

In particular, the symmetric contraction A is simple if and only if the qsc-extension T, or equivalently T^*, of A is $\mathfrak{N}$-minimal.

Proof. It follows from the Schur-Frobenius formula (2.9) that

$$(T-z)^{-1}\mathfrak{N} = \begin{pmatrix} -(A_0-z)^{-1}D_{A_0}K_0^*\mathfrak{N} \\ \mathfrak{N} \end{pmatrix}, \quad |z|>1,$$

which implies that

$$\begin{aligned} &\overline{\operatorname{span}}\,\{\,(T-zI)^{-1}\mathfrak{N} : |z|>1\,\} \\ &\quad = \overline{\operatorname{span}}\,\{\,(A_0-zI)^{-1}D_{A_0}K_0^*\mathfrak{N} : z\in\rho(A_0)\,\}\oplus\mathfrak{N} \\ &\quad = (\operatorname{clos} D_{A_0}\overline{\operatorname{span}}\,\{\,(A_0-zI)^{-1}K_0^*\mathfrak{N} : z\in\rho(A_0)\,\})\oplus\mathfrak{N}. \end{aligned}$$

This shows that

$$\mathfrak{H}_T' = (\operatorname{clos} D_{A_0}\mathfrak{H}_0^s)\oplus\mathfrak{N}. \tag{3.21}$$

Since $\operatorname{ran} K_0^*\subset\mathfrak{D}_{A_0}$ and $\mathfrak{D}_{A_0}$ is invariant under A_0 one has $\mathfrak{H}_0^s\subset\mathfrak{D}_{A_0}$. In particular, $\mathfrak{H}_0^s\cap\ker D_{A_0}=\{0\}$, which together with $D_{A_0}\mathfrak{H}_0^s\subset\mathfrak{H}_0^s$ implies that $\operatorname{clos} D_{A_0}\mathfrak{H}_0^s=\mathfrak{H}_0^s$. Hence, (3.21) implies the equality $\mathfrak{H}_T'=\mathfrak{H}_0^s\oplus\mathfrak{N}$. It follows from

$$(T^*-zI)^{-1}-(T-zI)^{-1} = (T-zI)^{-1}[T-T^*](T^*-zI)^{-1}, \quad |z|>1,$$

and the inclusion $\operatorname{ran}(T-T^*)\subset\mathfrak{N}$ that

$$(T^*-zI)^{-1}\mathfrak{N}\subset(T-zI)^{-1}\mathfrak{N}\subset\mathfrak{H}_T', \quad |z|>1.$$

Therefore, $\mathfrak{H}_{T^*}'\subset\mathfrak{H}_T'$ and the reverse inclusion follows by symmetry. This completes the proof of (3.20).

The last statement is clear from (3.20). □

For selfadjoint extensions of A the result in Proposition 3.3 has been given in [32]. In the case of closed densely defined symmetric operators A there is an equivalent criterion for the simplicity of A due to M.G. Kreĭn based on the defect elements:

$$\overline{\operatorname{span}}\,\{\,\ker(A^*-\lambda) : \lambda\in\mathbb{C}\setminus\mathbb{R}\,\} = \mathfrak{H},$$

cf. Lemma 3.2. This characterization has been extended to non-densely defined symmetric operators in [36].

4. Q-functions of qsc-operators

Let T be a qsc-operator in a separable Hilbert space $\mathfrak{H}$ and let $\mathfrak{N}$ be a subspace of $\mathfrak{H}$ such that $\mathfrak{N}\supset\operatorname{ran}(T-T^*)$. The operator-valued function

$$Q_T(z) = P_{\mathfrak{N}}(T-zI)^{-1}\upharpoonright\mathfrak{N}, \quad |z|>1, \tag{4.1}$$

where $P_{\mathfrak{N}}$ is the orthogonal projection in $\mathfrak{H}$ onto $\mathfrak{N}$, is said to be a *Q-function* associated with T and the subspace $\mathfrak{N}$. Clearly, it has the limit value $Q_T(\infty)=0$ and the Q-functions of T and T^* in $\mathfrak{N}$ are connected by

$$Q_{T^*}(z) = Q_T(\bar z)^*, \quad |z|>1. \tag{4.2}$$

If T is a selfadjoint contraction then the Q-function $Q_T(z)$ in (4.1) is a Nevanlinna function of the class $\mathbf{N}_{\mathfrak{N}}[-1,1]$. The next result contains some basic properties for the Q-function $Q_T(z)$ of a *qsc*-operator T as defined in (4.1).

Proposition 4.1. *Let $Q_T(z)$ be a Q-function of a qsc-operator T as defined in* (4.1). *Then:*

(i) *$Q_T(z)$ has the following asymptotic expansion:*

$$Q_T(z) = -\frac{1}{z}I + \frac{1}{z^2}F + o\left(\frac{1}{z^2}\right), \quad z \to \infty, \tag{4.3}$$

where $F = -P_{\mathfrak{N}}T{\upharpoonright}\mathfrak{N}$;

(ii) *$Q_T^{-1}(z) \in \mathbf{L}(\mathfrak{N})$ for all $|z|>1$;*

(iii) *$Q_T^{-1}(z)$ has strong limit values $Q_T^{-1}(\pm 1)$:*

$$Q_T^{-1}(-1) = \lim_{x\uparrow -1} Q_T^{-1}(x), \quad Q_T^{-1}(1) = \lim_{x\downarrow 1} Q_T^{-1}(x);$$

(iv) *for all $f,g \in \mathfrak{N}$ the following inequality holds:*

$$\begin{aligned}&\left|\left(\left(Q_T^{-1}(-1)+Q_T^{-1}(1)\right)f,g\right)\right|^2\\ &\quad\le \left(\left(Q_T^{-1}(-1)-Q_T^{-1}(1)\right)f,f\right)\left(\left(Q_T^{-1}(-1)-Q_T^{-1}(1)\right)g,g\right);\end{aligned}$$

(v) *the function $-Q_T^{-1}(z) - F - zI$ is an operator-valued Nevanlinna function;*

(vi) *$Q_T(z) \in \mathbf{N}_{\mathfrak{N}}[-1,1]$ if and only if $F = F^*$.*

Moreover, if T is decomposed as in (3.2) *with $\mathfrak{H}_0 = \mathfrak{H} \ominus \mathfrak{N}$ and $A = T{\upharpoonright}\mathfrak{H}_0$, then*

$$F = K_0A_0K_0^* - D_{K_0^*}XD_{K_0^*}, \tag{4.4}$$

$$Q_T^{-1}(-1) = D_{K_0^*}(X+I)D_{K_0^*}, \quad Q_T^{-1}(1) = D_{K_0^*}(X-I)D_{K_0^*}, \tag{4.5}$$

$$-Q_T^{-1}(z) - F - zI = K_0(I-A_0^2)(A_0-zI)^{-1}K_0^*. \tag{4.6}$$

Proof. (i) Clearly, $\lim_{z\to\infty} zQ_T(z)h = \lim_{z\to\infty} zP_{\mathfrak{N}}(T-zI)^{-1}h = -P_{\mathfrak{N}}h$ for all $h \in \mathfrak{N}$. Moreover, for all $h \in \mathfrak{N}$

$$\lim_{z\to\infty} z(I+zQ_T(z))h = \lim_{z\to\infty} zP_{\mathfrak{N}}T(T-zI)^{-1}h = -P_{\mathfrak{N}}Th. \tag{4.7}$$

Hence, $Q_T(z)$ admits the asymptotic expansion (4.3).

(ii) Let $|z|>1$, let $f \in \mathfrak{N}$, and let $\varphi = (T-zI)^{-1}f$. Then $\|f\| \le (1+|z|)\|\varphi\|$ and

$$\begin{aligned}|(Q_T(z)f,f)| &= \left|\left((T-zI)^{-1}f,f\right)\right| = |(\varphi,(T-zI)\varphi)|\\ &= \left|(\varphi,T\varphi) - \bar z\|\varphi\|^2\right| \ge \frac{|z|-1}{(|z|+1)^2}\|f\|^2.\end{aligned}$$

Since $|(Q_T(z)f,f)| = |(Q_T(z)^*f,f)|$, this implies that

$$\|Q_T(z)f\| \ge \frac{|z|-1}{(|z|+1)^2}\|f\|, \quad \|Q_T(z)^*f\| \ge \frac{|z|-1}{(|z|+1)^2}\|f\|.$$

Therefore, $Q_T^{-1}(z) \in \mathbf{L}(\mathfrak{N})$ for all $|z|>1$.

(iii) Decompose $\mathfrak{H} = \mathfrak{H}_0 \oplus \mathfrak{N}$ and write T in block form as in (3.2), where $\mathfrak{H}_0 = \mathfrak{H} \ominus \mathfrak{N}$, $A = T\upharpoonright \mathfrak{H}_0$, $A_0 = P_0 A$ is a selfadjoint contraction in $\mathfrak{H}_0$, $D_{A_0} = \left(I - A_0^2\right)^{1/2}$, $K_0 \in \mathbf{L}\left(\mathfrak{D}_{A_0}, \mathfrak{N}\right)$ is a contraction, and X is a contraction in the subspace $\mathfrak{D}_{K_0^*} \subset \mathfrak{N}$. The formula (4.4) for F is immediate from (3.2). Write $Q_T^{-1}(z)$ as in (2.10),

$$Q_T^{-1}(z) = -V_T(z) - zI, \quad |z| > 1,$$

where

$$V_T(z) = K_0\left[A_0 + (A_0 - zI)^{-1}(I - A_0^2)\right] K_0^* - D_{K_0^*} X D_{K_0^*}. \tag{4.8}$$

This shows that the limit values $Q_T^{-1}(\pm 1)$ exist and that they are given by (4.5).

(iv) It follows from (4.5) that

$$\begin{aligned} \frac{Q_T^{-1}(-1) + Q_T^{-1}(1)}{2} &= D_{K_0^*} X D_{K_0^*}, \\ \frac{Q_T^{-1}(-1) - Q_T^{-1}(1)}{2} &= D_{K_0^*}^2 = I - K_0 K_0^*. \end{aligned} \tag{4.9}$$

It remains to apply the criterion (2.3) with $S_0 = 0$ and $R_l = R_r = D_{K_0^*}^2$.

(v) It follows from (4.4) and (4.8) that (4.6) holds. Clearly, the function in (4.6) is a Nevanlinna function.

(vi) If $Q_T(z) \in \mathbf{N}_{\mathfrak{N}}[-1,1]$, then $-Q_T(z)^{-1}$ is a Nevanlinna function and now part (v) implies that $F = F^*$. Conversely, if $F = F^*$ then the function $V_T(z)$ in (4.8) and $-Q_T(z)^{-1} = V_T(z) + zI$ are Nevanlinna functions. Therefore, $Q_T(z) \in \mathbf{N}_{\mathfrak{N}}[-1,1]$. $\square$

Let T be a qsc-operator, let $Q_T(z)$ be defined by (4.1), and let F be defined by $F = -P_{\mathfrak{N}} T\upharpoonright \mathfrak{N}$. Associate with $Q_T(z)$ the following kernels:

$$\mathsf{G}_T(z,\xi) := \frac{Q_T(z) - Q_T(\xi)^* - Q_T(z)(F - F^*)Q_T(\xi)^*}{z - \bar{\xi}}, \tag{4.10}$$

$$\mathsf{M}_T(z,\xi) := I + zQ_T(z) + \bar{\xi}Q_T(\xi)^* + z\,\bar{\xi}\mathsf{G}_T(z,\xi), \tag{4.11}$$

$$\mathsf{L}_T(z,\xi) := \mathsf{G}_T(z,\xi) - \mathsf{M}_T(z,\xi), \tag{4.12}$$

and

$$\mathsf{K}_T(z,\xi) := \mathsf{L}_T(z,\xi) - Q_T(z)(F - F^*)Q_T(\xi)^*, \tag{4.13}$$

with $z \neq \bar{\xi}$, $|z|, |\xi| < 1$. The insertion of the definition of $\mathsf{G}_T(z,\xi)$ in $\mathsf{L}_T(z,\xi)$ and $\mathsf{K}_T(z,\xi)$ leads to the identities

$$\begin{aligned} (z - \bar{\xi})\mathsf{L}_T(z,\xi) = (1 - z^2)Q_T(z) - (1 - \bar{\xi}^2)Q_T(\xi)^* \\ - (1 - z\bar{\xi})Q_T(z)(F - F^*)Q_T(\xi)^* - (z - \bar{\xi})I, \end{aligned}$$

and

$$\begin{aligned} (z - \bar{\xi})\mathsf{K}_T(z,\xi) = (1 - z^2)Q_T(z) - (1 - \bar{\xi}^2)Q_T(\xi)^* \\ - (1 + z)(1 - \bar{\xi})Q_T(z)(F - F^*)Q_T(\xi)^* - (z - \bar{\xi})I. \end{aligned}$$

Proposition 4.2. *Let T be a qsc-operator, let $Q_T(z)$ be defined by (4.1), and let F be defined by $F = -P_{\mathfrak{N}}T\restriction\mathfrak{N}$. Let the kernels associated with $Q_T(z)$ be given by (4.10), (4.11), (4.12), and (4.13). Then the following equalities hold for every $z \neq \bar{\xi}$, $|z|, |\xi| > 1$:*

$$\mathsf{G}_T(z,\xi) = P_{\mathfrak{N}}(T - zI)^{-1}(T^* - \bar{\xi}I)^{-1}\restriction\mathfrak{N}, \tag{4.14}$$

$$\mathsf{M}_T(z,\xi) = P_{\mathfrak{N}}(T - zI)^{-1}TT^*(T^* - \bar{\xi}I)^{-1}\restriction\mathfrak{N}, \tag{4.15}$$

and

$$\mathsf{L}_T(z,\xi) = P_{\mathfrak{N}}(T - zI)^{-1}(I - TT^*)(T^* - \bar{\xi}I)^{-1}\restriction\mathfrak{N}. \tag{4.16}$$

The operator-valued functions $\mathsf{G}_T(z,\xi)$, $\mathsf{M}_T(z,\xi)$, and $\mathsf{L}_T(z,\xi)$ are nonnegative kernels. If, in addition, the operator T belongs to the class $C(\alpha)$ then the function

$$\mathsf{K}_T(z,\xi) = P_{\mathfrak{N}}(T - zI)^{-1}(I + T)(I - T^*)(T^* - \bar{\xi}I)^{-1}\restriction\mathfrak{N} \tag{4.17}$$

with $|z|$, $|\xi| > 1$ is an α-sectorial kernel.

Proof. Note that $\operatorname{ran}(T - T^*) \subset \mathfrak{N}$ implies that $\mathfrak{N}^\perp \subset \ker(T - T^*)$, and hence $T - T^* = P_{\mathfrak{N}}(T - T^*)P_{\mathfrak{N}}$. Therefore, for every $f, g \in \mathfrak{N}$,

$$\begin{aligned}((Q_T(z) - Q_T^*(\xi))f, g) &= \left(P_{\mathfrak{N}}(T - zI)^{-1}f - P_{\mathfrak{N}}(T^* - \bar{\xi}I)^{-1}f, g\right)\\ &= \left(P_{\mathfrak{N}}(T - zI)^{-1}(T^* - T)(T^* - \bar{\xi}I)^{-1}f, g\right)\\ &\quad + (z - \bar{\xi})\left(P_{\mathfrak{N}}(T - zI)^{-1}(T^* - \bar{\xi}I)^{-1}f, g\right)\\ &= (Q_T(z)(F - F^*)Q_T(\xi)^*f, g)\\ &\quad + (z - \bar{\xi})\left(P_{\mathfrak{N}}(T - zI)^{-1}(T^* - \bar{\xi}I)^{-1}f, g\right).\end{aligned}$$

Hence, it follows that

$$Q_T(z) - Q_T^*(\xi) = Q_T(z)(F - F^*)Q_T^*(\xi) + (z - \bar{\xi})P_{\mathfrak{N}}(T - zI)^{-1}(T^* - \bar{\xi}I)^{-1}\restriction\mathfrak{N},$$

and this proves (4.14). The identity (4.15) follows now from

$$\begin{aligned}&\left(T^*(T^* - \bar{\xi}I)^{-1}f, T^*(T^* - \bar{z}I)^{-1}g\right) = \left(f + \bar{\xi}(T^* - \bar{\xi}I)^{-1}f, g + \bar{z}(T^* - \bar{z}I)^{-1}g\right)\\ &= (f, g) + z(Q_T(z)f, g) + \bar{\xi}(Q_T^*(\xi)f, g) + z\bar{\xi}(\mathsf{G}_T(z,\xi)f, g), \quad f, g \in \mathfrak{N}.\end{aligned}$$

Subtracting (4.15) from (4.14) gives immediately the identity (4.16).

It is clear from the given formulas (4.14), (4.15), and (4.16), that the functions $\mathsf{G}_T(z,\xi)$, $\mathsf{M}_T(z,\xi)$, and $\mathsf{L}_T(z,\xi)$ are nonnegative kernels.

Since $T - T^* = P_{\mathfrak{N}}(T - T^*)P_{\mathfrak{N}}$, the definitions of $Q_T(z)$ and F in (4.1), (4.7) show that

$$-Q_T(z)(F - F^*)Q_T^*(\xi) = P_{\mathfrak{N}}(T - zI)^{-1}(T - T^*)(T^* - \bar{\xi}I)^{-1}.$$

Combining this identity with (4.16) leads to (4.17).

It is a consequence of (2.6) that $\mathsf{K}_T(z,\xi)$ is an α-sectorial kernel. □

Proposition 4.3. *Let T be a qsc-operator in a Hilbert space $\mathfrak{H}$, $\mathfrak{N} \supset \operatorname{ran}(T - T^*)$. Suppose that T is $\mathfrak{N}$-minimal, i.e., $\mathfrak{H} = \overline{\operatorname{span}}\{(T - z)^{-1}\mathfrak{N} : |z| > 1\}$. Then the following conditions are equivalent:*

(i) *$\mathfrak{N} = \mathfrak{H}$;*

(ii) $\mathsf{G}_T(z,z) = Q_T(z)Q_T(z)^*$ *for at least one (and equivalently for every)* z *with* $|z| > 1$*, where* $Q_T(z)$ *is Q-function of* T *defined by* (4.1) *and* $\mathsf{G}_T(z,\xi)$ *is defined by* (4.10)*;*

(iii) *the operator-valued function* $Q_T^{-1}(z) + zI$ *is constant.*

Proof. (i) $\Rightarrow$ (ii) & (iii) If $\mathfrak{N} = \mathfrak{H}$ then $Q_T(z) = (T - zI)^{-1}$ and the equality $\mathsf{G}_T(z,z) = Q_T(z)Q_T(z)^*$ for all z, $|z| > 1$, follows immediately from (4.14). Besides, $Q_T^{-1}(z) + zI = T$ for all z, $|z| > 1$.

(ii) $\Rightarrow$ (i) Now suppose that $\mathsf{G}_T(z,z) = Q_T(z)Q_T(z)^*$ for some z, $|z| > 1$. Then (4.1) and (4.14) yield

$$\|(T^* - \bar{z}I)^{-1}f\| = \|P_{\mathfrak{N}}(T^* - \bar{z}I)^{-1}f\| \quad \text{for every } f \in \mathfrak{N}.$$

Therefore, $(T^* - \bar{z}I)^{-1}\mathfrak{N} \subset \mathfrak{N}$ which implies that the subspace $\mathfrak{N}$ is invariant under T^*, and hence also under T, since $\operatorname{ran}(T - T^*) \subset \mathfrak{N}$. Because T is $\mathfrak{N}$-minimal, this leads to $\mathfrak{N} = \mathfrak{H}$.

(iii) $\Rightarrow$ (ii) Suppose that $Q_T^{-1}(z) + zI$ is constant for $|z| > 1$. According to Proposition 4.1 the function $-Q_T^{-1}(z) - zI - F$ has a holomorphic continuation onto $\operatorname{Ext}[-1,1]$ as a Nevanlinna function. Since $-Q_T^{-1}(z) - zI - F$ is constant for $|z| > 1$, one has

$$-Q_T^{-1}(z) - zI - F + Q_T(z)^{-*} + \bar{z}I + F^* = 0, \quad |z| > 1.$$

It follows that

$$\frac{-Q_T^{-1}(z) + Q_T(z)^{-*} - (F - F^*)}{z - \bar{z}} = I, \quad |z| > 1,$$

and thus

$$\frac{Q_T(z)\left(-Q_T^{-1}(z) + Q_T(z)^{-*} - (F - F^*)\right)Q_T(z)^*}{z - \bar{z}} = Q_T(z)Q_T(z)^*, \quad |z| > 1.$$

Therefore $\mathsf{G}(z,z) = Q_T(z)Q_T(z)^*$ for all z, $|z| > 1$. □

Observe, that the equality (4.14) can be rewritten in the following two equivalent forms:

$$\begin{aligned} &\frac{-Q_T(z)^{-1} - F - (-Q_T(\xi)^{-1} - F)^*}{z - \bar{\xi}} \\ &\qquad = Q_T(z)^{-1}P_{\mathfrak{N}}(T - zI)^{-1}(T^* - \bar{\xi}I)^{-1}Q_T(\xi)^{-*}, \end{aligned} \tag{4.18}$$

and

$$\begin{aligned} &\frac{-Q_T(z)^{-1} - F - zI - (-Q_T(\xi)^{-1} - F - \xi I)^*}{z - \bar{\xi}} \\ &\qquad = Q_T(z)^{-1}P_{\mathfrak{N}}(T - zI)^{-1}(I - P_{\mathfrak{N}})(T^* - \bar{\xi}I)^{-1}Q_T(\xi)^{-*}. \end{aligned} \tag{4.19}$$

These formulas show that $-Q_T(z)^{-1} - F$ and $-Q_T(z)^{-1} - F - zI$ indeed are Nevanlinna functions. In particular, the conditions (i)–(iii) in Proposition 4.3 are equivalent to the right side of (4.19) to vanish.

Remark 4.4. The Q-function $Q_T(z)$ as defined in (4.1) can be interpreted as the Weyl function for a special kind of boundary value space of a dual pair of operators, cf. [37], [39], [40]. To explain this, let $A = A_0 + K_0 D_{A_0}$ be a Hermitian contraction and let T be a *qsc*-extension of A, i.e., T is a contractive extension of a dual pair $\{A, A\}$. Let A^* be the adjoint linear relation of A in the Cartesian product $\mathfrak{H} \times \mathfrak{H}$. Then A^* can be represented as follows:

$$A^* = \left\{ \{f, Tf + \varphi\} : f \in \mathfrak{H}, \quad \varphi \in \mathfrak{N} \right\} = \left\{ \{f, T^* f + \psi\} : f \in \mathfrak{H}, \quad \psi \in \mathfrak{N} \right\}.$$

Define the following bounded linear operators acting from A^* into $\mathfrak{N}$:

$$\Gamma_0\{f, f'\} = P_{\mathfrak{N}} f, \quad \Gamma_1\{f, f'\} = P_{\mathfrak{N}} T^* f - P_{\mathfrak{N}} f', \quad \Gamma_2\{f, f'\} = P_{\mathfrak{N}} T f - P_{\mathfrak{N}} f',$$

where $\{f, f'\} \in A^*$. Then $\{\mathfrak{N}, \Gamma_0, \Gamma_1, \Gamma_2\}$ forms a boundary value space for A^*. In particular, for all $\widehat{f} = \{f, f'\}$, $\widehat{g} = \{g, g'\} \in A^*$ the following identity holds

$$(f', g) - (f, g') = (\Gamma_0 \widehat{f}, \Gamma_1 \widehat{g}) - (\Gamma_2 \widehat{f}, \Gamma_0 \widehat{g}),$$

and moreover $\ker \Gamma_1 = T^*$, $\ker \Gamma_2 = T$, and

$$\ker \Gamma_0 = \left\{ \{h, A_0 h + \varphi\} : h \in \mathfrak{H}_0, \quad \varphi \in \mathfrak{N} \right\}.$$

The corresponding γ-fields are the following operator functions

$$\begin{cases} \gamma_0(z)\varphi = -(A_0 - zI)^{-1} K_0^* D_{A_0} \varphi, \\ \gamma_1(z)\varphi = (T^* - zI)^{-1} \varphi, \\ \gamma_2(z)\varphi = (T - zI)^{-1} \varphi, \end{cases}$$

where $\varphi \in \mathfrak{N}$ and $|z| > 1$. It follows that $Q_T(z) = \Gamma_0 \gamma_2(z)$ is given by

$$Q_T(z) = P_{\mathfrak{N}} (T - zI)^{-1} \upharpoonright \mathfrak{N},$$

and that $-Q_T^{-1}(z) = \Gamma_2 \gamma_0(z)$ is given by

$$-Q_T^{-1}(z) = \left(K_0 \left[A_0 + (A_0 - zI)^{-1} (I - A_0^2) \right] K_0^* - D_{K_0^*} X D_{K_0^*} + zI \right) \upharpoonright \mathfrak{N},$$

where T is decomposed as in (3.2); see also Proposition 4.1. In particular, this means that $Q_T(z)$ can be interpreted as the Weyl function corresponding to the boundary value space $\{\mathfrak{N}, \Gamma_0, \Gamma_1, \Gamma_2\}$ in the sense of [39], [40].

5. The resolvent formula for *qsc*-extensions

Let $A = A_0 + K_0 D_{A_0}$ be a closed symmetric contraction in $\mathfrak{H}$ and let T be a *qsc*-extension of A given by the block matrix (3.2). If $Q_T(z) = P_{\mathfrak{N}} (T - zI)^{-1} \upharpoonright \mathfrak{N}$ is the Q-function of T, then by (4.9) the operator $(Q_T^{-1}(-1) - Q_T^{-1}(1))/2$ is nonnegative on $\mathfrak{N}$. Let

$$\mathbf{B}_{Q_T} := \mathbf{B} \left(-\frac{Q_T^{-1}(-1) + Q_T^{-1}(1)}{2}, \frac{Q_T^{-1}(-1) - Q_T^{-1}(1)}{2} \right) \tag{5.1}$$

be the operator ball in $\mathbf{L}(\mathfrak{N})$ with center

$$-(Q_T^{-1}(-1)+Q_T^{-1}(1))/2=-D_{K_0^*}XD_{K_0^*}$$

and equal left and right radii

$$(Q_T^{-1}(-1)-Q_T^{-1}(1))/2=D_{K_0^*}^2.$$

Recall that it is the set of all operators in $\mathfrak{N}$ of the form

$$-\frac{Q_T^{-1}(-1)+Q_T^{-1}(1)}{2}+\left(\frac{Q_T^{-1}(-1)-Q_T^{-1}(1)}{2}\right)^{1/2}Y\left(\frac{Q_T^{-1}(-1)-Q_T^{-1}(1)}{2}\right)^{1/2},$$

where $\|Y\|\leq 1$.

Theorem 5.1. *Let A be a closed symmetric operator in a Hilbert space $\mathfrak{H}$. Then the formula*

$$(\widetilde{T}-zI)^{-1}=(T-zI)^{-1}-(T-zI)^{-1}\widetilde{B}\left(I+Q_T(z)\widetilde{B}\right)^{-1}P_{\mathfrak{N}}(T-zI)^{-1} \quad (5.2)$$

with $|z|>1$ gives a one-to-one correspondence between the resolvents of all qsc-extensions $\widetilde{T}$ of A and all operators $\widetilde{B}$ belonging to the operator ball $\mathbf{B}_{Q_T}$ in (5.1).

Proof. By Theorem 3.1 every *qsc*-extension $\widetilde{T}$ of A can be written in the block form

$$\widetilde{T}=\begin{pmatrix} A_0 & D_{A_0}K_0^* \\ K_0D_{A_0} & -K_0A_0K_0^*+D_{K_0^*}\widetilde{Y}D_{K_0^*}\end{pmatrix}, \quad (5.3)$$

where $\|\widetilde{Y}\|\leq 1$. This together with (3.2) gives

$$\widetilde{B}:=\left(\widetilde{T}-T\right)\upharpoonright\mathfrak{N}=-D_{K_0^*}XD_{K_0^*}+D_{K_0^*}\widetilde{Y}D_{K_0^*} \quad (5.4)$$

which in view of (4.9) this means that $\widetilde{B}\in\mathbf{B}_{Q_T}$. It follows from

$$\widetilde{T}-zI=T-zI+\widetilde{B}P_{\mathfrak{N}} \quad (5.5)$$

that

$$(T-z)^{-1}=(\widetilde{T}-z)^{-1}+(T-z)^{-1}\widetilde{B}P_{\mathfrak{N}}(\widetilde{T}-z)^{-1}, \quad |z|>1,$$

and compression to $\mathfrak{N}$ leads to

$$Q(z)=\widetilde{Q}(z)+Q(z)\widetilde{B}\widetilde{Q}(z).$$

Since $Q(z)$ and $\widetilde{Q}(z)$ are invertible by part (ii) of Proposition 4.1, one obtains

$$\widetilde{Q}(z)^{-1}=Q(z)^{-1}+\widetilde{B}=Q(z)^{-1}(I+Q(z)\widetilde{B})=(I+\widetilde{B}Q(z))Q(z)^{-1}.$$

Therefore, the operators

$$I+Q(z)\widetilde{B}\quad\text{and}\quad I+\widetilde{B}Q(z),\quad |z|>1,$$

are invertible in $\mathfrak{N}$, too. Furthermore, by rewriting (5.5) in the form

$$\widetilde{T}-zI=\left(I+\widetilde{B}P_{\mathfrak{N}}(T-zI)^{-1}\right)(T-zI),$$

it is clear that $\left(I+\widetilde{B}P_{\mathfrak{N}}(T-zI)^{-1}\right)^{-1}\in\mathbf{L}(\mathfrak{H})$ for every $|z|>1$ and

$$(\widetilde{T}-zI)^{-1}=(T-zI)^{-1}\left(I+\widetilde{B}P_{\mathfrak{N}}(T-zI)^{-1}\right)^{-1},\quad |z|>1. \tag{5.6}$$

It also follows from (5.5) that

$$(\widetilde{T}-zI)^{-1}-(T-zI)^{-1}=-(\widetilde{T}-zI)^{-1}\widetilde{B}P_{\mathfrak{N}}(T-zI)^{-1}. \tag{5.7}$$

Now using the identities (5.6), (5.7), and

$$\left(I+\widetilde{B}P_{\mathfrak{N}}(T-zI)^{-1}\right)^{-1}\widetilde{B}P_{\mathfrak{N}}=\widetilde{B}P_{\mathfrak{N}}\left(I+P_{\mathfrak{N}}(T-zI)^{-1}\widetilde{B}P_{\mathfrak{N}}\right)^{-1},$$

one obtains

$$\begin{aligned}(\widetilde{T}-zI)^{-1}&-(T-zI)^{-1}\\ &=-(T-zI)^{-1}\left(I+\widetilde{B}P_{\mathfrak{N}}(T-zI)^{-1}\right)^{-1}\widetilde{B}P_{\mathfrak{N}}(T-zI)^{-1}\\ &=-(T-zI)^{-1}\widetilde{B}P_{\mathfrak{N}}\left(I+P_{\mathfrak{N}}(T-zI)^{-1}\widetilde{B}P_{\mathfrak{N}}\right)^{-1}P_{\mathfrak{N}}(T-zI)^{-1}\\ &=-(T-zI)^{-1}\widetilde{B}\left(I+Q_T(z)\widetilde{B}\right)^{-1}P_{\mathfrak{N}}(T-zI)^{-1},\end{aligned}$$

which gives the required identity (5.2).

Conversely, assume that $\widetilde{B}\in\mathbf{B}_{Q_T}$, i.e., that $\widetilde{B}$ is given by

$$-\frac{Q_T^{-1}(-1)+Q_T^{-1}(1)}{2}+\left(\frac{Q_T^{-1}(-1)-Q_T^{-1}(1)}{2}\right)^{1/2}\widetilde{Y}\left(\frac{Q_T^{-1}(-1)-Q_T^{-1}(1)}{2}\right)^{1/2}$$

for some $\|\widetilde{Y}\|\leq 1$. By (4.9) one has $\widetilde{B}=-D_{K_0^*}XD_{K_0^*}+D_{K_0^*}\widetilde{Y}D_{K_0^*}$. Consider the *qsc*-extension $\widetilde{T}$ of A given by the block operator $\widetilde{T}$ of the form (5.3) which is determined by $\widetilde{Y}$. Then clearly $\widetilde{B}=\left(\widetilde{T}-T\right)\upharpoonright\mathfrak{N}$. As was shown above, the operator $I+Q_T(z)\widetilde{B}$ is invertible for all $|z|>1$ and the resolvent of $\widetilde{T}$ takes the form (5.2).

The one-to-one correspondence is clear from the given arguments. □

Observe, that the Q-function $Q_{\widetilde{T}}(z)$ of the operator $\widetilde{T}$ in (5.2) and the Q-function $Q_T(z)$ of T are connected via

$$\begin{aligned}Q_{\widetilde{T}}(z)&=P_{\mathfrak{N}}(\widetilde{T}-zI)^{-1}\upharpoonright\mathfrak{N}\\ &=(I+Q_T(z)\widetilde{B})^{-1}Q_T(z)=Q_T(z)(I+\widetilde{B}Q_T(z))^{-1}\\ &=(\widetilde{B}+Q_T^{-1}(z))^{-1}.\end{aligned} \tag{5.8}$$

Remark 5.2. The resolvent formula (5.2) established in the theorem for the *qsc*-extensions $\widetilde{T}$ of A remains true for all quasi-selfadjoint extensions of A, i.e., T and $\widetilde{T}$ in Theorem 5.1 can be taken to be bounded, not necessarily contractive, extensions of A. Indeed, by defining the functions $Q_T(z)$, $|z|>\|T\|$, and $Q_{\widetilde{T}}(z)$, $|z|>\|\widetilde{T}\|$ as in (4.1) then they are bounded and boundedly invertible (cf., e.g., the proof presented for part (ii) of Proposition 4.1). It remains to repeat the proof of Theorem 5.1 to obtain the resolvent formula (5.2) for two arbitrary bounded quasi-selfadjoint extensions T and $\widetilde{T}$ of A in the domain $|z| > \max\{\|T\|, \|\widetilde{T}\|\}$. The parameter $\widetilde{B}$ is still given by (5.4), but now X and $\widetilde{Y}$ are not in general contractive and even if T is a qsc-extension of A ($\|X\| \leq 1$), $\widetilde{B}$ need not belong to the operator ball $\mathbf{B}_{Q_T}$ in (5.1). Of course, $\widetilde{T}$ can be interpreted as a bounded perturbation of T by the parameter $\widetilde{B}$ in (5.4). This allows one to apply the resolvent formula (5.2) to study, for instance, the behavior of the resolvents under arbitrary small perturbations $\|\widetilde{B}\| \leq \varepsilon$ with $\widetilde{B}$ not necessarily belonging to the operator ball $\mathbf{B}_{Q_T}$ in (5.1).

Descriptions of canonical and generalized resolvents of selfadjoint contractive extensions of a non-densely defined symmetric contraction A were given by M.G. Kreĭn and I.E. Ovcharenko in [32] by means of Q-functions of the selfadjoint extremal extensions A_μ and A_M. Later B. Textorius [47] described this set using an arbitrary selfadjoint contractive extension $\widetilde{A}$ of A and the corresponding special form of its Q-function. In [10], [12], [11] all canonical and generalized resolvents of qsc-extensions were parametrized by means of Q_μ- and Q_M-functions in the sense of Kreĭn-Ovcharenko in [32]. The proof of the parametrization formula (5.2), which is presented here, is kept very elementary, along the lines of a similar presentation in [9]. An abstract formula for canonical and generalized resolvents of extensions of a dual pair of linear relations in terms of a boundary value space and its Weyl function is derived in [39], [40].

6. A reproducing kernel space model for Q-functions of quasi-selfadjoint contractions

Let $\mathfrak{N}$ be a Hilbert space. An operator-valued function $Q(z)$ with values in $\mathbf{L}(\mathfrak{N})$ and holomorphic outside the unit disk is said to belong to the class $\mathbf{Q}(\mathfrak{N})$ if:

(i) $Q(z)$ has the expansion

$$Q(z) = -\frac{1}{z}I + \frac{1}{z^2}F + o\left(\frac{1}{z^2}\right), \quad z \to \infty; \tag{6.1}$$

(ii) the $\mathbf{L}(\mathfrak{N})$-valued function

$$\mathsf{G}(z,\xi) = \frac{Q(z) - Q(\xi)^* - Q(z)(F - F^*)Q(\xi)^*}{z - \bar{\xi}}, \quad z \neq \bar{\xi},$$

with $|z|, |\xi| > 1$, is a nonnegative kernel;

(iii) the $\mathbf{L}(\mathfrak{N})$-valued function

$$\mathsf{L}(z,\xi) = \frac{(1-z^2)Q(z) - (1-\bar{\xi}^2)Q(\xi)^* - (1-z\bar{\xi})Q(z)(F-F^*)Q(\xi)^* - (z-\bar{\xi})I}{z-\bar{\xi}},$$

with $z \neq \bar{\xi}$, $|z|, |\xi| > 1$, is a nonnegative kernel;

(iv) there exist a complex number z_0, $|z_0| > 1$, and a vector $f \in \mathfrak{N}$, such that $\mathsf{G}(z_0, z_0)f \neq Q(z_0)Q(z_0)^* f$.

If T is a qsc-operator in the Hilbert space $\mathfrak{H}$, $\mathfrak{N}$ is a subspace of $\mathfrak{H}$ such that $\mathfrak{N} \neq \mathfrak{H}$ and $\operatorname{ran}(T - T^*) \subset \mathfrak{N}$, and $Q_T(z)$ is its Q-function defined by (4.1), then according to Propositions 4.1, 4.2, and 4.3 the function $Q_T(z)$ belongs to the class $\mathbf{Q}(\mathfrak{N})$. The converse statement is also true.

Theorem 6.1. *Let $Q(z)$ be a function of the class $\mathbf{Q}(\mathfrak{N})$. Then there exist a Hilbert space $\mathfrak{H} \supset \mathfrak{N}$, $\mathfrak{N} \neq \mathfrak{H}$, and an $\mathfrak{N}$-minimal qsc-operator T in $\mathfrak{H}$, such that $\mathfrak{N} \supset \operatorname{ran}(T - T^*)$ and*

$$Q(z) = P_{\mathfrak{N}}(T - zI)^{-1} \upharpoonright \mathfrak{N}, \quad \textit{for all } |z| > 1. \tag{6.2}$$

If, in addition, the $\mathbf{L}(\mathfrak{N})$-valued function

$$\begin{aligned}\mathsf{K}(z,\xi) &:= \mathsf{L}(z,\xi) - Q(z)(F - F^*)Q(\xi)^* \\ &= \frac{(1 - z^2)Q(z) - (1 - \bar{\xi}^2)Q(\xi)^* - (1 + z)(1 - \bar{\xi})Q(z)(F - F^*)Q(\xi)^* - (z - \bar{\xi})I}{z - \bar{\xi}}\end{aligned}$$

with $z \neq \bar{\xi}$, $|z|, |\xi| > 1$, where F is given by (6.1), is an α-sectorial kernel with $\alpha \in [0, \pi/2)$, then the corresponding operator T belongs to the class $C(\alpha)$.

Proof. Step 1. Let $\widetilde{\mathfrak{H}}$ be the reproducing kernel Hilbert space associated with the nonnegative kernel $\mathsf{G}(z, \xi)$, i.e., $\widetilde{\mathfrak{H}}$ is the completion of

$$\operatorname{span}\{\, \mathsf{G}(\cdot, w)f \,:\, f \in \mathfrak{N}, \ |w| > 1 \,\}$$

with respect to the norm determined by the inner product

$$(\mathsf{G}(\cdot, w)f, \mathsf{G}(\cdot, \mu)g)_{\widetilde{\mathfrak{H}}} = (\mathsf{G}(\mu, w)f, g)_{\mathfrak{N}}.$$

For all $f \in \mathfrak{N}$ and $|w|, |\mu| > 1$,

$$\begin{aligned}\|\overline{w}\mathsf{G}(\cdot, w)f - \bar{\mu}\mathsf{G}(\cdot, \mu)f\|^2_{\widetilde{\mathfrak{H}}} &= |w|^2 \left(\mathsf{G}(w, w)f, f\right)_{\mathfrak{N}} + |\mu|^2 \left(\mathsf{G}(\mu, \mu)f, f\right)_{\mathfrak{N}} \\ &\quad - \mu\overline{w} \left(\mathsf{G}(\mu, w)f, f\right)_{\mathfrak{N}} - \bar{\mu}w \left(\mathsf{G}(w, \mu)f, f\right)_{\mathfrak{N}}.\end{aligned} \tag{6.3}$$

In view of (6.1) one has $Q(z) = (-1/z)I + o(1/z)$ as $z \to \infty$, which implies that

$$\lim_{w \to \infty} \overline{w}\mathsf{G}(z, w)f = -Q(z)f, \quad |z| > 1, \tag{6.4}$$

and moreover that

$$\lim_{\mu, w \widehat{\to} \infty} \mu\overline{w}\mathsf{G}(\mu, w)f = f, \quad f \in \mathfrak{N}. \tag{6.5}$$

(Here $\widehat{\to}$ stands for the nontangential limit in a sector $|\arg(z) - \pi/2| \leq \alpha < \pi/2$.) Hence (6.3) and (6.5) imply that the following limit exists in $\widetilde{\mathfrak{H}}$

$$Kf := -\lim_{w \widehat{\to} \infty} \overline{w}\mathsf{G}(z, w)f \tag{6.6}$$

and defines a linear operator $K : \mathfrak{N} \to \widetilde{\mathfrak{H}}$ for which

$$\|Kf\|^2_{\widetilde{\mathfrak{H}}} = \lim_{w \widehat{\to} \infty} \|\overline{w}\mathsf{G}(\cdot, w)f\|^2_{\widetilde{\mathfrak{H}}} = \lim_{w \widehat{\to} \infty} |w|^2 \left(\mathsf{G}(w, w)f, f\right)_{\mathfrak{N}} = \|f\|^2_{\mathfrak{N}}. \tag{6.7}$$

Thus, K is isometric. It follows from (6.4) that

$$\begin{aligned}(Kf, \mathsf{G}(\cdot,\mu)g)_{\widetilde{\mathfrak{H}}} &= -\lim_{w\widehat{\to}\infty} \overline{w}\,(\mathsf{G}(\cdot,w)f, \mathsf{G}(\cdot,\mu)g)_{\widetilde{\mathfrak{H}}} \\ &= -\lim_{w\widehat{\to}\infty} \overline{w}\,(\mathsf{G}(\mu,w)f, g)_{\mathfrak{N}} = (Q(\mu)f, g)_{\mathfrak{N}},\end{aligned}$$

which shows that

$$K^*\mathsf{G}(\cdot,\mu)g = Q(\mu)^*g, \quad g \in \mathfrak{N}. \tag{6.8}$$

Step 2. Define the linear relation $\mathcal{S}$ in $\widetilde{\mathfrak{H}}$ by

$$\mathcal{S} = \left\{ \left\{ \sum_{i=1}^{n} \mathsf{G}(\cdot,w_i)f_i, \sum_{i=1}^{n} Kf_i + \sum_{i=1}^{n} \overline{w}_i \mathsf{G}(\cdot,w_i)f_i \right\} : \, f_i \in \mathfrak{N}, \ |w_i| > 1 \right\}. \tag{6.9}$$

By definition the domain of $\mathcal{S}$ is dense in $\widetilde{\mathfrak{H}}$. In fact, $\mathcal{S}$ is a contractive linear operator in $\widetilde{\mathfrak{H}}$, since

$$\begin{aligned}&\left\| \sum_{i=1}^{n} \mathsf{G}(\cdot,w_i)f_i \right\|_{\widetilde{\mathfrak{H}}}^2 - \left\| \sum_{i=1}^{n} Kf_i + \sum_{i=1}^{n} \overline{w}_i \mathsf{G}(\cdot,w_i)f_i \right\|_{\widetilde{\mathfrak{H}}}^2 \\ &= \sum_{i,j=1}^{n} (\mathsf{G}(w_j,w_i)f_i, f_j)_{\mathfrak{N}} - \sum_{i,j=1}^{n} (f_i, f_j)_{\mathfrak{N}} - \sum_{i,j=1}^{n} \overline{w}_i (Q(w_i)^* f_i, f_j)_{\mathfrak{N}} \\ &\quad - \sum_{i,j=1}^{n} w_j (Q(w_j)f_i, f_j)_{\mathfrak{N}} - \sum_{i,j=1}^{n} w_j \overline{w}_i (\mathsf{G}(w_j,w_i)f_i, f_j)_{\mathfrak{N}} \\ &= \sum_{i,j=1}^{n} (\mathsf{L}(w_j,w_i)f_i, f_j)_{\mathfrak{N}} \geq 0,\end{aligned}$$

where (6.7) and (6.8) have been used. Therefore, the operator $\mathcal{S}$ has a unique contractive continuation which is defined everywhere on $\widetilde{\mathfrak{H}}$ and for which the same notation $\mathcal{S}$ is preserved.

Step 3. To calculate the imaginary part of $\mathcal{S}$ note that for $h = \sum_{i=1}^{n} \mathsf{G}(\cdot,w_i)f_i$ the following identities holds

$$\begin{aligned}(\mathcal{S}h, h) &= \left(\sum_{i=1}^{n} Kf_i + \sum_{i=1}^{n} \overline{w}_i \mathsf{G}(\cdot,w_i)f_i, \sum_{j=1}^{n} \mathsf{G}(\cdot,w_j)f_j \right)_{\widetilde{\mathfrak{H}}} \\ &= \sum_{i,j=1}^{n} (Q(w_j)f_i + \overline{w}_i \mathsf{G}(w_j,w_i)f_i, f_j)_{\mathfrak{N}}.\end{aligned}$$

Similarly one obtains

$$(h, \mathcal{S}h) = \sum_{i,j=1}^{n} (Q(w_i)^* f_i + w_j \mathsf{G}(w_j,w_i)f_i, f_j)_{\mathfrak{N}}.$$

Since $Q(w_j) - Q(w_i)^* + (\overline{w}_i - w_j)\mathsf{G}(w_j, w_i) = Q(w_j)(F - F^*)Q(w_i)^*$, one obtains

$$\begin{aligned}
&\left((\mathcal{S} - \mathcal{S}^*)\left(\sum_{i=1}^n \mathsf{G}(\cdot, w_i)f_i\right), \sum_{j=1}^n \mathsf{G}(\cdot, w_j)f_j\right)_{\widetilde{\mathfrak{H}}} \\
&\qquad = \sum_{i,j=1}^n (Q(w_j)(F - F^*)Q(w_i)^* f_i, f_j)_{\mathfrak{N}} \\
&\qquad = \sum_{i,j=1}^n ((F - F^*)K^*\mathsf{G}(\cdot, w_i)f_i, K^*\mathsf{G}(\cdot, w_j)f_j)_{\mathfrak{N}} \\
&\qquad = \left(K(F - F^*)K^*\left(\sum_{i=1}^n \mathsf{G}(\cdot, w_i)f_i\right), \sum_{j=1}^n \mathsf{G}(\cdot, w_j)f_j\right)_{\widetilde{\mathfrak{H}}}.
\end{aligned}$$

This implies that

$$\mathcal{S} - \mathcal{S}^* = K(F - F^*)K^*. \tag{6.10}$$

By the definition of $\mathcal{S}$ in (6.9) one has $(\mathcal{S} - \overline{w}I)\mathsf{G}(\cdot, w)f = Kf$, so that

$$(\mathcal{S} - \overline{w}I)^{-1}Kf = \mathsf{G}(\cdot, w)f, \quad f \in \mathfrak{N}, \quad |w| > 1. \tag{6.11}$$

Step 4. Since K is isometric, $\operatorname{ran} K$ is closed. Let $\mathfrak{H}_0 := \ker K^*$ and define $\mathfrak{H} := \mathfrak{H}_0 \oplus \mathfrak{N}$. Observe, that according to (6.8) $h = \sum_{i=1}^n \mathsf{G}(\cdot, w_i)f_i$ belongs to the subspace $\mathfrak{H}_0$ of $\widetilde{\mathfrak{H}}$ if and only if $\sum_{i=1}^n Q(w_i)^* f_i = 0$. Now decompose $\widetilde{\mathfrak{H}} = \mathfrak{H}_0 \oplus \operatorname{ran} K$ and define the operator $\mathfrak{U} : \widetilde{\mathfrak{H}} \to \mathfrak{H}$ by

$$\mathfrak{U}(x + y) = x + K^*y, \quad x \in \mathfrak{H}_0, \quad y \in \operatorname{ran} K.$$

Then $\mathfrak{U}$ maps $\widetilde{\mathfrak{H}}$ onto $\mathcal{H}$ and it is unitary. Hence, the operator T defined by $T := \mathfrak{U}\mathcal{S}^*\mathfrak{U}^{-1}$ is contractive in $\mathfrak{H}$ and (6.10) shows that $\operatorname{ran}(T - T^*) \subset \mathfrak{U}(\operatorname{ran} K) = \mathfrak{N}$. Furthermore, for $f, g \in \mathfrak{N}$ and $|z| > 1$ the identities (6.8) and (6.11) yield

$$\begin{aligned}
\left((T - z)^{-1}f, g\right)_{\mathfrak{H}} &= \left((\mathcal{S}^* - zI)^{-1}\mathfrak{U}^{-1}f, \mathfrak{U}^{-1}g\right)_{\widetilde{\mathfrak{H}}} \\
&= \left((\mathcal{S}^* - zI)^{-1}Kf, Kg\right)_{\widetilde{\mathfrak{H}}} \\
&= \left(Kf, (\mathcal{S} - \bar{z}I)^{-1}Kg\right)_{\widetilde{\mathfrak{H}}} \\
&= (Kf, \mathsf{G}(\cdot, z)g)_{\widetilde{\mathfrak{H}}} \\
&= (Q(z)f, g)_{\mathfrak{N}}.
\end{aligned}$$

Thus,

$$Q(z) = P_{\mathfrak{N}}(T - zI)^{-1}\upharpoonright \mathfrak{N}, \quad |z| > 1.$$

Moreover, it follows from (6.11) that the operator T is $\mathfrak{N}$-minimal.

Step 5. Finally it is shown that $\mathfrak{H}_0 \neq \{0\}$. If $\mathfrak{H}_0 = \{0\}$ then $\mathfrak{N} = \mathfrak{H}$ and by Proposition 4.3 the equality $\mathsf{G}(z, z) = Q(z)Q(z)^*$ holds for all $|z| > 1$. But this is impossible due to the condition (iv) of the definition of the class $\mathbf{Q}(\mathfrak{N})$. Therefore $\mathfrak{H}_0 \neq 0$, $\mathfrak{N} \neq \mathfrak{H}$, and T is a *qsc*-operator whose Q-function $Q_T(z)$ coincides with $Q(z)$.

As to the last statement observe, that since $Q(z)$ is of the form (6.2), the kernel $\mathsf{K}(z,\xi)$ admits the operator representation (4.17) in Proposition 4.2. Since T is $\mathfrak{N}$-minimal, it follows from (4.17) and (2.6) that $T \in C(\alpha)$. □

The *qsc*-operator T constructed in Theorem 6.1 is $\mathfrak{N}$-minimal. The next result shows that this model for functions $Q(z)$ belonging to the class $\mathbf{Q}(\mathfrak{N})$ is essentially unique. Namely, the $\mathfrak{N}$-minimal part of a *qsc*-operator T (and hence also of T^*) is up to unitary equivalence uniquely determined by its Q-function; a fact which is well known in the selfadjoint case.

Theorem 6.2. *Let $\mathfrak{H}_1 = \mathfrak{H}_{01} \oplus \mathfrak{N}$ and $\mathfrak{H}_2 = \mathfrak{H}_{02} \oplus \mathfrak{N}$ be two Hilbert spaces, and let T_1 and T_2 be qsc-operators in $\mathfrak{H}_1$ and $\mathfrak{H}_2$, respectively, such that* $\operatorname{ran}(T_1 - T_1^*) \subset \mathfrak{N}$ *and* $\operatorname{ran}(T_2 - T_2^*) \subset \mathfrak{N}$. *If $Q_{T_1}(z) = Q_{T_2}(z)$ in some neighborhood of infinity then the $\mathfrak{N}$-minimal parts of T_1 and T_2 are unitarily equivalent.*

Proof. Assume that $Q_{T_1}(z) = Q_{T_2}(z)$ holds in some neighborhood of infinity, say, for $|z| > R > 1$. Then these functions coincide everywhere outside the unit disk. It follows from (4.3) and (4.7) that $F_1 = F_2$, while (4.14) implies that

$$P_{\mathfrak{N}}(T_1^* - \bar{\xi}I)^{-1}(T_1 - zI)^{-1}\restriction\mathfrak{N} = P_{\mathfrak{N}}(T_2^* - \bar{\xi}I)^{-1}(T_2 - zI)^{-1}\restriction\mathfrak{N},$$

for all $|z|, |\xi| > 1$; cf. (4.2). Hence, for all $f, g \in \mathfrak{N}$

$$\left((T_1 - zI)^{-1}f, (T_1 - \xi I)^{-1}g\right) = \left((T_2 - zI)^{-1}f, (T_2 - \xi I)^{-1}g\right). \tag{6.12}$$

Now define the linear relation U from $\mathfrak{H}_1' = \overline{\operatorname{span}}\left\{(T_1 - zI)^{-1}\mathfrak{N} : |z| > 1\right\}$ into $\mathfrak{H}_2' = \overline{\operatorname{span}}\left\{(T_2 - zI)^{-1}\mathfrak{N} : |z| > 1\right\}$ by the formula

$$U = \left\{ \sum_{k=1}^{n}(T_1 - z_kI)^{-1}f_k, \sum_{k=1}^{n}(T_2 - z_kI)^{-1}f_k \right\}.$$

Then the identity (6.12) implies that U is a unitary operator from $\mathfrak{H}_1'$ onto $\mathfrak{H}_2'$. In addition, $Uf = f$ for all $f \in \mathfrak{N}$, and

$$\begin{aligned} UT_1\left(\sum_{k=1}^{n}(T_1 - z_kI)^{-1}f_k\right) &= \sum_{k=1}^{n} f_k + U\left(\sum_{k=1}^{n} z_k\,(T_1 - z_kI)^{-1}f_k\right) \\ &= \sum_{k=1}^{n} f_k + \sum_{k=1}^{n} z_k(T_2 - z_kI)^{-1}f_k = T_2U\left(\sum_{k=1}^{n}(T_1 - z_kI)^{-1}f_k\right). \end{aligned}$$

Therefore, the simple parts of T_1 and T_2 are unitarily equivalent. □

7. Characteristic properties of Q-functions of quasi-selfadjoint contractions

The definition of the class $\mathbf{Q}(\mathfrak{N})$ given in Section 6 can be seen as an analytical characterization for Q-functions of *qsc*-operators T as defined in (4.1). Another characterization is established in the next theorem.

Theorem 7.1. *Let $\mathfrak{N}$ be a Hilbert space. The following conditions are equivalent:*

(i) *the function $Q(z)$ belongs to the class $\mathbf{Q}(\mathfrak{N})$;*

(ii) (a) *$Q(z) \in \mathbf{L}(\mathfrak{N})$ is holomorphic in the domain $|z| > 1$ and with $F \in \mathbf{L}(\mathfrak{N})$ it has the asymptotic expansion*

$$Q(z) = -\frac{1}{z}I + \frac{1}{z^2}F + o\left(\frac{1}{z^2}\right), \qquad z \to \infty;$$

(b) *the function*

$$-Q^{-1}(z) - zI - F$$

is not constant, it has a holomorphic continuation onto $\operatorname{Ext}[-1,1]$ *as a bounded Nevanlinna function, and the strong limits $Q^{-1}(\pm 1)$ exist;*

(c) *$Q^{-1}(-1) - Q^{-1}(1) \geq 0$ and for all $f, g \in \mathfrak{N}$ the following inequality holds:*

$$\begin{aligned} &\left|\left(\left(Q^{-1}(-1) + Q^{-1}(1)\right) f, g\right)\right|^2 \\ &\quad \leq \left(\left(Q^{-1}(-1) - Q^{-1}(1)\right) f, f\right) \left(\left(Q^{-1}(-1) - Q^{-1}(1)\right) g, g\right). \end{aligned}$$

Proof. (i) $\Rightarrow$ (ii) Let the function $Q(z)$ belong to the class $\mathbf{Q}(\mathfrak{N})$. Then (a) holds by definition, see (6.1). By Theorem 6.1 the function $Q(z)$ has the operator representation $Q(z) = P_{\mathfrak{N}}(T - zI)^{-1}{\upharpoonright}\,\mathfrak{N}$, where T is a *qsc*-operator in a Hilbert space $\mathfrak{H} \supset \mathfrak{N}$, such that $\operatorname{ran}(T - T^*) \subset \mathfrak{N}$. Now (b) follows from parts (ii) and (v) of Proposition 4.1 and Proposition 4.3, see also the identity (4.19). The inequality in (c) is obtained from part (iv) of Proposition 4.1.

(ii) $\Rightarrow$ (i) Now assume that the function $Q(z)$ has the properties (a)–(c). It follows from (a) and (b) that

$$Q^{-1}(z) = -zI - F - G(z), \quad G(z) = o(1), \quad z \to \infty.$$

Here $G(z) \in \mathbf{N}_{\mathfrak{N}}[-1,1]$ and $G(\infty) = 0$. Now it follows from Theorem 2.3 that $G(z)$ has the representation

$$G(z) = K_0(A_0 - zI)^{-1}(I - A_0^2)K_0^*,$$

where A_0 is a selfadjoint contraction in some Hilbert space $\mathfrak{H}_0$ and $K_0 \in \mathbf{L}(\mathfrak{H}_0, \mathfrak{N})$. Moreover, according to (2.14)

$$\begin{aligned} G(-1) &= -Q^{-1}(-1) + I - F = K_0(I - A_0)K_0^*, \\ G(1) &= -Q^{-1}(1) - I - F = -K_0(I + A_0)K_0^*. \end{aligned}$$

This gives

$$\begin{cases} \dfrac{Q^{-1}(-1) - Q^{-1}(1)}{2} = I - K_0K_0^*, \\[2ex] \dfrac{Q^{-1}(-1) + Q^{-1}(1)}{2} = K_0A_0K_0^* - F. \end{cases} \tag{7.1}$$

Now the assumption (c) implies that $I - K_0K_0^* \geq 0$ and

$$|((K_0A_0K_0^* - F)f, g)| \leq \|D_{K_0^*}f\|\,\|D_{K_0^*}g\|, \quad f, g \in \mathfrak{N}.$$

By (2.3) there exists a contraction X in $\mathfrak{D}_{K_0^*}$ such that

$$-F = -K_0 A_0 K_0^* + D_{K_0^*} X D_{K_0^*}. \tag{7.2}$$

Consider the Hilbert space $\mathfrak{H} = \mathfrak{H}_0 \oplus \mathfrak{N}$ and let the operator T in $\mathfrak{H}$ be given by the block form (3.2). Then T is a contraction and, in fact, a *qsc*-extension of the closed symmetric contraction $A = A_0 + K_0 D_{A_0}$ defined on $\mathfrak{H}_0$. According to Schur-Frobenius formula (see (2.9), (2.10))

$$P_{\mathfrak{N}}(T - zI)^{-1}{\restriction}\mathfrak{N} = -(G(z) + zI + F)^{-1} = Q(z), \quad |z| > 1,$$

i.e., $Q(z)$ is the Q-function of T. Therefore, $Q(z)$ belongs to the class $\mathbf{Q}(\mathfrak{N})$. □

The model established in Theorem 6.1 yields the following simple characterizations of Q-functions corresponding to the extreme selfadjoint contractive extensions A_μ and A_M of A within the class $\mathbf{Q}(\mathfrak{N})$. Recall, that all *sc*-extensions $T = T^*$ of A can be described as the operator interval $[A_\mu, A_M]$, see [27].

Proposition 7.2. *Let $Q(z)$ belong to the class $\mathbf{Q}(\mathfrak{N})$ and suppose that*

$$\liminf_{x \uparrow -1} |(Q(x)f, f)| = \infty, \quad \text{for all } f \in \mathfrak{N} \setminus \{0\}, \tag{7.3}$$

or

$$\liminf_{x \downarrow 1} |(Q(x)f, f)| = \infty, \quad \text{for all } f \in \mathfrak{N} \setminus \{0\}. \tag{7.4}$$

Then $Q(z)$ is a Nevanlinna function in $\mathbf{N}_{\mathfrak{N}}[-1,1]$ and it can be represented in the form $Q(z) = P_{\mathfrak{N}}(A_\mu - zI)^{-1}{\restriction}\mathfrak{N}$ or $Q(z) = P_{\mathfrak{N}}(A_M - zI)^{-1}{\restriction}\mathfrak{N}$, $z \in \operatorname{Ext}[-1,1]$, respectively, where A_μ and A_M are the left and right extreme sc-extension of some symmetric contraction A.

Proof. According to Theorem 6.1 the function $Q(z)$ has the operator representation $Q(z) = P_{\mathfrak{N}}(T - zI)^{-1}{\restriction}\mathfrak{N}$, where T is a *qsc*-operator in a Hilbert space $\mathfrak{H} \supset \mathfrak{N}$, such that $\operatorname{ran}(T - T^*) \subset \mathfrak{N}$. Moreover, T is a *qsc*-extension of the closed symmetric contraction A defined by $A = T{\restriction}\operatorname{dom} A$ with $\operatorname{dom} A = \mathfrak{H} \ominus \mathfrak{N}$. Let $T_R = (T + T^*)/2$ and $T_I = (T - T^*)/2$ be the real and the imaginary part of T, respectively, so that $T = T_R + iT_I$. Then for $|x| > 1$,

$$(T - xI)^{-1} = (T_R - xI)^{-1/2}\,(I + iB)^{-1}\,(T_R - xI)^{-1/2},$$

where

$$B = (T_R - xI)^{-1/2} T_I (T_R - xI)^{-1/2}$$

is a bounded selfadjoint operator. This shows that for all $f \in \mathfrak{N}$

$$(Q(x)f, f) = \left((I + iB)^{-1}(T_R - xI)^{-1/2} f, (T_R - xI)^{-1/2} f\right).$$

Since $\|(I + iB)^{-1}\| \le 1$, one obtains

$$|(Q(x)f, f)| \le \left\|(T_R - xI)^{-1/2} f\right\|^2.$$

Now the assumption (7.3) implies that

$$\liminf_{x \uparrow -1} \left\|(T_R - xI)^{-1/2} f\right\|^2 = \infty, \quad \text{for all } f \in \mathfrak{N} \setminus \{0\}.$$

This means that ran $(I+T_R)^{1/2} \cap \mathfrak{N} = \{0\}$, cf., e.g., [7]. Since T_R is a *sc*-extension of A, one concludes from the characterizations in (3.15) that $T_R = A_\mu$, cf. [27], [7], [22]. Now, in view of (3.14) $T_I = 0$ and $T = A_\mu$. The proof of the other statement is similar. □

Some further characteristic properties of Q-functions in the selfadjoint case, in particular, of Q_μ- and Q_M-functions corresponding to the *sc*-extensions A_μ and A_M have been established in [7], including some corrections to the results stated in [32].

8. Linear fractional transformations of the class $\mathbf{Q}(\mathfrak{N})$

The Kreĭn formula (5.2) and the discussion following it concerning the formulas in (5.8) give rise to a linear fractional transformation of Q-functions.

Theorem 8.1. *Let $Q(z)$ belong to the class $\mathbf{Q}(\mathfrak{N})$. Then the function*

$$Q(z)\,(I+BQ(z))^{-1}, \quad |z|>1,$$

belongs to the class $\mathbf{Q}(\mathfrak{N})$ if and only if

$$B \in \mathbf{B}\left(-\frac{Q^{-1}(-1)+Q^{-1}(1)}{2}, \frac{Q^{-1}(-1)-Q^{-1}(1)}{2}\right). \tag{8.1}$$

Moreover, $Q(z)\,(I+BQ(z))^{-1}$ is a Nevanlinna function of the class $\mathbf{N}_{\mathfrak{N}}[-1,1]$ if and only if B satisfies the conditions

$$B+Q^{-1}(1) \le 0, \quad B+Q^{-1}(-1) \ge 0. \tag{8.2}$$

Proof. First observe that, if $B \in \mathbf{L}(\mathfrak{N})$ and $(I+BQ(z))^{-1} \in \mathbf{L}(\mathfrak{N})$ for all $|z|>1$, then it follows from (6.1) that

$$\widetilde{Q}(z) := Q(z)\,(I+BQ(z))^{-1} = -\frac{1}{z}I + \frac{1}{z^2}(F-B) + o\left(\frac{1}{z^2}\right), \quad z \to \infty,$$

and clearly $\widetilde{Q}^{-1}(z) = Q^{-1}(z) + B$.

Now assume that $\widetilde{Q}(z) \in \mathbf{Q}(\mathfrak{N})$. Then $\widetilde{Q}(z) \in \mathbf{L}(\mathfrak{N})$, $|z|>1$, and since by Theorem 7.1 $Q(z)^{-1}, \widetilde{Q}(z)^{-1} \in \mathbf{L}(\mathfrak{N})$, $|z|>1$, one has $B, (I+BQ(z))^{-1} \in \mathbf{L}(\mathfrak{N})$ for all $|z|>1$. Moreover, the limit values $\widetilde{Q}^{-1}(\pm 1)$ exist and satisfy

$$\widetilde{Q}^{-1}(-1) = Q^{-1}(-1) + B, \quad \widetilde{Q}^{-1}(1) = Q^{-1}(1) + B.$$

Now part (c) of Theorem 7.1 implies that

$$\begin{aligned} &\left|\left(\left(B + \frac{Q^{-1}(-1)+Q^{-1}(1)}{2}\right) f, g\right)\right|^2 \\ &\quad \le \left(\frac{Q^{-1}(-1)-Q^{-1}(1)}{2} f, f\right)\left(\frac{Q^{-1}(-1)-Q^{-1}(1)}{2} g, g\right) \end{aligned} \tag{8.3}$$

holds for all $f, g \in \mathfrak{N}$. Therefore, the condition (8.1) is satisfied.

Conversely, let the operator $B \in \mathbf{L}(\mathfrak{N})$ satisfy the condition (8.1). By assumption $Q(z)$ belongs to $\mathbf{Q}(\mathfrak{N})$ and Theorem 6.1 shows that $Q(z) = P_{\mathfrak{N}}(T-z)^{-1}\upharpoonright\mathfrak{N}$, where T is qsc-operator in some Hilbert space $\mathfrak{H} \supset \mathfrak{N}$. Moreover, T is a qsc-extension of the symmetric contraction $A = T\upharpoonright\mathfrak{H}_0$, $\mathfrak{H}_0 = \mathfrak{H}\ominus\mathfrak{N}$. Now by Theorem 5.1 the assumption (8.1) means that B defines a qsc-extension $\widetilde{T}$ of A whose resolvent is given by (5.2) with $\widetilde{B} = B$. According to (5.8) the Q-function $Q_{\widetilde{T}}(z)$ is of the form $Q_{\widetilde{T}}(z) = Q(z)(I + BQ(z))^{-1}$, $|z| > 1$, and as a Q-function belongs to the class $\mathbf{Q}(\mathfrak{N})$; see the discussion preceding Theorem 6.1.

To prove the second part of the theorem, observe that in view of (4.5)

$$Q_{\widetilde{T}}^{-1}(-1) = B + Q^{-1}(-1) = D_{K_0^*}(Y+I)D_{K_0^*},$$

and

$$Q_{\widetilde{T}}^{-1}(1) = B + Q^{-1}(1) = D_{K_0^*}(Y-I)D_{K_0^*},$$

where Y is a contraction in the subspace $\mathfrak{D}_{K_0^*} = \overline{\operatorname{ran}}\, D_{K_0^*}$. By Theorem 3.1 $\widetilde{T}$ is a selfadjoint contraction if and only if Y is a selfadjoint contraction in $\mathfrak{D}_{K_0^*}$, or equivalently, B satisfies the conditions (8.2). Now, if (8.2) holds then $\widetilde{T}$ is selfadjoint and $Q(z)(I + BQ(z))^{-1} = Q_{\widetilde{T}}(z) \in \mathbf{N}_{\mathfrak{N}}[-1,1]$.

Conversely, if $Q_{\widetilde{T}}(z) \in \mathbf{N}_{\mathfrak{N}}[-1,1]$ then by part (vi) of Proposition 4.1 one has $\widetilde{F} = \widetilde{F}^*$ and consequently $\widetilde{T} = \widetilde{T}^*$, i.e., the conditions (8.2) are satisfied. □

References

[1] D. Alpay, A. Dijksma, J. Rovnyak, and H.S.V. de Snoo, *Schur functions, operator colligations, and Pontryagin spaces*, Oper. Theory: Adv. Appl., 96, Birkhäuser Verlag, Basel-Boston, 1997.

[2] Yu.M. Arlinskiĭ, *Contractive extensions of a dual pair of contractions and their resolvents*, Ukrain. Mat. Zh., 37 (1985), 247–250.

[3] Yu.M. Arlinskiĭ, *A class of contractions in a Hilbert space*, Ukrain. Mat. Zh., 39 (1987), 691–696.

[4] Yu.M. Arlinskiĭ, *Characteristic functions of operators of the class $C(\alpha)$*, Izv. Vuzov, Mat., 2 (1991), 13–21.

[5] Yu.M. Arlinskiĭ, *On functions connected with sectorial operators and their extensions*, Int. Equat. Oper. Theory. 33 (1999), 125–152.

[6] Yu.M. Arlinskiĭ, S. Hassi, Z. Sebestyén, and H.S.V. de Snoo, *On the class of extremal extensions of a nonnegative operator*, Oper. Theory: Adv. Appl. (B. Sz.-Nagy memorial volume), 127 (2001), 41–81.

[7] Yu.M. Arlinskiĭ, S. Hassi, and H.S.V. de Snoo, *Q-functions of Hermitian contractions of Kreĭn–Ovcharenko type*, Integral Equations Operator Theory, 53 (2005), 153–189.

[8] Yu.M. Arlinskiĭ, S. Hassi, and H.S.V. de Snoo, *Parametrization of contractive block-operator matrices and passive discrete-time systems*, in preparation.

[9] Yu.M. Arlinskiĭ, S. Hassi, H.S.V. de Snoo, and E.R. Tsekanovskiĭ, *One-dimensional perturbations of selfadjoint operators with finite or discrete spectrum*, Contemporary Mathematics, 323 (2003), 419–433.

[10] Yu.M. Arlinskiĭ and E.R. Tsekanovskiĭ, *Nonselfadjoint contractive extensions of Hermitian contractions and M.G. Kreĭn's theorems*, Usp. Mat. Nauk, 37, (1982) 131–132.

[11] Yu.M. Arlinskiĭ and E.R. Tsekanovskiĭ, *Generalized resolvents of quasi-selfadjoint contracting extensions of a Hermitian contraction*, Ukrain. Mat. Zh., 35 (1983), 601–603.

[12] Yu.M. Arlinskiĭ and E.R. Tsekanovskiĭ, *Quasi-selfadjoint contractive extensions of Hermitian contractions*, Teor. Funkts., Funkts. Anal. Prilozhen, 50 (1988), 9–16.

[13] N. Aronszajn, *Theory of reproducing kernels*, Trans. Amer. Math. Soc., 68 (1950), 337–404.

[14] Gr. Arsene and A. Gheondea, *Completing matrix contractions*, J. Operator Theory, 7 (1982), 179–189.

[15] M.S. Brodskiĭ, *Triangular and Jordan representations of linear operators*, Nauka, Moscow 1968 (Russian).

[16] M.G. Crandall, *Norm preserving extensions of linear transformations on Hilbert spaces*, Proc. Amer. Math. Soc., 21 (1969), 335–340.

[17] Ch. Davis, W.M. Kahan, and H.F. Weinberger, *Norm preserving dilations and their applications to optimal error bounds*, SIAM J. Numer. Anal., 19 (1982), 445–469.

[18] V.A. Derkach, S. Hassi, M.M. Malamud, and H.S.V. de Snoo, *Boundary relations and their Weyl families*, Trans. Amer. Math. Soc. (to appear).

[19] V.A. Derkach and M.M. Malamud, *The extension theory of hermitian operators and the moment problem*, J. Math. Sciences, 73 (1995), 141–242.

[20] R.G. Douglas, *On majorization, factorization and range inclusion of operators in Hilbert space*, Proc. Amer. Math. Soc., 17 (1966), 413–416.

[21] V.I. Gorbachuk and M.L. Gorbachuk, *Boundary value problems for operator differential equations*, Kluwer Acad. Publ., Dordrecht–Boston–London, 1991 (Russian edition: Naukova Dumka, Kiev, 1984).

[22] S. Hassi, M.M. Malamud, and H.S.V. de Snoo, *On Kreĭn's extension theory of nonnegative operators*, Math. Nachr., 274/275 (2004), 40–73.

[23] P.A. Fillmore and J.P. Williams, *On operator ranges*, Advances in Math., 7 (1971), 254–281.

[24] I.S. Kac and M.G. Kreĭn, *R-functions – analytic functions mapping the upper half-plane into itself*, Supplement to the Russian edition of F.V. Atkinson, *Discrete and continuous boundary problems*, Mir, Moscow 1968 (Russian) (English translation: Amer. Math. Soc. Transl. Ser. 2, 103 (1974), 1–18).

[25] T. Kato, *Perturbation theory for linear operators*, Springer-Verlag, Berlin, Heidelberg, 1966.

[26] M.G. Kreĭn, *On Hermitian operators whose deficiency indices are 1*, Doklady Acad. Sci. SSSR, 43 (1944), 323–326.

[27] M.G. Kreĭn, *The theory of selfadjoint extensions of semibounded Hermitian operators and its applications*, Mat. Sb., 20 (1947) 431–495; 21 (1947), 365–404.

[28] M.G. Kreĭn, *On the resolvents of an Hermitian operator with deficiency index* (m, m), Doklady Acad. Sci. SSSR, 52 (1946), 657–660.

[29] M.G. Kreĭn and H. Langer, *The defect subspaces and generalized resolvents of a Hermitian operator in the space* Π_κ, Funkcional. Anal. i Priložen, 5 (1971), No. 2, 59–71 (English translation: Funct. Anal. Appl., 5 (1971), 136–146).

[30] M.G. Kreĭn and H. Langer, *The defect subspaces and generalized resolvents of a Hermitian operator in the space* Π_κ, Funkcional. Anal. i Priložen, 5 (1971), No. 3, 54–69 (English translation: Funct. Anal. Appl., 5 (1971), 217–228).

[31] M.G. Kreĭn and H. Langer, *Über die Q-function eines π-hermiteschen Operators in Raume* Π_κ, Acta Sci. Math. (Szeged), 34 (1973), 191–230.

[32] M.G. Kreĭn and I.E. Ovcharenko, *On the Q-functions and sc-extensions of a Hermitian contraction with nondense domain*, Sibirsk. Mat. J., 18 (1977), 1032–1056.

[33] M.G. Kreĭn and Sh.N. Saakyan, *On some new results in the theory of the resolvent of Hermitian operators*, Doklady Acad. Sci. SSSR, 169 (1966), 1269–1271.

[34] H. Langer and B. Textorius, *Generalized resolvents of contractions*, Acta Sci. Math. (Szeged), 44 (1982), 125–131.

[35] H. Langer and B. Textorius, *Generalized resolvents of dual pairs of contractions*, Operator Theory: Adv. Appl., 6 (1982), 103–118.

[36] H. Langer and B. Textorius, *On generalized resolvents and Q-functions of symmetric linear relations (subspaces) in Hilbert space*, Pacific J. Math., 72 (1977), 135–165.

[37] V.E. Lyantse and O.G. Storozh, *Methods of the theory of unbounded operators*, Naukova Dumka, Kiev, 1983.

[38] M.M. Malamud, *On extensions of sectorial operators and dual pairs of contractions*, Sov. Math. Dokl., 39 (1989), 252–259.

[39] M.M. Malamud and V.I. Mogilevskiĭ, *On extensions of dual pairs of operators*, Reports of National Academy of Sciences of Ukraine, (1997), 30–37.

[40] M.M. Malamud and V.I. Mogilevskiĭ, *Kreĭn type formula for canonical resolvents of dual pairs of linear relations*, Methods of Functional Analysis and Topology, 8 (2002), 72–100.

[41] M.A. Naimark, *On spectral functions of a symmetric operator*, Izv. Akad. Nauk SSSR, Ser. Matem., 7 (1943), 285–296.

[42] S. Saitoh, *Theory of reproducing kernels and its applications*, Pitman Research Notes in Mathematics Series, 189. Longman Scientific & Technical, Harlow; copublished in the United States with John Wiley & Sons, Inc., New York, 1988.

[43] Yu.L. Shmul'yan, *Two-sided division in the ring of operators*, Mat. Zametki, 1 (1967), 605–610.

[44] Yu.L. Shmul'yan, *Operator balls*, Teor. Funkciĭ Funkcional. Anal. i Priložen, 6 (1968), 68–81 (English translation: Integral Equations Operator Theory, 13 (1990), 864–882).

[45] Yu.L. Shmul'yan and R.N. Yanovskaya, *Blocks of a contractive operator matrix*, Izv. Vuzov, Mat., 7 (1981), 72–75.

[46] B. Sz.-Nagy and C. Foias, *Harmonic analysis of operators on Hilbert space*, North-Holland, New York, 1970.

[47] B. Textorius, *On generalized resolvents of nondensely defined symmetric contractions*, Acta Sci. Math. (Szeged), 49 (1985), 329–338.

Yury Arlinskiĭ
Department of Mathematical Analysis
East Ukrainian National University
Kvartal Molodyozhny 20-A
Lugansk 91034
Ukraine
e-mail: yma@snu.edu.ua

Seppo Hassi
Department of Mathematics and Statistics
University of Vaasa
P.O. Box 700
65101 Vaasa
Finland
e-mail: sha@uwasa.fi

Henk de Snoo
Department of Mathematics and Computing Science
University of Groningen
P.O. Box 800
9700 AV Groningen
Nederland
e-mail: desnoo@math.rug.nl

Operator Theory:
Advances and Applications, Vol. 163, 55–73

A Class of Abstract Boundary Value Problems with Locally Definitizable Functions in the Boundary Condition

Jussi Behrndt

Dedicated to Professor Heinz Langer

Abstract. For a class of boundary value problems where the spectral parameter appears in the boundary condition in the form of a locally definitizable function linearizations are constructed and their local spectral properties are investigated.

Mathematics Subject Classification (2000). Primary: 47B50, 34B07; Secondary: 46C20, 47A06, 47B40.

Keywords. Boundary value problems, locally definitizable operators and relations in Krein spaces, locally definitizable functions, boundary value spaces.

1. Introduction

In this paper we study a class of boundary value problems with eigenvalue dependent boundary conditions. Let A be a closed symmetric operator or relation of defect one in a separable Krein space $\mathcal{K}$, let $\{\mathbb{C}, \Gamma_0, \Gamma_1\}$ be a boundary value space for the adjoint A^+ and let τ be a function locally holomorphic in some open subset of the extended complex plane which is symmetric with respect to the real line such that $\tau(\overline{\lambda}) = \overline{\tau(\lambda)}$ holds. We investigate boundary value problems of the following form: For a given $k \in \mathcal{K}$ find a vector $\hat{f} = \binom{f}{f'} \in A^+$ such that

$$f' - \lambda f = k \quad \text{and} \quad \tau(\lambda)\Gamma_0 \hat{f} + \Gamma_1 \hat{f} = 0 \tag{1.1}$$

holds. Under additional assumptions on τ and A a solution of this problem can be obtained with the help of the compressed resolvent of a selfadjoint extension $\widetilde{A}$ of A which acts in a larger Krein space. Making use of the coupling method from [6] we construct this so-called linearization $\widetilde{A}$ and we study its local spectral properties, which are closely connected with the solvability of (1.1).

More precisely, let Ω be some domain in $\overline{\mathbb{C}}$ symmetric with respect to the real line such that $\Omega \cap \overline{\mathbb{R}} \neq \emptyset$ and the intersections of Ω with the upper and lower open half-planes are simply connected. We will assume that the selfadjoint extension $A_0 := \ker \Gamma_0$ of A is definitizable over Ω, i.e., for every subdomain Ω' of Ω with the same properties as Ω, $\overline{\Omega'} \subset \Omega$, there exists a selfadjoint projection E which reduces A_0 such that $A_0 \cap (E\mathcal{K})^2$ is definitizable in the Krein space $E\mathcal{K}$ and Ω' belongs to the resolvent set of $A_0 \cap ((1-E)\mathcal{K})^2$. With the help of approximative eigensequences or the local spectral function of A_0 the spectral points of A_0 in $\Omega \cap \overline{\mathbb{R}}$ can be classified in points of positive and negative type and critical points (cf. [13], [16]).

Further we assume that τ is a function which is definitizable in Ω, that is, for every domain Ω' as Ω, $\overline{\Omega'} \subset \Omega$, the function τ can be written as the sum of a definitizable function (cf. [14], [15]) and a function holomorphic on Ω'. Similarly to selfadjoint operators and relations definitizable over Ω the points in $\Omega \cap \overline{\mathbb{R}}$ can be classified in points of positive and negative type and critical points. It was shown in [17] that τ can be represented with a selfadjoint relation T_0 definitizable over Ω' in some Krein space $\mathcal{H}$ such that the sign types of τ and T_0 coincide in $\Omega' \cap \overline{\mathbb{R}}$.

If, in addition, the sign types of A_0 and τ are "compatible" in $\Omega \cap \overline{\mathbb{R}}$ (see Definition 2.8), then the selfadjoint relation $A_0 \times T_0$ in the Krein space $\mathcal{K} \times \mathcal{H}$ is definitizable over Ω'. The linearization $\widetilde{A}$ of the boundary value problem (1.1) turns out to be a two-dimensional perturbation in resolvent sense of $A_0 \times T_0$. Under some additional minimality assumptions on the selfadjoint relations A_0 and T_0 and with the help of a recent result of T.Ya. Azizov and P. Jonas which states that the inverse of a matrix-valued locally definitizable function is again locally definitizable we prove in Theorem 3.6 that $\widetilde{A}$ is definitizable over Ω'.

The paper is organized as follows. In Section 2 we introduce the necessary notations and we recall the definitions of locally definitizable operators and relations and locally definitizable functions which can be found in, e.g., [13], [16] and [17]. Section 3 deals with boundary value problems of the form (1.1). After some preparatory work in Section 3.1 and Section 3.2 we formulate and prove the main result in Section 3.3: The linearization $\widetilde{A}$ of the eigenvalue dependent boundary value problem (1.1) is locally definitizable.

I thank P. Jonas for encouragement and critical help in the preparation of the manuscript.

2. Locally definitizable selfadjoint relations and locally definitizable functions

2.1. Notations and definitions

Let $(\mathcal{K}, [\cdot,\cdot])$ be a separable Krein space. We study linear relations in $\mathcal{K}$, that is, linear subspaces of $\mathcal{K}^2$. The set of all closed linear relations in $\mathcal{K}$ is denoted by $\widetilde{\mathcal{C}}(\mathcal{K})$. Linear operators in $\mathcal{K}$ are viewed as linear relations via their graphs. For the usual definitions of the linear operations with relations, the inverse etc., we refer to [9]. We denote the sum (direct sum) of subspaces in $\mathcal{K}^2$ by $\,\hat{+}\,$ (resp. $\,\hat{\dot{+}}\,$). The linear space of bounded linear operators defined on a Krein space $\mathcal{K}_1$ with values in a Krein space $\mathcal{K}_2$ is denoted by $\mathcal{L}(\mathcal{K}_1, \mathcal{K}_2)$. In the case $\mathcal{K} := \mathcal{K}_1 = \mathcal{K}_2$ we simply write $\mathcal{L}(\mathcal{K})$.

If $(\mathcal{H}, [\cdot,\cdot]_{\mathcal{H}})$ is another separable Krein space the elements of $\mathcal{K} \times \mathcal{H}$ will be written in the form $\{k, h\}$, $k \in \mathcal{K}$, $h \in \mathcal{H}$. $\mathcal{K} \times \mathcal{H}$ equipped with the inner product $[\cdot,\cdot]_{\mathcal{K}\times\mathcal{H}}$ defined by

$$[\{k,h\},\{k',h'\}]_{\mathcal{K}\times\mathcal{H}} := [k,k'] + [h,h']_{\mathcal{H}}, \quad k,k' \in \mathcal{K}, \quad h,h' \in \mathcal{H},$$

is also a Krein space. If S is a relation in $\mathcal{K}$ and T is a relation in $\mathcal{H}$ we shall write $S \times T$ for the direct product of S and T which is a relation in $\mathcal{K} \times \mathcal{H}$,

$$S \times T = \left\{ \begin{pmatrix} \{s,t\} \\ \{s',t'\} \end{pmatrix} \,\Big|\, \begin{pmatrix} s \\ s' \end{pmatrix} \in S,\ \begin{pmatrix} t \\ t' \end{pmatrix} \in T \right\}. \tag{2.1}$$

For the pair $\left(\begin{smallmatrix} \{s,t\} \\ \{s',t'\} \end{smallmatrix}\right)$ on the right-hand side of (2.1) we shall also write $\{\hat{s}, \hat{t}\}$, where $\hat{s} = \left(\begin{smallmatrix} s \\ s' \end{smallmatrix}\right)$, $\hat{t} = \left(\begin{smallmatrix} t \\ t' \end{smallmatrix}\right)$.

Let S be a closed linear relation in $\mathcal{K}$. The resolvent set $\rho(S)$ of S is the set of all $\lambda \in \mathbb{C}$ such that $(S-\lambda)^{-1} \in \mathcal{L}(\mathcal{K})$, the spectrum $\sigma(S)$ of S is the complement of $\rho(S)$ in $\mathbb{C}$. The extended spectrum $\widetilde{\sigma}(S)$ of S is defined by $\widetilde{\sigma}(S) = \sigma(S)$ if $S \in \mathcal{L}(\mathcal{K})$ and $\widetilde{\sigma}(S) = \sigma(S) \cup \{\infty\}$ otherwise. We say that $\lambda \in \mathbb{C}$ belongs to the *approximate point spectrum of* S, denoted by $\sigma_{ap}(S)$, if there exists a sequence $\left(\begin{smallmatrix} x_n \\ y_n \end{smallmatrix}\right) \in S$, $n = 1, 2, \ldots$, such that $\|x_n\| = 1$ and $\lim_{n\to\infty} \|y_n - \lambda x_n\| = 0$. The *extended approximate point spectrum* $\widetilde{\sigma}_{ap}(S)$ *of* S is defined by

$$\widetilde{\sigma}_{ap}(S) := \begin{cases} \sigma_{ap}(S) \cup \{\infty\} & \text{if} \quad 0 \in \sigma_{ap}(S^{-1}) \\ \sigma_{ap}(S) & \text{if} \quad 0 \notin \sigma_{ap}(S^{-1}) \end{cases}.$$

We remark, that the boundary points of $\widetilde{\sigma}(S)$ in $\overline{\mathbb{C}}$ belong to $\widetilde{\sigma}_{ap}(S)$.

Next we recall the definitions of the spectra of positive and negative type of a closed linear relation (see [16], [19]). For equivalent descriptions of the spectra of positive and negative type we refer to [16, Theorem 3.18].

Definition 2.1. Let S be a closed linear relation in $\mathcal{K}$. A point $\lambda \in \sigma_{ap}(S)$ is said to be of *positive type* (*negative type*) with respect to S, if for every sequence $\left(\begin{smallmatrix} x_n \\ y_n \end{smallmatrix}\right) \in S$, $n = 1, 2 \ldots$, with $\|x_n\| = 1$, $\lim_{n\to\infty} \|y_n - \lambda x_n\| = 0$ we have

$$\liminf_{n\to\infty}\, [x_n, x_n] > 0 \quad \big(\text{resp. } \limsup_{n\to\infty}\, [x_n, x_n] < 0\big).$$

If $\infty \in \widetilde{\sigma}_{ap}(S)$, ∞ is said to be of *positive type* (*negative type*) with respect to S if 0 is of positive (resp. negative) type with respect to S^{-1}.

An open subset Δ of $\overline{\mathbb{R}}$ is said to be of *positive type* (*negative type*) with respect to S if each point $\lambda \in \Delta \cap \widetilde{\sigma}(S)$ is of positive (resp. negative) type with respect to S. Δ is called of *definite type* with respect to S if Δ is either of positive or negative type with respect to S.

Let A be a linear relation in $\mathcal{K}$. The *adjoint relation* $A^+ \in \widetilde{\mathcal{C}}(\mathcal{K})$ is defined as

$$A^+ := \left\{ \begin{pmatrix} h \\ h' \end{pmatrix} \,\middle|\, [g', h] = [g, h'] \quad \text{for all} \quad \begin{pmatrix} g \\ g' \end{pmatrix} \in A \right\}.$$

A is said to be *symmetric* (*selfadjoint*) if $A \subset A^+$ (resp. $A = A^+$).

For a selfadjoint relation A in $\mathcal{K}$ the points of definite type introduced in Definition 2.1 belong to $\overline{\mathbb{R}}$. In fact, if, e.g., $\lambda \neq \infty$ is of positive type with respect to A, and $\binom{x_n}{y_n} \in A$ is a sequence with $\|x_n\| = 1$ and $\lim_{n\to\infty} \|y_n - \lambda x_n\| = 0$, then

$$|\mathrm{Im}\,\lambda| \liminf_{n\to\infty} [x_n, x_n] = \liminf_{n\to\infty} \; |\mathrm{Im}\,[y_n - \lambda x_n, x_n]| \leq \lim_{n\to\infty} \; \|y_n - \lambda x_n\| = 0$$

implies $\lambda \in \mathbb{R}$.

2.2. Locally definitizable selfadjoint relations in Krein spaces

Let Ω be a domain in $\overline{\mathbb{C}}$ symmetric with respect to the real axis such that $\Omega \cap \overline{\mathbb{R}} \neq \emptyset$ and the intersections of Ω with the upper and lower open half-planes are simply connected.

Let A_0 be a selfadjoint relation in the Krein space $\mathcal{K}$ such that $\sigma(A_0) \cap (\Omega \backslash \overline{\mathbb{R}})$ consists of isolated points which are poles of the resolvent of A_0, and no point of $\Omega \cap \overline{\mathbb{R}}$ is an accumulation point of the non-real spectrum of A_0. Let Δ be an open subset of $\Omega \cap \overline{\mathbb{R}}$. We say that A_0 belongs to the class $S^\infty(\Delta)$, if for every finite union Δ' of open connected subsets, $\overline{\Delta'} \subset \Delta$, there exists $m \geq 1$, $M > 0$ and an open neighborhood $\mathcal{U}$ of $\overline{\Delta'}$ in $\overline{\mathbb{C}}$ such that

$$\|(A_0 - \lambda)^{-1}\| \leq M(1 + |\lambda|)^{2m-2} \, |\mathrm{Im}\,\lambda|^{-m} \tag{2.2}$$

holds for all $\lambda \in \mathcal{U} \backslash \overline{\mathbb{R}}$. We remark, that for an open subset Δ of $\Omega \cap \overline{\mathbb{R}}$ which is of positive type with respect to A_0 the estimate (2.2) holds with $m = 1$ (see [16, Theorem 3.18] and [19] for the case of a bounded operator A_0).

Definition 2.2. Let Ω be a domain as above and let A_0 be a selfadjoint relation in $\mathcal{K}$ such that $\sigma(A_0) \cap (\Omega \backslash \overline{\mathbb{R}})$ consists of isolated points which are poles of the resolvent of A_0 and no point of $\Omega \cap \overline{\mathbb{R}}$ is an accumulation point of the non-real spectrum of A_0 in Ω. The relation A_0 is said to be *definitizable over* Ω, if $A_0 \in S^\infty(\Omega \cap \overline{\mathbb{R}})$ and every point $\mu \in \Omega \cap \overline{\mathbb{R}}$ has an open connected neighborhood I_μ in $\overline{\mathbb{R}}$ such that both components of $I_\mu \backslash \{\mu\}$ are of definite type with respect to A_0.

The next theorem is a variant of [16, Theorem 4.8]. The simple modification of the proof is left to the reader.

Theorem 2.3. *Let A_0 be a selfadjoint relation in $\mathcal{K}$ and let Ω be a domain as above. A_0 is definitizable over Ω if and only if for every domain Ω' with the same properties as Ω, $\overline{\Omega'} \subset \Omega$, there exists a selfadjoint projection E in $\mathcal{K}$ such that A_0 can be decomposed in*

$$A_0 = \left(A_0 \cap (E\mathcal{K})^2\right) \dotplus \left(A_0 \cap ((1-E)\mathcal{K})^2\right)$$

and the following holds:

(i) $A_0 \cap (E\mathcal{K})^2$ *is a definitizable relation in the Krein space* $E\mathcal{K}$.

(ii) $\widetilde{\sigma}\big(A_0 \cap ((1-E)\mathcal{K})^2\big) \cap \Omega' = \emptyset$.

Let A_0 be a selfadjoint relation in $\mathcal{K}$ which is definitizable over Ω, let Ω' be a domain with the same properties as Ω, $\overline{\Omega'} \subset \Omega$, and let E be a selfadjoint projection with the properties as in Theorem 2.3. If E' is the spectral function of the definitizable selfadjoint relation $A_0 \cap (E\mathcal{K})^2$ in the Krein space $E\mathcal{K}$ (cf. [15, page 71], [10] and [18]), then the mapping

$$\delta \mapsto E'(\delta)E =: E_{A_0}(\delta) \tag{2.3}$$

defined for all finite unions δ of connected subsets of $\Omega' \cap \overline{\mathbb{R}}$ the endpoints of which belong to $\Omega' \cap \overline{\mathbb{R}}$ and are of definite type with respect to $A_0 \cap (E\mathcal{K})^2$, is the spectral function of A_0 on $\Omega' \cap \overline{\mathbb{R}}$ (see [16, Section 3.4 and Remark 4.9]).

2.3. Locally definitizable functions

Let Ω be a domain as in the beginning of Section 2.2 and let τ be an $\mathcal{L}(\mathbb{C}^n)$-valued piecewise meromorphic function in $\Omega\backslash\overline{\mathbb{R}}$ which is symmetric with respect to the real line, that is $\tau(\overline{\lambda}) = \tau(\lambda)^*$ for all points λ of holomorphy of τ. If, in addition, no point of $\Omega \cap \overline{\mathbb{R}}$ is an accumulation point of non-real poles of τ we write $\tau \in M^{n\times n}(\Omega)$. The set of the points of holomorphy of τ in $\Omega\backslash\overline{\mathbb{R}}$ and all points $\mu \in \Omega \cap \mathbb{R}$ such that τ can be analytically continued to μ and the continuations from $\Omega \cap \mathbb{C}^+$ and $\Omega \cap \mathbb{C}^-$ coincide, is denoted by $\mathfrak{h}(\tau)$.

In the next definition we introduce the sign type of open subsets in $\Omega \cap \overline{\mathbb{R}}$ with respect to functions from the class $M^{n\times n}(\Omega)$ (see [17]).

Definition 2.4. Let $\tau \in M^{n\times n}(\Omega)$. An open subset $\Delta \subset \Omega \cap \overline{\mathbb{R}}$ is said to be of *positive type* with respect to τ if for every $x \in \mathbb{C}^n$ and every sequence (μ_k) of points in $\Omega \cap \mathbb{C}^+ \cap \mathfrak{h}(\tau)$ which converges in $\overline{\mathbb{C}}$ to a point of Δ we have

$$\liminf_{k\to\infty} \operatorname{Im}\,(\tau(\mu_k)x, x) \geq 0.$$

An open subset $\Delta \subset \Omega \cap \overline{\mathbb{R}}$ is said to be of *negative type* with respect to τ if Δ is of positive type with respect to $-\tau$. Δ is said to be of *definite type* with respect to τ if Δ is of positive or negative type with respect to τ.

Definition 2.5. A function $\tau \in M^{n\times n}(\Omega)$ is called *definitizable in* Ω if the following holds.

(i) Every point $\mu \in \Omega \cap \overline{\mathbb{R}}$ has an open connected neighborhood I_μ in $\overline{\mathbb{R}}$ such that both components of $I_\mu\backslash\{\mu\}$ are of definite type with respect to τ.

(ii) For every open subset Δ in $\overline{\mathbb{R}}$, $\overline{\Delta} \subset \Omega \cap \overline{\mathbb{R}}$, there exists $m \geq 1$, $M > 0$ and an open neighborhood $\mathcal{U}$ of $\overline{\Delta}$ in $\overline{\mathbb{C}}$ such that

$$\|\tau(\lambda)\| \leq M(1 + |\lambda|)^{2m}\, |\operatorname{Im}\lambda|^{-m}$$

holds for all $\lambda \in \mathcal{U}\backslash\overline{\mathbb{R}}$.

In [17] it is shown that a function $\tau \in M^{n\times n}(\Omega)$ is definitizable in Ω if and only if for every finite union Δ of open connected subsets of $\overline{\mathbb{R}}$ such that $\overline{\Delta} \subset \Omega \cap \overline{\mathbb{R}}$, τ can be written as the sum of an $\mathcal{L}(\mathbb{C}^n)$-valued definitizable function (see [14], [15]) and an $\mathcal{L}(\mathbb{C}^n)$-valued function which is locally holomorphic on $\overline{\Delta}$.

The following theorem will be used in Section 3.2 and Section 3.3. It states that a locally definitizable function can be represented with a locally definitizable selfadjoint relation. A proof can be found in [17].

Theorem 2.6. *Let τ be an $\mathcal{L}(\mathbb{C}^n)$-valued locally definitizable function in Ω and let Ω' be a domain with the same properties as Ω, $\overline{\Omega'} \subset \Omega$. Then there exists a Krein space $\mathcal{H}$, a selfadjoint relation T_0 in $\mathcal{H}$ definitizable over Ω' and a mapping $\gamma' \in \mathcal{L}(\mathbb{C}^n, \mathcal{H})$ with the following properties.*

(a) $\rho(T_0) \cap \Omega' = \mathfrak{h}(\tau) \cap \Omega'$.

(b) *For a fixed $\lambda_0 \in \rho(T_0) \cap \Omega'$ and all $\lambda \in \rho(T_0) \cap \Omega'$*

$$\tau(\lambda) = \mathrm{Re}\, \tau(\lambda_0) + \gamma'^{+}\big((\lambda - \mathrm{Re}\, \lambda_0) + (\lambda - \lambda_0)(\lambda - \overline{\lambda}_0)(T_0 - \lambda)^{-1}\big)\gamma'$$

holds.

(c) *For any finite union Δ of open connected subsets of $\overline{\mathbb{R}}$, $\overline{\Delta} \subset \Omega' \cap \overline{\mathbb{R}}$, such that the boundary points of Δ are of definite type with respect to τ the spectral projection $E_{T_0}(\Delta)$ is defined. If Ω'' is a domain with the same properties as Ω and Ω', $\overline{\Omega''} \subset \Omega'$, and if we set $E := E_{T_0}(\Delta) + E_{T_0}(\overline{\Omega''} \backslash \overline{\mathbb{R}})$, then the minimality condition*

$$E\mathcal{H} = \mathrm{clsp}\,\big\{\big(1 + (\lambda - \lambda_0)(T_0 - \lambda)^{-1}\big)E\gamma' x \,|\, \lambda \in \rho(T_0) \cap \Omega',\, x \in \mathbb{C}^n\big\}$$

is fulfilled.

(d) *Any finite union Δ of open connected subsets of $\overline{\mathbb{R}}$, $\overline{\Delta} \subset \Omega' \cap \overline{\mathbb{R}}$, is of positive (negative) type with respect to τ if and only if Δ is of positive (resp. negative) type with respect to T_0.*

If τ and T_0 are as in Theorem 2.6 we shall say that T_0 is an *Ω'-minimal representing relation* for τ.

Remark 2.7. Let τ be an $\mathcal{L}(\mathbb{C}^n)$-valued locally definitizable function in Ω and let Ω' be a domain with the same properties as Ω, $\overline{\Omega'} \subset \Omega$. If, in addition, τ is the restriction of a definitizable function (see [14], [15]) or if, in addition, the boundary of Ω' is contained in $\mathfrak{h}(\tau)$, then the selfadjoint relation T_0 in Theorem 2.6 can be chosen such that the minimality condition

$$\mathcal{H} = \mathrm{clsp}\,\big\{\big(1 + (\lambda - \lambda_0)(T_0 - \lambda)^{-1}\big)\gamma' x \,|\, \lambda \in \rho(T_0) \cap \Omega',\, x \in \mathbb{C}^n\big\}$$

holds.

The next definition connects sign types of locally definitizable functions and sign types of spectral points of locally definitizable relations.

Definition 2.8. Let τ be an $\mathcal{L}(\mathbb{C}^n)$-valued locally definitizable function in Ω and let A_0 be a selfadjoint relation in the Krein space $\mathcal{K}$ which is definitizable over Ω.

We say that *the sign types of τ and A_0 are d-compatible in* Ω if for every point $\mu \in \Omega \cap \overline{\mathbb{R}}$ there exists an open connected neighborhood $I_\mu \subset \Omega \cap \overline{\mathbb{R}}$ of μ such that each component of $I_\mu \backslash \{\mu\}$ is either of positive type with respect to τ and A_0 or of negative type with respect to τ and A_0.

If τ is a function which is definitizable in Ω, Ω' is a domain as Ω, $\overline{\Omega'} \subset \Omega$, and T_0 is an Ω'-minimal representing relation for τ (see Theorem 2.6), then the sign types of τ and T_0 are d-compatible in Ω'.

3. Boundary value problems with locally definitizable functions in the boundary condition

3.1. Boundary value spaces and Weyl functions associated with symmetric relations in Krein spaces

Let $(\mathcal{K}, [\cdot,\cdot])$ be a separable Krein space, let J be a corresponding fundamental symmetry and let $A \in \widetilde{\mathcal{C}}(\mathcal{K})$ be a closed symmetric relation in $\mathcal{K}$. We say that A is of *defect* $m \in \mathbb{N} \cup \{\infty\}$, if both deficiency indices

$$n_\pm(JA) = \dim \ker((JA)^* - \overline{\lambda}), \qquad \lambda \in \mathbb{C}^\pm,$$

of the symmetric relation JA in the Hilbert space $(\mathcal{K}, [J\cdot,\cdot])$ are equal to m. Here * denotes the Hilbert space adjoint. We remark, that this is equivalent to the fact that there exists a selfadjoint extension of A in $\mathcal{K}$ and that each selfadjoint extension $\hat{A}$ of A in $\mathcal{K}$ satisfies $\dim(\hat{A}/A) = m$.

We shall use the so-called boundary value spaces for the description of the selfadjoint extensions of closed symmetric relations in Krein spaces. The following definition is taken from [5].

Definition 3.1. Let A be a closed symmetric relation in the Krein space $(\mathcal{K}, [\cdot,\cdot])$. We say that $\{\mathcal{G}, \Gamma_0, \Gamma_1\}$ is a *boundary value space for* A^+ if $(\mathcal{G}, (\cdot,\cdot))$ is a Hilbert space and there exist mappings $\Gamma_0, \Gamma_1 : A^+ \to \mathcal{G}$ such that $\Gamma = \binom{\Gamma_0}{\Gamma_1} : A^+ \to \mathcal{G} \times \mathcal{G}$ is surjective, and the relation

$$[f', g] - [f, g'] = (\Gamma_1 \hat{f}, \Gamma_0 \hat{g}) - (\Gamma_0 \hat{f}, \Gamma_1 \hat{g})$$

holds for all $\hat{f} = \binom{f}{f'}, \hat{g} = \binom{g}{g'} \in A^+$.

In the following we recall some basic facts on boundary value spaces which can be found in, e.g., [4] and [5]. For the Hilbert space case we refer to [11], [7] and [8]. Let A be a closed symmetric relation in $\mathcal{K}$ and let λ be a point of regular type of A. We denote by

$$\mathcal{N}_{\lambda,A^+} := \ker(A^+ - \lambda) = \operatorname{ran}(A - \overline{\lambda})^{[\perp]}$$

the defect subspace of A at the point λ and we set

$$\hat{\mathcal{N}}_{\lambda,A^+} = \left\{ \binom{f_\lambda}{\lambda f_\lambda} \,\middle|\, f_\lambda \in \mathcal{N}_{\lambda,A^+} \right\}.$$

When no confusion can arise we write $\mathcal{N}_\lambda$ and $\hat{\mathcal{N}}_\lambda$ instead of $\mathcal{N}_{\lambda,A^+}$ and $\hat{\mathcal{N}}_{\lambda,A^+}$. If there exists a selfadjoint extension A' of A such that $\rho(A') \neq \emptyset$ then we have

$$A^+ = A' \dot{+} \hat{\mathcal{N}}_\lambda \quad \text{for all} \quad \lambda \in \rho(A').$$

In this case there exists a boundary value space $\{\mathcal{G}, \Gamma_0, \Gamma_1\}$ for A^+ such that $\ker \Gamma_0 = A'$ (cf. [5]).

Let in the following A, $\{\mathcal{G}, \Gamma_0, \Gamma_1\}$ and Γ be as in Definition 3.1. It follows that the mappings Γ_0 and Γ_1 are continuous. The selfadjoint extensions

$$A_0 := \ker \Gamma_0 \qquad \text{and} \qquad A_1 := \ker \Gamma_1$$

of A are transversal, i.e., $A_0 \cap A_1 = A$ and $A_0 \hat{+} A_1 = A^+$. The mapping Γ induces, via

$$A_\Theta := \Gamma^{-1}\Theta = \bigl\{\hat{f} \in A^+ \,|\, \Gamma\hat{f} \in \Theta\bigr\}, \quad \Theta \in \widetilde{\mathcal{C}}(\mathcal{G}), \tag{3.1}$$

a bijective correspondence $\Theta \mapsto A_\Theta$ between the set of all closed linear relations $\widetilde{\mathcal{C}}(\mathcal{G})$ in $\mathcal{G}$ and the set of closed extensions $A_\Theta \subset A^+$ of A. In particular (3.1) gives a one-to-one correspondence between the closed symmetric (selfadjoint) extensions of A and the closed symmetric (resp. selfadjoint) relations in $\mathcal{G}$. If Θ is a closed operator in $\mathcal{G}$, then the corresponding extension A_Θ of A is determined by

$$A_\Theta = \ker\bigl(\Gamma_1 - \Theta\Gamma_0\bigr). \tag{3.2}$$

Assume that $\rho(A_0) \neq \emptyset$ and denote by π_1 the orthogonal projection onto the first component of $\mathcal{K} \times \mathcal{K}$. For every $\lambda \in \rho(A_0)$ we define the operators

$$\gamma(\lambda) = \pi_1(\Gamma_0|\hat{\mathcal{N}}_\lambda)^{-1} \in \mathcal{L}(\mathcal{G}, \mathcal{K}) \quad \text{and} \quad M(\lambda) = \Gamma_1(\Gamma_0|\hat{\mathcal{N}}_\lambda)^{-1} \in \mathcal{L}(\mathcal{G}).$$

The functions $\lambda \mapsto \gamma(\lambda)$ and $\lambda \mapsto M(\lambda)$ are called the *γ-field* and *Weyl function* corresponding to A and $\{\mathcal{G}, \Gamma_0, \Gamma_1\}$. γ and M are holomorphic on $\rho(A_0)$ and the relations

$$\gamma(\zeta) = (1 + (\zeta - \lambda)(A_0 - \zeta)^{-1})\gamma(\lambda) \tag{3.3}$$

and

$$M(\lambda) - M(\zeta)^* = (\lambda - \overline{\zeta})\gamma(\zeta)^+\gamma(\lambda) \tag{3.4}$$

hold for all $\lambda, \zeta \in \rho(A_0)$ (cf. [5]).

If $\Theta \in \widetilde{\mathcal{C}}(\mathcal{G})$ and A_Θ is the corresponding extension of A (see (3.1)), then a point $\lambda \in \rho(A_0)$ belongs to $\rho(A_\Theta)$ if and only if 0 belongs to $\rho(\Theta - M(\lambda))$. For $\lambda \in \rho(A_\Theta) \cap \rho(A_0)$ the well-known resolvent formula

$$(A_\Theta - \lambda)^{-1} = (A_0 - \lambda)^{-1} + \gamma(\lambda)\bigl(\Theta - M(\lambda)\bigr)^{-1}\gamma(\overline{\lambda})^+ \tag{3.5}$$

holds (for a proof see, e.g., [5]).

3.2. Locally definitizable functions as Weyl functions of symmetric relations

Let, as in Section 2.2, Ω be a domain in $\overline{\mathbb{C}}$ symmetric with respect to the real axis such that $\Omega \cap \overline{\mathbb{R}} \neq \emptyset$ and the intersections of Ω with the upper and lower open half-planes are simply connected.

In the next proposition we consider boundary value spaces and Weyl functions associated with symmetric relations in Krein spaces which have the additional property that there exists a locally definitizable selfadjoint extension. We restrict ourselves to the case of defect one.

Proposition 3.2. *Let A be a closed symmetric relation of defect one in the Krein space $\mathcal{K}$ and assume that there exists a selfadjoint extension A_0 of A which is definitizable over Ω. Let $\{\mathbb{C}, \Gamma_0, \Gamma_1\}$ be a boundary value space for A^+ such that $A_0 = \ker \Gamma_0$. Then the corresponding Weyl function M is definitizable in Ω. If Δ is an open subset of $\Omega \cap \overline{\mathbb{R}}$ which is of positive (negative) type with respect to A_0, then Δ is of positive (resp. negative) type with respect to M.*

Proof. As A_0 is a selfadjoint relation which is definitizable over Ω it follows that the Weyl function M corresponding to A and $\{\mathbb{C}, \Gamma_0, \Gamma_1\}$ is piecewise meromorphic in $\Omega\backslash\overline{\mathbb{R}}$ and no point of $\Omega \cap \overline{\mathbb{R}}$ is an accumulation point of the non-real poles of M. Let $\lambda_0 \in \rho(A_0)$. Making use of (3.3) and (3.4) we obtain that M is symmetric with respect to the real axis, $\operatorname{Im} M(\lambda_0) = (\operatorname{Im} \lambda_0)\gamma(\lambda_0)^+\gamma(\lambda_0)$ and

$$\begin{aligned} M(\lambda) &= \overline{M(\lambda_0)} + (\lambda - \overline{\lambda}_0)\gamma(\lambda_0)^+\gamma(\lambda) \\ &= \operatorname{Re} M(\lambda_0) - i\,(\operatorname{Im}\lambda_0)\gamma(\lambda_0)^+\gamma(\lambda_0) \\ &\qquad + \gamma(\lambda_0)^+\big((\lambda - \overline{\lambda}_0) + (\lambda - \lambda_0)(\lambda - \overline{\lambda}_0)(A_0 - \lambda)^{-1}\big)\gamma(\lambda_0) \\ &= \operatorname{Re} M(\lambda_0) + \gamma(\lambda_0)^+\big((\lambda - \operatorname{Re}\lambda_0) + (\lambda - \lambda_0)(\lambda - \overline{\lambda}_0)(A_0 - \lambda)^{-1}\big)\gamma(\lambda_0) \end{aligned} \tag{3.6}$$

holds for all $\lambda \in \rho(A_0)$. Since A_0 belongs to the class $S^\infty(\Omega \cap \overline{\mathbb{R}})$ it follows that M fulfils the second condition in Definition 2.5. Let $\mu \in \Omega \cap \overline{\mathbb{R}}$ and let $I_\mu \subset \Omega \cap \overline{\mathbb{R}}$ be an open connected neighborhood of μ in $\overline{\mathbb{R}}$ such that both components of $I_\mu\backslash\{\mu\}$ are of definite type with respect to A_0. By [16, Theorem 3.18] a component of $I_\mu\backslash\{\mu\}$ is of positive (negative) type with respect to A_0 if and only if it is of positive (resp. negative) type with respect to the function

$$\lambda \mapsto (\lambda - \operatorname{Re} \lambda_0) + (\lambda - \lambda_0)(\lambda - \overline{\lambda}_0)(A_0 - \lambda)^{-1}.$$

Now it follows from (3.6) that both components of $I_\mu\backslash\{\mu\}$ are of the same sign type with respect to A_0 and M and therefore the Weyl function M is definitizable in Ω. The same argument shows that an open subset $\Delta \subset \Omega \cap \overline{\mathbb{R}}$ which is of positive (negative) type with respect to A_0 is also of positive (negative) type with respect to M. □

The next theorem is a variant of [3, Theorem 3.3]. For the convenience of the reader we sketch the proof.

Theorem 3.3. *Let τ be a complex-valued locally definitizable function in Ω and assume that τ is not identically equal to a constant. Let Ω' be a domain with the* *same properties as Ω and $\overline{\Omega'} \subset \Omega$. Then there exists a Krein space $\mathcal{H}$, a closed symmetric relation T of defect one in $\mathcal{H}$ and a boundary value space $\{\mathbb{C}, \Gamma_0', \Gamma_1'\}$ for T^+ such that τ coincides with the corresponding Weyl function on Ω' and $T_0 := \ker \Gamma_0'$ is an Ω'-minimal representing relation for τ.*

Sketch of the proof of Theorem 3.3. Let τ be represented with an Ω'-minimal selfadjoint relation T_0 in a Krein space $\mathcal{H}$ as in Theorem 2.6. Let $\gamma' \in \mathcal{L}(\mathbb{C}, \mathcal{H})$ be as in Theorem 2.6 and fix some $\lambda_0 \in \mathfrak{h}(\tau) \cap \Omega'$. For all $\lambda \in \mathfrak{h}(\tau) \cap \Omega'$ we define

$$\gamma'(\lambda) := \big(1 + (\lambda - \lambda_0)(T_0 - \lambda)^{-1}\big)\gamma' \in \mathcal{L}(\mathbb{C}, \mathcal{H}).$$

The linear functional $\gamma'(\lambda)c \mapsto c$ defined on $\operatorname{ran} \gamma'(\lambda)$ is denoted by $\gamma'(\lambda)^{(-1)}$. The closed symmetric relation

$$T := \left\{ \begin{pmatrix} f \\ g \end{pmatrix} \in T_0 \,\middle|\, [g - \overline{\mu} f, \gamma'(\mu)\, 1] = 0 \right\}$$

has defect one and does not depend on the choice of $\mu \in \mathfrak{h}(\tau) \cap \Omega'$. For some fixed $\lambda \in \mathfrak{h}(\tau) \cap \Omega'$ we write the elements $\hat{f} \in T^+$ in the form $\hat{f} = \binom{f_0}{f_0'} + \binom{f_\lambda}{\lambda f_\lambda}$, where $\binom{f_0}{f_0'} \in T_0$ and $f_\lambda \in \operatorname{ran} \gamma'(\lambda) = \mathcal{N}_{\lambda, T^+}$. As in the proof of [3, Theorem 3.3] (see also [7, Theorem 1]) one verifies that $\{\mathbb{C}, \Gamma_0', \Gamma_1'\}$, where

$$\begin{aligned} \Gamma_0' \hat{f} &:= \gamma'(\lambda)^{(-1)} f_\lambda, \\ \Gamma_1' \hat{f} &:= \gamma'(\lambda)^+ (f_0' - \overline{\lambda} f_0) + \tau(\lambda)\gamma'(\lambda)^{(-1)} f_\lambda, \end{aligned}$$

is a boundary value space for T^+ and the corresponding Weyl function coincides with τ on Ω'. □

Remark 3.4. Let the function τ be definitizable in Ω and assume that τ is not identically equal to a constant. Let Ω' be a domain with the same properties as Ω, $\overline{\Omega'} \subset \Omega$, and assume that τ is the restriction of a definitizable function or that the boundary of Ω' is contained in $\mathfrak{h}(\tau)$. If we choose T_0 as in Remark 2.7 and $T \subset T_0 \subset T^+$ as in Theorem 3.3, then the condition

$$\mathcal{H} = \operatorname{clsp}\{\operatorname{ran} \gamma'(\lambda) \,|\, \lambda \in \rho(T_0) \cap \Omega'\} = \operatorname{clsp}\{\mathcal{N}_{\lambda, T^+} \,|\, \lambda \in \rho(T_0) \cap \Omega'\}$$

is fulfilled. In this case T is an operator.

In the following proposition we use Definition 2.8. The statements will be useful in the proof of our main result in Section 3.3.

Proposition 3.5. *Let A be a closed symmetric relation of defect one in the Krein space $\mathcal{K}$, let $\{\mathbb{C}, \Gamma_0, \Gamma_1\}$ be a boundary value space for A^+ and denote by M the corresponding Weyl function. Assume that the selfadjoint relation $A_0 = \ker \Gamma_0$ is definitizable over Ω and let τ be a complex-valued function which is definitizable in Ω such that the sign types of τ and A_0 are d-compatible in Ω.*
Let Ω' be a domain with the same properties as Ω, $\overline{\Omega'} \subset \Omega$, and let T_0 be an Ω'-minimal representing relation for τ in some Krein space $\mathcal{H}$ (see Theorem 2.6). Then the following holds:

(i) *The selfadjoint relation* $A_0 \times T_0 \in \widetilde{\mathcal{C}}(\mathcal{K} \times \mathcal{H})$ *is definitizable over* Ω' *and the sign types of* $A_0 \times T_0$ *and the functions* τ *and* M *are d-compatible in* Ω'.

(ii) *The function* $M + \tau$ *is definitizable in* Ω.

(iii) *If* $\tau(\eta) \neq 0$ *and* $(M + \tau)(\eta') \neq 0$ *for some* $\eta, \eta' \in \Omega$, *then the functions*

$$\lambda \mapsto \begin{pmatrix} M(\lambda) & 0 \\ 0 & -\tau(\lambda)^{-1} \end{pmatrix} \quad \text{and} \quad \lambda \mapsto -\begin{pmatrix} M(\lambda) & -1 \\ -1 & -\tau(\lambda)^{-1} \end{pmatrix}^{-1} \tag{3.7}$$

are definitizable in Ω *and their sign types are d-compatible with the sign types of the selfadjoint relation* $A_0 \times T_0$ *in* Ω'.

Proof. (i) Since A_0 and T_0 belong to $S^\infty(\Omega \cap \overline{\mathbb{R}})$ and $S^\infty(\Omega' \cap \overline{\mathbb{R}})$, respectively, we conclude that $A_0 \times T_0$ belongs to $S^\infty(\Omega' \cap \overline{\mathbb{R}})$. Let $\mu \in \Omega' \cap \overline{\mathbb{R}}$ and let $I_\mu \subset \Omega' \cap \overline{\mathbb{R}}$ be an open connected neighborhood of μ in $\overline{\mathbb{R}}$ such that each component of $I_\mu \backslash \{\mu\}$ is of the same sign type with respect to A_0 and τ. As T_0 is an Ω'-minimal representing relation for τ both components of $I_\mu \backslash \{\mu\}$ are of definite type with respect to $A_0 \times T_0$ and it follows that $A_0 \times T_0$ is definitizable over Ω'.

The assumption that the sign types of τ and A_0 are d-compatible in Ω implies that the sign types of τ and $A_0 \times T_0$ as well as the sign types of M and $A_0 \times T_0$ are d-compatible in Ω.

(ii) For $\mu \in \Omega \cap \overline{\mathbb{R}}$ we choose an open connected neighborhood $I_\mu \subset \Omega \cap \overline{\mathbb{R}}$ of μ such that both components of $I_\mu \backslash \{\mu\}$ are of the same sign type with respect to A_0 and τ. By Proposition 3.2 the sign types of M are the same as of A_0 and therefore both components of $I_\mu \backslash \{\mu\}$ are of the same sign type with respect to $M + \tau$. The growth properties of M and τ imply that $M + \tau$ fulfils the second condition in Definition 2.5 and therefore $M + \tau$ is definitizable in Ω.

(iii) By [1, Theorem 2.5] the function $-\tau^{-1}$ is definitizable in Ω and it follows from the proof of [1, Theorem 2.5] that each point $\mu \in \Omega \cap \overline{\mathbb{R}}$ has an open connected neighborhood $I_\mu \subset \Omega \cap \overline{\mathbb{R}}$ such that both components of $I_\mu \backslash \{\mu\}$ are of the same sign type with respect to $-\tau^{-1}$ and τ. Therefore the sign types of $-\tau^{-1}$ and A_0 are d-compatible in Ω. Now it is easy to see that the first function in (3.7) is definitizable in Ω and its sign types are d-compatible with the sign types of $A_0 \times T_0$ in Ω'.

As the function

$$\lambda \mapsto \begin{pmatrix} M(\lambda) & -1 \\ -1 & -\tau(\lambda)^{-1} \end{pmatrix}$$

is also definitizable in Ω another application of [1, Theorem 2.5] shows that the second function in (3.7) is definitizable in Ω and its sign types are d-compatible with the sign types of $A_0 \times T_0$ in Ω'. □

3.3. The main result

In this section we investigate the spectral properties of linearizations of a class of abstract eigenvalue dependent boundary value problems with locally definitizable functions in the boundary condition. Similar problems with a local variant of generalized Nevanlinna functions in the boundary condition have been considered in [3]. The main feature in Theorem 3.6 below is that the linearization turns out to be locally definitizable.

Theorem 3.6. *Let Ω be a domain as in the beginning of Section 3.2 and let A be a closed symmetric operator of defect one in the Krein space $\mathcal{K}$ such that $\mathcal{K} = \operatorname{clsp}\{\mathcal{N}_{\lambda,A^+} \mid \lambda \in \Omega\}$ holds. Assume that there exists a selfadjoint extension A_0 of A which is definitizable over Ω. Let $\{\mathbb{C}, \Gamma_0, \Gamma_1\}$ be a boundary value space for A^+, $A_0 = \ker \Gamma_0$, and denote by γ and M the corresponding γ-field and Weyl function, respectively.*
Let τ be a nonconstant function which is definitizable in Ω, let Ω' be a domain as Ω, $\overline{\Omega'} \subset \Omega$, choose $\mathcal{H}$, $T \subset T_0 \subset T^+$ and $\{\mathbb{C}, \Gamma_0', \Gamma_1'\}$ as in Theorem 3.3 and assume that the condition $\mathcal{H} = \operatorname{clsp}\{\mathcal{N}_{\lambda,T^+} \mid \lambda \in \Omega'\}$ is fulfilled.
Let the sign types of τ and A_0 be d-compatible in Ω, assume that the function $M + \tau$ is not identically equal to zero and define

$$\mathfrak{h}_0 := \mathfrak{h}(M) \cap \mathfrak{h}(\tau) \cap \mathfrak{h}(\tau^{-1}) \cap \mathfrak{h}\big((M+\tau)^{-1}\big).$$

Then the relation

$$\widetilde{A} = \Big\{\{\hat{f}_1, \hat{f}_2\} \in A^+ \times T^+ \,\big|\, \Gamma_1 \hat{f}_1 - \Gamma_1' \hat{f}_2 = \Gamma_0 \hat{f}_1 + \Gamma_0' \hat{f}_2 = 0\Big\} \tag{3.8}$$

is a selfadjoint extension of A in $\mathcal{K} \times \mathcal{H}$ which is definitizable over Ω' and the sign types of $\widetilde{A}$ are d-compatible with the sign types of τ and M in Ω'. The set $\Omega' \backslash (\overline{\mathbb{R}} \cup \mathfrak{h}_0)$ is finite. For every $k \in \mathcal{K}$ and every $\lambda \in \mathfrak{h}_0 \cap \Omega'$ the unique solution of the eigenvalue dependent boundary value problem

$$f_1' - \lambda f_1 = k, \quad \tau(\lambda)\Gamma_0 \hat{f}_1 + \Gamma_1 \hat{f}_1 = 0, \quad \hat{f}_1 = \begin{pmatrix} f_1 \\ f_1' \end{pmatrix} \in A^+, \tag{3.9}$$

is given by

$$\begin{aligned} f_1 &= P_{\mathcal{K}}(\widetilde{A} - \lambda)^{-1}\{k, 0\} = (A_0 - \lambda)^{-1}k - \gamma(\lambda)\big(M(\lambda) + \tau(\lambda)\big)^{-1}\gamma(\overline{\lambda})^+ k, \\ f_1' &= \lambda f_1 + k. \end{aligned} \tag{3.10}$$

Proof. **1.** In this step we construct the selfadjoint relation $\widetilde{A}$ in $\mathcal{K} \times \mathcal{H}$ with the help of the coupling method from [6, §5.2], we show that $\mathfrak{h}_0 \cap \Omega'$ belongs to $\rho(\widetilde{A})$ and that (3.10) is the unique solution of the boundary value problem (3.9). We follow the lines of [3, Proof of Theorem 4.1].

As the functions M, τ and $M + \tau$ are definitizable in Ω (see Proposition 3.5) [1, Theorem 2.3] implies that $-\tau^{-1}$ and $-(M+\tau)^{-1}$ are also definitizable in Ω. Let Ω' and $\mathfrak{h}_0$ be as in the assumptions of the theorem. As the non-real poles of the functions M, τ, τ^{-1} and $(M+\tau)^{-1}$ in Ω do not accumulate to $\Omega \cap \overline{\mathbb{R}}$ we conclude from $\overline{\Omega'} \subset \Omega$ that the set $\Omega' \backslash (\overline{\mathbb{R}} \cup \mathfrak{h}_0)$ is finite.

Let $\mathcal{H}$, $T \subset T^+$ and $\{\mathbb{C}, \Gamma_0', \Gamma_1'\}$ be as in Theorem 3.3. Then τ is the corresponding Weyl function and the selfadjoint relation $T_0 = \ker \Gamma_0'$ is definitizable over Ω'. We denote the γ-field corresponding to $\{\mathbb{C}, \Gamma_0', \Gamma_1'\}$ by γ' and we set

$T_1 := \ker \Gamma_1'$. As $\{\mathbb{C}, \Gamma_1', -\Gamma_0'\}$ is a boundary value space for T^+ with corresponding γ-field and Weyl function

$$\lambda \mapsto \gamma'(\lambda)\tau(\lambda)^{-1} \quad \text{and} \quad \lambda \mapsto -\tau(\lambda)^{-1}, \quad \lambda \in \mathfrak{h}(\tau) \cap \mathfrak{h}(\tau^{-1}) \cap \Omega', \tag{3.11}$$

respectively, it follows without difficulty that $\{\mathbb{C}^2, \widetilde{\Gamma}_0, \widetilde{\Gamma}_1\}$, where $\widetilde{\Gamma}_0$ and $\widetilde{\Gamma}_1$ are mappings from $A^+ \times T^+$ into $\mathbb{C}^2$ defined by

$$\widetilde{\Gamma}_0\{\hat{f}_1, \hat{f}_2\} := \begin{pmatrix} \Gamma_0 \hat{f}_1 \\ \Gamma_1' \hat{f}_2 \end{pmatrix} \quad \text{and} \quad \widetilde{\Gamma}_1\{\hat{f}_1, \hat{f}_2\} := \begin{pmatrix} \Gamma_1 \hat{f}_1 \\ -\Gamma_0' \hat{f}_2 \end{pmatrix},$$

$\{\hat{f}_1, \hat{f}_2\} \in A^+ \times T^+$, is a boundary value space for $A^+ \times T^+$ with corresponding γ-field

$$\lambda \mapsto \widetilde{\gamma}(\lambda) = \begin{pmatrix} \gamma(\lambda) & 0 \\ 0 & \gamma'(\lambda)\tau(\lambda)^{-1} \end{pmatrix}, \quad \lambda \in \mathfrak{h}(M) \cap \mathfrak{h}(\tau) \cap \mathfrak{h}(\tau^{-1}) \cap \Omega', \tag{3.12}$$

and Weyl function

$$\lambda \mapsto \widetilde{M}(\lambda) = \begin{pmatrix} M(\lambda) & 0 \\ 0 & -\tau(\lambda)^{-1} \end{pmatrix}, \quad \lambda \in \mathfrak{h}(M) \cap \mathfrak{h}(\tau) \cap \mathfrak{h}(\tau^{-1}) \cap \Omega'.$$

The selfadjoint relation $\widetilde{A}$ in $\mathcal{K} \times \mathcal{H}$ corresponding to $\Theta = \left(\begin{smallmatrix} 0 & 1 \\ 1 & 0 \end{smallmatrix}\right) \in \mathcal{L}(\mathbb{C}^2)$ via (3.1) and (3.2) is given by

$$\begin{aligned} \widetilde{A} &= \ker(\widetilde{\Gamma}_1 - \Theta \widetilde{\Gamma}_0) \\ &= \Big\{ \{\hat{f}_1, \hat{f}_2\} \in A^+ \times T^+ \,\Big|\, \Gamma_1 \hat{f}_1 - \Gamma_1' \hat{f}_2 = \Gamma_0 \hat{f}_1 + \Gamma_0' \hat{f}_2 = 0 \Big\}. \end{aligned} \tag{3.13}$$

For $\lambda \in \mathfrak{h}_0 \cap \Omega'$ the resolvent of $\widetilde{A}$ can be written as

$$(\widetilde{A} - \lambda)^{-1} = \begin{pmatrix} (A_0 - \lambda)^{-1} & 0 \\ 0 & (T_1 - \lambda)^{-1} \end{pmatrix} + \widetilde{\gamma}(\lambda)\big(\Theta - \widetilde{M}(\lambda)\big)^{-1} \widetilde{\gamma}(\overline{\lambda})^+, \tag{3.14}$$

(see (3.5)). Calculating $(\Theta - \widetilde{M}(\lambda))^{-1}$ one verifies that the compressed resolvent of $\widetilde{A}$ onto $\mathcal{K}$ is given by

$$P_{\mathcal{K}}(\widetilde{A} - \lambda)^{-1}|\mathcal{K} = (A_0 - \lambda)^{-1} - \gamma(\lambda)\big(M(\lambda) + \tau(\lambda)\big)^{-1} \gamma(\overline{\lambda})^+, \quad \lambda \in \mathfrak{h}_0 \cap \Omega'.$$

For $k \in \mathcal{K}$ we set $f_1 := P_{\mathcal{K}}(\widetilde{A} - \lambda)^{-1}\{k, 0\}$ and $f_2 := P_{\mathcal{H}}(\widetilde{A} - \lambda)^{-1}\{k, 0\}$. Then

$$\begin{pmatrix} \{f_1, f_2\} \\ \{\lambda f_1 + k, \lambda f_2\} \end{pmatrix} \in \widetilde{A} \subset A^+ \times T^+$$

and $\hat{f}_1 := \left(\begin{smallmatrix} f_1 \\ \lambda f_1 + k \end{smallmatrix}\right) \in A^+$, $\hat{f}_2 := \left(\begin{smallmatrix} f_2 \\ \lambda f_2 \end{smallmatrix}\right) \in \hat{\mathcal{N}}_{\lambda, T^+}$. From (3.13) and since τ is the Weyl function corresponding to $\{\mathbb{C}, \Gamma_0', \Gamma_1'\}$ we get

$$\Gamma_1 \hat{f}_1 = \Gamma_1' \hat{f}_2 = \tau(\lambda)\Gamma_0' \hat{f}_2 = -\tau(\lambda)\Gamma_0 \hat{f}_1, \quad \lambda \in \mathfrak{h}_0 \cap \Omega',$$

and it follows that $\hat{f}_1 \in A^+$ is a solution of (3.9).

Let us verify that this solution $\hat{f}_1 \in A^+$ is unique. Assume that the vector $\hat{g}_1 = \binom{g_1}{\lambda g_1 + k} \in A^+$ is also a solution of (3.9), $\lambda \in \mathfrak{h}_0 \cap \Omega'$. Then $\hat{f}_1 - \hat{g}_1$ belongs to $\hat{\mathcal{N}}_{\lambda, A^+}$ and

$$0 = \tau(\lambda)\Gamma_0(\hat{f}_1 - \hat{g}_1) + \Gamma_1(\hat{f}_1 - \hat{g}_1) = \big(\tau(\lambda) + M(\lambda)\big)\Gamma_0(\hat{f}_1 - \hat{g}_1)$$

implies $\hat{f}_1 - \hat{g}_1 \in A_0 \cap \hat{\mathcal{N}}_{\lambda, A^+}$ as $\tau(\lambda) + M(\lambda) \neq 0$. Therefore $\hat{f}_1 = \hat{g}_1$ since $\lambda \in \mathfrak{h}(M)$.

2. It remains to prove that $\widetilde{A}$ is definitizable over Ω' and that the sign types of $\widetilde{A}$ are d-compatible with the sign types of the functions τ and M in Ω'. In this step we show that for every point $\mu \in \Omega' \cap \overline{\mathbb{R}}$ there exists an open connected neighborhood I_μ of μ in $\Omega' \cap \overline{\mathbb{R}}$ such that both components of $I_\mu \backslash \{\mu\}$ are of definite type with respect to $\widetilde{A}$.

As the sign types of τ and A_0 are d-compatible in Ω, Proposition 3.5 implies that the selfadjoint relation $A_0 \times T_0$ is definitizable over Ω'. It is straightforward to check that $\{\mathbb{C}^2, \widehat{\Gamma}_0, \widehat{\Gamma}_1\}$, where

$$\widehat{\Gamma}_0 := \widetilde{\Gamma}_1 - \Theta\widetilde{\Gamma}_0, \qquad \widehat{\Gamma}_1 := -\widetilde{\Gamma}_0,$$

is a boundary value space for $A^+ \times T^+$ with $\ker \widehat{\Gamma}_0 = \widetilde{A}$. The corresponding γ-field $\widehat{\gamma}$ and Weyl function $\widehat{M}$ are defined on $\rho(\widetilde{A})$ and for $\lambda \in \mathfrak{h}_0 \cap \Omega'$ they are given by

$$\lambda \mapsto \widehat{\gamma}(\lambda) = \widetilde{\gamma}(\lambda)\big(\widetilde{M}(\lambda) - \Theta\big)^{-1} \tag{3.15}$$

and

$$\lambda \mapsto \widehat{M}(\lambda) = -(\widetilde{M}(\lambda) - \Theta)^{-1} = -\begin{pmatrix} M(\lambda) & -1 \\ -1 & -\tau(\lambda)^{-1} \end{pmatrix}^{-1},$$

respectively. In particular

$$\begin{aligned} \widehat{M}(\lambda) = \operatorname{Re} \widehat{M}(\lambda_0) + \widehat{\gamma}(\lambda_0)^+\big((\lambda - \operatorname{Re} \lambda_0) \\ + (\lambda - \lambda_0)(\lambda - \overline{\lambda}_0)(\widetilde{A} - \lambda)^{-1}\big)\widehat{\gamma}(\lambda_0) \end{aligned} \tag{3.16}$$

holds for a fixed $\lambda_0 \in \mathfrak{h}_0 \cap \Omega'$ and all $\lambda \in \mathfrak{h}_0 \cap \Omega'$ (see the proof of Proposition 3.2). By Proposition 3.5 the function $\widehat{M}$ is definitizable in Ω and the sign types of $\widehat{M}$ and $A_0 \times T_0$ are d-compatible in Ω'.

Let $\mu \in \Omega' \cap \overline{\mathbb{R}}$ and assume, e.g., that a one-sided open connected neighborhood Δ_+ of μ in $\overline{\mathbb{R}}$, $\overline{\Delta_+} \subset \Omega' \cap \overline{\mathbb{R}}$, is of positive type with respect to $A_0 \times T_0$. As the sign types of $\widehat{M}$ and $A_0 \times T_0$ are d-compatible in Ω', it is no restriction to assume that Δ_+ is also of positive type with respect to $\widehat{M}$. Since $A_0 \times T_0$ and $\widetilde{A}$ are both selfadjoint extensions of the symmetric relation $A \times T$ in $\mathcal{K} \times \mathcal{H}$ we have

$$\dim\big(\operatorname{ran}\big((\widetilde{A} - \lambda)^{-1} - ((A_0 \times T_0) - \lambda)^{-1}\big)\big) \leq 2$$

for all $\lambda \in \mathfrak{h}_0 \cap \Omega'$. Let Ω_{Δ_+} be a domain with the same properties as Ω, $\overline{\Omega}_{\Delta_+} \subset \Omega'$, such that $\Omega_{\Delta_+} \cap \overline{\mathbb{R}} = \Delta_+$. It follows from [2, Corollary 2.5] that $\widetilde{A}$ is definitizable over Ω_{Δ_+} and that for every finite union δ of open connected subsets in Δ_+, $\overline{\delta} \subset \Delta_+$, such that the spectral projection $E_{\widetilde{A}}(\delta)$ is defined the space $E_{\widetilde{A}}(\delta)(\mathcal{K} \times \mathcal{H})$

equipped with the inner product $[\cdot,\cdot]_{\mathcal{K}\times\mathcal{H}}$ is a Pontryagin space with finite rank of negativity.

Let Ω'' be a domain with the same properties as Ω such that $\overline{\Omega''} \subset \Omega'$ and $\overline{\Delta_+} \subset \Omega'' \cap \overline{\mathbb{R}}$. By Theorem 2.6 there exists an Ω''-minimal representing relation S for $\widehat{M}$, that is, S is a selfadjoint relation in some Krein space $\mathcal{G}$ which is definitizable over Ω'', $\rho(S) \cap \Omega'' = \mathfrak{h}(\widehat{M}) \cap \Omega''$, and with a suitable $\Lambda \in \mathcal{L}(\mathbb{C}^2, \mathcal{G})$ we have

$$\widehat{M}(\lambda) = \operatorname{Re}\widehat{M}(\lambda_0) + \Lambda^+\big((\lambda - \operatorname{Re}\lambda_0) + (\lambda-\lambda_0)(\lambda-\overline{\lambda}_0)(S-\lambda)^{-1}\big)\Lambda \tag{3.17}$$

for a fixed $\lambda_0 \in \rho(S) \cap \Omega''$ and all $\lambda \in \rho(S) \cap \Omega''$. The spectral function of S on $\Omega'' \cap \overline{\mathbb{R}}$ will be denoted by E_S (comp. (2.3)).

In the following we will assume that the point λ_0 in (3.16) and (3.17) belongs to $\rho(\widetilde{A}) \cap \rho(S) \cap \Omega''$. This is no restriction. From (3.16) and (3.17) we obtain

$$\widehat{\gamma}(\lambda_0)^+\widehat{\gamma}(\lambda_0) = \Lambda^+\Lambda \quad\text{and}\quad \widehat{\gamma}(\lambda_0)^+(\widetilde{A}-\lambda)^{-1}\widehat{\gamma}(\lambda_0) = \Lambda^+(S-\lambda)^{-1}\Lambda$$

for all $\lambda \in \rho(\widetilde{A}) \cap \rho(S) \cap \Omega''$. Therefore the relation

$$V := \left\{ \begin{pmatrix} \sum_{k=1}^n (1 + (\lambda_k - \lambda_0)(S-\lambda_k)^{-1})\Lambda x_k \\ \sum_{k=1}^n (1 + (\lambda_k - \lambda_0)(\widetilde{A}-\lambda_k)^{-1})\widehat{\gamma}(\lambda_0)x_k \end{pmatrix} \,\middle|\, \begin{matrix} \lambda_k \in \rho(S) \cap \rho(\widetilde{A}) \cap \Omega'' \\ x_k \in \mathbb{C}^2,\ k = 1,\dots,n \end{matrix} \right\}$$

is isometric. The assumptions

$$\mathcal{K} = \operatorname{clsp}\{\mathcal{N}_{\lambda,A^+} \mid \lambda \in \Omega\} \quad\text{and}\quad \mathcal{H} = \operatorname{clsp}\{\mathcal{N}_{\lambda,T^+} \mid \lambda \in \Omega'\}$$

imply

$$\mathcal{K} = \operatorname{clsp}\left\{\operatorname{ran}\gamma(\lambda) \mid \lambda \in \rho(S) \cap \rho(\widetilde{A}) \cap \Omega''\right\}$$

and

$$\mathcal{H} = \operatorname{clsp}\left\{\operatorname{ran}\gamma'(\lambda) \mid \lambda \in \rho(S) \cap \rho(\widetilde{A}) \cap \Omega''\right\},$$

respectively. From (3.12) and (3.15) we obtain

$$\mathcal{K} \times \mathcal{H} = \operatorname{clsp}\left\{\widehat{\gamma}(\lambda)x \mid \lambda \in \rho(S) \cap \rho(\widetilde{A}) \cap \Omega'',\ x \in \mathbb{C}^2\right\}$$

and therefore $\operatorname{ran} V$ is dense in $\mathcal{K}\times\mathcal{H}$. This implies that V is an isometric operator and the same holds for its closure $\overline{V}$.

Let δ be a finite union of open connected subsets in Δ_+, $\overline{\delta} \subset \Delta_+$, such that the boundary points of δ in $\overline{\mathbb{R}}$ are of definite type with respect to $\widetilde{A}$. Then $(E_{\widetilde{A}}(\delta)(\mathcal{K}\times\mathcal{H}), [\cdot,\cdot]_{\mathcal{K}\times\mathcal{H}})$ is a Pontryagin space with finite rank of negativity. As Δ_+ is of positive type with respect to S the spectral projection $E_S(\delta)$ is defined and $E_S(\delta)\mathcal{G}$ equipped with the inner product from $\mathcal{G}$ is a Hilbert space. Writing $E_S(\delta)$ and $E_{\widetilde{A}}(\delta)$ as strong limits of the resolvents of S and $\widetilde{A}$, respectively, one verifies that $\overline{V}$ is reduced by $E_S(\delta)\mathcal{G} \times E_{\widetilde{A}}(\delta)(\mathcal{K}\times\mathcal{H})$. Then

$$\overline{V}_\delta := \overline{V} \cap \big(E_S(\delta)\mathcal{G} \times E_{\widetilde{A}}(\delta)(\mathcal{K}\times\mathcal{H})\big)$$

is a closed isometric operator from the Hilbert space $E_S(\delta)\mathcal{G}$ with dense range in the Pontryagin space $E_{\widetilde{A}}(\delta)(\mathcal{K}\times\mathcal{H})$. As in the proof of [12, Theorem 6.2] one verifies that $\overline{V}_\delta$ is bounded and from this we conclude $\operatorname{dom}\overline{V}_\delta = E_S(\delta)\mathcal{G}$ and $\operatorname{ran}\overline{V}_\delta = E_{\widetilde{A}}(\delta)(\mathcal{K}\times\mathcal{H})$. The isometry of $\overline{V}_\delta$ implies that $E_{\widetilde{A}}(\delta)(\mathcal{K}\times\mathcal{H})$ is a Hilbert space.

Let $\xi \in \delta \cap \sigma(\widetilde{A})$ and choose a sequence $\binom{u_n}{v_n} \in \widetilde{A}$ with $\|u_n\| = 1$ and $\|v_n - \xi u_n\| \to 0$ for $n \to \infty$. From

$$\left(\widetilde{A} \cap \left((I - E_{\widetilde{A}}(\delta))(\mathcal{K} \times \mathcal{H})\right)^2 - \xi\right)^{-1} \in \mathcal{L}\left((I - E_{\widetilde{A}}(\delta))(\mathcal{K} \times \mathcal{H})\right)$$

and

$$\lim_{n\to\infty} \|(I - E_{\widetilde{A}}(\delta))(v_n - \xi u_n)\| = 0$$

we obtain $\|(I - E_{\widetilde{A}}(\delta))u_n\| \to 0$ and $\|E_{\widetilde{A}}(\delta)u_n\| \to 1$ for $n \to \infty$. As $E_{\widetilde{A}}(\delta)(\mathcal{K} \times \mathcal{H})$ is a Hilbert space we have

$$\liminf_{n\to\infty} [u_n, u_n]_{\mathcal{K}\times\mathcal{H}} = \liminf_{n\to\infty} [E_{\widetilde{A}}(\delta)u_n, E_{\widetilde{A}}(\delta)u_n]_{\mathcal{K}\times\mathcal{H}} > 0,$$

that is, ξ is of positive type with respect to $\widetilde{A}$. If ∞ belongs to $\delta \cap \widetilde{\sigma}(\widetilde{A})$ a similar reasoning shows that ∞ is of positive type with respect to $\widetilde{A}$. Therefore δ is of positive type with respect to $\widetilde{A}$. As this is true for every finite union δ of open connected subsets in Δ_+, $\overline{\delta} \subset \Delta_+$, such that the boundary points of δ in $\overline{\mathbb{R}}$ are of definite type with respect to $\widetilde{A}$ we conclude that Δ_+ is also of positive type with respect to $\widetilde{A}$.

Analogously one verifies that a one-sided open connected neighborhood Δ_- of μ in $\overline{\mathbb{R}}$, $\overline{\Delta_-} \subset \Omega' \cap \overline{\mathbb{R}}$, which is of negative type with respect to $A_0 \times T_0$ and $\widehat{M}$ is of negative type with respect to $\widetilde{A}$. We have shown that for every point $\mu \in \Omega' \cap \overline{\mathbb{R}}$ there is an open connected neighborhood I_μ in $\overline{\mathbb{R}}$ such that both components of $I_\mu \backslash \{\mu\}$ are of the same sign type with respect to $A_0 \times T_0$ and $\widetilde{A}$.

3. It remains to verify that $\widetilde{A}$ belongs to $S^\infty(\Omega' \cap \overline{\mathbb{R}})$. For this we use the relation (3.14). We show first that the selfadjoint relation $T_1 = \ker \Gamma_1'$ in $\mathcal{H}$ is definitizable over Ω'. As the function

$$\begin{aligned} -\tau(\lambda)^{-1} = \mathrm{Re}\,\left(-\tau(\lambda_0)^{-1}\right) + \tau(\overline{\lambda}_0)^{-1}\gamma'(\lambda_0)^+\big((\lambda - \mathrm{Re}\,\lambda_0) \\ + (\lambda - \lambda_0)(\lambda - \overline{\lambda}_0)(T_1 - \lambda)^{-1}\big)\gamma'(\lambda_0)\tau(\lambda_0)^{-1} \end{aligned}$$

(see (3.11) and the proof of Proposition 3.2) is definitizable in Ω and the selfadjoint relation T_0 is definitizable over Ω' the same considerations as in step 2 of the proof applied to $-\tau^{-1}$, T_0 and T_1 instead of $\widehat{M}$, $A_0 \times T_0$ and $\widetilde{A}$ show that every point $\mu \in \Omega' \cap \overline{\mathbb{R}}$ has an open connected neighborhood I_μ in $\overline{\mathbb{R}}$ such that both components of $I_\mu \backslash \{\mu\}$ are of definite type with respect to T_1. By (3.5) we have

$$(T_1 - \lambda)^{-1} = (T_0 - \lambda)^{-1} + \gamma'(\lambda)\left(-\tau(\lambda)^{-1}\right)\gamma'(\overline{\lambda})^+$$

for all $\lambda \in \mathfrak{h}(\tau) \cap \mathfrak{h}(\tau^{-1}) \cap \Omega'$. Since $-\tau^{-1}$ is definitizable in Ω the non-real spectrum of T_1 in Ω' does not accumulate to points in $\Omega' \cap \overline{\mathbb{R}}$. The growth properties of $-\tau^{-1}$ (see Definition 2.5) and the resolvent of T_0 imply $T_1 \in S^\infty(\Omega' \cap \overline{\mathbb{R}})$ and therefore T_1 is definitizable over Ω'.

As the sign types of $-\tau^{-1}$ and A_0 are d-compatible in Ω (see the proof of Proposition 3.5 (iii)) it follows that $A_0 \times T_1$ is definitizable over Ω'. The relation

$$(\widetilde{A} - \lambda)^{-1} = (A_0 \times T_1 - \lambda)^{-1} + \widetilde{\gamma}(\lambda)\widehat{M}(\lambda)\widetilde{\gamma}(\overline{\lambda})^+, \quad \lambda \in \mathfrak{h}_0 \cap \Omega',$$

(cf. (3.14)) and the growth properties of $\widehat{M}$ and the resolvent of $A_0 \times T_0$ show $\widetilde{A} \in S^\infty(\Omega' \cap \overline{\mathbb{R}})$. This completes the proof of Theorem 3.6. □

Remark 3.7. Let Ω, Ω', $A \subset A_0$, τ and T_0 be as in Theorem 3.6 and let Δ be an open connected subset in $\overline{\mathbb{R}}$, $\overline{\Delta} \subset \Omega' \cap \overline{\mathbb{R}}$, which is of positive (negative) type with respect to A_0 and τ. As T_0 is an Ω'-minimal representing relation for τ it follows that Δ is of positive (negative) type with respect to the selfadjoint relation $A_0 \times T_0$ in $\mathcal{K} \times \mathcal{H}$. From [2, Corollary 2.5] we obtain that for every finite union δ of open connected subsets in $\overline{\mathbb{R}}$, $\overline{\delta} \subset \Delta$, such that the spectral projection $E_{\widetilde{A}}(\delta)$ corresponding to $\widetilde{A}$ in (3.8) and the set δ is defined, $(E_{\widetilde{A}}(\delta)(\mathcal{K} \times \mathcal{H}), [\cdot,\cdot]_{\mathcal{K}\times\mathcal{H}})$ is a Pontryagin space with finite rank of negativity (positivity). This can also be deduced from [3, Theorem 4.1].

Remark 3.8. The assumption $\mathcal{H} = \operatorname{clsp}\{\mathcal{N}_{\lambda,T^+} \mid \lambda \in \Omega'\}$ in Theorem 3.6 implies that the selfadjoint extension $\widetilde{A}$ in (3.8) satisfies the minimality condition

$$\mathcal{K} \times \mathcal{H} = \operatorname{clsp}\left\{(1 + (\lambda - \lambda_0)(\widetilde{A} - \lambda)^{-1})\{k, 0\} \mid k \in \mathcal{K},\, \lambda \in \rho(\widetilde{A}) \cap \Omega'\right\} \tag{3.18}$$

for some fixed $\lambda_0 \in \rho(\widetilde{A}) \cap \Omega'$. This can be verified as in [3, Proof of Theorem 4.1]. Let $A \in \widetilde{\mathcal{C}}(\mathcal{K})$ be as in Theorem 3.6 and assume that $\widetilde{B}$ is a selfadjoint extension of A in some Krein space $\mathcal{K} \times \mathcal{H}'$ which is definitizable over Ω' such that the compressed resolvent of $\widetilde{B}$ onto $\mathcal{K}$ yields a solution of (3.9). Then $P_{\mathcal{K}}(\widetilde{B} - \lambda)^{-1}|_{\mathcal{K}}$ and $P_{\mathcal{K}}(\widetilde{A} - \lambda)^{-1}|_{\mathcal{K}}$ coincide. If $\widetilde{B}$ fulfils the minimality condition (3.18) with $\mathcal{K} \times \mathcal{H}$ and $\rho(\widetilde{A}) \cap \Omega'$ replaced by $\mathcal{K} \times \mathcal{H}'$ and $\rho(\widetilde{B}) \cap \Omega'$, respectively, and we choose $\lambda_0 \in \rho(\widetilde{A}) \cap \rho(\widetilde{B}) \cap \Omega'$, then

$$W := \left\{ \begin{pmatrix} \sum_{i=1}^n (1 + (\lambda_i - \lambda_0)(\widetilde{A} - \lambda_i)^{-1})\{k_i, 0\} \\ \sum_{i=1}^n (1 + (\lambda_i - \lambda_0)(\widetilde{B} - \lambda_i)^{-1})\{k_i, 0\} \end{pmatrix} \,\middle|\, \begin{matrix} \lambda_i \in \rho(\widetilde{A}) \cap \rho(\widetilde{B}) \cap \Omega', \\ k_i \in \mathcal{K},\, i = 1, 2, \ldots, n \end{matrix} \right\}$$

is a densely defined isometric operator in $\mathcal{K}\times\mathcal{H}$ with dense range in $\mathcal{K}\times\mathcal{H}'$ and the same holds for its closure $\overline{W}$. We denote the local spectral functions of $\widetilde{A}$ and $\widetilde{B}$ by $E_{\widetilde{A}}$ and $E_{\widetilde{B}}$, respectively. Let Δ be an open connected subset in $\overline{\mathbb{R}}$, $\overline{\Delta} \subset \Omega' \cap \overline{\mathbb{R}}$, such that $E_{\widetilde{A}}(\Delta)$ is defined. Then also $E_{\widetilde{B}}(\Delta)$ is defined and $\overline{W}$ is reduced by

$$E_{\widetilde{A}}(\Delta)(\mathcal{K} \times \mathcal{H}) \times E_{\widetilde{B}}(\Delta)(\mathcal{K} \times \mathcal{H}').$$

The closed isometric operator

$$\overline{W}_\Delta := \overline{W} \cap \left(E_{\widetilde{A}}(\Delta)(\mathcal{K} \times \mathcal{H}) \times E_{\widetilde{B}}(\Delta)(\mathcal{K} \times \mathcal{H}')\right)$$

intertwines the resolvents of

$$\widetilde{A}_1 := \widetilde{A} \cap \left(E_{\widetilde{A}}(\Delta)(\mathcal{K} \times \mathcal{H})\right)^2 \quad \text{and} \quad \widetilde{B}_1 := \widetilde{B} \cap \left(E_{\widetilde{B}}(\Delta)(\mathcal{K} \times \mathcal{H}')\right)^2,$$

i.e., for $\lambda \in \rho(\widetilde{A}_1) \cap \rho(\widetilde{B}_1) \cap \Omega'$ and $x \in \operatorname{dom} \overline{W}_\Delta$ we have

$$\overline{W}_\Delta(\widetilde{A}_1 - \lambda)^{-1}x = (\widetilde{B}_1 - \lambda)^{-1}\overline{W}_\Delta x.$$

In particular, the ranks of positivity and negativity of the inner products on the subspaces $E_{\widetilde{A}}(\Delta)(\mathcal{K} \times \mathcal{H})$ and $E_{\widetilde{B}}(\Delta)(\mathcal{K} \times \mathcal{H}')$ coincide.

If, in addition to the assumptions above, $(E_{\widetilde{A}}(\Delta)(\mathcal{K}\times\mathcal{H}),[\cdot,\cdot]_{\mathcal{K}\times\mathcal{H}})$ is a Pontryagin space, then $E_{\widetilde{B}}(\Delta)(\mathcal{K}\times\mathcal{H}')$ equipped with the inner product from $\mathcal{K}\times\mathcal{H}'$ is also a Pontryagin space and by [12, Theorem 6.2] the operator $\overline{W_\Delta}$ is an isometric isomorphism of $E_{\widetilde{A}}(\Delta)(\mathcal{K}\times\mathcal{H})$ onto $E_{\widetilde{B}}(\Delta)(\mathcal{K}\times\mathcal{H}')$, i.e., $\widetilde{A}_1$ and $\widetilde{B}_1$ are isometrically equivalent.

The case that the function τ is a real constant is excluded in Theorem 3.6. In this case we have the following theorem.

Theorem 3.9. *Let Ω be a domain as in the beginning of Section 3.2 and let A be a closed symmetric operator of defect one in the Krein space $\mathcal{K}$ such that $\mathcal{K}=\operatorname{clsp}\{\mathcal{N}_{\lambda,A^+}\,|\,\lambda\in\Omega\}$ holds. Assume that there exists a selfadjoint extension A_0 of A which is definitizable over Ω. Let $\{\mathbb{C},\Gamma_0,\Gamma_1\}$ be a boundary value space for A^+, $A_0=\ker\Gamma_0$, denote by γ and M the corresponding γ-field and Weyl function, respectively, and let τ be a real constant.*

Then the relation $A_{-\tau}=\ker\big(\Gamma_1+\tau\Gamma_0\big)$ is a selfadjoint extension of A in $\mathcal{K}$ which is definitizable over Ω. The sign types of M and $A_{-\tau}$ are d-compatible in Ω. For every $k\in\mathcal{K}$ and every $\lambda\in\mathfrak{h}(M)\cap\mathfrak{h}((M+\tau)^{-1})\cap\Omega$ a solution of the boundary value problem

$$f_1'-\lambda f_1=k,\quad \tau\Gamma_0\hat{f}_1+\Gamma_1\hat{f}_1=0,\quad \hat{f}_1=\begin{pmatrix}f_1\\f_1'\end{pmatrix}\in A^+, \tag{3.19}$$

is given by

$$f_1=(A_{-\tau}-\lambda)^{-1}k=(A_0-\lambda)^{-1}k-\gamma(\lambda)\big(M(\lambda)+\tau\big)^{-1}\gamma(\overline{\lambda})^+k,\quad f_1'=\lambda f_1+k.$$

Proof. The proof of Theorem 3.9 is a modification of the proof of Theorem 3.6. Note first that the relation (3.4) and $\mathcal{K}=\operatorname{clsp}\{\operatorname{ran}\gamma(\lambda)\,|\,\lambda\in\Omega\cap\rho(A_0)\}$ imply that the Weyl function M is not identically equal to a constant. Here it is obvious that the resolvent of $A_{-\tau}$ yields a solution of the boundary value problem (3.19) (compare (3.1), (3.2) and (3.5)). As in step 2 and step 3 of the proof of Theorem 3.6 the function $\lambda\mapsto-(M(\lambda)+\tau)^{-1}$, which by [1, Theorem 2.3] is definitizable in Ω, can be regarded as the Weyl function corresponding to a boundary value space $\{\mathbb{C},\hat{\Gamma}_0,\hat{\Gamma}_1\}$ for A^+ with $A_{-\tau}=\ker\hat{\Gamma}_0$. Now the same arguments as in the proof of Theorem 3.6 show that $A_{-\tau}$ is definitizable over Ω. The details are left to the reader. □

References

[1] T.Ya. Azizov, P. Jonas, On Locally Definitizable Matrix Functions, Preprint 21-2005, Institute of Mathematics, Technische Universität Berlin.

[2] J. Behrndt, P. Jonas, On Compact Perturbations of Locally Definitizable Selfadjoint Relations in Krein Spaces, *Integral Equations Operator Theory* **52** (2005), 17–44.

[3] J. Behrndt, P. Jonas, Boundary Value Problems with Local Generalized Nevanlinna Functions in the Boundary Condition, to appear in *Integral Equations Operator Theory.*

[4] V.A. Derkach, On Weyl Function and Generalized Resolvents of a Hermitian Operator in a Krein Space. *Integral Equations Operator Theory* **23** (1995), 387–415.

[5] V.A. Derkach, On Generalized Resolvents of Hermitian Relations in Krein Spaces. *J. Math. Sci.* (New York) **97** (1999), 4420–4460.

[6] V.A. Derkach, S. Hassi, M.M. Malamud, H.S.V. de Snoo, Generalized Resolvents of Symmetric Operators and Admissibility. *Methods Funct. Anal. Topology* **6** (2000), 24–53.

[7] V.A. Derkach, M.M. Malamud, Generalized Resolvents and the Boundary Value Problems for Hermitian Operators with Gaps. *J. Funct. Anal.* **95** (1991), 1–95.

[8] V.A. Derkach, M.M. Malamud, The Extension Theory of Hermitian Operators and the Moment Problem. *J. Math. Sci. (New York)* **73** (1995), 141–242.

[9] A. Dijksma, H.S.V. de Snoo, Symmetric and Selfadjoint Relations in Krein Spaces I. *Oper. Theory Adv. Appl.* **24**, Birkhäuser Verlag Basel (1987), 145–166.

[10] A. Dijksma, H.S.V. de Snoo, Symmetric and Selfadjoint Relations in Krein Spaces II, *Ann. Acad. Sci. Fenn. Math.* **12**, 1987, 199–216.

[11] V.I. Gorbachuk, M.L. Gorbachuk, *Boundary Value Problems for Operator Differential Equations.* Kluwer Academic Publishers, Dordrecht (1991).

[12] I.S. Iohvidov, M.G. Krein, H. Langer, *Introduction to the Spectral Theory of Operators in Spaces with an Indefinite Metric.* Mathematical Research **9** (1982), Akademie-Verlag Berlin.

[13] P. Jonas, On a Class of Unitary Operators in Krein Space. *Oper. Theory Adv. Appl.* **17**, Birkhäuser Verlag Basel (1986), 151–172.

[14] P. Jonas, A Class of Operator-Valued Meromorphic Functions on the Unit Disc, Ann. Acad. Sci. Fenn. Math. **17** (1992), 257-284.

[15] P. Jonas, Operator Representations of Definitizable Functions. *Ann. Acad. Sci. Fenn. Math.* **25** (2000), 41–72.

[16] P. Jonas, On Locally Definite Operators in Krein Spaces. in: Spectral Theory and Applications, Theta Foundation, (2003).

[17] P. Jonas, On Operator Representations of Locally Definitizable Functions, *Oper. Theory Adv. Appl.* **162**, Birkhäuser Verlag, Basel (2005), 165–190.

[18] H. Langer, Spectral Functions of Definitizable Operators in Krein Spaces. Functional Analysis Proceedings of a Conference held at Dubrovnik, Yugoslavia, November 2-14 (1981), *Lecture Notes in Mathematics* **948**, Springer Verlag Berlin-Heidelberg-New York (1982), 1–46.

[19] H. Langer, A. Markus, V. Matsaev, Locally Definite Operators in Indefinite Inner Product Spaces, *Math. Ann.* **308** (1997), 405–424.

Jussi Behrndt
Institut für Mathematik, MA 6-4
Technische Universität Berlin
Straße des 17. Juni 136
D-10623 Berlin
Germany
e-mail: behrndt@math.tu-berlin.de

Operator Theory:
Advances and Applications, Vol. 163, 75–95

Riesz Bases of Root Vectors of Indefinite Sturm-Liouville Problems with Eigenparameter Dependent Boundary Conditions, I

Paul Binding and Branko Ćurgus

Abstract. We consider a regular indefinite Sturm-Liouville problem with two self-adjoint boundary conditions, one being affinely dependent on the eigenparameter. We give sufficient conditions under which a basis of each root subspace for this Sturm-Liouville problem can be selected so that the union of all these bases constitutes a Riesz basis of a corresponding weighted Hilbert space.

Mathematics Subject Classification (2000). Primary 34L10, 34B24, 34B09, 47B50.

Keywords. Sturm-Liouville equations, indefinite weight functions, Riesz bases.

1. Introduction

We consider a regular indefinite Sturm-Liouville boundary eigenvalue problem of the form

$$-(p\,f')' + q\,f = \lambda\, r\, f \quad \text{on} \quad [-1,1]. \tag{1.1}$$

The coefficients $1/p, q, r$ in (1.1) are assumed to be real and integrable over $[-1,1]$, $p(x) > 0$, and $x\,r(x) > 0$ for almost all $x \in [-1,1]$. We impose two boundary conditions on (1.1) (only one of which is λ-dependent):

$$\mathsf{L}\,\mathbf{b}(f) = 0, \qquad \mathsf{M}\mathbf{b}(f) = \lambda\,\mathsf{N}\mathbf{b}(f). \tag{1.2}$$

where L, M and N are 1×4 non-zero (row) matrices and the boundary mapping $\mathbf{b}$ is defined for all f in the domain of (1.1) by

$$\mathbf{b}(f) = \begin{bmatrix} f(-1) & f(1) & (pf')(-1) & (pf')(1) \end{bmatrix}^T.$$

We shall utilize an operator theoretic framework developed in [3]. Under Condition 2.1 below, a self-adjoint operator A in the Krein space $L_{2,r}(-1,1)\oplus\mathbb{C}_\Delta$ can be associated with the eigenvalue problem (1.1), (1.2). Here Δ is a non-zero real number which is determined by M and N – see Section 2 for details. We remark that the topology of this Krein space is that of the corresponding Hilbert space $L_{2,|r|}(-1,1) \oplus \mathbb{C}_{|\Delta|}$. (In the rest of the paper we abbreviate $L_{2,r}(-1,1)$ to $L_{2,r}$ and $L_{2,|r|}(-1,1)$ to $L_{2,|r|}$.) Our main goal in this paper is to provide sufficient conditions on the coefficients in (1.1), (1.2) under which there is a Riesz basis of the above Hilbert space consisting of the union of bases for all the root subspaces of the above operator A. This will be referred to for the remainder of this section as the *Riesz-basis property of* A.

Completeness and expansion theorems with a stronger topology, but in a smaller space corresponding to the form domain of the operator A, have been considered by many authors – see [3] (and the references there) and [12]. Although the topology of the Krein space $L_{2,r} \oplus \mathbb{C}_{\Delta}$ is weaker than the topology of the form domain, which in our case is a Pontryagin space, the expansion question turns out to be much more challenging mathematically.

Indeed, even for the case when the boundary conditions are λ-independent this problem is nontrivial. In our notation, this case corresponds to L being a nonsingular 2×4 matrix, with the second equation in (1.2) suppressed. The Riesz-basis property of the operator corresponding to A, now defined in $L_{2,r}$, has been discussed by several authors, e.g., in [2, 6, 9, 14, 15]. The first general sufficient condition for this was given by Beals [2], who required the weight function r to behave like a power of the independent variable x in an open neighborhood of the turning point $x = 0$, although his method does allow more general weight functions. Refinements of Beals's method in [9] and [15] show that a "one-sided" condition on r (i.e., in only a half-neighborhood of $x = 0$) is enough to guarantee the Riesz-basis property. That some extra condition on r is indeed necessary follows from [15] where Volkmer showed that weight functions r exist for which the corresponding Sturm-Liouville problem (1.1), under the conditions used here, does not have the Riesz-basis property. Explicit examples of such weight functions were given in [1] and [10]. Recently, Parfyonov [13] has given an explicit necessary and sufficient condition for the Riesz basis property in the case $p = 1, q = 0$ with odd weight function r. Here, and in most of the above references, Dirichlet boundary conditions were imposed.

General self-adjoint (perhaps non-separated, but still λ-independent) boundary conditions were treated by Ćurgus and Langer [6]. They showed that if the *essential* boundary conditions, i.e., those not including derivatives, were separated, then a Beals-type condition in a neighborhood of $x = 0$ was sufficient for the Riesz-basis property. But if some of the non-separated boundary conditions were essential then [6] established the Riesz-basis property only by imposing extra restrictions on the weight function in (half-)neighborhoods of *both* endpoints of the interval $[-1,1]$. Again, some extra restriction is necessary, since in [4] we gave an explicit example of (1.1) under the conditions used here, satisfying a Beals-type condition at $x=0$, but without the Riesz-basis property. Of course at least one (in fact one, in this antiperiodic case) boundary condition was essential and non-separated. In some sense, then, the boundary ± 1 behaves as a turning point under such boundary conditions.

In summary, the Riesz-basis property is quite subtle, and depends significantly on the nature of the boundary conditions even when they are independent of λ. In this paper and its sequel, we shall examine the analogous situation for the cases of one and two λ-dependent boundary conditions, where the possibilities for the (λ-dependent) boundary conditions are much greater. As in the λ-independent case, a condition on the weight function is needed near the turning point $x = 0$ to ensure the Riesz-basis property of A. We shall develop such a condition (which is implied by the ones discussed above) in Section 4. Depending on the nature of the boundary conditions (1.2), we may also need a condition near the boundary, and this is discussed in Section 5. It should be remarked that for the case of exactly one λ-dependent boundary condition treated here we need only one such condition, near either $x = -1$ or $x = 1$, and this can be viewed as a "one-sided" condition at ± 1. In the case of two λ-dependent boundary conditions we shall also need a condition involving both boundary points $x = -1$ and $x = 1$.

It turns out that all the above conditions have a common core. This is not immediately obvious, since there are differences between the "turning points" 0 and ± 1. For example, when the boundary conditions are separated, the values of f and f' are equal at 0 but are independent at -1 and 1. The common core, which will also be needed in Part II, involves the notion of smoothly connected half-neighborhoods, and this is defined and studied in Section 3.

In order to apply the above conditions, we use a criterion in Theorem 2.2, equivalent to the Riesz-basis property of A, involving a positive homeomorphism of $L_{2,r} \oplus \mathbb{C}_\Delta$ with the form domain of A as an invariant subspace. This, together with certain mollification arguments, is used for our main results, which are detailed in Section 6. To paraphrase these, we recall that a λ-independent boundary condition is *essential* if it does not include derivatives. Similarly, a λ-dependent boundary condition will be called *essential* if it does not include derivatives in the λ-terms.

In Theorem 6.1 we discuss situations when a condition on r near $x = 0$ suffices for the Riesz-basis property of A. For example this holds when the first (λ-independent) boundary condition in (1.2) is either non-essential, or essential and separated, and the second (λ-dependent) one is non-essential. If the latter condition is essential instead, then the same result holds if a sign condition is also satisfied, and this includes a result of Fleige [11], which is the only reference we know where the Riesz-basis property of A has been studied for λ-dependent boundary conditions.

In Theorems 6.2 and 6.3 we consider those cases of (1.2) which are not covered by Theorem 6.1. Then we require a condition near just one of the boundary points ± 1, not both as in [6]. The choice of the boundary point is arbitrary in Theorem 6.2 which deals with the case when the boundary conditions in (1.2) are, respectively, essential non-separated and non-essential. In Theorem 6.3, however, this choice is not arbitrary but depends on the sign of the number Δ used in defining the inner product on $L_{2,r} \oplus \mathbb{C}_\Delta$.

2. Operators associated with the eigenvalue problem

The maximal operator $S_{\max}$ in $L_{2,r}(-1,1) = L_{2,r}$ associated with (1.1) is defined by

$$S_{\max} : f \mapsto \ell(f) := \frac{1}{r}\big(-(pf')' + qf\big), \quad f \in \mathcal{D}(S_{\max}),$$

where

$$\mathcal{D}(S_{\max}) = \mathcal{D}_{\max} = \big\{ f \in L_{2,r} : f, pf' \in AC[0,1],\ \ell(f) \in L_{2,r} \big\}.$$

We define the boundary mapping $\mathbf{b}$ by

$$\mathbf{b}(f) := \begin{bmatrix} f(-1) & f(1) & (pf')(-1) & (pf')(1) \end{bmatrix}^T, \quad f \in \mathcal{D}(S_{\max}),$$

and the concomitant matrix Q corresponding to $\mathbf{b}$ by

$$\mathsf{Q} := i \begin{bmatrix} 0 & 0 & -1 & 0 \\ 0 & 0 & 0 & 1 \\ 1 & 0 & 0 & 0 \\ 0 & -1 & 0 & 0 \end{bmatrix}.$$

We notice that $\mathsf{Q} = \mathsf{Q}^{-1}$. Integrating by parts we easily calculate that

$$\int_{-1}^{1} S_{\max} f\, \overline{g}\, r - \int_{-1}^{1} f\, S_{\max} \overline{g}\, r = i\, \mathbf{b}(g)^* \mathsf{Q} \mathbf{b}(f), \quad f, g \in \mathcal{D}(S_{\max}).$$

Throughout, we shall impose the following non-degeneracy and self-adjointness conditions on the boundary data.

Condition 2.1. The row vectors L, M and N in (1.2) satisfy:

(1) the 3×4 matrix $\begin{bmatrix} \mathsf{L} \\ \mathsf{M} \\ \mathsf{N} \end{bmatrix}$ has rank 3,

(2) $\mathsf{LQL}^* = \mathsf{MQM}^* = \mathsf{NQN}^* = \mathsf{LQM}^* = \mathsf{LQN}^* = 0$,

(3) $i\,\mathsf{MQ}^{-1}\mathsf{N}^*$ is a non-zero real number and we define

$$\Delta = -\frac{i}{\mathsf{MQ}^{-1}\mathsf{N}^*}. \tag{2.1}$$

Clearly the boundary value problem (1.1)–(1.2) will not change if row reduction is applied to the coefficient matrix

$$\begin{bmatrix} \mathsf{L} & 0 \\ \mathsf{M} & \mathsf{N} \end{bmatrix}. \tag{2.2}$$

In what follows we will assume that the 2×8 matrix in (2.2) is row reduced to row echelon form (starting the reduction at the bottom right corner). After the row reduction, we write the row vectors L and N as

$$\mathsf{L} = \begin{bmatrix} \mathsf{L}_e & \mathsf{L}_n \end{bmatrix}, \qquad \mathsf{N} = \begin{bmatrix} \mathsf{N}_e & \mathsf{N}_n \end{bmatrix}. \tag{2.3}$$

If either of the 1×2 matrices $\mathsf{L}_n, \mathsf{N}_n$ is non-zero, the corresponding boundary condition is called "non-essential". In any case these matrices do not appear in the representation of the form domain of A, discussed below, but they will play an important role in our conditions for Riesz bases in Section 6. The 1×2 matrices L_e and N_e represent the "essential" boundary conditions if the non-essential parts L_n and N_n are zero matrices.

Next we define a Krein space operator associated with the problem (1.1)–(1.2). We consider the linear space $L_{2,|r|} \oplus \mathbb{C}$, equipped with the inner product

$$\left[\begin{pmatrix} f \\ z \end{pmatrix}, \begin{pmatrix} g \\ w \end{pmatrix}\right] := \int_{-1}^{1} f\overline{g}r + \overline{w}\Delta z, \quad f, g \in L_{2,|r|},\ z, w \in \mathbb{C}.$$

Then $\left(L_{2,|r|} \oplus \mathbb{C}, [\,\cdot\,,\cdot\,]\right)$ is a Krein space, which we denote by $L_{2,r} \oplus \mathbb{C}_\Delta$. A fundamental symmetry on this Krein space is given by

$$J := \begin{bmatrix} J_0 & 0 \\ 0 & \operatorname{sgn}\Delta \end{bmatrix}, \tag{2.4}$$

where $\operatorname{sgn}\Delta \in \{-1, 1\}$ and $J_0 : L_{2,r} \to L_{2,r}$ is defined by

$$(J_0 f)(x) := f(x)\operatorname{sgn}(r(x)), \quad x \in [-1, 1].$$

Then $[J\,\cdot\,,\cdot\,]$ is a positive definite inner product which turns $L_{2,r} \oplus \mathbb{C}_\Delta$ into a Hilbert space $L_{2,|r|} \oplus \mathbb{C}_{|\Delta|}$. The topology of $L_{2,r} \oplus \mathbb{C}_\Delta$ is defined to be that of $L_{2,|r|} \oplus \mathbb{C}_{|\Delta|}$, and a Riesz basis of $L_{2,r} \oplus \mathbb{C}_\Delta$ is defined as a homeomorphic image of an orthonormal basis of $L_{2,|r|} \oplus \mathbb{C}_{|\Delta|}$.

We define the operator A in the Krein space $L_{2,r} \oplus \mathbb{C}_\Delta$ on the domain

$$\mathcal{D}(A) = \left\{ \begin{bmatrix} f \\ z \end{bmatrix} \in \begin{array}{c} L_{2,r} \\ \oplus \\ \mathbb{C}_\Delta \end{array} : f \in \mathcal{D}(S_{\max}),\ \mathsf{L}\,\mathbf{b}(f) = 0,\ \ z = \mathsf{N}\mathbf{b}(f) \right\} \tag{2.5}$$

by

$$A \begin{bmatrix} f \\ \mathsf{N}\mathbf{b}(f) \end{bmatrix} := \begin{bmatrix} S_{\max} f \\ \mathsf{M}\mathbf{b}(f) \end{bmatrix}, \qquad \begin{bmatrix} f \\ \mathsf{N}\mathbf{b}(f) \end{bmatrix} \in \mathcal{D}(A). \tag{2.6}$$

Using [3, Theorems 3.3 and 4.1] we see that this operator is self-adjoint in $L_{2,r} \oplus \mathbb{C}_\Delta$ and in particular:

(i) A is quasi-uniformly positive [7] (and therefore definitizable) in $L_{2,r} \oplus \mathbb{C}_\Delta$.
(ii) A has a discrete spectrum.
(iii) The root subspaces corresponding to real distinct eigenvalues of A are mutually orthogonal in the Krein space $L_{2,r} \oplus \mathbb{C}_\Delta$.
(iv) All but finitely many eigenvalues of A are semisimple and real.

For further properties of A, we refer the reader to [3, Theorem 3.3]. From (i), (ii) and the characterization of the regularity of the critical point infinity for definitizable operators in Krein spaces given in [5, Theorem 3.2], we then obtain the following, which is our central tool.

Theorem 2.2. *Let $\mathcal{F}(A)$ denote the form domain of A. There exists a basis for each root subspace of A, so that the union of all these bases is a Riesz basis of $L_{2,|r|} \oplus \mathbb{C}_{|\Delta|}$ if and only if there exists a bounded, boundedly invertible, positive operator W in $L_{2,r} \oplus \mathbb{C}_\Delta$ such that $W\,\mathcal{F}(A) \subset \mathcal{F}(A)$.*

In order to apply this result, we need to characterize the form domain $\mathcal{F}(A)$. To this end, let $\mathcal{F}_{\max}$ be the set of all functions f in $L_{2,r}$, absolutely continuous on $[-1,1]$, such that $\int_{-1}^{1} p\,|f'|^2 < +\infty$. On $\mathcal{F}_{\max}$ we define the *essential boundary mapping* $\mathbf{b}_e : \mathcal{F}_{\max} \to \mathbb{C}^2$ by

$$\mathbf{b}_e(f) := \begin{bmatrix} f(-1) & f(1) \end{bmatrix}^T, \quad f \in \mathcal{F}_{\max}.$$

Clearly $\mathbf{b}_e$ is surjective.

By [3, Theorem 4.2], there are four possible cases for the form domain $\mathcal{F}(A)$ of A: If $\mathsf{L}_n \neq 0$ and $\mathsf{N}_n \neq 0$, then

$$\mathcal{F}(A) = \left\{ \begin{bmatrix} f \\ z \end{bmatrix} \in \begin{matrix} L_{2,r} \\ \oplus \\ \mathbb{C}_\Delta \end{matrix} : f \in \mathcal{F}_{\max},\ z \in \mathbb{C} \right\}. \tag{2.7}$$

If $\mathsf{L}_n = 0$ and $\mathsf{N}_n \neq 0$, then

$$\mathcal{F}(A) = \left\{ \begin{bmatrix} f \\ z \end{bmatrix} \in \begin{matrix} L_{2,r} \\ \oplus \\ \mathbb{C}_\Delta \end{matrix} : f \in \mathcal{F}_{\max}, \mathsf{L}_e\mathbf{b}_e(f) = 0,\ z \in \mathbb{C} \right\}. \tag{2.8}$$

If $\mathsf{L}_n \neq 0$ and $\mathsf{N}_n = 0$, then

$$\mathcal{F}(A) = \left\{ \begin{bmatrix} f \\ \mathsf{N}_e\mathbf{b}_e(f) \end{bmatrix} \in \begin{matrix} L_{2,r} \\ \oplus \\ \mathbb{C}_\Delta \end{matrix} : f \in \mathcal{F}_{\max} \right\}. \tag{2.9}$$

If $\mathsf{L}_n = 0$ and $\mathsf{N}_n = 0$, then

$$\mathcal{F}(A) = \left\{ \begin{bmatrix} f \\ \mathsf{N}_e\mathbf{b}_e(f) \end{bmatrix} \in \begin{matrix} L_{2,r} \\ \oplus \\ \mathbb{C}_\Delta \end{matrix} : f \in \mathcal{F}_{\max}, \mathsf{L}_e\mathbf{b}_e(f) = 0 \right\}. \tag{2.10}$$

To construct an operator W as in Theorem 2.2 we need to impose conditions on the coefficients p and r in (1.1). In all cases we need Condition 4.1 in a neighborhood of 0, and in some cases we also need one of two conditions, 5.1 or 5.2, on r in neighborhoods of -1 or 1. These will be discussed in Sections 4 and 5 respectively.

3. Smooth connection and associated operator

To prepare the ground for the conditions mentioned above (and in Part II), we develop the concept of smoothly connected half-neighborhoods. A closed interval of non-zero length is said to be a *left half-neighborhood* of its right endpoint and a *right half-neighborhood* of its left endpoint.

Let $\imath$ be a closed subinterval of $[-1,1]$. By $\mathcal{F}_{\max}(\imath)$ we denote the set of all functions f in $L_{2,r}(\imath)$ which are absolutely continuous on $\imath$ and such that $\int_\imath p\,|f'|^2 <$

$+\infty$. Note that $\mathcal{F}_{\max}[-1,1]$ is the space $\mathcal{F}_{\max}$ defined below Theorem 2.2. In the next definition affine function α means $\alpha(t) = a + \alpha' t$ where $a, \alpha', t \in \mathbb{R}$.

Definition 3.1. Let p and r be the coefficients in (1.1). Let $a, b \in [-1,1]$ and let h_a and h_b, respectively, be half-neighborhoods of a and b which are contained in $[-1,1]$. We say that the ordered pair (h_a, h_b) is *smoothly connected* if there exist

(a) positive real numbers ϵ and τ,
(b) non-constant affine functions $\alpha : [0,\epsilon] \to h_a$ and $\beta : [0,\epsilon] \to h_b$,
(c) non-negative real functions ρ and ϖ defined on $[0,\epsilon]$,

such that

(i) $\alpha(0) = a$ and $\beta(0) = b$,
(ii) $p \circ \alpha$ and $p \circ \beta$ are locally integrable on the interval $(0,\epsilon]$,
(iii) $\rho \circ \alpha^{-1} \in \mathcal{F}_{\max}\big(\alpha([0,\epsilon])\big)$,
(iv) $1/\tau < \varpi < \tau$ a.e. on $[0,\epsilon]$,
(v) $\rho(t) = \dfrac{\big|r\big(\beta(t)\big)\big|}{\big|r\big(\alpha(t)\big)\big|}$, and $\varpi(t) = \dfrac{p\big(\beta(t)\big)}{p\big(\alpha(t)\big)}$, for $t \in (0,\epsilon]$.

The numbers α', β', (the slopes of α, β, respectively) and $\rho(0)$ are called the *parameters* of the smooth connection.

Remark 3.2. Since the function α in Definition 3.1 is affine, the condition $\rho \circ \alpha^{-1} \in \mathcal{F}_{\max}\big(\alpha([0,\epsilon])\big)$ in (iii) is equivalent to

$$\rho \in AC[0,\epsilon] \quad \text{and} \quad \int_0^\epsilon |\rho'(t)|^2 p(\alpha(t))dt < +\infty. \tag{3.1}$$

Under the assumption that $1/\tau < \varpi < \tau$ a.e. on $[0,\epsilon]$, it follows that property (3.1) is equivalent to

$$\rho \in AC[0,\epsilon] \quad \text{and} \quad \int_0^\epsilon |\rho'(t)|^2 p(\beta(t))dt < +\infty.$$

To illustrate Definition 3.1, we make the following

Definition 3.3. Let ν and a be real numbers and let h_a be a half-neighborhood of a. Let g be a function defined on h_a. Then g is called *of order ν on h_a* if there exists $g_1 \in C^1(h_a)$ such that

$$g(x) = (x-a)^\nu g_1(x) \quad \text{and} \quad g_1(x) \neq 0, \quad x \in h_a.$$

Example 3.4. Let $a, b \in [-1,1]$. Let h_a and h_b, respectively, be half-neighborhoods of a and b contained in $[-1,1]$. Assume that the coefficient r in (1.1) is of order ν on both half-neighborhoods h_a and h_b. Assume also that the functions p and $1/p$ are bounded on h_a and h_b (or, alternatively, that p is of order μ on both half-neighborhoods h_a and h_b.) Then lengthy, but straightforward, reasoning shows that the half-neighborhoods h_a and h_b are smoothly connected. Moreover the parameters of the smooth connection are non-zero numbers.

Remark 3.5. Throughout the paper we use the following convention: A product of functions is defined to have value 0 whenever one of its terms has value zero, even if some other terms are not defined.

Theorem 3.6. *Let $\imath$ and $\jmath$ be closed intervals, $\imath, \jmath \in \{[-1,0],[0,1]\}$. Let a be an endpoint of $\imath$ and let b be an endpoint of $\jmath$. Denote by a_1 and b_1, respectively, the remaining endpoints. Assume that the half-neighborhoods $\imath$ of a and $\jmath$ of b are smoothly connected with parameters α', β' and $\rho(0)$. Then there exists an operator*

$$S : L_{2,|r|}(\imath) \to L_{2,|r|}(\jmath)$$

such that:

(S-1) $S \in \mathcal{L}\big(L_{2,|r|}(\imath), L_{2,|r|}(\jmath)\big),\ S^* \in \mathcal{L}\big(L_{2,|r|}(\jmath), L_{2,|r|}(\imath)\big)$.
(S-2) $(Sf)(x) = 0,\ |x - b_1| \leq \frac{1}{2}$ *for all* $f \in L_{2,|r|}(\imath)$, *and* $(S^*g)(x) = 0,\ |x - a_1| \leq \frac{1}{2}$ *for all* $g \in L_{2,|r|}(\jmath)$.
(S-3) $S\mathcal{F}_{\max}(\imath) \subset \mathcal{F}_{\max}(\jmath),\quad S^*\mathcal{F}_{\max}(\jmath) \subset \mathcal{F}_{\max}(\imath)$.
(S-4) *For all* $f \in \mathcal{F}_{\max}(\imath)$ *and all* $g \in \mathcal{F}_{\max}(\jmath)$ *we have*

$$\lim_{\substack{y \to b\\ y \in \jmath}} (Sf)(y) = |\alpha'| \lim_{\substack{x \to a\\ x \in \imath}} f(x), \qquad \lim_{\substack{x \to a\\ x \in \imath}} (S^*g)(x) = |\beta'|\rho(0) \lim_{\substack{y \to b\\ y \in \jmath}} g(y).$$

Proof. Let $\epsilon > 0$ be the real number and α and β the affine functions introduced in Definition 3.1. Thus $\alpha(0) = a$ and $\beta(0) = b$. It is no loss of generality to assume that each of the intervals $\alpha([0,\epsilon])$ and $\beta([0,\epsilon])$ has a length $< 1/2$. Let $\alpha_1 : [0,1] \to \imath$ and $\beta_1 : [0,1] \to \jmath$ be strictly monotonic and continuously differentiable bijections such that $\alpha_1(x) = \alpha(x)$ and $\beta_1(x) = \beta(x)$ for all $x \in [0,\epsilon]$. Then $\alpha_1(1) = a_1$ and $\beta_1(1) = b_1$.

Let $\phi : [0,1] \to [0,1]$, $\phi \in C^1[0,1]$, be such that

$$\phi(t) = 1, \quad 0 \leq t \leq \epsilon/2, \qquad \phi(t) = 0, \quad \epsilon \leq t \leq 1. \tag{3.2}$$

Define the operator $S : L_{2,|r|}(\imath) \to L_{2,|r|}(\jmath)$ by

$$(Sf)\big(\beta_1(t)\big) := |\alpha'| f\big(\alpha_1(t)\big)\phi(t), \quad f \in L_{2,|r|}(\imath), \quad t \in [0,1]. \tag{3.3}$$

Clearly S is linear.

In what follows we shall use the combination of property (3.2) and Remark 3.5 to simplify the notation and calculations. For example these imply that in the definition (3.3) of S we could use β and α instead of β_1 and α_1 without changing the substance of the definition.

At various points of the proof we shall employ the monotonic (increasing or decreasing) substitutions

$$x = \beta(t), \qquad \alpha(t) = \xi.$$

To prove that S is bounded we let $f \in L_{2,|r|}(\imath)$ and calculate

$$\begin{aligned}\int_\jmath |(Sf)(x)|^2|r(x)|dx &= \operatorname{sgn}(\beta')\int_0^\epsilon |(Sf)(\beta(t))|^2|r(\beta(t))|\,\beta'\,dt\\ &= |\alpha'|^2|\beta'|\int_0^\epsilon |f(\alpha(t))|^2\,|\phi(t)|^2\,\rho(t)\,|r(\alpha(t))|dt\\ &\le |\alpha'|^2|\beta'|R\int_0^\epsilon |f(\alpha(t))|^2\,|\phi(t)|^2\,|r(\alpha(t))|dx\\ &\le |\alpha'||\beta'|R\int_\imath |f(\xi)|^2\,|r(\xi)|d\xi,\end{aligned}$$

where R is an upper bound of the function ρ. The above calculation proves that S is bounded and $\|S\| \le |\alpha'||\beta'|R$.

To verify the first claim in (S-2), let $|x - b_1| < 1/2$ and $f \in L_{2,|r|}(\imath)$. Note that the length of $\jmath$ is 1, the endpoints of $\jmath$ are b, b_1, and $\beta(0) = b$. Since $\beta\big([0,\epsilon]\big)$ has the length $< 1/2$ and since β_1 is strictly monotonic we conclude that $t = \beta_1^{-1}(x) > \epsilon$. Therefore, by (3.2),

$$(Sf)(x) = |\alpha'| f\big(\alpha(t)\big)\phi(t) = 0.$$

This proves the first claim in (S-2).

To prove $S\mathcal{F}_{\max}(\imath) \subset \mathcal{F}_{\max}(\jmath)$, let $f \in \mathcal{F}_{\max}(\imath)$. By definition (3.3), since f is absolutely continuous on $\imath$ and $\phi \in C^1[0,1]$, the function Sf is absolutely continuous on $\jmath$ and for almost all $t \in [0,1]$ we have

$$\beta'(Sf)'(\beta(t)) = |\alpha'|\big(\alpha'\, f'(\alpha(t))\phi(t) + f(\alpha(t))\phi'(t)\big). \tag{3.4}$$

To prove that $Sf \in \mathcal{F}_{\max}(\jmath)$ we need to show that $(Sf)' \in L_{2,p}(\jmath)$, that is

$$\int_\jmath |(Sf)'(x)|^2 p(x)dx = |\beta'|\int_0^\epsilon |(Sf)'(\beta(t))|^2 p(\beta(t))dt < +\infty. \tag{3.5}$$

We consider each summand in (3.4) separately. By (3.2), the second function in the sum in (3.4) is a continuous function which vanishes outside of the interval $[\epsilon/2, \epsilon]$. Since by assumption $p \circ \beta$ is an integrable function on $[\epsilon/2, \epsilon]$, it follows that

$$\int_0^1 |f(\alpha(t))\phi'(t)|^2 p(\beta(t))dt < +\infty. \tag{3.6}$$

Using the notation and assumptions from Definition 3.1, for the first function in the sum in (3.4) we have

$$\begin{aligned}\int_0^1 |f'(\alpha(t))|^2|\phi(t)|^2 p(\beta(t))dt &= \int_0^\epsilon |f'(\alpha(t))|^2\,|\phi(t)|^2\,\varpi(t)\,p(\alpha(t))dt\\ &\le \tau\int_0^\epsilon |f'(\alpha(t))|^2\,|\phi(t)|^2\,p(\alpha(t))dt \qquad (3.7)\\ &\le \frac{\tau}{|\alpha'|}\int_\imath |f'(\xi)|^2\,p(\xi)d\xi.\end{aligned}$$

Since $f' \in L_{2,p}(\imath)$ the last expression is finite. Based on (3.4), (3.5), (3.6) and (3.7) we conclude that $(Sf)' \in L_{2,p}(\jmath)$ and consequently $Sf \in \mathcal{F}_{\max}(\jmath)$.

The next step in the proof is to calculate

$$S^* : L_{2,|r|}(\jmath) \to L_{2,|r|}(\imath).$$

Note that S^* is calculated with respect to the Hilbert space inner products on the underlying spaces. Property (3.2) allows us to consider only affine changes of variable in the integrals below. Let $f \in L_{2,|r|}(\imath)$ and $g \in L_{2,|r|}(\jmath)$. Then

$$\begin{aligned}\int_\jmath (Sf)(x)\,\overline{g(x)}\,|r(x)|dx \\ &= |\beta'| \int_0^\epsilon (Sf)(\beta(t))\,\overline{g(\beta(t))}\,|r(\beta(t))|\,dt \\ &= |\beta'||\alpha'| \int_0^\epsilon f(\alpha(t))\,\phi(t)\,\overline{g(\beta(t))}\,\rho(t)\,|r(\alpha(t))|\,dt \\ &= |\beta'| \int_\imath f(\xi)\,\phi(\alpha^{-1}(\xi))\,\overline{g(\beta(\alpha^{-1}(\xi)))}\,\rho(\alpha^{-1}(\xi))\,|r(\xi)|\,d\xi.\end{aligned}$$

Therefore for $g \in L_{2,|r|}(\jmath)$ we have

$$(S^*g)\big(x\big) := |\beta'|\big(\rho\,\phi\,(g\circ\beta)\big)\big(\alpha^{-1}(x)\big), \quad x \in \imath.$$

Thus

$$(S^*g)\big(\alpha(t)\big) = |\beta'|g\big(\beta(t)\big)\rho(t)\,\phi(t), \quad g \in L_{2,|r|}(\jmath), \quad t \in [0,1]. \tag{3.8}$$

As the adjoint of a bounded operator, the operator S^* is bounded. To verify the second part of (S-2) let $|x - a_1| < 1/2$ and $g \in L_{2,|r|}(\jmath)$. Note that the length of $\imath$ is 1 and $\alpha(0) = a$. Since $\alpha\big([0,\epsilon]\big)$ has length $< 1/2$ and since α_1 is strictly monotonic we conclude that $t = \alpha_1^{-1}(x) > \epsilon$. Therefore, by (3.2),

$$(S^*g)(x) = |\beta'|g\big(\beta(t)\big)\rho(t)\phi(t) = 0.$$

To prove $S^*\mathcal{F}_{\max}(\jmath) \subset \mathcal{F}_{\max}(\imath)$, let $g \in \mathcal{F}_{\max}(\jmath)$. Since g and ρ are absolutely continuous and $\phi \in C^1[0,1]$, the function $(S^*g)\circ\alpha$ is absolutely continuous on $[0,1]$. Differentiation of (3.8) yields

$$\begin{aligned}\alpha'\,(S^*g)'\big(\alpha(t)\big) \\ &= |\beta'|\,\Big(\big(\rho'\,\phi\,(g\circ\beta)\big)(t) + \big(\rho\,\phi'\,(g\circ\beta)\big)(t) + \beta'\big(\rho\,\phi\,(g'\circ\beta)\big)(t)\Big),\end{aligned} \tag{3.9}$$

for almost all $t \in [0,1]$. To prove that $S^*f \in \mathcal{F}_{\max}(\imath)$ we need to show that $(S^*f)' \in L_{2,p}(\imath)$, that is

$$\int_\imath |(S^*f)'(\xi)|^2 p(\xi)d\xi = |\alpha'| \int_0^\epsilon |(S^*f)'(\alpha(t))|^2 p(\alpha(t))dt < +\infty. \tag{3.10}$$

We prove that each summand on the right-hand side of (3.9) belongs to $L_{2,p}(\imath)$. By (3.2), the second summand is a continuous function which vanishes outside of the interval $[\epsilon/2, \epsilon]$. Since $p \circ \alpha$ is an integrable function on $[\epsilon/2, \epsilon]$, it follows that

$$\int_0^1 |g(\beta(t))\phi'(t)\rho(t)|^2 p(\alpha(t))dt < +\infty. \tag{3.11}$$

Next, we consider the third summand in (3.9). Since ρ is continuous on $[0,1]$ we can consider only $\phi\,(g' \circ \beta)$:

$$\begin{aligned}\int_0^1 \left|g'(\beta(t))\,\phi(t)\right|^2 p(\alpha(t))dt &= \int_0^\epsilon \left|g'(\beta(t))\,\phi(t)\right|^2 \frac{1}{\varpi(t)} p(\beta(t))dt \\ &\le \tau \int_0^\epsilon \left|g'(\beta(t))\,\phi(t)\right|^2 p(\beta(t))dt \\ &= \frac{\tau}{|\beta'|} \int_\jmath |g'(x)|^2 p(x)dx < +\infty.\end{aligned} \tag{3.12}$$

Finally, for the first summand in (3.9), it is sufficient to consider $\rho'\phi$, since $g \circ \beta$ is absolutely continuous. By (3.1)

$$\int_0^1 \left|\rho'(t)\,\phi(t)\right|^2 p(\alpha(t))dt \le \int_0^\epsilon \left|\rho'(t)\right|^2 p(\alpha(t))dt < +\infty. \tag{3.13}$$

Based on (3.9), (3.10), (3.11), (3.12) and (3.13) we conclude that $(S^* f)' \in L_{2,p}(\imath)$ and consequently $S^* f \in \mathcal{F}_{\max}(\imath)$.

Thus we have verified the properties (S-1), (S-2), (S-3). Since (S-4) is clear the theorem is proved. □

4. Condition at 0 and associated operator

Condition 4.1 (Condition at 0). Let p and r be coefficients in (1.1). Denote by 0_- a generic left and by 0_+ a generic right half-neighborhood of 0. We assume that at least one of the four ordered pairs of half-neighborhoods

$$(0_-, 0_-), \qquad (0_-, 0_+), \qquad (0_+, 0_-), \qquad (0_+, 0_+), \tag{4.1}$$

is smoothly connected with connection parameters α_0', β_0' and $\rho_0(0)$ such that $|\alpha_0'| \neq |\beta_0'|\rho_0(0)$.

Theorem 4.2. *Assume that the coefficients p and r satisfy Condition 4.1. Then there exists an operator $W_0 : L_{2,r} \to L_{2,r}$ such that*

(a) *W_0 is bounded on $L_{2,|r|}$.*
(b) *The operator $J_0 W_0 - I$ is nonnegative on the Hilbert space $L_{2,|r|}$. In particular W_0^{-1} is bounded and W_0 is positive on the Krein space $L_{2,r}$.*
(c) *$(W_0 f)(x) = (Jf)(x)$, $\frac{1}{2} \le |x| \le 1$, $f \in L_{2,r}$.*
(d) *$W_0 \mathcal{F}_{\max}[-1,1] \subset \mathcal{F}_{\max}[-1,1]$.*

Proof. Let α_0', β_0', and $\rho_0(0)$ be given by Condition 4.1. Recall that $|\alpha_0'| \neq |\beta_0'|\rho_0(0)$. Let $\phi_0 : [-1,1] \to [0,1]$, $\phi_0 \in C^1[-1,1]$ be an even function such that

$$\phi_0(0) = 1 \qquad \text{and} \qquad \phi_0(x) = 0 \quad \text{for} \quad 1/2 \le |x| \le 1.$$

Define the operators

$$P_{0,-} : L_{2,|r|}(-1,0) \to L_{2,|r|}(-1,0) \quad \text{and} \quad P_{0,+} : L_{2,|r|}(0,1) \to L_{2,|r|}(0,1)$$

by

$$\begin{aligned} (P_{0,-}f)(x) &= f(x)\,\phi_0(x), \quad f \in L_{2,|r|}(-1,0), \quad x \in [-1,0], \\ (P_{0,+}f)(x) &= f(x)\,\phi_0(x), \quad f \in L_{2,|r|}(0,1), \quad x \in [0,1]. \end{aligned}$$

Then $P_{0,-}$ and $P_{0,+}$ are self-adjoint operators with the following properties:

$$(P_{0,-}f)(x) = 0, \quad f \in L_{2,|r|}(-1,0), \quad -1 \le x \le -\tfrac{1}{2}, \tag{4.2}$$

$$(P_{0,+}f)(x) = 0, \quad f \in L_{2,|r|}(0,1), \quad \tfrac{1}{2} \le x \le 1, \tag{4.3}$$

$$P_{0,-}\mathcal{F}_{\max}[-1,0] \subset \mathcal{F}_{\max}[-1,0], \qquad P_{0,+}\mathcal{F}_{\max}[0,1] \subset \mathcal{F}_{\max}[0,1], \tag{4.4}$$

and

$$(P_{0,-}f)(0-) = f(0-), \quad f \in \mathcal{F}_{\max}[-1,0], \tag{4.5}$$

$$(P_{0,+}f)(0+) = f(0+), \quad f \in \mathcal{F}_{\max}[0,1]. \tag{4.6}$$

Here, the value of a function at $0\pm$ represents its one sided limit.

Condition 4.1 requires that one of the four ordered pairs of half neighborhoods is smoothly connected. For such a pair, Theorem 3.6 guarantees the existence of a specific operator which we denote by S_0. For each of the four pairs we shall use different combinations of scaled operators $P_{0,-}, P_{0,+}, S_0$ and S_0^* to define a bounded block operator

$$X_0 \;:\; \begin{matrix} L_{2,|r|}(-1,0) \\ \oplus \\ L_{2,|r|}(0,1) \end{matrix} \;\longrightarrow\; \begin{matrix} L_{2,|r|}(-1,0) \\ \oplus \\ L_{2,|r|}(0,1) \end{matrix}$$

with the following properties

$$(X_0^* f)(x) = 0, \quad 1/2 \le |x| \le 1, \tag{4.7}$$

$$X_0\mathcal{F}_{\max}[-1,1] \subset \mathcal{F}_{\max}[-1,1], \tag{4.8}$$

$$(X_0 f)(0) = f(0), \quad f \in \mathcal{F}_{\max}[-1,1], \tag{4.9}$$

$$X_0^*\mathcal{F}_{\max}[-1,1] \subset \mathcal{F}_{\max}[-1,0] \oplus \mathcal{F}_{\max}[0,1], \tag{4.10}$$

$$(X_0^* f)(0+) + (X_0^* f)(0-) = -2f(0), \quad f \in \mathcal{F}_{\max}[-1,1]. \tag{4.11}$$

These properties of X_0 and X_0^* imply that the operator

$$W_0 = J\left(X_0^* X_0 + I\right)$$

has all the properties stated in the theorem.

Since we assume that $|\alpha_0'| \neq |\beta_0'|\rho_0(0)$, the system

$$\gamma_1|\alpha_0'| + \gamma_2 = 1, \qquad \gamma_1|\beta_0'|\rho_0(0) + \gamma_2 = -3$$

has a nontrivial real solution γ_1, γ_2. We use this solution in the definitions below.

Case 1. Assume that the half-neighborhoods $0_-, 0_-$ in (4.1) are smoothly connected. Then by Theorem 3.6 there exists an operator

$$S_0 : L_{2,|r|}(-1,0) \to L_{2,|r|}(-1,0)$$

which satisfies (S-1)–(S-4) in Theorem 3.6 with $\imath = \jmath = [-1,0]$, $a = b = 0$. In particular, for $f \in \mathcal{F}_{\max}[-1,0]$,

$$(S_0 f)(0-) = |\alpha_0'|\, f(0-), \qquad (S_0^* f)(0-) = |\beta_0'|\, \rho_0(0)\, f(0-).$$

We define X_0 and calculate X_0^* as

$$X_0 = \begin{bmatrix} \gamma_1 S_0 + \gamma_2 P_{0,-} & 0 \\ 0 & P_{0,+} \end{bmatrix}, \qquad X_0^* = \begin{bmatrix} \gamma_1 S_0^* + \gamma_2 P_{0,-} & 0 \\ 0 & P_{0,+} \end{bmatrix}.$$

Case 2. Assume that the half-neighborhoods $0_-, 0_+$ in (4.1) are smoothly connected. Then by Theorem 3.6 there exists an operator

$$S_0 : L_{2,|r|}(-1,0) \to L_{2,|r|}(0,1)$$

which satisfies (S-1)–(S-4) in Theorem 3.6 with $\imath = [-1,0]$, $\jmath = [0,1]$, $a = b = 0$. In particular, for $f \in \mathcal{F}_{\max}[-1,0]$,

$$(S_0 f)(0+) = |\alpha_0'|\, f(0-), \qquad (S_0^* f)(0-) = |\beta_0'|\, \rho_0(0)\, f(0+).$$

We define X_0 and calculate X_0^* as

$$X_0 = \begin{bmatrix} P_{0,-} & 0 \\ \gamma_1 S_0 & \gamma_2 P_{0,+} \end{bmatrix}, \qquad X_0^* = \begin{bmatrix} P_{0,-} & \gamma_1 S_0^* \\ 0 & \gamma_2 P_{0,+} \end{bmatrix}.$$

Case 3. Assume that the half-neighborhoods $0_+, 0_-$ in (4.1) are smoothly connected. Then by Theorem 3.6 there exists an operator

$$S_0 : L_{2,|r|}(0,1) \to L_{2,|r|}(-1,0)$$

which satisfies (S-1)–(S-4) in Theorem 3.6 with $\imath = [0,1]$, $\jmath = [-1,0]$, $a = b = 0$. In particular, for $f \in \mathcal{F}_{\max}[0,1]$,

$$(S_0 f)(0-) = |\alpha_0'|\, f(0+), \qquad (S_0^* f)(0+) = |\beta_0'|\, \rho_0(0)\, f(0-).$$

We define X_0 and calculate X_0^* as

$$X_0 = \begin{bmatrix} \gamma_2 P_{0,-} & \gamma_1 S_0 \\ 0 & P_{0,+} \end{bmatrix}, \qquad X_0^* = \begin{bmatrix} \gamma_2 P_{0,-} & 0 \\ \gamma_1 S_0^* & P_{0,+} \end{bmatrix}.$$

Case 4. Assume that the half-neighborhoods $0_+, 0_+$ in (4.1) are smoothly connected. Then by Theorem 3.6 there exists an operator

$$S_0 : L_{2,|r|}(0,1) \to L_{2,|r|}(0,1)$$

which satisfies (S-1)–(S-4) in Theorem 3.6 with $\imath = [0,1]$, $\jmath = [0,1]$, $a = b = 0$. In particular, for $f \in \mathcal{F}_{\max}[0,1]$,

$$(S_0 f)(0+) = |\alpha_0'|\, f(0+), \qquad (S_0^* f)(0+) = |\beta_0'|\, \rho_0(0)\, f(0+).$$

We define X_0 and calculate X_0^* as

$$X_0 = \begin{bmatrix} P_{0,-} & 0 \\ 0 & \gamma_1 S_0 + \gamma_2 P_{0,+} \end{bmatrix}, \qquad X_0^* = \begin{bmatrix} P_{0,-} & 0 \\ 0 & \gamma_1 S_0^* + \gamma_2 P_{0,+} \end{bmatrix}.$$

First note that in each of the four cases above, the operator X_0 is bounded since each of its components is bounded.

In each of the four cases above, the property (4.7) follows from (S-2) in Theorem 3.6, and properties (4.2) and (4.3).

Let $f \in \mathcal{F}_{\max}[-1,1]$. Since $\gamma_1|\alpha_0'| + \gamma_2 = 1$, the function $X_0 f$ is continuous in each case and $(X_0 f)(0) = f(0)$. This, (4.4), (4.5), (4.6), (S-3) and (S-4) in Theorem 3.6 imply (4.8). Inclusion (4.10) follows similarly.

In each of the above cases, equation (4.11) is a consequence of

$$\gamma_1|\beta_0'|\rho_0(0) + \gamma_2 = -3,$$

(4.5), (4.6) and (S-4) in Theorem 3.6. This proves the theorem. □

Remark 4.3. Note the behavior of the operator W_0 in Theorem 4.2 at the boundary of the interval $[-1,1]$:

$$\begin{bmatrix} (W_0 f)(-1) \\ (W_0 f)(1) \end{bmatrix} = \begin{bmatrix} -f(-1) \\ f(1) \end{bmatrix}, \qquad f \in \mathcal{F}_{\max}[-1,1].$$

This property of W_0 will be used in Section 6. In the next section, under additional assumptions on the coefficients p and r in a neighborhood of -1 and 1, we shall construct operators W with specified behaviors at -1 and 1.

5. Conditions at –1 and 1, and associated operators

In this section we show that under additional assumptions on the coefficients p and r near -1 we can construct an operator W_{-1} with prescribed behavior at -1 and under additional assumptions near 1 we can construct an operator W_{+1} with prescribed behavior at 1.

Condition 5.1 (Condition at -1). Let p and r be coefficients in (1.1). We assume that a right half-neighborhood of -1 is smoothly connected to a right half-neighborhood of -1 with the connection parameters $\alpha_{-1}', \beta_{-1}'$ and $\rho_{-1}(0)$ such that $|\alpha_{-1}'| \neq |\beta_{-1}'|\rho_{-1}(0)$.

Condition 5.2 (Condition at 1). Let p and r be coefficients in (1.1). We assume that a left half-neighborhood of 1 is smoothly connected to a left half-neighborhood of 1 with the connection parameters $\alpha_{+1}', \beta_{+1}'$ and $\rho_{+1}(0)$ such that $|\alpha_{+1}'| \neq |\beta_{+1}'|\rho_{+1}(0)$.

In the rest of this section we shall need two operators analogous to $P_{0,-}$ and $P_{0,+}$ introduced in Section 4. Let $\phi_1 : [-1,1] \to [0,1]$ be a smooth even function such that

$$\phi_1(-1) = 1, \qquad \phi_1(x) = 0 \quad \text{for} \quad 0 \le |x| \le 1/2, \qquad \phi_1(1) = 1. \tag{5.1}$$

Define the operators

$$P_{1,-} : L_{2,|r|}(-1,0) \to L_{2,|r|}(-1,0) \quad \text{and} \quad P_{1,+} : L_{2,|r|}(0,1) \to L_{2,|r|}(0,1)$$

by

$$(P_{1,-}f)(x) = f(x)\,\phi_1(x), \quad f \in L_{2,|r|}(-1,0), \quad x \in [-1,0], \tag{5.2}$$

$$(P_{1,+}f)(x) = f(x)\,\phi_1(x), \quad f \in L_{2,|r|}(0,1), \quad x \in [0,1]. \tag{5.3}$$

Then $P_{1,-}$ and $P_{1,+}$ are self-adjoint operators with the following properties:

$$(P_{1,-}f)(x) = 0, \quad f \in L_{2,|r|}(-1,0), \quad -\tfrac{1}{2} \le x \le 0, \tag{5.4}$$

$$(P_{1,+}f)(x) = 0, \quad f \in L_{2,|r|}(0,1), \quad 0 \le x \le \tfrac{1}{2}, \tag{5.5}$$

$$P_{1,-}\mathcal{F}_{\max}[-1,0] \subset \mathcal{F}_{\max}[-1,0], \qquad P_{1,+}\mathcal{F}_{\max}[0,1] \subset \mathcal{F}_{\max}[0,1], \tag{5.6}$$

and

$$(P_{1,-}f)(-1+) = f(-1+), \quad f \in \mathcal{F}_{\max}[-1,0], \tag{5.7}$$

$$(P_{1,+}f)(1-) = f(1-), \quad f \in \mathcal{F}_{\max}[0,1]. \tag{5.8}$$

Proposition 5.3. *Assume that the coefficients p and r satisfy Condition* 5.1. *Let μ be an arbitrary complex number. Then there exists an operator $W_{-1} : L_{2,r} \to L_{2,r}$ such that*

(a) *W_{-1} is bounded on $L_{2,|r|}$.*
(b) *The operator $J_0W_{-1} - I$ is nonnegative on the Hilbert space $L_{2,|r|}$. In particular $(W_{-1})^{-1}$ is bounded and W_{-1} is positive on the Krein space $L_{2,r}$.*
(c) $(W_{-1}f)(x) = (Jf)(x), \quad -\frac{1}{2} \le x \le 1, \quad f \in L_{2,r}$.
(d) $W_{-1}\mathcal{F}_{\max}[-1,1] \subset \mathcal{F}_{\max}[-1,0] \oplus \mathcal{F}_{\max}[0,1]$.
(e) $(W_{-1}f)(-1) = \mu f(-1)$ *for all* $f \in \mathcal{F}_{\max}[-1,1]$.

Proof. We use the notation introduced in Condition 5.1. By Theorem 3.6 there exists a bounded operator $S_{-1} : L_{2,|r|}(-1,0) \to L_{2,|r|}(-1,0)$ such that

$$S_{-1}\mathcal{F}_{\max}[-1,0] \subset \mathcal{F}_{\max}[-1,0] \quad \text{and} \quad S_{-1}^*\mathcal{F}_{\max}[-1,0] \subset \mathcal{F}_{\max}[-1,0],$$

and, for all $f \in \mathcal{F}_{\max}[-1,0]$,

$$(S_{-1}f)(-1) = |\alpha'_{-1}|\, f(-1), \qquad (S_{-1}^*f)(-1) = |\beta'_{-1}|\,\rho_{-1}(0)f(-1).$$

Let μ be an arbitrary complex number. Since we assume that $|\alpha'_{-1}| \ne |\beta'_{-1}|\,\rho_{-1}(0)$, the complex numbers γ_1 and γ_2 can be chosen such that

$$\gamma_1|\alpha'_{-1}| + \gamma_2 = 1, \qquad \overline{\gamma}_1|\beta'_{-1}|\,\rho_{-1}(0) + \overline{\gamma}_2 = -\mu - 1.$$

We define X_{-1} and calculate X_{-1}^* as

$$X_{-1} = \begin{bmatrix} \gamma_1 S_{-1} + \gamma_2 P_{1,-} & 0 \\ 0 & 0 \end{bmatrix}, \qquad X_{-1}^* = \begin{bmatrix} \overline{\gamma}_1 S_{-1}^* + \overline{\gamma}_2 P_{1,-} & 0 \\ 0 & 0 \end{bmatrix}.$$

Then for all $f \in \mathcal{F}_{\max}[-1,1]$ we have

$$(X_{-1}f)(-1) = f(-1) \quad \text{and} \quad (X_{-1}^*f)(-1) = (-\mu - 1)f(-1).$$

Therefore

$$W_{-1} = J\big(X_{-1}^*X_{-1} + I\big)$$

has all the properties stated in the proposition. □

The proof of the next proposition is very similar to the preceding proof, and will be omitted.

Proposition 5.4. *Assume that the coefficients p and r satisfy Condition* 5.2. *Let* μ *be an arbitrary complex number. Then there exists an operator*

$$W_{+1} : L_{2,r} \to L_{2,r}$$

such that

(a) W_{+1} *is bounded on* $L_{2,|r|}$.
(b) *The operator* $J_0 W_{+1} - I$ *is nonnegative on the Hilbert space* $L_{2,|r|}$. *In particular* $(W_{+1})^{-1}$ *is bounded and* W_{+1} *is positive on the Krein space* $L_{2,r}$.
(c) $(W_{+1} f)(x) = (Jf)(x), \quad -1 \leq x \leq \frac{1}{2}, \quad f \in L_{2,r}$.
(d) $W_{+1} \mathcal{F}_{\max}[-1,1] \subset \mathcal{F}_{\max}[-1,0] \oplus \mathcal{F}_{\max}[0,1]$.
(e) $(W_{+1} f)(1) = \mu f(1)$ *for all* $f \in \mathcal{F}_{\max}[-1,1]$.

6. Riesz basis of root vectors

In this section we return to the eigenvalue problem (1.1)–(1.2) and the operator A associated with it. We start with cases when the conditions in Section 5 are not needed. We remark that the notation of Section 2 is used extensively in the rest of this section.

Theorem 6.1. *Assume that the following three conditions are satisfied.*

(a) *The coefficients p and r satisfy Condition* 4.1.
(b) *One of the following is true:*
  (i) $\mathsf{L}_n \neq 0$,
  (ii) $\mathsf{L} = [1 \;\; 0 \;\; 0 \;\; 0]$,
  (iii) $\mathsf{L} = [0 \;\; 1 \;\; 0 \;\; 0]$.
(c) *One of the following is true:*
  (i) $\mathsf{N}_n \neq 0$,
  (ii) $\mathsf{N} = [1 \;\; 0 \;\; 0 \;\; 0]$ *and* $\Delta < 0$,
  (iii) $\mathsf{N} = [0 \;\; 1 \;\; 0 \;\; 0]$ *and* $\Delta > 0$.

Then there is a basis for each root subspace of A, so that the union of all these bases is a Riesz basis of $L_{2,|r|} \oplus \mathbb{C}_{|\Delta|}$.

Proof. Assume first that $\mathsf{N}_n \neq 0$. By (2.7), the form domain of A when $\mathsf{L}_n \neq 0$ is given by

$$\mathcal{F}(A) = \left\{ \begin{bmatrix} f \\ z \end{bmatrix} \in \begin{array}{c} L_{2,r} \\ \oplus \\ \mathbb{C}_\Delta \end{array} : f \in \mathcal{F}_{\max}, \; z \in \mathbb{C} \right\},$$

and in the other two cases in (b), (2.8) gives

$$\mathcal{F}(A) = \left\{ \begin{bmatrix} f \\ z \end{bmatrix} \in \begin{array}{c} L_{2,r} \\ \oplus \\ \mathbb{C}_\Delta \end{array} : f \in \mathcal{F}_{\max},\ z \in \mathbb{C},\ \mathsf{L}_e \mathbf{b}_e(f) = 0 \right\},$$

where $\mathsf{L}_e \mathbf{b}_e(f) = f(-1)$ in case (b-ii) and $\mathsf{L}_e \mathbf{b}_e(f) = f(1)$ in case (b-iii).

Next assume (c-ii). Then $\mathsf{N}_e \mathbf{b}_e(f) = f(-1)$ and (2.9) shows that

$$\mathcal{F}(A) = \left\{ \begin{bmatrix} f \\ f(-1) \end{bmatrix} \in \begin{array}{c} L_{2,r} \\ \oplus \\ \mathbb{C}_\Delta \end{array} : f \in \mathcal{F}_{\max} \right\}$$

when $\mathsf{L}_n \neq \mathsf{0}$, and in the other cases in (b), (2.10) gives

$$\mathcal{F}(A) = \left\{ \begin{bmatrix} f \\ f(-1) \end{bmatrix} \in \begin{array}{c} L_{2,r} \\ \oplus \\ \mathbb{C}_\Delta \end{array} : f \in \mathcal{F}_{\max},\ \mathsf{L}_e \mathbf{b}_e(f) = 0 \right\}.$$

Finally, assume (c-iii). Then $\mathsf{N}_e \mathbf{b}_e(f) = f(1)$ and (2.9) shows that

$$\mathcal{F}(A) = \left\{ \begin{bmatrix} f \\ f(1) \end{bmatrix} \in \begin{array}{c} L_{2,r} \\ \oplus \\ \mathbb{C}_\Delta \end{array} : f \in \mathcal{F}_{\max} \right\}$$

when $\mathsf{L}_n \neq \mathsf{0}$, and in the other cases in (b), (2.10) gives

$$\mathcal{F}(A) = \left\{ \begin{bmatrix} f \\ f(1) \end{bmatrix} \in \begin{array}{c} L_{2,r} \\ \oplus \\ \mathbb{C}_\Delta \end{array} : f \in \mathcal{F}_{\max},\ \mathsf{L}_e \mathbf{b}_e(f) = 0 \right\}.$$

Let W_0 be the operator constructed in Theorem 4.2, and let

$$W = \begin{bmatrix} W_0 & 0 \\ 0 & \operatorname{sgn}(\Delta) \end{bmatrix} : \begin{array}{c} L_{2,r} \\ \oplus \\ \mathbb{C}_\Delta \end{array} \to \begin{array}{c} L_{2,r} \\ \oplus \\ \mathbb{C}_\Delta \end{array}.$$

A straightforward verification shows that W is a bounded, boundedly invertible, positive operator in the Krein space $L_{2,r} \oplus \mathbb{C}_\Delta$ and $W\,\mathcal{F}(A) \subset \mathcal{F}(A)$ in each of the above listed cases. Consequently, the theorem follows from Theorem 2.2. □

In the next result we shall assume that one of the conditions from Section 5 is satisfied.

Theorem 6.2. *Assume that the following three conditions are satisfied.*

(a) *The coefficients* p *and* r *satisfy Condition* 4.1, *and one of Conditions* 5.1, 5.2.
(b) $\mathsf{N}_n \neq \mathsf{0}$.
(c) $\mathsf{L} = [u \quad v \quad 0 \quad 0]$ *with* $uv \neq 0$.

Then there is a basis for each root subspace of A, *so that the union of all these bases is a Riesz basis of* $L_{2,|r|} \oplus \mathbb{C}_{|\Delta|}$.

Proof. Under the assumptions of the theorem, (2.8) shows that the form domain of A is given by

$$\mathcal{F}(A) = \left\{ \begin{bmatrix} f \\ z \end{bmatrix} \in \begin{matrix} L_{2,r} \\ \oplus \\ \mathbb{C}_\Delta \end{matrix} : f \in \mathcal{F}_{\max},\ z \in \mathbb{C},\ uf(-1) + vf(1) = 0 \right\}.$$

Define the following two Krein spaces:

$$\mathcal{K}_0 := L_{2,r}\left(-\tfrac{1}{2}, \tfrac{1}{2}\right), \qquad \mathcal{K}_1 := L_{2,r}(-1, -\tfrac{1}{2})[\dot{+}]L_{2,r}(\tfrac{1}{2}, 1). \tag{6.1}$$

Extending the functions in $\mathcal{K}_0$ and $\mathcal{K}_1$ by 0 onto the rest of $[-1, 1]$, we can consider the spaces $\mathcal{K}_0$ and $\mathcal{K}_1$ as subspaces of $L_{2,r}$.

Then

$$L_{2,r} = \mathcal{K}_0[\dot{+}]\mathcal{K}_1.$$

Assume that the functions p and r satisfy Conditions 4.1 and 5.1. Let W_0 be the operator constructed in Theorem 4.2 and let W_{-1} be the operator constructed in Proposition 5.3 with $\mu = 1$. Then properties (c) in Theorem 4.2 and Proposition 5.3, imply that $\mathcal{K}_0$ and $\mathcal{K}_1$ are invariant under W_0 and W_{-1}. As we chose $\mu = 1$, we have $(W_{-1}f)(-1) = f(-1)$ and $(W_{-1}f)(1) = f(1)$. Define

$$W_{01} := W_0|_{\mathcal{K}_0}[\dot{+}]W_{-1}|_{\mathcal{K}_1}. \tag{6.2}$$

Since W_0 and W_{-1} are bounded, boundedly invertible and positive in the Krein space $L_{2,r}$, so is the the operator W_{01}. Also, $W_{01}\mathcal{F}_{\max}[-1, 1] \subset \mathcal{F}_{\max}[-1, 1]$ and

$$\begin{bmatrix} (W_{01}f)(-1) \\ (W_{01}f)(1) \end{bmatrix} = \begin{bmatrix} f(-1) \\ f(1) \end{bmatrix}. \tag{6.3}$$

If the functions p and r satisfy Conditions 4.1 and 5.2, then, instead of W_{-1}, we use the operator W_{+1} constructed in Proposition 5.4 with $\mu = -1$. Redefining the operator W_{01} as

$$W_{01} := W_0|_{\mathcal{K}_0}[\dot{+}]W_{+1}|_{\mathcal{K}_1} \tag{6.4}$$

we see that it is again bounded, boundedly invertible, and positive in the Krein space $L_{2,r}$, $W_{01}\mathcal{F}_{\max}[-1, 1] \subset \mathcal{F}_{\max}[-1, 1]$ and (since we use $\mu = -1$)

$$\begin{bmatrix} (W_{01}f)(-1) \\ (W_{01}f)(1) \end{bmatrix} = -\begin{bmatrix} f(-1) \\ f(1) \end{bmatrix}. \tag{6.5}$$

Now a simple inspection shows that, in both above cases, the operator

$$W = \begin{bmatrix} W_{01} & 0 \\ 0 & \Delta \end{bmatrix} : \begin{matrix} L_{2,r} \\ \oplus \\ \mathbb{C}_\Delta \end{matrix} \to \begin{matrix} L_{2,r} \\ \oplus \\ \mathbb{C}_\Delta \end{matrix}$$

is bounded, boundedly invertible and positive in the Krein space $L_{2,r} \oplus \mathbb{C}_\Delta$ and $W\mathcal{F}(A) \subset \mathcal{F}(A)$. Thus the theorem again follows from Theorem 2.2. □

Our final result covers the remaining cases, but in the interests of presentation we shall impose no conditions on L. Of course, there is some overlap with Theorem 6.1.

Theorem 6.3. *Assume that the following two conditions are satisfied.*

(a) *The coefficients p and r satisfy Condition* 4.1, *and*
 (i) *Condition* 5.1 *if* $\Delta > 0$,
 (ii) *Condition* 5.2 *if* $\Delta < 0$.

(b) $\mathsf{N}_n = 0$.

Then there is a basis for each root subspace of A, so that the union of all these bases is a Riesz basis of $L_{2,|r|} \oplus \mathbb{C}_{|\Delta|}$.

Proof. In this case (2.9) shows that the form domain of A is

$$\mathcal{F}(A) = \left\{ \begin{bmatrix} f \\ \mathsf{N}_e \mathbf{b}_e(f) \end{bmatrix} \in \begin{matrix} L_{2,r} \\ \oplus \\ \mathbb{C}_\Delta \end{matrix} : f \in \mathcal{F}_{\max} \right\}$$

if $\mathsf{L}_n \neq 0$, and

$$\mathcal{F}(A) = \left\{ \begin{bmatrix} f \\ \mathsf{N}_e \mathbf{b}_e(f) \end{bmatrix} \in \begin{matrix} L_{2,r} \\ \oplus \\ \mathbb{C}_\Delta \end{matrix} : f \in \mathcal{F}_{\max},\ \mathsf{L}_e \mathbf{b}_e(f) = 0 \right\}$$

by (2.10) if $\mathsf{L}_n = 0$.

Let $W_{01} : L_{2,r} \to L_{2,r}$ be the operator constructed in the proof of Theorem 6.2, and define $W : L_{2,r} \oplus \mathbb{C}_\Delta \to L_{2,r} \oplus \mathbb{C}_\Delta$ by

$$W = \begin{bmatrix} W_{01} & 0 \\ 0 & \operatorname{sgn}(\Delta) \end{bmatrix}.$$

As before, in both cases the properties of W_{01} imply that W is a bounded, boundedly invertible, positive operator in the Krein space $L_{2,r} \oplus \mathbb{C}_\Delta$ and $W\mathcal{F}(A) \subset \mathcal{F}(A)$. Now the theorem follows from Theorem 2.2. □

Remark 6.4. It is instructive to look at the above results from the viewpoint of non-essential boundary conditions (L_n, $\mathsf{N}_n \neq 0$) and essential ones (whose essential parts L_e, N_e can be separated or not). Let us call a boundary condition essentially separated if it is either non-essential, or else its essential part is separated. Theorem 6.1 states that if both boundary conditions are essentially separated, then subject to the sign conditions in (c-ii) and (c-iii), Condition 4.1 suffices for the existence of a Riesz basis of root vectors.

If any of these assumptions fail, then we impose conditions from Section 5. In particular, if the λ-dependent boundary condition is non-essential, then either of these conditions suffice, but in other cases the choice is governed by the sign of Δ.

We conclude with a simple example.

Example 6.5. We suppose that

$$\mathsf{L} = [d_1 \;\; d_2 \;\; d_3 \;\; d_4], \quad \mathsf{M} = [m_1 \;\; m_2 \;\; m_3 \;\; m_4], \quad \mathsf{N} = [0 \;\; \gamma \;\; 0 \;\; 0],$$

where $(d_3, d_4) \neq (0,0)$ and $\gamma m_4 > 0$. Note that the only λ-dependent term in (1.2) involves $f(1)$.

We calculate

$$\mathsf{M}\mathsf{Q}^{-1}\mathsf{N}^* = -i\gamma m_4,$$

so by (2.1), $\Delta > 0$. It follows from Theorem 6.1 with condition (b-i) that Condition 4.1 suffices for the existence of a Riesz basis of root vectors.

This example overlaps with [11, Corollary 3.8], where separated boundary conditions, also satisfying $d_2 = d_4 = m_1 = m_3 = 0$, $d_3 = m_4 = 1$, $\gamma > 0$, were considered by Fleige for a Krein-Feller equation instead of (1.1).

Acknowledgment

We thank a referee for a very careful reading of the submitted version of this article. This has led to the correction of a number of inaccuracies.

References

[1] N.L. Abasheeva, S.G. Pyatkov, *Counterexamples in indefinite Sturm-Liouville problems.* Siberian Advances in Mathematics. Siberian Adv. Math. **7** (1997), 1–8.

[2] R. Beals, *Indefinite Sturm-Liouville problems and half-range completeness.* J. Differential Equations **56** (1985), 391–407.

[3] P.A. Binding, B. Ćurgus, *Form domains and eigenfunction expansions for differential equations with eigenparameter dependent boundary conditions.* Canad. J. Math. **54** (2002), 1142–1164.

[4] P.A. Binding, B. Ćurgus, *A counterexample in Sturm-Liouville completeness theory.* Proc. Roy. Soc. Edinburgh Sect. A **134** (2004), 244–248.

[5] B. Ćurgus, *On the regularity of the critical point infinity of definitizable operators.* Integral Equations Operator Theory **8** (1985), 462–488.

[6] B. Ćurgus, H. Langer, *A Krein space approach to symmetric ordinary differential operators with an indefinite weight function.* J. Differential Equations **79** (1989), 31–61.

[7] B. Ćurgus, B. Najman, *Quasi-uniformly positive operators in Krein space.* Operator theory and boundary eigenvalue problems (Vienna, 1993), 90–99, Oper. Theory Adv. Appl., **80**, Birkhäuser, 1995.

[8] A. Dijksma, *Eigenfunction expansions for a class of J-self-adjoint ordinary differential operators with boundary conditions containing the eigenvalue parameter.* Proc. Roy. Soc. Edinburgh Sect. A **86** (1980), 1–27.

[9] A. Fleige, *The "turning point condition" of Beals for indefinite Sturm-Liouville problems.* Math. Nachr. **172** (1995), 109–112.

[10] A. Fleige, *A counterexample to completeness properties for indefinite Sturm-Liouville problems.* Math. Nachr. **190** (1998), 123–128.

[11] A. Fleige, *Spectral theory of indefinite Krein-Feller differential operators.* Mathematical Research, **98** Akademie Verlag, 1996.

[12] H. Langer, A. Schneider, *On spectral properties of regular quasidefinite pencils $F - \lambda G$.* Results Math. **19** (1991), 89–109.

[13] A.I. Parfyonov, *On an embedding criterion for interpolation spaces and application to indefinite spectral problems.* Siberian Math. J. **44** (2003), 638–644.

[14] S.G. Pyatkov, *Interpolation of some function spaces and indefinite Sturm-Liouville problems.* Differential and integral operators (Regensburg, 1995), 179–200, Oper. Theory Adv. Appl. **102**, Birkhäuser, (1998).

[15] H. Volkmer, *Sturm-Liouville problems with indefinite weights and Everitt's inequality.* Proc. Roy. Soc. Edinburgh Sect. A **126** (1996), 1097–1112.

Paul Binding
Department of Mathematics and Statistics
University of Calgary
Calgary
Alberta, T2N 1N4
Canada
e-mail: binding@ucalgary.ca

Branko Ćurgus
Department of Mathematics
Western Washington University
Bellingham
WA 98225
USA
e-mail: curgus@cc.wwu.edu

Operator Theory:
Advances and Applications, Vol. 163, 97–134

Minimal Models for $\mathcal{N}_\kappa^\infty$-functions

Aad Dijksma, Annemarie Luger and Yuri Shondin

To Heinz Langer, wishing him a happy retirement

Abstract. We present explicit realizations in terms of self-adjoint operators and linear relations for a non-zero scalar generalized Nevanlinna function $N(z)$ and the function $\widehat{N}(z) = -1/N(z)$ under the assumption that $\widehat{N}(z)$ has exactly one generalized pole which is not of positive type namely at $z = \infty$. The key tool we use to obtain these models is reproducing kernel Pontryagin spaces.

Mathematics Subject Classification (2000). Primary 47B25, 47B50, 47B32; Secondary 47A06.

Keywords. Generalized Nevanlinna function, generalized pole, realization, model, reproducing kernel spaces, Pontryagin spaces, self-adjoint operator, symmetric operator, linear relation, block operator matrix.

1. Introduction

An $n \times n$ matrix function N is called a generalized Nevanlinna function with κ negative squares if (i) it is defined and meromorphic on $\mathbb{C} \setminus \mathbb{R}$, (ii) it satisfies $N(z) = N(z^*)^*$ for all $z \in \mathcal{D}(N)$, the domain of holomorphy of N, and (iii) the kernel

$$K_N(\zeta, z) = \frac{N(\zeta) - N(z)^*}{\zeta - z^*}, \quad \zeta, z \in \mathcal{D}(N),$$

has κ negative squares. Here the expression on the right-hand side for $\zeta = z^*$ is to be understood as $N'(\zeta)$. If $\kappa = 0$, the function N is called a Nevanlinna function;

The authors gratefully acknowledge support from the "Fond zur Förderung der wissenschaftlichen Forschung" (FWF, Austria, grant number P15540-N05), the Netherlands Organization for Scientific Research NWO (grant NWO 047-008-008), and the Research Training Network HPRN-CT-2000-00116 of the European Union.

in this case N is holomorphic on $\mathbb{C}\setminus\mathbb{R}$, satisfies $N(z) = N(z^*)^*$ there and the kernel condition is equivalent to the condition

$$\frac{\operatorname{Im} N(z)}{\operatorname{Im} z} \geq 0, \quad \operatorname{Im} z \neq 0.$$

The class of $n \times n$ matrix functions with κ negative squares is denoted by $\mathcal{N}_\kappa^{n\times n}$ and by $\mathcal{N}_\kappa$ when the functions are scalar.

A realization for a function $N \in \mathcal{N}_\kappa^{n\times n}$ in some Pontryagin space $\mathcal{P}$ is a pair (A, Γ_z) consisting of a self-adjoint relation A in $\mathcal{P}$ with a nonempty resolvent set $\rho(A)$ and a corresponding Γ-field Γ_z, that is, a family of mappings $\Gamma_z : \mathbb{C}^n \to \mathcal{P}$, $z \in \rho(A)$, which satisfy

$$\Gamma_z = (I_{\mathcal{P}} + (z-\zeta)(A-z)^{-1})\Gamma_\zeta, \quad \zeta,\, z \in \rho(A),$$

and

$$\frac{N(\zeta) - N(z)^*}{\zeta - z^*} = \Gamma_z^*\Gamma_\zeta, \quad \zeta,\, z \in \rho(A),\ z \neq \zeta^*.$$

If a point $z_0 \in \rho(A)$ is fixed this implies the following representation of N:

$$N(z) = N(z_0)^* + (z - z_0^*)\Gamma_{z_0}^*\big(I_{\mathcal{P}} + (z - z_0)(A-z)^{-1}\big)\Gamma_{z_0}, \qquad z \in \mathcal{D}(N).$$

The function N is determined by the self-adjoint relation A in $\mathcal{P}$ and the Γ-field Γ_z up to an additive constant hermitian $n \times n$ matrix. The space $\mathcal{P}$ is called the state space of the realization (A, Γ_z). The realization (A, Γ_z) can always be chosen minimal which means that

$$\overline{\operatorname{span}}\,\{\Gamma_z \mathbf{c} \mid z \in \varrho(A), \mathbf{c} \in \mathbb{C}^n\} = \mathcal{P}.$$

In that case the negative index of the state space $\mathcal{P}$ is equal to the number of negative squares of the kernel $K_N(\zeta, z)$ and $\mathcal{D}(N) = \rho(A)$; see [16, Theorem 1.1]. Two minimal realizations of N are unitarily equivalent. With a minimal realization (A, Γ_z) often a symmetric restriction S of the relation A is associated and defined by

$$S = \{\{f, g\} \in A \,|\, \Gamma_{z_0}^*(g - z_0^* f) = 0\}.$$

This definition is independent of $z_0 \in \mathcal{D}(N)$, S is an operator, and Γ_z maps $\mathbb{C}^n$ onto the defect subspace $\operatorname{ran}(S - z^*)^\perp$ of S at z. The triplet (A, Γ_z, S) is called a model in $\mathcal{P}$ for the realization of N or, for short, a model for the function N in $\mathcal{P}$. The model will be called minimal if the realization is minimal.

If $n = 1$ the function

$$\varphi(z) = \Gamma_z 1 = (I_{\mathcal{P}} + (z - z_0)(A - z)^{-1})\varphi(z_0),$$

called a defect function for S and A, spans the defect subspace of S at z and the representation of N takes the form

$$N(z) = N(z_0)^* + (z - z_0^*)\langle\varphi(z), \varphi(z_0)\rangle_{\mathcal{P}}.$$

Every $N \in \mathcal{N}_\kappa$ admits a basic factorization of the form

$$N(z) = r^{\#}(z) N_1(z) r(z), \tag{1.1}$$

where $N_1 \in \mathcal{N}_0$ and r is a rational function whose zeros (poles) are the generalized zeros (poles) of N in $\mathbb{C}^+ \cup \mathbb{R}$ ($\mathbb{C}^- \cup \mathbb{R}$, respectively) which are not of positive type; for definitions and a proof of (1.1), see, for example, [10] and [9]. Here and in the sequel for a vector function f we denote by $f^\#$ the function $f^\#(z) = f(z^*)^*$. If κ_1 is the number of zeros of r and κ_2 is the number of poles of r (counted according to their multiplicities), then $\kappa = \max\{\kappa_1, \kappa_2\}$. If $\tau = \kappa_1 - \kappa_2$ is positive (negative) then $z = \infty$ is a generalized pole (zero) of N which is not of positive type and with degree of non-positivity $|\tau|$. In particular, if r is a polynomial (necessarily of degree κ), then $z = \infty$ is the only generalized pole of N and not of positive type; if on the other hand $\kappa_1 = 0$ (so that $\kappa_2 = \kappa$), then $z = \infty$ is the only generalized zero of N and not of positive type.

In this paper we are describing minimal models for functions $0 \neq N \in \mathcal{N}_\kappa$ and $\widehat{N} = -N^{-1}$ (which also belongs to $\mathcal{N}_\kappa$) under the assumption that the latter belongs to the class $\mathcal{N}_\kappa^\infty$ considered in [12]. By definition, a function $\widehat{N}$ belongs to the class $\mathcal{N}_\kappa^\infty$ if and only if it belongs to $\mathcal{N}_\kappa$ and has a representation of the form

$$\widehat{N}(z) = c^\#(z) N_0(z) c(z) + p(z), \tag{1.2}$$

where $N_0(z)$ is a Nevanlinna function with the properties

$$\lim_{y\to\infty} y \operatorname{Im} N_0(iy) = +\infty, \qquad \lim_{y\to\infty} y^{-1} N_0(iy) = 0, \qquad \operatorname{Re} N_0(i) = 0, \tag{1.3}$$

$c(z) = (z - z_0)^m$ with $m \in \mathbb{N}_0$ and $z_0 \in \mathcal{D}(\widehat{N})$, and p is some real polynomial. As explained in [12], the representation (1.2) (with (1.3)) is irreducible and implies that $z = \infty$ is the only generalized pole of non-positive type of the function $\widehat{N}(z)$. The first two conditions in (1.3) are equivalent to the fact that in the minimal model for N_0 the symmetric operator is densely defined in the state space. The third condition is simply a normalization. In the definition of the class $\mathcal{N}_\kappa^\infty$ given in [12] it was required that the point z_0 belongs to the set $\mathbb{C} \setminus \mathbb{R}$, but in view of [12, Remark 1.3] z_0 may belong to the possibly larger set $\mathcal{D}(\widehat{N})$ and the definition is independent of the choice of $z_0 \in \mathcal{D}(\widehat{N})$. The minimal models, which we obtain for N and $\widehat{N}$ and which are related to the irreducible representation (1.2) of $\widehat{N}$, have a state space of the form $\mathcal{K} = \mathcal{H}_0 \oplus \mathbb{C}^n \oplus \mathbb{C}^m \oplus \mathbb{C}^m$, $n = \max\{\deg p - 2m, 0\}$, equipped with the indefinite inner product $\langle G\,\cdot\,,\,\cdot\,\rangle_{\mathcal{K}}$, where $\mathcal{H}_0$ is the state space for a minimal model of the function N_0 and the Gram matrix G is the 4×4 block matrix given by (6.1) with blocks determined by the polynomials p and q from the realization (4.1) of $\widehat{N}(z)$. In [5], [6], [10], and [25] the minimal models for N related to the basic factorization (1.1) are studied. The model considered in [25] has a state space which is a subspace with finite co-dimension of $\mathcal{L} = \mathcal{H}_1 \oplus \mathbb{C}^\kappa \oplus \mathbb{C}^\kappa$ equipped with the indefinite inner product $\langle G_{\mathcal{L}}\,\cdot\,,\,\cdot\,\rangle_{\mathcal{L}}$, where $\mathcal{H}_1$ is the state space for a minimal model of the function N_1 and the Gram matrix is given by

$$G_{\mathcal{L}} = \begin{pmatrix} I_{\mathcal{H}_1} & 0 & 0 \\ 0 & 0 & I_{\mathbb{C}^\kappa} \\ 0 & I_{\mathbb{C}^\kappa} & 0 \end{pmatrix}.$$

The model in the present paper is more detailed than this model because we consider a more special class of generalized Nevanlinna functions.

To motivate our study of the model problem we list some applications where functions of the form (1.2) play a role. First we note that Nevanlinna functions $N_0(z)$ satisfying the asymptotic conditions in (1.3) (in the following we disregard the normalization condition) appear naturally (i) as a Q-function of the minimal operator associated with a self-adjoint boundary value problem for a formally symmetric ordinary differential expression (Titchmarsh-Weyl coefficient) and (ii) as the main ingredient in the formula of the resolvent for the singular perturbation

$$A_0 + \alpha^{-1}\chi\langle\,\cdot\,,\chi\rangle_0 \tag{1.4}$$

of an unbounded self-adjoint operator A_0 in a Hilbert space $\mathcal{H}_0$ with inner product $\langle\,\cdot\,,\,\cdot\,\rangle_0$, generated by a generalized element $\chi \in \mathcal{H}_{-1} \setminus \mathcal{H}_0$:

$$N_0(z) = (z - z_0^*)\langle\varphi_0(z), \varphi_0(z_0)\rangle_0 - i\mathrm{Im}\, z_0\langle\varphi_0(z_0), \varphi_0(z_0)\rangle_0 + c,$$

where $\varphi(z) = (A_0 - z)^{-1}\chi$ and c is a real number; see [1, 2]. Here, in the scale of Hilbert spaces associated with $\mathcal{H}_0$ and A_0, the space $\mathcal{H}_{-m}$ is the dual of the space $H_m = \mathrm{dom}\,|A_0|^m$ equipped with the inner product $\langle(|A_0|+1)^m\,\cdot\,,(|A_0|+1)^m\,\cdot\,\rangle_0$, $m = 1, 2 \ldots$. As explained in [13], generalized Nevanlinna functions of the form (1.2) with $\deg p \leq 2m$ play a similar role as in (ii) but now for the strongly singular perturbation (1.4) with $\chi \in \mathcal{H}_{-m-1} \setminus \mathcal{H}_{-m}$. Furthermore, in [19] and [24] point-like perturbations of the Laplacian in $\mathbb{R}^3$ were constructed to describe the low energy asymptotic behavior

$$k \cot \delta_0(k) = \sum_{j=1}^{n} a_j k^{2j} + o(k^{2n})$$

of the quantum mechanical scattering data at zero orbital momenta, where $E = k^2$ is the energy of scattering particle and $\delta_0(k)$ is the scattering phase. This construction amounts to building a model of the generalized Nevanlinna function of the form (1.2) with $c(z) = 1$, $\mathrm{deq}\; p > 0$, and $N_0(z) = -\sqrt{-z}$. In the two papers just mentioned two different models in Pontryagin spaces were given. To describe a given truncated series of low energy scattering with non-zero angular momentum models for generalized Nevanlinna functions (1.2) with arbitrary $\deg c(z)$ and $\deg p(z)$ are needed. Some models of this kind where considered in [8]. As a further motivation for the models in this paper, we discuss in Section 8 an approximation problem where generalized Nevanlinna functions of the form (1.2) with various values of $\deg c(z)$ and $\deg p(z)$ appear.

We summarize the contents of the seven sections which come after this introduction. In Section 2 we recall the main theorem from [12] which characterizes realizations of the functions N and $\widehat{N} = -1/N$ under the assumption that $\widehat{N}$ belongs to the class $\mathcal{N}_\kappa^\infty$. The self-adjoint operator A and the self-adjoint relation in the models for N and $\widehat{N}$ are related via infinite coupling. This notion from [20] is explained after Theorem 2.1. In the sequel we make it a point to indicate this connection between various versions of the two models. To do this, we also consider minimal models for certain one-dimensional perturbations $A^{\langle\alpha\rangle}$ of $A = A^{\langle 0\rangle}$, where α is a real number. The key tool in the further analysis of the realizations in Section 2 is the theory of reproducing kernel Pontryagin spaces and in Section 3 we collect some theorems from this theory which will be used in the sequel. In particular, we recall the so-called canonical models. The irreducible representation (1.2) induces a canonical model for the generalized Nevanlinna matrix functions

$$\widetilde{N}(z) = \begin{pmatrix} N_0(z) & 0 & 0 & 0 \\ 0 & q(z) & 0 & 0 \\ 0 & 0 & p_0(z) & c^\#(z) \\ 0 & 0 & c(z) & 0 \end{pmatrix}, \quad M(z) = \begin{pmatrix} p_0(z) & c^\#(z) \\ c(z) & 0 \end{pmatrix},$$

where the real polynomials q and p_0 are uniquely determined by the polynomial p in (1.2) via the equality $p(z) = c^\#(z)q(z)c(z) + p_0(z)$ and the requirement that $\deg p_0 \le 2m - 1$. In Section 4 we present models for N and $\widehat{N}$ in which the reproducing kernel space $\mathcal{L}(\widetilde{N})$ is the state space. See Theorem 4.1, where, as in all our theorems (unless stated otherwise), the case $n = \deg q > 0$ and $m > 0$ is considered. The resolvents of the corresponding self-adjoint operators/relation are given in Corollary 4.5. The cases where $n = 0$ or $m = 0$ are considered separately in Theorem 4.6 and Theorem 4.7; these cases are important in our examples. The space $\mathcal{L}(\widetilde{N})$ admits the decomposition $\mathcal{L}(\widetilde{N}) = \mathcal{L}(N_0) \oplus \mathcal{L}(q) \oplus \mathcal{L}(M)$, where the direct summands are the reproducing kernel spaces associated with the functions N_0, q and M. In Section 5 we study special bases for the last two summands and the associated Gram matrices (see Lemma 5.1). These bases allow us to identify $\mathcal{L}(\widetilde{N})$ with $\widetilde{\mathcal{L}} = \mathcal{L}(N_0) \oplus \mathbb{C}^n \oplus \mathbb{C}^m \oplus \mathbb{C}^m$. The corresponding matrix representations of the models in Theorems 4.1, 4.6, and 4.7 are given in Theorems 6.1, 6.3, and 6.4 in Section 6. In that section we also determine formulas for the compressions of the resolvents of the self-adjoint operators/relation in the models and the compressions of the operators/relation themselves to the subspaces $\mathcal{L}(N_0)$ and $\mathcal{L}(N_0) \oplus \mathbb{C}^n$; see Theorem 6.5 and Theorem 6.6. By changing the bases slightly, the self-adjoint operator in the model for N can be given in a block operator matrix form, and this result is shown in Section 7. In Section 8, the last section of this paper, we give some examples and discuss an approximation problem associated with the Bessel differential expression.

We thank the referees for their useful comments.

2. Characterization of the class $\mathcal{N}_\kappa^\infty$

In [12] the following characterization of the class $\mathcal{N}_\kappa^\infty$ was established. We recall that if A is a self-adjoint operator or a self-adjoint relation in some Pontryagin space $\mathcal{P}$ and w an element in $\mathcal{P}$, then w is called cyclic for A if

$$\overline{\operatorname{span}}\,\{w,\, (A - z)^{-1}w \,|\, z \in \rho(A)\} = \mathcal{P},$$

or, equivalently, if for some (and then for every) $z_0 \in \rho(A)$, the function

$$\varphi(z) = w + (z - z_0)(A - z)^{-1} w$$

generates the space $\mathcal{P}$, that is,

$$\overline{\operatorname{span}}\{\varphi(z) \,|\, z \in \rho(A)\} = \mathcal{P}.$$

If A is an operator then w is cyclic for A if and only if

$$\overline{\operatorname{span}}\{(A - z)^{-1} w \,|\, z \in \rho(A)\} = \mathcal{P}.$$

Theorem 2.1. *For the functions $N(z)$ and $\widehat{N}(z) = -N(z)^{-1}$, the following four assertions are equivalent.*

(i) *$N(z)$ has a representation :*

$$N(z) = \left\langle (A - z)^{-1} w, w \right\rangle_{\mathcal{P}}, \quad z \in \mathcal{D}(N), \tag{2.1}$$

where A is a self-adjoint operator in a Pontryagin space $\mathcal{P}$ with negative index κ, $w \in \mathcal{P}$ is a cyclic element for A with the property

$$w \in \operatorname{dom} A^{m+n-1} \setminus \operatorname{dom} A^{m+n}$$

for some integers $m, n \in \mathbb{N}_0$, $m + n > 0$, the subspace

$$\mathcal{L} = \operatorname{span}\{w, Aw, \ldots, A^{m-1}w, A^m w, \ldots, A^{m+n-1}w\}$$

has index of non-positivity κ, and

$$\begin{aligned} \langle A^j w, A^k w\rangle &= 0, \qquad 0 \le j, k \le m + n - 1,\ j + k \le 2m + n - 2, \\ \langle A^j w, A^k w\rangle &\ne 0, \qquad 0 \le j, k \le m + n - 1,\ j + k = 2m + n - 1. \end{aligned}$$

(ii) *$N(z) \in \mathcal{N}_\kappa$, $z = \infty$ is the only generalized zero of non-positive type of $N(z)$, and $N(z)$ has a representation*

$$N(z) = -\sum_{j=2m+n}^{2m+2n-1} \frac{s_{j-1}}{z^j} + \frac{1}{z^{2m+2n-1}} M(z) \tag{2.2}$$

with $m, n \in \mathbb{N}_0$, $m + n > 0$, real numbers s_j, $j = 2m + n - 1, \ldots, 2m + 2n - 2$, $s_{2m+n-1} \ne 0$ if $n > 0$, and a function $M(z)$ with the properties

$$\lim_{y\to\infty} M(iy) = 0, \quad \lim_{y\to\infty} y^2 \operatorname{Re} M(iy) = +\infty.$$

(iii) *$\widehat{N}(z)$ has a representation*

$$\widehat{N}(z) = \widehat{N}(z_0^*) + (z - z_0^*)\langle (I_{\widehat{\mathcal{P}}} + (z - z_0)(\widehat{A} - z)^{-1})u, u\rangle_{\widehat{\mathcal{P}}}, \quad z \in \mathcal{D}(\widehat{N}),$$

where $z_0 \in \mathcal{D}(\widehat{N})$, $\widehat{A}$ is a self-adjoint relation in a Pontryagin space $\widehat{\mathcal{P}}$ with negative index κ, $\rho(\widehat{A}) \ne \emptyset$, $u \in \widehat{\mathcal{P}}$ is a cyclic element for $\widehat{A}$, the root space $\widehat{\mathcal{L}}$ of $\widehat{A}$ at $z = \infty$ is spanned by $m + n$ vectors $w_1, w_2, \ldots, w_{m+n}$, which form a Jordan chain of $\widehat{A}$ at ∞, $\widehat{\mathcal{L}}$ has index of non-positivity κ and $\operatorname{span}\{w_1, w_2, \ldots, w_m\}$ is its isotropic subspace. If $m = 0$ and P_0 denotes the orthogonal projection onto $\mathcal{H}_0 = \widehat{\mathcal{P}} \ominus \widehat{\mathcal{L}}$, which is a uniformly positive subspace of $\widehat{\mathcal{P}}$, then $P_0 u \notin \operatorname{dom} \widehat{A}$.

(iv) $\widehat{N}(z) \in \mathcal{N}_\kappa^\infty$, *the irreducible representation of* $\widehat{N}(z)$ *being*

$$\widehat{N}(z) = c(z)^\# N_0(z)c(z) + p_\ell(z), \quad c(z) = (z - z_0)^m,$$

where $N_0 \in \mathcal{N}_0$ *has the properties*

$$\lim_{y\to\infty} y \operatorname{Im} N_0(iy) = \infty, \qquad \lim_{y\to\infty} y^{-1} N_0(iy) = 0, \qquad \operatorname{Re} N_0(i) = 0,$$

$m \in \mathbb{N}_0$, $z_0 \in \mathcal{D}(\widehat{N})$, $p_\ell(z) = \sum_{k=0}^{\ell} a_k z^k$ *is a real polynomial of degree* ℓ, *and we set* $n = \max\{\ell - 2m, 0\}$.

The Pontryagin spaces in (i) *and* (iii) *can be chosen the same and then the element* w_1 *in* (iii) *can be chosen to coincide with* w *in* (i) *and to satisfy* $\langle w_1, u\rangle = 1$; *in this case* $w_j = A^{j-1}w$, $j = 1, \ldots, m+n$, *and* $\mathcal{L} = \widehat{\mathcal{L}}$. *With* A *and* w *from* (i) *and the coefficients* s_j, $2m+n-1 \le j \le 2m+2n-2$ *in* (2.2), $s_j = 0$ *if* $0 \le j \le 2m+n-2$, *it holds*

$$s_j = \langle A^r w, A^s w\rangle \quad \text{if} \quad r + s = j,\ 0 \le r, s \le m + n - 1,$$

and, if $n > 0$, *then* $s_{2m+n-1} = 1/a_{2m+n} = 1/a_\ell$, *where* a_ℓ *is the leading coefficient of the polynomial* p *in* (iv). *The relation between the negative index* κ *of the Pontryagin spaces in* (i) *and* (iii) *and the integers* m, *and* n *is given by*

$$\kappa = \begin{cases} m & \text{if } n = 0, \\ m + \frac{n+1}{2} & \text{if } n > 0,\ n \text{ odd },\ a_\ell < 0, \\ m + \left[\frac{n}{2}\right] & \text{otherwise }. \end{cases}$$

Note that in case $m = 0$, the root space $\widehat{\mathcal{L}}$ of $\widehat{A}$ at ∞ in part (iii) is a regular subspace, whereas if $m > 0$ it is degenerate with an m-dimensional isotropic part. In the first case ∞ is called a critical singular point and in the second case it is called a singular critical point of $\widehat{A}$.

To the last part of the theorem can be added that $\widehat{A} = A^\infty$, where A^∞ is defined through infinite coupling of A and w. This means that it is obtained as the limit in the resolvent sense of the self-adjoint operator

$$A^{\langle\alpha\rangle} = A + \alpha\langle\,\cdot\,, w\rangle w \tag{2.3}$$

by letting $\alpha \to \infty$: Since for $\alpha \in \mathbb{R} \setminus \{0\}$,

$$(A^{\langle\alpha\rangle} - z)^{-1} = (A - z)^{-1} - \frac{\langle\,\cdot\,, \varphi(z^*)\rangle_{\mathcal{P}}}{N(z) + 1/\alpha}\,\varphi(z), \quad \varphi(z) = (A - z)^{-1}w, \tag{2.4}$$

we have

$$(A^\infty - z)^{-1} = (A - z)^{-1} - \frac{\langle\,\cdot\,, \varphi(z^*)\rangle_{\mathcal{P}}}{N(z)}\,\varphi(z). \tag{2.5}$$

For later reference we note that (2.4) implies

$$\langle (A^{\langle\alpha\rangle} - z)^{-1}w, w\rangle = \frac{N(z)}{1 + \alpha N(z)}, \tag{2.6}$$

which for $\alpha = 0$ is consistent with (2.1) and for $\alpha = \infty$ with $(A^\infty - z)^{-1}w = 0$, which follows from (2.5).

Formula (2.2) is related to the moment problem for generalized Nevanlinna functions, the numbers s_j being the moments; see, for example, [23] and [3]. The purpose of this paper is to provide some explicit minimal models for the operator $A = A^{\langle 0 \rangle}$ and the relation $\widehat{A} = A^{\infty}$. To derive these models we use the theory of reproducing kernels.

3. Reproducing kernel Pontryagin spaces and canonical models

A by now well-known model for $N \in \mathcal{N}_\kappa^{n\times n}$ is described in the following theorem. Here the state space is the reproducing kernel space $\mathcal{L}(N)$ associated with the kernel $K_N(\zeta, z)$. Recall that the elements of this space are n-vector functions defined and holomorphic on $\mathcal{D}(N)$, that the functions $K_N(\,\cdot\,, z)\mathbf{c}$, where z runs through $\mathcal{D}(N)$ and $\mathbf{c}$ runs through $\mathbb{C}^n$, are dense in $\mathcal{L}(N)$, and that the kernel has the reproducing property:

$$\langle f, K_N(\,\cdot\,, z)\mathbf{c}\rangle_{\mathcal{L}(N)} = \mathbf{c}^* f(z), \quad f \in \mathcal{L}(N), \mathbf{c} \in \mathbb{C}^n.$$

Whenever defined we denote by R_z the difference-quotient operator and, for later use, by E_ζ the operator of evaluation at the point ζ, that is,

$$R_z f(\zeta) = \frac{f(\zeta) - f(z)}{\zeta - z}, \quad E_\zeta f = f(\zeta), \tag{3.1}$$

where f is a vector function.

Theorem 3.1. *Let $N \in \mathcal{N}_\kappa^{n\times n}$ be given. Then:*

(i) $A = \big\{\{f, g\} \in \mathcal{L}(N)^2 \,|\, \exists\, \mathbf{c} \in \mathbb{C}^n : \; g(\zeta) - \zeta f(\zeta) \equiv \mathbf{c}\big\}$ *is a self-adjoint relation in $\mathcal{L}(N)$ with $\rho(A) \neq \emptyset$, and*

$$\big(\Gamma_z \mathbf{c}\big)(\zeta) = K_N(\zeta, z^*)\mathbf{c} = \frac{N(\zeta) - N(z)}{\zeta - z}\mathbf{c}, \quad \mathbf{c} \in \mathbb{C}^n,$$

is a corresponding Γ-field. The pair (A, Γ_z) is a minimal realization of N.

(ii) *The resolvent of A is the difference-quotient operator in $\mathcal{L}(N)$:*

$$(A - z)^{-1} = R_z, \quad z \in \rho(A).$$

(iii) $S = \big\{\{f, g\} \in \mathcal{L}(N)^2 \,|\, g(\zeta) - \zeta f(\zeta) \equiv 0\big\}$ *is a symmetric operator in the space $\mathcal{L}(N)$ with equal defect indices $n - d$, where $d = \dim\ker\Gamma_z$. Moreover, $\sigma_p(S) = \emptyset$ and the adjoint of S is given by*

$$\begin{aligned} S^* &= \overline{\operatorname{span}}\,\big\{\{\Gamma_z \mathbf{h}, z\Gamma_z \mathbf{h}\} \;|\; \mathbf{h} \in \mathbb{C}^n, z \in \mathcal{D}(N)\big\} \\ &= \big\{\{f, g\} \in \mathcal{L}(N)^2 \,|\, \exists\, \mathbf{c}, \mathbf{d} \in \mathbb{C}^n : \; g(\zeta) - \zeta f(\zeta) \equiv \mathbf{c} - N(\zeta)\mathbf{d}\big\}. \end{aligned}$$

For the proof of this theorem and remarks concerning its origin we refer to [10, Theorem 2.1]. The minimal realization of N described here is called the canonical realization of N and the triplet (A, Γ_z, S) is called the canonical model. For these canonical models only, to denote the dependence on N we often write A_N, Γ_{Nz}, S_N etc. instead of A, Γ_z, S, etc.

By $\mathbf{N}$ we denote the operator of multiplication by the function N:

$$(\mathbf{N}f)(\zeta) = N(\zeta)f(\zeta). \tag{3.2}$$

Theorem 3.2. *Let $N \in \mathcal{N}_\kappa^{n\times n}$, assume that $N(z)$ is invertible for some point $z \in \mathcal{D}(N)$, and set $\widehat{N} = -N^{-1}$. Then $\widehat{N} \in \mathcal{N}_\kappa^{n\times n}$ and the following statements hold.*

(i) $\mathbf{N}$ *as a mapping from $\mathcal{L}(\widehat{N})$ to $\mathcal{L}(N)$ is unitary; its inverse is the operator of multiplication by N^{-1}.*

(ii) *We have*

$$\widehat{A}_{\widehat{N}} = \mathbf{N}^{-1}A_N\mathbf{N} = \Big\{\{f,g\} \in \mathcal{L}(\widehat{N})^2 \,\big|\, \exists\, \mathbf{d} \in \mathbb{C}^n : \, g(\zeta) - \zeta f(\zeta) \equiv \widehat{N}(\zeta)\mathbf{d}\Big\}$$

and hence $\rho(\widehat{A}_{\widehat{N}}) \neq \emptyset$.

(iii) *For $0 \neq \mathbf{c} \in \mathbb{C}^n$ and $j = 0, 1, \ldots,$ we have $\zeta^j N(\zeta)\mathbf{c} \in \mathcal{L}(N)$ if and only if $\zeta^j \mathbf{c} \in \mathcal{L}(\widehat{N})$.*

The theorem coincides in part with [10, Corollary 2.3]. Part (i) follows from the kernel identity

$$N(\zeta)K_{\widehat{N}}(\zeta, z)N(z)^* = K_N(\zeta, z),$$

part (ii) from (i) and Theorem 3.1 (i), and part (iii) follows from (i). The inclusions in (iii) hold for $j = 0$ if and only if $z = \infty$ is a generalized zero of N or, equivalently, $z = \infty$ is a generalized pole of $\widehat{N}$. That we use the notation $\widehat{A}_{\widehat{N}}$ for the operator/relation $\mathbf{N}^{-1}A_N\mathbf{N}$ in part (ii) of Theorem 3.2 comes from applying our convention that if A is the self-adjoint operator/relation in a model for N then we write $\widehat{A}$ for the corresponding operator/relation associated with $\widehat{N}$: If $N \in \mathcal{N}_\kappa^{n\times n}$ is invertible at some point in $\mathcal{D}(N)$, then the triplet $(A_{\widehat{N}}, K_{\widehat{N}}(\,\cdot\,, z^*), S_{\widehat{N}})$ is the canonical model for $\widehat{N}$ and the triplet

$$(\widehat{A}_{\widehat{N}}, K_{\widehat{N}}(\,\cdot\,, z^*)N(z), S_{\widehat{N}}) \tag{3.3}$$

is a minimal model for the function $N = \widehat{\widehat{N}}$ in $\mathcal{L}(\widehat{N})$, because it is isomorphic under $\mathbf{N}$ with the canonical model (A_N, Γ_{Nz}, S_N) for N in $\mathcal{L}(N)$.

For use in the next section we recall the following theorem (see [10, Theorem 2.4]). A function $N \in \mathcal{N}_\kappa^{n\times n}$ is called strict if for some non-real point $z_0 \in \mathcal{D}(N)$ it holds

$$\bigcap_{\zeta \in \mathcal{D}(N)} \ker K_N(\zeta, z_0) = \{0\}.$$

Theorem 3.3. *Suppose that $N \in \mathcal{N}_\kappa^{n\times n}$ is strict and let $(A,\, \Gamma_z,\, S)$ be the canonical model for N. Then:*

(i) *A relation is a canonical self-adjoint extension of S if and only if it is of the form*

$$A_{\mathcal{A},\mathcal{B}} = \big\{\{f,g\} \in \mathcal{L}(N)^2 \,\big|\, \exists\, \mathbf{h} \in \mathbb{C}^n : \; g(\zeta) - \zeta f(\zeta) \equiv (\mathcal{A} + N(\zeta)\mathcal{B})\mathbf{h}\big\}$$

with $n \times n$ matrices $\mathcal{A}$ and $\mathcal{B}$ satisfying the relations

$$\operatorname{rank}\begin{pmatrix}\mathcal{A}\\ \mathcal{B}\end{pmatrix} = n, \qquad \mathcal{A}^*\mathcal{B} - \mathcal{B}^*\mathcal{A} = 0.$$

If $A_{\mathcal{A},\mathcal{B}}$ and $A_{\mathcal{A}',\mathcal{B}'}$ are two such canonical self-adjoint extensions of S then $A_{\mathcal{A}',\mathcal{B}'} = A_{\mathcal{A},\mathcal{B}}$ if and only if $\mathcal{A}' = \mathcal{A}\mathcal{C}$ and $\mathcal{B}' = \mathcal{B}\mathcal{C}$ for some invertible $n\times n$ matrix $\mathcal{C}$.

(ii) *$\rho\,(A_{\mathcal{A},\mathcal{B}}) \neq \emptyset$ if and only if for some non-real point $z_0 \in \mathcal{D}(N)$ the matrices $\mathcal{A} + N(z_0)\mathcal{B}$ and $\mathcal{A} + N(z_0)^*\mathcal{B}$ are invertible. In this case for $z \in \rho\,(A_{\mathcal{A},\mathcal{B}}) \cap \rho(A)$:*

$$(A_{\mathcal{A},\mathcal{B}} - z)^{-1} = (A - z)^{-1} - \Gamma_z\mathcal{B}\,(\mathcal{A} + N(z)\mathcal{B})^{-1}\,\Gamma_{z^*}^*. \tag{3.4}$$

We specialize to case $n = 1$ and assume $0 \neq N \in \mathcal{L}(N)$. Then on account of Theorem 3.1 (i) and (ii), $w = N$ is a cyclic element of $\mathcal{L}(N)$ for A_N and

$$N(z) = \langle (A_N - z)^{-1} w, w\rangle_{\mathcal{L}(N)}.$$

The operator

$$A_N^{\langle\alpha\rangle} = A_N + \alpha\langle\,\cdot\,, w\rangle_{\mathcal{L}(N)} w, \qquad \alpha \in \mathbb{R},$$

can also be written as

$$A_N^{\langle\alpha\rangle} = \{\{f,g\} \in \mathcal{L}(N)^2 | \exists c \in \mathbb{C} : g(\zeta) - \zeta f(\zeta) = (1 + \alpha N(\zeta))c\}. \tag{3.5}$$

This can be seen by comparing the resolvents (2.4) and (3.4) applied to this situation. If we let $\alpha \to \infty$, then in the resolvent sense $A_N^{\langle\alpha\rangle}$ converges to

$$\begin{aligned} A_N^{\infty} &= \{\{f,g\} \in \mathcal{L}(N)^2 | \exists c \in \mathbb{C} : g(\zeta) - \zeta f(\zeta) = N(\zeta)c\} \\ &= S_N + \{\{0, cw\} \mid c \in \mathbb{C}\} = S_N + \{0\} \times (\operatorname{dom} S_N)^{\perp}. \end{aligned}$$

The set on the right-hand side after the first equality can at least formally be obtained from the set on the right-hand side of (3.5) by replacing $\{f,g\}$ by $\{\alpha f, \alpha g\}$ and letting $\alpha \to \infty$. Now we apply the unitary map $\mathbf{N}$ and find that (the constant function) $1 \in \mathcal{L}(\widehat{N})$, 1 is a cyclic element for $\widehat{A}_{\widehat{N}}$, and

$$\begin{aligned} (\widehat{A}_{\widehat{N}})^{\langle\alpha\rangle} &= \mathbf{N}^{-1} A_N^{\langle\alpha\rangle}\mathbf{N} = \widehat{A}_{\widehat{N}} + \alpha\langle\,\cdot\,, 1\rangle_{\mathcal{L}(\widehat{N})} \\ &= \{\{f,g\} \in \mathcal{L}(\widehat{N})^2 | \exists c \in \mathbb{C} : g(\zeta) - \zeta f(\zeta) = (\alpha - \widehat{N}(\zeta))c\}. \end{aligned} \tag{3.6}$$

Hence in the infinite coupling of $\widehat{A}_{\widehat{N}}$ and 1 we have

$$(\widehat{A}_{\widehat{N}})^{\infty} = A_{\widehat{N}} = \mathbf{N}^{-1} A_N^{\infty}\mathbf{N}. \tag{3.7}$$

4. Minimal models in the space $\mathcal{L}(\widetilde{N})$

From now on we assume that

(1) N is a non-zero scalar generalized Nevanlinna function in $\mathcal{N}_\kappa$,

(2) $\widehat{N} = -1/N \in \mathcal{N}_\kappa^{\infty}$,

(3) $\widehat{N}$ has representation (1.2), and
(4) $z_0 \in \mathcal{D}(\widehat{N})$ belongs to the possibly smaller set $\mathcal{D}(N_0)$.

We rewrite the irreducible representation (1.2) of $\widehat{N}$ in the form

$$\widehat{N}(z) = c^\#(z)(N_0(z) + q(z))c(z) + p_0(z), \quad c(z) = (z - z_0)^m, \tag{4.1}$$

where q and p_0 are real polynomials such that $c^\#(z)q(z)c(z) + p_0(z) = p(z)$, $\ell_0 = \deg p_0 < 2m$, and $n = \deg q$ is the number appearing in Theorem 2.1. With the decomposition (4.1) we associate the generalized Nevanlinna matrix functions

$$\widetilde{N}(z) = \begin{pmatrix} N_0(z) & 0 & 0 & 0 \\ 0 & q(z) & 0 & 0 \\ 0 & 0 & p_0(z) & c^\#(z) \\ 0 & 0 & c(z) & 0 \end{pmatrix}, \quad M(z) = \begin{pmatrix} p_0(z) & c^\#(z) \\ c(z) & 0 \end{pmatrix}.$$

It follows that the reproducing kernel space $\mathcal{L}(\widetilde{N})$ with kernel $K_{\widetilde{N}}(\,\cdot\,,\,\cdot\,)$ can be decomposed as the orthogonal sum $\mathcal{L}(\widetilde{N}) = \mathcal{L}(N_0) \oplus \mathcal{L}(q) \oplus \mathcal{L}(M)$. If $n > 0$ and $m > 0$, then the elements of $\mathcal{L}(q)$ are the polynomials of degree $< n$ and the elements of $\mathcal{L}(M)$ are 2-vector functions with polynomial entries. Unless stated otherwise we assume $n > 0$ and $m > 0$. If $n = 0$ or $m = 0$, then $\mathcal{L}(q) = \{0\}$ or $\mathcal{L}(M) = \{0\}$ and the formulas simplify; we consider these cases separately.

In this section we give minimal models for N and $\widehat{N}$ in the space $\mathcal{L}(\widetilde{N})$. For this we introduce the vector function

$$v(z) = \begin{pmatrix} c(z) & c(z) & 1 & c(z)(N_0(z) + q(z)) \end{pmatrix}^\top$$

and the following 4×4 matrices

$$\mathcal{A}_\alpha = \begin{pmatrix} 1 & 0 & 0 & 0 \\ 0 & 1 & 0 & 0 \\ 0 & 0 & \alpha & 0 \\ 0 & 0 & 0 & 1 \end{pmatrix}, \quad \mathcal{B} = -\begin{pmatrix} 0 & 0 & 0 & 1 \\ 0 & 0 & 0 & 1 \\ 0 & 0 & 1 & 0 \\ 1 & 1 & 0 & 0 \end{pmatrix}, \quad \widehat{\mathcal{B}} = -\begin{pmatrix} 0 & 0 & 0 & 1 \\ 0 & 0 & 0 & 1 \\ 0 & 0 & 0 & 0 \\ 1 & 1 & 0 & 0 \end{pmatrix}.$$

Theorem 4.1. *Assume the conditions* (1)–(4) *hold.*

(i) *The minimal models of N and $\widehat{N}$ in $\mathcal{L}(\widetilde{N})$ are given by the triplets*

$$(B, K_{\widetilde{N}}(\,\cdot\,, z^*)v(z)N(z), \widetilde{S}) \quad \text{and} \quad (\widehat{B}, K_{\widetilde{N}}(\,\cdot\,, z^*)v(z), \widetilde{S}),$$

where

$$B = \{\{\widetilde{f}, \widetilde{g}\} \in \mathcal{L}(\widetilde{N}) \mid \exists \mathbf{h} \in \mathbb{C}^4 : \widetilde{g}(\zeta) - \zeta\widetilde{f}(\zeta) = (\mathcal{A}_0 + \widetilde{N}(\zeta)\mathcal{B})\mathbf{h}\}, \tag{4.2}$$

$$\widehat{B} = \{\{\widetilde{f}, \widetilde{g}\} \in \mathcal{L}(\widetilde{N}) \mid \exists \mathbf{h} \in \mathbb{C}^4 : \widetilde{g}(\zeta) - \zeta\widetilde{f}(\zeta) = (I_{\mathbb{C}^4} + \widetilde{N}(\zeta)\widehat{\mathcal{B}})\mathbf{h}\}, \tag{4.3}$$

and

$$\widetilde{S} = \{\{\widetilde{f}, \widetilde{g}\} \in \mathcal{L}(\widetilde{N}) \mid \exists \mathbf{h} \in \mathbb{C}^4 \text{ with } h_3 = 0 : \widetilde{g}(\zeta) - \zeta\widetilde{f}(\zeta) = (\mathcal{A}_0 + \widetilde{N}(\zeta)\widehat{\mathcal{B}})\mathbf{h}\}.$$

(ii) $\widetilde{S}$ *has defect* $(1,1)$ *and the family of all its self-adjoint extensions in* $\mathcal{L}(\widetilde{N})$ *is given by* $\widehat{B}$ *and* $B^{\langle\alpha\rangle}$, $\alpha \in \mathbb{R}$, *where*

$$\begin{aligned} B^{\langle\alpha\rangle} &= B + \alpha\langle\,\cdot\,,\widetilde{w}\rangle_{\mathcal{L}(\widetilde{N})}\widetilde{w} \\ &= \big\{\{\widetilde{f},\widetilde{g}\} \in \mathcal{L}(\widetilde{N}) \mid \exists \mathbf{h} \in \mathbb{C}^4 : \widetilde{g}(\zeta) - \zeta\widetilde{f}(\zeta) = (\mathcal{A}_\alpha + \widetilde{N}(\zeta)\mathcal{B})\mathbf{h}\big\} \end{aligned} \tag{4.4}$$

with

$$\widetilde{w} = \begin{pmatrix} 0 & 0 & 1 & 0 \end{pmatrix}^\top \in \mathcal{L}(\widetilde{N}).$$

Moreover, $B = B^{\langle 0\rangle}$ *and* $\widehat{B} = B^{\infty}$, *the limit of* $B^{\langle\alpha\rangle}$ *in the resolvent sense by letting* $\alpha \to \infty$.

Note that $\widetilde{S}$ has defect $(1,1)$ shows that it does not coincide with $S_{\widetilde{N}}$ in the canonical representation of $\widetilde{N}$ in $\mathcal{L}(\widetilde{N})$, which has defect $(4,4)$ if $n \neq 0$ and $(3,3)$ if $n = 0$.

On account of (1)–(4) the four equivalent statements in Theorem 2.1 hold for N and $\widehat{N}$. For the Pontryagin space $\widehat{\mathcal{P}}$ we take the reproducing kernel Pontryagin space $\mathcal{L}(\widehat{N})$. For $\mathcal{P}$ we take $\mathcal{L}(N)$ but we identify it with $\mathcal{L}(\widehat{N})$ via the unitary map $\mathbf{N}$ defined by (3.2). Since ∞ is a generalized zero of N, we have $N \in \mathcal{L}(N)$ (see [10, Corollary 2.3(iii)]). It follows that (2.1) holds with $w = N$ and $A = A_N$ in $\mathcal{P}$ and in the identification of $\mathcal{P}$ with $\mathcal{L}(\widehat{N})$ we have $w = w_1 = 1$ and $A = \widehat{A}_{\widehat{N}}$ and so

$$N(z) = \langle(\widehat{A}_{\widehat{N}} - z)^{-1}1, 1\rangle_{\mathcal{L}(\widehat{N})}.$$

The expansion (2.2) for N implies that the functions $w_1, w_2, \ldots, w_{m+n}$ (or, what amounts to the same, $w, Aw, \ldots, A^{m+n-1}w$) are given by $1, \zeta, \ldots, \zeta^{m+n-1}$ (see [3, Lemma 5.2]). Here we use that the moments s_j are zero for $0 \leq j \leq 2m+n-2$. The representation for $\widehat{N}$ in statement (iii) of Theorem 2.1 holds with $\widehat{A} = A_{\widehat{N}}$ and

$$u = K_{\widehat{N}}(\,\cdot\,, z_0^*). \tag{4.5}$$

Notice that $\langle u, w_1\rangle_{\mathcal{L}(\widehat{N})} = 1$ by the reproducing property of the kernel. Since $(A_{\widehat{N}} - z)^{-1}1 = R_z 1 = 0$, we see directly that $A_{\widehat{N}}$ is a relation with a nontrivial multi-valued part: $1 \in A_{\widehat{N}}(0)$. In Section 3 we showed that the triplets

$$(\widehat{A}_{\widehat{N}}, K_{\widehat{N}}(\,\cdot\,,z^*)N(z), S_{\widehat{N}}) \quad \text{and} \quad (A_{\widehat{N}}, K_{\widehat{N}}(\,\cdot\,,z^*), S_{\widehat{N}})$$

are minimal models of N and $\widehat{N}$ in $\mathcal{L}(\widehat{N})$ and that $A_{\widehat{N}}$ is the limit in the resolvent sense of $\widehat{A}_{\widehat{N}}^{\langle\alpha\rangle}$; see (3.3), Theorem 3.1 applied to $\widehat{N}$, and (3.6) and (3.7). The main idea of the proof of the theorem is that $B^{\langle\alpha\rangle}$ and $\widehat{B}$ are isomorphic copies of $\widehat{A}_{\widehat{N}}^{\langle\alpha\rangle}$ and $A_{\widehat{N}}$. The isomorphism is given in Lemma 4.4 below. We begin with two technical lemmas.

Lemma 4.2. *If* $\begin{pmatrix} h_1 \\ h_2 \end{pmatrix} \in \mathcal{L}(M)$ *and* $h_2 = 0$, *then* $\deg h_1 < m$.

Proof. Since $\deg p_0 < 2m$, the space $\mathcal{L}(M)$ is spanned by the $2m$ linearly independent 2-vector functions

$$R_0^j \begin{pmatrix} p_0 \\ c \end{pmatrix}, \quad R_0^j \begin{pmatrix} c^\# \\ 0 \end{pmatrix}, \qquad j = 1, 2, \ldots, m,$$

where R_0 is the difference-quotient operator given by (3.1) with $z = 0$. If $h_2 = 0$, then

$$\begin{pmatrix} h_1 \\ h_2 \end{pmatrix} \in \operatorname{span}\left\{R_0^j \begin{pmatrix} c^\# \\ 0 \end{pmatrix} \mid j = 1, 2, \ldots, m\right\},$$

which implies $\deg h_1 < m$. □

Lemma 4.3. *Assume $N_0 \in \mathcal{N}_0$ satisfies the conditions in* (1.3). *If h_1 and h_2 are polynomials and $N_0 h_1 + h_2 \in \mathcal{L}(N_0)$, then $h_1 = h_2 = 0$.*

Proof. Whenever defined we have

$$R_{w_0}(fg)(\zeta) = R_{w_0}(f)(\zeta)g(w_0) + f(\zeta)R_{w_0}(g)(\zeta).$$

Let h_1 and h_2 be polynomials such that $N_0 h_1 + h_2 \in \mathcal{L}(N_0)$. Assume that h_1 and h_2 are not both identically equal to 0. If we apply R_{w_0} with $w_0 \in \mathcal{D}(N_0)$ a number of times to the function $N_0 h_1 + h_2 \in \mathcal{L}(N_0)$ and use the above formula and that $R_{w_0}(N_0)(\zeta)c \in \mathcal{L}(N_0)$, $c \in \mathbb{C}$, we find that there is pair of complex numbers $(c_1, c_2) \neq (0, 0)$ such that $N_0 c_1 + c_2 \in \mathcal{L}(N_0)$. If $N_0 c_1 + c_2 = 0$, then N_0 is a real constant and therefore, on account of the last equality in (1.3) equal to zero. But this is in contradiction with the first equality in (1.3). If $N_0 c_1 + c_2 \neq 0$, then, by Theorem 3.1(iii) applied to N_0,

$$\{0, 0\} \neq \{0, N_0 c_1 + c_2\} \in S_{N_0}^*,$$

which implies that the minimal operator S_{N_0} in $\mathcal{L}(N_0)$ is not densely defined. This is in contradiction with the first two equalities in (1.3). These contradictions imply that h_1 and h_2 are identically equal to 0. □

Lemma 4.4. *The mapping $\mathbf{V} : \mathcal{L}(\widetilde{N}) \to \mathcal{L}(\widehat{N})$ defined by $(\mathbf{V}\widetilde{f})(\zeta) = v^\#(\zeta)\,\widetilde{f}(\zeta)$ is unitary.*

The corresponding mapping in [10, Lemma 3.1], where realizations of N related to its basic factorization (1.1) are considered, is a partial isometry but not necessarily injective.

Proof of Lemma 4.4. A straightforward calculation shows

$$v^\#(\zeta)K_{\widetilde{N}}(\zeta, z)v(z^*) = K_{\widehat{N}}(\zeta, z). \tag{4.6}$$

We claim that the number of negative squares of the kernels $K_{\widetilde{N}}$ and $K_{\widehat{N}}$ coincide. Indeed, the first number is the sum of the number of negative squares of the scalar function q and the matrix function M. The first of which is equal to $(n+1)/2$ if n is odd and the leading coefficient of the polynomial q is negative, and otherwise it is $\left[n/2\right]$. The kernel of M has m negative squares since z_0 is the only zero of M in the closed upper half-plane and its multiplicity is m. By Theorem 2.1, the sum of these numbers equals κ, which is the number of negative squares of $K_{\widehat{N}}$. Hence

[4, Theorem 1.5.7.] implies that $\mathbf{V}$ is a surjective partial isometry. We show that it is in fact a unitary mapping, that is, $\ker \mathbf{V}$ is trivial. Assume there is an element $\widetilde{f}(\zeta) = \begin{pmatrix} f(\zeta) & a(\zeta) & b(\zeta) & d(\zeta) \end{pmatrix}^\top \in \ker \mathbf{V}$. Then

$$c^\#(\zeta)f(\zeta) + c^\#(\zeta)a(\zeta) + b(\zeta) + c^\#(\zeta)(N_0(\zeta) + q(\zeta))d(\zeta) \equiv 0. \tag{4.7}$$

Since N_0 is holomorphic at the point z_0^*, we have that $f \in \mathcal{L}(N_0)$ is holomorphic at z_0^* also and the equality (4.7) implies that the polynomial b has a zero of order at least m at z_0^*. So we have $b(\zeta) = (\zeta - z_0^*)^m b_1(\zeta)$ for some polynomial b_1. Equality (4.7) implies $N_0 h_1 + h_2 = -f \in \mathcal{L}(N_0)$ with polynomials $h_1 = d$ and $h_2 = a + b_1 + qd$. By Lemma 4.3, $h_1 = h_2 = 0$, that is, $d = 0$ and $a + b_1 = 0$. Now we use Lemma 4.2: Since $\begin{pmatrix} b & d \end{pmatrix}^\top \in \mathcal{L}(M)$ and $d = 0$, the lemma yields that $\deg b < m$, which implies $b_1 = 0$ and hence $b = 0$ and $a = 0$. Finally, on account of (4.7), we have $f = 0$. We conclude that $\widetilde{f} = 0$, that is, $\ker \mathbf{V} = \{0\}$. □

Proof of Theorem 4.1. We first define B, $\widehat{B}$, and $\widetilde{S}$ by the formulas

$$B = \mathbf{V}^{-1}\widehat{A}_{\widehat{N}}\mathbf{V}, \quad \widehat{B} = \mathbf{V}^{-1}A_{\widehat{N}}\mathbf{V}, \quad \widetilde{S} = B \cap \widehat{B} = \mathbf{V}^{-1}S_{\widehat{N}}\mathbf{V}$$

and claim that they coincide with the relations in part (i) of the theorem. Assuming the claim is true, we have, according to (4.6), that under the unitary mapping $\mathbf{V}$ the element $K_{\widetilde{N}}(\,\cdot\,, z^*)v(z)$ in $\mathcal{L}(\widetilde{N})$ is the isomorphic copy of the element $K_{\widehat{N}}(\,\cdot\,, z^*)$ in $\mathcal{L}(\widehat{N})$. Hence the triplets in (i) are isomorphic copies of the minimal models (3.3) and $(A_{\widehat{N}}, K_{\widehat{N}}(\,\cdot\,, z^*), S_{\widehat{N}})$ for N and $\widehat{N}$ in $\mathcal{L}(\widetilde{N})$. It remains to prove the claim. It is easy to see that

$$\begin{aligned} B &= \{\{\widetilde{f}, \widetilde{g}\} \in \mathcal{L}(\widetilde{N})^2 \mid \exists d \in \mathbb{C} : v^\#(\zeta)(\widetilde{g}(\zeta) - \zeta\widetilde{f}(\zeta)) = \widehat{N}(\zeta)d\}, \\ \widehat{B} &= \{\{\widetilde{f}, \widetilde{g}\} \in \mathcal{L}(\widetilde{N})^2 \mid \exists c \in \mathbb{C} : v^\#(\zeta)(\widetilde{g}(\zeta) - \zeta\widetilde{f}(\zeta)) = c\}. \end{aligned}$$

First we prove formula (4.2). Denote by $B_{\mathcal{A}_0,\mathcal{B}}$ the relation defined by the right-hand side of (4.2). If $\{\widetilde{f}, \widetilde{g}\} \in B_{\mathcal{A}_0,\mathcal{B}}$ and $\mathbf{h} = \begin{pmatrix} h_1 & h_2 & h_3 & h_4 \end{pmatrix}^\top$ then

$$v^\#(\zeta)\big(\widetilde{g}(\zeta) - \zeta\widetilde{f}(\zeta)\big) = v^\#(\zeta)(\mathcal{A}_0 + \widetilde{N}(\zeta)\mathcal{B})\mathbf{h} = -\widehat{N}(\zeta)h_3, \tag{4.8}$$

hence $B_{\mathcal{A}_0,\mathcal{B}} \subset B$. Since $B_{\mathcal{A}_0,\mathcal{B}}$ and B are self-adjoint operators, see Theorem 3.3, equality holds. In the same way, if $\{\widetilde{f}, \widetilde{g}\} \in B_{I_{\mathbb{C}^4},\widehat{\mathcal{B}}}$, the operator defined by the right-hand side of (4.3), then

$$v^\#(\zeta)(\widetilde{g}(\zeta) - \zeta\widetilde{f}(\zeta)) = v^\#(\zeta)(I_{\mathbb{C}^4} + \widetilde{N}(\zeta)\widehat{\mathcal{B}})\mathbf{h} = h_3, \tag{4.9}$$

hence $B_{I_{\mathbb{C}^4},\widehat{\mathcal{B}}} \subset \widehat{B}$ and equality prevails because both self-adjoint relations have nonempty resolvent sets. The formula for $\widetilde{S}$ follows from (4.8) and (4.9). This completes the proof of part (i).

As to (ii) we first define the operators $B^{\langle\alpha\rangle}$ by

$$B^{\langle\alpha\rangle} = \mathbf{V}^{-1}(\widehat{A}_{\widehat{N}})^{\langle\alpha\rangle}\mathbf{V}, \quad \alpha \in \mathbb{R}.$$

On account of (3.6), we have

$$B^{\langle\alpha\rangle} = B + \alpha\langle\,\cdot\,,\widetilde{w}\rangle_{\mathcal{L}(\widetilde{N})}\widetilde{w},$$

where

$$\widetilde{w} = \mathbf{V}^{-1}1 = \begin{pmatrix}0 & 0 & 1 & 0\end{pmatrix}^\top. \tag{4.10}$$

We now show (4.4). From (2.4), applied to this situation, we have

$$(B^{\langle\alpha\rangle} - z)^{-1} = (B-z)^{-1} - \frac{\langle\,\cdot\,,(B-z^*)^{-1}\widetilde{w}\rangle_{\mathcal{L}(\widetilde{N})}}{N(z)+1/\alpha}(B-z)^{-1}\widetilde{w}$$

and, by Theorem 3.3,

$$(B_{\mathcal{A}_\alpha,\mathcal{B}} - z)^{-1} = (A_{\widetilde{N}} - z)^{-1} - \Gamma_{\widetilde{N}z}\mathcal{B}(\mathcal{A}_\alpha + \widetilde{N}(\zeta)\mathcal{B})^{-1}E_z,$$

where $E_z = \Gamma^*_{\widetilde{N}z^*}$ is the operator of evaluation at the point z. For $\alpha = 0$, the last equality yields

$$(B-z)^{-1} = (A_{\widetilde{N}} - z)^{-1} - \Gamma_{\widetilde{N}z}\mathcal{B}(\mathcal{A}_0 + \widetilde{N}(\zeta)\mathcal{B})^{-1}E_z$$

and, on account of (4.10), $(B-z)^{-1}\widetilde{w} = N(z)\Gamma_{\widetilde{N}z}v(z)$. Combining these relations we find

$$\begin{aligned}&(B^{\langle\alpha\rangle} - z)^{-1} - (B_{\mathcal{A}_\alpha,\mathcal{B}} - z)^{-1}\\ &= \Gamma_{\widetilde{N}z}\left[-\mathcal{B}(\mathcal{A}_0 + \widetilde{N}(\zeta)\mathcal{B})^{-1} - \frac{N(z)^2}{N(z)+\frac{1}{\alpha}}v(z)v^{\#}(z) + \mathcal{B}(\mathcal{A}_\alpha + \widetilde{N}(\zeta)\mathcal{B})^{-1}\right]E_z.\end{aligned}$$

A straightforward calculation shows that the expression in square brackets vanishes, which implies (4.4) for all $\alpha \in \mathbb{R}$.

Clearly, $B = B^{\langle 0\rangle}$ and, because of (3.7), $\widehat{B} = B^\infty$, where the relation on the right-hand side is obtained via infinite coupling of B and $\widetilde{w}$, that is, by taking the limit of $B^{\langle\alpha\rangle}$ in the resolvent sense by letting $\alpha \to \infty$. Finally, the statement concerning $\widetilde{S}$ and its extensions follow from the corresponding results for $S_{\widehat{N}}$ and the unitarity of $\mathbf{V}$. □

The following corollary is an immediate consequence of Theorem 3.3. We set

$$K_\alpha(z) = \begin{pmatrix} c(z)c^{\#}(z) & c(z)c^{\#}(z) & c(z) & \alpha - p_0(z)\\ c(z)c^{\#}(z) & c(z)c^{\#}(z) & c(z) & \alpha - p_0(z)\\ c^{\#}(z) & c^{\#}(z) & 1 & (N_0(z)+q(z))c^{\#}(z)\\ \alpha - p_0(z) & \alpha - p_0(z) & (N_0(z)+q(z))c(z) & (N_0(z)+q(z))(\alpha - p_0(z))\end{pmatrix}$$

and

$$\widehat{K}(z) = \lim_{\alpha\to\infty}\frac{1}{\alpha}K_\alpha(z) = \begin{pmatrix}0&0&0&1\\0&0&0&1\\0&0&0&0\\1&1&0&N_0(z)+q(z)\end{pmatrix}.$$

Corollary 4.5. *The resolvents of* $B^{\langle\alpha\rangle}$ *and* $\widehat{B}$ *are given by*

$$(B^{\langle\alpha\rangle}-z)^{-1}=(A_{\widetilde{N}}-z)^{-1}+\frac{N(z)}{1+\alpha N(z)}\Gamma_{\widetilde{N}z}K_\alpha(z)E_z, \tag{4.11}$$

$$(\widehat{B}-z)^{-1}=(A_{\widetilde{N}}-z)^{-1}+\Gamma_{\widetilde{N}z}\widehat{K}(z)E_z, \tag{4.12}$$

where $(A_{\widetilde{N}}-z)^{-1}$ *is the difference-quotient operator in the space* $\mathcal{L}(\widetilde{N})$,

$$\Gamma_{\widetilde{N}z}=\operatorname{diag}\{K_{N_0}(\,\cdot\,,z^*),K_q(\,\cdot\,,z^*),K_M(\,\cdot\,,z^*)\},$$

and $E_z=(\Gamma_{\widetilde{N}z^*})^*$ *is the evaluation operator at the point* z.

From (4.11) it readily follows that

$$\langle (B^{\langle\alpha\rangle}-z)^{-1}\widetilde{w},\widetilde{w}\rangle=\frac{N(z)}{1+\alpha N(z)},\quad \alpha\in\mathbb{R}\cup\{\infty\},$$

which is consistent with (2.6).

It remains to discuss the simplifications if $n=0$ or $m=0$.

The case $n=0$ *and* $m>0$: Then $q(z)=q_0$ with $q_0\in\mathbb{R}$, hence $K_q(\zeta,z)=0$ and $\mathcal{L}(q)=\{0\}$; as to the 2×2 matrix function $M(z)$: $\deg p_0(z)<2m$ and the space $\mathcal{L}(M)$ is nontrivial. Now $\widetilde{N}(z)$ becomes the 3×3 matrix function

$$\widetilde{N}(z)=\operatorname{diag}\{N_0(z)+q_0,M(z)\}\quad\text{and}\quad \mathcal{L}(\widetilde{N})=\mathcal{L}(N_0)\oplus\mathcal{L}(M).$$

With $v(z)=\begin{pmatrix}c(z) & 1 & c(z)(N_0(z)+q_0)\end{pmatrix}^\top$ the operator $\mathbf{V}:\mathcal{L}(\widetilde{N})\to\mathcal{L}(\widehat{N})$ of multiplication by $v^\#(z)=\begin{pmatrix}c^\#(z) & 1 & c^\#(z)(N_0(z)+q_0)\end{pmatrix}$ is unitary and $\widetilde{w}$ in (4.10) becomes $\widetilde{w}:=\mathbf{V}^{-1}1=\begin{pmatrix}0 & 1 & 0\end{pmatrix}^\top$.

Theorem 4.6. *Assume* $n=0$ *and* $m>0$. *Then Theorem* 4.1 *and Corollary* 4.5 *remain true provided in all* 4×4 *matrices the* 2*-nd row and the* 2*-nd column are deleted and in the formulas for* B, $\widehat{B}$, $\widetilde{S}$, *and* $B^{\langle\alpha\rangle}$ *the space* $\mathbb{C}^4$ *and the entry* h_3 *are replaced by* $\mathbb{C}^3$ *and* h_2.

The case $n>0$ *and* $m=0$: Now $c(z)=1$, $q(z)$ is a nonconstant real polynomial, and the irreducible representation (1.2) becomes $\widehat{N}(z)=N_0(z)+q(z)$. From $\widetilde{N}(z)=\operatorname{diag}\{N_0(z),q(z)\}$ it follows that $\mathcal{L}(\widetilde{N})=\mathcal{L}(N_0)\oplus\mathcal{L}(q)$. With the vector function $v(z)=\begin{pmatrix}1 & 1\end{pmatrix}^\top$ the mapping $V:\mathcal{L}(\widetilde{N})\to\mathcal{L}(\widehat{N})$ of multiplication by $v^\#(z)=\begin{pmatrix}1 & 1\end{pmatrix}$ is unitary.

Theorem 4.7. *Assume* $n>0$ *and* $m=0$ *and for* $\alpha\in\mathbb{R}$ *define the operators*

$$B^{\langle\alpha\rangle}=\{\{\widetilde{f},\widetilde{g}\}\in\mathcal{L}(\widetilde{N})\mid \exists\mathbf{h}\in\mathbb{C}^2:\widetilde{g}(\zeta)-\zeta\widetilde{f}(\zeta)=(\mathcal{A}_\alpha+\widetilde{N}(\zeta)\mathcal{B})\mathbf{h}\},$$

where

$$\mathcal{A}_\alpha=\begin{pmatrix}1 & 0\\ -1 & \alpha\end{pmatrix},\quad \mathcal{B}=-\begin{pmatrix}0 & 1\\ 0 & 1\end{pmatrix}.$$

(i) *Theorem* 4.1 *holds provided* $B=B^{\langle 0\rangle}$, $\widehat{B}=A_{\widetilde{N}}=B^\infty$, *and* $\widetilde{S}=B\cap\widehat{B}$, *which takes the form*

$$\widetilde{S}=\{\{\widetilde{f},\widetilde{g}\}\in A_{\widetilde{N}}\mid \exists h\in\mathbb{C}:(\widetilde{g}(\zeta)-\zeta\widetilde{f}(\zeta))=h\begin{pmatrix}1 & -1\end{pmatrix}^\top\}.$$

(ii) *Corollary* 4.5 *becomes: For* $\alpha \in \mathbb{R}$,

$$(B^{\langle\alpha\rangle} - z)^{-1} = (A_{\widetilde{N}} - z)^{-1} - \frac{1}{N_0(z) + q(z) - \alpha}\, \Gamma_{\widetilde{N}z} \begin{pmatrix} 1 & 1 \\ 1 & 1 \end{pmatrix} E_z,$$

where $(A_{\widetilde{N}} - z)^{-1}$ *is the difference-quotient operator in the space* $\mathcal{L}(\widetilde{N})$,

$$\Gamma_{\widetilde{N}z} = \operatorname{diag}\{K_{N_0}(\,\cdot\,, z^*), K_q(\,\cdot\,, z^*)\},$$

and $E_z = (\Gamma_{\widetilde{N}z^*})^*$ *is point evaluation at* z.

5. A decomposition of $\mathcal{L}(\widetilde{N})$

In this section we choose a basis in $\mathcal{L}(q) \oplus \mathcal{L}(M)$ and determine the associated Gram matrix $\widetilde{G}$; see Lemma 5.1 below. In the next section we identify $\mathcal{L}(q) \oplus \mathcal{L}(M)$ with $\mathbb{C}^n \oplus \mathbb{C}^m \oplus \mathbb{C}^m$ equipped with an inner product determined by $\widetilde{G}$ and exhibit the matrix representations of the operators $B^{\langle\alpha\rangle}$ and the relation $\widehat{B}$. As in [12] we define in $\mathcal{L}(\widetilde{N})$ the linearly independent elements

$$v_j = (B - z_0^*)^{j-1}\widetilde{w}, \quad j = 1, \ldots, m, m+1, \ldots, m+n,$$

and with $\varphi(z) = \varphi(\,\cdot\,, z) = K_{\widetilde{N}}(\,\cdot\,, z^*)v(z)$

$$u_j = (\widehat{B} - z_0)^{-j+1}\varphi(z_0) = \frac{1}{(j-1)!}\left(\frac{d}{dz}\right)^{j-1}\varphi(z)\,|_{z=z_0}, \quad j = 1, \ldots, m.$$

Here, on account of (4.6), $u_1 = \mathbf{V}^{-1}u$, where u is given by (4.5). Moreover, we introduce the three subspaces

$$\mathcal{L}^0 = \operatorname{span}\{v_1, \ldots, v_m\},\ \mathcal{L}' = \operatorname{span}\{v_{m+1}, \ldots, v_{m+n}\},\ \mathcal{M} = \operatorname{span}\{u_1, \ldots, u_m\}.$$

According to Theorem 2.1, the root space of $\widehat{B}$ at ∞ is the direct sum $\mathcal{L}' \dot{+} \mathcal{L}^0$ and $\mathcal{L}^0$ is its isotropic part.

Lemma 5.1. *We have* $\mathcal{L}(q) = \mathcal{L}'$ *and* $\mathcal{L}(M) = \mathcal{L}^0 \dot{+} \mathcal{M}$. *The basis elements for* $\mathcal{L}(q)$ *and* $\mathcal{L}(M)$ *can be written more explicitly as*

$$v_{m+j}(\zeta) = \begin{pmatrix} 0 & (\zeta - z_0^*)^{j-1} & 0 & 0 \end{pmatrix}^\top, \quad j = 1, \ldots, n,$$

and

$$\begin{aligned} v_j(\zeta) &= \begin{pmatrix} 0 & 0 & (\zeta - z_0^*)^{j-1} & 0 \end{pmatrix}^\top, & j &= 1, \ldots, m, \\ u_j(\zeta) &= \begin{pmatrix} 0 & 0 & R_{z_0}^j\, p_0(\zeta) & (\zeta - z_0)^{m-j} \end{pmatrix}^\top, & j &= 1, \ldots, m, \end{aligned}$$

where R_z *stands for the difference-quotient operator at* z. *The Gram matrix associated with this basis for the space* $\mathcal{L}(q) \oplus (\mathcal{L}^0 \dot{+} \mathcal{M})$ *is given by*

$$\widetilde{G} = \begin{pmatrix} G_q & 0 & 0 \\ 0 & 0 & I_{\mathbb{C}^m} \\ 0 & I_{\mathbb{C}^m} & G_{p_0} \end{pmatrix},$$

in which $G_q = (q_{i,j})_{i,j=1}^n$ *has entries*

$$\begin{aligned} q_{i,j} &= \langle v_{m+j}, v_{m+i}\rangle_{\mathcal{L}(\tilde{N})} \\ &= \sum_{k=0}^{m+j-1}\sum_{l=0}^{m+i-1}\binom{m+j-1}{k}\binom{m+i-1}{l}(-z_0^*)^{m+j-1-k}(-z_0)^{m+i-1-l}s_{k+l}, \end{aligned}$$

where s_{k+l} *are the moments of* N *in* (2.2), *and* $G_{p_0} = (p_{i,j})_{i,j=1}^m$ *has entries*

$$\begin{aligned} p_{i,j} &= \langle u_j, u_i\rangle_{\mathcal{L}(\tilde{N})} \\ &= \frac{1}{(j-1)!}\frac{1}{(i-1)!}\left(\frac{d}{dz}\right)^{j-1}\left(\frac{d}{dw^*}\right)^{i-1}\frac{p_0(z)-p_0(w^*)}{z-w^*}\bigg|_{z=w=z_0}. \end{aligned} \tag{5.1}$$

The formulas for the basis element $u_j(\zeta)$ and the entry $p_{i,j}$ of the Gram matrix G_{p_0} in this lemma are independent of the way the polynomial $p_0(z)$ is written. If we write $p_0(\zeta) = \sum_{k=0}^{\ell_0} p_k(\zeta - z_0)^k$ and set $p_k = 0$ if $k > \ell_0$, then $R_{z_0}^j\, p_0(\zeta) = \sum_{k=j}^{\ell_0} p_k(\zeta - z_0)^{k-j}$ and straightforward calculations yield

$$p_{i,j} = \sum_{k=i}^{2m-j}\binom{k-1}{i-1}p_{k+j-1}(z_0^* - z_0)^{k-i};$$

so, in particular, if $z_0 \in \mathbb{R}$ then $p_{i,j} = p_{i+j-1}$. After the proof of the lemma we give some other formulas for the Gram matrix G_q as well.

Proof of Lemma 5.1. Since $v_1 = \widetilde{w}$, the element v_1 is of the given form. We calculate $v_j = (B - z_0^*)v_{j-1}$ for $j = 2, \ldots, m+n$. Write Bv_{j-1} as

$$Bv_{j-1}(\zeta) = \begin{pmatrix} f(\zeta) & a(\zeta) & b(\zeta) & d(\zeta)\end{pmatrix}^\top \in \mathcal{L}(N_0)\oplus\mathcal{L}(q)\oplus\mathcal{L}(M).$$

Then, by (4.4), there exists a vector $\mathbf{h} = \begin{pmatrix} h_1 & h_2 & h_3 & h_4\end{pmatrix}^\top \in \mathbb{C}^4$ such that

$$\begin{pmatrix} f(\zeta) \\ a(\zeta) \\ b(\zeta) - \zeta(\zeta - z_0^*)^{j-2} \\ d(\zeta)\end{pmatrix} = \begin{pmatrix} h_1 - N_0(\zeta)h_4 \\ h_2 - q(\zeta)h_4 \\ -c^\#(\zeta)(h_1+h_2) - p_0(\zeta)h_3 \\ -c(\zeta)h_3 + h_4\end{pmatrix}. \tag{5.2}$$

By Lemma 4.3, $h_1 = h_4 = 0$. Since d is a polynomial of degree less than m also $h_3 = 0$. Thus b is the first component of an element in $\mathcal{L}(M)$ whose second component $d = 0$, therefore, see Lemma 4.2, the degree of b is less than m. The equality between the third components of the vectors in (5.2) now reads as

$$b(\zeta) - \zeta(\zeta - z_0^*)^{j-2} = -c^\#(\zeta)h_2. \tag{5.3}$$

For $2 \le j \le m$, a comparison of the degrees of the polynomials on both sides, yields $h_2 = 0$. Thus $b(\zeta) = (\zeta - z_0^*)^{j-1} + z_0^*(\zeta - z_0^*)^{j-2}$ and hence we have

$$v_j(\zeta) = \begin{pmatrix} 0 & 0 & (\zeta - z_0^*)^{j-1} & 0\end{pmatrix}^\top, \quad j = 1, \ldots, m.$$

If $j = m + 1$ then (5.3) implies $h_2 = 1$ and hence we find

$$v_{m+1}(\zeta) = \begin{pmatrix} 0 & 1 & 0 & 0 \end{pmatrix}^\top.$$

Now the formula for v_j, $j = m+2, \dots, n$, can be checked in a similar way as above. It is easy to see that the element $u_1 = K_{\widetilde{N}}(\,\cdot\,, z_0^*)v(z_0)$ has the stated form. By (4.12), we have for $2 \le j \le m$,

$$u_j(\zeta) = (\widehat{B} - z_0)^{-1} u_{j-1}(\zeta) = (A_{\widetilde{N}} - z_0)^{-1} \begin{pmatrix} 0 \\ 0 \\ R_{z_0}^{j-1} p_0(\zeta) \\ (\zeta - z_0)^{m-j+1} \end{pmatrix} = \begin{pmatrix} 0 \\ 0 \\ R_{z_0}^{j} p_0(\zeta) \\ (\zeta - z_0)^{m-j} \end{pmatrix}.$$

As to the Gram matrix $\widetilde{G}$, the zeros come from the facts that $\mathcal{L}(q) \perp \mathcal{L}(M)$ and $\mathcal{L}^0$ is neutral. The formula for G_q follows from expanding

$$q_{i,j} = \langle (B - z_0^*)^{m+j-1} \widetilde{w}, (B - z_0^*)^{m+i-1} \widetilde{w} \rangle_{\mathcal{L}(\widetilde{N})}$$

in terms of

$$\begin{aligned} &\langle B^l \widetilde{w}, B^k \widetilde{w} \rangle_{\mathcal{L}(\widetilde{N})} \\ &\quad = \langle (\widehat{A}_{\widehat{N}})^l 1\,, (\widehat{A}_N)^k\, 1 \rangle_{\mathcal{L}(\widehat{N})} = \langle A_N^l N, A_N^k N \rangle_{\mathcal{L}(N)} = \langle A^l w, A^k w \rangle_{\mathcal{P}} = s_{k+l}. \end{aligned}$$

The entries $I_{\mathbb{C}^m}$ in $\widetilde{G}$ are obtained from the reproducing kernel property of $K_{\widetilde{N}}(\zeta, z)$:

$$\begin{aligned} \langle u_j, v_i \rangle_{\mathcal{L}(\widetilde{N})} &= \frac{1}{(j-1)!} \left(\frac{d}{dz}\right)^{j-1} \langle K_{\widetilde{N}}(\,\cdot\,, z^*)v(z), v_i \rangle_{\mathcal{L}(\widetilde{N})} \Big|_{z=z_0} \\ &= \frac{1}{(j-1)!} \left(\frac{d}{dz}\right)^{j-1} (z - z_0)^{i-1} \Big|_{z=z_0} = \delta_{ij}, \quad 1 \le i, j \le m. \end{aligned}$$

Finally, the formula for $p_{i,j}$ in the lemma readily follows from

$$\begin{aligned} \langle u_j, u_i \rangle_{\mathcal{L}(\widetilde{N})} &= \frac{1}{(j-1)!} \frac{1}{(i-1)!} \\ &\times \left(\frac{d}{dz}\right)^{j-1} \left(\frac{d}{dw^*}\right)^{i-1} \langle K_{\widetilde{N}}(\,\cdot\,, z^*)v(z), K_{\widetilde{N}}(\,\cdot\,, w^*)v(w) \rangle_{\mathcal{L}(\widetilde{N})} \Big|_{z=w=z_0}. \end{aligned}$$ □

We claim that the Gram matrix $G_q = (\langle v_{m+j}, v_{m+i} \rangle_{\mathcal{L}(\widetilde{N})})_{i,j=1}^n$ in Lemma 5.1 is lower diagonal with respect to the second diagonal. To see this we use the equality

$$\langle v_{m+j}, v_{m+i} \rangle_{\mathcal{L}(\widetilde{N})} = \langle v_{m+j+1}, v_{m+i-1} \rangle_{\mathcal{L}(\widetilde{N})} + (z_0^* - z_0) \langle v_{m+j}, v_{m+i-1} \rangle_{\mathcal{L}(\widetilde{N})}, \tag{5.4}$$

which readily follows from the relation $v_{m+i} = (B - z_0^*) v_{m+i-1}$. Since $\mathcal{L}(q)$ is orthogonal to $\mathcal{L}^0$ the recurrence relation (5.4) implies

$$\langle v_{m+j}, v_{m+1} \rangle_{\mathcal{L}(\widetilde{N})} = 0, \quad j = 1, \dots, n-1,$$

and hence, again with (5.4), also the lower triangular form of G_q. Furthermore, (5.4) also implies that the entries on the second diagonal: $\langle v_{m+d}, v_{m+n+1-d} \rangle_{\mathcal{L}(\widetilde{N})}$

are independent of $d = 1, \ldots, n$. Note that G_q is not a Hankel matrix in general, however, it is if $z_0 = z_0^*$.

In the following two propositions we present two other formulas for G_q. The first one is in terms of the real coefficients τ_j of q:

$$q(z) = \tau_n z^n + \tau_{n-1} z^{n-1} + \cdots + \tau_1 z + \tau_0, \quad \tau_n \neq 0.$$

The Gram matrix S_q associated with the standard basis $\{1, \zeta, \ldots, \zeta^{n-1}\}$ in $\mathcal{L}(q)$ is equal to

$$S_q = \begin{pmatrix} \tau_1 & \tau_2 & \cdots & \tau_{n-1} & \tau_n \\ \tau_2 & \tau_3 & \cdots & \tau_n & 0 \\ \vdots & \vdots & & \vdots & \vdots \\ \tau_n & 0 & \cdots & 0 & 0 \end{pmatrix}^{-1} = \begin{pmatrix} 0 & 0 & \cdots & 0 & \sigma_{n-1} \\ 0 & 0 & \cdots & \sigma_{n-1} & \sigma_n \\ \vdots & \vdots & & \vdots & \vdots \\ \sigma_{n-1} & \sigma_n & \cdots & \sigma_{2n-3} & \sigma_{2n-2} \end{pmatrix}$$

where the real numbers σ_j are defined by the expansion

$$\frac{1}{q(z)} = \frac{\sigma_{n-1}}{z^n} + \cdots + \frac{\sigma_{2n-2}}{z^{2n-1}} + O\left(\frac{1}{z^{2n}}\right), \quad z = iy, \ y \uparrow \infty \tag{5.5}$$

(see, for example, [10, Theorem 3.4]). Notice that τ_0 does not play a role. Since G_q is the Gram matrix of the basis $v_{m+1}, v_{m+2}, \ldots, v_{m+n}$, we express this basis in terms of the standard basis via the $n \times n$ matrix $\mathbb{H}$:

$$\begin{pmatrix} 1 & \zeta - z_0^* & (\zeta - z_0^*)^2 & \ldots & (\zeta - z_0^*)^{n-1} \end{pmatrix} = \begin{pmatrix} 1 & \zeta & \zeta^2 & \ldots & \zeta^{n-1} \end{pmatrix} \mathbb{H}.$$

and then we have $G_q = \mathbb{H}^* S_q \mathbb{H}$. If $\mathbb{H} = (h_{i,j})_{i,j=1}^n$ then the entries in the jth column are given by

$$h_{i,j} = \begin{cases} \binom{j-1}{i-1} (-z_0^*)^{j-i}, & i = 1, \ldots j, \\ 0, & i = j+1, \ldots n. \end{cases}$$

The connection with the moments s_j for N given by (2.2) can be obtained from the fact that the asymptotics of $N(z) = -1/\widehat{N}(z)$ in Theorem 2.1(ii) is the same as the asymptotics of the function $-1/(c^{\#}(z)q(z)c(z))$: Write for $|z| > |z_0|$,

$$\frac{1}{c(z)} = \sum_{k=0}^{\infty} \frac{t_{m+k-1}}{z^{m+k}}, \quad t_{m+k-1} = z_0^k \binom{m+k-1}{m-1} \tag{5.6}$$

(note that $t_{m-1} = 1$) and set

$$\mathbb{T} = \begin{pmatrix} t_{m-1} & t_m & \cdots & t_{m+n-3} & t_{m+n-2} \\ 0 & t_{m-1} & \cdots & t_{m+n-4} & t_{m+n-3} \\ \vdots & \vdots & & \vdots & \vdots \\ 0 & 0 & \cdots & 0 & t_{m-1} \end{pmatrix}.$$

Using (5.5) and (5.6) to calculate the asymptotics of $-1/(c^\#(z)q(z)c(z))$ and comparing it with (2.2) we find that

$$M_N = \begin{pmatrix} 0 & 0 & \cdots & 0 & s_{2m+n-1} \\ 0 & 0 & \cdots & s_{2m+n-1} & s_{2m+n} \\ \vdots & \vdots & & \vdots & \vdots \\ s_{2m+n-1} & s_{2m+n} & \cdots & s_{2m+2n-3} & s_{2m+2n-2} \end{pmatrix} = \mathbb{T}^* S_q \mathbb{T}.$$

Hence we have proved the following proposition.

Proposition 5.2. *With $\mathbb{H}$, S_q, $\mathbb{T}$ and M_N as defined above we have*

$$G_q = \mathbb{H}^*\mathbb{T}^{-*} M_N \mathbb{T}^{-1}\mathbb{H} = \mathbb{H}^* S_q \mathbb{H}.$$

The first equality follows from the formula for G_q given in Lemma 5.1. The triangular forms of the matrices $\mathbb{H}$ with 1 on the diagonal and S_q with $\sigma_{n-1} = 1/\tau_n$ on the second diagonal yield the triangular form of G_q with $1/\tau_n$ on the second diagonal.

To derive yet another formula for G_q, we identify the elements $v_{m+j} \in \mathcal{L}(\widetilde{N})$ with the functions $(\zeta - z_0^*)^{j-1} \in \mathcal{L}(q)$, $j = 1, \ldots, n$. Also for later use, we introduce the vector polynomial

$$\mathbf{s}_q(z) = (s_i(z))_{i=1}^n, \quad s_i(z) = R_{z_0^*}^i\, q(z), \tag{5.7}$$

where R_z is the difference-quotient operator at z. The kernel $K_q(\,\cdot\,, z)$ can be expressed in the basis $\{v_{m+1}, \ldots, v_{m+n}\}$ as

$$K_q(\,\cdot\,, z) = \sum_{i=1}^n v_{m+i}\, s_i(z^*). \tag{5.8}$$

For this, write q as $q(z) = \sum\limits_{k=0}^{n} q_k(z-z_0^*)^k$, then $s_i(z) = \sum_{k=i}^n q_k(z-z_0^*)^{k-i}$ and hence

$$\begin{aligned} K_q(\zeta, z) &= \sum_{k=1}^n q_k \frac{(\zeta - z_0^*)^k - (z^* - z_0^*)^k}{\zeta - z^*} = \sum_{k=1}^n q_k \sum_{i=1}^k (\zeta - z_0^*)^{i-1}(z^* - z_0^*)^{k-i} \\ &= \sum_{i=1}^n (\zeta - z_0^*)^{i-1} \sum_{k=i}^n q_k (z^* - z_0^*)^{k-i} = \sum_{i=1}^n v_{m+i}(\zeta)\, s_i(z^*). \end{aligned}$$

For the next proposition, we choose n distinct points $z_1, \ldots, z_n \in \mathbb{C}$, denote by $\mathbb{V}$ the $n \times n$ Vandermonde matrix

$$\mathbb{V} = \begin{pmatrix} 1 & z_1 - z_0^* & \ldots & (z_1 - z_0^*)^{n-1} \\ 1 & z_2 - z_0^* & \ldots & (z_2 - z_0^*)^{n-1} \\ \vdots & \vdots & & \vdots \\ 1 & z_n - z_0^* & \ldots & (z_n - z_0^*)^{n-1} \end{pmatrix},$$

and define the $n \times n$ matrix $\mathbb{S} = (s_{i,j})_{i,j=1}^n$ by $s_{i,j} = s_i(z_j^*)$.

Proposition 5.3. *With the above notation $G_q = \mathbb{S}^{-*}\mathbb{V}$.*

Proof. If $\mathbb{S}^{-1} = (t_{i,j})_{i,j=1}^n$, then on account of (5.8), $v_{m+i} = \sum_{k=1}^n K_q(\,\cdot\,, z_k)t_{k,i}$, $i, j = 1, \dots, n$, and by the reproducing kernel property,

$$\begin{aligned}\langle v_{m+j}, v_{m+i}\rangle_{\mathcal{L}(q)} &= \sum_{k=1}^n t_{k,i}^* \langle v_{m+j}, K_q(\,\cdot\,, z_k)\rangle_{\mathcal{L}(q)} = \sum_{k=1}^n t_{k,i}^* v_{m+j}(z_k) \\ &= \sum_{k=1}^n (\mathbb{S}^{-*})_{i,k}(z_k - z_0^*)^{j-1} = (\mathbb{S}^{-*}\mathbb{V})_{i,j}. \qquad \square\end{aligned}$$

6. Minimal models in the space $(\mathcal{K}; G)$

In this section we construct minimal models for the functions N and $\widehat{N}$ in the orthogonal sum $\mathcal{K} = \mathcal{L}(N_0) \oplus \mathbb{C}^n \oplus \mathbb{C}^m \oplus \mathbb{C}^m$, where $\mathcal{L}(N_0)$ is the reproducing kernel Hilbert space associated with the Nevanlinna function N_0 in the representation (4.1) of $\widehat{N}$. The inner product on $\mathbb{C}^m$ will be denoted by $(\mathbf{x}, \mathbf{y})_m = \mathbf{y}^*\mathbf{x}$, $\mathbf{x}, \mathbf{y} \in \mathbb{C}^m$; the index m in the inner product will be omitted when it is clear from the context. We denote by $(\mathcal{K}; G)$ the linear space $\mathcal{K}$ equipped with the indefinite inner product $\langle G\,\cdot\,, \,\cdot\,\rangle_{\mathcal{K}}$ defined by the Gram matrix

$$G = \begin{pmatrix} I_{\mathcal{L}(N_0)} & 0 & 0 & 0 \\ 0 & G_q & 0 & 0 \\ 0 & 0 & 0 & I_{\mathbb{C}^m} \\ 0 & 0 & I_{\mathbb{C}^m} & G_{p_0} \end{pmatrix}, \tag{6.1}$$

where G_q and G_{p_0} are given in Lemma 5.1.

Because of Lemma 4.3 we have $N_0 \notin \mathcal{L}(N_0)$. But since the element

$$R_{w_0}N_0 = K_{N_0}(\,\cdot\,, w_0^*), \quad w_0 \in \mathcal{D}(N_0),$$

belongs to $\mathcal{L}(N_0)$, we see that N_0 is a generalized element belonging to the space $\mathcal{L}(N_0)_{-1}$ defined in the Introduction. Thus the pairing $\langle f_0, N_0\rangle$ between an element $f_0 \in \operatorname{dom} A_{N_0}$ and N_0 is well defined:

$$\langle f_0, N_0\rangle = \langle (A_{N_0} - w_0)f_0, K_{N_0}(\,\cdot\,, w_0)\rangle_{\mathcal{L}(N_0)} = g_0(w_0) - w_0 f_0(w_0), \tag{6.2}$$

where $g_0 = A_{N_0}f_0$ and the right-hand side is independent of $w_0 \in \mathcal{D}(N_0)$. In this connection we write χ_{-1} for $N_0 \in \mathcal{L}(N_0)_{-1}$ and we define

$$\varphi_0(z) = (A_{N_0} - z)^{-1}\chi_{-1} = K_{N_0}(\,\cdot\,, z^*), \quad z \in \mathcal{D}(N_0).$$

By Theorem 3.1, the triplet $(A_{N_0}, \varphi_0(z), S_{N_0})$ is a minimal model for N_0 in $\mathcal{L}(N_0)$.

To keep the formulation of the next theorem short, we introduce the following notation. We write

$$q(z) = \sum_{j=0}^n q_j(z - z_0^*)^j, \quad p_0(z) = \sum_{j=0}^{\ell_0} p_k(z - z_0)^j, \tag{6.3}$$

set $p_k = 0$ for $k > \ell_0$, and define the column vectors

$$\mathbf{q} = \begin{pmatrix} q_0 \cdots q_{n-1}\end{pmatrix}^\top \in \mathbb{C}^n, \quad \mathbf{p} = \begin{pmatrix} p_0 \cdots p_{m-1}\end{pmatrix}^* \in \mathbb{C}^m.$$

We use the column vector polynomials

$\mathbf{s}_q(z)=(s_j(z))_{j=1}^n$, $\mathbf{t}_{p_0}(z)=(t_j(z))_{j=1}^m$, $\mathbf{r}_{1p_0}(z)=(r_{1j}(z))_{j=1}^m$, $\mathbf{r}_{2p_0}(z)=(r_{1j}(z))_{j=1}^m$.

Here $\mathbf{s}_q(z)$ is as in (5.7), $t_j(z) = R_{z_0}^j p_0(z)$, and the entries of the last two vectors are the coefficients in the expansions

$$R_z R_{z_0}^m p_0(\zeta) = \sum_{j=1}^m r_{1j}(z)(\zeta - z_0^*)^{j-1},$$

and

$$\sum_{j=m+1}^{\ell_0} R_{z_0}^j p_0(\zeta)(z-z_0)^{j-1} = \sum_{j=1}^m r_{2j}(z)(\zeta - z_0^*)^{j-1}.$$

Furthermore, we write

$$\mathbf{b}(z) = \begin{pmatrix}(z-z_0^*)^{m-1} & (z-z_0^*)^{m-2} & \cdots & 1\end{pmatrix}^\top,$$
$$\mathbf{d}(z) = \begin{pmatrix}1 & (z-z_0) & \cdots & (z-z_0)^{m-1}\end{pmatrix}^\top.$$

We denote by $J_m(z_0)$ the $m \times m$ Jordan block matrix at z_0 with $J_m(z_0)\mathbf{e}_{m,1} = z_0\mathbf{e}_{m,1}$, where for $j = 1, 2, \ldots, m$, $\mathbf{e}_{m,j}$ stands for the jth element in the standard orthogonal basis of $\mathbb{C}^m$.

Theorem 6.1. *Assume the conditions* (1)–(4).

(i) *For $\alpha \in \mathbb{R}$, let $C^{\langle\alpha\rangle}$ be the set of all pairs of the form*

$$\left\{\begin{pmatrix} f_0+\lambda\chi_0 \\ \mathbf{a} \\ \mathbf{b} \\ \mathbf{d}\end{pmatrix}, \begin{pmatrix} g_0 + w_0\lambda\chi_0 \\ -\langle f_0,\chi_{-1}\rangle\mathbf{e}_{n,1} - \lambda(N_0(w_0)\mathbf{e}_{n,1}+\mathbf{q}) + J_n(z_0)^*\mathbf{a} + (\mathbf{b},\mathbf{e}_{m,m})\mathbf{e}_{n,1} \\ -\lambda G_{p_0}\mathbf{e}_{m,m} + J_m(z_0)^*\mathbf{b} + (\mathbf{d}, \alpha\mathbf{e}_{m,1} - \mathbf{p})\mathbf{e}_{m,1} \\ \lambda\mathbf{e}_{m,m} + J_m(z_0)\mathbf{d}\end{pmatrix}\right\},$$

with $\chi_0 = \varphi_0(w_0)$, $\{f_0, g_0\} \in A_{N_0}$, $\lambda \in \mathbb{C}$, $\mathbf{a} \in \mathbb{C}^n$ such that $a_n = \lambda q_n$, and $\mathbf{b}$, $\mathbf{d} \in \mathbb{C}^m$, where w_0 is a fixed point in $\mathcal{D}(N_0)$.
Then $C^{\langle\alpha\rangle}$ is the graph of a self-adjoint operator (also denoted by $C^{\langle\alpha\rangle}$) in the space $(\mathcal{K}; G)$.

(ii) *Let $\widehat{C}$ be the set of all pairs of the form*

$$\left\{\begin{pmatrix} f_0+\lambda\chi_0 \\ \mathbf{a} \\ \mathbf{b} \\ \mathbf{d}\end{pmatrix}, \begin{pmatrix} g_0 + w_0\lambda\chi_0 \\ -\langle f_0,\chi_{-1}\rangle\mathbf{e}_{n,1} - \lambda(N_0(w_0)\mathbf{e}_{n,1}+\mathbf{q}) + J_n(z_0)^*\mathbf{a} + (\mathbf{b},\mathbf{e}_{m,m})\mathbf{e}_{n,1} \\ -\lambda G_{p_0}\mathbf{e}_{m,m} + J_m(z_0)^*\mathbf{b} + \mu\mathbf{e}_{m,1} \\ \lambda\mathbf{e}_{m,m} + J_m(z_0)\mathbf{d}\end{pmatrix}\right\},$$

with $\chi_0 = \varphi_0(w_0)$, $\{f_0, g_0\} \in A_{N_0}$, $\lambda, \mu \in \mathbb{C}$, $\mathbf{a} \in \mathbb{C}^n$ such that $a_n = \lambda q_n$, $\mathbf{b} \in \mathbb{C}^m$, and $\mathbf{d} \in \mathbb{C}^m$ such that $d_1 = 0$, where w_0 is a fixed point in $\mathcal{D}(N_0)$.
Then $\widehat{C}$ is a self-adjoint relation in $(\mathcal{K}; G)$.

(iii) *The minimal models of N and $\widehat{N}$ in the space $(\mathcal{K}; G)$ are given by the triplets*

$$(C,\ N(z)\Gamma_z,\ S) \quad \textit{and} \quad (\widehat{C},\ \Gamma_z,\ S),$$

where $C = C^{\langle 0 \rangle}$, $S = C \cap \widehat{C}$, *and*

$$\Gamma_z = \begin{pmatrix} \varphi_0(z)c(z) \\ \mathbf{s}_q(z)c(z) \\ \mathbf{r}_{2p_0}(z) + \mathbf{b}(z)c(z)(N_0(z) + q(z)) \\ \mathbf{d}(z) \end{pmatrix}.$$

(iv) *The family of all self-adjoint extensions of* S *in* $(\mathcal{K}; G)$ *is given by* $C^{\langle \alpha \rangle}$, $\alpha \in \mathbb{R}$, *and* $\widehat{C}$. *Moreover,* $\widehat{C} = C^{\infty}$, *the limit in the resolvent sense of* $C^{\langle \alpha \rangle}$ *as* $\alpha \to \infty$.

Note that $C^{\langle \alpha \rangle}$ is of the form (2.3):

$$C^{\langle \alpha \rangle} = C + \alpha \left\langle \cdot, \begin{pmatrix} 0 \\ 0 \\ \mathbf{e}_{m,1} \\ 0 \end{pmatrix} \right\rangle \begin{pmatrix} 0 \\ 0 \\ \mathbf{e}_{m,1} \\ 0 \end{pmatrix}.$$

Note also $\widehat{C}$ is multi-valued: $\begin{pmatrix} 0 & 0 & \mathbf{e}_{m,1} & 0 \end{pmatrix}^{\top} \in \widehat{C}(0)$, S can also be written as

$$S = C^{\langle \alpha \rangle} \cap \left\{ \left\{ \begin{pmatrix} 0 \\ 0 \\ 0 \\ 0 \end{pmatrix}, \begin{pmatrix} 0 \\ 0 \\ \mathbf{e}_{m,1} \\ 0 \end{pmatrix} \right\} \right\}^{*},$$

and that

$$C^{\langle \alpha \rangle} = S + \operatorname{span} \left\{ \left\{ \begin{pmatrix} 0 \\ 0 \\ 0 \\ \mathbf{e}_{m,1} \end{pmatrix}, \begin{pmatrix} 0 \\ 0 \\ (\alpha - \widehat{N}_0(z_0))\mathbf{e}_{m,1} \\ z_0 \mathbf{e}_{m,1} \end{pmatrix} \right\} \right\},$$

in particular, the domain of $C^{\langle \alpha \rangle}$ is dense and independent of $\alpha \in \mathbb{R}$.

In the proof of Theorem 6.1 we identify – according to the basis discussed in Section 5 – the space $\mathcal{L}(\widetilde{N})$ with $\mathcal{K} = \mathcal{L}(N_0) \oplus \mathbb{C}^n \oplus \mathbb{C}^m \oplus \mathbb{C}^m$ and the relations $B^{\langle \alpha \rangle}$, $\widehat{B}$, and $\widetilde{S}$ with $C^{\langle \alpha \rangle}$, $\widehat{C}$, and S defined in the theorem. To explain the identification, let the vector function $\widetilde{f} \in \mathcal{L}(\widetilde{N})$ be given as

$$\widetilde{f}(\zeta) = \begin{pmatrix} f(\zeta) & a(\zeta) & b(\zeta) & d(\zeta) \end{pmatrix}^{\top},$$

where

$$\mathcal{L}(q) \ni a(\zeta) = \sum_{j=1}^{n} a_j \big(v_{m+j}(\zeta)\big)_2 = \sum_{j=1}^{n} a_j (\zeta - z_0^*)^{j-1} \tag{6.4}$$

and

$$\mathcal{L}^0 \dot{+} \mathcal{M} \ni \begin{pmatrix} b(\zeta) \\ d(\zeta) \end{pmatrix}$$

$$= \sum_{j=1}^{m} b_j \big(v_j(\zeta)\big)_{3,4} + d_j \big(u_j(\zeta)\big)_{3,4} = \begin{pmatrix} \sum\limits_{j=1}^{m} d_j R_{z_0}^j p_0(\zeta) + b_j (\zeta - z_0^*)^{j-1} \\ \sum\limits_{j=1}^{m} d_j (\zeta - z_0)^{m-j} \end{pmatrix}.$$

Then $\widetilde{f}$ will be identified with the element

$$\begin{pmatrix} f \\ \mathbf{a} \\ \mathbf{b} \\ \mathbf{d} \end{pmatrix} \in \mathcal{K},$$

where $\mathbf{a} = \begin{pmatrix} a_1 & \dots & a_n \end{pmatrix}^\top$, etc. We write $\widetilde{f} \simeq \begin{pmatrix} f & \mathbf{a} & \mathbf{b} & \mathbf{d} \end{pmatrix}^\top$. For example, for $\widetilde{w}$ in (4.10) we have $\widetilde{w} \simeq \begin{pmatrix} 0 & 0 & \mathbf{e}_{m,1} & 0 \end{pmatrix}^\top$.

Proof of Theorem 6.1. Identify the vector functions $\widetilde{f}$, $\widetilde{g} \in \mathcal{L}(\widetilde{N})$ given by

$$\widetilde{f}(\zeta) = \begin{pmatrix} f(\zeta) & a_1(\zeta) & b_1(\zeta) & d_1(\zeta) \end{pmatrix}^\top, \quad \widetilde{g}(\zeta) = \begin{pmatrix} g(\zeta) & a_2(\zeta) & b_2(\zeta) & d_2(\zeta) \end{pmatrix}^\top$$

with the elements

$$\widetilde{f} \simeq \begin{pmatrix} f & \mathbf{a}_1 & \mathbf{b}_1 & \mathbf{d}_1 \end{pmatrix}^\top, \quad \widetilde{g} \simeq \begin{pmatrix} g & \mathbf{a}_2 & \mathbf{b}_2 & \mathbf{d}_2 \end{pmatrix}^\top$$

in $\mathcal{K}$. Here for $i = 1, 2$, the entries of the vectors $\mathbf{a}_i = \begin{pmatrix} a_{i,1} & \dots & a_{i,n} \end{pmatrix}^\top$, etc., appear as coefficients in the representations

$$a_i(\zeta) = \sum_{j=1}^{n} a_{i,j} (\zeta - z_0^*)^{j-1}, \quad \begin{pmatrix} b_i(\zeta) \\ d_i(\zeta) \end{pmatrix} = \begin{pmatrix} \sum\limits_{j=1}^{m} \Big(d_{i,j} R_{z_0}^j p_0(\zeta) + b_{i,j} (\zeta - z_0^*)^{j-1} \Big) \\ \sum\limits_{j=1}^{m} d_{i,j} (\zeta - z_0)^{m-j} \end{pmatrix}.$$

According to Theorem 4.1 we have

$$\{\widetilde{f}(\zeta), \widetilde{g}(\zeta)\} \in B^{\langle \alpha \rangle} \iff \exists \begin{pmatrix} h_1 & h_2 & h_3 & h_4 \end{pmatrix}^\top \in \mathbb{C}^4 :$$

$$\widetilde{g}(\zeta) - \zeta \widetilde{f}(\zeta) = \begin{pmatrix} h_1 - N_0(\zeta) h_4 \\ h_2 - q(\zeta) h_4 \\ -c^\#(\zeta)(h_1 + h_2) + (\alpha - p_0(\zeta)) h_3 \\ -c(\zeta) h_3 + h_4 \end{pmatrix}. \tag{6.5}$$

Comparison of the fourth components on both sides of this equality yields $h_3 = d_{1,1}$ and

$$d_{2,j} = z_0\, d_{1,j} + d_{1,j+1}, \; j = 1, \dots, m-1; \quad d_{2,m} = z_0\, d_{1,m} + h_4.$$

In the same way the second components in (6.5) yield $h_4 = a_{1,n}/q_n$ and

$$a_{2,1} = z_0^* a_{1,1} + h_2 - q_0 \frac{a_{1,n}}{q_n},$$

$$a_{2,j} = z_0^* a_{1,j} + a_{1,j-1} - q_{j-1} \frac{a_{1,n}}{q_n}, \qquad j = 2, \ldots, n.$$

We now consider the third components in (6.5). Inserting the expressions for $d_{2,j}$ already obtained we find that

$$\sum_{j=1}^{m} d_{2,j} \sum_{k=j}^{\ell_0} p_k (\zeta - z_0)^{k-j} - \sum_{j=1}^{m} \sum_{k=j}^{\ell_0} d_{1,j} p_k \Big((\zeta - z_0)^{k-j+1} + z_0 (\zeta - z_0)^{k-j} \Big)$$

reduces to

$$\sum_{j=1}^{m} d_{1,j} p_{j-1} - h_3 p_0(\zeta) + h_4 \sum_{k=m}^{\ell_0} p_k (\zeta - z_0)^{k-m}.$$

In the last term we may replace ℓ_0 by $2m-1$, because $\ell_0 < 2m$ and $p_k = 0$ if $k > \ell_0$. Using the relation (5.1) the last term can now be rewritten as $h_4 \sum_{j=1}^{m} \langle u_m, u_j \rangle_{\mathcal{L}(\widetilde{N})} (\zeta - z_0^*)^{j-1}$. Then in the third components there only remain powers of $\zeta - z_0^*$ and comparing these we find $h_1 + h_2 = b_{1,m}$ and

$$b_{2,1} = z_0^* b_{1,1} + \alpha d_{1,1} - \sum_{j=1}^{m} p_{j-1} d_{1,j} - h_4 \langle u_m, u_1 \rangle_{\mathcal{L}(\widetilde{N})},$$

$$b_{2,j} = z_0^* b_{1,j} + b_{1,j-1} - h_4 \langle u_m, u_j \rangle_{\mathcal{L}(\widetilde{N})}, \quad j = 2, \ldots, m.$$

If we rewrite $N_0(\zeta)$ in the first component of (6.5) as $(\zeta - w_0) K_{N_0}(\zeta, w_0^*) + N_0(w_0)$ we find

$$\underbrace{\big(g(\zeta) - w_0 h_4 K_{N_0}(\zeta, w_0^*) \big)}_{=: g_0(\zeta)} - \zeta \underbrace{\big(f(\zeta) - h_4 K_{N_0}(\zeta, w_0^*) \big)}_{=: f_0(\zeta)} = h_1 - h_4 N_0(w_0).$$

So $\{f_0, g_0\} \in A_{N_0}$ and with $\lambda = a_{1,n}/q_n = h_4$ we have

$$\left\{ \begin{array}{lcl} f & = & f_0 + \lambda K_{N_0}(\,\cdot\,, w_0^*), \\ g & = & g_0 + w_0 \lambda K_{N_0}(\,\cdot\,, w_0^*). \end{array} \right. \tag{6.6}$$

Because $N_0 \notin \mathcal{L}(N_0)$, we have that $K_{N_0}(\,\cdot\,, w_0^*) \notin \operatorname{dom} A_{N_0}$, hence the decomposition (6.6) is unique. Furthermore, we have that $h_1 = \langle f_0, N_0 \rangle + \lambda N_0(w_0)$. Indeed, since $\{f_0, g_0\} \in A_{N_0}$, the difference $g_0(\zeta) - \zeta f_0(\zeta)$ is identically equal to a constant and hence, on account of (6.2),

$$\begin{aligned} h_1 - \lambda N_0(w_0) &= h_1 - h_4 N_0(w_0) \\ &= g_0(\zeta) - \zeta f_0(\zeta) = g_0(w_0) - w_0 f_0(w_0) = \langle f_0, N_0 \rangle. \end{aligned}$$

Hence $h_2 = b_{1,m} - h_1 = b_{1,m} - \langle f_0, N_0 \rangle - \lambda N_0(w_0)$. Together these formulas show that $\{\widetilde{f}(\zeta), \widetilde{g}(\zeta)\} \in B^{\langle \alpha \rangle}$ can be identified with a pair of elements in $\mathcal{K}$ of the form described in the theorem. Hence under the identification $B^{\langle \alpha \rangle}$ coincides with $C^{\langle \alpha \rangle}$.

(ii) That $\widehat{C}$ can be identified with $\widehat{B}$ can be proved in a similar way and therefore the details are omitted.
(iii) In the identification between $\mathcal{L}(\widetilde{N})$ and $\mathcal{K}$, the Γ-field $K_{\widetilde{N}}(\,\cdot\,,z^*)v(z)$ in Theorem 4.1 coincides with the Γ-field Γ_z in (iii). $\square$

Example. Consider $\widehat{N}(z) = (z^2+1)N_0(z) + \gamma_3 z^3 + \gamma_2 z^2 + \gamma_1 z + \gamma_0$, where N_0 is a Nevanlinna function satisfying (1.3) and the γ_j's are real numbers with $\gamma_3 \neq 0$. We rewrite $\widehat{N}$ in the form (4.1) and (6.3):

$$\widehat{N}(z) = (z+i)\{N_0(z) + q_1(z+i) + q_0\}(z-i) + p_1(z-i) + p_0$$

with $q_1 = \gamma_3$, $q_0 = \gamma_2 - i\gamma_3$, $p_1 = \gamma_1 - \gamma_3$ and $p_0 = \widehat{N}(i) = \gamma_0 - \gamma_2 + i(\gamma_1 - \gamma_3)$. Then $m = 1$, $n = 1$, $\widehat{N} \in \mathcal{N}_\kappa^\infty$, where $\kappa = 2$ if $\gamma_3 < 0$ and $\kappa = 1$ if $\gamma_3 > 0$. The state space for N and $\widehat{N}$ is $\mathcal{K} = \mathcal{L}(N_0) \oplus \mathbb{C} \oplus \mathbb{C}^2$ equipped with the indefinite inner product $\langle G\,\cdot\,,\,\cdot\,\rangle_{\mathcal{K}}$ in which the Gram matrix is given by

$$G = \operatorname{diag}\left(I_{\mathcal{L}(N_0)},\, \frac{1}{\gamma_3},\, \begin{pmatrix} 0 & 1 \\ 1 & \gamma_1 - \gamma_3 \end{pmatrix}\right).$$

We take $w_0 = i$, and recall $\chi_0(z) = K_{N_0}(\,\cdot\,,z^*) \in \mathcal{L}(N_0)$ and $\chi_{-1} = N_0 \in \mathcal{L}(N_0)_{-1}$. We find that the graph of the self-adjoint operator $C^{\langle\alpha\rangle}$ is the set of all elements of the form

$$\left\{ \begin{pmatrix} f_0 + \lambda\chi_0 \\ \lambda\gamma_3 \\ b \\ d \end{pmatrix}, \begin{pmatrix} g_0 + i\lambda\chi_0 \\ -\langle f_0, \chi_{-1}\rangle - \lambda(N_0(i) + \gamma_2) + b \\ -\lambda(\gamma_1 - \gamma_3) - ib + (\alpha - \widehat{N}(i))d \\ \lambda + id \end{pmatrix} \right\}$$

with $\{f_0, g_0\} \in A_{N_0}$, $\lambda, b, d \in \mathbb{C}$. The self-adjoint relation $\widehat{C}$ is the set of all elements of the form

$$\left\{ \begin{pmatrix} f_0 + \lambda\chi_0 \\ \lambda\gamma_3 \\ b \\ 0 \end{pmatrix}, \begin{pmatrix} g_0 + i\lambda\chi_0 \\ -\langle f_0, \chi_{-1}\rangle - \lambda(N_0(i) + \gamma_2) + b \\ \mu \\ \lambda \end{pmatrix} \right\}$$

with $\{f_0, g_0\} \in A_{N_0}$, $\lambda, b, \mu \in \mathbb{C}$.

From Corollary 4.5 we obtain the following theorem. We set $S_1(z; z_0) = 0$ and if $n \geq 2$,

$$S_n(z; z_0) = \begin{pmatrix} 0 & 1 & (z-z_0) & \dots & (z-z_0)^{n-2} \\ & \ddots & \ddots & & \vdots \\ & & \ddots & \ddots & (z-z_0) \\ & & & \ddots & 1 \\ & & & & 0 \end{pmatrix}$$

and we recall that the definitions of the matrix functions $K_\alpha(z)$ and $\widehat{K}(z)$ are given just before Corollary 4.5.

Theorem 6.2. *The resolvents of the operators* $C^{\langle\alpha\rangle}$, $\alpha \in \mathbb{R}$, *and the relation* $\widehat{C}$ *in* $\mathcal{K}$ *are given by*

$$(C^{\langle\alpha\rangle}-z)^{-1} = \operatorname{diag}\left((A_{N_0}-z)^{-1},\ S_n(z;z_0^*),\ \begin{pmatrix} S_m(z;z_0^*) & \mathbf{r}_{1p_0}(z)\mathbf{b}^{\#}(z) \\ 0 & S_m^{\#}(z;z_0^*)\end{pmatrix}\right)$$
$$+\frac{N(z)}{1+\alpha N(z)}\operatorname{diag}\left(\varphi_0(z), \mathbf{s}_q(z), \begin{pmatrix}\mathbf{r}_{2p_0}(z) & \mathbf{b}(z)\\ \mathbf{d}(z) & 0\end{pmatrix}\right) K_\alpha(z)$$
$$\times \operatorname{diag}\left(\langle\,\cdot\,,\varphi_0(z^*)\rangle_{\mathcal{L}(N_0)},\ \mathbf{d}^{\#}(z),\ \begin{pmatrix}\mathbf{d}^{\#}(z) & \mathbf{t}_{p_0}(z)^\top \\ 0 & \mathbf{b}^{\#}(z)\end{pmatrix}\right),$$

where $\varphi_0(z) = K_{N_0}(\,\cdot\,,z^*)$, *and*

$$(\widehat{C}-z)^{-1} = \operatorname{diag}\left((A_{N_0}-z)^{-1},\ S_n(z;z_0^*),\ \begin{pmatrix} S_m(z;z_0^*) & \mathbf{r}_{1p_0}(z)\mathbf{b}^{\#}(z) \\ 0 & S_m^{\#}(z;z_0^*)\end{pmatrix}\right)$$
$$+\operatorname{diag}\left(\varphi_0(z), \mathbf{s}_q(z), \begin{pmatrix}\mathbf{r}_{2p_0}(z) & \mathbf{b}(z)\\ \mathbf{d}(z) & 0\end{pmatrix}\right) \widehat{K}(z)$$
$$\times \operatorname{diag}\left(\langle\,\cdot\,,\varphi_0(z^*)\rangle_{\mathcal{L}(N_0)},\ \mathbf{d}^{\#}(z),\ \begin{pmatrix}\mathbf{d}^{\#}(z) & \mathbf{t}_{p_0}(z)^\top \\ 0 & \mathbf{b}^{\#}(z)\end{pmatrix}\right).$$

As to the proof of the theorem we only mention the following identifications between the elements and operators in $\mathcal{L}(q)$ and $\mathbb{C}^n$:

$$\begin{array}{ccccc} K_q(\,\cdot\,,z^*) & \simeq & \mathbf{s}_q(z) & = & (J_n(z_0)^*-z)^{-1}(\boldsymbol{q}-q(z)\boldsymbol{e}_{n,1}), \\ \langle\,\cdot\,,K_q(\,\cdot\,,z)\rangle & \simeq & (G_q\,\cdot\,,\mathbf{s}_q(z^*)) & = & \mathbf{d}^{\#}(z), \\ (A_q-z)^{-1} & \simeq & S_n(z;z_0^*) & = & (I+(z_0^*-z)J_n(0))^{-1}J_n(0). \end{array}$$

The first identification follows from (5.7) and (5.8) which show that $\mathbf{s}_q(z)$ is the vector representation of the element $K_q(\,\cdot\,,z^*)\in\mathcal{L}(q)$ relative to the basis $\{v_{m+i}\}_{i=1}^n$. The linear functional $\langle\,\cdot\,,K_q(\,\cdot\,,z^*)\rangle_{\mathcal{L}(q)}$ on $\mathcal{L}(q)$ can be identified with $(G_q\,\cdot\,,s_q(z^*)) = \mathbf{d}^{\#}(z)$ viewed as a mapping from $\mathbb{C}^n$ to $\mathbb{C}$. This also follows from (6.4). In the theorem $\mathbf{r}_{1p_0}(z)\mathbf{b}^{\#}(z)$ is an $m\times m$ matrix polynomial of rank 1. Also, the matrix

$$\begin{pmatrix}\mathbf{r}_{2p_0}(z) & \mathbf{b}(z)\\ \mathbf{d}(z) & 0\end{pmatrix}$$

whose entries are column m-vectors should be viewed as a mapping from $\mathbb{C}^2$ to $\mathbb{C}^{2m}$, whereas the matrix

$$\begin{pmatrix}\mathbf{d}^{\#}(z) & \mathbf{t}_{p_0}(z)^\top \\ 0 & \mathbf{b}^{\#}(z)\end{pmatrix}$$

whose entries are row m-vectors should be seen as a mapping from $\mathbb{C}^{2m}$ to $\mathbb{C}^2$. The matrices are related via the formula

$$\begin{pmatrix}\mathbf{r}_{2p_0}(z) & \mathbf{b}(z)\\ \mathbf{d}(z) & 0\end{pmatrix}^{\#}\begin{pmatrix}0 & I_{\mathbb{C}^m}\\ I_{\mathbb{C}^m} & G_{p_0}\end{pmatrix} = \begin{pmatrix}\mathbf{d}^{\#}(z) & \mathbf{t}_{p_0}(z)^\top \\ 0 & b^{\#}(z)\end{pmatrix},$$

which corresponds to a part of the identity $\Gamma^*_{\widetilde{N}z^*} = E_z$.

We now consider the analogs of Theorem 6.1 in the cases $n = 0$ and $m = 0$. *The case $n = 0$ and $m > 0$:* Here $\mathcal{K} = \mathcal{L}(N_0) \oplus \mathbb{C}^m \oplus \mathbb{C}^m$ and the Gram matrix takes the form

$$G = \begin{pmatrix} I_{\mathcal{L}(N_0)} & 0 & 0 \\ 0 & 0 & I_{\mathbb{C}^m} \\ 0 & I_{\mathbb{C}^m} & G_{p_0} \end{pmatrix}.$$

Theorem 6.3. *Assume $n = 0$ and $m > 0$. Then Theorem 6.1 holds if we delete the second component in all 4-vectors in the formulas, omit in (i) and (ii) the statement "$\mathbf{a} \in \mathbb{C}^n$ such that $a_n = \lambda q_n$", add in (i) and (ii) the statement "$\mathfrak{b}_m = \langle f_0, \chi_{-1}\rangle + \lambda(N_0(w_0) + q_0)$", and set $q(z) = q_0$ in Γ_z in (iii).*

The case $n > 0$ and $m = 0$: Now $\mathcal{K} = \mathcal{L}(N_0) \oplus \mathbb{C}^n$ and $G = \operatorname{diag}\{I_{\mathcal{L}(N_0)}, G_q\}$.

Theorem 6.4. *Assume $n > 0$ and $m = 0$. Then Theorem 6.1 holds if $C^{\langle\alpha\rangle}$ is the set of all pairs of the form*

$$\left\{ \begin{pmatrix} f_0 + \lambda\chi_0 \\ \mathbf{a} \end{pmatrix}, \begin{pmatrix} g_0 + w_0\lambda\chi_0 \\ -\langle f_0, \chi_{-1}\rangle \mathbf{e}_{n,1} - \lambda(N_0(w_0)\mathbf{e}_{n,1} + \mathbf{q} + \alpha\mathbf{e}_{n,1}) + J_n(z_0)^*\mathbf{a} \end{pmatrix} \right\} \tag{6.7}$$

with $\chi_0 = \varphi_0(w_0)$, $\{f_0, g_0\} \in A_{N_0}$, $\lambda \in \mathbb{C}$, and $\mathbf{a} \in \mathbb{C}^n$ such that $a_n = \lambda q_n$, where w_0 is a fixed point in $\mathcal{D}(N_0)$, if

$$\widehat{C} = A_{N_0} \oplus \{\{\mathbf{a}, J_n(z_0)^*\mathbf{a} + \mu\mathbf{e}_{n,1}\} \mid \mathbf{a} \in \mathbb{C}^n, a_n = 0, \mu \in \mathbb{C}\},$$

and if $\Gamma_z = \big(\varphi_0(z), \mathbf{s}_q(z)\big)^\top$.

Next we give the formulas for the compressions of the resolvents $(C^{\langle\alpha\rangle} - z)^{-1}$, $\alpha \in \mathbb{R}$, and $(\widehat{C} - z)^{-1}$ to the subspaces $\mathcal{L}(N_0)$ and $\mathcal{L}(N_0) \oplus \mathbb{C}^n$ of $(\mathcal{K}; G)$ in the case $n > 0$ and $m > 0$; similar formulas can be obtained in the other two cases. We denote by P_0 and P_1 the orthogonal projections in $(\mathcal{K}; G)$ onto $\mathcal{L}(N_0)$ and $\mathcal{L}(N_0) \oplus \mathbb{C}^n$.

Theorem 6.5.

(i) *For $\alpha \in \mathbb{R}$,*

$$P_0(C^{\langle\alpha\rangle} - z)^{-1}|_{\mathcal{L}(N_0)} = (A_{N_0} - z)^{-1} - \frac{1}{N_0(z) + T_\alpha(z)}\langle\,\cdot\,, \varphi_0(z^*)\rangle_{\mathcal{L}(N_0)}\varphi_0(z)$$

with parameter $T_\alpha(z) = q(z) + \dfrac{p_0(z) - \alpha}{c^\#(z)c(z)}$, and for $\alpha = \infty$,

$$P_0(\widehat{C} - z)^{-1}|_{\mathcal{L}(N_0)} = (A_{N_0} - z)^{-1}.$$

(ii) *For $\alpha \in \mathbb{R}$,*

$$P_1(C^{\langle\alpha\rangle} - z)^{-1}|_{\mathcal{L}(N_0)\oplus\mathbb{C}^n} = \begin{pmatrix} (A_{N_0} - z)^{-1} & 0 \\ 0 & S_n(z; z_0^*) \end{pmatrix}$$

$$- \frac{1}{N_0(z) + T_\alpha(z)} \begin{pmatrix} \langle\,\cdot\,, \varphi_0(z^*)\rangle_{\mathcal{L}(N_0)}\varphi_0(z) & \varphi_0(z)\mathbf{d}^\#(z) \\ \langle\,\cdot\,, \varphi_0(z^*)\rangle_{\mathcal{L}(N_0)}\mathbf{s}_q(z) & \mathbf{s}_q(z)\mathbf{d}^\#(z) \end{pmatrix},$$

and for $\alpha = \infty$,

$$P_1(C^\infty - z)^{-1}|_{\mathcal{L}(N_0)\oplus\mathbb{C}^n} = \begin{pmatrix} (A_{N_0} - z)^{-1} & 0 \\ 0 & S_n(z; z_0^*) \end{pmatrix}.$$

The resolvent formula

$$(A_\tau - z)^{-1} = (A_{N_0} - z)^{-1} - \frac{1}{N_0(z) + \tau}\langle\,\cdot\,, \varphi_0(z^*)\rangle\varphi_0(z)$$

describes the family of all self-adjoint extensions A_τ of S_{N_0} in $\mathcal{L}(N_0)$ in terms of the parameter $\tau \in \mathbb{R} \cup \{\infty\}$. If the number τ is replaced by a generalized Nevanlinna function $\tau(z)$ the formula describes the minimal self-adjoint extensions which act in spaces containing $\mathcal{L}(N_0)$. This is called Krein's resolvent formula. In part (i) the parameter describing $C^{\langle\alpha\rangle}$ with $\alpha \in \mathbb{R}$ is explicitly given by $\tau(z) = T_\alpha(z) \in \mathcal{N}_\kappa$. Related to Krein's formula here are the references [21, Theorem 4.7], [13, Theorem 4.2], and [7, Theorem 3.5].

It is of interest to compare the compressed resolvents of $C^{\langle\alpha\rangle}$ and $\widehat{C}$ with the compressions of these operators/relation themselves. Recall Stenger's lemma (see [15, Theorem 3.3 and a remark after the theorem]) that if A is a self-adjoint relation with $\rho(A) \neq \emptyset$ in a Pontryagin space $\widetilde{\mathcal{H}}$ and $\mathcal{H}$ is a Hilbert or Pontryagin subspace of $\widetilde{\mathcal{H}}$, such that $\dim \widetilde{\mathcal{H}} \ominus \mathcal{H} < \infty$, then the compression of A to $\mathcal{H}$, that is, the linear relation

$$P_{\mathcal{H}}A\,|_{\mathcal{H}} = \{\{f, P_{\mathcal{H}}g\} \mid \{f, g\} \in A,\ f \in \mathcal{H}\},$$

where $P_{\mathcal{H}}$ is the orthogonal projection in $\widetilde{\mathcal{H}}$ onto $\mathcal{H}$, is self-adjoint in $\mathcal{H}$. In our case $\mathcal{L}(N_0)$ is a Hilbert subspace and $\mathcal{L}(N_0) \oplus \mathbb{C}^n$ is a Pontryagin subspace of the Pontryagin space $\mathcal{K}$ and both have a finite codimension. Therefore the compressions just mentioned are self-adjoint. The following theorem follows directly from Theorem 6.1 and its versions for the special cases $n = 0$ or $m = 0$.

Theorem 6.6.

(i) *If* $n > 0$ *and* $m \geq 0$, *then*

$$P_0 C^{\langle\alpha\rangle}|_{\mathcal{L}(N_0)} = P_0\widehat{C}|_{\mathcal{L}(N_0)} = A_{N_0},$$

and if $n = 0$ *and* $m > 0$, *then (in graph notation)*

$$\begin{aligned} P_0 C^{\langle\alpha\rangle}|_{\mathcal{L}(N_0)} &= P_0\widehat{C}|_{\mathcal{L}(N_0)} \\ &= \{\{f_0 + \lambda\chi_0, g_0 + w_0\lambda\chi_0\} | \{f_0, g_0\} \in A_{N_0}, \langle f_0, \chi_{-1}\rangle + \lambda(N_0(w_0) + q_0) = 0\}. \end{aligned}$$

(ii) *If* $n > 0$ *and* $m > 0$, *then*

$$P_1 C^{\langle\alpha\rangle}|_{\mathcal{L}(N_0)\oplus\mathbb{C}^n} = P_1\widehat{C}|_{\mathcal{L}(N_0)\oplus\mathbb{C}^n}$$

and their graphs coincide with the set of all pairs of the form (6.7) *with* $\alpha = 0$.

7. Block operator matrix models in the space $(\mathcal{K}; H)$

Changing the basis we have considered in the previous sections we can write $C^{\langle\alpha\rangle}$ and $\widehat{C}$ in Theorem 6.1 with $w_0 = z_0^*$ in a block operator matrix form, which is not possible with the basis used so far. Set

$$T = \begin{pmatrix} I_{\mathcal{L}(N_0)} & \frac{1}{q_n}(\,\cdot\,,\mathbf{e}_{n,n})\chi_0 & 0 & 0 \\ 0 & I_{\mathbb{C}^n} & 0 & 0 \\ 0 & 0 & I_{\mathbb{C}^m} & 0 \\ 0 & 0 & 0 & I_{\mathbb{C}^m} \end{pmatrix},$$

where $\chi_0 = \varphi_0(z_0^*)$, and define the operators $D^{\langle\alpha\rangle} = T^{-1}C^{\langle\alpha\rangle}T$, $\alpha \in \mathbb{R}$, the linear relation $\widehat{D} = T^{-1}\widehat{C}T$, and the Gram matrix

$$H = T^*GT = \begin{pmatrix} I_{\mathcal{L}(N_0)} & \frac{1}{q_n}(\,\cdot\,,\mathbf{e}_{n,n})\chi_0 & 0 & 0 \\ \frac{1}{q_n}\langle\,\cdot\,,\chi_0\rangle_{\mathcal{L}(N_0)}\mathbf{e}_{n,n} & \frac{h_0}{q_n^2}(\,\cdot\,,\mathbf{e}_{n,n})\mathbf{e}_{n,n} + G_q & 0 & 0 \\ 0 & 0 & 0 & I_{\mathbb{C}^m} \\ 0 & 0 & I_{\mathbf{C}^m} & G_{p_0} \end{pmatrix},$$

where $h_0 = \langle\chi_0,\chi_0\rangle_{\mathcal{L}(N_0)} = K_{N_0}(z_0,z_0)$. Since $T^{-1}\widetilde{w} = \widetilde{w}$, we have

$$D^{\langle\alpha\rangle} = D + \alpha\langle H\,\cdot\,,\widetilde{w}\rangle_{\mathcal{K}}\widetilde{w}, \quad D = T^{-1}CT.$$

The space $\mathcal{K}$ equipped with the indefinite inner product $\langle H\,\cdot\,,\,\cdot\,\rangle_{\mathcal{K}}$ will be denoted by $(\mathcal{K}; H)$. Clearly, $D^{\langle\alpha\rangle}$ and $\widehat{D}$ are self-adjoint in $(\mathcal{K}; H)$. The relation $\widehat{D}$ can be obtained via infinite coupling of D and $\widetilde{w}$, that is, as limit of $D^{\langle\alpha\rangle}$ in the resolvent sense by letting $\alpha \to \infty$. The following theorem shows that $D^{\langle\alpha\rangle}$ and $\widehat{D}$ can be expressed by means of block operator matrices. We use the notation explained directly above Theorem 6.1.

Theorem 7.1. *Let $D^{\langle\alpha\rangle}$, $\alpha \in \mathbb{R}$, and $\widehat{D}$ be as defined above. Then:*

(i) $\operatorname{dom} D^{\langle\alpha\rangle} = (\operatorname{dom} A_{N_0}) \oplus \mathbb{C}^n \oplus \mathbb{C}^m \oplus \mathbb{C}^m$ *and on this domain $D^{\langle\alpha\rangle}$ has the block matrix form*

$$D^{\langle\alpha\rangle} = \begin{pmatrix} A_{N_0} & D_{12} & 0 & 0 \\ D_{21} & D_{22} & (\,\cdot\,,\mathbf{e}_{m,m})\mathbf{e}_{n,1} & 0 \\ 0 & -\frac{1}{q_n}(\,\cdot\,,\mathbf{e}_{n,n})G_q\mathbf{e}_{m,m} & J_m(z_0)^* & (\,\cdot\,,\alpha\mathbf{e}_{m,1} - \mathbf{p})\mathbf{e}_{m,1} \\ 0 & \frac{1}{q_n}(\,\cdot\,,\mathbf{e}_{n,n})\mathbf{e}_{m,m} & 0 & J_m(z_0) \end{pmatrix},$$

where with $\chi_0 = \varphi_0(z_0)$

$$D_{12} = \Big(\frac{q_{n-1}}{q_n^2}(\,\cdot\,,\mathbf{e}_{n,n}) - \frac{1}{q_n}(\,\cdot\,,\mathbf{e}_{n,n-1})\Big)\chi_0, \quad D_{21} = -\langle\,\cdot\,,\chi_{-1}\rangle\mathbf{e}_{n,1},$$

and

$$D_{22} = J_n(z_0)^* - \frac{1}{q_n}(\,\cdot\,, \mathbf{e}_{n,n})\big(N_0(z_0^*)\mathbf{e}_{n,1} + \mathbf{q}\big).$$

(ii) $\operatorname{dom} \widehat{D} = (\operatorname{dom} A_{N_0}) \oplus \mathbb{C}^n \oplus \mathbb{C}^m \oplus (\mathbb{C}^m \ominus \{\mathbf{e}_{m,1}\})$ *and*

$$\widehat{D} = \{\{\widetilde{f}, \widetilde{g}\} | \widetilde{f} \in \operatorname{dom} \widehat{D},\ \exists \mu \in \mathbb{C} : \widetilde{g} = D\widetilde{f} + \begin{pmatrix} 0 & 0 & \mu\mathbf{e}_{m,1} & 0 \end{pmatrix}^\top\},$$

where $D = D^{\langle 0 \rangle}$.

(iii) *The triplets* $(D, N(z)\widetilde{\Gamma}_z, \widetilde{S})$, $(\widehat{D}, \widetilde{\Gamma}_z, \widetilde{S})$ *are minimal models for* N *and* $\widehat{N}$ *in* $(\mathcal{K}; H)$, *where* $\widetilde{S} = D \cap \widehat{D}$ *and*

$$\widetilde{\Gamma}_z = \begin{pmatrix} (\varphi_0(z) - \varphi_0(z_0))c(z) \\ \mathbf{s}_q(z)c(z) \\ \mathbf{r}_{2p_0}(z) + \mathbf{b}(z)c(z)(N_0(z) + q(z)) \\ \mathbf{d}(z) \end{pmatrix}.$$

The theorem follows directly from Theorem 6.1 with $w_0 = z_0^*$, the definitions of $D^{\langle \alpha \rangle}$ and $\widehat{D}$, and $\widetilde{\Gamma}_z = T^{-1}\Gamma_z$. Note that $\widetilde{\Gamma}_{z_0} = \begin{pmatrix} 0 & 0 & 0 & \mathbf{e}_{m,1} \end{pmatrix}^\top$.

From the theorems in Section 6 one can easily obtain formulas for the resolvents, the compressions of the resolvent and the compressions of $D^{\langle \alpha \rangle}$ and $\widehat{D}$. We leave the details to the reader.

8. Examples

We give two examples and discuss an approximation problem taken from [11], [17], and [14] to which we refer for details and proofs. They are related to the Bessel differential expression

$$\ell_\nu\, y(x) = -y''(x) + \frac{\nu^2 - 1/4}{x^2} y(x)$$

on $(0, 1]$ with a self-adjoint boundary condition at the regular endpoint $x = 1$ and on $(0, \infty)$, which is limit point at $x = \infty$. We recall the series expansion of the Bessel function

$$J_\nu(z) = \left(\frac{z}{2}\right)^\nu \sum_{k=0}^{\infty} \frac{(-1)^k}{k!\Gamma(k + \nu + 1)} \left(\frac{z}{2}\right)^{2k}, \tag{8.1}$$

where the series on the right converges absolutely, and uniformly in any bounded domain of z and ν. We denote by

$$\begin{aligned} Y_\nu(z) &= \frac{1}{\sin(\nu\pi)}(J_\nu(z)\cos(\nu\pi) - J_{-\nu}(z)), \\ H_\nu^{(1)}(z) &= \frac{1}{i\sin(\nu\pi)}(J_{-\nu}(z) - J_\nu(z)\mathrm{e}^{-i\nu\pi}), \\ K_\nu(z) &= i\tfrac{\pi}{2}\mathrm{e}^{i\frac{\pi}{2}\nu} H_\nu^{(1)}(iz) \end{aligned}$$

the Neumann function of order ν, the first Hankel function of order ν, and the Basset (or MacDonald) function of order ν, respectively; see, for example, [18].

Example. See [11]. We consider ℓ_ν on the interval $(0,1]$ and impose the boundary condition $y(1)=0$ at the regular endpoint $x=1$ on all functions y.
(i) First assume $0<\nu<1$. Then the minimal realization S of ℓ_ν in $\mathcal{H}_0=L^2(0,1)$ is a symmetric operator with defect indices $(1,1)$. We denote by A_0 the self-adjoint operator extension of S with graph

$$A_0=\{\{y,\ell_\nu y\}\in S^*|\lim_{x\downarrow 0} x^{\nu-1/2}y(x)=0\}. \tag{8.2}$$

The function

$$\widehat{\varphi}(x,z)=-\frac{\pi}{2\sin\pi\nu}z^{\nu/2}x^{1/2}\left(\frac{J_{-\nu}(\sqrt{z})}{J_\nu(\sqrt{z})}J_\nu(x\sqrt{z})-J_{-\nu}(x\sqrt{z})\right) \tag{8.3}$$

belongs to $\ker(S^*-z)$ and is a defect function for S and A_0 with corresponding Q-function

$$\widehat{N}(z)=-\frac{\pi}{2\sin\pi\nu}z^\nu\frac{J_{-\nu}(\sqrt{z})}{J_\nu(\sqrt{z})}. \tag{8.4}$$

Thus the following relations hold:

$$\widehat{\varphi}(z)=(I+(z-z_0)(A_0-z)^{-1})\widehat{\varphi}(z_0),\qquad \frac{\widehat{N}(z)-\widehat{N}(w)^*}{z-w^*}=\langle\widehat{\varphi}(z),\widehat{\varphi}(w)\rangle_0. \tag{8.5}$$

(ii) Now assume $\nu>1$, $\nu\neq 2,3,\ldots$. Then the results are quite different from those in (i): The minimal realization of ℓ_ν in $\mathcal{H}_0$ is self-adjoint, the function $\widehat{\varphi}(\,\cdot\,,z)$ in (8.3) is well defined but it does not belong to $\mathcal{H}_0$, and the function $\widehat{N}$ in (8.4) is now a generalized Nevanlinna function with $\kappa=[(\nu+1)/2]$ negative squares. Thus the model for this function involves a self-adjoint operator or relation in a Pontryagin space with κ negative squares. In [11] we show that $\widehat{N}\in\mathcal{N}_\kappa^\infty$: Let z_n, $n=1,2\ldots$, be the enumeration of the zeros of the function $z^{-\nu/2}J_\nu(\sqrt{z})$ in increasing order. (The zeros are positive and, since the function is entire, countable.) Then $\widehat{N}$ admits the decomposition

$$\widehat{N}(z)=z^{2\kappa}(N_0(z)+q_0)+p_0(z),$$

where

$$N_0(z)=\sum_{n=1}^{\infty}\left(\frac{1}{z_n-z}-\frac{z_n}{z_n^2+1}\right)\frac{2z_n^{\nu-2\kappa}}{J_\nu'(z_n)^2},$$

$$q_0=\frac{1}{(2\kappa)!}\widehat{N}^{(2\kappa)}(0)-\sum_{n=1}^{\infty}\frac{2z_n^{\nu-2\kappa-1}}{(z_n^2+1)J_\nu'(z_n)^2},$$

and

$$p_0(z)=\sum_{j=0}^{2\kappa-1}p_j\,z^j,\quad p_j=\frac{1}{j!}\widehat{N}^{(j)}(0).$$

The Nevanlinna function N_0 satisfies the relations (1.3) and hence $\widehat{N}\in\mathcal{N}_\kappa^\infty$. Theorem 6.3 with $m=\kappa>0$, $z_0=0$, and $G_{p_0}=(p_{i+j-1})_{i,j=1}^{\kappa}$ yields the description of the models for $\widehat{N}$ and $N=-1/\widehat{N}$.

Example. See [17]. We now consider ℓ_ν on $(0,\infty)$. The endpoint $x=\infty$ is limit point so we do not need to impose a condition at this endpoint. Where possible we use the same notation as in the previous example.
(i) First assume $0<\nu<1$. Then the minimal realization S of ℓ_v in $\mathcal{H}_0 = L^2(0,\infty)$ is a symmetric operator with defect indices $(1,1)$. The self-adjoint extension A_0 of S defined by formula (8.2) is uniquely determined by the facts that its spectrum $\sigma(A_0)=[0,\infty)$ is absolutely continuous and that the functions $y(x,\lambda)=c(\lambda)x^{1/2}J_\nu(x\sqrt{\lambda})$, $\lambda\in[0,\infty)$, form a complete set of generalized eigenfunctions of A_0, where $c(\lambda)$ is some normalizing factor. The function

$$\widehat{\varphi}(x,z)=\sqrt{x}(-z)^{\frac{\nu}{2}}K_\nu(x\sqrt{-z}) \tag{8.6}$$

belongs to $\ker(S^*-z)$ and is a defect function for S and A_0 with corresponding Q-function

$$\widehat{N}(z)=-\frac{\pi}{2\sin\pi\nu}(-z)^\nu. \tag{8.7}$$

Thus the relations (8.5) are also valid in this case. It follows from (8.7) that $\widehat{N}$ is a Nevanlinna function, which satisfies the limit conditions in (1.3).
(ii) Now assume $\nu>1$ and $\nu\neq 2,3,\ldots$. Then, as in the previous example, the minimal realization of ℓ_ν in $\mathcal{H}_0$ is self-adjoint, the function $\widehat{\varphi}(\,\cdot\,,z)$ in (8.6) does not belong to $\mathcal{H}_0$, and the function $\widehat{N}$ in (8.7) is a generalized Nevanlinna function with $\kappa=[(\nu+1)/2]$ negative squares. Here the branch of $(-z)^\nu$ is chosen so that $(-z)^\nu=r^\nu\mathrm{e}^{i\nu(\theta-\pi)}$ if $z=r\mathrm{e}^{i\theta}$, $0<\theta<2\pi$. For any $z_0\in(-\infty,0)$ the function $\widehat{N}$ admits the decomposition

$$\widehat{N}(z)=(z-z_0)^{2\kappa}(N_0(z)+q_0)+p_0(z),$$

where

$$N_0(z)=\int_0^\infty\left(\frac{1}{t-z}-\frac{t}{t^2+1}\right)\frac{t^\nu}{2(t-z_0)^{2\kappa}}\,dt,\quad q_0=-\frac{\pi}{4\sin\frac{\pi\nu}{2}},$$

and

$$p_0(z)=\sum_{j=0}^{2\kappa-1}p_j(z-z_0)^j,\quad p_j=\frac{1}{j!}\widehat{N}^{(j)}(z_0)=\frac{\pi(-1)^{j+1}}{2\sin\pi\nu}\binom{\nu}{j}(-z_0)^{\nu-j}.$$

Since N_0 is a Nevanlinna function which satisfies the relations (1.3), we have that $\widehat{N}\in\mathcal{N}_\kappa^\infty$. Hence Theorem 6.3 with $m=\kappa>0$ applies and provides the description of the models for $\widehat{N}(z)$ and $N(z)=-1/\widehat{N}(z)$. Since z_0 is real, the Gram matrix G_{p_0} is given by $G_{p_0}=(p_{i+j-1})_{i,j=1}^\kappa$.

Inspired by [29] and [28], we discuss an approximation problem in $\mathcal{N}_\kappa^\infty$; for details we refer to the paper [14] in preparation. In the context of the discussion around (1.4), the problem is to approximate strongly singular perturbations by smoother perturbations. Consider a function $\widehat{N}\in\mathcal{N}_\kappa^\infty$ $(\kappa>0)$ with irreducible representation (4.1):

$$\widehat{N}(z)=(z-z_0)^{2\kappa}(N_0(z)+q_0)+p_0(z),$$

and a sequence of functions $\widehat{N}_j \in \mathcal{N}_\kappa^\infty$ with irreducible representation (4.1):

$$\widehat{N}_j(z) - N_{0j}(z) + q_j(z), \quad j = 1, 2, \dots,$$

where N_0 and all N_{0j} are Nevanlinna functions satisfying (1.3), z_0 is a real number belonging to their common domain $\mathcal{D}$ of holomorphy, $q_0 \in \mathbb{R}$, $p_0(z)$ and $q_j(z)$ are real polynomials with $\deg p_0 \leq 2\kappa - 1$, and $n = \deg q_j$ is either 2κ or $2\kappa \pm 1$: if n is odd and the leading coefficient of $q_j(z)$ is negative, then $n = 2\kappa - 1$; if n is odd and the leading coefficient of $q_j(z)$ is positive, then $n = 2\kappa + 1$. Assume that, as $j \to \infty$, $\widehat{N}_j$ converges to $\widehat{N}$ uniformly on compact subsets of $\mathcal{D}$. The approximation problem with variable spaces then is to describe this convergence in terms of the models of $\widehat{N}_j$ and $\widehat{N}$ and of the corresponding state spaces. We rewrite $\widehat{N}_j$ in the following form

$$\widehat{N}_j(z) = (z - z_0)^{2\kappa}(M_{0j}(z) + q_{0j}) + p_{0j}(z) + q_{j,2\kappa+1}(z - z_0)^{2\kappa+1}, \tag{8.8}$$

where $M_{0j}(z)$ is a Nevanlinna function, $p_{0j}(z)$ is a real polynomial with $\deg p_{0j} \leq 2\kappa - 1$, and $q_{0j}, q_{j,2\kappa+1} \in \mathbb{R}$ with $q_{j,2\kappa+1} \geq 0$ (if $q_{j,2\kappa+1} > 0$ then it is the leading coefficient of $q_j(z)$). The convergence assumption is equivalent to the convergence of $M_{0j}(z) + q_{0j}$ to $N_0(z) + q_0$ uniformly on compact subsets of $\mathcal{D}$, the pointwise convergence of the polynomials $p_{0j}(z)$ to $p_0(z)$, and the convergence $q_{j,2\kappa+1} \to 0$, as $j \to \infty$. The representation (8.8) of $\widehat{N}_j(z)$ need not be irreducible, and so models will have to be constructed, which fall outside the scope of this paper. Approximation of operators with variation of the space in which they act has been considered in [22, pp. 512, 513]; for such approximations in an indefinite setting, see [27] and [26].

The application we have in mind is related to ℓ_ν and the last example. In [14] we show that the function $\widehat{N}$ in (8.7) with $\nu > 1$, $\nu \neq 2, 3, \dots$, can be approximated by functions of the form

$$\widehat{N}^\delta(z) = N_0^\delta(z) + q^\delta(z) \tag{8.9}$$

by letting $\delta \downarrow 0$. Here $q^\delta(z)$ is some real polynomial of degree $[\nu]$ with coefficients depending on δ, which we will not further specify here, and the function N_0^δ is obtained as follows. Consider the family of regularized differential expressions

$$l_{\nu,\delta} y(x) = -y''(x) + \frac{\nu^2 - 1/4}{(x+\delta)^2} y(x)$$

on $(0, \infty)$, where the parameter δ varies over some interval $(0, \delta_0)$, $\delta_0 > 0$. Let S_δ be the minimal operator associated with $l_{\nu,\delta}$ in the Hilbert space $\mathcal{H}_0 = L^2(0, \infty)$; it is symmetric and its defect indices are $(1, 1)$. Each self-adjoint extension of S_δ can be obtained as the restriction of the maximal operator $S_\delta{}^*$ by the boundary condition $y'(0) = \alpha y(0)$ with $\alpha \in \mathbb{R} \cup \{\infty\}$. We denote by A_δ the extension corresponding to $\alpha = \infty$. The function

$$\varphi_\delta(x, z) = \gamma(x+\delta)^{1/2} \frac{K_\nu((x+\delta)\sqrt{-z})}{\delta^\nu K_\nu(\delta\sqrt{-z})}, \quad \gamma = 2^{\nu-1}\Gamma(\nu),$$

is a defect function for S_δ and A_δ. The function considered in (8.9) is by definition the function

$$N_0^\delta(z) = \gamma^2 \delta^{1-2\nu} \sqrt{-z}\, \frac{K_\nu'(\delta\sqrt{-z})}{K_\nu(\delta\sqrt{-z})}. \tag{8.10}$$

It satisfies the relation

$$\frac{N_0^\delta(z) - N_0^\delta(w)^*}{z - w^*} = \langle \varphi_\delta(z), \varphi_\delta(w)\rangle_0$$

and hence is a Q-function for S_δ and A_δ. It follows that N_0^δ is a Nevanlinna function with integral representation

$$N_0^\delta(z) = \int_0^\infty \left(\frac{1}{t-z} - \frac{t}{t^2+1}\right) d\sigma_\delta(t) + \operatorname{Re} N_0^\delta(i),$$

where, for $t \geq 0$,

$$d\sigma_\delta(t) = \frac{1}{\pi}\operatorname{Im} N_0^\delta(t+i0)\,dt = \frac{2\gamma^2}{\pi^2\delta^{2\nu}} \frac{1}{J_\nu^2(\delta\sqrt{t}) + Y_\nu^2(\delta\sqrt{t})}\,dt. \tag{8.11}$$

It will be shown (in [14]) that the function $\widehat{N}^\delta$ in (8.9) belongs to $\mathcal{N}_\kappa^\infty$ with $\kappa = [(\nu+1)/2]$ and, if $\delta \downarrow 0$, converges to $\widehat{N}$ in (8.7) uniformly on compact subsets of $\mathcal{D}(\widehat{N})$. Note that the representation (8.9) of $\widehat{N}^\delta$ is irreducible and corresponds to (4.1) with $m = 0$. This in contrast with the limit function $\widehat{N}$ whose irreducible representation corresponds to (4.1) with $m = \kappa > 0$.

We conclude the paper with a final remark.

Remark 8.1. From the beginning up to and including Section 7 we may replace the factor $c(z) = (z - z_0)^m$ by $c(z) = (z - z_1)\cdots(z - z_m)$ with $z_j \in \mathcal{D}(N_0)$ to obtain similar but more general models as in [13, Sections 6 and 7].

References

[1] S. Albeverio, F. Gesztesy, R. Høegh-Krohn, and H. Holden, *Solvable models in quantum mechanics*, Springer, 1988.

[2] S. Albeverio and P. Kurasov, *Singular perturbations of differential operators*, London Mathematical Society Lecture Notes 271, Cambridge Univ. Press, 2000.

[3] D. Alpay, A. Dijksma, and H. Langer, *Factorization of J-unitary matrix polynomials on the line and a Schur algorithm for generalized Nevanlinna functions*, Linear Algebra Appl. **387C** (2004), 313–342.

[4] D. Alpay, A. Dijksma, J. Rovnyak, and H. de Snoo, *Schur functions, operator colligations, and reproducing kernel Pontryagin spaces*, Operator Theory: Adv. Appl. **96**, Birkhäuser Verlag, Basel, 1997.

[5] V.A. Derkach and S. Hassi, *A reproducing kernel space model for N_κ-functions*, Proc. Amer. Math. Soc. **131** (12) (2003), 3795–3806.

[6] V. Derkach, S. Hassi, and H. de Snoo, *Operator models associated with singular perturbations*, Methods Funct. Anal. Topology **7** (3) (2001), 1–21.

[7] V. Derkach, S. Hassi, and H. de Snoo, *Singular perturbations of self-adjoint operators*, Mathematical Physics, Analysis and Geometry **6** (2003), 349–384.

[8] J.F. van Diejen and A. Tip, *Scattering from generalized point interaction using self-adjoint extensions in Pontryagin spaces*, J. Math. Phys. **32** (3) (1991), 630–641.

[9] A. Dijksma, H. Langer, A. Luger, and Yu. Shondin, *A factorization result for generalized Nevanlinna functions of the class* $\mathcal{N}_\kappa$, Integral Equations Operator Theory **36** (2000), 121–125.

[10] A. Dijksma, H. Langer, A. Luger, and Yu. Shondin, *Minimal realizations of scalar generalized Nevanlinna functions related to their basic factorization*, Operator Theory: Adv. Appl., **154**, Birkhäuser Verlag, Basel, 2004, 69–90.

[11] A. Dijksma, H. Langer, A. Luger, and Yu. Shondin, *Singular and regular point-like perturbations of the Bessel operator on* $(0, 1]$ *in a Pontryagin space*, in preparation.

[12] A. Dijksma, H. Langer, and Yu. Shondin, *Rank one perturbations at infinite coupling in Pontryagin spaces*, J. Funct. Anal. **209** (2004), 206–246.

[13] A. Dijksma, H. Langer, Y. Shondin, and C. Zeinstra, *Self-adjoint operators with inner singularities and Pontryagin spaces*, Operator Theory: Adv., Appl., **118**, Birkhäuser Verlag, Basel, 2000, 105–175.

[14] A. Dijksma, A. Luger, and Yu. Shondin, *Approximation in* $\mathcal{N}_\kappa^\infty$ *with variable state spaces* (tentative title), in preparation.

[15] A. Dijksma, H. Langer, and H.S.V. de Snoo, *Unitary colligations in* Π_κ*-spaces, characteristic functions and Straus extensions*, Pacific J. Math. **125** (2) (1986), 347–362.

[16] A. Dijksma, H. Langer, and H.S.V. de Snoo, *Eigenvalues and pole functions of Hamiltonian systems with eigenvalue depending boundary conditions*, Math. Nachr. **161** (1993), 107–154.

[17] A. Dijksma and Yu. Shondin, *Singular point-like perturbations of the Bessel operator in a Pontryagin space*, J. Differential Equations **164** (2000), 49–91.

[18] A. Erdélyi, *Higher transcendental functions*, vol. ii, Mcgraw-Hill, New York, 1953.

[19] C.J. Fewster, *Generalized point interactions for the radial Schrodinger equation via unitary dilation*, J. Phys. A: **28** (1995), 1107–1127.

[20] F. Gesztesy and B. Simon, *Rank one perturbation at infinite coupling*, J. Funct. Anal. **128** (1995), 245–252.

[21] S. Hassi, M. Kaltenbäck, and H.S.V. de Snoo, *The sum of matrix Nevanlinna functions and self-adjoint extensions in exit spaces*, Operator Theory: Adv., Appl., **103**, Birkhäuser Verlag, Basel, 1998, 137–154.

[22] T. Kato, *Perturbation theory for linear operators*, Die Grundlehren der mathematischen Wissenschaften, Band 132, Springer-Verlag, Heidelberg, 1966.

[23] M.G. Krein and H. Langer, *Über einige Fortsetzungsprobleme, die eng mit der Theorie hermitescher Operatoren im Raume* Π_κ *zusammenhängen. I. Einige Funktionenklassen und ihre Darstellungen*, Math. Nachr. **77** (1977), 187–236.

[24] V.I. Kruglov and B.S. Pavlov, *Zero-range potentials with inner structure: fitting parameters for resonance scattering*, Preprint arXiv.org: quant-ph/0306150.

[25] M. Langer and A. Luger, *Scalar generalized Nevanlinna functions: realizations with block operator matrices*, Operator Theory: Adv. Appl. **162**, Birkhäuser Verlag, Basel, 2005, 253–267.

[26] H. Langer and B. Najman, *Perturbation theory for definizable operators in Krein spaces*, J. Operator Theory **9**(1983), 297–317.

[27] B. Najman, *Perturbation theory for selfadjoint operators in Pontrjagin spaces*, Glasnik Mat. **15**(35) (1980), 351–371.

[28] Yu. Shondin, *On approximation of high order singular perturbations*, J. Phys. A: Math. Gen. **38**(2005), 5023–5039.

[29] O.Yu. Shvedov, *Approximations for strongly singular evolution equations*, J. Funct. Anal. **210**(2) (2004), 259–294.

Aad Dijksma
Department of Mathematics
University of Groningen
P.O. Box 800
9700 AV Groningen
The Netherlands
e-mail: dijksma@math.rug.nl

Annemarie Luger
Institute for Analysis and Scientific Computing
Vienna University of Technology
Wiedner Hauptstrasse 8–10
A-1040 Vienna
Austria
e-mail: aluger@mail.zserv.tuwien.ac.at

Yuri Shondin
Department of theoretical Physics
State Pedagogical University
Str. Ulyanova 1
Nizhny Novgorod
603950 Russia
e-mail: shondin@sinn.ru

Operator Theory:
Advances and Applications, Vol. 163, 135–145

Generalized Friedrichs Extensions Associated with Interface Conditions for Sturm-Liouville Operators

Andreas Fleige, Seppo Hassi, Henk de Snoo and Henrik Winkler

Dedicated to Heinz Langer on the occasion of his retirement

Abstract. For a class of Sturm-Liouville operators with an interface condition at an interior point all selfadjoint realizations are determined. This result is obtained via a description of the selfadjoint extensions of the coupling of two symmetric operators. The (generalized) Friedrichs extension, when it exists, is determined. Sufficient conditions for the (generalized) Friedrichs extension to exist are given.

Mathematics Subject Classification (2000). Primary 47A10, 47B25; Secondary 34B05, 34B24.

Keywords. Symmetric operator, selfadjoint extension, (generalized) Friedrichs extension, boundary triplet, Weyl function, interface condition.

1. Introduction

Let $-DpD+q$ be a Sturm-Liouville expression with real coefficients on the subset $[-b,0)\cup(0,b]$ of $\mathbb{R}$. It is assumed that there are fixed separated boundary conditions at $-b$ and b and that there is a selfadjoint interface condition at 0 of the form

$$(pu')(0+)=\tau(u(0+)-u(0-)),\quad (pu')(0+)=(pu')(0-),$$

when $\tau\in\mathbb{R}$, and of the form

$$u(0+)=u(0-),\quad (pu')(0+)=(pu')(0-),$$

when $\tau=\infty$. For $\tau\in\mathbb{R}\cup\{\infty\}$ these interface conditions describe all selfadjoint extensions of the symmetric operator which is given by the interface condition

$$u(0+)=u(0-),\quad (pu')(0+)=(pu')(0-)=0,$$

together with the fixed separated boundary conditions at $-b$ and b. The purpose of this note is to indicate conditions which show the existence of a generalized Friedrichs extension (cf. [5], [6]) and to determine the corresponding value of $\tau \in \mathbb{R} \cup \{\infty\}$ under mild conditions on the singularity at 0. Note that the conditions on the coefficients p and q at 0 can be further relaxed along the lines of, e.g., [10].

It is clear that interface conditions at an interior point can be easily described via the orthogonal coupling of symmetric operators. The selfadjoint extensions for such orthogonal couplings are best expressed in terms of the corresponding boundary triplets; see the extension results in [1]. The description of generalized Friedrichs extensions can also be given in terms of sesquilinear forms along the lines of [3], which will be done in [4].

2. Preliminaries

2.1. Boundary triplets

Let S be a closed symmetric operator (or relation) with equal defect numbers in a Hilbert space $\mathfrak{H}$ with inner product $(\cdot,\cdot)$. Recall that a relation in $\mathfrak{H}$ is a linear subspace of the Cartesian product $\mathfrak{H} \times \mathfrak{H}$, and its elements are denoted as ordered pairs. Let $\mathfrak{N}_\lambda = \ker(S^* - \lambda)$ be the defect subspace of S, and denote

$$\widehat{\mathfrak{N}}_\lambda := \{\, \widehat{f}_\lambda = \{f_\lambda, \lambda f_\lambda\} : f_\lambda \in \mathfrak{N}_\lambda \,\}, \quad \lambda \in \mathbb{C},$$

so that $\widehat{\mathfrak{N}}_\lambda \subset S^*$. A boundary triplet $\Pi = \{\mathcal{H}, \Gamma_0, \Gamma_1\}$ of S^* consists of a Hilbert space $\mathcal{H}$ and the boundary mappings Γ_j, $j = 0, 1$, from S^* to $\mathcal{H}$ such that $\Gamma : \widehat{f} \to \{\, \Gamma_0 \widehat{f}, \Gamma_1 \widehat{f} \,\}$ from S^* into $\mathcal{H} \times \mathcal{H}$ is surjective and such that the identity

$$(f', g) - (f, g') = \left(\Gamma_1 \widehat{f}, \Gamma_0 \widehat{g} \right)_{\mathcal{H}} - \left(\Gamma_0 \widehat{f}, \Gamma_1 \widehat{g} \right)_{\mathcal{H}} \tag{2.1}$$

holds for all $\widehat{f} = \{f, f'\}$, $\widehat{g} = \{g, g'\} \in S^*$. The mappings Γ_i define two selfadjoint extensions A_i of S via $A_i = \ker \Gamma_i$, $i = 0, 1$. The mapping $\Gamma : \widehat{f} \to \{\Gamma_0 \widehat{f}, \Gamma_1 \widehat{f}\}$ induces a one-to-one correspondence between the closed extensions H of S which are intermediate (i.e., which satisfy $S \subset H \subset S^*$) and the closed linear relations τ in $\mathcal{H}$, via

$$H := \{\, \widehat{f} \in S^* : \Gamma \widehat{f} \in -\tau^{-1} \,\}. \tag{2.2}$$

In particular, this correspondence (2.2) is one-to-one between all selfadjoint extensions H of S and all selfadjoint relations τ in $\mathcal{H}$, cf. [1], [2]. When S is densely defined, then often the mappings Γ_0 and Γ_1 are interpreted as being defined on $\operatorname{dom} S^*$ instead of on S^*, i.e., in that case one speaks of $\Gamma_0 f$ and $\Gamma_1 f$ when $\widehat{f} = \{f, f'\} \in S^*$. This convention will be followed in the present note. Associated to the boundary triplet Π are two operator functions: the Weyl function $M(\lambda)$, defined by

$$M(\lambda) = \{\, \{\Gamma_0 \widehat{f}_\lambda, \Gamma_1 \widehat{f}_\lambda\} : \widehat{f}_\lambda \in \widehat{\mathfrak{N}}_\lambda \,\} \quad \lambda \in \rho(A_0), \tag{2.3}$$

the graph of a bounded linear operator in $\mathcal{H}$, and the γ-field $\gamma(\lambda)$ defined by

$$\gamma(\lambda) = \{ \{\Gamma_0 \widehat{f}_\lambda, f_\lambda\} : \ \widehat{f}_\lambda \in \widehat{\mathfrak{N}}_\lambda \}, \quad \lambda \in \rho(A_0), \tag{2.4}$$

the graph of a bounded linear operator from $\mathcal{H}$ to $\mathfrak{N}_\lambda$. Both functions are holomorphic on $\rho(A_0)$. The relation between the Weyl function $M(\lambda)$ and the γ-field $\gamma(\lambda)$ is given by

$$\frac{M(\lambda) - M(\mu)^*}{\lambda - \bar{\mu}} = \gamma(\mu)^* \gamma(\lambda). \tag{2.5}$$

The identity (2.5) implies that $M(\lambda)$ is the Q-function of the pair $\{S, A_0\}$. Each Weyl function belongs to the class $\mathbf{N}$ of Nevanlinna functions, i.e., is holomorphic on $\mathbb{C} \setminus \mathbb{R}$, and satisfies $M(\lambda)^* = M(\bar{\lambda})$, and $\operatorname{Im} M(\lambda) \geq 0$ for $\lambda \in \mathbb{C}_+$. Moreover, $M(\lambda)$ is strict, i.e., $0 \in \rho(\operatorname{Im} M(\lambda))$ for all $\lambda \in \mathbb{C} \setminus \mathbb{R}$. In general, the Weyl function $M(\lambda)$ determines up to unitary isomorphisms, a model for the symmetric operator S and its selfadjoint extension A_0.

2.2. Generalized Friedrichs extensions

For the purposes of this paper it is sufficient to consider the case of a symmetric operator S with defect numbers $(1,1)$, in which case the Weyl function $M(\lambda)$ is a scalar function. The Weyl function $M(\lambda)$ is said to belong to the Kac class $\mathbf{N}_1$ if

$$\int_1^\infty \frac{\operatorname{Im} M(iy)}{y} \, dy < \infty.$$

In this case there is a real limit

$$\gamma = \lim_{y \to \infty} M(iy) \in \mathbb{R}. \tag{2.6}$$

It follows from (2.2) (identifying the selfadjoint relations in $\mathbb{R}$ with $\mathbb{R} \cup \{\infty\}$) that there is a one-to-one correspondence between all selfadjoint extensions of S in $\mathfrak{H}$ and all numbers in $\mathbb{R} \cup \{\infty\}$ via

$$A(\tau) = \{ \widehat{f} \in S^* : \Gamma_0 \widehat{f} = -\tau \Gamma_1 \widehat{f} \}, \quad \tau \in \mathbb{R} \cup \{\infty\}. \tag{2.7}$$

Recall that the selfadjoint extension $A(\tau)$ is determined by $\tau \in \mathbb{R} \cup \{\infty\}$ via Kreĭn's formula:

$$(A(\tau) - z)^{-1} = (A_0 - z)^{-1} - \gamma(z)(M(z) + 1/\tau)^{-1}(\cdot, \gamma(z)), \quad z \in \mathbb{C} \setminus \mathbb{R}.$$

Note that $A(0)$ corresponds with $A_0 = \ker \Gamma_0$ and that $A(\infty)$ corresponds with $A_1 = \ker \Gamma_1$. The Weyl functions $M_\tau(\lambda)$ of the selfadjoint extensions $A(\tau)$ of S in (2.7) are related by

$$M_\tau(\lambda) = \frac{M(\lambda) - \tau}{1 + \tau M(\lambda)}, \quad \tau \in \mathbb{R} \cup \{\infty\}. \tag{2.8}$$

If $M(\lambda)$ belongs to the Kac class $\mathbf{N}_1$ then all $M_\tau(\lambda)$, $\tau \in \mathbb{R} \cup \{\infty\}$, belong to $\mathbf{N}_1$, except for $\tau = -1/\gamma$ where γ is given by (2.6). The value $\tau = -1/\gamma$ gives rise to an exceptional selfadjoint extension of S, the *generalized Friedrichs extension*, cf. [5], [6].

Closely related to the Kac class $\mathbf{N}_1$ is the class $\mathbf{S}$ of Stieltjes functions. A Nevanlinna function $M(\lambda)$ is said to belong to the class $\mathbf{S}$ if both $M(\lambda)$ and $\lambda M(\lambda)$ are Nevanlinna functions. Equivalently, $M(\lambda)$ belongs to $\mathbf{S}$ if and only if $M(\lambda)$ belongs to $\mathbf{N}_1$ and is analytic on $\mathbb{C} \setminus [0,\infty)$, such that

$$\lim_{x\to-\infty} M(x) \geq 0,$$

cf. [8]. Analogously, a Nevanlinna function $M(\lambda)$ is said to belong to the class $\mathbf{S}_-$ if both $M(\lambda)$ and $\lambda M(-\lambda)$ are Nevanlinna functions. Now $M(\lambda)$ belongs to $\mathbf{S}_-$ if and only if $M(\lambda)$ belongs to $\mathbf{N}_1$ and is analytic on $\mathbb{C} \setminus (-\infty,0]$, such that

$$\lim_{x\to\infty} M(x) \leq 0.$$

2.3. Coupling of symmetric operators with defect numbers $(1,1)$

Consider two closed symmetric operators S_+ and S_- in the Hilbert spaces $\mathfrak{H}_+$ and $\mathfrak{H}_-$ and assume that their defect numbers are $(1,1)$. Form the orthogonal sum $\mathfrak{H} = \mathfrak{H}_+ \oplus \mathfrak{H}_-$ and define in $\mathfrak{H}$ the closed symmetric operator $S = S_+ \oplus S_-$. Clearly, $S^* = S_+^* \oplus S_-^*$ and the defect numbers of S are $(2,2)$. Let $\Pi_+ = \{\mathcal{H}, \Gamma_0^+, \Gamma_1^+\}$ and $\Pi_- = \{\mathcal{H}, \Gamma_0^-, \Gamma_1^-\}$ be boundary triplets for S_+^* and S_-^* with γ-fields $\gamma_+(\lambda)$ and $\gamma_-(\lambda)$. Then

$$\Gamma_0 = \Gamma_0^+ \oplus \Gamma_0^-, \quad \Gamma_1 = \Gamma_1^+ \oplus \Gamma_1^-, \tag{2.9}$$

forms a boundary triplet for the orthogonal sum S^*. Observe that the corresponding γ-field and Weyl function are of the form

$$\gamma_+(\lambda) \oplus \gamma_-(\lambda), \quad M_+(\lambda) \oplus M_-(\lambda), \tag{2.10}$$

where $\gamma_\pm(\lambda)$ and $M_\pm(\lambda)$ correspond to the selfadjoint extensions $A_0^+ = \ker \Gamma_0^+$ and $A_0^- = \ker \Gamma_0^-$ of S_+ and S_-, respectively. It is also of interest to consider one-dimensional symmetric extensions $\widetilde{S}$ of S, cf. [1], [5].

Proposition 2.1. *Let $\Pi_+ = \{\mathcal{H}, \Gamma_0^+, \Gamma_1^+\}$ and $\Pi_- = \{\mathcal{H}, \Gamma_0^-, \Gamma_1^-\}$ be boundary triplets for S_+^* and S_-^* with γ-fields $\gamma_+(\lambda)$, $\gamma_-(\lambda)$ and Weyl functions $M_+(\lambda)$, $M_-(\lambda)$, respectively. Then the linear relation $\widetilde{S}$ defined by*

$$\widetilde{S} = \{\, \widehat{f} = \widehat{f}_+ \oplus \widehat{f}_- \in S_+^* \oplus S_-^* : \Gamma_0^+ \widehat{f}_+ = \Gamma_0^- \widehat{f}_- = \Gamma_1^+ \widehat{f}_+ + \Gamma_1^- \widehat{f}_- = 0 \,\},$$

is closed and symmetric in $\mathfrak{H} = \mathfrak{H}_+ \oplus \mathfrak{H}_-$ and has defect numbers $(1,1)$. Its adjoint $\widetilde{S}^*$ is given by

$$\widetilde{S}^* = \{\, \widehat{f} = \widehat{f}_+ \oplus \widehat{f}_- \in S_+^* \oplus S_-^* : \Gamma_0^+ \widehat{f}_+ = \Gamma_0^- \widehat{f}_-, \},$$

and the defect spaces of $\widetilde{S}$ are of the form

$$\mathfrak{N}_\lambda(\widetilde{S}^*) = \ker(\widetilde{S}^* - \lambda) = \{\, \gamma_+(\lambda) \oplus \gamma_-(\lambda)\, h : h \in \mathcal{H} \,\}.$$

A boundary triplet for $\widetilde{S}^*$ is given by

$$\Pi = \{\, \mathcal{H}, \Gamma_0^+ \upharpoonright \widetilde{S}^*, (\Gamma_1^+ + \Gamma_1^-) \upharpoonright \widetilde{S}^* \,\},$$

and the corresponding Weyl function is given by

$$M_+(\lambda) + M_-(\lambda).$$

All selfadjoint extensions $A(\tau)$ *of* $\widetilde{S}$ *in* $\mathfrak{H}$ *are in one-to-one correspondence with* $\tau \in \mathbb{R} \cup \{\infty\}$ *via*

$$A(\tau) = \{\, \widehat{f} = \widehat{f}_+ \oplus \widehat{f}_- \in S_+^* \oplus S_-^* : \Gamma_0^+ \widehat{f}_+ = \Gamma_0^- \widehat{f}_-,\ \Gamma_0^+ \widehat{f}_+ = -\tau(\Gamma_1^+ \widehat{f}_+ + \Gamma_1^- \widehat{f}_-) \,\}.$$

In particular,

$$A(\infty) = \{\, \widehat{f} = \widehat{f}_+ \oplus \widehat{f}_- \in S_+^* \oplus S_-^* : \Gamma_0^+ \widehat{f}_+ = \Gamma_0^- \widehat{f}_-,\ \Gamma_1^+ \widehat{f}_+ + \Gamma_1^- \widehat{f}_- = 0 \,\}.$$

Now assume that the Weyl functions $M_+(\lambda)$ and $M_-(\lambda)$ corresponding to the selfadjoint extensions $A_0^+ = \ker \Gamma_0^+$ and $A_0^- = \ker \Gamma_0^-$, respectively, each belong to $\mathbf{N}_1$. Clearly, the sum $M_+(\lambda) + M_-(\lambda)$ then also belongs to the class $\mathbf{N}_1$. Therefore, the symmetric extension $\widetilde{S}$ of $S_+ \oplus S_-$ has a generalized Friedrichs extension. In fact, if $\lim_{y\to\infty} M_+(iy) = 0$ and $\lim_{y\to\infty} M_-(iy) = 0$, then the value $\tau = \infty$ corresponds to the generalized Friedrichs extension.

3. Sturm-Liouville expressions and interface conditions

3.1. Sturm-Liouville operators

Let p and q be real-valued functions on an interval $(0, b]$, $b > 0$, such that $1/p$ and q are integrable on $(0, b]$, and consider the Sturm-Liouville expression $L_+ := -DpD + q$ on $(0, b]$. The following lemma can be checked directly.

Lemma 3.1. *Let* $L_{+,\max}$ *be the maximal differential operator in* $L^2(0, b)$ *associated with* L_+ *on* $(0, b]$, $b > 0$. *Then:*

(i) *the restriction* S_+ *of* $L_{+,\max}$ *defined by*

$$S_+ = \{\, u \in \operatorname{dom} L_{+,\max} : u(0) = (pu')(0) = u(b) = 0 \,\}$$

is a closed, densely defined, and symmetric operator with defect numbers $(1, 1)$;

(ii) *the adjoint of* S_+ *is given by*

$$S_+^* = \{\, u \in \operatorname{dom} L_{+,\max} : u(b) = 0 \,\};$$

(iii) *a boundary triplet for* S_+^* *is defined by the boundary mappings*

$$\Gamma_0^+ u = -(pu')(0), \quad \Gamma_1^+ u = u(0).$$

Now let p and q be real-valued functions on an interval $[-b, 0)$, $b > 0$, such that $1/p$ and q are integrable on $[-b, 0)$, and consider the Sturm-Liouville expression $L_- = -DpD + q$ on the interval $[-b, 0)$. The following analog of Lemma 3.1 is immediate.

Lemma 3.2. *Let* $L_{-,\max}$ *be the maximal differential operator in* $L^2(-b, 0)$ *associated with* (3.1) *on* $[-b, 0)$, $b > 0$. *Then:*

(i) *the restriction* S_- *of* $L_{-,\max}$ *defined by*

$$S_- = \{\, u \in \operatorname{dom} L_{-,\max} : u(0) = (pu')(0) = u(-b) = 0 \,\}$$

is a closed, densely defined, and symmetric operator with defect numbers $(1, 1)$;

(ii) *the adjoint of* S_- *is given by*

$$S_-^* = \{ u \in \operatorname{dom} L_{-,\max} : u(-b) = 0 \};$$

(iii) *a boundary triplet for* S_-^* *is defined by the boundary mappings*

$$\Gamma_0^- u = -(pu')(0), \quad \Gamma_1^- u = -u(0),$$

The corresponding Weyl functions can be calculated by fixing a fundamental system for the equation

$$-(pu')' + qu = \lambda u, \quad \lambda \in \mathbb{C}. \tag{3.1}$$

Let $\varphi(\cdot,\lambda)$ and $\psi(\cdot,\lambda)$ denote the fundamental solutions of (3.1) on $(0,b]$ which satisfy the initial conditions

$$\begin{cases} \varphi(0,\lambda) = 1, & (p\varphi')(0,\lambda) = 0, \\ \psi(0,\lambda) = 0, & (p\psi')(0,\lambda) = -1. \end{cases} \tag{3.2}$$

Introduce the function $M_+(\lambda)$ by

$$M_+(\lambda) = -\frac{\psi(b,\lambda)}{\varphi(b,\lambda)}, \tag{3.3}$$

and define a solution of $L_+u = \lambda u$ by $\chi_+(\cdot,\lambda) = \psi(\cdot,\lambda) + M_+(\lambda)\varphi(\cdot,\lambda)$. Then $\chi_+(x,\lambda)$ satisfies $\chi_+(b,\lambda) = 0$ and thus it spans $\ker(S_+^* - \lambda)$. Furthermore, $M_+(\lambda)$ is the Weyl function corresponding to the boundary triplet in (iii) of Lemma 3.1, since $\Gamma_0^+ \chi_+(\cdot,\lambda) = 1$, $\Gamma_1^+ \chi_+(\cdot,\lambda) = M_+(\lambda)$, cf. (2.3). Likewise it can be shown that $\chi_+(\cdot,\lambda)$ is the corresponding γ-field, cf. (2.4).

Observe that the functions $\varphi(x,\lambda)$ and $\psi(x,\lambda)$ in (3.2) can also be seen as a fundamental system for the equation (3.1) on the interval $[-b,0)$. Introduce the function $M_-(\lambda)$ by

$$M_-(\lambda) = \frac{\psi(-b,\lambda)}{\varphi(-b,\lambda)}, \tag{3.4}$$

and define a solution of $L_-u = \lambda u$ by $\chi_-(\cdot,\lambda) = \psi(\cdot,\lambda) - M_-(\lambda)\varphi(\cdot,\lambda)$. Then $\chi_-(x,\lambda)$ satisfies $\chi_-(-b,\lambda) = 0$ and thus it spans $\ker(S_-^* - \lambda)$. Completely analogous to the previous case, $M_-(\lambda)$ is the Weyl function corresponding to the boundary triplet in (iii) of Lemma 3.2, since $\Gamma_0^- \chi_-(\cdot,\lambda) = 1$, $\Gamma_1^- \chi_-(\cdot,\lambda) = M_-(\lambda)$. Furthermore, $\chi_-(\cdot,\lambda)$ is the corresponding γ-field.

3.2. Coupling of Sturm-Liouville operators

The two differential operators S_+^* in $L^2[0,b]$ and S_-^* in $L^2[-b,0]$ are used to define a differential operator in $L^2[-b,b]$ by means of an interface conditions at the origin. For this purpose define the orthogonal sum $S^* = S_+^* \oplus S_-^*$ of S_+^* and S_-^* in $L^2[-b,b]$, so that $S = S_+ \oplus S_-$ has defect numbers $(2,2)$. A boundary triplet for S^* is obtained from Lemmas 3.1 and 3.2 and the construction in (2.9), which means for the present situation

$$\Gamma_0 u = \begin{pmatrix} -(pu')(0+) \\ -(pu')(0-) \end{pmatrix}, \quad \Gamma_1 u = \begin{pmatrix} u(0+) \\ -u(0-) \end{pmatrix}, \quad u \in \operatorname{dom} S^*. \tag{3.5}$$

Here the domain $\operatorname{dom} S^*$ is given by

$$\operatorname{dom} S^* = \{ u \in \operatorname{dom} L_{+,\max} \oplus L_{-,\max} : u(-b) = u(b) = 0 \},$$

i.e., by the set of all $u \in L^2[-b,b]$ which satisfy

$$u, pu' \in AC([-b,b] \setminus \{0\}), \; (pu')' \in L^2[-b,b], \; u(-b) = u(b) = 0. \tag{3.6}$$

The notation $u(0\pm) = \lim_{t \to \pm 0} u(t)$ is used to indicate a reference to the underlying operator $S_\pm^*$. Observe that the corresponding γ-field and Weyl function are of the form (2.10), where $\gamma_\pm(\lambda)$ and $M_\pm(\lambda)$ correspond to the selfadjoint extensions A_0^+ and A_0^- determined by the boundary condition $(pu')(0+) = 0$ and $(pu')(0-) = 0$, respectively. The following proposition is an immediate consequence of Proposition 2.1 and Lemmas 3.1 and 3.2.

Proposition 3.3. *Let $S^* = S_+^* \oplus S_+^*$ be the differential operator in $L^2[-b,b]$, where S_+^* and S_-^* are as in Lemma 3.1 and Lemma 3.2, respectively, and let $\{\mathbb{C}^2, \Gamma_0, \Gamma_1\}$ be a boundary triplet for S^* determined by the boundary mappings Γ_0, Γ_1 in (3.5), so that the γ-field and the Weyl function are of the form (2.10). Then:*

(i) *the linear relation $\widetilde{S}$ defined by*

$$\widetilde{S} = \{ u \in \operatorname{dom} S^* : u(0+) = u(0-), \quad (pu')(0+) = (pu')(0-) = 0 \} \tag{3.7}$$

is a closed symmetric extension of S in $L^2[-b,b]$ with defect numbers $(1,1)$;

(ii) *the adjoint of $\widetilde{S}$ is given by*

$$\widetilde{S}^* = \{ u \in \operatorname{dom} S^* : (pu')(0+) = (pu')(0-) := (pu')(0) \}; \tag{3.8}$$

(iii) *a boundary triplet $\{\mathbb{C}, \widetilde{\Gamma}_0, \widetilde{\Gamma}_1\}$ for $\widetilde{S}^*$ is determined by*

$$\widetilde{\Gamma}_0 u = -(pu')(0), \quad \widetilde{\Gamma}_1 u = u(0+) - u(0-);$$

(iv) *the corresponding Weyl function is equal to $M_+(\lambda) + M_-(\lambda)$.*

Moreover, the selfadjoint extensions $A(\tau)$ of $\widetilde{S}$ in $L^2[-b,b]$ are given for $\tau \in \mathbb{R}$ by

$$\begin{aligned}&\operatorname{dom} A(\tau)\\ &\quad = \{ u \in \operatorname{dom} S^* : (pu')(0+) = \tau(u(0+) - u(0-)), \, (pu')(0+) = (pu')(0-) \},\end{aligned}$$

and for $\tau = \infty$ by

$$\operatorname{dom} A(\infty) = \{ u \in \operatorname{dom} S^* : u(0+) = u(0-), \, (pu')(0+) = (pu')(0-) \}.$$

4. Generalized Friedrichs extensions and interface conditions

The Sturm-Liouville expressions in Section 3 satisfy conditions which guarantee the interpretation as differential operators in the spaces $L^2[0,b]$ and $L^2[-b,0]$ with Weyl functions belonging to the general Nevanlinna class $\mathbf{N}$. Under additional conditions it can be shown that the Weyl functions belong to the Kac class $\mathbf{N}_1$.

Lemma 4.1. *Assume that the coefficient p in the Sturm-Liouville expression $L_+ = -DpD + q$ satisfies $p(x) > 0$ for $0 < x \leq b$ and*

$$\int_0^b \frac{dx}{p(x)} < \infty, \tag{4.1}$$

and that the coefficient $q(x) \geq c$, $c \in \mathbb{R}$, is bounded from below and integrable on $(0, b]$. Then the Weyl function $M_+(\lambda)$ in (3.3) belongs to the Kac class $\mathbf{N_1}$, it is analytic on $\mathbb{C} \setminus [c, \infty)$, and

$$\lim_{\lambda \to -\infty} M_+(\lambda) = 0. \tag{4.2}$$

In particular, if $q(x) \geq 0$ on $(0, b]$, then $M_+(\lambda)$ belongs to the class $\mathbf{S}$ of Stieltjes functions.

Proof. It suffices to show that the spectral measure σ of $M_+(\lambda)$ satisfies the relation $\operatorname{supp} \sigma \subset [c, \infty)$, and that (4.2) holds. It follows from the relation (3.1) that

$$\varphi(x, \lambda) = 1 + \int_0^x \varphi'(t, \lambda) dt, \quad p(x)\varphi'(x, \lambda) = \int_0^x (q(x) - \lambda)\varphi(t, z) dt, \tag{4.3}$$

and these relations imply for $0 < x \leq b$ and $\lambda \leq c$ that $\varphi'(x, \lambda) \geq 0$, and further that the functions $\varphi(x, \lambda)$ and $p(x)\varphi'(x, \lambda)$ are nondecreasing with respect to x. In particular, $\varphi(b, \lambda)$ has no zeros off $[c, \infty)$, and according to (3.3) the relation $\operatorname{supp} \sigma \subset [c, \infty)$ is shown. The Wronskian identity and the initial conditions (3.2) imply that

$$-\left(\frac{\psi(x, \lambda)}{\varphi(x, \lambda)}\right)' = \frac{1}{p(x)\varphi(x, \lambda)^2},$$

and hence, with the relation (3.3), that

$$M_+(\lambda) = \int_0^b \frac{1}{\varphi(t, \lambda)^2} \frac{dt}{p(t)}. \tag{4.4}$$

As $\varphi(x, \lambda) \geq 1$ for $0 < x \leq b$ and $\lambda \leq c$, it follows from the relations (4.3) that

$$\lim_{\lambda \to -\infty} p(x)\varphi'(x, \lambda) = \infty, \qquad \lim_{\lambda \to -\infty} \varphi(x, \lambda) = \infty.$$

By dominated convergence, the relation (4.4) implies the relation (4.2). If $q(x) \geq 0$ then $\operatorname{supp} \sigma \subset [0, \infty)$ and it follows from [8] that $M_+(\lambda) \in \mathbf{S}$. □

The following result can be established in a completely analogous fashion.

Lemma 4.2. *Assume that the coefficient p in the Sturm-Liouville expression $L_- = -DpD + q$ on $[-b, 0)$ satisfies $p(x) < 0$ for $-b \leq x < 0$ and*

$$-\infty < \int_{-b}^0 \frac{dx}{p(x)}, \tag{4.5}$$

and that the coefficient $q(x) \le c$, $c \in \mathbb{R}$, *is bounded from above and integrable on* $[-b, 0)$. *Then the Weyl function* $M_-(\lambda)$ *in* (3.3) *belongs to the Kac class* $\mathbf{N}_1$, *it is analytic on* $\mathbb{C} \setminus (-\infty, c]$, *and*

$$\lim_{\lambda \to \infty} M_-(\lambda) = 0. \tag{4.6}$$

In particular, if $q(x) \le 0$ *on* $[-b, 0)$, *then* $M_-(\lambda)$ *belongs to the class* $\mathbf{S}_-$.

A combination of Lemmas 4.1 and 4.2 in conjunction with the remarks following Proposition 2.1 now leads to a description of the generalized Friedrichs extension in the situation of Proposition 3.3.

Proposition 4.3. *Assume that the coefficients* p *and* q *of the Sturm-Liouville expression* $-DpD + q$ *on* $[-b, b]$ *satisfy the assumptions of Lemma* 4.1 *on* $(0, b]$, *and the assumptions of Lemma* 4.2 *on* $[-b, 0)$. *Then the symmetric operator* $\widetilde{S}$ *in* (3.7) *has a generalized Friedrichs extension and it corresponds to* $\tau = \infty$.

It is also possible to obtain a similar result for the case when the sign of the coefficient p is not fixed on $(0, b]$ or on $[-b, 0)$. Define for this purpose the following functions

$$P_0(x) = \int_0^x \frac{1}{|p(t)|}\, dt,$$
$$P_1(x) = \int_0^x \frac{|t|}{|p(t)|}\, dt,$$
$$P_{01}(x) = \int_0^x \frac{1}{P_0(t)}\, dt.$$

The following result is inspired by [7] (and could be stated in a slightly more general fashion, similar to the formulations in [7]).

Proposition 4.4. *Assume that one of the following functions*

$$\frac{P_{01}(\delta)}{P_1(\delta)|p(\delta)|} \quad \text{or} \quad \frac{P_0(\delta)}{\delta} \tag{4.7}$$

is locally integrable in a neighborhood of 0. *Then the symmetric operator* $\widetilde{S}$ *in* (3.7) *has a generalized Friedrichs extension corresponding to* $\tau = \infty$.

Proof. First consider the case of the interval $(0, b]$. It suffices to recall the inequality [7, Theorem 4.1 and (4.8)]:

$$|M_+(iy)| \le C \frac{1 + y \int_0^{\delta(y)} P_0(t)\, dt}{y\delta(y)}, \tag{4.8}$$

where $\delta(y)$ is a monotonically decreasing function with

$$\delta(y) \to 0, \quad y\delta(y) \to \infty, \quad y \to \infty,$$

which satisfies further conditions, cf. [7, Lemma 3.2]. Hence, if for some $y_0 > 0$, the following inequality

$$\int_{y_0}^{\infty} \frac{1 + y \int_0^{\delta(y)} P_0(t)\, dt}{y^2 \delta(y)}\, dy < \infty \tag{4.9}$$

is satisfied, then the function $M_+(\lambda) \in \mathbf{N}_1$ and $\lim_{y\to\infty} M_+(iy) = 0$. Now if one of the functions in (4.7) is locally integrable near 0 then there exists a monotonically decreasing, absolutely continuous function $\delta(y) : [y_0, \infty) \to (0, 1)$ with the required properties such that the inequality (4.9) holds, cf. [7]. Completely similar to the case of the interval $(0, b]$, the case of the interval $[-b, 0)$ can be treated. □

References

[1] V.A. Derkach, S. Hassi, M.M. Malamud, and H.S.V. de Snoo, *Generalized resolvents of symmetric operators and admissibility*, Methods of Functional Analysis and Topology, 6 (2000), 24–55.

[2] V.A. Derkach, S. Hassi, M.M. Malamud, and H.S.V. de Snoo, *Boundary relations and their Weyl families*, Trans. Amer. Math. Soc., to appear.

[3] A. Fleige, S. Hassi, and H.S.V. de Snoo, *A Kreĭn space approach to representation theorems and generalized Friedrichs extensions*, Acta Sci. Math. (Szeged), 66 (2000), 633–650.

[4] A. Fleige, S. Hassi, H.S.V. de Snoo, and H. Winkler, *Sesquilinear forms corresponding to a non-semibounded Sturm-Liouville operator*, in preparation.

[5] S. Hassi, M. Kaltenbäck, and H.S.V. de Snoo, *Triplets of Hilbert spaces and Friedrichs extensions associated with the subclass* $\mathbf{N}_1$ *of Nevanlinna functions*, J. Operator Theory, 37 (1997), 155–181.

[6] S. Hassi, H. Langer, and H.S.V. de Snoo, *Selfadjoint extensions for a class of symmetric operators with defect numbers* (1, 1), 15th OT Conference Proceedings, (1995), 115–145.

[7] S. Hassi, M. Möller, and H.S.V. de Snoo, *Sturm-Liouville operators and their spectral functions*, J. Math. Anal. Appl., 282 (2003), 584–602.

[8] I.S. Kac and M.G. Kreĭn, *R-functions-analytic functions mapping the upper halfplane into itself*, Supplement II to the Russian edition of F.V. Atkinson, *Discrete and continuous boundary problems*, Mir, Moscow, 1968 (Russian) (English translation: Amer. Math. Soc. Transl., (2) 103 (1974), 1–18.

[9] I.S. Kac and M.G. Kreĭn, *On the spectral functions of the string*, Supplement II to the Russian edition of F.V. Atkinson, *Discrete and continuous boundary problems*, Mir, Moscow, 1968 (Russian) (English translation: Amer. Math. Soc. Transl., (2) 103 (1974), 19–102).

[10] H.-D. Niessen and A. Zettl, *Singular Sturm-Liouville problems: The Friedrichs extension and comparison of eigenvalues*, Proc. London Math. Soc., 64 (1992), 545–578.

Andreas Fleige
Am Südwestfriedhof 27
44137 Dortmund
Deutschland
e-mail: andreas.fleige@lycos.de

Seppo Hassi
Department of Mathematics and Statistics
University of Vaasa
P.O. Box 700
65101 Vaasa
Finland
e-mail: sha@uwasa.fi

Henk de Snoo
Department of Mathematics and Computing Science
University of Groningen
P.O. Box 800
9700 AV Groningen
Nederland
e-mail: desnoo@math.rug.nl

Henrik Winkler
Department of Mathematics and Computing Science
University of Groningen
P.O. Box 800
9700 AV Groningen
Nederland
e-mail: winkler@math.rug.nl

Operator Theory:
Advances and Applications, Vol. 163, 147–162

Spectral Properties of Operator Polynomials with Nonnegative Coefficients

Karl-Heinz Förster and Béla Nagy

Dedicated to Professor Dr. Heinz Langer

Abstract. We study properties of the polynomial $Q(\lambda) = \lambda^m I - S(\lambda)$ where $S(\lambda) = \lambda^l A_l + \cdots + \lambda A_1 + A_0$ and $1 \leq m < l$. The coefficients $A_l, \ldots, A_0$ are in the positive cone of an ordered Banach algebra or are positive operators on a complex Banach lattice E and I is the identity. We study the properties of the spectral radius of $S(\lambda)$ if λ is a nonnegative real number, and its connection with the existence of spectral divisors with nonnegative coefficients in the considered sense. We prove factorization results for nonnegative elements in an ordered decomposing Banach algebra with closed normal algebra cone and in the Wiener algebra. Earlier results on monic (nonnegative) operator polynomials are applied to the operator polynomial class studied here.

Mathematics Subject Classification (2000). MSC(2000): Primary 47A56; Secondary 46H99, 47B65 .

Keywords. Operator polynomials, positive coefficients, factorization, ordered Banach algebras, spectral radius.

1. Introduction

In the present paper we consider the operator polynomial

$$Q(\lambda) = \lambda^m I - \lambda^l A_l - \cdots - \lambda A_1 - A_0,$$

where $A_l, \ldots, A_0$ are nonnegative (= positive) operators on a complex Banach lattice E, $A_l \neq 0$ and I is the identity operator on E. Most of our results will be valid and proved for polynomials of the type above with coefficients in the cone of an ordered Banach algebra.

This work was completed with partial support of the Hungarian National Science Grants OTKA Nos T-030042 and T- 047276 and partial support of the DAAD and the Technical University of Berlin.

Concerning the used concepts of Banach algebras we refer to the monographs T.W. Palmer[18] and I. Gohberg, S. Goldberg and M. A. Kaashoek [10], whereas concerning the spectral theory of operator polynomials we refer to the monographs A.S. Markus [16] and L. Rodman [19], concerning the theory of nonnegative operators on Banach lattices we refer to the monographs P. Meyer-Nieberg [17] and H.H. Schaefer [23]. We will adopt several results on operator polynomials for our more general case, if no essential differences in their proofs appear.

Monic operator polynomials (i.e., $m > l$) with nonnegative operator coefficients $A_l, \ldots, A_0$ have been considered in [4], [13], [15] and [19]. Here we investigate mainly the case $1 \leq m < l$.

We change the notation and consider in the following polynomials

$$q(\lambda) = \lambda^m e - \lambda^l a_l - \cdots - \lambda a_1 - a_0, \quad \lambda \in \mathbb{C}, \tag{1.1}$$

where e is the unit element of a complex Banach algebra A and the coefficients a_j belong to a cone C in A for $j = 0, 1, \ldots, l$; we say also, they are nonnegative (with respect to this cone). We define the polynomial

$$s(\lambda) = \lambda^l a_l + \cdots + \lambda a_1 + a_0. \tag{1.2}$$

By $\varrho(s(\lambda))$ we denote the spectral radius of $s(\lambda)$. One of our main results is the following (see Section 5) :
Let $1 \leq m < l$ and $\varrho(s(r_0)) < r_0^m$ for some $r_0 > 0$. Then $q(\cdot)$ has a monic (right) spectral divisor of degree m with nonnegative coefficients and a comonic (left) spectral divisor with nonnegative coefficients.
In Section 4 we prove factorization results for nonnegative elements in an abstract ordered decomposing Banach algebra with closed normal algebra cone. As corollaries we obtain results on factorizations in the Wiener algebra.
In Section 2 we recall the notation of an ordered Banach algebra with closed (normal) cone, [20], and collect some spectral properties of elements in such cones.
In Section 3 we study the function

$$\varrho_s : [0, \infty[\longrightarrow \mathbb{R}_+ \text{ with } \varrho_s(r) = \varrho(s(r)). \tag{1.3}$$

An important property of this function is its log-log convexity (for this notion see the text below).
In the last section we consider operator polynomials and apply known results on monic operator polynomials to our case here.

2. Ordered Banach algebras

In this section we consider ordered Banach algebras in the sense of [20]. We assume the reader to be familiar with the definition and elementary properties of Banach algebras.

Throughout this section A will denote a (real or complex) Banach algebra with unit e and zero element 0. As in [20, §3] we call a subset $C \subset A$ an *algebra cone* if C satisfies the following conditions:

1. $C + C \subset C$, 2. $\lambda\, C \subset C$ for all $\lambda \geq 0$,
3. $C \cdot C \subset C$, 4. $e \in C$.

C is called a *proper* cone, if $-C \cap C = \{0\}$. Any cone in A induces an ordering "$\leq$" on A in the following way:

$a \leq b$ if and only if $b - a \in C$ (for every $a, b \in A$).

It is well known that this is a partial ordering on A, i.e., $\leq$ is reflexive and transitive; $\leq$ is antisymmetric if and only if C is proper. If C is an algebra cone in A, then the induced partial ordering $\leq$ satisfies for $a, b \in A$ and a scalar λ :

$1'.\ 0 \leq a, 0 \leq b \Longrightarrow 0 \leq a + b$, $2'.\ 0 \leq a, 0 \leq \lambda \Longrightarrow 0 \leq \lambda a$
$3'.\ 0 \leq a, 0 \leq b \Longrightarrow 0 \leq a \cdot b$, $4'.\ 0 \leq e$.

Conversely, if $\leq$ is a partial ordering on A such that $1'$.–$4'$. hold, then $C = \{a \in A : 0 \leq a\}$ is an algebra cone which induces $\leq$. If A is ordered by an algebra cone, then we call A an *ordered Banach algebra.*

An algebra cone C of A is said to be *normal* if there exists a constant $\gamma > 0$ such that $0 \leq a \leq b$ in A implies $\|a\| \leq \gamma\ \|b\|$. It is easy to see that a normal algebra cone is proper.

In applications the Banach algebra $L(E)$ of all linear bounded operators in a (complex) Banach lattice E with cone $L(E)_+$ is an important example for an ordered Banach algebras with a closed normal algebra cone. Especially the algebra $\mathbb{C}^{n \times n}$ of all complex $n \times n$ matrices with the cone of all entrywise nonnegative matrices is such an algebra.
In the next proposition we collect some properties of an ordered Banach algebra with a normal cone.

Proposition 2.1. *Let A be an ordered Banach algebra with a closed normal algebra cone C. Then*

1. *The spectral radius ϱ is a monotone function on C; i.e., if $0 \leq a \leq b$, then $\varrho(a) \leq \varrho(b)$.*
2. *For all $a \in C$ its spectral radius $\varrho(a)$ belongs to its spectrum $\sigma(a)$; i.e., $\varrho(a) \in \sigma(a)$ for $a \in C$.*
3. *Let $a \in C$ and let $\lambda \in \mathbb{C}$. Then $\lambda \notin \sigma(a)$ and $(\lambda e - a)^{-1} \in C$ if and only if λ is real and $\varrho(a) < \lambda$.*

Proof. The first two assertion are proved in [20, Theorem 4.1.1 and Proposition 5.1] We will prove the third assertion. It is clear that $\varrho(a) < \lambda$ is sufficient for $\lambda \notin \sigma(a)$; $(\lambda e - a)^{-1} \in C$ follows from the expansion of $(\lambda e - a)^{-1}$ at ∞ (C. Neumann's series). Suppose that $(\lambda e - a)^{-1} \in C$ for some $\lambda \notin \sigma(a)$. For $n = 0, 1, 2 \ldots$ set $x_n = (\lambda e - a)^{-n}$, where $x_0 = (\lambda e - a)^0 = e$. Clearly $x_n \in C, x_n \neq 0$ and $\lambda x_n = a x_n + x_{n-1}$ for $n = 1, 2, \ldots$. By induction on n it follows from the last equality that $\lambda^n x_n \in C, \lambda^{n-1} x_n \in C$ and $\lambda^n x_n \geq \lambda^{n-1} x_{n-1} \geq x_0 = e$. Now we can proceed as in the proof of [23, Appendix 2.3, p. 264] to obtain that $\varrho(a) < \lambda$. □

3. The spectral radius of $s(\cdot)$ as a function on the nonnegative axis

In this section we investigate the function

$$\varrho_s : [0, r_0[\longrightarrow \mathbb{R}_+ \text{ with } \varrho_s(r) = \varrho(s(r)).$$

where $s(\cdot)$ is holomorphic on the open disc $\mathbb{D}_{r_0} = \{z \in \mathbb{C} \mid |z| < r_0\}$, so

$$s(\lambda) = \sum_{j=0}^{\infty} \lambda^j a_j \quad \text{for } |\lambda| < r_0,$$

and that the coefficients a_j are in the cone C of an ordered complex Banach algebra A.

The next lemma may be known. For sake of completeness we include a proof; it is an appropriate modification of the proof of [2, Lemma 3]

Lemma 3.1. *Let C be a normal closed cone in a complex Banach algebra. Then there exists a constant $\beta > 0$ such that*

$$\|s(\lambda)\| \le \beta \|s(|\lambda|)\|, \quad \varrho(s(\lambda)) \le \varrho(s(|\lambda|))$$

for all functions $s(\cdot)$ which are holomorphic in $\mathbb{D}_{r_0}$ with coefficients in C and for all $\lambda \in \mathbb{D}_{r_0}$.

Proof. C is normal; i.e., there exists a positive constant γ such that $x, y \in C$ and $x \le y$ imply $\| x \| \le \gamma \| y \|$.
Then $u \in C$, $v \in A$ and

$$-u \le v \le u \quad \text{imply} \quad \| v \| \le 2\gamma \| u \|.$$

Indeed, from $u - v \ge 0$ and $u + v \ge 0$ there follows $2\gamma \| u \| = \gamma \| u + v + (u - v) \| \ge \max\{\| u + v \|, \| u - v \|\} \ge \frac{1}{2}(\| u + v \| + \| u - v \|) \ge \| v \|$.

Let $\lambda = re^{i\phi}$ and $\theta \in [0, 2\pi]$. Then

$$\begin{aligned} \| s(\lambda) \| = \| e^{i\theta} s(\lambda) \| &= \| \sum_{k=0}^{\infty} r^k (\cos(\theta + k\phi) + i \sin(\theta + k\phi)) a_k \| \\ &\le 2 \sup_{0 \le \omega \le 2\pi} \| \sum_{k=0}^{\infty} r^k \cos(\omega + k\phi) a_k \|. \end{aligned}$$

Now $-r^k a_k \le r^k \cos(\omega + k\phi) a_k \le r^k a_k$, for $k = 0, 1, 2, \ldots$. Adding, we have

$$-\sum_{k=0}^{\infty} r^k a_k \le \sum_{k=0}^{\infty} r^k \cos(\omega + k\phi) a_k \le \sum_{k=0}^{\infty} r^k a_k \in C.$$

Therefore $\| s(\lambda) \| \le 4\gamma \| \sum_{k=0}^{\infty} r^k a_k \| = \|s(|\lambda|)\|$. The inequality of the spectral radii follows now from $\|(s(\lambda))^k\| \le \beta \|(s(|\lambda\|))^k\|$ for $k = 0, 1, 2, \ldots$ (note that $s^k(\cdot)$ is also holomorphic in $\mathbb{D}_{r_0}$ and has nonnegative coefficients) and the well-known formula for the spectral radius. □

By a result of E. Vesentini (see [1, p. 52]) the function $\lambda \longmapsto \log(\varrho(s(\lambda)))$ (and then also the function $\lambda \longmapsto \varrho(s(\lambda))$) is subharmonic on $\mathbb{D}_{r_0}$. Since the coefficients a_j are in C we obtain from the lemma above $\varrho_s(r) = \max\limits_{|\lambda|=r} \varrho(s(\lambda))$ for all $r \in [0, r_0[$. From the theory of subharmonic functions (see [14], Theorem 2.13) we obtain

Proposition 3.2. *Under the conditions above the function ϱ_s is continuous and not decreasing, and $\log(\varrho_s(\cdot))$ is convex in $\log(r)$ on $]0, r_0[$.*

$\log(\varrho_s(\cdot))$ is convex in $\log(r)$ on $]0, r_0[$ means that the function

$$\eta_s :]-\infty, \log(r_0)[\to \mathbb{R} \quad \text{with} \quad t \longmapsto \log(\varrho_s(e^t)) \tag{3.1}$$

is convex, or equivalently

$$\varrho_s({r_1}^{\tau} {r_2}^{1-\tau}) \leq \varrho_s(r_1)^{\tau}\ \varrho_S(r_2)^{1-\tau} \quad \text{for} \quad r_1, r_2 \in]0, r_0[, \quad \tau \in [0,1].$$

We call a function which satisfies the last functional inequality log-log convex. Fundamental properties of log-log convex functions give

Proposition 3.3. *Assume that the function $s(\cdot)$ satisfies the conditions above. Let $k \in \mathbb{N}_0$.*

1. *Let $0 < r_1 < r_2 < r_3 < r_0$ be such that $\varrho_s(r_j) = r_j^k$ for $j = 1, 2, 3$. Then $\varrho_s(r) = r^k$ for all $r \in [r_1, r_3]$.*
2. *Let $r_1, r_2 > 0$ be such that $r_1 \neq r_2$ and $\varrho_s(r_j) = r_j^k$ for $j = 1, 2$, let ϱ_s be differentiable in r_1, and let $\varrho_s'(r_1) = k r_1^{k-1}$. Then $\varrho_s(r) = r^k$ for all r in the closed interval with the endpoints r_1 and r_2.*

Proof. The function η_s is convex. Conditions 1 imply that the graphs of η_s and of the line $\mathbb{R} \to \mathbb{R}$ with $t \longmapsto kt$ have three different points in common. Therefore they coincide on the corresponding interval, this implies the assertion.
Conditions 2 imply that the graphs of η_s and of the line have two different points in common and in one of the points the line is the tangent of η_s in this point. Again, they coincide on the corresponding interval, and we obtain the assertion. □

Proposition 3.4. *Assume that the function $s(\cdot)$ satisfies the conditions above. Let $q(\lambda) = \lambda^m e - s(\lambda)$ for all $\lambda \in \mathbb{D}_{r_0}$ and some $m \in \mathbb{N}$. Assume that there exist positive numbers r_1 and r_2 such that $0 < r_1 < r_2 < r_0$ and $\varrho_s(r) < r^m$ for $r \in]r_1, r_2[$. Then*

$$\sigma(q(\cdot)) \cap \{\lambda \in \mathbb{C} \mid r_1 < |\lambda| < r_2\} = \emptyset.$$

Proof. For $\lambda \in \mathbb{C}$ with $r_1 < |\lambda| < r_2$ we obtain from Lemma 3.1 that

$$\varrho(s(\lambda)) \leq \varrho(s(|\lambda|)) = \varrho_s(|\lambda|) < |\lambda^m|.$$

Therefore $q(\lambda) = \lambda^m(e - \lambda^{-m} s(\lambda))$ is invertible. □

Example 3.5. For $n \in \mathbb{N}$ and $k_j \in \mathbb{N}_0 = \mathbb{N} \cup \{0\}$ for $j = 1, 2, \ldots n$ consider the $n \times n$ (weighted cyclic) matrix

$$S(\lambda) = \begin{pmatrix} 0 & \lambda^{k_1} & 0 & \cdots & 0 \\ 0 & 0 & \lambda^{k_2} & & \vdots \\ \vdots & \vdots & \ddots & \ddots & \vdots \\ 0 & 0 & \cdots & 0 & \lambda^{k_{n-1}} \\ \lambda^{k_n} & 0 & \cdots & 0 & 0 \end{pmatrix}, \quad \lambda \in \mathbb{C}.$$

We assume that $l := \max_{1 \le j \le n} k_j \ge 1$. Then $S(\cdot)$ is a matrix polynomial of degree l with entrywise nonnegative matrices as coefficients. These matrices are nonnegative operators on the Banach lattice $\mathbb{C}^n$ and elements in the normal cone ${\mathbb{R}_+}^{n\times n}$ of the nonnegative $n \times n$ square matrices in the Banach algebra $\mathbb{C}^{n\times n}$ of the complex $n \times n$ square matrices with the spectral norm.

Let $\lambda \neq 0$. Then $\sigma(S(\lambda)) = \{z \in \mathbb{C} : z^n = \lambda^k\}$ where $k = k_1 + \cdots + k_n$. The eigenvalues of $S(\lambda)$ have simple (geometric and algebraic) multiplicities, the corresponding eigenvectors of $S(\lambda)$ to $z \in \sigma(S(\lambda))$ are multiples of $(1, \lambda^{-k_1} z, \lambda^{-k_1-k_2} z^2, \ldots,$
$\lambda^{-k_1-\cdots-k_{n-1}} z^{n-1})$. For $r > 0$ the matrix $S(r)$ is nonnegative and irreducible. $S(1)$ is row-stochastic; i.e., $S(1)\mathbf{1} = \mathbf{1}$, where $\mathbf{1}$ is the vector in $\mathbb{C}^n$ which has all components equal to 1. We have $\varrho_S(r) = r^{\frac{k}{n}}$. Therefore ϱ_S is not convex if $k < n$. For $k = n$ we obtain $\varrho_S(r) = r$ for $r > 0$, therefore $\varrho_S(1) = \varrho_S'(1) = 1$ and $\varrho_S''(1) = 0$ in this case.

For $L(\lambda) = \lambda^m - S(\lambda)$ we have $\sigma(L) = \{z \in \mathbb{C} : z^{nm} = z^k\}$, especially $\sigma(L) = \mathbb{C}$ if and only if $mn = k$.

4. Factorization in ordered decomposing Banach algebras

In this section we introduce the concept of an ordered decomposing Banach algebra and prove a factorization result on elements in the algebra cone which will be applied in the next section to polynomials with nonnegative coefficients.

A Banach algebra A is called a *decomposing Banach algebra* if A is the direct sum of two closed subalgebras A_+ and A_-; see [3], [10], [11]. Let P_+ denote the bounded linear projection of A onto A_+ annihilating A_-. Then $P_- := I - P_+$ is the bounded linear projection of A onto A_- annihilating A_+. Let A be at the same time an ordered Banach algebra with algebra cone C and a decomposing Banach algebra (with the projections P_+ and P_-). We call A an *ordered decomposing Banach algebra*, if C is invariant under P_+ and under P_-; i.e., $P_\pm C \subset C$.

The following theorem is an order theoretic version of Lemma 5.1 in Chapter 1 of [8]; cf. also Theorem 23.3 in [16].

Theorem 4.1. *Let A be an ordered decomposing Banach algebra with closed normal algebra cone C. If*

$$a \in C \quad \text{and} \quad \varrho(a) < 1,$$

then $e - a$ *admits a factorization*

$$e - a = b_+(e - b_-)$$

with $b_+ \in A_+$, $e - b_+ \in A_+ \cap C$, $b_- \in A_- \cap C$, *the elements* b_+ *and* $e - b_-$ *are invertible, and* $b_+^{-1} \in A_+ \cap C$, $(e - b_-)^{-1} - e \in A_- \cap C$ *and* $(e - b_-)^{-1} \in C$.

Proof. Let $a \in C$ and $\varrho(a) < 1$. We define the bounded linear operator $T_+ : A \to A$ by $T_+x = P_+(xa)$ for all $x \in A$. Then $T_+C \subset A_+ \cap C$ and $0 \leq T_+e = P_+a \leq a$. Therefore $0 \leq {T_+}^2 e \leq T_+a = P_+a^2 \leq a^2$, and then $0 \leq {T_+}^n e \leq a^n$ for $n = 1, 2, \ldots$. The normality of the cone C implies $\|{T_+}^n e\| \leq \gamma \|a^n\|$ for $n = 1, 2, \ldots$ and some constant γ. Therefore from $\varrho(a) < 1$ it follows that $\hat{x} = \sum\limits_{n=0}^{\infty} {T_+}^n e$ converges, belongs to $A_+ \cap C$ and is a solution of the equation

$$x - P_+(xa) = x - T_+x = e.$$

Set $b_- = e - \hat{x} + \hat{x}a$, then $b_- = P_-(\hat{x}a) \in A_- \cap C$. Next we consider the operator $T_- : A \to A$ with $T_-y = P_-(ay)$ for all $y \in A$. As above, $T_-C \subset A_- \cap C$ and $\|{T_-}^n e\| \leq \gamma \|a^n\|$ for $n = 1, 2, \ldots$. Therefore $\hat{y} = \sum\limits_{n=0}^{\infty} {T_-}^n e$ converges, belongs to C, and is a solution of the equation

$$y - P_-(ay) = y - T_-y = e.$$

Set $b_+ = e - P_+(a\hat{y})$. Then $b_+ \in A_+$ (note that we assume $e \in A_+$), $e - b_+ = P_+(a\hat{y}) \in A_+ \cap C$ and $b_+ = (e - a)(e + P_-(a\hat{y}))$. The last equation implies $\hat{x}b_+ = (e - b_-)(e + P_-(a\hat{y}))$, and this is equivalent to

$$\hat{x}b_+ - e = -b_- + P_-(a\hat{y}) - b_-P_-(a\hat{y}).$$

In the last equation the left-hand side belongs to A_+, while the right-hand side belongs to A_-. Therefore both sides are equal to zero, consequently

$$\hat{x}b_+ = e = (e - b_-)(e + P_-(a\hat{y})).$$

Thus b_+ is left invertible, and $e - b_-$ is right invertible. We employ now a standard argument to prove that these elements are invertible. We replace in the argument above a by τa, where $\tau \in [0, 1]$. Then $\tau a \in C$ and $\varrho(\tau a) < 1$, and the formulae for $\hat{x}, \hat{y}, b_+$ and b_- show that we have to replace them by

$$\hat{x}(\tau) = \sum_{n=0}^{\infty} (\tau T_+)^m e, \qquad \hat{y}(\tau) = \sum_{n=0}^{\infty} (\tau T_-)^m e,$$

$$b_+(\tau) = e - \tau P_+(a\hat{y}(\tau)), \quad b_-(\tau) = \tau P_-(\hat{x}(\tau)a),$$

respectively. These functions are continuous on $[0, 1], b_+(\tau)$ is left invertible and $e - b_-(\tau)$ is right invertible for all $\tau \in [0, 1]$, and $b_+(0) = e - b_-(0) = e$ is invertible. By Lemma 23.2 in [16], $b_+ = b_+(1)$ and $b_- = b_-(1)$ are invertible and $b_+^{-1} = \hat{x} \in A_+ \cap C, (e - b_-)^{-1} = e + P_-(a\hat{y}) \in C$ and $(e - b_-)^{-1} - e = P_-(a\hat{y}) \in A_- \cap C$. We obtained above the equation $b_+ = (e - a)(e + P_-(a\hat{y}))$. Multiplying on the right by $e - b_-$, we obtain $b_+(e - b_-) = e - a$. □

We will apply Theorem 4.1 to the Wiener algebra $W(A;\mathbb{T})$, see [11], where A is an ordered Banach algebra and $\mathbb{T}=\mathbb{T}_1$ denotes the unit circle, i.e.,

$$W(A;\mathbb{T})=\{a(\cdot)\ :\ a(\lambda)=\sum_{k=-\infty}^{\infty}\lambda^k a_k, \lambda\in\mathbb{T},\ (a_n)\in l^1_{\mathbb{Z}}(A)\}$$

where $l^1_{\mathbb{Z}}(A)=\{(a_k)_{k\in\mathbb{Z}}\subset A|\ \sum_{k=-\infty}^{\infty}\|a_k\|$ is convergent$\}$. Clearly, a_k is the kth Fourier coefficient of $a(\cdot)\in W(A;\mathbb{T})$.

Let A be a Banach algebra. Then $W(A;\mathbb{T})$ is a Banach algebra with norm $\|a(\cdot)\|_W=\sum_{k=-\infty}^{\infty}\|a_k\|$. $W(A;\mathbb{T})$ is the direct sum of the closed subalgebras

$$\begin{aligned} W_+(A;\mathbb{T}) &= \{\, a(\cdot)\ :\ a_k=0 \text{ for } k<0\},\\ W_{-,0}(A;\mathbb{T}) &= \{\, a(\cdot)\ :\ a_k=0 \text{ for } k\geq 0\}. \end{aligned}$$

Obviously, $W_+(A;\mathbb{T})$ is the set of those functions in $W(A;\mathbb{T})$ that admit continuous extensions onto $\{\lambda\in\mathbb{C}:\ |\lambda|\leq 1\}$ and are analytic on $\mathbb{D}_1$, while $W_{-,0}(A;\mathbb{T})$ is the set of those functions in $W(A;\mathbb{T})$ that admit continuous extensions onto $\{\lambda\in\mathbb{C}\mid|\lambda|\geq 1\}$, are analytic on $\{\lambda\in\mathbb{C};\ |\lambda|>1\}$ and vanish at infinity. Clearly, the operator P_+ defined on $W(A;\mathbb{T})$ by

$$(P_+a(\cdot))(\lambda)=\sum_{k=0}^{\infty}\lambda^k a_k,\quad \lambda\in\mathbb{T},$$

is the projection of $W(A;\mathbb{T})$ onto $W_+(A;\mathbb{T})$ annihilating $W_{-,0}(A;\mathbb{T})$. $e_W(\cdot)$ with $e_W(\lambda)\equiv e$ for all $\lambda\in\mathbb{T}$ is the identity in $W(A;\mathbb{T})$. For $a(\cdot)\in W(A;\mathbb{T})$ we denote by $\varrho_W(a(\cdot))$ the spectral radius of $a(\cdot)$ in the Banach algebra $W(A;\mathbb{T})$.

Let A be an ordered Banach algebra with algebra cone C. Then $W(A;\mathbb{T})$ is an ordered Banach algebra with algebra cone

$$C_W=\{a(\cdot)\in W(A;\mathbb{T}):\ a_k\in C \text{ for all } k\in\mathbb{Z}\}.$$

If C is normal (or closed), then C_W is normal (or closed, respectively).

The following corollary is the reformulation of Theorem 4.1 for the Wiener algebra $W(A;\mathbb{T})$; see also Theorem 23.4 in [16].

Corollary 4.2. *Let A be an ordered Banach algebra with closed normal algebra cone C. Suppose $a(\cdot)\in W(A;\mathbb{T})$ such that*

$$a(\cdot)\in C_W \quad \text{and } \varrho_W(a(\cdot))<1$$

Then $e_W(\cdot)-a(\cdot)$ admits a canonical factorization with respect to the unit circle; i.e., there exist $b_+(\cdot)\in W_+(A;\mathbb{T})$ and $b_-(\cdot)\in W_{-,0}(A;\mathbb{T})$ such that

$$e-a(\lambda)=b_+(\lambda)(e-b_-(\lambda)) \quad \text{for all } \lambda\in\mathbb{T},$$

and $b_+(\cdot)\in W_+(A;\mathbb{T}), b_-(\cdot)\in W_-(A;\mathbb{T})\cap C_W, e_W(\cdot)-b_+(\cdot)\in W_+(A;\mathbb{T})\cap C_W, b_+(\cdot)^{-1}\in W_+(A;\mathbb{T})\cap C_W, (e_W(\cdot)-b_-(\cdot))^{-1}-e_W(\cdot)\in W_{-,0}(A;\mathbb{T})\cap C_W$ and $(e_W(\cdot)-b_-(\cdot))^{-1}\in C_W$.

For functions in $W(A;\mathbb{T})$ with only finitely many non-zero Fourier coefficients we obtain stronger results. Note that $W(A;\mathbb{T})$ is a subalgebra of $C(A;\mathbb{T}) := \{f \mid f : \mathbb{T} \to A \text{ is continuous}\}$ and $\| a(\cdot) \|_\infty := \max_{|\lambda|=1} \| a(\lambda) \|_A \leq \| a(\cdot) \|_W$ for all $a(\cdot) \in W(A;\mathbb{T})$. In the following we sometimes write for $a \in A$ for clarity $\|a\|_A$ and $\varrho_A(a)$ instead of $\|a\|$ and $\varrho(a)$, respectively.

Lemma 4.3. *Let A be a Banach algebra, and let $a(\cdot) \in W(A;\mathbb{T})$ be such that $a_k = 0$ for all $k \in \mathbb{Z}$ with $|k| \geq m$ for some $m \in \mathbb{N}$. Then*

$$\| a(\cdot) \|_W \leq (2m+1) \| a(\cdot) \|_\infty \quad \textit{and} \quad \varrho_W(a(\cdot)) = \varrho_\infty(a(\cdot)).$$

Proof. The inequality follows from Cauchy's formula for the coefficients a_k of $a(\cdot)$, see [16, p.127]. Applying this inequality to $a^n(\cdot)$, we obtain $\| a^n(\cdot) \|_W \leq (2nm + 1) \|a^n(\cdot) \|_\infty$ for $n = 1, 2 \dots$. Using the known formula for the spectral radius, we have that $\varrho_W(a(\cdot)) \leq \varrho_\infty(a(\cdot))$. The reverse inequality follows from $\| a^n(\cdot) \|_\infty \leq \| a^n(\cdot) \|_W$ for $n = 1, 2 \dots$. □

The next lemma is very similar to Lemma 3.1, and the proof of Lemma 3.1 needs only minor technical changes.

Lemma 4.4. *Let A be an ordered Banach algebra with closed normal algebra cone C. Then there exists a $\beta > 0$ such that*

$$\| a(\cdot) \|_\infty \leq \beta \| a(1) \|_A \quad \textit{and} \quad \varrho_\infty(a(\cdot)) = \varrho_A(a(1))$$

for all $a(\cdot) \in C_W$.

Theorem 4.5. *Let A be an ordered Banach algebra with closed normal algebra cone C. Suppose that $a(\cdot) \in W(A;\mathbb{T})$ and $m \in \mathbb{N}$ is such that*

$$a_k = 0 \textit{ for } |k| > m, \quad a(\cdot) \in C_W \quad \textit{and} \quad \varrho_A(a(1)) < 1.$$

Then $e_W(\cdot) - a(\cdot)$ admits a canonical factorization with respect to the unit circle and all assertions of Corollary 4.2 hold.

Proof. Lemmata 4.3 and 4.4 show that $\varrho_W(a(\cdot)) = \varrho_\infty(a(\cdot)) = \varrho_A(a(1)) < 1$. Therefore, the assumptions of Corollary 3.4 are satisfied. □

The results of this section can be used to study matrix functions having values which are (in a certain sense) entrywise nonnegative matrices; for the factorization theory of general matrix functions relative to a curve we refer to [11]. In the next section we will apply the last theorem to polynomials with coefficients in the cone of an ordered Banach algebra.

5. Factorization of polynomials with nonnegative coefficients

In this section we apply the results of the preceding section to obtain a factorization of polynomials with coefficients in a normal cone of an ordered Banach algebra; the next theorem is the main result of this paper.

Theorem 5.1. *Let A be an ordered Banach algebra with closed normal algebra cone C. Let $q(\cdot)$ be a polynomial $q(\lambda) = \lambda^m e - \sum_{j=0}^{l} \lambda^j a_j$ with $1 \leq m < l$ and $a_j \in C$ for $j = 0, 1 \ldots, l$, and $a_l \neq 0$. Suppose there exists a $r_0 > 0$ such that*

$$\varrho(s(r_0)) < {r_0}^m,$$

where $s(\lambda) = \sum_{j=0}^{l} \lambda^j a_j$.

Then

1. *The A-valued function $e - a(\cdot)$ with $a(\lambda) = \sum_{j=0}^{l} \lambda^{j-m} a_j = \lambda^{-m} s(\lambda)$ for $\lambda \neq 0$ admits a (right) canonical factorization with respect to the circle $\mathbb{T}_{r_0} = \{\lambda \in \mathbb{C} \mid |\lambda| = r_0\}$;*
2. *$q(\cdot)$ has a monic spectral (right) divisor $c(\cdot)$ of degree m with nonnegative coefficients, i.e.,*

$$c(\lambda) = \lambda^m e - \sum_{j=0}^{m-1} c_j \lambda^j, \quad c_j \in C \tag{5.1}$$

for $j = 0, 1, 2 \ldots, m-1$, and

$$\sigma(c(\cdot)) = \sigma(q(\cdot)) \cap \{\lambda \in \mathbb{C} \mid |\lambda| < r_0\}.$$

3. *Let $b(\cdot)$ be the (uniquely defined) polynomial of degree $l - m$ such that $q(\lambda) = b(\lambda)c(\lambda)$ for all $\lambda \in \mathbb{C}$. Then*

$$b(\lambda) = b_0(e - \sum_{j=1}^{l-m} \lambda^j b_j),\ b_0 \text{ is invertible, } b_0^{-1} \in C \text{ and } b_j \in C \tag{5.2}$$

for $j = 1, 2, \ldots, l-m$,

$$\sigma(b(\cdot)) = \sigma(q(\cdot)) \cap \{\lambda \in \mathbb{C} \,:\, |\lambda| > r_0\}.$$

4. *$c(t)^{-1} \in C$ for $r_0 \leq t$; $b(t)^{-1} \in C$ for $0 \leq t \leq r_0$.*

Proof. 1. The function $a_{r_0}(\cdot)$ with $a_{r_0}(\lambda) = a(r_0\lambda)$ for $\lambda \in \mathbb{T}_1$ satisfies the assumptions of Theorem 4.5. Therefore we have $e - a_{r_0}(\lambda) = b_+(\lambda)(e - b_-(\lambda))$ for $\lambda \in \mathbb{T}_1$ with some $b_+(\cdot) \in W_+(A; \mathbb{T})$ and some $b_-(\cdot) \in W_{-,0}(A; \mathbb{T})$.
2. Then $q(r_0\lambda) = r_0^m b_+(\lambda)(\lambda^m e - \lambda^m b_-(\lambda))$ for $\lambda \in \mathbb{T}_1$. Now $b_+(\cdot)$ and $b_+^{-1}(\cdot)$ have analytic extensions onto $\mathbb{D}_1$, and $b_-(\cdot)$ has an analytic extension onto $\{z \in \mathbb{C} : |z| > 1\}$ which vanishes at infinity. Then it follows (cf. 22.11 in [16]) that $c(\lambda) = \lambda^m e - \lambda^m b_-(r_0^{-1}\lambda)$ is a monic polynomial of degree m and $c(\cdot)$ is a spectral divisor of $q(\cdot)$ with $\sigma(c(\cdot)) = \sigma(q(\cdot)) \cap \{z \in \mathbb{C} : |z| < r_0\}$. Now $b_-(\cdot) \in C_W$, therefore $\lambda^m b_-(r_0^{-1}\lambda) = \lambda^{m-1} c_{m-1} + \cdots + c_0$ with $c_j \in C$.
3. That $b(\cdot)$ is uniquely defined follows from Lemma 22.8 in [16]; this Lemma is proved for an algebra of operators but its proof works in our more general situation. From the proof of part 2 it follows that $b(\lambda) = b_+(r_0^{-1}\lambda)$ for $\lambda \in \mathbb{C}$. $b_+(\cdot)^{-1} \in$

$C_W \cap W_+(A;\mathbb{T})$ implies $\sigma(b(\cdot)) \subset \mathbb{C}\backslash\overline{\mathbb{D}_{r_0}}$ and $b(\lambda)^{-1} = \sum_{k=0}^{\infty} \lambda^k f_k$ for $|\lambda| \le r_0$ with $f_k \in C$ for $k = 0, 1, 2 \ldots$. Therefore $b(t)^{-1} \in C$ for $0 \le t \le r_0$; especially, b_0 is invertible and $b_0^{-1} \in C$. From Corollary 4.2 we know that $e_W(\cdot) - b_+(\cdot) \in C_W$. Then $e - b_0 \in C$ and $b_j \in C$ for $j = 1, \ldots, l-m$.

4. From $t > \varrho(c(\cdot))$ we obtain

$$c(t)^{-1} = \left(t^m e - \sum_{j=0}^{m-1} t^j\, c_j\right)^{-1} = t^{-m} \sum_{k=0}^{\infty} \hat{c}^k \in C \quad \text{where} \quad \hat{c} = \sum_{j=0}^{m-1} t^{j-m} c_j \in C.$$

The proof for $b(t)^{-1} \in C$ is similar. □

Corollary 5.2. *Let A be an ordered Banach algebra with closed normal cone C and let $q(\cdot)$ be the polynomial $q(\lambda) = \lambda^m e - \sum_{j=0}^{l} \lambda^i a_j = \lambda^m e - s(\lambda)$ where $1 \le m < l$ and $a_j \in C, a_l \neq 0$. Then there exist polynomials $b(\cdot)$ and $c(\cdot)$ as in (5.1) and (5.2) with $q(\lambda) = b(\lambda)c(\lambda)$ for all $\lambda \in \mathbb{C}$ and an $r_0 > 0$ such that part 4 of Theorem 5.1 holds if and only if $\varrho(s(r_0)) < r_0^m$.*
Any factorization with these properties is unique.

Proof. We have to prove the "only if" part. Let ${r_0}^m e - s(r_0) = q(r_0) = b(r_0)c(r_0)$ be such that $b(r_0)^{-1}$ and $c(r_0)^{-1}$ exist and belong to C. Then ${r_0}^m e - s(r_0)$ has an inverse in C. It follows from Proposition 2.1.3 that $\varrho(s(r_0)) < {r_0}^m$. The last assertion follows from [16, Lemma 22.8]. □

Theorem 5.3. *Let A be an ordered complex Banach algebra with closed normal cone C and let the polynomials $q(\cdot)$ and $s(\cdot)$ satisfy the assumptions of the last theorem; hence $\varrho_S(r_0) < r_0^m$ for some $r_0 > 0$. Then*

1. *Either $\varrho_s(t) < t^m$ for all $t \in]0, r_0]$, or there exists exactly one $r_1 \in]0, r_0[$ such that $\varrho_S(r_1) = {r_1}^m$. In the second case we have*

$$\varrho_s(t) \;<\; t^m \quad \textit{for} \quad r_1 < t \le r_0, \tag{5.3}$$
$$\varrho_s(t) \;>\; t^m \quad \textit{for} \quad 0 < t < r_1, \tag{5.4}$$

$r_1 = \varrho(c(\cdot)) \in \sigma(c(\cdot))$ and ${r_1}^m = \varrho(c_+(r_1))$, here $c_+(\lambda) = \sum_{j=0}^{m-1} c_j\lambda^j$, see (5.1).

2. *Either $\varrho_s(t) < t^m$ for all $t \in [r_0, \infty[$, or there exists exactly one $r_2 \in]r_0, \infty[$ such that $\varrho_s(r_2) = {r_2}^m$. In the second case we have*

$$\varrho_s(t) \;<\; t^m \quad \textit{for} \quad r_0 \le t < r_2, \tag{5.5}$$
$$\varrho_s(t) \;>\; t^m \quad \textit{for} \quad r_2 < t, \tag{5.6}$$

$r_2 \in \sigma(b(\cdot)), \sigma(b(\cdot)) \subset \{\lambda \in \mathbb{C} : r_2 \le |\lambda|\}$ and $\varrho(b_+(r_2)) = 1$, here $b_+(\lambda)) = \sum_{j=1}^{l-m} \lambda^j b_j$, see (5.2).

Proof. 1. Assume that $\varrho_s(t) \geq t^m$ for some $t \in]0, r_0[$. The function $\varrho_s(\cdot)$ is continuous, by the intermediate-value theorem there exists an $r_1 \in [t, r_0[$ such that $\varrho_s(r_1) = r_1^m$. For any such r_1 we obtain from the convexity of $\eta_s(\cdot)$ (see Proposition 3.2) that $\eta_s(u) \leq \dfrac{\eta_s(u_0) - mu_1}{u_0 - u_1}(u - u_1) + mu_1$ for $u_1 \leq u \leq u_0$. Here $u_j = \log r_j$ for $j = 0, 1$. Clearly $\eta_s(u_0) < mu_0$ implies $\eta_s(u) < mu$ for $u_1 < u \leq u_0$. This is equivalent to $\varrho_s(t) < t^m$ for $t \in]r_1, r_0]$. Now it is clear there is exactly one $r \in]0, r_0]$ such that $\varrho_s(r) = r^m$. The inequalities (5.3) and (5.4) are now clear.
For this r_1 we have ${r_1}^m \in \sigma(s(r_1))$ by Proposition 2.1.1. Therefore $r_1 \in \sigma(q(\cdot)) \cap \mathbb{D}_{r_0} = \sigma(c(\cdot))$ by Theorem 5.1.2, thus $r_1 \leq \varrho(c(\cdot))$. Set $r = \varrho(c(\cdot))$. Then $r \in [r_1, r_0]$. Further $r \in$ boundary $\sigma(c(\cdot)) \subset \sigma(q(\cdot))$, see [16, Lemma 22.3, p.112], therefore $r^m \in \sigma(s(r))$, and then $r^m \leq \varrho(s(r))$. From (5.3) we obtain $r_1 = r = \varrho(c(\cdot))$. For monic polynomials of degree m with coefficients in C we have $\varrho(c_+(\cdot)) = \varrho(c(\cdot))^m = r^m$, see [19, Proposition 2.1].
2. Assume that $\varrho_s(t) \geq t^m$ for some $t \in]r_0, \infty[$. A similar argument as in the proof of part 1 shows that there exists exactly one $r_2 \in]r_0, \infty[$ with $\varrho_s(r_2) = {r_2}^m$ and $\varrho_s(t) < t^m$ for $t \in [r_0, r[$. The inequalities (5.5) and (5.6) are now clear.
For a proof of the other assertions we define $\tilde{b}(\cdot)$ by $\tilde{b}(\lambda) = \lambda^{l-m} {b_0}^{-1} b(1/\lambda)$ for $\lambda \neq 0$ and $\tilde{b}(0) = b_{l-m}$, here we use (5.2). Then $\tilde{b}(\cdot)$ is the monic polynomial with nonnegative coefficients

$$\tilde{b}(\lambda) = \lambda^{l-m} e - (\lambda^{l-m-1} b_1 + \cdots + b_{l-m}) = \lambda^{l-m} e - \tilde{b}_+(\lambda).$$

Note that $\sigma(\tilde{b}(\cdot)) \backslash \{0\} = \{\frac{1}{\lambda} : \lambda \in \sigma(b(\cdot))\} = \{\frac{1}{\lambda} : \lambda \in \sigma(q(\cdot)), r_0 \leq |\lambda|\}$ by Theorem 5.1.3. We will show that $\frac{1}{r_2} = \varrho(\tilde{b}(\cdot))$. Now $q(r_2) = {r_2}^{m-l} b_0 \tilde{b}(\frac{1}{r_2}) c(r_2)$ is singular but b_0 and $c(r_2)$ are invertible, thus $\tilde{b}(\frac{1}{r_2})$ is singular. Then $\frac{1}{r_2} \in \sigma(\tilde{b}(\cdot))$, and $\frac{1}{r_2} \leq \varrho(\tilde{b}(\cdot))$. Set $\tilde{r} = \varrho(\tilde{b}(\cdot))$. Then $\tilde{r} \in$ boundary $\sigma(\tilde{b}(\cdot))$, and $b(\frac{1}{\tilde{r}})$ is singular. Therefore $q(\frac{1}{\tilde{r}})$ is not invertible, see [19, Proposition 2.1], and $r_0 < \frac{1}{\tilde{r}} \leq r_2$. Thus $\frac{1}{\tilde{r}^m} \in \sigma(s(\frac{1}{\tilde{r}}))$, and then $\frac{1}{\tilde{r}^m} \leq \varrho(s(\frac{1}{\tilde{r}}))$. From (5.5) we obtain $\frac{1}{\tilde{r}} = r_2$.
Now $r_2 \in \{\frac{1}{\lambda} : \lambda \in \sigma(\tilde{b}(\cdot)), \lambda \neq 0\} = \sigma(b(\cdot)) \subset \{\lambda \in \mathbb{C} : r_2 \leq |\lambda|\}$. By [19, Proposition 2.1] we have $r_2^{m-l} = \varrho(\tilde{b}(\frac{1}{r_2}))$, and then $\varrho(b_+(r_2)) = \varrho({r_2}^{l-m} \tilde{b}_+(\frac{1}{r_2})) = 1$. □

6. Operator polynomials with nonnegative coefficients

In this section we consider the operator polynomial

$$Q(\lambda) = \lambda^m I - \lambda^l A_l - \cdots - \lambda A_1 - A_0 = \lambda^m I - S(\lambda) \tag{6.1}$$

where $A_0, \ldots, A_l$ are nonnegative (= positive) operators on a complex Banach lattice E with $A_l \neq 0$.
We assume that $1 \leq m < l$ and $\varrho(S(r_0)) < r_0^m$ for some positive number r_0. The results of the preceding section (Theorem 5.1) show that $Q(\cdot)$ can be factorized in a special way, namely

$$Q(\lambda) = B_0 B(\lambda) C(\lambda) \quad \text{for all } \lambda \in \mathbb{C}. \tag{6.2}$$

Here B_0 is invertible with $B_0^{-1} \in L(E)_+$, $B(\cdot)$ is a comonic operator polynomial of degree $l - m$ with nonnegative coefficients, i.e.,

$$B(\lambda) = I - \lambda B_1 - \cdots - \lambda^{l-m} B_{l-m} = I - B_+(\lambda), \tag{6.3}$$

where $B_j \in L(E)_+$ for $j = 1, \ldots, l-m$ and $B_{l-m} = B_0^{-1} A_l \neq 0$, and $C(\cdot)$ is a monic operator polynomial of degree m with nonnegative coefficients, i.e.,

$$C(\lambda) = \lambda^m I - \lambda^{m-1} C_{m-1} - \ldots - \lambda_1 C_1 - C_0 = \lambda^m I - C_+(\lambda) \tag{6.4}$$

where $C_j \in L(E)_+$ for $j = 0, 1, \ldots, m-1$.

We will use results on monic (and comonic) operator polynomials with nonnegative coefficients from [4], [5] and [19] to describe spectral properties of $Q(\cdot)$ on the circles with radius r such that $\varrho_S(r) = r^m$.

Theorem 6.1. *Let $E, Q(\cdot), S(\cdot), B(\cdot), C(\cdot), B_+(\cdot), C_+(\cdot), B_0$ and r_0 be as above, especially $\varrho(S(r_0)) < r_0{}^m$. Then for all $r \in [0, \infty[$ with $\varrho_S(r) = r^m$ the following statements hold:*

1. $r^m \in \sigma(S(r))$ *and* $r \in \sigma(Q(\cdot))$.
2. *Let $0 < r < r_0$. Then r is a pole of order k of $Q^{-1}(\cdot)$ and the space of all Jordan chains of $Q(\cdot)$ corresponding to r has a finite dimension h if and only if r is a pole of $R(\cdot, r^{-m+1} C_+(r))$ with residue of finite rank h.*
3. *Let $r_0 < r$. Then r is a pole of order k of $Q^{-1}(\cdot)$ and the space of all Jordan chains of $Q(\cdot)$ corresponding to r has a finite dimension h if and only if 1 is a pole of $R(\cdot, B_+(r))$ with residue of finite rank h.*
4. *If r is a pole of $Q^{-1}(\cdot)$ and the space of all Jordan chains of $Q(\cdot)$ corresponding to r is finite-dimensional, then*

$$\{\lambda \in \sigma(Q(\cdot)) \mid \mid \lambda \mid = r\} = r \cdot \mathbb{G}.$$

   here $\mathbb{G}$ is the union of finitely many (finite) groups of roots of unity and consists entirely of poles of $Q^{-1}(\cdot)$
5. *Let r and λ_0 be poles of $Q^{-1}(\cdot)$ with $\mid \lambda_0 \mid = r$. Then the following hold:*
   (a) *The order of the pole λ_0 of $Q^{-1}(\cdot)$ is not greater than the order of the pole r of $Q^{-1}(\cdot)$.*
   (b) *Every nontrivial Jordan chain of $Q(\cdot)$ corresponding to λ_0 is a linearly independent set.*

Proof. Let $0 < r < r_0$ be such that $\varrho_S(r) = r^m$. Then $B(\lambda_0)$ is invertible for all $\lambda_0 \in \mathbb{T}_r$ by Theorem 5.1.3. Therefore $Q^{-1}(\cdot)$ has a pole of order k at λ_0 if and only $C^{-1}(\cdot)$ has a pole of order k at λ_0. By [16, Lemma 22.5, p.123] the Jordan chains of $Q(\cdot)$ and $C(\cdot)$ corresponding to λ_0 coincide. Now the assertions of the theorem concerning the case $0 < r < r_0$ follow from the corresponding results of [19]; namely, assertion 2 follows from [19, Corollary 5.5], assertion 4 follows from [19, Corollary 4.7] and assertion 5(b) follows from [19, Corollary 5.4]. Assertion 5(a) is a consequence of the Pringsheim Theorem, see [22, p.262].
Let $r_0 < r$ be such that $\varrho_S(r) = r^m$. Then $C(\lambda_0)$ is invertible for all $\lambda_0 \in \mathbb{T}_r$ by Theorem 5.1.3. Therefore $Q^{-1}(\cdot)$ has a pole of order k at λ_0 if and only if $B^{-1}(\cdot)$

has a pole of order k at λ_0. Direct computation shows: x_j for $j = 0, 1, \ldots, p$ form a Jordan chain of $Q(\cdot)$ corresponding to λ_0 if and only if the vectors

$$y_j = \sum_{i=0}^{j} \binom{j}{i} C^{(j-i)}(\lambda_0) x_i \quad \text{for} \quad j = 0, 1, \ldots, p$$

form a Jordan chain of $B(\cdot)$ corresponding to λ_0. Therefore $Q^{-1}(\cdot)$ has a pole of order k at λ_0 if and only $B^{-1}(\cdot)$ has a pole of order k at λ_0; recall that the order of a pole is the length of the longest nontrivial Jordan chain corresponding to it and that a Jordan chain is called nontrivial if its first vector is non-zero. Further, it follows that the space of Jordan chains of $Q(\cdot)$ corresponding to λ_0 has a finite dimension h if and only if the space of Jordan chains of $B(\cdot)$ corresponding to λ_0 has a finite dimension h. We define now the monic operator polynomial $\tilde{B}(\cdot)$ by $\tilde{B}(\lambda) = \lambda^{l-m} B(1/\lambda)$ for $\lambda \neq 0$. Note that (see the proof of Theorem 5.3.2)

$$\sigma(\tilde{B}(\cdot))\backslash\{0\} = \{\frac{1}{\lambda} : \lambda \in \sigma(B(\cdot))\} \quad \text{and} \quad \tilde{r} = \frac{1}{r} = \varrho(\tilde{B}(\cdot)).$$

Then the companion operator $C_{\tilde{B}}$ of $\tilde{B}(\cdot)$ is the comonic companion operator of $B(\cdot)$, see [12, p. 187]; note that the results formulated there for the matrix case hold also for bounded linear operators in Banach spaces. Note also that in [16, §12] slightly different companion operators are defined, but the two variants are cogredient, i.e., they coincide after a permutation of some operator rows and corresponding operator columns. Therefore the results of [16, §12 and §13] hold for the companion operators defined above and the following sequence of equivalences holds.

$Q^{-1}(\cdot)$ has a pole of order k at λ_0 (and the space of Jordan chains of $Q(\cdot)$ corresponding to λ_0 has finite dimension h) if and only if

$B^{-1}(\cdot)$ has a pole of order k at λ_0 (and the space of Jordan chains of $B(\cdot)$ corresponding to λ_0 has a finite dimension h, respectively) if and only if

$(I - \cdot C_{\tilde{B}})^{-1}$ has a pole of order k at λ_0 (and the space of Jordan chains of $(I - \cdot C_{\tilde{B}})$ corresponding to λ_0 has finite a dimension h, respectively), see [16, Lemma 12.5 and 12.6 and the remarks on p. 63] if and only if

the resolvent $R(\cdot, C_{\tilde{B}}) = (\cdot I - C_{\tilde{B}})^{-1}$ has a pole at $\tilde{\lambda}_0 = \frac{1}{\lambda_0}$ of order k (and the residuum of this resolvent at λ_0 has finite rank h, respectively), see [16, Lemma 12.8] if and only if

$\tilde{B}^{-1}(\cdot)$ has a pole of order k at $\tilde{\lambda}_0$ and the space of Jordan chains of $\tilde{B}(\cdot)$ corresponding to $\tilde{\lambda}_0$ has finite a dimension h, respectively), see [15, 2.1, Hilfssatz 1].

Now assertion 4 follows from [19, Corollary 4.7]. Assertion 5(a) is a consequence of the Pringsheim Theorem, see [22, p. 262].

From the connection between the Jordan chains of $Q(\cdot)$ and $B(\cdot)$ given above it follows directly that the nontrivial Jordan chains of $Q(\cdot)$ corresponding to $\lambda_0 \neq 0$ are linearly independent if and only if the nontrivial Jordan chains of $B(\cdot)$ corresponding to $\lambda_0 \neq 0$ are linearly independent if and only if the nontrivial Jordan chains of $\tilde{B}(\cdot)$ corresponding to $\tilde{\lambda}_0 = \frac{1}{\lambda_0}$ are linearly independent, see [16,

Lemma 12.3 and 12.7 and remarks on p.63]. Assertion 5(b) follows now from [19, Corollary 5.4].
If $\lambda_0 = r$, we can continue the equivalences above by [19, Corollary 5.5], and obtain:
$Q^{-1}(\cdot)$ has a pole of order k at r and the space of Jordan chains of $Q(\cdot)$ corresponding to r has a finite dimension h if and only if
$\tilde{r}$ is a pole of the resolvent $R(\cdot, \tilde{r}^{-l+m+1})\tilde{B}_+(\tilde{r})$ and the residuum of this resolvent at $\tilde{r}$ has finite rank h, and this assertion is equivalent to
1 is a pole of the resolvent $R(\cdot, B_+(r))$ and the residuum of this resolvent at 1 has finite rank h (since $\tilde{r}^{-l+m+1}\tilde{B}_+(\tilde{r}) = \tilde{r}B_+(r)$). □

Acknowledgement

The authors wish to thank two referees for their careful work and constructive remarks.

References

[1] B. Aupetit, *A Primer on Spectral Theory.* Springer-Verlag, New York – Berlin – Heidelberg, 1991.

[2] F.F. Bonsall, *Endomorphismus of a Partially Ordered Vector Space without Order Unit.* J. London Math. Soc. **30** (1955), 144–153.

[3] K.F. Clancey and I. Gohberg, *Factorization of Matrix Functions and Singular Integral Operators.* Birkhäuser, Basel and Boston, 1981.

[4] K.-H. Förster and B. Nagy, *Some Properties of the Spectral Radius of a Monic Operator Polynomial with Nonnegative Compact Coefficients.* Integral Equations Operator Theory, **14** (1991), 794–805.

[5] K.H. Förster and B. Nagy, *On the Linear Independence of Jordan Chains.* Operator Theory: Advances and Applications, 122 (2001), 229–245.

[6] H.R. Gail, S.L. Hantler and B.A. Taylor. *Spectral Analysis of M/G/1 and G/M/1 Type Markov Chains.* Adv. Appl. Prob. **28** (1996), 114–165.

[7] H.R. Gail, S.L. Hantler and B.A. Taylor. *Matrix-Geometric Invariant Measures for G/M/1 Type Markov Chains.* Comm. Statist. Stochastic Models **14** (3) (1998), 537–569.

[8] I. Gohberg and I. Feldman. *Convolution Equations and Projection Methods for their Solution.* Amer. Math. Soc., Providence, R. I., 1974.

[9] I. Gohberg, S. Goldberg and M.A. Kaashoek, *Classes of Linear Operators.* I.OT49. Birkhäuser Verlag, 1990.

[10] I. Gohberg, S. Goldberg and M.A. Kaashoek, *Classes of Linear Operators,* Vol. II. Birkhäuser Verlag, Basel and Boston, 1993.

[11] I. Gohberg, M.A. Kaashoek and I.M. Spitkovsky, *An Overview of Matrix Factorization Theory and Operator Applications,* in: Operator Theory: Advances and Applications, vol. 141, pp. 1–102, Birkhäuser Verlag, Basel and Boston, 2003.

[12] I. Gohberg, P. Lancaster and L. Rodman, *Matrix Polynomials.* Academic Press, New York, 1982.

[13] K.P. Hadeler, *Eigenwerte von Operatorpolynomen.* Archive Rat. Mech. Appl. **20** (1965), 72–80.

[14] W.K. Hayman, P.B. Kennedy, *Subharmonic Functions.* Academic Press, London – New York – San Francisco, 1976.

[15] G. Maibaum, *Über Scharen positiver Operatoren.* Math. Ann. **184** (1970), 238–256.

[16] A.S. Markus, *Introduction to the Spectral Theory of Polynomial Operator Pencils.* Transl. of Math. Monographs, vol. 71. Amer. Math. Soc., Providence, 1988.

[17] P. Meyer-Nieberg, *Banach Lattices.* Springer-Verlag, New York, 1991.

[18] T.W. Palmer, *Banach Algebras and The General Theory of *-Algebras.* Cambridge University Press, Cambridge, 1994.

[19] R.T. Rau, *On the Peripheral Spectrum of a Monic Operator Polynomial with Positive Coefficients.* Integral Equations Operator Theory **15** (1992), 479–495.

[20] H. Raubenheimer and S. Rode, *Cones in Banach Algebras.* Indag. Math. (N.S.) **7** (4) (1996), 489–502.

[21] L. Rodman, *An Introduction to Operator Polynomials.* Operator Theory: Advances and Applications, vol. 38, Birkhäuser: Basel-Boston-Berlin, 1989.

[22] H.H. Schaefer, *Banach Lattices and Positive Operators.* Springer Verlag, New York, 1980.

[23] H.H. Schaefer, *Topological Vector Spaces.* Springer Verlag, New York, 1971.

Karl-Heinz Förster
Technische Universität Berlin
Fakultät II
Institut für Mathematik, MA 6-4
Straße des 17. Juni 136
D-10623 Berlin
Germany
e-mail: `foerster@math.tu-berlin.de`

Béla Nagy
Department of Analysis
Institute of Mathematics
Technical University of Budapest
H-1521 Budapest
Hungary
e-mail: `bnagy@math.bme.hu`

Operator Theory:
Advances and Applications, Vol. 163, 163–189

Szegő Pairs of Orthogonal Rational Matrix-valued Functions on the Unit Circle

Bernd Fritzsche, Bernd Kirstein and Andreas Lasarow

Dedicated to Professor Heinz Langer

Abstract. We study distinguished pairs of orthonormal systems of rational matrix-valued functions on the unit circle, namely the so-called Szegő pairs. These pairs are determined by an initial condition and a sequence of strictly contractive $q \times q$ matrices, which is called the sequence of Szegő parameters. The Szegő parameters contain essential information on the underlying $q \times q$ nonnegative Hermitian Borel measure on the unit circle.

Mathematics Subject Classification (2000). Primary 42C05, 47A56.

Keywords. Orthogonal rational matrix-valued functions, Recurrence relations of Szegő-type, Szegő parameters, Favard-type theorem.

0. Introduction

This paper continues the line of investigations started in [FKL1]–[FKL3] where we realized first steps towards a Szegő theory of orthogonal rational matrix-valued functions on the unit circle $\mathbb{T}$. The main feature of our conception of Szegő theory is the distinguished role of the Christoffel-Darboux formulas (see [FKL2], [FKL3]). As a cornerstone of our approach we introduced in [FKL2, Section 6] the notions of left and right Christoffel-Darboux pairs for rational matrix-valued functions. The recursion formulas for left and right Christoffel-Darboux pairs which were obtained in Section 2 of [FKL3] realize a crucial step in constructing our Szegő theory. Namely, these recursion formulas indicated that there is a one-to-one correspondence between Christoffel-Darboux pairs and sequences of matrices which are connected to particular signature matrices.

The work of the third author of the present paper was supported by the German Academy of Natural Scientists Leopoldina by means of the Federal Ministry of Education and Research on badge BMBF-LPD 9901/8-88.

The main topic of this paper is to discuss those pairs of orthogonal systems of rational matrix-valued functions which are direct generalizations of the Szegő pairs of orthonormal systems of matrix polynomials which were discussed by Delsarte, Genin, and Kamp in [DGK] (cf. [DFK, Section 3.6]), who developed the theory along the classical case of orthogonal polynomials of Szegő [S] (see also [GS], [G1], and [G2]). The present paper is also guided by the work of Bultheel, González-Vera, Hendriksen, and Njåstad on scalar orthogonal rational functions (see, e.g., [BGHN1], [BGHN2], and the probably first work on this topic by Djrbashian [Dj]). Note that, in the case of matrix polynomials and in the scalar case of rational functions, Szegő recursions for orthogonal functions were derived in [DGK, Section V] and in [BGHN2, Chapter 4], respectively, using a different way as demonstrated below.

Szegő pairs of rational matrix-valued functions on the unit circle $\mathbb{T}$ are special pairs of orthogonal systems which are determined by an initial condition and a sequence of strictly contractive $q \times q$ matrices which are called the Szegő parameters. In [FKL3] it is already shown that, in the generic case, each pair of orthonormal systems of rational matrix-valued functions is characterized by an initial condition and a sequence of $\mathbf{j}_{qq}$-unitary matrices. Roughly speaking, Szegő pairs of orthonormal systems of rational matrix-valued functions correspond to sequences of positive Hermitian $\mathbf{j}_{qq}$-unitary matrices. Since there is a one-to-one correspondence between $\mathbf{j}_{qq}$-unitary matrices and strictly contractive $q \times q$ matrices (see, e.g., [D, Theorem 1.2]), Szegő pairs of rational matrix-valued functions on the unit circle $\mathbb{T}$ can be described by an initial condition and a sequence of strictly contractive $q \times q$ matrices.

A brief synopsis is as follows. In addition to several notations we will use, we explain in Section 1 some basics on rational matrix-valued functions and, especially, we recall the Christoffel-Darboux formulas which every pair of orthonormal rational matrix-valued functions satisfies. In Section 2, we introduce the notion Szegő pair of orthonormal systems in the context of rational matrix-valued functions and, by an application of the Christoffel-Darboux formulas, we will see that such pairs of orthonormal systems are determined by an initial condition and a sequence of strictly contractive $q \times q$ matrices via certain recurrence relations. Then, in Section 3, we study an inverse question about this and we prove a result which is often called Favard-type theorem. Our Favard-type theorem says that if we have a pair of rational matrix-valued functions which satisfies the initial condition and which is recursively connected by some strictly contractive $q \times q$ matrices based on the recurrence relations in Section 2, then there is a (not necessarily unique) nonnegative Hermitian $q \times q$ matrix measure on the Borelian σ-algebra on the unit circle with respect to which the sequences of rational matrix-valued functions form left and right orthonormal systems, respectively. Finally, we will see in Section 4 that the Szegő parameter $\mathbf{E}_n$, which determines the elements X_n and Y_n of a Szegő pair of rational matrix-valued functions by the recurrence relations assuming the knowledge of the previous elements X_{n-1} and Y_{n-1}, can be computed by some integral formulas and, in particular, in terms of X_{n-1} and Y_{n-1}.

1. Notation and preliminaries

Throughout this paper, let p and q belong to the set $\mathbb{N}$ of all positive integers. Let us use $\mathbb{C}, \mathbb{Z}$, and $\mathbb{N}_0$ to denote the sets of all complex numbers, of all integers, and of all nonnegative integers, respectively. If $m \in \mathbb{Z}$ and if $n \in \mathbb{Z}$ or $n = +\infty$, then we will write $\mathbb{N}_{m,n}$ for the set of all integers k which satisfy $m \leq k \leq n$. If $\mathfrak{X}$ is a nonempty set, then let $\mathfrak{X}^{p\times q}$ be the set of $p \times q$ matrices each entry of which belongs to $\mathfrak{X}$. The notation $\mathbf{0}_{p\times q}$ stands for the null matrix that belongs to the set $\mathbb{C}^{p\times q}$ of all $p \times q$ complex matrices, and the identity matrix which belongs to $\mathbb{C}^{q\times q}$ will be denoted by $\mathbf{I}_q$. If the size of a null matrix or a identity matrix is obvious, we will omit the indices. If $\mathbf{A} \in \mathbb{C}^{q\times q}$, then rank $\mathbf{A}$ and det $\mathbf{A}$ indicate the rank of $\mathbf{A}$ and the determinant of $\mathbf{A}$, respectively. Furthermore, if $\mathbf{A}$ is a nonnegative Hermitian $q \times q$ matrix, then $\sqrt{\mathbf{A}}$ stands for the (unique) nonnegative Hermitian square root of $\mathbf{A}$.

The symbol $\mathbb{T}$ stands for the unit circle, $\mathbb{D}$ for its interior, and $\mathbb{E}$ for its exterior with respect to the extended complex plane $\mathbb{C}_0 := \mathbb{C} \cup \{\infty\}$, i.e., let $\mathbb{T} := \{z \in \mathbb{C} : |z| = 1\}$, $\mathbb{D} := \{z \in \mathbb{C} : |z| < 1\}$, and $\mathbb{E} := \mathbb{C}_0 \setminus (\mathbb{D} \cup \mathbb{T})$. We will use the notation $\mathcal{T}_1$ to designate the set of all sequences $(\alpha_j)_{j=1}^{\infty}$ of complex numbers which satisfy $\overline{\alpha_j}\alpha_k \neq 1$ for all positive integers j and k. Obviously, if $(\alpha_j)_{j=1}^{\infty} \in \mathcal{T}_1$, then $\alpha_j \notin \mathbb{T}$ for all $j \in \mathbb{N}$. For technical reasons, throughout this paper, we set

$$\alpha_0 := 0.$$

Let $\mathfrak{B}_{p\times q}$ (respectively, $\mathfrak{B}_1$) be the σ-algebra of all Borel subsets of $\mathbb{C}^{p\times q}$ (respectively, $\mathbb{C}$), and let $\mathfrak{B}_{\mathbb{T}} := \mathfrak{B}_1 \cap \mathbb{T}$. The linear Lebesgue measure on $(\mathbb{T}, \mathfrak{B}_{\mathbb{T}})$ is designated by $\underline{\lambda}$. If f is a matrix-valued function defined on a subset $\mathcal{M}$ of $\mathbb{C}_0$ with $\mathbb{T} \subseteq \mathcal{M}$, then we will write $\underline{f}$ for the restriction of f to $\mathbb{T}$. If $\mathcal{G}$ is a nonempty subset of $\mathbb{C}_0$ and if f is a matrix-valued function defined on $\mathcal{G}$, then $f^*(z)$ is short for $\big(f(z)\big)^*$, $z \in \mathcal{G}$.

If $\alpha \in \mathbb{C} \setminus (\mathbb{T} \cup \{0\})$, then b_α denotes the function $b_\alpha : \mathbb{C}_0 \setminus \{\frac{1}{\overline{\alpha}}\} \to \mathbb{C}$ given by

$$b_\alpha(w) := \begin{cases} \frac{\overline{\alpha}}{|\alpha|} \frac{\alpha - w}{1 - \overline{\alpha} w} & \text{for } w \in \mathbb{C} \setminus \{\frac{1}{\overline{\alpha}}\} \\ \frac{1}{|\alpha|} & \text{for } w = \infty \end{cases}. \tag{1.1}$$

Further, let $b_0 : \mathbb{C} \to \mathbb{C}$ be defined by $b_0(w) := w$ for each $w \in \mathbb{C}$. Clearly, if $\alpha \in \mathbb{D}$, then the function b_α is exactly the elementary Blaschke factor corresponding to α.

Now we turn our attention to left and right $\mathbb{C}^{q\times q}$-modules of rational matrix-valued functions with prescribed pole structure. Let $\tau \in \mathbb{N}$ or $\tau = +\infty$, and let $(\alpha_j)_{j=1}^{\tau}$ be a sequence of complex numbers. Further, let $\pi_{\alpha,0} : \mathbb{C}_0 \to \mathbb{C}$ be the constant function with value 1 and let $\mathcal{R}_{\alpha,0}$ denote the set of all constant complex-valued functions defined on $\mathbb{C}_0$. For each $n \in \mathbb{N}_{1,\tau}$, let $\pi_{\alpha,n} : \mathbb{C} \to \mathbb{C}$ be defined by

$$\pi_{\alpha,n}(w) := \prod_{j=1}^{n} (1 - \overline{\alpha_j} w)$$

and let $\mathcal{R}_{\alpha,n}$ designate the set of all complex-valued functions f which are rational and which admit a representation $f = \frac{1}{\pi_{\alpha,n}} P$ with some complex polynomial P of degree not greater than n. Further, for each $n \in \mathbb{N}_{0,\tau}$, let

$$\mathbb{P}_{\alpha,n} := \bigcup_{j=1}^{n} \left\{ \frac{1}{\overline{\alpha_j}} \right\} \quad \text{and} \quad \mathbb{Z}_{\alpha,n} := \bigcup_{j=1}^{n} \{\alpha_j\},$$

where we use the conventions $\frac{1}{0} := \infty$, $\frac{1}{\infty} := 0$, and $\bigcup_{j=1}^{0} \left\{ \frac{1}{\overline{\alpha_j}} \right\} := \emptyset$. For each $n \in \mathbb{N}_{0,\tau}$, every function which belongs to $\mathcal{R}_{\alpha,n}$ is holomorphic in $\mathbb{C}_0 \setminus \mathbb{P}_{\alpha,n}$. Obviously, if $0 \leq n < \tau$, then $\mathcal{R}_{\alpha,n}$ does not depend on the numbers $\alpha_j, j \in \mathbb{N}_{n+1,\tau}$. Identifying a constant complex-valued function defined on $\mathbb{C}_0$ with its restriction to $\mathbb{C}$, one can easily see that in the case $\alpha_j = 0$ for all $j \in \mathbb{N}_{1,n}$ the class $\mathcal{R}_{\alpha,n}$ coincides with the set $\mathcal{P}_n$ of all complex-valued polynomials of degree not greater than n. If n_1 and n_2 are integers with $0 \leq n_1 \leq n_2 \leq \tau$, then $\mathcal{R}_{\alpha,n_1} \subseteq \mathcal{R}_{\alpha,n_2}$. If a sequence $(\alpha_j)_{j=1}^{\infty}$ of complex numbers is given, then let

$$\mathcal{R}_{\alpha,\infty} := \bigcup_{n=0}^{\infty} \mathcal{R}_{\alpha,n}.$$

Every function $f \in \mathcal{R}_{\alpha,\infty}$ is holomorphic in $\mathbb{C}_0 \setminus \left(\bigcup_{j=1}^{\infty} \left\{ \frac{1}{\overline{\alpha_j}} \right\} \right)$.

For each $n \in \mathbb{N}_0$, the class $\mathcal{R}_{\alpha,n}^{p \times q}$ can be considered as a right $\mathbb{C}^{q \times q}$-submodule of the right $\mathbb{C}^{q \times q}$-module $\mathcal{R}_{\alpha,\infty}^{p \times q}$. On the other hand, for each $n \in \mathbb{N}_0$, the class $\mathcal{R}_{\alpha,n}^{p \times q}$ is also a left $\mathbb{C}^{p \times p}$-submodule of the left $\mathbb{C}^{p \times p}$-module $\mathcal{R}_{\alpha,\infty}^{p \times q}$.

Now let $(\alpha_j)_{j=1}^{\infty} \in \mathcal{T}_1$. We will use $B_{\alpha,0}^{(q)} : \mathbb{C}_0 \to \mathbb{C}^{q \times q}$ to denote the constant matrix-valued function with value $\mathbf{I}_q$ and, for each $n \in \mathbb{N}$, let the function $B_{\alpha,n}^{(q)} : \mathbb{C}_0 \setminus \mathbb{P}_{\alpha,n} \to \mathbb{C}^{q \times q}$ be defined by

$$B_{\alpha,n}^{(q)}(w) := \left(\prod_{j=1}^{n} b_{\alpha_j}(w) \right) \mathbf{I}_q.$$

If $\tau \in \mathbb{N}_0$ or $\tau = +\infty$, then $\left\{ B_{\alpha,k}^{(q)} : k \in \mathbb{N}_{0,\tau} \right\}$ is both a basis of the right $\mathbb{C}^{q \times q}$-module $\mathcal{R}_{\alpha,\tau}^{q \times q}$ and a basis of the left $\mathbb{C}^{q \times q}$-module $\mathcal{R}_{\alpha,\tau}^{q \times q}$. (For a more detailed discussion of these modules we refer to [FKL1].) In particular, if $n \in \mathbb{N}_0$ and if $X \in \mathcal{R}_{\alpha,n}^{q \times q}$, then there are unique matrices $\mathbf{A}_0, \mathbf{A}_1, \ldots, \mathbf{A}_n \in \mathbb{C}^{q \times q}$ such that $X = \sum_{j=0}^{n} \mathbf{A}_j B_{\alpha,j}^{(q)}$ and the *reciprocal rational (matrix-valued) function* $X^{[\alpha,n]}$ *of* X *with respect to* $(\alpha_j)_{j=1}^{\infty}$ *and* n is given by $X^{[\alpha,n]} := \sum_{j=0}^{n} \mathbf{A}_{n-j}^{*} B_{\beta,j}^{(q)}$ where the sequence $(\beta_j)_{j=1}^{\infty}$ is defined by $\beta_k := \alpha_{n+1-k}$ for each $k \in \mathbb{N}_{1,n}$ and $\beta_j := \alpha_j$ for each integer j with $j \geq n+1$. In particular,

$$X^{[\alpha,n]} \in \mathcal{R}_{\alpha,n}^{q \times q} \tag{1.2}$$

and

$$X^{[\alpha,n]}(\alpha_n) = \mathbf{A}_n^{*}. \tag{1.3}$$

If $n \in \mathbb{N}_0$ and if $\alpha_j = 0$ for each $j \in \mathbb{N}_{1,n}$, then every $X \in \mathcal{R}^{q\times q}_{\alpha,n}$ is a $q \times q$ matrix polynomial of degree not greater than n and $X^{[\alpha,n]}$ is exactly the reciprocal matrix polynomial $\tilde{X}^{[n]}$ of X with respect to the unit circle $\mathbb{T}$ and the formal degree n (cf. [DGK] or also [DFK, Section 1.2]).

A mapping F whose domain is the σ-algebra $\mathfrak{B}_{\mathbb{T}}$ of all Borelian subsets of $\mathbb{T}$ and whose values are nonnegative Hermitian complex $q \times q$ matrices is called nonnegative Hermitian $q \times q$ Borel measure on $\mathbb{T}$ if it is countably additive, i.e., if F satisfies $F\left(\bigcup_{k=1}^{\infty} A_k\right) = \sum_{k=1}^{\infty} F(A_k)$ for every infinite sequence $(A_k)_{k=1}^{\infty}$ of pairwise disjoint sets which belong to $\mathfrak{B}_{\mathbb{T}}$. We will use $\mathcal{M}^q_{\geq}(\mathbb{T}, \mathfrak{B}_{\mathbb{T}})$ to denote the set of all nonnegative Hermitian $q \times q$ Borel measures on $\mathbb{T}$. Let $F \in \mathcal{M}^q_{\geq}(\mathbb{T}, \mathfrak{B}_{\mathbb{T}})$. For each $j \in \mathbb{N}_{1,q}$ and each $k \in \mathbb{N}_{1,q}$, the entry function F_{jk} of F in the jth row and kth column is a complex-valued measure. One can easily see that F is absolutely continuous with respect to the trace measure $\tau F := \sum_{j=1}^{q} F_{jj}$ of F. An ordered pair $[\Phi, \Psi]$ consisting of Borel measurable matrix-valued functions $\Phi : \mathbb{T} \to \mathbb{C}^{p\times q}$ and $\Psi : \mathbb{T} \to \mathbb{C}^{s\times q}$ is called left-integrable with respect to F if each entry of the matrix-valued function $\Phi F'_\tau \Psi^*$ is integrable with respect to τF and the corresponding integral is defined by

$$\int_{\mathbb{T}} \Phi dF \Psi^* := \int_{\mathbb{T}} \Phi F'_\tau \Psi^* d\tau F. \tag{1.4}$$

We will also write $\int_{\mathbb{T}} \Phi(z) F(dz) \Psi^*(z)$ for the integral given in (1.4). An ordered pair $[\Phi, \Psi]$ consisting of Borel measurable matrix-valued functions $\Phi : \mathbb{T} \to \mathbb{C}^{q\times p}$ and $\Psi : \mathbb{T} \to \mathbb{C}^{q\times s}$ is said to be right-integrable with respect to F if $[\Phi^*, \Psi^*]$ is left-integrable with respect to F. It is known that the set $p \times q$-$\mathcal{L}^2_l(\mathbb{T}, \mathfrak{B}_{\mathbb{T}}, F)$ $\big($respectively, $q \times p - \mathcal{L}^2_r(\mathbb{T}, \mathfrak{B}_{\mathbb{T}}, F)\big)$ of all matrix-valued functions $\Phi : \mathbb{T} \to \mathbb{C}^{p\times q}$ (respectively, $\Phi : \mathbb{T} \to \mathbb{C}^{q\times p}$) for which $[\Phi, \Phi]$ is left-integrable (respectively, right-integrable) with respect to F is a left (respectively, right) $\mathbb{C}^{p\times p}$-semi Hilbert module, where the Gramian structure is given by

$$(\Phi, \Psi)_{F,l} := \int_{\mathbb{T}} \Phi dF \Psi^* \quad \left(\text{respectively, } (\Phi, \Psi)_{F,r} := \int_{\mathbb{T}} \Phi^* dF \Psi\right). \tag{1.5}$$

For more details on the integration theory with respect to nonnegative Hermitian $q \times q$ measures, we refer to Kats [Kt] and Rosenberg [R].

Now let a sequence $(\alpha_j)_{j=1}^{\infty}$ of numbers which belong to $\mathbb{C} \setminus \mathbb{T}$ and a nonnegative Hermitian $q \times q$ Borel measure F on $\mathbb{T}$ be given. Then one can immediately see that $\{\underline{X} : X \in \mathcal{R}^{p\times q}_{\alpha,\infty}\} \subseteq p \times q - \mathcal{L}^2_l(\mathbb{T}, \mathfrak{B}_{\mathbb{T}}, F)$ and $\{\underline{X} : X \in \mathcal{R}^{q\times p}_{\alpha,\infty}\} \subseteq q \times p - \mathcal{L}^2_r(\mathbb{T}, \mathfrak{B}_{\mathbb{T}}, F)$ hold. If X and Y belong to $\mathcal{R}^{p\times q}_{\alpha,\infty}$ (respectively, $\mathcal{R}^{q\times p}_{\alpha,\infty}$), then $(X, Y)_{F,l}$ (respectively, $(X, Y)_{F,r}$) is short for $(\underline{X}, \underline{Y})_{F,l}$ (respectively, $(\underline{X}, \underline{Y})_{F,r}$). For each $n \in \mathbb{N}_0$, it is readily checked that $\mathcal{R}^{q\times q}_{\alpha,n}$ is both a left $\mathbb{C}^{q\times q}$-semi Hilbert submodule of $q \times q - \mathcal{L}^2_l(\mathbb{T}, \mathfrak{B}_{\mathbb{T}}, F)$ and a right $\mathbb{C}^{q\times q}$-semi Hilbert submodule of $q \times q - \mathcal{L}^2_r(\mathbb{T}, \mathfrak{B}_{\mathbb{T}}, F)$. Under a certain assumption on the measure F the space $\mathcal{R}^{q\times q}_{\alpha,n}$ is a $\mathbb{C}^{q\times q}$-Hilbert module. This is based on a non-degeneracy concept which is studied in [FKL1, Section 5]. If $n \in \mathbb{N}_0$, then a nonnegative Hermitian $q \times q$

Borel measure F on $\mathbb{T}$ is called *non-degenerate of order* n if the block Toeplitz matrix $\mathbf{T}_n^{(F)} := (\mathbf{\Gamma}_{j-k}^{(F)})_{j,k=0}^n$ is nonsingular where

$$\mathbf{\Gamma}_j^{(F)} := \int_{\mathbb{T}} z^{-j} F(dz), \quad j \in \mathbb{Z},$$

are the Fourier coefficients of F. For each $n \in \mathbb{N}_0$, we will use $\mathcal{M}_{\geq}^{q,n}(\mathbb{T}, \mathfrak{B}_{\mathbb{T}})$ to denote the set of all $F \in \mathcal{M}_{\geq}^{q}(\mathbb{T}, \mathfrak{B}_{\mathbb{T}})$ which are non-degenerate of order n. Clearly, $\mathcal{M}_{\geq}^{q,n+1}(\mathbb{T}, \mathfrak{B}_{\mathbb{T}}) \subseteq \mathcal{M}_{\geq}^{q,n}(\mathbb{T}, \mathfrak{B}_{\mathbb{T}})$ holds for all $n \in \mathbb{N}_0$. A nonnegative Hermitian $q \times q$ Borel measure F on $\mathbb{T}$ is said to be *non-degenerate of order* ∞ if F belongs to

$$\mathcal{M}_{\geq}^{q,\infty}(\mathbb{T}, \mathfrak{B}_{\mathbb{T}}) := \bigcap_{n=0}^{\infty} \mathcal{M}_{\geq}^{q,n}(\mathbb{T}, \mathfrak{B}_{\mathbb{T}}).$$

Observe that in [FKL1, Section 5] several characterizations of the set $\mathcal{M}_{\geq}^{q,n}(\mathbb{T}, \mathfrak{B}_{\mathbb{T}})$ are given. If $n \in \mathbb{N}_0$ and if $F \in \mathcal{M}_{\geq}^{q,n}(\mathbb{T}, \mathfrak{B}_{\mathbb{T}})$, in view of [FKL1, Theorem 5.8] one can easily check that $\mathcal{R}_{\alpha,n}^{q\times q}$ (with the Gramian structure given in (1.5)) is both a left and a right $\mathbb{C}^{q\times q}$-Hilbert module. If F belongs to $\mathcal{M}_{\geq}^{q,\infty}(\mathbb{T}, \mathfrak{B}_{\mathbb{T}})$, the space $\mathcal{R}_{\alpha,\infty}^{q\times q}$ turns out to be both a left and a right $\mathbb{C}^{q\times q}$-pre-Hilbert module.

Let $(\alpha_j)_{j=1}^{\infty} \in \mathcal{T}_1$ and let $F \in \mathcal{M}_{\geq}^{q}(\mathbb{T}, \mathfrak{B}_{\mathbb{T}})$. Further, let $\tau \in \mathbb{N}_0$ or let $\tau = +\infty$. A sequence $(X_k)_{k=0}^{\tau}$ of matrix-valued functions which belong to $\mathcal{R}_{\alpha,\infty}^{q\times q}$ is called a *left* (respectively, *right*) *orthonormal system corresponding to* $(\alpha_j)_{j=1}^{\infty}$ *and* F if the following two conditions are satisfied:

(i) For each $k \in \mathbb{N}_{0,\tau}$, the function X_k belongs to $\mathcal{R}_{\alpha,k}^{q\times q}$.
(ii) For each $j \in \mathbb{N}_{0,\tau}$ and each $k \in \mathbb{N}_{0,\tau}$,

$$(X_j, X_k)_{F,l} = \delta_{jk}\mathbf{I} \quad \Big(\text{respectively,} \quad (X_j, X_k)_{F,r} = \delta_{jk}\mathbf{I}\Big).$$

We call a pair $[(X_k)_{k=0}^{\tau}, (Y_k)_{k=0}^{\tau}]$ consisting of a left orthonormal system $(X_k)_{k=0}^{\tau}$ corresponding to $(\alpha_j)_{j=1}^{\infty}$ and F and a right orthonormal system $(Y_k)_{k=0}^{\tau}$ corresponding to $(\alpha_j)_{j=1}^{\infty}$ and F a *pair of orthonormal systems corresponding to* $(\alpha_j)_{j=1}^{\infty}$ *and* F. In [FKL2, Corollary 4.4] it is verified that if $(\alpha_j)_{j=1}^{\infty} \in \mathcal{T}_1$, if $\tau \in \mathbb{N}_0$ or $\tau = +\infty$, and if $F \in \mathcal{M}_{\geq}^{q}(\mathbb{T}, \mathfrak{B}_{\mathbb{T}})$, then there exists a pair $[(X_k)_{k=0}^{\tau}, (Y_k)_{k=0}^{\tau}]$ of orthonormal systems corresponding to $(\alpha_j)_{j=1}^{\infty}$ and F if and only if the underlying measure F is non-degenerate of order τ. Pairs of orthonormal systems satisfy the following Christoffel-Darboux formulas.

Theorem 1.1. *Let* $(\alpha_j)_{j=1}^{\infty} \in \mathcal{T}_1$, *let* $\tau \in \mathbb{N}$ *or* $\tau = \infty$, *and let* $F \in \mathcal{M}_{\geq}^{q,\tau}(\mathbb{T}, \mathfrak{B}_{\mathbb{T}})$. *Further, let* $[(X_k)_{k=0}^{\tau}, (Y_k)_{k=0}^{\tau}]$ *be a pair of orthonormal systems corresponding to* $(\alpha_j)_{j=1}^{\infty}$ *and* F. *For all* $n \in \mathbb{N}_{1,\tau}$ *and every choice of* v *and* w *in* $\mathbb{C}_0 \setminus \mathbb{P}_{\alpha,n}$,

$$\big(1 - \overline{b_{\alpha_n}(v)b_{\alpha_n}(w)}\big)\sum_{k=0}^{n-1} X_k^*(v)X_k(w) = \big(Y_n^{[\alpha,n]}(v)\big)^* Y_n^{[\alpha,n]}(w) - X_n^*(v)X_n(w),$$

$$\big(1 - b_{\alpha_n}(v)\overline{b_{\alpha_n}(w)}\big)\sum_{k=0}^{n-1} Y_k(v)Y_k^*(w) = X_n^{[\alpha,n]}(v)\big(X_n^{[\alpha,n]}(w)\big)^* - Y_n(v)Y_n^*(w),$$

$$\left(1-\overline{b_{\alpha_n}(v)}b_{\alpha_n}(w)\right)\sum_{k=0}^{n}X_k^*(v)X_k(w)$$
$$=\left(Y_n^{[\alpha,n]}(v)\right)^*Y_n^{[\alpha,n]}(w)-\overline{b_{\alpha_n}(v)}b_{\alpha_n}(w)X_n^*(v)X_n(w),$$

and

$$\left(1-b_{\alpha_n}(v)\overline{b_{\alpha_n}(w)}\right)\sum_{k=0}^{n}Y_k(v)Y_k^*(w)$$
$$=X_n^{[\alpha,n]}(v)\left(X_n^{[\alpha,n]}(w)\right)^*-b_{\alpha_n}(v)\overline{b_{\alpha_n}(w)}Y_n(v)Y_n^*(w).$$

A proof of Theorem 1.1 is given in [FKL2, Theorem 5.4 and Corollary 5.5]. Note that, conversely, systems of rational matrix-valued functions satisfying identities of the type given in Theorem 1.1 are necessarily orthonormal systems corresponding to some nonnegative Hermitian measure (see [FKL3, Theorems 3.10 and 4.12]).

Remark 1.2. Let $(\alpha_j)_{j=1}^{\infty}\in\mathcal{T}_1$, let $\tau\in\mathbb{N}$ or $\tau=\infty$, and let $F\in\mathcal{M}_{\geq}^{q,\tau}(\mathbb{T},\mathfrak{B}_{\mathbb{T}})$. Further, let $[(X_k)_{k=0}^{\tau},(Y_k)_{k=0}^{\tau}]$ be a pair of orthonormal systems corresponding to $(\alpha_j)_{j=1}^{\infty}$ and F. For each $n\in\mathbb{N}_{1,\tau}$, the following statements are satisfied (see [FKL2, Remark 6.2, Theorems 6.7, 6.9, and 6.10]):

(a) There is a number $z_n\in\mathbb{T}$ such that
$$z_n\cdot\det X_n(w)=\det Y_n(w)\quad\text{and}\quad\det X_n^{[\alpha,n]}(w)=z_n\cdot\det Y_n^{[\alpha,n]}(w)$$
are satisfied for every choice of $w\in\mathbb{C}_0\setminus\mathbb{P}_{\alpha,n}$.

(b) If $|\alpha_n|<1$, then $\det X_n$ vanishes nowhere in $\mathbb{E}\setminus\mathbb{P}_{\alpha,n}$ and $\det X_n^{[\alpha,n]}$ vanishes nowhere in $\mathbb{D}\setminus\mathbb{P}_{\alpha,n-1}$.

(c) If $|\alpha_n|>1$, then $\det X_n$ vanishes nowhere in $\mathbb{D}\setminus\mathbb{P}_{\alpha,n}$ and $\det X_n^{[\alpha,n]}$ vanishes nowhere in $\mathbb{E}\setminus\mathbb{P}_{\alpha,n-1}$.

2. Szegő pairs of orthonormal systems

Inspired by the case of matrix polynomials we turn our attention to distinguished pairs of orthonormal systems of rational matrix-valued functions.

Definition 2.1. Let $(\alpha_j)_{j=1}^{\infty}\in\mathcal{T}_1$, let $\tau\in\mathbb{N}$ or $\tau=\infty$, and let $F\in\mathcal{M}_{\geq}^{q,\tau}(\mathbb{T},\mathfrak{B}_{\mathbb{T}})$. For each $k\in\mathbb{N}_{0,\tau}$, let

$$\eta_k:=\begin{cases}\frac{\overline{\alpha_k}}{|\alpha_k|} & \text{if } \alpha_k\neq 0\\ -1 & \text{if } \alpha_k=0\,.\end{cases}\tag{2.1}$$

Let $[(X_k)_{k=0}^{\tau},(Y_k)_{k=0}^{\tau}]$ be a pair of orthonormal systems corresponding to $(\alpha_j)_{j=1}^{\infty}$ and F. Then $[(X_k)_{k=0}^{\tau},(Y_k)_{k=0}^{\tau}]$ is called a *Szegő pair of orthonormal systems corresponding to* $(\alpha_j)_{j=1}^{\infty}$ *and* F if for each $n\in\mathbb{N}_{1,\tau}$ the following statements holds:

(i) If $(1-|\alpha_n|)(1-|\alpha_{n-1}|)>0$, then the matrices

$$\frac{\eta_n\overline{\eta_{n-1}}(1-|\alpha_{n-1}|^2)}{1-\overline{\alpha_n}\alpha_{n-1}}Y_{n-1}^{[\alpha,n-1]}(\alpha_{n-1})\left(Y_n^{[\alpha,n]}(\alpha_{n-1})\right)^{-1}\tag{2.2}$$

and

$$\frac{\eta_n \overline{\eta_{n-1}}(1-|\alpha_{n-1}|^2)}{1-\overline{\alpha_n}\alpha_{n-1}} \left(X_n^{[\alpha,n]}(\alpha_{n-1})\right)^{-1} X_{n-1}^{[\alpha,n-1]}(\alpha_{n-1}) \tag{2.3}$$

are both positive Hermitian.

(ii) If $(1-|\alpha_n|)(1-|\alpha_{n-1}|) < 0$, then the matrices

$$\frac{|\alpha_{n-1}|^2 - 1}{1-\overline{\alpha_n}\alpha_{n-1}} \left(Y_n(\alpha_{n-1})\right)^{-1} X_{n-1}^{[\alpha,n-1]}(\alpha_{n-1}) \tag{2.4}$$

and

$$\frac{|\alpha_{n-1}|^2 - 1}{1-\overline{\alpha_n}\alpha_{n-1}} Y_{n-1}^{[\alpha,n-1]}(\alpha_{n-1}) \left(X_n(\alpha_{n-1})\right)^{-1} \tag{2.5}$$

are both positive Hermitian.

Observe that, in view of Remark 1.2, the inverse matrices occurring in the formulae (2.2)–(2.5) are well defined. The notion introduced in Definition 2.1 is a generalization of the corresponding notion in the matrix polynomial case, i.e., in the case that $\alpha_j = 0$ for all $j \in \mathbb{N}$ (see [DGK] and [DFK, Definition 3.6.5]).

Remark 2.2. Let $(\alpha_j)_{j=1}^\infty \in \mathcal{T}_1$, let $\tau \in \mathbb{N}$ or $\tau = \infty$, and let $F \in \mathcal{M}_{\geq}^{q,\tau}(\mathbb{T}, \mathfrak{B}_{\mathbb{T}})$. From [FKL2, Proposition 3.7 and Corollary 4.4] and the polar decomposition of matrices one can immediately see that there exists a Szegő pair of orthonormal systems corresponding to $(\alpha_j)_{j=1}^\infty$ and F.

Remark 2.3. Let $(\alpha_j)_{j=1}^\infty \in \mathcal{T}_1$, let $\tau \in \mathbb{N}$ or $\tau = \infty$, and let $F \in \mathcal{M}_{\geq}^{q,\tau}(\mathbb{T}, \mathfrak{B}_{\mathbb{T}})$. From [FKL2, Proposition 3.7] and the polar decomposition of matrices one can also see that the following statements hold:

(a) If $[(X_k)_{k=0}^\tau, (Y_k)_{k=0}^\tau]$ is a Szegő pair of orthonormal systems corresponding to $(\alpha_j)_{j=1}^\infty$ and F, then, for all unitary complex $q\times q$ matrices $\mathbf{U}$ and $\mathbf{V}$, the pair $[(\tilde{X}_k)_{k=0}^\tau, (\tilde{Y}_k)_{k=0}^\tau]$ given for all $k \in \mathbb{N}_{0,\tau}$ by

$$\tilde{X}_k := \begin{cases} \mathbf{U}X_k & \text{if } \alpha_k \in \mathbb{D} \\ \mathbf{V}^* X_k & \text{if } \alpha_k \in \mathbb{E} \end{cases} \quad \text{and} \quad \tilde{Y}_k := \begin{cases} Y_k\mathbf{V} & \text{if } \alpha_k \in \mathbb{D} \\ Y_k\mathbf{U}^* & \text{if } \alpha_k \in \mathbb{E} \end{cases}$$

is also a Szegő pair of orthonormal systems corresponding to $(\alpha_j)_{j=1}^\infty$ and F.

(b) If $[(X_k)_{k=0}^\tau, (Y_k)_{k=0}^\tau]$ and $[(\tilde{X}_k)_{k=0}^\tau, (\tilde{Y}_k)_{k=0}^\tau]$ are Szegő pairs of orthonormal systems corresponding to $(\alpha_j)_{j=1}^\infty$ and F, then $\mathbf{U} := X_0(0)\left(\tilde{X}_0(0)\right)^{-1}$ and $\mathbf{V} := \left(\tilde{Y}_0(0)\right)^{-1} Y_0(0)$ are unitary complex $q\times q$ matrices and for all $k \in \mathbb{N}_{0,\tau}$ the identities

$$\tilde{X}_k = \begin{cases} \mathbf{U}X_k & \text{if } \alpha_k \in \mathbb{D} \\ \mathbf{V}^* X_k & \text{if } \alpha_k \in \mathbb{E} \end{cases} \quad \text{and} \quad \tilde{Y}_k = \begin{cases} Y_k\mathbf{V} & \text{if } \alpha_k \in \mathbb{D} \\ Y_k\mathbf{U}^* & \text{if } \alpha_k \in \mathbb{E} \end{cases}$$

are satisfied.

Remark 2.4. Let $(\alpha_j)_{j=1}^{\infty} \in \mathcal{T}_1$, let $\tau \in \mathbb{N}$ or $\tau = \infty$, and let $F \in \mathcal{M}_{\geq}^{q,\tau}(\mathbb{T}, \mathfrak{B}_{\mathbb{T}})$. Further, let $[(X_k)_{k=0}^{\tau}, (Y_k)_{k=0}^{\tau}]$ be a Szegő pair of orthonormal systems corresponding to $(\alpha_j)_{j=1}^{\infty}$ and F. In view of Remark 1.2 and Definition 2.1, one can get that there exists a $z \in \mathbb{T}$ such that for all $n \in \mathbb{N}_{1,\tau}$ and all $w \in \mathbb{C}_0 \setminus \mathbb{P}_{\alpha,n}$ the equalities

$$z \cdot \det X_n(w) = \det Y_n(w) \quad \text{and} \quad \det X_n^{[\alpha,n]}(w) = z \cdot \det Y_n^{[\alpha,n]}(w)$$

are satisfied.

Definition 2.5. Let $(\alpha_j)_{j=1}^{\infty} \in \mathcal{T}_1$, let $\tau \in \mathbb{N}$ or $\tau = \infty$, and let $F \in \mathcal{M}_{\geq}^{q,\tau}(\mathbb{T}, \mathfrak{B}_{\mathbb{T}})$. For each $k \in \mathbb{N}_{0,\tau}$, let η_k be given by (2.1). Further, let $[(X_k)_{k=0}^{\tau}, (Y_k)_{k=0}^{\tau}]$ be a Szegő pair of orthonormal systems corresponding to $(\alpha_j)_{j=1}^{\infty}$ and F. Then $(\mathbf{E}_n)_{n=1}^{\tau}$ given by

$$\mathbf{E}_n := \begin{cases} \eta_n \overline{\eta_{n-1}} X_n(\alpha_{n-1}) \big(Y_n^{[\alpha,n]}(\alpha_{n-1})\big)^{-1} & \text{if } (1-|\alpha_n|)(1-|\alpha_{n-1}|) > 0 \\ \eta_n \overline{\eta_{n-1}} \Big(Y_n^{[\alpha,n]}(\alpha_{n-1}) \big(X_n(\alpha_{n-1})\big)^{-1}\Big)^* & \text{if } (1-|\alpha_n|)(1-|\alpha_{n-1}|) < 0 . \end{cases} \tag{2.6}$$

is said to be *the sequence of Szegő parameters corresponding to* $[(X_k)_{k=0}^{\tau}, (Y_k)_{k=0}^{\tau}]$.

Note that, in view of Remark 1.2, the inverses occurring in (2.6) are well defined. The notion introduced in Definition 2.5 is a generalization of the corresponding notion in the matrix polynomials case (see [DGK] and [DFK, Definition 3.6.6]). This can be seen from [DFK, Lemma 3.6.14 and Proposition 3.6.5]. Moreover, under the assumptions of Definition 2.5 from [FKL2, Remark 6.2 and Lemma 6.5] and Remark 1.2, for each $n \in \mathbb{N}_{1,\tau}$, it follows that

$$\mathbf{E}_n = \begin{cases} \eta_n \overline{\eta_{n-1}} \big(X_n^{[\alpha,n]}(\alpha_{n-1})\big)^{-1} Y_n(\alpha_{n-1}) & \text{if } (1-|\alpha_n|)(1-|\alpha_{n-1}|) > 0 \\ \eta_n \overline{\eta_{n-1}} \Big(\big(Y_n(\alpha_{n-1})\big)^{-1} X_n^{[\alpha,n]}(\alpha_{n-1})\Big)^* & \text{if } (1-|\alpha_n|)(1-|\alpha_{n-1}|) < 0 . \end{cases} \tag{2.7}$$

In the matrix polynomial case the Szegő parameters can be used to obtain recursion formulas for Szegő pairs of orthonormal systems (of matrix polynomials). We will see that these formulas can be generalized to the rational case.

Remark 2.6. Let $(\alpha_j)_{j=1}^{\infty} \in \mathcal{T}_1$, let $\tau \in \mathbb{N}$ or $\tau = \infty$, and let $F \in \mathcal{M}_{\geq}^{q,\tau}(\mathbb{T}, \mathfrak{B}_{\mathbb{T}})$. Let $[(X_k)_{k=0}^{\tau}, (Y_k)_{k=0}^{\tau}]$ be a Szegő pair of orthonormal systems corresponding to $(\alpha_j)_{j=1}^{\infty}$ and F, and let $(\mathbf{E}_n)_{n=1}^{\tau}$ be the sequence of its Szegő parameters. Then F^T belongs to $\mathcal{M}_{\geq}^{q,\tau}(\mathbb{T}, \mathfrak{B}_{\mathbb{T}})$. Moreover, $[(Y_k^T)_{k=0}^{\tau}, (X_k^T)_{k=0}^{\tau}]$ is a Szegő pair of orthonormal systems corresponding to $(\alpha_j)_{j=1}^{\infty}$ and F^T, and $(\mathbf{E}_n^T)_{n=1}^{\tau}$ is the sequence of its Szegő parameters (see [FKL1, Remark 5.11], [FKL2, Remark 2.10 and Remark 3.5], and (2.7)).

Remark 2.7. Let $(\alpha_j)_{j=1}^{\infty} \in \mathcal{T}_1$, let $\tau \in \mathbb{N}$ or $\tau = \infty$, and let $F \in \mathcal{M}_{\geq}^{q,\tau}(\mathbb{T}, \mathfrak{B}_{\mathbb{T}})$. If $\mathbf{A}$ is a positive Hermitian $q \times q$ matrix, then [FKL1, Remarks 5.3, 3.10, and 5.13] yield that $H : \mathfrak{B}_{\mathbb{T}} \to \mathbb{C}^{q \times q}$ given by $H(B) := \sqrt{\mathbf{A}}\sqrt{F(\mathbb{T})}^{-1} F(B) \sqrt{F(\mathbb{T})}^{-1} \sqrt{\mathbf{A}}$ belongs also to $\mathcal{M}_{\geq}^{q,\tau}(\mathbb{T}, \mathfrak{B}_{\mathbb{T}})$ and satisfies $H(\mathbb{T}) = \mathbf{A}$. Moreover, if $[(X_k)_{k=0}^{\tau}, (Y_k)_{k=0}^{\tau}]$ is a Szegő pair of orthonormal systems corresponding to $(\alpha_j)_{j=1}^{\infty}$ and F, and if $(\mathbf{E}_n)_{n=1}^{\tau}$

is the sequence of its Szegő parameters, then, in view of [FKL2, Remark 2.8], it is not hard to see that $\left[\left(X_k\sqrt{F(\mathbb{T})}\sqrt{\mathbf{A}}^{-1}\right)_{k=0}^{\tau}, \left(\sqrt{\mathbf{A}}^{-1}\sqrt{F(\mathbb{T})}Y_k\right)_{k=0}^{\tau}\right]$ is a Szegő pair of orthonormal systems corresponding to $(\alpha_j)_{j=1}^{\infty}$ and H, and $(\mathbf{E}_n)_{n=1}^{\tau}$ is also the sequence of the Szegő parameters corresponding to this Szegő pair.

Lemma 2.8. *Let $(\alpha_j)_{j=1}^{\infty} \in \mathcal{T}_1$, let $\tau \in \mathbb{N}$ or $\tau = \infty$, and let $F \in \mathcal{M}_{\geq}^{q,\tau}(\mathbb{T}, \mathfrak{B}_{\mathbb{T}})$. Let $[(X_k)_{k=0}^{\tau}, (Y_k)_{k=0}^{\tau}]$ be a pair of orthonormal systems corresponding to $(\alpha_j)_{j=1}^{\infty}$ and F. For each $n \in \mathbb{N}_{1,\tau}$ and each $w \in \mathbb{C} \setminus \mathbb{P}_{\alpha,n}$, then*

$$\begin{aligned}
&\left(Y_n^{[\alpha,n]}(\alpha_{n-1})\right)^* Y_n^{[\alpha,n]}(w) - \left(X_n(\alpha_{n-1})\right)^* X_n(w) \\
&\quad = \left(1 - \overline{b_{\alpha_n}(\alpha_{n-1})}b_{\alpha_n}(w)\right)\left(Y_{n-1}^{[\alpha,n-1]}(\alpha_{n-1})\right)^* Y_{n-1}^{[\alpha,n-1]}(w)
\end{aligned}$$

and

$$\begin{aligned}
&X_n^{[\alpha,n]}(w)\left(X_n^{[\alpha,n]}(\alpha_{n-1})\right)^* - Y_n(w)\left(Y_n(\alpha_{n-1})\right)^* \\
&\quad = \left(1 - b_{\alpha_n}(w)\overline{b_{\alpha_n}(\alpha_{n-1})}\right)X_{n-1}^{[\alpha,n-1]}(w)\left(X_{n-1}^{[\alpha,n-1]}(\alpha_{n-1})\right)^*.
\end{aligned}$$

Proof. Apply Theorem 1.1 and use (1.1). □

Proposition 2.9. *Let $(\alpha_j)_{j=1}^{\infty} \in \mathcal{T}_1$, let $\tau \in \mathbb{N}$ or $\tau = \infty$, and let $F \in \mathcal{M}_{\geq}^{q,\tau}(\mathbb{T}, \mathfrak{B}_{\mathbb{T}})$. For each $k \in \mathbb{N}_{0,\tau}$, let η_k be given by (2.1). For each $n \in \mathbb{N}_{1,\tau}$, let*

$$\rho_n := \begin{cases} \sqrt{\frac{1-|\alpha_n|^2}{1-|\alpha_{n-1}|^2}} & \text{if } (1-|\alpha_n|)(1-|\alpha_{n-1}|) > 0 \\ -\sqrt{\frac{|\alpha_n|^2-1}{1-|\alpha_{n-1}|^2}} & \text{if } (1-|\alpha_n|)(1-|\alpha_{n-1}|) < 0\,. \end{cases} \tag{2.8}$$

Let $[(X_k)_{k=0}^{\tau}, (Y_k)_{k=0}^{\tau}]$ be a Szegő pair of orthonormal systems corresponding to $(\alpha_j)_{j=1}^{\infty}$ and F, and let $(\mathbf{E}_n)_{n=1}^{\tau}$ be the sequence of its Szegő parameters. For each $n \in \mathbb{N}_{1,\tau}$, the following statements hold:

(a) *If $(1-|\alpha_n|)(1-|\alpha_{n-1}|) > 0$, then*

$$\mathbf{I} - \mathbf{E}_n^*\mathbf{E}_n = \rho_n^2\left(\frac{\eta_n\overline{\eta_{n-1}}(1-|\alpha_{n-1}|^2)}{1-\overline{\alpha_n}\alpha_{n-1}}Y_{n-1}^{[\alpha,n-1]}(\alpha_{n-1})\left(Y_n^{[\alpha,n]}(\alpha_{n-1})\right)^{-1}\right)^2$$

and

$$\mathbf{I} - \mathbf{E}_n\mathbf{E}_n^* = \rho_n^2\left(\frac{\eta_n\overline{\eta_{n-1}}(1-|\alpha_{n-1}|^2)}{1-\overline{\alpha_n}\alpha_{n-1}}\left(X_n^{[\alpha,n]}(\alpha_{n-1})\right)^{-1}X_{n-1}^{[\alpha,n-1]}(\alpha_{n-1})\right)^2.$$

(b) *If $(1-|\alpha_n|)(1-|\alpha_{n-1}|) < 0$, then*

$$\mathbf{I} - \mathbf{E}_n^*\mathbf{E}_n = \rho_n^2\left(\frac{|\alpha_{n-1}|^2-1}{1-\overline{\alpha_n}\alpha_{n-1}}\left(Y_n(\alpha_{n-1})\right)^{-1}X_{n-1}^{[\alpha,n-1]}(\alpha_{n-1})\right)^2$$

and

$$\mathbf{I} - \mathbf{E}_n\mathbf{E}_n^* = \rho_n^2\left(\frac{|\alpha_{n-1}|^2-1}{1-\overline{\alpha_n}\alpha_{n-1}}Y_{n-1}^{[\alpha,n-1]}(\alpha_{n-1})\left(X_n(\alpha_{n-1})\right)^{-1}\right)^2.$$

(c) *The Szegő parameter $\mathbf{E}_n$ is a strictly contractive $q \times q$ matrix.*

Proof. Let $n \in \mathbb{N}_{1,\tau}$. First we consider the case $(1-|\alpha_n|)(1-|\alpha_{n-1}|) > 0$. Hence, Remark 1.2 implies

$$\det Y_n^{[\alpha,n]}(\alpha_{n-1}) \neq 0. \tag{2.9}$$

Using (2.9), (2.6), and Lemma 2.8 we obtain

$$\begin{aligned}
&\big(Y_n^{[\alpha,n]}(\alpha_{n-1})\big)^*\big(\mathbf{I}-\mathbf{E}_n^*\mathbf{E}_n\big)Y_n^{[\alpha,n]}(\alpha_{n-1})\\
&\quad=\big(Y_n^{[\alpha,n]}(\alpha_{n-1})\big)^*Y_n^{[\alpha,n]}(\alpha_{n-1})-\big(X_n(\alpha_{n-1})\big)^*X_n(\alpha_{n-1})\\
&\quad=\big(1-|b_{\alpha_n}(\alpha_{n-1})|^2\big)\big(Y_{n-1}^{[\alpha,n-1]}(\alpha_{n-1})\big)^*Y_{n-1}^{[\alpha,n-1]}(\alpha_{n-1}).
\end{aligned}\tag{2.10}$$

From (1.1) we see that

$$1-|b_{\alpha_n}(\alpha_{n-1})|^2=\frac{(1-|\alpha_n|^2)(1-|\alpha_{n-1}|^2)}{(1-\alpha_n\overline{\alpha_{n-1}})(1-\overline{\alpha_n}\alpha_{n-1})}. \tag{2.11}$$

Since η_n and η_{n-1} are unimodular complex numbers from (2.8), (2.10), (2.9), and (2.11), we can conclude that, in the considered case, the first equation in (a) is fulfilled. In view of (2.7), in the case $(1-|\alpha_n|)(1-|\alpha_{n-1}|) < 0$, the first formula in (b) can be verified similarly. Using Remark 2.6 the second equation in (a) and (b) follows from the first one, respectively. Thus, parts (a) and (b) are proved. In view of Definition 2.1 and (2.8), part (c) is a consequence of parts (a) and (b). □

In the following, if a sequence $(\alpha_j)_{j=1}^{\infty} \in \mathcal{T}_1$ is given, we will use the notations η_k and ρ_n which are defined by (2.1) for all $k \in \mathbb{N}_0$ and by (2.8) for all $n \in \mathbb{N}$, respectively.

Proposition 2.10. *Let $(\alpha_j)_{j=1}^{\infty} \in \mathcal{T}_1$, let $\tau \in \mathbb{N}$ or $\tau = \infty$, and let $F \in \mathcal{M}_{\geq}^{q,\tau}(\mathbb{T},\mathfrak{B}_{\mathbb{T}})$. Let $[(X_k)_{k=0}^{\tau},(Y_k)_{k=0}^{\tau}]$ be a Szegő pair of orthonormal systems corresponding to $(\alpha_j)_{j=1}^{\infty}$ and F, and let $(\mathbf{E}_n)_{n=1}^{\tau}$ be the sequence of its Szegő parameters. For each $n \in \mathbb{N}_{1,\tau}$ and each $w \in \mathbb{C}\setminus\mathbb{P}_{\alpha,n}$, then the following statements hold:*

(a) *If $(1-|\alpha_n|)(1-|\alpha_{n-1}|) > 0$, then*

$$Y_n(w)-\overline{\eta_n}\eta_{n-1}X_n^{[\alpha,n]}(w)\mathbf{E}_n=\rho_n\frac{1-\overline{\alpha_{n-1}}w}{1-\overline{\alpha_n}w}b_{\alpha_{n-1}}(w)Y_{n-1}(w)\sqrt{\mathbf{I}-\mathbf{E}_n^*\mathbf{E}_n}$$

and

$$X_n(w)-\overline{\eta_n}\eta_{n-1}\mathbf{E}_nY_n^{[\alpha,n]}(w)=\rho_n\frac{1-\overline{\alpha_{n-1}}w}{1-\overline{\alpha_n}w}b_{\alpha_{n-1}}(w)\sqrt{\mathbf{I}-\mathbf{E}_n\mathbf{E}_n^*}X_{n-1}(w).$$

(b) *If $(1-|\alpha_n|)(1-|\alpha_{n-1}|) < 0$, then*

$$Y_n(w)-\overline{\eta_n}\eta_{n-1}X_n^{[\alpha,n]}(w)\mathbf{E}_n=\rho_n\frac{1-\overline{\alpha_{n-1}}w}{1-\overline{\alpha_n}w}X_{n-1}^{[\alpha,n-1]}(w)\sqrt{\mathbf{I}-\mathbf{E}_n^*\mathbf{E}_n}$$

and

$$X_n(w)-\overline{\eta_n}\eta_{n-1}\mathbf{E}_nY_n^{[\alpha,n]}(w)=\rho_n\frac{1-\overline{\alpha_{n-1}}w}{1-\overline{\alpha_n}w}\sqrt{\mathbf{I}-\mathbf{E}_n\mathbf{E}_n^*}Y_{n-1}^{[\alpha,n-1]}(w).$$

Proof. Let $n \in \mathbb{N}_{1,\tau}$ and $w \in \mathbb{C} \setminus \mathbb{P}_{\alpha,n}$. At first we shall prove the first equation in (a) and hence we consider in the following the case $(1-|\alpha_n|)(1-|\alpha_{n-1}|) > 0$. Using (2.6) and Lemma 2.8 we get

$$\begin{aligned}
&Y_n^{[\alpha,n]}(w) - \eta_n\overline{\eta_{n-1}}\,\mathbf{E}_n^* X_n(w)\\
&\quad= \left(Y_n^{[\alpha,n]}(\alpha_{n-1})\right)^{-*}\left(\left(Y_n^{[\alpha,n]}(\alpha_{n-1})\right)^* Y_n^{[\alpha,n]}(w) - \left(X_n(\alpha_{n-1})\right)^* X_n(w)\right)\\
&\quad= \left(1-\overline{b_{\alpha_n}(\alpha_{n-1})}b_{\alpha_n}(w)\right)\left(Y_n^{[\alpha,n]}(\alpha_{n-1})\right)^{-*}\left(Y_{n-1}^{[\alpha,n-1]}(\alpha_{n-1})\right)^* Y_{n-1}^{[\alpha,n-1]}(w)
\end{aligned}$$

and therefore, in view of [FKL2, Remarks 2.4 and 2.8] and Proposition 2.9,

$$\begin{aligned}
&Y_n(w) - \overline{\eta_n}\eta_{n-1}X_n^{[\alpha,n]}(w)\mathbf{E}_n\\
&\quad= \left(b_{\alpha_n}(w) - b_{\alpha_n}(\alpha_{n-1})\right)Y_{n-1}(w)Y_n^{[\alpha,n-1]}(\alpha_{n-1})\left(Y_n^{[\alpha,n]}(\alpha_{n-1})\right)^{-1}\\
&\quad= \frac{\left(b_{\alpha_n}(w) - b_{\alpha_n}(\alpha_{n-1})\right)\overline{\eta_n}\eta_{n-1}(1-\overline{\alpha_n}\alpha_{n-1})}{(1-|\alpha_{n-1}|^2)\rho_n}Y_{n-1}(w)\sqrt{\mathbf{I}-\mathbf{E}_n^*\mathbf{E}_n}.
\end{aligned}$$

Taking into account the identity

$$\frac{1-\overline{\alpha_{n-1}}w}{1-\overline{\alpha_n}w}b_{\alpha_{n-1}}(w) = \frac{\left(b_{\alpha_n}(w) - b_{\alpha_n}(\alpha_{n-1})\right)\overline{\eta_n}\eta_{n-1}(1-\overline{\alpha_n}\alpha_{n-1})}{1-|\alpha_n|^2}$$

it follows the first identity in (a). Now let $(1-|\alpha_n|)(1-|\alpha_{n-1}|) < 0$. Using (2.7), Lemma 2.8, (1.1), (2.8), and Proposition 2.9 we can conclude

$$\begin{aligned}
&Y_n(w) - \overline{\eta_n}\eta_{n-1}X_n^{[\alpha,n]}(w)\mathbf{E}_n\\
&\quad= \left(Y_n(w)\left(Y_n(\alpha_{n-1})\right)^* - X_n^{[\alpha,n]}(w)\left(X_n^{[\alpha,n]}(\alpha_{n-1})\right)^*\right)\left(Y_n(\alpha_{n-1})\right)^{-*}\\
&\quad= \left(b_{\alpha_n}(w)\overline{b_{\alpha_n}(\alpha_{n-1})} - 1\right)X_{n-1}^{[\alpha,n-1]}(w)\left(X_{n-1}^{[\alpha,n-1]}(\alpha_{n-1})\right)^*\left(Y_n(\alpha_{n-1})\right)^{-*}\\
&\quad= \frac{(|\alpha_n|^2-1)(1-\overline{\alpha_{n-1}}w)}{(1-\alpha_n\overline{\alpha_{n-1}})(1-\overline{\alpha_n}w)}X_{n-1}^{[\alpha,n-1]}(w)\left(\left(Y_n(\alpha_{n-1})\right)^{-1}X_{n-1}^{[\alpha,n-1]}(\alpha_{n-1})\right)^*\\
&\quad= \rho_n\frac{1-\overline{\alpha_{n-1}}w}{1-\overline{\alpha_n}w}X_{n-1}^{[\alpha,n-1]}(w)\sqrt{\mathbf{I}-\mathbf{E}_n^*\mathbf{E}_n}.
\end{aligned}$$

Hence, the first identity in (b) is also proved. The second equation in (a) and (b) can be obtained using Remark 2.6, [FKL2, Remark 2.11], and the first one, respectively. □

For each strictly contractive complex $q \times q$ matrix $\mathbf{E}$, let $\mathbf{H}(\mathbf{E})$ be the so-called *Halmos extension of* $\mathbf{E}$, i.e., let

$$\mathbf{H}(\mathbf{E}) := \begin{pmatrix} \sqrt{\mathbf{I}-\mathbf{E}\mathbf{E}^*}^{\,-1} & \mathbf{E}\sqrt{\mathbf{I}-\mathbf{E}^*\mathbf{E}}^{\,-1} \\ \mathbf{E}^*\sqrt{\mathbf{I}-\mathbf{E}\mathbf{E}^*}^{\,-1} & \sqrt{\mathbf{I}-\mathbf{E}^*\mathbf{E}}^{\,-1} \end{pmatrix}.$$

Furthermore, in view of Definition 2.1, Definition 2.5, and part (c) of Proposition 2.9, if $(\alpha_j)_{j=1}^\infty \in \mathcal{T}_1$ and if $(\mathbf{E}_n)_{n=1}^\tau$ is a sequence of strictly contractive complex $q \times q$ matrices where $\tau \in \mathbb{N}$ or $\tau = \infty$, then for $n \in \mathbb{N}_{1,\tau}$ and $w \in \mathbb{C} \setminus \mathbb{P}_{\alpha,n}$ we set

$$H_{\alpha,\mathbf{E};n}(w) := \begin{pmatrix} \mathbf{I}_q & \mathbf{0} \\ \mathbf{0} & \eta_n\overline{\eta_{n-1}}\,\mathbf{I}_q \end{pmatrix}\mathbf{H}(\mathbf{E}_n)\begin{pmatrix} b_{\alpha_{n-1}}(w)\mathbf{I}_q & \mathbf{0} \\ \mathbf{0} & \mathbf{I}_q \end{pmatrix}$$

and

$$G_{\alpha,\mathbf{E};n}(w) := \begin{pmatrix} b_{\alpha_{n-1}}(w)\mathbf{I}_q & \mathbf{0} \\ \mathbf{0} & \mathbf{I}_q \end{pmatrix} \mathbf{H}(\mathbf{E}_n^*) \begin{pmatrix} \mathbf{I}_q & \mathbf{0} \\ \mathbf{0} & \eta_n \overline{\eta_{n-1}}\, \mathbf{I}_q \end{pmatrix}$$

if $(1-|\alpha_n|)(1-|\alpha_{n-1}|) > 0$ and in the case $(1-|\alpha_n|)(1-|\alpha_{n-1}|) < 0$ we put

$$H_{\alpha,\mathbf{E};n}(w) := \begin{pmatrix} \mathbf{I}_q & \mathbf{0} \\ \mathbf{0} & \eta_n \overline{\eta_{n-1}}\, \mathbf{I}_q \end{pmatrix} \mathbf{H}(\mathbf{E}_n) \begin{pmatrix} \mathbf{0} & \mathbf{I}_q \\ \mathbf{I}_q & \mathbf{0} \end{pmatrix} \begin{pmatrix} b_{\alpha_{n-1}}(w)\mathbf{I}_q & \mathbf{0} \\ \mathbf{0} & \mathbf{I}_q \end{pmatrix}$$

and

$$G_{\alpha,\mathbf{E};n}(w) := \begin{pmatrix} b_{\alpha_{n-1}}(w)\mathbf{I}_q & \mathbf{0} \\ \mathbf{0} & \mathbf{I}_q \end{pmatrix} \begin{pmatrix} \mathbf{0} & \mathbf{I}_q \\ \mathbf{I}_q & \mathbf{0} \end{pmatrix} \mathbf{H}(\mathbf{E}_n^*) \begin{pmatrix} \mathbf{I}_q & \mathbf{0} \\ \mathbf{0} & \eta_n \overline{\eta_{n-1}}\, \mathbf{I}_q \end{pmatrix}.$$

Theorem 2.11. *Let $(\alpha_j)_{j=1}^{\infty} \in \mathcal{T}_1$, let $\tau \in \mathbb{N}$ or $\tau = \infty$, and let $F \in \mathcal{M}_{\geq}^{q,\tau}(\mathbb{T}, \mathfrak{B}_{\mathbb{T}})$. Let $[(X_k)_{k=0}^{\tau}, (Y_k)_{k=0}^{\tau}]$ be a Szegő pair of orthonormal systems corresponding to $(\alpha_j)_{j=1}^{\infty}$ and F, and let $(\mathbf{E}_n)_{n=1}^{\tau}$ be the sequence of its Szegő parameters. For all $n \in \mathbb{N}_{1,\tau}$ and all $w \in \mathbb{C} \setminus \mathbb{P}_{\alpha,n}$, the following recurrence relations are satisfied:*

$$\begin{pmatrix} X_n(w) \\ Y_n^{[\alpha,n]}(w) \end{pmatrix} = \rho_n \frac{1-\overline{\alpha_{n-1}}w}{1-\overline{\alpha_n}w} H_{\alpha,\mathbf{E};n}(w) \begin{pmatrix} X_{n-1}(w) \\ Y_{n-1}^{[\alpha,n-1]}(w) \end{pmatrix}$$

and

$$\left(Y_n(w),\ X_n^{[\alpha,n]}(w)\right) = \rho_n \frac{1-\overline{\alpha_{n-1}}w}{1-\overline{\alpha_n}w} \left(Y_{n-1}(w),\ X_{n-1}^{[\alpha,n-1]}(w)\right) G_{\alpha,\mathbf{E};n}(w).$$

Proof. Let $n \in \mathbb{N}_{1,\tau}$ and $w \in \mathbb{C} \setminus \mathbb{P}_{\alpha,n}$. Suppose $(1-|\alpha_n|)(1-|\alpha_{n-1}|) > 0$. We have

$$\begin{aligned} &\mathbf{H}(-\mathbf{E}_n) \begin{pmatrix} \mathbf{I}_q & \mathbf{0} \\ \mathbf{0} & \overline{\eta_n}\eta_{n-1}\mathbf{I}_q \end{pmatrix} \begin{pmatrix} X_n(w) \\ Y_n^{[\alpha,n]}(w) \end{pmatrix} \\ &= \begin{pmatrix} \sqrt{\mathbf{I}-\mathbf{E}_n\mathbf{E}_n^*}^{\,-1} X_n(w) - \overline{\eta_n}\eta_{n-1}\mathbf{E}_n \sqrt{\mathbf{I}-\mathbf{E}_n^*\mathbf{E}_n}^{\,-1} Y_n^{[\alpha,n]}(w) \\ -\mathbf{E}_n^* \sqrt{\mathbf{I}-\mathbf{E}_n\mathbf{E}_n^*}^{\,-1} X_n(w) + \overline{\eta_n}\eta_{n-1} \sqrt{\mathbf{I}-\mathbf{E}_n^*\mathbf{E}_n}^{\,-1} Y_n^{[\alpha,n]}(w) \end{pmatrix} \\ &= \begin{pmatrix} \sqrt{\mathbf{I}-\mathbf{E}_n\mathbf{E}_n^*}^{\,-1} \left(X_n(w) - \overline{\eta_n}\eta_{n-1}\mathbf{E}_n Y_n^{[\alpha,n]}(w)\right) \\ \sqrt{\mathbf{I}-\mathbf{E}_n^*\mathbf{E}_n}^{\,-1} \left(-\mathbf{E}_n^* X_n(w) + \overline{\eta_n}\eta_{n-1} Y_n^{[\alpha,n]}(w)\right) \end{pmatrix}. \end{aligned} \tag{2.12}$$

From Proposition 2.10, [FKL2, Remark 2.8 and Proposition 2.13], and the identity

$$\frac{1-\overline{\alpha_{n-1}}w}{1-\overline{\alpha_n}w} b_{\alpha_{n-1}}(w) = \frac{\eta_{n-1}(\alpha_{n-1}-w)}{1-\overline{\alpha_n}w}$$

we get

$$\begin{aligned} &Y_n^{[\alpha,n]}(w) - \eta_n\overline{\eta_{n-1}}\, \mathbf{E}_n^* X_n(w) \\ &= \rho_n \frac{-\eta_n(\overline{\eta_{n-1}}\ \overline{\alpha_{n-1}}\, w - \overline{\eta_{n-1}})}{1-\overline{\alpha_n}w} \sqrt{\mathbf{I}-\mathbf{E}_n^*\mathbf{E}_n} Y_{n-1}^{[\alpha,n-1]}(w) \\ &= \rho_n \eta_n \overline{\eta_{n-1}} \frac{1-\overline{\alpha_{n-1}}w}{1-\overline{\alpha_n}w} \sqrt{\mathbf{I}-\mathbf{E}_n^*\mathbf{E}_n} Y_{n-1}^{[\alpha,n-1]}(w). \end{aligned} \tag{2.13}$$

Using (2.13) and again Proposition 2.10 we can see that the right-hand side of (2.12) is equal to

$$\rho_n \frac{1-\overline{\alpha_{n-1}}w}{1-\overline{\alpha_n}w} \begin{pmatrix} b_{\alpha_{n-1}}(w) X_{n-1}(w) \\ Y_{n-1}^{[\alpha,n-1]}(w) \end{pmatrix}.$$

Taking into account $\mathbf{H}(\mathbf{E}_n)\mathbf{H}(-\mathbf{E}_n) = \mathbf{I}$ thus the first identity for the considered case $(1-|\alpha_n|)(1-|\alpha_{n-1}|)>0$ follows. Analogously, one can prove that this identity is satisfied if $(1-|\alpha_n|)(1-|\alpha_{n-1}|)<0$. In view of [FKL2, Remark 2.8 and Proposition 2.13], the second recurrence formula is a consequence of the first one. □

Corollary 2.12. *Let* $(\alpha_j)_{j=1}^{\infty} \in \mathcal{T}_1$, *let* $\tau \in \mathbb{N}$ *or* $\tau = \infty$, *and let* $F \in \mathcal{M}_{\geq}^{q,\tau}(\mathbb{T}, \mathfrak{B}_{\mathbb{T}})$. *Let* $[(X_k)_{k=0}^{\tau}, (Y_k)_{k=0}^{\tau}]$ *be a Szegő pair of orthonormal systems corresponding to* $(\alpha_j)_{j=1}^{\infty}$ *and* F, *and let* $(\mathbf{E}_n)_{n=1}^{\tau}$ *be the sequence of its Szegő parameters. For every choice of* $n \in \mathbb{N}_{1,\tau}$ *and* $w \in \mathbb{C} \setminus \mathbb{P}_{\alpha,n}$, *the following statements hold:*

(a) *If* $(1-|\alpha_n|)(1-|\alpha_{n-1}|) > 0$, *then*

$$X_n(w) = \rho_n \frac{1-\overline{\alpha_{n-1}}w}{1-\overline{\alpha_n}w} \sqrt{\mathbf{I}-\mathbf{E}_n\mathbf{E}_n^*}^{\,-1} \left(b_{\alpha_{n-1}}(w)X_{n-1}(w) + \mathbf{E}_n Y_{n-1}^{[\alpha,n-1]}(w)\right)$$

and

$$Y_n(w) = \rho_n \frac{1-\overline{\alpha_{n-1}}w}{1-\overline{\alpha_n}w} \left(b_{\alpha_{n-1}}(w)Y_{n-1}(w) + X_{n-1}^{[\alpha,n-1]}(w)\mathbf{E}_n\right) \sqrt{\mathbf{I}-\mathbf{E}_n^*\mathbf{E}_n}^{\,-1}.$$

(b) *If* $(1-|\alpha_n|)(1-|\alpha_{n-1}|) < 0$, *then*

$$X_n(w) = \rho_n \frac{1-\overline{\alpha_{n-1}}w}{1-\overline{\alpha_n}w} \sqrt{\mathbf{I}-\mathbf{E}_n\mathbf{E}_n^*}^{\,-1} \left(b_{\alpha_{n-1}}(w)\mathbf{E}_n X_{n-1}(w) + Y_{n-1}^{[\alpha,n-1]}(w)\right)$$

and

$$Y_n(w) = \rho_n \frac{1-\overline{\alpha_{n-1}}w}{1-\overline{\alpha_n}w} \left(b_{\alpha_{n-1}}(w)Y_{n-1}(w)\mathbf{E}_n + X_{n-1}^{[\alpha,n-1]}(w)\right) \sqrt{\mathbf{I}-\mathbf{E}_n^*\mathbf{E}_n}^{\,-1}.$$

Proof. Apply Theorem 2.11. □

At the end of this section we note again that the particular case of $q \times q$ matrix polynomials is considered in [DGK] and [DFK, Section 3.6]. In this special case the recurrence relations stated in Corollary 2.12 coincide with those given in [DFK, Lemma 3.6.12].

3. Szegő pairs of rational matrix-valued functions

Motivated by the matrix polynomial case and Corollary 2.12 we introduce the following notion.

Definition 3.1. Let $(\alpha_j)_{j=1}^{\infty} \in \mathcal{T}_1$, let $\tau \in \mathbb{N}$ or $\tau = \infty$, and let $(\mathbf{E}_n)_{n=1}^{\tau}$ be a sequence of strictly contractive $q \times q$ matrices. Further, let $\mathbf{X}_0$ and $\mathbf{Y}_0$ be nonsingular complex $q\times q$ matrices such that $\mathbf{X}_0^*\mathbf{X}_0 = \mathbf{Y}_0\mathbf{Y}_0^*$, and let X_0 and Y_0 be the constant matrix-valued functions (defined on $\mathbb{C}_0$) with values $\mathbf{X}_0$ and $\mathbf{Y}_0$, respectively. For all $n \in \mathbb{N}_{1,\tau}$ such that $(1-|\alpha_n|)(1-|\alpha_{n-1}|) > 0$, let X_n and Y_n be the matrix-valued functions which are given for all $w \in \mathbb{C} \setminus \mathbb{P}_{\alpha,n}$ by

$$X_n(w) := \rho_n \frac{1-\overline{\alpha_{n-1}}w}{1-\overline{\alpha_n}w} \sqrt{\mathbf{I}-\mathbf{E}_n\mathbf{E}_n^*}^{\,-1} \left(b_{\alpha_{n-1}}(w)X_{n-1}(w) + \mathbf{E}_n Y_{n-1}^{[\alpha,n-1]}(w)\right)$$

and

$$Y_n(w) := \rho_n \frac{1-\overline{\alpha_{n-1}}w}{1-\overline{\alpha_n}w}\left(b_{\alpha_{n-1}}(w)Y_{n-1}(w) + X_{n-1}^{[\alpha,n-1]}(w)\mathbf{E}_n\right)\sqrt{\mathbf{I}-\mathbf{E}_n^*\mathbf{E}_n}^{\,-1}.$$

For all $n \in \mathbb{N}_{1,\tau}$ such that $(1-|\alpha_n|)(1-|\alpha_{n-1}|) < 0$, let X_n and Y_n be the matrix-valued functions which are defined for all $w \in \mathbb{C} \setminus \mathbb{P}_{\alpha,n}$ by

$$X_n(w) := \rho_n \frac{1-\overline{\alpha_{n-1}}w}{1-\overline{\alpha_n}w}\sqrt{\mathbf{I}-\mathbf{E}_n\mathbf{E}_n^*}^{\,-1}\left(b_{\alpha_{n-1}}(w)\mathbf{E}_n X_{n-1}(w) + Y_{n-1}^{[\alpha,n-1]}(w)\right)$$

and

$$Y_n(w) := \rho_n \frac{1-\overline{\alpha_{n-1}}w}{1-\overline{\alpha_n}w}\left(b_{\alpha_{n-1}}(w)Y_{n-1}(w)\mathbf{E}_n + X_{n-1}^{[\alpha,n-1]}(w)\right)\sqrt{\mathbf{I}-\mathbf{E}_n^*\mathbf{E}_n}^{\,-1}.$$

Then the pair $[(X_k)_{k=0}^{\tau}, (Y_k)_{k=0}^{\tau}]$ is called *the Szegő pair of rational matrix-valued functions generated by* $[(\alpha_j)_{j=1}^{\infty}; (\mathbf{E}_n)_{n=1}^{\tau}; \mathbf{X}_0, \mathbf{Y}_0]$.

Remark 3.2. Let $(\alpha_j)_{j=1}^{\infty} \in \mathcal{T}_1$, let $\tau \in \mathbb{N}$ or $\tau = \infty$, and let $F \in \mathcal{M}_{\geq}^{q,\tau}(\mathbb{T}, \mathfrak{B}_{\mathbb{T}})$. Let $[(X_k)_{k=0}^{\tau}, (Y_k)_{k=0}^{\tau}]$ be a Szegő pair of orthonormal systems corresponding to $(\alpha_j)_{j=1}^{\infty}$ and F, and let $(\mathbf{E}_n)_{n=1}^{\tau}$ be a sequence of Szegő parameters. Further, let $\mathbf{X}_0 := X_0(0)$ and $\mathbf{Y}_0 := Y_0(0)$. From [FKL2, Remark 5.3] we know that $\mathbf{X}_0$ and $\mathbf{Y}_0$ are nonsingular matrices such that $\mathbf{X}_0^*\mathbf{X}_0 = \mathbf{Y}_0\mathbf{Y}_0^*$ holds. Hence, Corollary 2.12 shows that $[(X_k)_{k=0}^{\tau}, (Y_k)_{k=0}^{\tau}]$ is the Szegő pair of rational matrix-valued functions generated by $[(\alpha_j)_{j=1}^{\infty}; (\mathbf{E}_n)_{n=1}^{\tau}; \mathbf{X}_0, \mathbf{Y}_0]$.

Remark 3.3. Let $(\alpha_j)_{j=1}^{\infty} \in \mathcal{T}_1$, let $\tau \in \mathbb{N}$ or $\tau = \infty$, and let $(\mathbf{E}_n)_{n=1}^{\tau}$ be a sequence of strictly contractive $q \times q$ matrices. Further, let $\mathbf{X}_0$ and $\mathbf{Y}_0$ be nonsingular complex $q \times q$ matrices such that $\mathbf{X}_0^*\mathbf{X}_0 = \mathbf{Y}_0\mathbf{Y}_0^*$. Let $[(X_k)_{k=0}^{\tau}, (Y_k)_{k=0}^{\tau}]$ be the Szegő pair of rational matrix-valued functions generated by $[(\alpha_j)_{j=1}^{\infty}; (\mathbf{E}_n)_{n=1}^{\tau}; \mathbf{X}_0, \mathbf{Y}_0]$. For each $k \in \mathbb{N}_{0,\tau}$, from Definition 3.1, (1.1), and (1.2) one can easily see that the matrix-valued functions X_k and Y_k both belong to $\mathcal{R}_{\alpha,k}^{q\times q}$.

Remark 3.4. Let $(\alpha_j)_{j=1}^{\infty} \in \mathcal{T}_1$, let $\tau \in \mathbb{N}$ or $\tau = \infty$, and let $(\mathbf{E}_n)_{n=1}^{\tau}$ be a sequence of strictly contractive complex $q \times q$ matrices. Furthermore, let $\mathbf{X}_0$ and $\mathbf{Y}_0$ be nonsingular complex $q \times q$ matrices such that $\mathbf{X}_0^*\mathbf{X}_0 = \mathbf{Y}_0\mathbf{Y}_0^*$, let X_0 and Y_0 be the constant matrix-valued functions with values $\mathbf{X}_0$ and $\mathbf{Y}_0$ respectively, and for each $n \in \mathbb{N}_{1,\tau}$ let X_n and Y_n be a matrix-valued function belonging to $\mathcal{R}_{\alpha,n}^{q\times q}$. By using the same arguments as in the proof of Theorem 2.11, one can verify that $[(X_k)_{k=0}^{\tau}, (Y_k)_{k=0}^{\tau}]$ is the Szegő pair of rational matrix-valued functions generated by $[(\alpha_j)_{j=1}^{\infty}; (\mathbf{E}_n)_{n=1}^{\tau}; \mathbf{X}_0, \mathbf{Y}_0]$ if and only if for each $n \in \mathbb{N}_{1,\tau}$ and each $w \in \mathbb{C}\setminus\mathbb{P}_{\alpha,n}$ the recurrence relations stated in Proposition 2.10 are fulfilled.

As already in the introduction announced, we are going now to prove some results with respect to the pairs of sequences of rational matrix-valued functions defined by Definition 3.1, which are often called Favard-type theorems (cf. [BGHN2, Chapter 8]). At first, we study the case that a finite sequence of strictly contractive complex $q \times q$ matrices forms the basis of the pairs.

Theorem 3.5. *Let* $(\alpha_j)_{j=1}^{\infty} \in \mathcal{T}_1$ *and let* $m \in \mathbb{N}$. *Further, let* $(\mathbf{E}_n)_{n=1}^{m}$ *be a sequence of strictly contractive complex* $q \times q$ *matrices, let* $\mathbf{X}_0$ *and* $\mathbf{Y}_0$ *be nonsingular complex* $q \times q$ *matrices such that* $\mathbf{X}_0^* \mathbf{X}_0 = \mathbf{Y}_0 \mathbf{Y}_0^*$, *and let* $[(X_k)_{k=0}^m, (Y_k)_{k=0}^m]$ *be the Szegő pair of rational matrix-valued functions generated by* $[(\alpha_j)_{j=1}^{\infty}; (\mathbf{E}_n)_{n=1}^m; \mathbf{X}_0, \mathbf{Y}_0]$. *Then* $F_m : \mathfrak{B}_{\mathbb{T}} \to \mathbb{C}^{q\times q}$ *defined by*

$$F_m(B) := \frac{1}{2\pi} \int_B \frac{\left|1 - |\alpha_m|^2\right|}{|z-\alpha_m|^2} \big(X_m(z)\big)^{-1} \big(X_m(z)\big)^{-*} \underline{\lambda}(dz) \tag{3.1}$$

is a (well-defined) nonnegative Hermitian $q \times q$ *measure which belongs to the set* $\mathcal{M}_{\geq}^{q,\infty}(\mathbb{T}, \mathfrak{B}_{\mathbb{T}})$ *and* $[(X_k)_{k=0}^m, (Y_k)_{k=0}^m]$ *is a Szegő pair of orthonormal systems corresponding to* $(\alpha_j)_{j=1}^{\infty}$ *and* F_m. *Moreover,* $(\mathbf{E}_n)_{n=1}^m$ *is exactly the sequence of Szegő parameters corresponding to* $[(X_k)_{k=0}^m, (Y_k)_{k=0}^m]$.

Proof. From Remark 3.3 we know that, for each $k \in \mathbb{N}_{0,m}$, the matrix-valued functions X_k and Y_k both belong to $\mathcal{R}_{\alpha,k}^{q\times q}$. The matrix

$$\mathbf{j}_{qq} := \begin{pmatrix} \mathbf{I}_q & \mathbf{0} \\ \mathbf{0} & -\mathbf{I}_q \end{pmatrix}$$

is obviously a $2q \times 2q$ signature matrix, i.e., $\mathbf{j}_{qq}^2 = \mathbf{j}_{qq}$ and $\mathbf{j}_{qq}^* = \mathbf{j}_{qq}$ hold. We are going to verify that $[(X_k)_{k=0}^m, (Y_k)_{k=0}^m]$ is a pair of $\mathcal{R}_{\alpha,\infty}^{q\times q}$-sequences which is $\mathbf{j}_{qq}$-recursively connected, i.e., that X_0 and Y_0 are constant matrix-valued functions which satisfy $X_0^* X_0 = Y_0 Y_0^*$, that there is a sequence $(\mathbf{U}_n)_{n=1}^m$ of complex $2q \times 2q$ matrices such that

$$\mathbf{U}_n^* \, \mathbf{j}_{qq} \mathbf{U}_n = \begin{cases} \mathbf{j}_{qq} & \text{if } (1-|\alpha_n|)(1-|\alpha_{n-1}|) > 0 \\ -\mathbf{j}_{qq} & \text{if } (1-|\alpha_n|)(1-|\alpha_{n-1}|) < 0 \end{cases} \tag{3.2}$$

for all $n \in \mathbb{N}_{1,m}$ and that

$$\begin{pmatrix} X_n(w) \\ Y_n^{[\alpha,n]}(w) \end{pmatrix} = \rho_n \frac{1-\overline{\alpha_{n-1}}w}{1-\overline{\alpha_n}} \mathbf{U}_n \begin{pmatrix} b_{\alpha_{n-1}}(w)\mathbf{I}_q & \mathbf{0} \\ \mathbf{0} & \mathbf{I}_q \end{pmatrix} \begin{pmatrix} X_{n-1}(w) \\ Y_{n-1}^{[\alpha,n-1]}(w) \end{pmatrix} \tag{3.3}$$

for all $n \in \mathbb{N}_{1,m}$ and all $w \in \mathbb{C} \setminus \mathbb{P}_{\alpha,n}$. Obviously, X_0 and Y_0 are constant matrix-valued functions which satisfy $X_0^* X_0 = Y_0 Y_0^*$. For each $n \in \mathbb{N}_{1,m}$, the matrix

$$\mathbf{U}_n := \begin{cases} \begin{pmatrix} \mathbf{I}_q & \mathbf{0} \\ \mathbf{0} & \eta_n \overline{\eta_{n-1}} \mathbf{I}_q \end{pmatrix} \mathbf{H}(\mathbf{E}_n) & \text{if } (1-|\alpha_n|)(1-|\alpha_{n-1}|) > 0 \\ \begin{pmatrix} \mathbf{I}_q & \mathbf{0} \\ \mathbf{0} & \eta_n \overline{\eta_{n-1}} \mathbf{I}_q \end{pmatrix} \mathbf{H}(\mathbf{E}_n) \begin{pmatrix} \mathbf{0} & \mathbf{I}_q \\ \mathbf{I}_q & \mathbf{0} \end{pmatrix} & \text{if } (1-|\alpha_n|)(1-|\alpha_{n-1}|) < 0 \end{cases}$$

fulfills (3.2). Let $n \in \mathbb{N}_{1,m}$ and let $w \in \mathbb{C} \setminus \mathbb{P}_{\alpha,n}$. In view of Definition 3.1 and [FKL2, Remark 2.8 and Proposition 2.13], if $(1-|\alpha_n|)(1-|\alpha_{n-1}|) > 0$ then

$$\begin{aligned} &Y_n^{[\alpha,n]}(w) \\ &= \rho_n \eta_n \overline{\eta_{n-1}} \frac{1-\overline{\alpha_{n-1}}w}{1-\overline{\alpha_n}w} \sqrt{\mathbf{I}-\mathbf{E}_n^*\mathbf{E}_n}^{\,-1} \Big(Y_{n-1}^{[\alpha,n-1]}(w) + b_{\alpha_{n-1}}(w) \mathbf{E}_n^* X_{n-1}(w)\Big). \end{aligned} \tag{3.4}$$

Analogously, in the case $(1-|\alpha_n|)(1-|\alpha_{n-1}|)<0$, we have

$$Y_n^{[\alpha,n]}(w) = \rho_n\eta_n\overline{\eta_{n-1}}\frac{1-\overline{\alpha_{n-1}}w}{1-\overline{\alpha_n}w}\sqrt{\mathbf{I}-\mathbf{E}_n^*\mathbf{E}_n}^{-1}\left(\mathbf{E}_n^* Y_{n-1}^{[\alpha,n-1]}(w)+b_{\alpha_{n-1}}(w)X_{n-1}(w)\right). \quad (3.5)$$

Consequently, (3.3) is valid. By virtue of [FKL3, Lemma 3.11], for all $z\in\mathbb{T}$, the matrix $X_m(z)$ is nonsingular. From [FKL1, Example 5.4] we know then that F_m is a well-defined nonnegative Hermitian measure which belongs to $\mathcal{M}_{\geq}^{q,\infty}(\mathbb{T},\mathfrak{B}_{\mathbb{T}})$. Applying [FKL3, Theorem 4.4] we obtain that $[(X_k)_{k=0}^m,(Y_k)_{k=0}^m]$ is an orthonormal system corresponding to $(\alpha_j)_{j=1}^\infty$ and F_m. Let $n\in\mathbb{N}_{1,m}$. We consider the case that the inequality $(1-|\alpha_n|)(1-|\alpha_{n-1}|)>0$ holds. From (3.4) we see that

$$\sqrt{\mathbf{I}-\mathbf{E}_n^*\mathbf{E}_n}Y_n^{[\alpha,n]}(\alpha_{n-1}) = \rho_n\eta_n\overline{\eta_{n-1}}\frac{1-|\alpha_{n-1}|^2}{1-\overline{\alpha_n}\alpha_{n-1}}Y_{n-1}^{[\alpha,n-1]}(\alpha_{n-1})$$

and, in view of [FKL3, Lemma 3.11],

$$\sqrt{\mathbf{I}-\mathbf{E}_n^*\mathbf{E}_n} = \rho_n\eta_n\overline{\eta_{n-1}}\frac{1-|\alpha_{n-1}|^2}{1-\overline{\alpha_n}\alpha_{n-1}}Y_{n-1}^{[\alpha,n-1]}(\alpha_{n-1})\left(Y_n^{[\alpha,n]}(\alpha_{n-1})\right)^{-1}. \quad (3.6)$$

In particular, the matrix stated in (2.2) is positive Hermitian. For each $w\in\mathbb{C}\setminus\mathbb{P}_{\alpha,n}$, from Definition 3.1 we get

$$X_n^{[\alpha,n]}(w)\sqrt{\mathbf{I}-\mathbf{E}_n\mathbf{E}_n^*} = \rho_n\eta_n\overline{\eta_{n-1}}\frac{1-\overline{\alpha_{n-1}}w}{1-\overline{\alpha_n}w}\left(X_{n-1}^{[\alpha,n-1]}(w)+b_{\alpha_{n-1}}(w)Y_{n-1}(w)\mathbf{E}_n^*\right)$$

and consequently, in view of [FKL3, Lemma 3.12],

$$\sqrt{\mathbf{I}-\mathbf{E}_n\mathbf{E}_n^*} = \rho_n\eta_n\overline{\eta_{n-1}}\frac{1-|\alpha_{n-1}|^2}{1-\overline{\alpha_n}\alpha_{n-1}}\left(X_n^{[\alpha,n]}(\alpha_{n-1})\right)^{-1}X_{n-1}^{[\alpha,n-1]}(\alpha_{n-1}).$$

Therefore the matrix stated in (2.3) is also positive Hermitian. In the case that $(1-|\alpha_n|)(1-|\alpha_{n-1}|)<0$ holds one can analogously check that the matrices stated in (2.4) and (2.5) are both positive Hermitian. If $(1-|\alpha_n|)(1-|\alpha_{n-1}|)>0$, then Definition 3.1 and (3.6) provide us

$$\eta_n\overline{\eta_{n-1}}X_n(\alpha_{n-1})\left(Y_n^{[\alpha,n]}(\alpha_{n-1})\right)^{-1} = \eta_n\overline{\eta_{n-1}}\rho_n\frac{1-|\alpha_{n-1}|^2}{1-\overline{\alpha_n}\alpha_{n-1}}\sqrt{\mathbf{I}-\mathbf{E}_n\mathbf{E}_n^*}^{-1}\mathbf{E}_nY_{n-1}^{[\alpha,n-1]}(\alpha_{n-1})\left(Y_n^{[\alpha,n]}(\alpha_{n-1})\right)^{-1}=\mathbf{E}_n.$$

If $(1-|\alpha_n|)(1-|\alpha_{n-1}|)<0$, from (3.5) and Definition 3.1 we obtain similarly

$$\eta_n\overline{\eta_{n-1}}\left(Y_n^{[\alpha,n]}(\alpha_{n-1})\left(X_n(\alpha_{n-1})\right)^{-1}\right)^* = \rho_n\left(\frac{1-|\alpha_{n-1}|^2}{1-\overline{\alpha_n}\alpha_{n-1}}\sqrt{\mathbf{I}-\mathbf{E}_n^*\mathbf{E}_n}^{-1}\mathbf{E}_n^*Y_{n-1}^{[\alpha,n-1]}(\alpha_{n-1})\left(X_n(\alpha_{n-1})\right)^{-1}\right)^*=\mathbf{E}_n.$$

The proof is complete. □

Corollary 3.6. *Let $(\alpha_j)_{j=1}^{\infty} \in \mathcal{T}_1$, let $\tau \in \mathbb{N}$, let $(\mathbf{E}_n)_{n=1}^{\tau}$ be a sequence of strictly contractive complex $q \times q$ matrices, let $\mathbf{X}_0$ and $\mathbf{Y}_0$ be nonsingular complex $q \times q$ matrices such that $\mathbf{X}_0^*\mathbf{X}_0 = \mathbf{Y}_0\mathbf{Y}_0^*$, and let $[(X_k)_{k=0}^{\tau}, (Y_k)_{k=0}^{\tau}]$ be the Szegő pair of rational matrix-valued functions generated by $[(\alpha_j)_{j=1}^{\infty}; (\mathbf{E}_n)_{n=1}^{\tau}; \mathbf{X}_0, \mathbf{Y}_0]$. Further, let the sequences $(X_m)_{m=\tau+1}^{\infty}$ and $(Y_m)_{m=\tau+1}^{\infty}$ of rational matrix-valued functions be defined for all $m \in \mathbb{N}_{\tau+1,\infty}$ and $w \in \mathbb{C}_0 \setminus \mathbb{P}_{\alpha,m}$ by*

$$X_m(w) := \begin{cases} \eta_{m-1}\sqrt{\frac{1-|\alpha_m|^2}{1-|\alpha_{m-1}|^2}}\,\frac{\alpha_{m-1}-w}{1-\overline{\alpha_m}w}X_{m-1}(w) & \text{if } (1-|\alpha_m|)(1-|\alpha_{m-1}|) > 0 \\ -\sqrt{\frac{|\alpha_m|^2-1}{1-|\alpha_{m-1}|^2}}\,\frac{1-\overline{\alpha_{m-1}}w}{1-\overline{\alpha_m}w}Y_{m-1}^{[\alpha,m-1]}(w) & \text{if } (1-|\alpha_m|)(1-|\alpha_{m-1}|) < 0 \end{cases}$$

and

$$Y_m(w) := \begin{cases} \eta_{m-1}\sqrt{\frac{1-|\alpha_m|^2}{1-|\alpha_{m-1}|^2}}\,\frac{\alpha_{m-1}-w}{1-\overline{\alpha_m}w}Y_{m-1}(w) & \text{if } (1-|\alpha_m|)(1-|\alpha_{m-1}|) > 0 \\ -\sqrt{\frac{|\alpha_m|^2-1}{1-|\alpha_{m-1}|^2}}\,\frac{1-\overline{\alpha_{m-1}}w}{1-\overline{\alpha_m}w}X_{m-1}^{[\alpha,m-1]}(w) & \text{if } (1-|\alpha_m|)(1-|\alpha_{m-1}|) < 0\,, \end{cases}$$

and let $\mathbf{E}_m := \mathbf{0}_{q\times q}$ for all $m \in \mathbb{N}_{\tau+1,\infty}$. Then $[(X_k)_{k=0}^{\infty}, (Y_k)_{k=0}^{\infty}]$ is the Szegő pair of rational matrix-valued functions generated by $[(\alpha_j)_{j=1}^{\infty}; (\mathbf{E}_n)_{n=1}^{\infty}; \mathbf{X}_0, \mathbf{Y}_0]$. Moreover, for each $m \in \mathbb{N}_{\tau,\infty}$, if the nonnegative Hermitian $q \times q$ Borel measure F_m on $\mathbb{T}$ is given as in (3.1) then $F_m = F_\tau$, $[(X_k)_{k=0}^{\infty}, (Y_k)_{k=0}^{\infty}]$ is the Szegő pair of orthonormal systems corresponding to $(\alpha_j)_{j=1}^{\infty}$ and F_τ, and $(\mathbf{E}_n)_{n=1}^{\infty}$ is exactly the sequence of Szegő parameters corresponding to $[(X_k)_{k=0}^{\infty}, (Y_k)_{k=0}^{\infty}]$.

Proof. Since $\mathbf{0}_{q\times q}$ is a strictly contractive $q \times q$ matrix, from Definition 3.1, (2.8), (1.1), and (2.1) it follows immediately that $[(X_k)_{k=0}^{\infty}, (Y_k)_{k=0}^{\infty}]$ is the Szegő pair of rational matrix-valued functions generated by $[(\alpha_j)_{j=1}^{\infty}; (\mathbf{E}_n)_{n=1}^{\infty}; \mathbf{X}_0, \mathbf{Y}_0]$. Furthermore, an application of Theorem 3.5 and [FKL2, Remark 2.5, Remark 6.2, and Lemma 6.5] implies for all $m \in \mathbb{N}_{\tau+1,\infty}$ and $z \in \mathbb{T}$ the identity

$$\big(X_m(z)\big)^* X_m(z) = \left|\frac{1-|\alpha_m|^2}{1-|\alpha_\tau|^2}\right| \frac{|z-\alpha_\tau|^2}{|z-\alpha_m|^2}\big(X_\tau(z)\big)^* X_\tau(z).$$

Therefore, we obtain for each $m \in \mathbb{N}_{\tau+1,\infty}$ the equality $F_m = F_\tau$. Thus, we can finally conclude from Theorem 3.5 that $[(X_k)_{k=0}^{\infty}, (Y_k)_{k=0}^{\infty}]$ is the Szegő pair of orthonormal systems corresponding to $(\alpha_j)_{j=1}^{\infty}$ and F_τ, and that $(\mathbf{E}_n)_{n=1}^{\infty}$ is exactly the sequence of Szegő parameters corresponding to $[(X_k)_{k=0}^{\infty}, (Y_k)_{k=0}^{\infty}]$. □

Theorem 3.7. *Let $(\alpha_j)_{j=1}^{\infty} \in \mathcal{T}_1$, let $(\mathbf{E}_n)_{n=1}^{\infty}$ be a sequence of strictly contractive complex $q \times q$ matrices, and let $\mathbf{X}_0$ and $\mathbf{Y}_0$ be nonsingular complex $q \times q$ matrices such that $\mathbf{X}_0^*\mathbf{X}_0 = \mathbf{Y}_0\mathbf{Y}_0^*$. Let $[(X_k)_{k=0}^{\infty}, (Y_k)_{k=0}^{\infty}]$ be a Szegő pair of rational matrix-valued functions generated by $[(\alpha_j)_{j=1}^{\infty}; (\mathbf{E}_n)_{n=1}^{\infty}; \mathbf{X}_0, \mathbf{Y}_0]$. Then there is an $F \in \mathcal{M}_{\geq}^{q,\infty}(\mathbb{T}, \mathfrak{B}_{\mathbb{T}})$ such that $[(X_k)_{k=0}^{\infty}, (Y_k)_{k=0}^{\infty}]$ is a Szegő pair of orthonormal systems corresponding to $(\alpha_j)_{j=1}^{\infty}$ and F. Moreover, $(\mathbf{E}_n)_{n=1}^{\infty}$ is the sequence of Szegő parameters corresponding to $[(X_k)_{k=0}^{\infty}, (Y_k)_{k=0}^{\infty}]$.*

Proof. According to Theorem 3.5, for each $m \in \mathbb{N}$, $[(X_k)_{k=0}^m, (Y_k)_{k=0}^m]$ is a Szegő pair of orthonormal systems corresponding to $(\alpha_j)_{j=1}^\infty$ and the nonnegative Hermitian $q \times q$ measure F_m which is given by (3.1). From [FKL2, Remark 5.3] we know that $F_m(\mathbb{T}) = (\mathbf{X}_0^*\mathbf{X}_0)^{-1}$ for all $m \in \mathbb{N}$. By virtue of [FKL3, Lemma 4.7], there are an $F \in \mathcal{M}_\geq^q(\mathbb{T}, \mathfrak{B}_\mathbb{T})$ and a subsequence $(F_{m_n})_{n=1}^\infty$ of $(F_m)_{m=1}^\infty$ which converges weakly to F (cf. [FK] or [FKL3, Section 4]). Applying [FKL3, Lemma 4.6] (see also [FK, Satz 3]), for all nonnegative integers j and k, we obtain

$$\begin{aligned}(X_j, X_k)_{F,l} &= \int_\mathbb{T} \underline{X_j} dF \underline{X_k^*} = \lim_{n\to\infty} \int_\mathbb{T} \underline{X_j} dF_{m_n} \underline{X_k^*} \\ &= \lim_{n\to\infty} (X_j, X_k)_{F_{m_n},l} = \delta_{j,k}\mathbf{I}_q\end{aligned}$$

and analogously

$$(Y_j, Y_k)_{F,r} = \delta_{jk}\mathbf{I}_q.$$

Hence, in view of Remark 3.3, $[(X_k)_{k=0}^\infty, (Y_k)_{k=0}^\infty]$ is a pair of orthonormal systems corresponding to $(\alpha_j)_{j=1}^\infty$ and F. The rest of the assertion follows immediately from [FKL2, Corollary 4.4] and Theorem 3.5. □

4. Integral representations of Szegő parameters

In the following, for each $w \in \mathbb{C}$, let $f_w : \mathbb{C} \to \mathbb{C}$ and $g_w : \mathbb{C} \to \mathbb{C}$ be defined by

$$f_w(z) := z - w \quad \text{and} \quad g_w(z) := 1 - \overline{w}z, \tag{4.1}$$

respectively. If a sequence $(\alpha_j)_{j=1}^\infty \in \mathcal{T}_1$ is given, we will again use the notation η_k given by (2.2).

Remark 4.1. Let $(\alpha_j)_{j=1}^\infty \in \mathcal{T}_1$, let $n \in \mathbb{N}$, and let $F \in \mathcal{M}_\geq^q(\mathbb{T}, \mathfrak{B}_\mathbb{T})$. Let X and Y belong to $\mathcal{R}_{\alpha,n-1}^{q\times q}$. Further, let P be a polynomial over $\mathbb{C}$ of degree one. In view of [FKL2, Remark 2.5], one obtains that $\frac{1}{g_{\alpha_n}} PX^{[\alpha,n-1]}$ and $\frac{1}{g_{\alpha_n}} PY$ belong to $\mathcal{R}_{\alpha,n}^{q\times q}$ and that

$$\left(\frac{1}{g_{\alpha_n}} PX^{[\alpha,n-1]}, Y^{[\alpha,n-1]}\right)_{F,l} = \left(X, \frac{1}{g_{\alpha_n}} PY\right)_{F,r}.$$

Lemma 4.2. *Let $(\alpha_j)_{j=1}^\infty \in \mathcal{T}_1$, let $\tau \in \mathbb{N}$ or $\tau = \infty$, and let $F \in \mathcal{M}_\geq^{q,\tau}(\mathbb{T}, \mathfrak{B}_\mathbb{T})$. Let $[(X_k)_{k=0}^\tau, (Y_k)_{k=0}^\tau]$ be a Szegő pair of orthonormal systems corresponding to $(\alpha_j)_{j=1}^\infty$ and F, and let $(\mathbf{E}_n)_{n=1}^\tau$ be the sequence of its Szegő parameters. For each $n \in \mathbb{N}_{1,\tau}$, the following statements hold:*

(a) *If $(1 - |\alpha_n|)(1 - |\alpha_{n-1}|) > 0$, then for all $Z \in \mathcal{R}_{\alpha,n-1}^{q\times q}$, we have*

$$\mathbf{E}_n \left(\frac{g_{\alpha_{n-1}}}{g_{\alpha_n}} Y_{n-1}^{[\alpha,n-1]}, Z\right)_{F,l} = \eta_{n-1} \left(\frac{f_{\alpha_{n-1}}}{g_{\alpha_n}} X_{n-1}, Z\right)_{F,l},$$

$$\left(Z, \frac{g_{\alpha_{n-1}}}{g_{\alpha_n}} X_{n-1}^{[\alpha,n-1]}\right)_{F,r} \mathbf{E}_n = \eta_{n-1} \left(Z, \frac{f_{\alpha_{n-1}}}{g_{\alpha_n}} Y_{n-1}\right)_{F,r},$$

$$\left(\frac{g_{\alpha_{n-1}}}{g_{\alpha_n}}Z, X_{n-1}\right)_{F,l} \mathbf{E}_n = \eta_{n-1}\left(\frac{f_{\alpha_{n-1}}}{g_{\alpha_n}}Z, Y_{n-1}^{[\alpha,n-1]}\right)_{F,l},$$

and

$$\mathbf{E}_n\left(Y_{n-1}, \frac{g_{\alpha_{n-1}}}{g_{\alpha_n}}Z\right)_{F,r} = \eta_{n-1}\left(X_{n-1}^{[\alpha,n-1]}, \frac{f_{\alpha_{n-1}}}{g_{\alpha_n}}Z\right)_{F,r}.$$

(b) *If* $(1-|\alpha_n|)(1-|\alpha_{n-1}|) < 0$, *then*

$$\mathbf{E}_n\left(\frac{f_{\alpha_{n-1}}}{g_{\alpha_n}}X_{n-1}, Z\right)_{F,l} = \overline{\eta_{n-1}}\left(\frac{g_{\alpha_{n-1}}}{g_{\alpha_n}}Y_{n-1}^{[\alpha,n-1]}, Z\right)_{F,l},$$

$$\left(Z, \frac{f_{\alpha_{n-1}}}{g_{\alpha_n}}Y_{n-1}\right)_{F,r} \mathbf{E}_n = \overline{\eta_{n-1}}\left(Z, \frac{g_{\alpha_{n-1}}}{g_{\alpha_n}}X_{n-1}^{[\alpha,n-1]}\right)_{F,r},$$

$$\left(\frac{f_{\alpha_{n-1}}}{g_{\alpha_n}}Z, Y_{n-1}^{[\alpha,n-1]}\right)_{F,l} \mathbf{E}_n = \overline{\eta_{n-1}}\left(\frac{g_{\alpha_{n-1}}}{g_{\alpha_n}}Z, X_{n-1}\right)_{F,l},$$

and

$$\mathbf{E}_n\left(X_{n-1}^{[\alpha,n-1]}, \frac{f_{\alpha_{n-1}}}{g_{\alpha_n}}Z\right)_{F,r} = \overline{\eta_{n-1}}\left(Y_{n-1}, \frac{g_{\alpha_{n-1}}}{g_{\alpha_n}}Z\right)_{F,r}.$$

Proof. Let $n \in \mathbb{N}_{1,\tau}$ and $Z \in \mathcal{R}_{\alpha,n-1}^{q\times q}$. From [FKL2, Remark 2.1 and Lemma 3.6] we get that

$$(X_n, Z)_{F,l} = \mathbf{0} \quad \text{and} \quad (Z, Y_n)_{F,r} = \mathbf{0}.$$

Application of Corollary 2.12 yields the first and the second identity of (a) if $(1-|\alpha_n|)(1-|\alpha_{n-1}|) > 0$ and in the case $(1-|\alpha_n|)(1-|\alpha_{n-1}|) < 0$ the first and the second identity of (b). The other identities follow then from Remark 4.1. □

Lemma 4.3. *Let* $(\alpha_j)_{j=1}^{\infty} \in \mathcal{T}_1$, *let* $\tau \in \mathbb{N}$ *or* $\tau = \infty$, *and let* $F \in \mathcal{M}_{\geq}^{q,\tau}(\mathbb{T}, \mathfrak{B}_{\mathbb{T}})$. *Let* $[(X_k)_{k=0}^{\tau}, (Y_k)_{k=0}^{\tau}]$ *be a pair of orthonormal systems corresponding to* $(\alpha_j)_{j=1}^{\infty}$ *and* F. *Further, let* $n \in \mathbb{N}_{1,\tau}$ *and let* $Z \in \mathcal{R}_{\alpha,n-1}^{q\times q}$.

(a) *If* $(1-|\alpha_n|)(1-|\alpha_{n-1}|) > 0$, *then the following statements are equivalent:*
  - (i) $\det\left(\frac{g_{\alpha_{n-1}}}{g_{\alpha_n}}Z, X_{n-1}\right)_{F,l} \neq 0$.
  - (ii) $\det\left(Y_{n-1}, \frac{g_{\alpha_{n-1}}}{g_{\alpha_n}}Z\right)_{F,r} \neq 0$.
  - (iii) $\det Z^{[\alpha,n-1]}(\alpha_n) \neq 0$.

(b) *If* $(1-|\alpha_n|)(1-|\alpha_{n-1}|) < 0$, *then the following statements are equivalent:*
  - (iv) $\det\left(\frac{f_{\alpha_{n-1}}}{g_{\alpha_n}}Z, Y_{n-1}^{[\alpha,n-1]}\right)_{F,l} \neq 0$.
  - (v) $\det\left(X_{n-1}^{[\alpha,n-1]}, \frac{f_{\alpha_{n-1}}}{g_{\alpha_n}}Z\right)_{F,r} \neq 0$.
  - (vi) $\det Z^{[\alpha,n-1]}(\alpha_n) \neq 0$.

Proof. (a) Suppose $(1-|\alpha_n|)(1-|\alpha_{n-1}|)>0$. Obviously, $W := \frac{g_{\alpha_{n-1}}}{g_{\alpha_n}} Z$ belongs to $\mathcal{R}^{q\times q}_{\alpha,n}$. Using [FKL2, Remark 2.8 and Proposition 2.13] we obtain

$$W^{[\alpha,n]} = -\eta_n \frac{f_{\alpha_{n-1}}}{g_{\alpha_n}} Z^{[\alpha,n-1]}. \tag{4.2}$$

From $f_{\alpha_{n-1}}(\alpha_{n-1})=0$ it follows

$$W^{[\alpha,n]}(\alpha_{n-1}) = \mathbf{0}. \tag{4.3}$$

In view of [FKL2, Remark 2.1], let $(\mathbf{W}_k)_{k=0}^n$ be the unique sequence of complex $q\times q$ matrices such that

$$W = \sum_{k=0}^n \mathbf{W}_k B^{(q)}_{\alpha,k}$$

and for each $j\in\mathbb{N}_{0,n}$ let $(\mathbf{X}_{jk})_{k=0}^n$ be the sequence of complex $q\times q$ matrices such that

$$X_j = \sum_{k=0}^n \mathbf{X}_{jk} B^{(q)}_{\alpha,k}. \tag{4.4}$$

We have

$$X_n^{[\alpha,n]}(\alpha_{n-1}) = \mathbf{X}^*_{nn} + b_{\alpha_n}(\alpha_{n-1})\mathbf{X}^*_{n,n-1} \tag{4.5}$$

and

$$W^{[\alpha,n]}(\alpha_{n-1}) = \mathbf{W}^*_n + b_{\alpha_n}(\alpha_{n-1})\mathbf{W}^*_{n-1}.$$

From (4.3) it follows

$$\mathbf{W}_n = -\overline{b_{\alpha,n}(\alpha_{n-1})}\mathbf{W}_{n-1}. \tag{4.6}$$

Using (1.3) and (4.2) we get

$$\begin{aligned}\mathbf{W}^*_n &= W^{[\alpha,n]}(\alpha_n) = -\eta_n \frac{f_{\alpha_{n-1}}(\alpha_n)}{g_{\alpha_n}(\alpha_n)} Z^{[\alpha,n-1]}(\alpha_n)\\ &= -\frac{1-\overline{\alpha_n}\alpha_{n-1}}{1-|\alpha_n|^2} b_{\alpha_n}(\alpha_{n-1}) Z^{[\alpha,n-1]}(\alpha_n).\end{aligned}$$

If $\alpha_n\neq\alpha_{n-1}$, then comparing this with (4.6) yields

$$\frac{1-\overline{\alpha_n}\alpha_{n-1}}{1-|\alpha_n|^2} Z^{[\alpha,n-1]}(\alpha_n) = \mathbf{W}^*_{n-1}. \tag{4.7}$$

If $\alpha_n=\alpha_{n-1}$ we have $W=Z$ and hence, in view of (1.3), equation (4.7) also holds in this case. From (4.4) and [FKL2, Lemma 3.6] we obtain

$$\begin{aligned}\mathbf{0} &= (X_n, X_{n-1})_{F,l} = \mathbf{X}_{nn}(B^{(q)}_{\alpha,n}, X_{n-1})_{F,l} + \left(\sum_{k=0}^{n-1} \mathbf{X}_{nk} B^{(q)}_{\alpha,k}, X_{n-1}\right)_{F,l}\\ &= \mathbf{X}_{nn}(B^{(q)}_{\alpha,n}, X_{n-1})_{F,l} + \mathbf{X}_{n,n-1}\mathbf{X}^{-1}_{n-1,n-1}. \end{aligned} \tag{4.8}$$

Analogously, we get

$$(W, X_{n-1})_{F,l} = \mathbf{W}_n(B^{(q)}_{\alpha,n}, X_{n-1})_{F,l} + \mathbf{W}_{n-1}\mathbf{X}^{-1}_{n-1,n-1}.$$

Hence from (4.8), (4.6), (4.5), and (4.7) we can conclude

$$\begin{aligned}(W,X_{n-1})_{F,l} &= \big(-\mathbf{W}_n\mathbf{X}_{nn}^{-1}\mathbf{X}_{n,n-1}+\mathbf{W}_{n-1}\big)\mathbf{X}_{n-1,n-1}^{-1}\\ &= \mathbf{W}_{n-1}\mathbf{X}_{nn}^{-1}\big(\overline{b_{\alpha_n}(\alpha_{n-1})}\mathbf{X}_{n,n-1}+\mathbf{X}_{nn}\big)\mathbf{X}_{n-1,n-1}^{-1}\\ &= \mathbf{W}_{n-1}\mathbf{X}_{nn}^{-1}\big(X_n^{[\alpha,n]}(\alpha_{n-1})\big)^*\mathbf{X}_{n-1,n-1}^{-1}\\ &= \frac{1-\overline{\alpha_{n-1}}\alpha_n}{1-|\alpha_n|^2}\big(Z^{[\alpha,n-1]}(\alpha_n)\big)^*\mathbf{X}_{nn}^{-1}\big(X_n^{[\alpha,n]}(\alpha_{n-1})\big)^*\mathbf{X}_{n-1,n-1}^{-1}. \qquad (4.9)\end{aligned}$$

Since we know from Remark 1.2 that $\det X_n^{[\alpha,n]}(\alpha_{n-1}) \neq 0$ holds we see then from (4.9) that (i) and (iii) are equivalent. In view of [FKL2, Remark 2.11 and Remark 3.5], we also obtain that (ii) and (iii) are equivalent.

(b) Now let $(1-|\alpha_n|)(1-|\alpha_{n-1}|)<0$. Let $V := \frac{f_{\alpha_{n-1}}}{g_{\alpha_n}}Z$, let $(\mathbf{V}_k)_{k=0}^n$ be the unique sequence of complex $q\times q$ matrices such that

$$V=\sum_{k=0}^n \mathbf{V}_k B_{\alpha,k}^{(q)}$$

and let $(\mathbf{Y}_{nk})_{k=0}^n$ be the sequence of complex $q\times q$ matrices such that

$$Y_n=\sum_{k=0}^n \mathbf{Y}_{nk} B_{\alpha,k}^{(q)}.$$

Obviously, $R := \sum_{k=0}^{n-1}\mathbf{Y}_{nk}B_{\alpha,k}^{(q)}$ belongs to $\mathcal{R}_{\alpha,n-1}^{q\times q}$. From [FKL2, Remark 2.8 and Proposition 2.13] we get

$$V^{[\alpha,n]} = -\eta_n\frac{g_{\alpha_{n-1}}}{g_{\alpha_n}}Z^{[\alpha,n-1]}$$

and (1.3) yields then

$$\mathbf{V}_n=\big(V^{[\alpha,n]}(\alpha_n)\big)^* = -\overline{\eta_n}\frac{1-\overline{\alpha_n}\alpha_{n-1}}{1-|\alpha_n|^2}\big(Z^{[\alpha,n-1]}(\alpha_n)\big)^*. \qquad (4.10)$$

Since $X_{n-1}^{[\alpha,n-1]}$ belongs to $\mathcal{R}_{\alpha,n-1}^{q\times q}$, from [FKL2, Lemma 3.6 and Remark 4.2] and (1.3) we obtain

$$\begin{aligned}\mathbf{0} &= \Big(X_{n-1}^{[\alpha,n-1]},Y_n\Big)_{F,r} = \Big(X_{n-1}^{[\alpha,n-1]},B_{\alpha,n}^{(q)}\Big)_{F,r}\mathbf{Y}_{nn}+\Big(R^{[\alpha,n-1]},X_{n-1}\Big)_{F,l}^*\\ &= \Big(X_{n-1}^{[\alpha,n-1]},B_{\alpha,n}^{(q)}\Big)_{F,r}\mathbf{Y}_{nn}+\Big(\big(R(\alpha_{n-1})\big)^*\mathbf{X}_{n-1,n-1}^{-1}\Big)^*. \qquad (4.11)\end{aligned}$$

Furthermore, we have

$$R(\alpha_{n-1}) = Y_n(\alpha_{n-1})-\mathbf{Y}_{nn}B_{\alpha,n}^{(q)}(\alpha_{n-1}) = Y_n(\alpha_{n-1}). \qquad (4.12)$$

The matrix-valued function

$$H := \sum_{k=0}^{n-1}\mathbf{V}_kB_{\alpha,k}^{(q)}$$

belongs to $\mathcal{R}^{q\times q}_{\alpha,n-1}$ and satisfies

$$H(\alpha_{n-1}) = V(\alpha_{n-1}) - \mathbf{V}_n B^{(q)}_{\alpha,n}(\alpha_{n-1}) = V(\alpha_{n-1}) = \frac{f_{\alpha_{n-1}}(\alpha_{n-1})}{g_{\alpha_n}(\alpha_{n-1})} Z(\alpha_{n-1}) = \mathbf{0}.$$

By virtue of [FKL2, Lemma 3.6 and Remark 4.2], (1.3), (4.10), (4.11), and (4.12) it follows

$$\begin{aligned}\left(X^{[\alpha,n-1]}_{n-1}, V\right)_{F,r} &= \left(X^{[\alpha,n-1]}_{n-1}, B^{(q)}_{\alpha,n}\right)_{F,r} \mathbf{V}_n + \left(\left(H(\alpha_{n-1})\right)^* \mathbf{X}^{-1}_{n-1,n-1}\right)^* \\ &= \left(X^{[\alpha,n-1]}_{n-1}, B^{(q)}_{\alpha,n}\right)_{F,r} \mathbf{V}_n = -\left(\left(R(\alpha_{n-1})\right)^* \mathbf{X}^{-1}_{n-1,n-1}\right)^* \mathbf{Y}^{-1}_{nn} \mathbf{V}_n \\ &= -\overline{\eta_n} \frac{1-\overline{\alpha_n}\alpha_{n-1}}{1-|\alpha_n|^2} \left(\left(Y_n(\alpha_{n-1})\right)^* \mathbf{X}^{-1}_{n-1,n-1}\right)^* \mathbf{Y}^{-1}_{nn} \left(Z^{[\alpha,n-1]}(\alpha_n)\right)^*.\end{aligned}$$

Since Remark 1.2 yields that $\det Y_n(\alpha_{n-1}) \neq 0$ is fulfilled we see that (v) and (vi) are equivalent. The equivalence of (iv) and (vi) follows then by application of [FKL2, Remark 2.11 and Remark 3.5]. □

Theorem 4.4. *Let $(\alpha_j)_{j=1}^\infty \in \mathcal{T}_1$, let $\tau \in \mathbb{N}$ or $\tau = \infty$, and let $F \in \mathcal{M}^{q,\tau}_{\geq}(\mathbb{T}, \mathfrak{B}_\mathbb{T})$. Let $[(X_k)_{k=0}^\tau, (Y_k)_{k=0}^\tau]$ be a Szegő pair of orthonormal systems corresponding to $(\alpha_j)_{j=1}^\infty$ and F, and let $(\mathbf{E}_n)_{n=1}^\tau$ be the sequence of its Szegő parameters. For each $n \in \mathbb{N}_{1,\tau}$, the following statements hold:*

(a) *If $(1-|\alpha_n|)(1-|\alpha_{n-1}|) > 0$, then*

$$\mathbf{E}_n = \eta_{n-1} \left(\int_\mathbb{T} \frac{g_{\alpha_{n-1}}}{g_{\alpha_n}} Z dF X^*_{n-1}\right)^{-1} \int_\mathbb{T} \frac{f_{\alpha_{n-1}}}{g_{\alpha_n}} Z dF \left(Y^{[\alpha,n-1]}_{n-1}\right)^*$$

and

$$\mathbf{E}_n = \eta_{n-1} \int_\mathbb{T} \left(X^{[\alpha,n-1]}_{n-1}\right)^* dF \left(\frac{f_{\alpha_{n-1}}}{g_{\alpha_n}} Z\right) \left(\int_\mathbb{T} Y^*_{n-1} dF \left(\frac{g_{\alpha_{n-1}}}{g_{\alpha_n}} Z\right)\right)^{-1}$$

for all $Z \in \mathcal{R}^{q\times q}_{\alpha,n-1}$ which satisfy

$$\det Z^{[\alpha,n-1]}(\alpha_n) \neq 0. \tag{4.13}$$

(b) *If $(1-|\alpha_n|)(1-|\alpha_{n-1}|) < 0$, then*

$$\mathbf{E}_n = \overline{\eta_{n-1}} \left(\int_\mathbb{T} \frac{f_{\alpha_{n-1}}}{g_{\alpha_n}} Z dF \left(Y^{[\alpha,n-1]}_{n-1}\right)^*\right)^{-1} \int_\mathbb{T} \frac{g_{\alpha_{n-1}}}{g_{\alpha_n}} Z dF X^*_{n-1}$$

and

$$\mathbf{E}_n = \overline{\eta_{n-1}} \int_\mathbb{T} Y^*_{n-1} dF \left(\frac{g_{\alpha_{n-1}}}{g_{\alpha_n}} Z\right) \left(\int_\mathbb{T} \left(X^{[\alpha,n-1]}_{n-1}\right)^* dF \left(\frac{f_{\alpha_{n-1}}}{g_{\alpha_n}} Z\right)\right)^{-1}$$

for all $Z \in \mathcal{R}^{q\times q}_{\alpha,n-1}$ which satisfy (4.13).

(c) *If $(1-|\alpha_n|)(1-|\alpha_{n-1}|) > 0$, then*

$$\mathbf{E}_n = \eta_{n-1} \int_\mathbb{T} \frac{f_{\alpha_{n-1}}}{g_{\alpha_n}} X_{n-1} dF Z^* \left(\int_\mathbb{T} \frac{g_{\alpha_{n-1}}}{g_{\alpha_n}} Y^{[\alpha,n-1]}_{n-1} dF Z^*\right)^{-1}$$

and

$$\mathbf{E}_n = \eta_{n-1}\left(\int_{\mathbb{T}} Z^* dF\left(\frac{g_{\alpha_{n-1}}}{g_{\alpha_n}} X_{n-1}^{[\alpha,n-1]}\right)\right)^{-1} \int_{\mathbb{T}} Z^* dF\left(\frac{f_{\alpha_{n-1}}}{g_{\alpha_n}} Y_{n-1}\right)$$

for all $Z \in \mathcal{R}_{\alpha,n-1}^{q\times q}$ *which satisfy*

$$\det Z(\alpha_n) \neq 0. \tag{4.14}$$

(d) *If* $(1-|\alpha_n|)(1-|\alpha_{n-1}|) < 0$, *then*

$$\mathbf{E}_n = \overline{\eta_{n-1}} \int_{\mathbb{T}} \frac{g_{\alpha_{n-1}}}{g_{\alpha_n}} Y_{n-1}^{[\alpha,n-1]} dF Z^* \left(\int_{\mathbb{T}} \frac{f_{\alpha_{n-1}}}{g_{\alpha_n}} X_{n-1} dF Z^*\right)^{-1}$$

and

$$\mathbf{E}_n = \overline{\eta_{n-1}}\left(\int_{\mathbb{T}} Z^* dF\left(\frac{f_{\alpha_{n-1}}}{g_{\alpha_n}} Y_{n-1}\right)\right)^{-1} \int_{\mathbb{T}} Z^* dF\left(\frac{g_{\alpha_{n-1}}}{g_{\alpha_n}} X_{n-1}^{[\alpha,n-1]}\right)$$

for all $Z \in \mathcal{R}_{\alpha,n-1}^{q\times q}$ *which satisfy* (4.14).

Proof. Application of Lemma 4.2 and Lemma 4.3 yields the proof of parts (a) and (b). Let $Z \in \mathcal{R}_{\alpha,n-1}^{q\times q}$. In view of Remark 4.1 and [FKL2, Remark 2.4], we have

$$\left(\frac{g_{\alpha_{n-1}}}{g_{\alpha_n}} Z^{[\alpha,n-1]}, X_{n-1}\right)_{F,l} = \left(Z, \frac{g_{\alpha_{n-1}}}{g_{\alpha_n}} X_{n-1}^{[\alpha,n-1]}\right)_{F,r},$$

$$\left(Y_{n-1}, \frac{g_{\alpha_{n-1}}}{g_{\alpha_n}} Z^{[\alpha,n-1]}\right)_{F,r} = \left(\frac{g_{\alpha_{n-1}}}{g_{\alpha_n}} Y_{n-1}^{[\alpha,n-1]}, Z\right)_{F,l},$$

$$\left(\frac{f_{\alpha_{n-1}}}{g_{\alpha_n}} Z^{[\alpha,n-1]}, Y_{n-1}^{[\alpha,n-1]}\right)_{F,l} = \left(Z, \frac{f_{\alpha_{n-1}}}{g_{\alpha_n}} Y_{n-1}\right)_{F,r},$$

and

$$\left(X_{n-1}^{[\alpha,n-1]}, \frac{f_{\alpha_{n-1}}}{g_{\alpha_n}} Z^{[\alpha,n-1]}\right)_{F,r} = \left(\frac{f_{\alpha_{n-1}}}{g_{\alpha_n}} X_{n-1}, Z\right)_{F,l}.$$

All the matrices stated in the left-hand sides of these equalities are in view of (4.14), [FKL2, Remark 2.4], and Lemma 4.3 nonsingular. Application of Lemma 4.2 completes the proof of parts (c) and (d). □

The following remark presents particular matrix-valued functions Z which satisfy (4.13) and (4.14), respectively.

Remark 4.5. Let $(\alpha_j)_{j=1}^{\infty} \in \mathcal{T}_1$, let $\tau \in \mathbb{N}$ or $\tau = \infty$, and let $F \in \mathcal{M}_{\geq}^{q,\tau}(\mathbb{T}, \mathfrak{B}_{\mathbb{T}})$. Let $[(X_k)_{k=0}^{\tau}, (Y_k)_{k=0}^{\tau}]$ be a Szegő pair of orthonormal systems corresponding to $(\alpha_j)_{j=1}^{\infty}$ and F, and let $(\mathbf{E}_n)_{n=1}^{\tau}$ be the sequence of its Szegő parameters. From Remark 1.2 one can see that for all $n \in \mathbb{N}_{1,\tau}$, the following statements hold:

(a) If $(1-|\alpha_n|)(1-|\alpha_{n-1}|) > 0$, then the relations $\det X_{n-1}^{[\alpha,n-1]}(\alpha_n) \neq 0$ and $\det Y_{n-1}^{[\alpha,n-1]}(\alpha_n) \neq 0$ are satisfied.
(b) If $(1-|\alpha_n|)(1-|\alpha_{n-1}|) < 0$, then $\det X_{n-1}(\alpha_n) \neq 0$ and $\det Y_{n-1}(\alpha_n) \neq 0$.

We want to draw the attention of the reader to the special case $\alpha_{n-1} = \alpha_n$ where the integral representations of the Szegő parameters stated in Theorem 4.4 can be simplified. In particular, this situation will be met if one studies orthogonal $q \times q$ matrix polynomials.

Corollary 4.6. *Let $(\alpha_j)_{j=1}^{\infty} \in \mathcal{T}_1$, let $\tau \in \mathbb{N}$ or $\tau = \infty$, and let $F \in \mathcal{M}_{\geq}^{q,\tau}(\mathbb{T}, \mathfrak{B}_{\mathbb{T}})$. Let $[(X_k)_{k=0}^{\tau}, (Y_k)_{k=0}^{\tau}]$ be a Szegő pair of orthonormal systems corresponding to $(\alpha_j)_{j=1}^{\infty}$ and F, and let $(\mathbf{E}_n)_{n=1}^{\tau}$ be the sequence of its Szegő parameters. Let $n \in \mathbb{N}_{1,\tau}$ and let $\alpha_{n-1} = \alpha_n$. For all $Z \in \mathcal{R}_{\alpha,n-1}^{q\times q}$ which satisfy (4.13),*

$$\mathbf{E}_n = -\left(\left(Z^{[\alpha,n-1]}(\alpha_{n-1})\right)^{-1} X_{n-1}^{[\alpha,n-1]}(\alpha_{n-1})\right)^* \int_{\mathbb{T}} b_{\alpha_n} Z dF \left(Y_{n-1}^{[\alpha,n-1]}\right)^* \tag{4.15}$$

and

$$\mathbf{E}_n = -\int_{\mathbb{T}} \left(X_{n-1}^{[\alpha,n-1]}\right)^* dF\left(b_{\alpha_n} Z\right)\left(Y_{n-1}^{[\alpha,n-1]}(\alpha_{n-1})\left(Z^{[\alpha,n-1]}(\alpha_{n-1})\right)^{-1}\right)^*. \tag{4.16}$$

In particular,

$$\mathbf{E}_n = -\int_{\mathbb{T}} b_{\alpha_n} X_{n-1} dF\left(Y_{n-1}^{[\alpha,n-1]}\right)^* \tag{4.17}$$

and

$$\mathbf{E}_n = -\int_{\mathbb{T}} \left(X_{n-1}^{[\alpha,n-1]}\right)^* dF\left(b_{\alpha_n} Y_{n-1}\right). \tag{4.18}$$

Proof. In view of (1.1), (2.1), and (4.1), we have

$$\eta_{n-1} \frac{f_{\alpha_{n-1}}}{g_{\alpha_n}} = \eta_{n-1} \frac{f_{\alpha_{n-1}}}{g_{\alpha_{n-1}}} = -b_{\alpha_{n-1}} = -b_{\alpha_n}.$$

Further, using [FKL2, Remark 2.1 and Lemma 3.6] and (1.3), we obtain

$$\int_{\mathbb{T}} \frac{g_{\alpha_{n-1}}}{g_{\alpha_n}} Z dF X_{n-1}^* = \int_{\mathbb{T}} Z dF X_{n-1}^* = \left(Z^{[\alpha,n-1]}(\alpha_{n-1})\right)^* \left(X_{n-1}^{[\alpha,n-1]}(\alpha_{n-1})\right)^{-*}.$$

Consequently, part (a) of Theorem 4.4 yields (4.15). Equation (4.16) follows analogously. By virtue of part (a) of Remark 4.5, choosing $Z = X_{n-1}$ (respectively, $Z = Y_{n-1}$) from (4.15) and (4.16) we get (4.17) and (4.18). □

Note that, similar as in Corollary 4.6, starting from part (c) of Theorem 4.4 one can also obtain simpler expressions for the Szegő parameter $\mathbf{E}_n$ in the special case $\alpha_{n-1} = \alpha_n$.

In view of Corollary 3.6, we consider finally the situation that the Szegő parameter $\mathbf{E}_n$ is equal to zero for some $n \in \mathbb{N}_{1,\tau}$.

Corollary 4.7. *Let $(\alpha_j)_{j=1}^{\infty} \in \mathcal{T}_1$, let $\tau \in \mathbb{N}$ or $\tau = \infty$, and let $F \in \mathcal{M}_{\geq}^{q,\tau}(\mathbb{T}, \mathfrak{B}_{\mathbb{T}})$. Let $[(X_k)_{k=0}^{\tau}, (Y_k)_{k=0}^{\tau}]$ be a Szegő pair of orthonormal systems corresponding to $(\alpha_j)_{j=1}^{\infty}$ and F, and let $(\mathbf{E}_n)_{n=1}^{\tau}$ be the sequence of its Szegő parameters. Further, let $n \in \mathbb{N}_{1,\tau}$ and let $(1-|\alpha_n|)(1-|\alpha_{n-1}|) > 0$. The following statements are equivalent:*

(i) $\mathbf{E}_n = \mathbf{0}$.

(ii) $X_n(\alpha_{n-1}) = \mathbf{0}$.

(iii) $X_n = -\eta_{n-1}\sqrt{\frac{1-|\alpha_n|^2}{1-|\alpha_{n-1}|^2}}\frac{f_{\alpha_{n-1}}}{g_{\alpha_n}}X_{n-1}$.

(iv) *For all* $Z \in \mathcal{R}_{\alpha,n-1}^{q\times q}$,

$$\int_{\mathbb{T}} \frac{f_{\alpha_{n-1}}}{g_{\alpha_n}} X_{n-1} dF Z^* = \mathbf{0}. \tag{4.19}$$

(v) *There exists a* $Z \in \mathcal{R}_{\alpha,n-1}^{q\times q}$ *such that* (4.14) *and* (4.19) *hold.*

(vi) $Y_n(\alpha_{n-1}) = \mathbf{0}$.

(vii) $Y_n = -\eta_{n-1}\sqrt{\frac{1-|\alpha_n|^2}{1-|\alpha_{n-1}|^2}}\frac{f_{\alpha_{n-1}}}{g_{\alpha_n}}Y_{n-1}$.

(viii) *For all* $Z \in \mathcal{R}_{\alpha,n-1}^{q\times q}$,

$$\int_{\mathbb{T}} Z^* dF\Big(\frac{f_{\alpha_{n-1}}}{g_{\alpha_n}} Y_{n-1}\Big) = \mathbf{0}. \tag{4.20}$$

(ix) *There exists a* $Z \in \mathcal{R}_{\alpha,n-1}^{q\times q}$ *such that* (4.14) *and* (4.20) *hold.*

Proof. (i) $\Leftrightarrow$ (ii): This equivalence follows immediately from (2.6) and (2.1).
(i) $\Rightarrow$ (iii): Use part (a) of Corollary 2.12, (1.1), (2.1), (2.8), and (4.1).
(iii) $\Rightarrow$ (iv): This implication is an easy consequence of [FKL2, Remark 2.1 and Lemma 3.6].
(iv) $\Rightarrow$ (v): From [FKL2, Remark 2.4] and part (a) of Remark 4.5 it follows that (v) is necessary for (iv).
(v) $\Rightarrow$ (i): Apply part (c) of Theorem 4.4.
Analogously, in view of (2.7), one can verify that (i), (vi), (vii), (viii), and (ix) are equivalent. □

Observe that if $(1-|\alpha_n|)(1-|\alpha_{n-1}|) < 0$ is fulfilled, the case $\mathbf{E}_n = \mathbf{0}$ can be similarly characterized as in Corollary 4.7. We omit the details.

References

[BGHN1] A. Bultheel, P. González-Vera, E. Hendriksen, O. Njåstad, *A Szegő Theory for Rational Functions.* Report TW 131, Department of Computer Science, K.U. Leuven 1990.

[BGHN2] A. Bultheel, P. González-Vera, E. Hendriksen, O. Njåstad, *Orthogonal Rational Functions.* Cambridge Monographs on Applied and Comput. Math. 5, Cambridge University Press, Cambridge 1999.

[DGK] P. Delsarte, Y. Genin, Y. Kamp, *Orthogonal polynomial matrices on the unit circle.* IEEE Trans. Circuits and Systems CAS **25** (1978), 145–160.

[Dj] M.M. Djrbashian, *Orthogonal systems of rational functions on the circle with given set of poles* (in Russian). Dokl. Akad. Nauk SSSR **147** (1962), 1278–1281.

[DFK] V.K. Dubovoj, B. Fritzsche, B. Kirstein, *Matricial Version of the Classical Schur Problem.* Teubner-Texte zur Mathematik 129, Teubner, Leipzig 1992.

[D] H. Dym, *J-contractive Matrix Functions, Reproducing Kernel Hilbert Spaces and Interpolation.* CBMS Regional Conf. Ser. Math. 71, Amer. Math. Soc., Providence, R. I. 1989.

[FK] B. Fritzsche, B. Kirstein, *Schwache Konvergenz nichtnegativ hermitescher Borelmaße*. Wiss. Z. Karl-Marx-Univ. Leipzig, Math.-Naturwiss. R. **37** (1988), 375–398.

[FKL1] B. Fritzsche, B. Kirstein, A. Lasarow, *On rank invariance of moment matrices of nonnegative Hermitian-valued Borel measures on the unit circle*. Math. Nachr. **263–264** (2004), 103–132.

[FKL2] B. Fritzsche, B. Kirstein, A. Lasarow, *Orthogonal rational matrix-valued functions on the unit circle*. Math. Nachr. **278** (2005), 525–553.

[FKL3] B. Fritzsche, B. Kirstein, A. Lasarow, *Orthogonal rational matrix-valued functions on the unit circle: Recurrence relations and a Favard-type theorem*. to appear in: Math. Nachr.

[G1] Ja. L. Geronimus, *On polynomials orthogonal on the unit circle, on the trigonometric moment problem and on associated functions of classes of Carathéodory-Schur* (in Russian). Mat. USSR-Sb. **15** (1944), 99–130.

[G2] Ja. L. Geronimus, *Orthogonal Polynomials* (in Russian). Fizmatgiz, Moskva 1958.

[GS] U. Grenander, G. Szegő, *Toeplitz Forms and Their Applications*. University of California Press, Berkeley-Los Angeles 1958.

[Kt] I. S. Kats, *On Hilbert spaces generated by Hermitian monotone matrix functions* (in Russian). Zupiski Nauc.-issled. Inst. Mat. i Mech. i Kharkov. Mat. Obsh. **22** (1950), 95–113.

[R] M. Rosenberg, *The square integrability of matrix-valued functions with respect to a non-negative Hermitian measure*. Duke Math. J. **31** (1964), 291–298.

[S] G. Szegő, *Orthogonal Polynomials*. Amer. Math. Soc. Coll. Publ. 23, Providence, R. I. 1939.

Bernd Fritzsche, Bernd Kirstein
Fakultät für Mathematik und Informatik
Universität Leipzig
Augustusplatz 10
D-04109 Leipzig
Germany
e-mail: {fritzsche,kirstein}@mathematik.uni-leipzig.de

Andreas Lasarow
Katholieke Universiteit Leuven
Departement Computerwetenschappen
Celestijnenlaan 200A
B-3001 Heverlee (Leuven)
Belgium
e-mail: Andreas.Lasarow@cs.kuleuven.be

Operator Theory:
Advances and Applications, Vol. 163, 191–248

Singularities of Generalized Strings

Michael Kaltenbäck, Henrik Winkler and Harald Woracek

Abstract. We investigate the structure of a maximal chain of matrix functions whose Weyl coefficient belongs to $\mathcal{N}_\kappa^+$. It is shown that its singularities must be of a very particular type. As an application we obtain detailed results on the structure of the singularities of a generalized string which are explicitly stated in terms of the mass function and the dipole function. The main tool is a transformation of matrices, the construction of which is based on the theory of symmetric and semibounded de Branges spaces of entire functions. As byproducts we obtain inverse spectral results for the classes of symmetric and essentially positive generalized Nevanlinna functions.

Mathematics Subject Classification (2000). 46C20, 34A55, 30H05.

Keywords. generalized string, chain of matrices, de Branges space.

1. Introduction

A vibrating string $S[L, m]$ with inhomogeneous mass distribution is given by its length $L > 0$ and a function $m : [0, L) \to [0, \infty)$ which is nondecreasing and continuous from the left. The function m measures the total mass of the part of the string between 0 and x. In the description of the motion of the string the following boundary value problem appears:

$$y'(x) + z \int_0^x y(t)dm(t) = 0 \,,$$

$$y'(0) = 0, \ y(L) = 0 \text{ if } L + m(L) < \infty \,.$$

Thereby z is a complex parameter. The concept of the principal Titchmarsh-Weyl coefficient q_S of a string S was introduced by I.S.Kac and M.G.Krein, cf. [KaK1]. It turned out to be of fundamental importance. The principal Titchmarsh-Weyl coefficient belongs to the Stieltjes class $\mathcal{S}$, i.e., it is analytic in the open upper half-plane $\mathbb{C}^+$ and satisfies

$$\operatorname{Im} q_S(z) \geq 0, \operatorname{Im} zq_S(z) \geq 0, \ z \in \mathbb{C}^+ \,.$$

A basic inverse result states that to each function $q \in \mathcal{S}$ there exists a unique string $S[L,m]$, such that q is the principal Titchmarsh-Weyl coefficient of $S[L,m]$.

A canonical system of differential equations, or one-dimensional Hamiltonian system, is a 2×2-system of differential equations of the form

$$y'(x) = zJH(x)y(x),\ \ x \in [0, l_H)\,, \tag{1.1}$$

where H is a locally integrable 2×2-matrix-valued function on $[0,l_H)$ whose values are real and nonnegative matrices. Moreover, J is the matrix

$$J := \begin{pmatrix} 0 & -1 \\ 1 & 0 \end{pmatrix} \quad .$$

If x is interpreted as time parameter, it models the motion of a particle under the influence of a time-dependent potential. The function H is called the Hamiltonian of the system under consideration and describes its total energy. To a canonical system there is associated its Weyl coefficient q_H, which is a function belonging to the Nevanlinna class $\mathcal{N}$, i.e., is analytic in $\mathbb{C}^+$ and satisfies $\operatorname{Im} q_H(z) \geq 0$, $z \in \mathbb{C}^+$. A basic inverse result of L. de Branges, cf. [dB1] (a proper formulation can be found, e.g., in [W1]), states that to each function $q \in \mathcal{N}$ there exists an essentially unique Hamiltonian which has q as its Weyl coefficient.

The notions of strings and canonical systems are closely related. In fact, in view of the above inverse results, we know that to each string $S[L,m]$ there exists a unique Hamiltonian H_s such that $q_S = q_{H_s}$, and that the behavior of the string is completely determined by the behavior of the canonical system with Hamiltonian H_s.

There are also other ways to relate strings and canonical systems. Since $q_S \in \mathcal{S}$ we know that also $zq_S(z) \in \mathcal{N}$, and therefore that there exists a Hamiltonian H_0 such that $q_{H_0} = zq_S(z)$. Moreover, it is known that, if $q \in \mathcal{S}$, then also $zq(z^2) \in \mathcal{N}$. Thus we have naturally associated yet another Hamiltonian H_d, namely such that $q_{H_d} = zq_S(z^2)$. Each of the Hamiltonians H_s, H_0 and H_d fully describes the string, the most natural choice is H_s. Each of H_s, H_0 and H_d can be determined explicitly in terms of the string $S[L,m]$. A detailed exposition of these topics is given in [KWW3].

During the last decades a theory was developed which deals with generalizations of the notions and theorems mentioned above to an indefinite setting. The class $\mathcal{N}_{<\infty}$ of generalized Nevanlinna functions is defined by substituting the positivity condition in the definition of $\mathcal{N}$ by the requirement that a certain kernel function has only a finite number of negative squares. For the exact definition see §2.3. This class of functions was intensively studied, for the basic results we refer to [KL1]. Moreover, a function q is said to belong to the class $\mathcal{N}^+_{<\infty}$, if $q \in \mathcal{N}_{<\infty}$ and $zq(z) \in \mathcal{N}$. This can be viewed as a generalization of the Stieltjes class. For example it is known that, if $q \in \mathcal{N}^+_{<\infty}$, then $zq(z^2) \in \mathcal{N}_{<\infty}$.

The theory of canonical systems and the inverse spectral theorem of L. de Branges was generalized to an indefinite setting in [KW1, KW2, KW3]. Thereby the differential equation $y'(x) = zJH(x)y$ is substituted by the family of its fundamental matrices $\omega(x)$, which form a so-called chain of matrices. The notion of chains of matrices can be axiomatically accessed and generalized to the indefinite situation. It is then proved that to each chain of matrices a function $q_\infty(\omega) \in \mathcal{N}_{<\infty}$ is associated, which is called the Weyl coefficient of the chain, and that conversely to each $q \in \mathcal{N}_{<\infty}$ there exists an essentially unique chain of matrices such that q is its Weyl coefficient. We will recall these notions and results in more detail in §4. The interpretation of an indefinite chain of matrices as the family of fundamental matrices of an indefinite canonical system is work in progress, cf. [KW4]. In contrast to the classical situation, a chain of matrices in the indefinite case has singularities. The peculiarities of indefiniteness are reflected in the structure of these singularities.

A generalization of the notion of a string to the indefinite setting was proposed in [LW]. A generalized string is a triple $S[L,m,D]$ where $L > 0$, m is a locally square integrable function defined on $[0,L)$ which is nondecreasing and continuous from the left with possible exception of a finite number of points, and where D is a step function defined on $[0,L)$ which has only a finite number of points of increase, is nondecreasing and continuous from the left. A point $x_e \in [0,L)$ is called critical, if either $D(x_e+) - D(x_e) > 0$, $m(x_e+) - m(x_e) < 0$ or $\limsup_{x\to x_e} |m(x)| = \infty$. The relation

$$f'(x) + z\int_{[0,x]} f(x)\,dm(x) + z^2\int_{[0,x]} f(x)\,dD(x) = 0, \quad f'(0-) = 0\,.$$

is called the differential equation of the generalized string. Of course, this equation requires an appropriate interpretation. Also to a generalized string $S[L,m,D]$ a function q_S is associated and again called the principal Titchmarsh-Weyl coefficient of $S[L,m,D]$. It belongs to the class $\mathcal{N}^+_{<\infty}$. An inverse theorem is established which states that this notion induces a bijective correspondence between the set of all generalized strings and $\mathcal{N}^+_{<\infty}$.

By the above inverse results we know that, given a generalized string $S[L,m,D]$, there exist chains of matrices ω_s, ω_0 and ω_d, such that $q_\infty(\omega_s)(z) = q_S(z)$, $q_\infty(\omega_0)(z) = zq_S(z)$ and $q_\infty(\omega_d)(z) = zq_S(z^2)$. Thereby ω_0 will be positive definite, since $zq_S(z) \in \mathcal{N}$, whereas ω_s and ω_d will in general be indefinite. The behavior of the generalized string is fully determined by each of these chains.

It is the aim of this paper to describe the chain ω_s, in particular the structure of its singularities, in terms of the generalized string $S[L,m,D]$. It turns out that the singularities of ω_s correspond exactly to the critical points of $S[L,m,D]$, and that their structure can be explicitly read off from the behavior of the mass function m and the possible presence of dipoles. We obtain a noteworthy inverse result, which states that the singularities of a chain whose Weyl coefficient belongs to $\mathcal{N}^+_{<\infty}$ are of a very special kind. In fact, there are just five different types which can occur.

We will obtain these results by explicit construction. Assume that $q \in \mathcal{N}_{<\infty}$ such that also $zq(z), zq(z^2) \in \mathcal{N}_{<\infty}$. Let $\omega_s, \omega_0, \omega_d$ be the chains of matrices with

$$q_\infty(\omega_s)(z) = q(z),\ q_\infty(\omega_0)(z) = zq(z),\ q_\infty(\omega_d)(z) = zq(z^2)\,.$$

We will give a method how to construct all three of the chains $\omega_s, \omega_0, \omega_d$, once one of them is known. From this we deduce results on the structure of singularities of either of these chains. In the situation that $S[L, m, D]$ is a generalized string and $q = q_S$ we can apply this knowledge to obtain what we were aiming for.

We are also led to the conclusion that indeed the principal Titchmarsh-Weyl coefficient q_S, whose definition in [LW] might seem to be a bit 'ad hoc', is the most natural object to describe the structure of a generalized string $S[L, m, D]$. Although, of course, $S[L, m, D]$ is also determined by either of $zq_S(z)$ or $zq_S(z^2)$. However, looking at the chains ω_0 and ω_d, we see that ω_0 does not have any singularities and it will be apparent from our results that the singularities of ω_d can be of a much more complicated type as those of ω_s. Hence the information on $S[L, m, D]$ in ω_0 is somewhat hidden and ω_d is simply to big to describe $S[L, m, D]$ in a neat way.

The present work is divided into three parts. The first part consists of Sections 2 and 3. In Section 2 we introduce some classes of functions which are of importance in our investigations, study the relationship between those classes as well as the reproducing kernel spaces generated by such type of functions. These are first of all the $\mathcal{M}$-classes of 2×2-matrix functions, in particular the subclasses of symmetric and essentially positive matrix functions, cf. Definition 2.1,Definition 2.2. Secondly we recall the notions of generalized Nevanlinna functions and of Hermite-Biehler functions, and the appropriate analogues of symmetry and essential positivity on the level of these functions. Moreover, we recall the definition of a de Branges space of entire functions, cf. Definition 2.12, and the relation of those spaces to the introduced classes of functions. Some of these results are well known, however, we wish to set up these notations in sufficient generality and to collect what is needed in the sequel. In Section 3 we deal with a transformation of matrix functions and its converse. This transformation relates the $\mathcal{M}$-classes of symmetric and essentially positive matrix functions, cf. Theorem 3.2. Although the methods employed in these investigations are mostly elementary, they lead to two striking results on the structure of de Branges spaces, Proposition 3.12 and Proposition 3.14.

The second part of this paper consists of Sections 4, 5 and 6. In this part we lift the above-mentioned transformations to the level of chains of matrices and investigate the evolution of singularities. In Section 4 we give the definition of a chain of matrices, cf. Definition 4.1, and of symmetric and essentially positive chains of matrices, cf. Definition 4.3. Moreover, we recall some basic facts concerning these notions. A noteworthy inverse result, which is proved in this section, states that a chain of matrices is symmetric if and only if its Weyl coefficient is symmetric, cf. Proposition 4.4. Section 5 deals with the proper lifting of the transformations of matrix functions to whole chains of matrices. It contains Theorem 5.1 which can be viewed as the core of our present work. It shows explicitly how we can obtain the chain ϖ whose Weyl coefficient is $q(z)$ from the chain ω whose Weyl coefficient is $zq(z^2)$. As a corollary we obtain another inverse result which states that a chain of matrices is essentially positive if and only if its Weyl coefficient has this property, cf. Proposition 5.6. Also a first result on the structure of essentially positive chains is deduced, cf. Corollary 5.8. Moreover, we consider the inverse transformation, cf. Theorem 5.10 where we give an explicit construction of the chain whose Weyl coefficient is $zq(z^2)$ assuming that the chain with Weyl coefficient $q(z)$ is known. These results also lead to a construction of the chain with Weyl coefficient $zq(z)$ out of the one with Weyl coefficient q as indicated in Remark 5.11. The statement of Remark 5.11 is of fundamental importance, since it tells us how to proceed in order to reach our aim stated above. In Section 6 we recall the classification of singularities of a chain of matrices and investigate how singularities appear and transform when switching from the chain with Weyl coefficient q to those whose Weyl coefficients are $zq(z^2)$ and $-(zq(z))^{-1}$, respectively. In view of our needs in the discussion of generalized strings we restrict ourselves to the case that $q \in \mathcal{N}$. However, it is obvious how a more general discussion can be carried out (and how tedious this might be).

Finally, in the third part, Section 7, we prove the results on generalized strings we were aiming for, cf. Theorem 7.3. Following the idea which was made explicit in Remark 5.11, they are deduced from the results of Sections 5 and 6.

2. Some classes of functions and their interrelation

In this preliminary we set up our notation and collect some results on various classes of functions and their interrelation. Only some of these results are new, some are well known, some are taken from previous work. However, we feel that it is a benefit for the reader to have this collection of preliminaries at hand.

2.1. The class $\mathcal{M}$ of matrix functions

Main objects of our studies in the present paper are 2×2-matrix-valued entire functions of a particular kind. For a function $f : \mathbb{C} \to \mathbb{C}$ we denote by $f^{\#}$ the function

$$f^{\#}(z) := \overline{f(\overline{z})}. \tag{2.1}$$

If $f = f^{\#}$, we call f *real.*

2.1. Definition. Let $\mathcal{M}$ be the set of all 2×2-matrix-valued functions

$$W = (w_{ij})_{i,j=1}^{2} : \mathbb{C} \to \mathbb{C}^{2\times 2}$$

such that the entries w_{ij} are real and entire functions, $\det W \equiv 1$ and $W(0) = I$. Denote by $\mathcal{M}^{\text{sym}}$ the subset of $\mathcal{M}$ which consists of those functions W such that w_{11}, w_{22} are even and w_{12}, w_{21} are odd. Let $\mathcal{M}^{\text{ep}}$ be the set of all functions $W \in \mathcal{M}$ which have the property that each of their entries has only finitely many zeros off the positive real axis.

Let J be the matrix

$$J := \begin{pmatrix} 0 & -1 \\ 1 & 0 \end{pmatrix}.$$

2.2. Definition. Let $\kappa \in \mathbb{N} \cup \{0\}$. We write $W \in \mathcal{M}_\kappa$, if W belongs to $\mathcal{M}$ and if the 2×2-matrix-valued kernel

$$H_W(w,z) := \frac{W(z)JW(w)^* - J}{z - \overline{w}}$$

has κ negative squares on $\mathbb{C}$.

Throughout this paper we will use the notation

$$\mathcal{M}_{\leq\kappa} := \bigcup_{0 \leq \nu \leq \kappa} \mathcal{M}_\nu, \ \mathcal{M}_{<\infty} := \bigcup_{\nu \in \mathbb{N}\cup\{0\}} \mathcal{M}_\nu ,$$

and write $\operatorname{ind}_- W = \kappa$ to express the fact that $W \in \mathcal{M}$ belongs to $\mathcal{M}_\kappa$. Moreover,

$$\mathcal{M}_\kappa^{\mathrm{sym}} := \mathcal{M}^{\mathrm{sym}} \cap \mathcal{M}_\kappa, \ \mathcal{M}_\kappa^{\mathrm{ep}} := \mathcal{M}^{\mathrm{ep}} \cap \mathcal{M}_\kappa ,$$

and $\mathcal{M}_{\leq\kappa}^{\mathrm{sym}}$, $\mathcal{M}_{<\infty}^{\mathrm{sym}}$, $\mathcal{M}_{\leq\kappa}^{\mathrm{ep}}$, $\mathcal{M}_{<\infty}^{\mathrm{ep}}$ are defined correspondingly.

Similarly we can define classes ${}^+\mathcal{M}_\kappa$, ${}^+\mathcal{M}_\kappa^{\mathrm{sym}}$ etc., by imposing a restriction on the numbers of positive squares, instead of negative squares, of the kernel H_W.

For later reference let us explicitly state the following elementary and mostly well-known results. Put

$$V := \begin{pmatrix} 0 & 1 \\ 1 & 0 \end{pmatrix} .$$

2.3. Lemma. *The classes $\mathcal{M}$, $\mathcal{M}^{\mathrm{sym}}$, $\mathcal{M}_{<\infty}$, $\mathcal{M}_{<\infty}^{\mathrm{sym}}$, ${}^+\mathcal{M}_{<\infty}$, ${}^+\mathcal{M}_{<\infty}^{\mathrm{sym}}$ are closed with respect to multiplication. Each of the following transformations ϕ_j is an involution of $\mathcal{M}$. The subclasses $\mathcal{M}^{\mathrm{sym}}$, $\mathcal{M}^{\mathrm{ep}}$ remain invariant (symbolized by a ✓) or not and the class $\mathcal{M}_\kappa$ remains invariant or is mapped to the class ${}^+\mathcal{M}_\kappa$ according to the following scheme:*

Transformation	$\mathcal{M}^{\mathrm{sym}}$	$\mathcal{M}^{\mathrm{ep}}$	$\mathcal{M}_\kappa$
$\phi_1 : W(z) \mapsto W(-z)^{-1}$	✓		✓
$\phi_2 : W(z) \mapsto W(-z)$	✓		${}^+\mathcal{M}_\kappa$
$\phi_3 : W(z) \mapsto W(z)^{-1}$	✓	✓	${}^+\mathcal{M}_\kappa$
$\phi_4 : W(z) \mapsto -JW(z)J$	✓	✓	✓
$\phi_5 : W(z) \mapsto VW(z)^{-1}V$	✓	✓	✓
$\phi_6 : W(z) \mapsto W(z)^T$	✓	✓	${}^+\mathcal{M}_\kappa$

Proof. The fact that $\mathcal{M}$ and $\mathcal{M}^{\mathrm{sym}}$ are closed with respect to multiplication is obvious. The kernel relation

$$\frac{(W_1W_2)(z)J(W_1W_2)^*(w) - J}{z - \overline{w}} =$$

$$= W_1(z)\frac{W_2(z)JW_2^*(w) - J}{z - \overline{w}}W_1(w)^* + \frac{W_1(z)JW_1^*(w) - J}{z - \overline{w}}$$

shows that also $\mathcal{M}_{<\infty}$ and ${}^+\mathcal{M}_{<\infty}$ have this property.

Each of the transformations ϕ_j is an involution on $\mathcal{M}$. The facts that $\mathcal{M}^{\rm sym}$ is mapped into itself by all ϕ_j and that $\mathcal{M}^{\rm ep}$ is invariant under $\phi_3, \ldots, \phi_6$ is seen by explicitly writing down the matrix $\phi_j(W)$.

The fact that ϕ_1, ϕ_4 and ϕ_5 map $\mathcal{M}_\kappa$ into itself follows from the kernel relations

$$\frac{W(-z)^{-1}JW(-w)^{-*} - J}{z - \overline{w}} = W(-z)^{-1}\frac{-J + W(z)JW(w)^*}{(-z) - (-\overline{w})}W(w)^{-*},$$

$$\frac{(-JW(z)J)J(-JW(w)J)^* - J}{z - \overline{w}} = J\frac{W(z)JW(w)^* - J}{z - \overline{w}}J^*,$$

and

$$\frac{(VW^{-1}V)(z)J(VW^{-1}V)^*(w) - J}{z - \overline{w}} = VW(z)^{-1}\frac{W(z)JW^*(w) - J}{z - \overline{w}}W(w)^{-*}V.$$

Moreover, $\phi_2(\mathcal{M}_\kappa) = {}^+\mathcal{M}_\kappa$ since

$$H_{W(-z)}(w,z) = -H_W(-w,-z).$$

From the relations $\phi_3 = \phi_1 \circ \phi_2$ and $\phi_6 = \phi_3 \circ \phi_4$, we find that also $\phi_3(\mathcal{M}_\kappa) = {}^+\mathcal{M}_\kappa$ and $\phi_6(\mathcal{M}_\kappa) = {}^+\mathcal{M}_\kappa$. □

2.4. *Remark.* The simplest examples of matrices in $\mathcal{M}$, besides the constant I, are linear polynomials. An elementary argument shows that a nonconstant linear polynomial W belongs to $\mathcal{M}$ if and only it is of the form

$$W(z) = W_{(l,\phi)}(z) := \begin{pmatrix} 1 - lz\sin\phi\cos\phi & lz\cos^2\phi \\ -lz\sin^2\phi & 1 + lz\sin\phi\cos\phi \end{pmatrix} \tag{2.2}$$

for some $l \in \mathbb{R} \setminus \{0\}$ and $\phi \in [0,\pi)$. The numbers l and ϕ are uniquely determined by W.

Note that for all $l \in \mathbb{R} \setminus \{0\}$ and $\phi \in [0,\pi)$ we have $W_{(l,\phi)} \in \mathcal{M}^{\rm ep}$, that $W_{(l,\phi)} \in \mathcal{M}^{\rm sym}$ if and only if $\phi = 0$ or $\phi = \frac{\pi}{2}$, and that

$$W_{(l,\phi)} \in \begin{cases} \mathcal{M}_0 \cap {}^+\mathcal{M}_1 &, l > 0 \\ \mathcal{M}_1 \cap {}^+\mathcal{M}_0 &, l < 0 \end{cases}$$

Any function $W \in \mathcal{M}_\kappa$ generates a Pontryagin space $\mathfrak{K}(W)$ by means of the reproducing kernel H_W. Recall that this space is obtain as completion of the linear space

$$\operatorname{span}\left\{H_W(w,.)\begin{pmatrix} x \\ y \end{pmatrix} : w \in \mathbb{C}, \begin{pmatrix} x \\ y \end{pmatrix} \in \mathbb{C}^2\right\},$$

equipped with the inner product

$$\left[H_W(w_1,.)\begin{pmatrix} x_1 \\ y_1 \end{pmatrix}, H_W(w_2,.)\begin{pmatrix} x_2 \\ y_2 \end{pmatrix}\right] := \begin{pmatrix} x_2 \\ y_2 \end{pmatrix}^* H_W(w_1,w_2)\begin{pmatrix} x_1 \\ y_1 \end{pmatrix},$$

see for example [ADRS].

Let $W \in \mathcal{M}_{<\infty}$. It follows from the fact that the entries of W are real, i.e., satisfy $w_{ij}^{\#} = w_{ij}$, that the mapping

$$\begin{pmatrix} f \\ g \end{pmatrix} \longmapsto \begin{pmatrix} f \\ g \end{pmatrix}^{\#} := \begin{pmatrix} f^{\#} \\ g^{\#} \end{pmatrix}$$

is an anti-isometry of $\mathfrak{K}(W)$ onto itself. Moreover, cf. [KW1, Proposition 8.3], the space $\mathfrak{K}(W)$ is invariant under the difference quotient operator ($w \in \mathbb{C}$)

$$\mathcal{R}_w : \begin{pmatrix} f \\ g \end{pmatrix} \longmapsto \begin{pmatrix} \frac{f(z)-f(w)}{z-w} \\ \frac{g(z)-g(w)}{z-w} \end{pmatrix}.$$

We put

$$\mathfrak{K}_-(W) := \operatorname{cls}\left\{ H_W(w,z) \begin{pmatrix} 0 \\ 1 \end{pmatrix} : w \in \mathbb{C} \right\}, \ ,$$

and, similarly, $\mathfrak{K}_+(W) := \operatorname{cls}\{ H_W(w,z) \begin{pmatrix} 1 \\ 0 \end{pmatrix} : w \in \mathbb{C}\}$.

2.2. A characterization of $\mathcal{M}_{<\infty}^{\mathrm{sym}}$

The fact that $W \in \mathcal{M}_\kappa^{\mathrm{sym}}$ is reflected in a symmetry property of $\mathfrak{K}(W)$.

Denote by $\mathcal{O}(\mathbb{C})$ the set of all entire functions and consider the map

$$M : \begin{pmatrix} F(z) \\ G(z) \end{pmatrix} \mapsto \begin{pmatrix} -F(-z) \\ G(-z) \end{pmatrix} \tag{2.3}$$

This map is an involution of $\mathcal{O}(\mathbb{C})^2$.

2.5. Proposition. *Let $W \in \mathcal{M}_\kappa$. Then $W \in \mathcal{M}_\kappa^{\mathrm{sym}}$ if and only if $M|_{\mathfrak{K}(W)}$ is an isometry of $\mathfrak{K}(W)$ onto itself.*

Proof. Assume that $W \in \mathcal{M}_\kappa^{\mathrm{sym}}$. Then

$$W(-z) = \begin{pmatrix} -1 & 0 \\ 0 & 1 \end{pmatrix} W(z) \begin{pmatrix} -1 & 0 \\ 0 & 1 \end{pmatrix}$$

and hence

$$H_W(-w,-z) = \begin{pmatrix} -1 & 0 \\ 0 & 1 \end{pmatrix} H_W(w,z) \begin{pmatrix} -1 & 0 \\ 0 & 1 \end{pmatrix}.$$

Since

$$M \begin{pmatrix} F(z) \\ G(z) \end{pmatrix} = \begin{pmatrix} -1 & 0 \\ 0 & 1 \end{pmatrix} \begin{pmatrix} F(-z) \\ G(-z) \end{pmatrix},$$

we obtain

$$\begin{aligned} M\, H_W(w,z) \begin{pmatrix} \alpha \\ \beta \end{pmatrix} &= \begin{pmatrix} -1 & 0 \\ 0 & 1 \end{pmatrix} H_W(w,-z) \begin{pmatrix} \alpha \\ \beta \end{pmatrix} \\ &= H_W(-w,z) \begin{pmatrix} -\alpha \\ \beta \end{pmatrix} \in \mathfrak{K}(W). \end{aligned}$$

Moreover,

$$\left[M\, H_W(w_1,z) \begin{pmatrix} \alpha_1 \\ \beta_1 \end{pmatrix}, M\, H_W(w_2,z) \begin{pmatrix} \alpha_2 \\ \beta_2 \end{pmatrix} \right]_{\mathfrak{K}(W)}$$

$$= \left[H_W(-w_1, z) \begin{pmatrix} -\alpha_1 \\ \beta_1 \end{pmatrix}, H_W(-w_2, z) \begin{pmatrix} -\alpha_2 \\ \beta_2 \end{pmatrix} \right]_{\mathfrak{K}(W)}$$

$$= \begin{pmatrix} -\alpha_2 \\ \beta_2 \end{pmatrix}^* H_W(-w_1, -w_2) \begin{pmatrix} -\alpha_1 \\ \beta_1 \end{pmatrix} = \begin{pmatrix} \alpha_2 \\ \beta_2 \end{pmatrix}^* H_W(w_1, w_2) \begin{pmatrix} \alpha_1 \\ \beta_1 \end{pmatrix}$$

$$= \left[H_W(w_1, z) \begin{pmatrix} \alpha_1 \\ \beta_1 \end{pmatrix}, H_W(w_2, z) \begin{pmatrix} \alpha_2 \\ \beta_2 \end{pmatrix} \right]_{\mathfrak{K}(W)} .$$

Let

$$\mathcal{L} := \operatorname{span} \left\{ H_W(w, z) \begin{pmatrix} \alpha \\ \beta \end{pmatrix} : w, \alpha, \beta \in \mathbb{C} \right\} .$$

Then $\mathcal{L}$ is a dense linear subspace of $\mathfrak{K}(W)$ and $M|_{\mathcal{L}}$ maps $\mathcal{L}$ isometrically onto itself. Hence there exists an isometric continuation of $M|_{\mathcal{L}}$ to $\mathfrak{K}(W)$ which must be given by (2.3), since point evaluation is continuous.

Conversely, assume that (2.3) is an isometry of $\mathfrak{K}(W)$ onto itself. As $M^2 = \mathrm{id}$, we obtain

$$\left[\begin{pmatrix} F(z) \\ G(z) \end{pmatrix}, M\, H_W(w, z) \begin{pmatrix} \alpha \\ \beta \end{pmatrix} \right]_{\mathfrak{K}(W)} = \left[M \begin{pmatrix} F(z) \\ G(z) \end{pmatrix}, H_W(w, z) \begin{pmatrix} \alpha \\ \beta \end{pmatrix} \right]_{\mathfrak{K}(W)}$$

$$= \begin{pmatrix} \alpha \\ \beta \end{pmatrix}^* \begin{pmatrix} -F(-w) \\ G(-w) \end{pmatrix} = \begin{pmatrix} -\alpha \\ \beta \end{pmatrix}^* \begin{pmatrix} F(-w) \\ G(-w) \end{pmatrix}$$

$$= \left[\begin{pmatrix} F(z) \\ G(z) \end{pmatrix}, H_W(-w, z) \begin{pmatrix} -\alpha \\ \beta \end{pmatrix} \right]_{\mathfrak{K}(W)} .$$

It follows that

$$\begin{pmatrix} -1 & 0 \\ 0 & 1 \end{pmatrix} H_W(w, -z) \begin{pmatrix} \alpha \\ \beta \end{pmatrix} = M\, H_W(w, z) \begin{pmatrix} \alpha \\ \beta \end{pmatrix}$$

$$= H_W(-w, z) \begin{pmatrix} -\alpha \\ \beta \end{pmatrix} = H_W(-w, z) \begin{pmatrix} -1 & 0 \\ 0 & 1 \end{pmatrix} \begin{pmatrix} \alpha \\ \beta \end{pmatrix} .$$

Since α, β were arbitrary, we conclude that

$$\begin{pmatrix} -1 & 0 \\ 0 & 1 \end{pmatrix} H_W(w, -z) = H_W(-w, z) \begin{pmatrix} -1 & 0 \\ 0 & 1 \end{pmatrix} .$$

Substituting $w = 0$ in this relation yields

$$\begin{pmatrix} -1 & 0 \\ 0 & 1 \end{pmatrix} \frac{W(-z)J - J}{-z} = \frac{W(z)J - J}{z} \begin{pmatrix} -1 & 0 \\ 0 & 1 \end{pmatrix} ,$$

and since

$$\begin{pmatrix} -1 & 0 \\ 0 & 1 \end{pmatrix} J = -J \begin{pmatrix} -1 & 0 \\ 0 & 1 \end{pmatrix} ,$$

it follows that

$$\begin{pmatrix} -1 & 0 \\ 0 & 1 \end{pmatrix} W(-z) = W(z) \begin{pmatrix} -1 & 0 \\ 0 & 1 \end{pmatrix} ,$$

i.e., that $W \in \mathcal{M}^{\mathrm{sym}}$. □

This result puts us in position to apply [KWW2, Lemma 2.1], and to obtain a splitting of the space $\mathfrak{K}(W)$.

2.6. Corollary. *Define*

$$\mathfrak{K}(W)_e := \ker(I-M) = \left\{ \begin{pmatrix} F \\ G \end{pmatrix} \in \mathfrak{K}(W) : F \text{ odd}, G \text{ even} \right\},$$

$$\mathfrak{K}(W)_o := \ker(I+M) = \left\{ \begin{pmatrix} F \\ G \end{pmatrix} \in \mathfrak{K}(W) : F \text{ even}, G \text{ odd} \right\}.$$

Then $\mathfrak{K}(W)_e$ *and* $\mathfrak{K}(W)_o$ *are closed subspaces of* $\mathfrak{K}(W)$ *and*

$$\mathfrak{K}(W) = \mathfrak{K}(W)_e [\dot{+}] \mathfrak{K}(W)_o .$$

The reproducing kernels of $\mathfrak{K}(W)_e$ *and* $\mathfrak{K}(W)_o$ *are given by* $\frac{1}{2}(I+M)H_W(w,z)$ *and* $\frac{1}{2}(I-M)H_W(w,z)$*, respectively.*

Of particular interest is the situation when $\mathfrak{K}_-(W) = \mathfrak{K}(W)$.

2.7. Lemma. *Let* $W \in \mathcal{M}^{\mathrm{sym}}_{<\infty}$, $(F,G)^T \in \mathfrak{K}(W)$*, and assume that* $\mathfrak{K}_-(W) = \mathfrak{K}(W)$*. Then* $(F,G)^T \in \mathfrak{K}(W)_e$ *if and only if* G *is even, and* $(F,G)^T \in \mathfrak{K}(W)_o$ *if and only if* G *is odd. The analogous assertion holds when* $\mathfrak{K}_+(W) = \mathfrak{K}(W)$ *and* G *is replaced by* F.

Proof. Assume that $\mathfrak{K}_-(W) = \mathfrak{K}(W)$ and that $(F,G)^T \in \mathfrak{K}(W)$ is such that G is even. Then

$$(I-M)\begin{pmatrix} F \\ G \end{pmatrix} = \begin{pmatrix} F(z)+F(-z) \\ 0 \end{pmatrix}$$

and hence $(I-M)(F,G)^T = 0$. The other assertions follow similarly. □

2.3. The class $\mathcal{N}_{<\infty}$ of generalized Nevanlinna functions

Let us recall the notion of matrix-valued *generalized Nevanlinna functions*: A $n \times n$-matrix-valued function Q is said to belong to $\mathcal{N}_\kappa^{n\times n}$, if it is defined and meromorphic on $\mathbb{C} \setminus \mathbb{R}$, satisfies $Q(\overline{z}) = Q(z)^*$, and has the property that the $n \times n$-matrix-valued kernel

$$L_Q(w,z) := \frac{Q(z) - Q(w)^*}{z - \overline{w}}$$

has κ negative squares.

The following subclasses of generalized Nevanlinna functions were investigated in [KWW2]. A function $Q \in \mathcal{N}_\kappa^{n\times n}$ is said to be

(i) *symmetric*, if $Q(-z) = -Q(z)$, i.e., if Q is odd.
(ii) *essentially positive*, if Q is analytic on $\mathbb{C} \setminus [0,\infty)$ with possible exception of finitely many poles.

The subset of $\mathcal{N}_\kappa^{n\times n}$ which consists of all symmetric (essentially positive) functions will be denoted by $\mathcal{N}_\kappa^{n\times n,\mathrm{sym}}$ ($\mathcal{N}_\kappa^{n\times n,\mathrm{ep}}$, respectively). If we deal with scalar-valued functions, i.e., $n = 1$, then the upper index $n \times n$ will be suppressed. Moreover, the scalar function $q(z) \equiv \infty$ will be regarded as an element of $\mathcal{N}_0$. We will freely use self-explanatory notation like $\mathcal{N}^{\mathrm{ep}}_{\leq\kappa}$, $\mathcal{N}^{n\times n,\mathrm{sym}}_{<\infty}$ etc.

The $\mathcal{M}$-classes of matrix functions are related to the generalized Nevanlinna classes in several ways, two of them are of importance in the present context.

The first one we deal with is the Potapov-Ginzburg transform. If W is a 2×2-matrix function whose entries are real analytic functions such that $\det W \equiv 1$ and $w_{21} \not\equiv 0$, then the *Potapov-Ginzburg transform* $\Psi(W)$ is defined as (cf., e.g., [Br])

$$\Psi(W)(z) := \begin{pmatrix} \frac{w_{11}(z)}{w_{21}(z)} & \frac{1}{w_{21}(z)} \\ \frac{1}{w_{21}(z)} & \frac{w_{22}(z)}{w_{21}(z)} \end{pmatrix}.$$

2.8. Lemma. *We have $W \in \mathcal{M}_\kappa$ if and only if $\Psi(W) \in \mathcal{N}_\kappa^{2\times 2}$. Moreover, $W \in \mathcal{M}_\kappa^{\mathrm{sym}}$ ($W \in \mathcal{M}_\kappa^{\mathrm{ep}}$) if and only if $\Psi(W) \in \mathcal{N}_\kappa^{\mathrm{sym}}$ ($\Psi(W) \in \mathcal{N}_\kappa^{\mathrm{ep}}$, respectively).*

Proof. The assertion follows from the kernel relation

$$H_W(w,z) = \begin{pmatrix} -1 & w_{11}(z) \\ 0 & w_{21}(z) \end{pmatrix} \frac{\Psi(W)(z) - \Psi(W)(w)^*}{z - \overline{w}} \begin{pmatrix} -1 & w_{11}(w) \\ 0 & w_{21}(w) \end{pmatrix}^* \qquad \square$$

Secondly, matrix functions of the class $\mathcal{M}_{<\infty}$ operate on $\mathcal{N}_{<\infty}$ via fractional linear transformations: If $W \in \mathcal{M}$ and $\tau : \Omega \subseteq \mathbb{C} \to \mathbb{C}$ is an analytic function such that $w_{21}(z)\tau(z) + w_{22}(z) \not\equiv 0$, then we define

$$(W \star \tau)(z) := \frac{w_{11}(z)\tau(z) + w_{12}(z)}{w_{21}(z)\tau(z) + w_{22}(z)}.$$

Then $W \star \tau$ is a meromorphic function on Ω. If $w_{21}(z)\tau(z) + w_{22}(z) \equiv 0$ we set $W \star \tau \equiv \infty$. Moreover,

$$(W \star \infty)(z) := \frac{w_{11}(z)}{w_{21}(z)}.$$

2.9. Lemma. *If $W \in \mathcal{M}_\kappa$ and $\tau \in \mathcal{N}_\nu$, then $W \star \tau \in \mathcal{N}_{\leq \kappa+\nu}$. If $W \in {}^{+}\mathcal{M}_\kappa$ and $\tau \in \mathbb{R} \cup \{\infty\}$, then $-(W \star \tau) \in \mathcal{N}_{\leq\kappa}$.*

Proof. This assertion follows from the kernel relation

$$(w_{21}(z)\tau(z) + w_{22}(z)) \frac{(W \star \tau)(z) - (W \star \tau)(\overline{w})}{z - \overline{w}} (w_{21}(\overline{w})\tau(\overline{w}) + w_{22}(\overline{w}))$$
$$= \begin{pmatrix} -\tau(z) & 1 \end{pmatrix} \frac{(VW^{-1}V)(z)J(VW^{-1}V)^*(w) - J}{z - \overline{w}} \begin{pmatrix} -\tau(\overline{w}) \\ 1 \end{pmatrix} + \frac{\tau(z) - \tau(\overline{w})}{z - \overline{w}}. \qquad \square$$

2.10. Corollary.

(i) *If $W \in \mathcal{M}_\kappa$, then each of the functions*

$$\frac{w_{11}(z)}{w_{21}(z)}, \frac{w_{12}(z)}{w_{22}(z)}, \quad -\frac{w_{11}(z)}{w_{12}(z)}, -\frac{w_{21}(z)}{w_{22}(z)} \tag{2.4}$$

belongs to $\mathcal{N}_{\leq\kappa}$.

(ii) *A function $W \in \mathcal{M}_{<\infty}$ belongs to $\mathcal{M}_{<\infty}^{\mathrm{ep}}$ if and only if one of its entries has the property to have only finitely many zeros in $\mathbb{C}\setminus[0,\infty)$. In fact, if $W \in \mathcal{M}_\kappa$ and one entry has n zeros in $\mathbb{C}\setminus[0,\infty)$, then each other entry has at most $4n + 6 + 3\kappa$ zeros in this region.*

Proof. For the first two functions in (2.4) use the first part of Lemma 2.9 with the matrix W and the parameters $\tau = \infty$ and $\tau = 0$, respectively. For the second pair of functions use the second part of Lemma 2.9 with the matrix W^T and the parameters $\tau = 0, \infty$.

The second assertion follows from (2.4) by employing [KWW2, Lemma 4.5]. An inspection of its proof shows the explicit estimate. □

The estimate in Corollary 2.10, (ii), is very rough, but sufficient for our purposes.

2.4. de Branges spaces of entire functions. The class $\mathcal{H}B_{<\infty}$ of Hermite-Biehler functions

Let us recall the notion of dB-spaces and their connection to the class of Hermite-Biehler functions. To make the present work more self-contained, let us recall the notion of an almost Pontryagin space, [KWW1].

2.11. Definition. Let $\mathfrak{L}$ be a linear space, $[.,.]$ an inner product on $\mathfrak{L}$ and $\mathcal{O}$ a Hilbert space topology on $\mathfrak{L}$. The triplet $(\mathfrak{L}, [.,.], \mathcal{O})$ is called an *almost Pontryagin space*, if

(aPS1) $[.,.]$ is $\mathcal{O}$-continuous.

(aPS2) There exists a $\mathcal{O}$-closed linear subspace $\mathfrak{M}$ of $\mathfrak{L}$ with finite codimension such that $(\mathfrak{M}, [.,.])$ is a Hilbert space.

2.12. Definition. An inner product space $(\mathfrak{P}, [.,.])$ is called a *de Branges space* (*dB-space*, for short), if the following axioms hold true:

(dB1) $(\mathfrak{P}, [.,.])$ is a reproducing kernel almost Pontryagin space on $\mathbb{C}$ whose elements are entire functions.

(dB2) If $F \in \mathfrak{P}$, then also $F^\# \in \mathfrak{P}$. Moreover,

$$[F^\#, G^\#] = [G, F], \ \ F, G \in \mathfrak{P}.$$

(dB3) If $F \in \mathfrak{P}$ and $z_0 \in \mathbb{C} \setminus \mathbb{R}$ with $F(z_0) = 0$, then

$$\frac{z - \overline{z_0}}{z - z_0} F(z) \in \mathfrak{P}.$$

Moreover, if additionally $G \in \mathfrak{P}$ with $G(z_0) = 0$, then

$$\Big[\frac{z - \overline{z_0}}{z - z_0} F(z), \frac{z - \overline{z_0}}{z - z_0} G(z)\Big] = [F, G].$$

We will assume throughout this paper that also

(Z) For every $t \in \mathbb{R}$ there exists $F \in \mathfrak{P}$ with $F(t) \neq 0$.

Let $\mathfrak{P}$ be a dB-space. We define the set of *associated functions* as

$$\operatorname{Assoc} \mathfrak{P} := z\mathfrak{P} + \mathfrak{P}.$$

A function S belongs to $\operatorname{Assoc} \mathfrak{P}$ if and only if $\mathfrak{P}$ is closed with respect to the difference quotient operator ($w \in \mathbb{C}$)

$$F(z) \mapsto \frac{F(z)S(w) - F(w)S(z)}{z - w}.$$

The *Hermite-Biehler class* $\mathcal{HB}_\kappa$ with negative index $\kappa \in \mathbb{N} \cup \{0\}$ is defined as the set of all entire functions E, such that E and $E^\#$ have no common non-real zeros, $E^{-1}E^\#$ is not constant, and the kernel

$$S_{\frac{E^\#}{E}}(w,z) := i\frac{1 - \frac{E^\#(z)}{E(z)}\overline{\left(\frac{E^\#(w)}{E(w)}\right)}}{z - \overline{w}}$$

has κ negative squares on $\mathbb{C}^+$.

The Hermite-Biehler class is related to the notion of dB-Pontryagin spaces, i.e., non-degenerated dB-spaces, by the fact that, if $\mathfrak{P}$ is a dB-Pontryagin space, then its reproducing kernel K is of the form

$$K(w,z) = i\frac{E(z)\overline{E(w)} - E^\#(z)E(\overline{w})}{2\pi(z - \overline{w})}$$

for a (not necessarily unique) Hermite-Biehler function E. Conversely, every Hermite-Biehler E function generates in this way a dB-Pontryagin space which we will denote by $\mathfrak{P}(E)$, cf. [KW1]

The class of Hermite-Biehler functions is also connected with generalized Nevanlinna functions. Let E be an entire function and write $E = A - iB$ with A, B real entire functions. Then

$$E \in \mathcal{HB}_\kappa \iff \frac{A}{B} \in \mathcal{N}_\kappa\,,$$

$$E \in \mathcal{HB}_\kappa^{\mathrm{sym}} \iff \frac{A}{B} \in \mathcal{N}_\kappa^{\mathrm{sym}},\ E \in \mathcal{HB}_\kappa^{\mathrm{sb}} \iff \frac{A}{B} \in \mathcal{N}_\kappa^{\mathrm{ep}}\,.$$

The following two special classes of dB-spaces were investigated in [KWW4]. Let $M : \mathcal{O}(\mathbb{C}) \to \mathcal{O}(\mathbb{C})$ be defined as

$$M : \left\{ \begin{array}{ccc} \mathcal{O}(\mathbb{C}) & \to & \mathcal{O}(\mathbb{C}) \\ F(z) & \mapsto & F(-z) \end{array} \right.$$

The map M is a linear involution of $\mathcal{O}(\mathbb{C})$. A dB-space $(\mathfrak{P}, [.,.])$ is called *symmetric*, if M induces an isometric involution of $\mathfrak{P}$, i.e., if $M(\mathfrak{P}) \subseteq \mathfrak{P}$ and

$$[MF, MG] = [F, G],\ F, G \in \mathfrak{P}\,.$$

Let $(\mathfrak{P}, [.,.])$ be a dB-space, and denote by $\mathcal{S}$ the operator of multiplication with the independent variable in $\mathfrak{P}$. Then $\mathfrak{P}$ is called *semibounded* if the inner product

$$[F, G]_S := [\mathcal{S}F, G],\ F, G \in \operatorname{dom} \mathcal{S}\,,$$

has a finite number of negative squares on $\operatorname{dom} \mathcal{S}$.

The sets of Hermite-Biehler functions which correspond to these classes of dB-Pontryagin spaces are the following: Define $\mathcal{HB}_\kappa^{\mathrm{sym}}$ to be the subset of $\mathcal{HB}_\kappa$ consisting of all functions E which have the property that $E^\#(z) = E(-z)$. Then $\mathfrak{P}$ is a symmetric dB-Pontryagin space if and only $\mathfrak{P} = \mathfrak{P}(E)$ for some $E \in \mathcal{HB}_{<\infty}^{\mathrm{sym}}$. Moreover, we denote by $\mathcal{HB}_\kappa^{\mathrm{sb}}$ the set of all functions $E = A - iB \in \mathcal{HB}_\kappa$ such that B has only finitely many zeros in $\mathbb{C} \setminus [0, \infty)$. Then $\mathfrak{P}$ is a semibounded dB-Pontryagin space if and only $\mathfrak{P} = \mathfrak{P}(E)$ for some $E \in \mathcal{HB}_{<\infty}^{\mathrm{sb}}$.

We will need the following result which supplements our discussion of symmetric dB-spaces in [KWW4].

A subspace $\mathfrak{Q}$ of a dB-space $\mathfrak{P}$ is called a *dB-subspace* of $\mathfrak{P}$, if it is with the topology and inner product inherited from $\mathfrak{P}$ a dB-space. This is the case if and only if $F \in \mathfrak{Q}$ implies $F^\# \in \mathfrak{Q}$ and if $F \in \mathfrak{Q}$, $F(w) = 0$, implies $(z-w)^{-1}F(z) \in \mathfrak{Q}$. A main result in the theory of dB-spaces, cf. [dB1], [KW1], is the ordering theorem for subspaces of $\mathfrak{P}$. It states that the set of all dB-subspaces of a given dB-space $\mathfrak{P}$ is totally ordered with respect to inclusion.

2.13. Lemma. *Let $\mathfrak{P}$ be a symmetric a dB-space and let $\mathfrak{Q}$ be a dB-subspace of $\mathfrak{P}$. Then $\mathfrak{Q}$ is symmetric.*

Proof. Put $\tilde{\mathfrak{Q}} := \mathfrak{Q}$, we have to show that $\tilde{\mathfrak{Q}} \subseteq \mathfrak{Q}$. It is straightforward to check that $\tilde{\mathfrak{Q}}$ is a dB-subspace of $\mathfrak{P}$. By the ordering theorem for subspaces of $\mathfrak{P}$ we have either $\tilde{\mathfrak{Q}} \subseteq \mathfrak{Q}$ or $\mathfrak{Q} \subseteq \tilde{\mathfrak{Q}}$. In the first case we are already done. In the second case we have $M\tilde{\mathfrak{Q}} = \mathfrak{Q} \subseteq \tilde{\mathfrak{Q}}$. Since M is an involution, this implies $M\tilde{\mathfrak{Q}} = \tilde{\mathfrak{Q}}$. □

Matrices of the class $\mathcal{M}_{<\infty}$ give rise to Hermite-Biehler functions. If $W \in \mathcal{M}_\kappa$, then the function $E_W := w_{22} + iw_{21}$ belongs to $\mathcal{HB}_{\leq\kappa}$. It was shown in [KW1] that the projection $\pi_- : (F,G)^T \mapsto G$ is an isometric isomorphism of $\mathfrak{K}_-(W)/\mathfrak{K}_-(W)^\circ$ onto $\mathfrak{P}(E_W)$.

Similarly, if we put $\tilde{E}_W := w_{11} - iw_{12}$, then $\tilde{E}_W \in \mathcal{HB}_{\leq\kappa}$ and the projection $\pi_+ : (F,G)^T \mapsto F$ is an isometric isomorphism of $\mathfrak{K}_+(W)/\mathfrak{K}_+(W)^\circ$ onto $\mathfrak{P}(\tilde{E}_W)$.

It is an important result, cf. [KW1], that for a non-degenerated dB-space $\mathfrak{P}$ we have $1 \in \operatorname{Assoc}\mathfrak{P}$ if and only if there exists a matrix $W \in \mathcal{M}_{<\infty}$, $\mathfrak{K}_-(W) = \mathfrak{K}(W)$ such that $\mathfrak{P} = \mathfrak{P}(E_W)$.

If $\mathfrak{Q}$ is a dB-subspace of $\mathfrak{P}$, then trivially $\operatorname{Assoc}\mathfrak{Q} \subseteq \operatorname{Assoc}\mathfrak{P}$. It follows from [dB1] that, if $1 \in \operatorname{Assoc}\mathfrak{P}$, then also $1 \in \operatorname{Assoc}\mathfrak{Q}$.

The dB-subspaces of a $\mathfrak{P}(E_W)$ can be obtained from certain subspaces of $\mathfrak{K}(W)$. Note that, if we assume that $\mathfrak{K}_-(W) = \mathfrak{K}(W)$, then the projection π_- onto the second component is an isometry of $\mathfrak{K}(W)$ onto $\mathfrak{P}(E_W)$.

2.14. Lemma. *Let $W \in \mathcal{M}_{<\infty}$ and assume that $\mathfrak{K}_-(W) = \mathfrak{K}(W)$. Then the projection π_- induces an order-preserving bijection of the set of all closed subspaces $\mathcal{L}$ of $\mathfrak{K}(W)$ which are invariant under the mappings $.^\#$ (cf. (2.1)) and $\mathcal{R}_w$, $w \in \mathbb{C}$, and the set of all dB-subspaces of $\mathfrak{P}(E_W)$. Thereby $\operatorname{ind}_- \pi_-(\mathcal{L}) = \operatorname{ind}_- \mathcal{L}$ and $\dim \pi_-(\mathcal{L})^\circ = \dim \mathcal{L}^\circ$.*

Under the assumption that $\mathfrak{K}_+(W) = \mathfrak{K}(W)$ the same assertion holds with π_+ and $\mathfrak{P}(\tilde{E}_W)$.

Proof. Since $\mathfrak{K}_-(W) = \mathfrak{K}(W)$, the mapping π_- is an isometric isomorphism of $\mathfrak{K}(W)$ onto $\mathfrak{P}(E_W)$. Hence it induces an order preserving bijection of the set of all closed subspaces of $\mathfrak{K}(W)$ onto the set of all closed subspaces of $\mathfrak{P}(E_W)$, and hence leaves negative indices and degree of degeneracy invariant.

Assume that $\mathcal{L}$ is a closed subspace of $\mathfrak{K}(W)$ which is closed with respect to $.^\#$ and $\mathcal{R}_w$. Since $\pi_- \circ .^\# = .^\# \circ \pi_-$ and $\pi_- \circ \mathcal{R}_w = \mathcal{R}_w \circ \pi_-$, also $\pi(\mathcal{L})$ has this property. We see that $F \in \pi_-(\mathcal{L})$ implies $F^\# \in \pi_-(\mathcal{L})$ and that, if $F \in \pi_-(\mathcal{L})$ and $F(w) = 0$, then $(z-w)^{-1}F(z) \in \pi_-(\mathcal{L})$. Thus $\pi_-(\mathcal{L})$ is a dB-subspace of $\mathfrak{P}(E_W)$.

Conversely, let $\mathfrak{Q}$ be a dB-subspace of $\mathfrak{P}(E_W)$ and put $\mathcal{L} := \pi_-^{-1}(\mathfrak{Q})$. Then $F \in \mathfrak{Q}$ implies $F^\# \in \mathfrak{Q}$. Assume that $F = \pi_-(G,F)^T$, $F^\# = \pi_-(H,F^\#)^T$. It follows that $(G^\# - H, 0)^T \in \mathfrak{K}(W)$ and thus that $G^\# = H$. Thus also $\mathcal{L}$ is closed under $.^\#$. As $1 \in \operatorname{Assoc}\mathfrak{P}(E_W)$, also $1 \in \operatorname{Assoc}\mathfrak{Q}$, i.e., $\mathfrak{Q}$ is invariant under $\mathcal{R}_w$. Again using that $\mathfrak{K}_-(W) = \mathfrak{K}(W)$, we see that also $\mathcal{L}$ has this property. □

3. The square root transformation

In this section we investigate a transformation $\mathcal{T}_{\surd}$ which assigns to each matrix $W \in \mathcal{M}^{\text{sym}}$ an element of $\mathcal{M}$ and its converse transformations $\mathcal{T}_{2,\gamma}$. These results have consequences on the structure of symmetric and semibounded dB-spaces. Moreover, they are the basic tool for the subsequent sections.

3.1. The transformations $\mathcal{T}_{\surd}$ and $\mathcal{T}_{2,\gamma}$

3.1. Definition. Define a transformation $\mathcal{T}_{\surd} : \mathcal{M}^{\text{sym}} \to \mathcal{M}$ by

$$\mathcal{T}_{\surd}(W)(z^2) := \begin{pmatrix} w_{11}(z) & \frac{w_{12}(z)}{z} - w_{12}'(0)w_{11}(z) \\ zw_{21}(z) & w_{22}(z) - w_{12}'(0)zw_{21}(z) \end{pmatrix}.$$

Let $\gamma \in \mathbb{R}$. Define a transformation $\mathcal{T}_{2,\gamma} : \mathcal{M} \to \mathcal{M}^{\text{sym}}$ by

$$\mathcal{T}_{2,\gamma}(W)(z) := \begin{pmatrix} w_{11}(z^2) & z(w_{12}(z^2) + \gamma w_{11}(z^2)) \\ \frac{w_{21}(z^2)}{z} & w_{22}(z^2) + \gamma w_{21}(z^2) \end{pmatrix}.$$

The facts that $\mathcal{T}_{\surd}(W)$ is well defined and belongs to $\mathcal{M}$, and that $\mathcal{T}_{2,\gamma}(W) \in \mathcal{M}^{\text{sym}}$ follow on inspecting the defining formulas.

3.2. Theorem. *The transformation $\mathcal{T}_{\surd}$ maps $\mathcal{M}^{\text{sym}}$ surjectively onto $\mathcal{M}$. For each $W \in \mathcal{M}$ we have*

$$\mathcal{T}_{\surd}^{-1}(\{W\}) = \{\mathcal{T}_{2,\gamma}(W) : \gamma \in \mathbb{R}\}.$$

By $\mathcal{T}_{\surd}$ the class $\mathcal{M}_{<\infty}^{\text{sym}}$ is mapped onto $\mathcal{M}_{<\infty}^{\text{ep}}$. In fact, $\operatorname{ind}_- \mathcal{T}_{\surd}(W) \leq \operatorname{ind}_- W$. *Moreover, the map*

$$\Phi : \begin{pmatrix} f(z) \\ g(z) \end{pmatrix} \mapsto \begin{pmatrix} zf(z^2) \\ g(z^2) \end{pmatrix}$$

is an isometry of $\mathfrak{K}(\mathcal{T}_{\surd}(W))$ onto $\mathfrak{K}(W)_e$.

The proof of this theorem needs some preparation. First we list some elementary properties of $\mathcal{T}_{\surd}$ and $\mathcal{T}_{2,\gamma}$, and show that these transformations are in a way inverse to each other.

3.3. Lemma.

(i) *Let* $W \in \mathcal{M}^{\mathrm{sym}}$, *then*

$$\mathcal{T}_{\surd}(W)(z^2) = \begin{pmatrix} \frac{1}{z} & 0 \\ 0 & 1 \end{pmatrix} W(z) \begin{pmatrix} z & 0 \\ 0 & 1 \end{pmatrix} \begin{pmatrix} 1 & -w_{12}'(0) \\ 0 & 1 \end{pmatrix} . \tag{3.1}$$

If additionally $\hat{W} = (\hat{w}_{ij})_{i,j=1}^2 \in \mathcal{M}^{\mathrm{sym}}$, *then*

$$\mathcal{T}_{\surd}(W)^{-1}(z^2) \cdot \mathcal{T}_{\surd}(\hat{W})(z^2) = \\ = \begin{pmatrix} 1 & w_{12}'(0) \\ 0 & 1 \end{pmatrix} \begin{pmatrix} \frac{1}{z} & 0 \\ 0 & 1 \end{pmatrix} W^{-1}(z)\hat{W}(z) \begin{pmatrix} z & 0 \\ 0 & 1 \end{pmatrix} \begin{pmatrix} 1 & -\hat{w}_{12}'(0) \\ 0 & 1 \end{pmatrix} . \tag{3.2}$$

Moreover,

$$\operatorname{tr}\left(\mathcal{T}_{\surd}(W)'(0)J\right) = -w_{21}'(0) + \frac{w_{12}'''(0)}{6} - w_{12}'(0)\frac{w_{11}''(0)}{2} , \tag{3.3}$$

and we have

$$z\left(\mathcal{T}_{\surd}(W) \star \infty\right)(z^2) = (W \star \infty)(z) , \\ \frac{\mathcal{T}_{\surd}(W)_{22}(z^2)}{\mathcal{T}_{\surd}(W)_{21}(z^2)} = \frac{w_{22}(z)}{w_{21}(z)} \cdot \frac{1}{z} - w_{12}'(0) . \tag{3.4}$$

(ii) *Let* $W \in \mathcal{M}$, *then*

$$\mathcal{T}_{2,\gamma}(W)(z) = \begin{pmatrix} z & 0 \\ 0 & 1 \end{pmatrix} W(z^2) \begin{pmatrix} 1 & \gamma \\ 0 & 1 \end{pmatrix} \begin{pmatrix} \frac{1}{z} & 0 \\ 0 & 1 \end{pmatrix} . \tag{3.5}$$

If additionally $\hat{W} \in \mathcal{M}$, *then*

$$\mathcal{T}_{2,\gamma}(W)^{-1}(z)\mathcal{T}_{2,\hat{\gamma}}(\hat{W})(z) \\ = \begin{pmatrix} z & 0 \\ 0 & 1 \end{pmatrix} \begin{pmatrix} 1 & -\gamma \\ 0 & 1 \end{pmatrix} W(z^2)^{-1}\hat{W}(z^2) \begin{pmatrix} 1 & \hat{\gamma} \\ 0 & 1 \end{pmatrix} \begin{pmatrix} \frac{1}{z} & 0 \\ 0 & 1 \end{pmatrix} . \tag{3.6}$$

Moreover,

$$\operatorname{tr}\left(\mathcal{T}_{2,\gamma}(W)'(0)J\right) = \gamma - w_{21}'(0) . \tag{3.7}$$

(iii) *We have*

$$\left(\mathcal{T}_{\surd} \circ \mathcal{T}_{2,\gamma}\right)(W) = W,\ W \in \mathcal{M} , \\ \left(\mathcal{T}_{2,w_{12}'(0)} \circ \mathcal{T}_{\surd}\right)(W) = W,\ W \in \mathcal{M}^{\mathrm{sym}} .$$

Proof. The formulas (3.1), (3.5) and (3.7) are verified by straightforward computation. The relation (3.3) is proved by comparison of the Taylor coefficients in

$$\mathcal{T}_{\surd}(W)'(z^2) \cdot 2z = \begin{pmatrix} w_{11}'(z) & \frac{w_{12}'(z)}{z} - \frac{w_{12}(z)}{z^2} - w_{12}'(0)w_{11}'(z) \\ w_{21}(z) + zw_{21}'(z) & w_{22}'(z) - w_{12}'(0)w_{21}(z) - w_{12}'(0)zw_{21}'(z) \end{pmatrix}$$

The relation (3.2) is established by substituting the expression (3.1) for $\mathcal{T}_{\surd}(W)(z^2)$ and $\mathcal{T}_{\surd}(\hat{W})(z^2)$, respectively. The relation (3.6) follows from (3.5). Finally, (3.4) follows from the definition of $\mathcal{T}_{\surd}(W)$.

The second relation in (iii) follows from (3.5) and (3.1). Since $\mathcal{T}_{2,\gamma}(W)_{12}'(0) = \gamma$, the same source implies the validity of the first relation in (iii). □

The kernel of the map $\mathcal{T}_{\surd}$ can be determined explicitly.

3.4. Lemma. *Let $W, \hat{W} \in \mathcal{M}^{\mathrm{sym}}$. Then $\mathcal{T}_{\surd}(W) = \mathcal{T}_{\surd}(\hat{W})$ if and only if*

$$\hat{W} = W W_{(l,0)}\,,$$

for some $l \in \mathbb{R}$, where $W_{(l,0)}$ is as in (2.2).

Proof. Assume that $\hat{W} = W W_{(l,0)}$. Then

$$\hat{W}(z) = W(z)\begin{pmatrix}1 & lz\\ 0 & 1\end{pmatrix},\ \hat{w}_{12}'(0) = w_{12}'(0) + l\,,$$

and hence, by (3.2),

$$\begin{aligned}&\mathcal{T}_{\surd}(W)^{-1}(z^2)\mathcal{T}_{\surd}(\hat{W})(z^2)\\ &= \begin{pmatrix}1 & w_{12}'(0)\\ 0 & 1\end{pmatrix}\begin{pmatrix}\frac{1}{z} & 0\\ 0 & 1\end{pmatrix}\begin{pmatrix}1 & lz\\ 0 & 1\end{pmatrix}\begin{pmatrix}z & 0\\ 0 & 1\end{pmatrix}\begin{pmatrix}1 & -w_{12}'(0) - l\\ 0 & 1\end{pmatrix} =\\ &= \begin{pmatrix}1 & w_{12}'(0)\\ 0 & 1\end{pmatrix}\begin{pmatrix}1 & l\\ 0 & 1\end{pmatrix}\begin{pmatrix}1 & -w_{12}'(0) - l\\ 0 & 1\end{pmatrix} = \begin{pmatrix}1 & 0\\ 0 & 1\end{pmatrix}.\end{aligned}$$

Conversely, assume that $\mathcal{T}_{\surd}(W)^{-1}\mathcal{T}_{\surd}(\hat{W}) = I$. Then we obtain from (3.2) that

$$\begin{aligned}W(z)^{-1}\hat{W}(z) &= \begin{pmatrix}z & 0\\ 0 & 1\end{pmatrix}\begin{pmatrix}1 & \hat{w}_{12}'(0) - w_{12}'(0)\\ 0 & 1\end{pmatrix}\begin{pmatrix}\frac{1}{z} & 0\\ 0 & 1\end{pmatrix}\\ &= \begin{pmatrix}1 & (\hat{w}_{12}'(0) - w_{12}'(0))z\\ 0 & 1\end{pmatrix}.\end{aligned}$$

□

The relationship between the spaces $\mathfrak{K}(W)$ and $\mathfrak{K}(\mathcal{T}_{\surd}(W))$ is expressed by the following kernel relation.

3.5. Lemma. *For each $W \in \mathcal{M}^{\mathrm{sym}}$ we have*

$$\begin{aligned}&2\begin{pmatrix}z & 0\\ 0 & 1\end{pmatrix} H_{\mathcal{T}_{\surd}(W)}(w^2, z^2)\begin{pmatrix}\overline{w} & 0\\ 0 & 1\end{pmatrix}\\ &= H_W(w,z) + \begin{pmatrix}-1 & 0\\ 0 & 1\end{pmatrix} H_W(w,-z) = (I+M)H_W(w,z)\,.\end{aligned} \tag{3.8}$$

Proof. Since $w_{12}'(0) \in \mathbb{R}$ we have

$$\begin{pmatrix}1 & -w_{12}'(0)\\ 0 & 1\end{pmatrix} J \begin{pmatrix}1 & -w_{12}'(0)\\ 0 & 1\end{pmatrix}^* = J\,,$$

and hence we obtain, using (3.1),

$$\begin{aligned}&\begin{pmatrix}z & 0\\ 0 & 1\end{pmatrix}\frac{\mathcal{T}_{\surd}(W)(z^2)J\mathcal{T}_{\surd}(W)(w^2)^* - J}{z^2 - \overline{w}^2}\begin{pmatrix}\overline{w} & 0\\ 0 & 1\end{pmatrix}\\ &= \frac{1}{z^2 - \overline{w}^2}\Big[W(z)\begin{pmatrix}z & 0\\ 0 & 1\end{pmatrix} J \begin{pmatrix}\overline{w} & 0\\ 0 & 1\end{pmatrix} W(w)^* - \begin{pmatrix}z & 0\\ 0 & 1\end{pmatrix} J \begin{pmatrix}\overline{w} & 0\\ 0 & 1\end{pmatrix}\Big]\\ &\quad= \frac{1}{z^2 - \overline{w}^2}\Big[W(z)\begin{pmatrix}0 & -z\\ \overline{w} & 0\end{pmatrix} W(w)^* - \begin{pmatrix}0 & -z\\ \overline{w} & 0\end{pmatrix}\Big].\end{aligned}$$

We compute the expression on the right-hand side of (3.8).

$$H_W(w,z) + \begin{pmatrix} -1 & 0 \\ 0 & 1 \end{pmatrix} H_W(w,-z)$$

$$= \frac{W(z)JW(w)^* - J}{z-\overline{w}} + \begin{pmatrix} -1 & 0 \\ 0 & 1 \end{pmatrix} \frac{W(-z)JW(w)^* - J}{-z-\overline{w}}$$

$$= \frac{1}{z^2-\overline{w}^2}\Big[W(z)\Big((z+\overline{w})J - (z-\overline{w})\begin{pmatrix} -1 & 0 \\ 0 & 1 \end{pmatrix} J\Big)W(w)^*$$

$$-\Big((z+\overline{w})J - (z-\overline{w})\begin{pmatrix} -1 & 0 \\ 0 & 1 \end{pmatrix} J\Big)\Big].$$

Since

$$(z+\overline{w})J - (z-\overline{w})\begin{pmatrix} -1 & 0 \\ 0 & 1 \end{pmatrix} J = \begin{pmatrix} 0 & -2z \\ 2\overline{w} & 0 \end{pmatrix},$$

the equality (3.8) follows. □

We are now in position to prove Theorem 3.2.

Proof of Theorem 3.2. By the first relation in Lemma 3.3, (iii), it follows that $\mathcal{T}_{\surd} : \mathcal{M}^{\mathrm{sym}} \to \mathcal{M}$ is surjective. Moreover, for each $W \in \mathcal{M}$ and all $\gamma \in \mathbb{R}$, we have $\mathcal{T}_{2,\gamma}(W) \in \mathcal{T}_{\surd}^{-1}(\{W\})$. The relation (3.6) shows that

$$\mathcal{T}_{2,\hat{\gamma}}(W) = \mathcal{T}_{2,\gamma}(W)W_{(\hat{\gamma}-\gamma,0)},$$

hence we conclude from Lemma 3.4 that the matrices $\mathcal{T}_{2,\gamma}(W)$ exhaust $\mathcal{T}_{\surd}^{-1}(\{W\})$.

Let $W \in \mathcal{M}_\kappa^{\mathrm{sym}}$. Then, by the kernel relation (3.8), we have $\mathcal{T}_{\surd}(W) \in \mathcal{M}_{\leq\kappa}$. Corollary 2.6 together with (3.8) yields that the map Φ is an isometry of $\mathfrak{K}(\mathcal{T}_{\surd}(W))$ onto $\mathfrak{K}(W)_e$.

By (3.4) and Corollary 2.10, (i), the function $q(z) = \mathcal{T}_{\surd}(W) \star \infty$ has the property that $zq(z^2)$ belongs to $\mathcal{N}_{\leq\kappa}$. It follows from [KWW2, Theorem 4.1] that $q \in \mathcal{N}_{\leq\kappa}^{\mathrm{ep}}$. Since the entries of the first column of $\mathcal{T}_{\surd}(W)$ can not have common zeros, we see that the entry $\mathcal{T}_{\surd}(W)_{21}$ has only finitely many zeros in $\mathbb{C} \setminus [0,\infty)$. Thus, by Corollary 2.10, (ii), we have $\mathcal{T}_{\surd}(W) \in \mathcal{M}_{\leq\kappa}^{\mathrm{ep}}$.

Conversely, assume that $W \in \mathcal{M}_{<\infty}^{\mathrm{ep}}$. We show that $\mathcal{T}_{2,\gamma}(W) \in \mathcal{M}_{<\infty}^{\mathrm{sym}}$. Since $\mathcal{T}_{2,\gamma}(W) = \mathcal{T}_{2,0}(W)W_{(\gamma,0)}$, it suffices to consider the particular case $\gamma = 0$. In this case

$$\mathcal{T}_{2,0}(W)(z) = \begin{pmatrix} w_{11}(z^2) & zw_{12}(z^2) \\ \frac{w_{21}(z^2)}{z} & w_{22}(z^2) \end{pmatrix}.$$

Hence the Potapov-Ginzburg transform computes as

$$\Psi\big(\mathcal{T}_{2,0}(W)\big)(z) = \begin{pmatrix} z\frac{w_{11}(z^2)}{w_{21}(z^2)} & \frac{z}{w_{21}(z^2)} \\ \frac{z}{w_{21}(z^2)} & z\frac{w_{22}(z^2)}{w_{21}(z^2)} \end{pmatrix}$$

$$= z\Psi(W)(z^2).$$

Since $W \in \mathcal{M}^{\mathrm{ep}}_{<\infty}$ we have $\Psi(W) \in \mathcal{N}^{2\times 2,\mathrm{ep}}_{<\infty}$ and hence, by [KWW2, Theorem 4.1],

$$\Psi\big(\mathcal{T}_{2,0}(W)\big)(z) \in \mathcal{N}^{2\times 2,\mathrm{sym}}_{<\infty}\,.$$

This shows $\mathcal{T}_{2,0}(W) \in \mathcal{M}^{\mathrm{sym}}_{<\infty}$. □

The subject of the next proposition is to clarify how linear polynomials are transformed when either of $\mathcal{T}_{\surd}$ or $\mathcal{T}_{2,\gamma}$ is performed. This result is an important tool for our later investigation of transformations of matrix chains.

3.6. Proposition.

(i) *Let $W, \hat{W} \in \mathcal{M}^{\mathrm{sym}}$ and assume that $W^{-1}\hat{W} = W_{(l,\alpha)}$. Then $\alpha \in \{0, \frac{\pi}{2}\}$. If $\alpha = 0$, we have $\mathcal{T}_{\surd}(W)^{-1}\mathcal{T}_{\surd}(\hat{W}) = I$. If $\alpha = \frac{\pi}{2}$, then*

$$\mathcal{T}_{\surd}(W)^{-1}\mathcal{T}_{\surd}(\hat{W}) = W_{(l',\phi)}\,, \quad \textit{where } l' = l(1 + w_{12}'(0)^2),\ \phi = \operatorname{Arccot} w_{12}'(0)\,.$$

(ii) *Let $W, \hat{W} \in \mathcal{M}$, $\gamma, \hat{\gamma} \in \mathbb{R}$. If $W^{-1}\hat{W} = W_{(l,\phi)}$ for some $l \in \mathbb{R}$, $\phi \in [0,\pi)$, cf. (2.2), then*

$$\begin{aligned}&\mathcal{T}_{2,\gamma}(W)^{-1}(z)\mathcal{T}_{2,\hat{\gamma}}(\hat{W})(z)\\ &= \begin{pmatrix} 1 - z^2 l\sin\phi(\cos\phi - \gamma\sin\phi) & \begin{smallmatrix} z(\hat{\gamma}-\gamma)+ \\ +z^3 l(\cos\phi-\gamma\sin\phi)(\cos\phi-\hat{\gamma}\sin\phi)\end{smallmatrix} \\ -zl\sin^2\phi & 1 + z^2 l\sin\phi(\cos\phi - \hat{\gamma}\sin\phi)\end{pmatrix}.\end{aligned} \tag{3.9}$$

We have $\mathcal{T}_{2,\gamma}(W)^{-1}\mathcal{T}_{2,\hat{\gamma}}(\hat{W}) = W_{(L,\alpha)}$ if and only if $W^{-1}\hat{W} = W_{(l,\phi)}$ and either

$$l = 0, \quad \textit{in which case} \quad L = \hat{\gamma} - \gamma \quad \textit{and} \quad \alpha = 0\,,$$

$$\textit{or} \quad l \neq 0, \phi \neq 0, \gamma = \hat{\gamma} = \cot\phi, \quad \textit{in which case} \quad L = l\sin^2\phi, \alpha = \frac{\pi}{2}\,.$$

Proof. The matrix function $W_{(l,\alpha)}$ belongs to $\mathcal{M}^{\mathrm{sym}}$ if and only if $\alpha \in \{0, \frac{\pi}{2}\}$. This proves the first assertion in (i).

The case that $\alpha = 0$ was already treated in Lemma 3.4. Assume that $\alpha = \frac{\pi}{2}$. Then

$$\hat{W}(z) = W(z)\begin{pmatrix} 1 & 0 \\ -lz & 1\end{pmatrix}, \quad \hat{w}_{12}'(0) = w_{12}'(0)\,.$$

It follows that

$$\begin{aligned}&\mathcal{T}_{\surd}(W)^{-1}(z^2)\mathcal{T}_{\surd}(\hat{W})(z^2)\\ &= \begin{pmatrix} 1 & w_{12}'(0) \\ 0 & 1\end{pmatrix}\begin{pmatrix} \frac{1}{z} & 0 \\ 0 & 1\end{pmatrix}\begin{pmatrix} 1 & 0 \\ -lz & 1\end{pmatrix}\begin{pmatrix} z & 0 \\ 0 & 1\end{pmatrix}\begin{pmatrix} 1 & -w_{12}'(0) - l \\ 0 & 1\end{pmatrix}\\ &= \begin{pmatrix} 1 & w_{12}'(0) \\ 0 & 1\end{pmatrix}\begin{pmatrix} 1 & 0 \\ -lz^2 & 1\end{pmatrix}\begin{pmatrix} 1 & -w_{12}'(0) \\ 0 & 1\end{pmatrix}\\ &= \begin{pmatrix} 1 - lw_{12}'(0)z^2 & lw_{12}'(0)^2z^2 \\ -lz^2 & 1 + lw_{12}'(0)z^2\end{pmatrix} = W_{(l',\phi)}(z^2)\end{aligned}$$

when $l' = l(1 + w_{12}'(0))$ and $\cot\phi = w_{12}'(0)$.

We come to the proof of (ii). Assume that $W^{-1}\hat{W} = W_{(l,\phi)}$. Then

$$\mathcal{T}_{2,\gamma}(W)^{-1}(z)\mathcal{T}_{2,\hat{\gamma}}(\hat{W})(z) =$$
$$= \begin{pmatrix} z & -\gamma z \\ 0 & 1 \end{pmatrix} \begin{pmatrix} 1 - lz^2 \sin\phi\cos\phi & lz^2\cos^2\phi \\ -lz^2\sin^2\phi & 1 + lz^2\sin\phi\cos\phi \end{pmatrix} \begin{pmatrix} \frac{1}{z} & \hat{\gamma} \\ 0 & 1 \end{pmatrix} =$$
$$= \begin{pmatrix} 1 - z^2 l\sin\phi(\cos\phi - \gamma\sin\phi) & \begin{smallmatrix} z(\hat{\gamma}-\gamma)+ \\ +z^3 l(\cos^2\phi - \gamma\hat{\gamma}\sin^2\phi - \hat{\gamma}\sin\phi\cos\phi - \gamma\sin\phi\cos\phi) \end{smallmatrix} \\ -zl\sin^2\phi & 1 + z^2 l\sin\phi(\cos\phi - \hat{\gamma}\sin\phi) \end{pmatrix}.$$

Assume that $\mathcal{T}_{2,\gamma}(W)^{-1}(z)\mathcal{T}_{2,\hat{\gamma}}(\hat{W})(z) = W_{(L,\alpha)}$. By the already proved part (i) of the present proposition, we must have $\alpha \in \{0, \frac{\pi}{2}\}$ and $W^{-1}\hat{W} = W_{(l,\phi)}$ for certain $l \in \mathbb{R}$, $\phi \in [0, \pi)$. Considering the just proved formula (3.9) we see that either (i) or (ii) holds. The converse follows from (3.9). $\square$

3.7. *Remark.* Note that in case $\phi = 0$ the matrix (3.9) is equal to

$$\begin{pmatrix} 1 & z(\hat{\gamma} - \gamma) + lz^3 \\ 0 & 1 \end{pmatrix}.$$

If $\phi \neq 0, l \neq 0$ we can decompose (3.9) as

$$\begin{pmatrix} 1 & (\cot\phi - \gamma)z \\ 0 & 1 \end{pmatrix} \begin{pmatrix} 1 & 0 \\ -lz\sin^2\phi & 1 \end{pmatrix} \begin{pmatrix} 1 & -(\cot\phi - \hat{\gamma})z \\ 0 & 1 \end{pmatrix}.$$

Then its Potapov-Ginzburg transform is equal to

$$-\frac{1}{z}\frac{1}{l\sin^2\phi}\begin{pmatrix} 1 & 1 \\ 1 & 1 \end{pmatrix} + z\begin{pmatrix} \cot\phi - \gamma & 0 \\ 0 & -(\cot\phi - \hat{\gamma}) \end{pmatrix}.$$

From the isomorphy of $\mathfrak{K}(W)_e$ and $\mathfrak{K}(\mathcal{T}_{\surd}(W))$ we also obtain a relation between the spaces $\mathfrak{K}_-(W)$ and $\mathfrak{K}_-(\mathcal{T}_{\surd}(W))$.

3.8. Proposition. *Let $W \in \mathcal{M}_\kappa^{\text{sym}}$, then*

$$\dim \mathfrak{K}_-\big(\mathcal{T}_{\surd}(W)\big)^\perp = \big[\frac{1}{2}\dim \mathfrak{K}_-(W)^\perp\big],$$

and

$$\dim \mathfrak{K}_-\big(\mathcal{T}_{\surd}(W)\big)^\circ = \big[\frac{1}{2}\dim \mathfrak{K}_-(W)^\circ\big].$$

Proof. By [KW1, Corollary 9.7,Proposition 8.3] we have for any matrix W

$$\mathfrak{K}_-(W)^\perp = \Big\{ \begin{pmatrix} p(z) \\ 0 \end{pmatrix} \in \mathfrak{K}(W) : p(z) \text{ polynomial}\Big\}$$
$$= \Big\{ \begin{pmatrix} p(z) \\ 0 \end{pmatrix} : p(z) \text{ polynomial}, \deg p < \dim \mathfrak{K}_-(W)^\perp\Big\}.$$

Let Φ be defined as in Theorem 3.2. Then we conclude that Φ maps $\mathfrak{K}_-(\mathcal{T}_{\surd}(W))^\perp$ into $\mathfrak{K}_-(W)^\perp$, and

$$\dim \mathfrak{K}_-(W)^\perp \geq 2\dim \mathfrak{K}_-\big(\mathcal{T}_{\surd}(W)\big)^\perp.$$

Conversely, if l is the largest odd number $\leq \dim \mathfrak{K}_-(W)^\perp - 1$, then

$$\begin{pmatrix} z^l \\ 0 \end{pmatrix} \in \mathfrak{K}_-(W)^\perp, \quad \text{and hence} \quad \begin{pmatrix} z^{\frac{l-1}{2}} \\ 0 \end{pmatrix} \in \mathfrak{K}_-\big(\mathcal{T}_{\surd}(W)\big)^\perp.$$

If $\dim \mathfrak{K}_-(W)^\perp \equiv 0 \mod 2$, then

$$\frac{l-1}{2} = \frac{(\dim \mathfrak{K}_-(W)^\perp - 1) - 1}{2} = \frac{\dim \mathfrak{K}_-(W)^\perp}{2} - 1,$$

and hence $\dim \mathfrak{K}_-(\mathcal{T}_{\surd}(W))^\perp \geq \frac{1}{2} \dim \mathfrak{K}_-(W)^\perp$. If $\dim \mathfrak{K}_-(W)^\perp \equiv 1 \mod 2$, then

$$\frac{l-1}{2} = \frac{(\dim \mathfrak{K}_-(W)^\perp - 2) - 1}{2} = \frac{\dim \mathfrak{K}_-(W)^\perp - 1}{2} - 1,$$

and hence $\dim \mathfrak{K}_-(\mathcal{T}_{\surd}(W))^\perp \geq \frac{1}{2}(\dim \mathfrak{K}_-(W)^\perp - 1)$. This yields the first equality.

For any matrix W we have

$$\mathfrak{K}_-(W)^\circ = \big(\mathfrak{K}_-(W)^\perp\big)^\circ = \Big\{ \begin{pmatrix} p(z) \\ 0 \end{pmatrix} \in \mathfrak{K}(W) : p(z) \text{ polynomial} \Big\}^\circ,$$

and it follows from [KW1, Proposition 8.3] that the space $\mathfrak{K}_-(W)^\circ$ is invariant with respect the difference quotient operator. Hence

$$\mathfrak{K}_-(W)^\circ = \Big\{ \begin{pmatrix} p(z) \\ 0 \end{pmatrix} : p(z) \text{ polynomial}, \deg p < \dim \mathfrak{K}_-(W)^\circ \Big\}.$$

Using the fact that

$$\begin{aligned} \mathfrak{K}_-(W)^\perp &= \big(\mathfrak{K}_-(W)^\perp \cap \mathfrak{K}_-(W)_e\big)[\dot{+}]\big(\mathfrak{K}_-(W)^\perp \cap \mathfrak{K}_-(W)_o\big) \\ &= \operatorname{span}\Big\{ \begin{pmatrix} z^l \\ 0 \end{pmatrix} : l \text{ odd}, l < \dim \mathfrak{K}_-(W)^\perp \Big\}[\dot{+}] \\ &\quad [\dot{+}] \operatorname{span}\Big\{ \begin{pmatrix} z^l \\ 0 \end{pmatrix} : l \text{ even}, l < \dim \mathfrak{K}_-(W)^\perp \Big\}, \end{aligned}$$

the same argument as in the previous paragraph shows that

$$\dim \mathfrak{K}_-\big(\mathcal{T}_{\surd}(W)\big)^\circ = \Big[\frac{1}{2} \dim \mathfrak{K}_-(W)^\circ\Big].$$

□

3.2. Two characteristic values of a matrix $W \in \mathcal{M}_{<\infty}^{\text{ep}}$

It was shown in [KWW2, Proposition 4.9] that for a function $q \in \mathcal{N}_{<\infty}^{\text{ep}}$ the limit $\lim_{t\to-\infty} q(t)$ exists in $\mathbb{R} \cup \{\pm\infty\}$. If $W \in \mathcal{M}_{<\infty}^{\text{ep}}$, we obtain two essentially positive generalized Nevanlinna functions, namely $W \star \infty = w_{21}^{-1} w_{11}$ and $w_{21}^{-1} w_{22}$. We denote the respective limits by

$$m(W) := \lim_{t\to-\infty} (W \star \infty)(t), \;\; \lambda(W) := \lim_{t\to-\infty} \frac{w_{22}(t)}{w_{21}(t)}. \tag{3.10}$$

These numbers have interpretations in terms of the transformation $\mathcal{T}_{2,\gamma}$ and the corresponding Pontryagin spaces.

3.9. Proposition. *Let* $W \in \mathcal{M}^{\mathrm{ep}}_{<\infty}$ *and* $\gamma \in \mathbb{R}$. *Then* $m(W) \in \mathbb{R}$ *if and only if* $\mathfrak{K}_-(W) = \mathfrak{K}(W)$ *and* $\mathfrak{K}_-(\mathcal{T}_{2,\gamma}(W))^\circ = \{0\}$. *Thereby* $\mathfrak{K}_-(\mathcal{T}_{2,\gamma}(W)) = \mathfrak{K}(\mathcal{T}_{2,\gamma}(W))$ *if and only if* $m(W) = 0$.

Otherwise, if $m(W) \in \mathbb{R} \setminus \{0\}$, *we have* $\dim \mathfrak{K}_-(\mathcal{T}_{2,\gamma}(W))^\perp = 1$.

Proof. Since $W \in \mathcal{M}^{\mathrm{ep}}_{<\infty}$ we have $\hat{W} := \mathcal{T}_{2,\gamma}(W) \in \mathcal{M}^{\mathrm{sym}}_{<\infty}$, and, hence, may consider the space $\mathfrak{K}(\hat{W})$. Keep in mind that $W = \mathcal{T}_{\surd}(\hat{W})$.

Assume first that $m(W) \in \mathbb{R}$. By (3.4) we have

$$m(W) = \lim_{y \to +\infty} \big(W \star \infty\big)\big((iy)^2\big) = \lim_{y \to +\infty} \frac{1}{iy} \big(\hat{W} \star \infty\big)(iy) .$$

We conclude from [KW2, Theorem 5.7] that

$$\tilde{W}(z) := \begin{pmatrix} 1 & -m(W)z \\ 0 & 1 \end{pmatrix} \hat{W}(z) .$$

satisfies $\mathfrak{K}_-(\tilde{W}) = \mathfrak{K}(\tilde{W})$. In case $m(W) \neq 0$ the space generated by the linear matrix in the above relation is one-dimensional and spanned by the constant $(1,0)^T$, in fact $\mathfrak{K}(\hat{W}) = \mathfrak{K}(\tilde{W}) \oplus \mathrm{span}\{(1,0)^T\}$. Hence $\mathfrak{K}_-(\hat{W})$ is orthocomplemented in $\mathfrak{K}(\hat{W})$ and $\dim \mathfrak{K}_-(\hat{W})^\perp = 1$. We conclude from Proposition 3.8 that $\mathfrak{K}_-(W) = \mathfrak{K}(W)$. In the case $m(W) = 0$ we have $\dim \mathfrak{K}_-(\hat{W})^\perp = 0$.

Conversely, assume that $\mathfrak{K}_-(W) = \mathfrak{K}(W)$ and $\mathfrak{K}(\hat{W})^\circ = \{0\}$. Again appealing to [KW2, Theorem 5.7] we find that there exists a polynomial $p(z)$ such that

$$\tilde{W}(z) := \begin{pmatrix} 1 & p(z) \\ 0 & 1 \end{pmatrix} \hat{W}(z)$$

satisfies $\mathfrak{K}_-(\tilde{W}) = \mathfrak{K}(\tilde{W})$. Thereby $\dim \mathfrak{K}_-(\hat{W})^\perp = \deg p$. Since, by Proposition 3.8, $\dim \mathfrak{K}_-(\hat{W})^\perp \leq 1$ we can choose $p(z) = az$ and thus

$$\begin{aligned} \lim_{y \to +\infty} \big(W \star \infty\big)\big((iy)^2\big) &= \lim_{y \to +\infty} \frac{1}{iy} \big(\hat{W} \star \infty\big)(iy) \\ &= \lim_{y \to +\infty} \frac{1}{iy} \big(\tilde{W} \star \infty\big)(iy) - a = -a . \end{aligned}$$

□

3.10. Proposition. *Let* $W \in \mathcal{M}^{\mathrm{ep}}_{<\infty}$ *and assume that* $m(W) = 0$. *Then* $\lambda(W) \notin \mathbb{R}$ *if and only if* $\mathrm{ind}_- \mathcal{T}_{2,\gamma}(W) \star \infty < \mathrm{ind}_- \mathcal{T}_{2,\gamma}(W)$ *for all* $\gamma \in \mathbb{R}$. *If* $\lambda(W) \in \mathbb{R}$, *we have*

$$\lambda(W) = -\sup\big\{\gamma \in \mathbb{R} : \mathrm{ind}_- \mathcal{T}_{2,\gamma}(W) \star \infty < \mathrm{ind}_- \mathcal{T}_{2,\gamma}(W)\big\} .$$

If $\gamma, \hat{\gamma} \in \mathbb{R}$, *then we have*

$$\mathrm{ind}_- \mathcal{T}_{2,\hat{\gamma}}(W) = \mathrm{ind}_- \mathcal{T}_{2,\gamma}(W) + \begin{cases} -1 & , \ \gamma < -\lambda(W) \leq \hat{\gamma} \\ 0 & , \ \gamma, \hat{\gamma} \geq -\lambda(W) \text{ or } \gamma, \hat{\gamma} < -\lambda(W) \\ 1 & , \ \hat{\gamma} < -\lambda(W) \leq \gamma \end{cases} .$$

Proof. According to Proposition 3.9 our assumption $m(W) = 0$ implies that $\mathfrak{K}_-(W) = \mathfrak{K}(W)$ and that $\mathfrak{K}_-(\mathcal{T}_{2,\gamma}(W)) = \mathfrak{K}(\mathcal{T}_{2,\gamma}(W))$ for all $\gamma \in \mathbb{R}$. Put $W_\gamma := \mathcal{T}_{2,\gamma}(W)$ and $E_\gamma := w_{\gamma,22} + iw_{\gamma,21}$. Note that the function

$$q(z) := \big(W_\gamma \star \infty\big)(z) = z\big(W \star \infty\big)(z^2)$$

does not depend on γ. By (3.4) we have

$$\frac{w_{22}(-y^2)}{w_{21}(-y^2)} = \frac{1}{iy}\frac{w_{\gamma,22}(iy)}{w_{\gamma,21}(iy)} - \gamma\,.$$

There are four possibilities:

$$\lambda(W) + \gamma = \lim_{y\to+\infty} \frac{1}{iy}\frac{w_{\gamma,22}(iy)}{w_{\gamma,21}(iy)} = \begin{cases} = 0 & , \ w_{\gamma,21} \notin \mathfrak{P}(E_\gamma) \\ > 0 & , \ w_{\gamma,21} \in \mathfrak{P}(E_\gamma),\ [w_{\gamma,21}, w_{\gamma,21}] > 0 \\ < 0 & , \ w_{\gamma,21} \in \mathfrak{P}(E_\gamma),\ [w_{\gamma,21}, w_{\gamma,21}] < 0 \\ \nexists & , \ w_{\gamma,21} \in \mathfrak{P}(E_\gamma),\ [w_{\gamma,21}, w_{\gamma,21}] = 0 \end{cases} \tag{3.11}$$

In the first two cases (cf. [KW2, Lemma 5.12])

$$\operatorname{ind}_- W_\gamma = \operatorname{ind}_- W_\gamma \star \infty = \operatorname{ind}_- q\,,$$

and in the third

$$\operatorname{ind}_- W_\gamma = \operatorname{ind}_- W_\gamma \star \infty + 1 = \operatorname{ind}_- q + 1\,.$$

Assume now that $\lambda(W) \in \mathbb{R}$, i.e., we are in one of the first three cases of (3.11). Then

$$\operatorname{ind}_- W_\gamma = \begin{cases} \operatorname{ind}_- q & , \ \gamma \geq -\lambda(W) \\ \operatorname{ind}_- q + 1 & , \ \gamma < -\lambda(W) \end{cases}$$

Hence, in this case, the assertion of the lemma follows.

Consider the case that $\lambda(W) \notin \mathbb{R}$, i.e., that $w_{\gamma,21} \in \mathfrak{P}(E_\gamma)$ and $[w_{\gamma,21}, w_{\gamma,21}] = 0$. Then, by the proof of [KW2, Theorem 7.1], for all $l < 0$

$$\operatorname{ind}_- W_\gamma W_{(l,0)} = \operatorname{ind}_- W_\gamma\,.$$

It follows that $\operatorname{ind}_- W_\gamma = \operatorname{ind}_- W_{\hat\gamma}$ and that $\operatorname{ind}_- W_\gamma \star \infty < \operatorname{ind}_- W_\gamma$ for all $\gamma, \hat\gamma \in \mathbb{R}$ (cf. [KW3]). □

3.3. Structure of symmetric and semibounded dB-spaces

The above Proposition 3.9 has two consequences on the structure of symmetric and semibounded dB-Pontryagin spaces, which we shall elaborate in the following. The first one gives a growth restriction on the elements of a semibounded dB-space. For the proof we use a lemma which supplements [KW5, Theorem 3.17, Corollary 3.18].

3.11. Lemma. *Let $q \in \mathcal{N}_{<\infty}$ and let B be an entire function of finite order ρ. If $A(z) := q(z)B(z)$ is entire, then the order of A is also equal to ρ. Moreover, B is of finite type if and only if A is. If ρ is not an integer, then B being of minimal type is equivalent to A possessing the same property.*

Proof. By [DLLS] we can write $q(z) = r(z) \cdot q_1(z)$ with $q_1 \in \mathcal{N}_0$ and a rational function r. This shows that it suffices to prove the assertion for the case $\kappa = 0$. By our assumption the function q must be meromorphic in the whole plane and hence we may write according to [L, VII. Lehrsatz 1]

$$q(z) = c\frac{z - a_0}{z - b_0} \prod_{k \neq 0} \Big(1 - \frac{z}{a_k}\Big)\Big(1 - \frac{z}{b_k}\Big)^{-1}. \tag{3.12}$$

Denote the zeros of B by x_k. Since $q(z) \cdot B(z)$ is entire, the zero set of B splits up as $\{x_k\} = \{b_k\} \cup X$. Here and for the rest of this proof we understand that a zero is listed as often as its multiplicity states, and also understand set theoretic notations as including multiplicities. It follows that the zeros $\{y_k\}$ of A are given by $\{y_k\} = \{a_k\} \cup X$. From [KWW2, Lemma 4.5] we conclude that the convergence exponents of $\{x_k\}$ and $\{y_k\}$ are equal and that the upper densities of these sequences coincide:

$$\limsup_{r \to \infty} \frac{1}{r^\rho} \big|\{x_k : |x_k| \leq r\}\big| = \limsup_{r \to \infty} \frac{1}{r^\rho} \big|\{y_k : |y_k| \leq r\}\big|.$$

Moreover, the values

$$\limsup_{r \to \infty} \sum_{|x_k| \leq r} \frac{1}{x_k^\rho}, \ \limsup_{r \to \infty} \sum_{|x_k| \leq r} \frac{1}{y_k^\rho}$$

are together finite or infinite. If we arrange the sequences (a_k) and (b_k) so that $a_k < b_k < a_{k+1}$, for all $l \in \mathbb{N}$ the series

$$\sum_k \Big[\Big(\frac{1}{a_k}\Big)^l - \Big(\frac{1}{b_k}\Big)^l\Big] \tag{3.13}$$

converges. Consider the Hadamard factorization of B:

$$\begin{aligned} B(z) &= e^{P(z)} \prod \Big(1 - \frac{z}{x_k}\Big) \exp\Big[\sum_{l=1}^p \frac{1}{l}\Big(\frac{z}{x_k}\Big)^l\Big] \\ &= e^{P(z)} \prod \Big(1 - \frac{z}{b_k}\Big) \exp\Big[\sum_{l=1}^p \frac{1}{l}\Big(\frac{z}{b_k}\Big)^l\Big] \prod_X \Big(1 - \frac{z}{x_k}\Big) \exp\Big[\sum_{l=1}^p \frac{1}{l}\Big(\frac{z}{x_k}\Big)^l\Big]. \end{aligned}$$

Here P is a polynomial of degree at most ρ and p is the genus of the zeros of B. From $A = qB$, (3.12) and the convergence of the series (3.13) it follows that the entire function $\tilde{P}$ in the product representation

$$A(z) = e^{\tilde{P}(z)} \prod \Big(1 - \frac{z}{a_k}\Big) \exp\Big[\sum_{l=1}^p \frac{1}{l}\Big(\frac{z}{a_k}\Big)^l\Big] \prod_X \Big(1 - \frac{z}{x_k}\Big) \exp\Big[\sum_{l=1}^p \frac{1}{l}\Big(\frac{z}{x_k}\Big)^l\Big]$$

of A must be equal to

$$\tilde{P}(z) = \log\Big(c\frac{a_0}{b_0}\Big) \cdot P(z) \cdot \sum_{l=1}^p \frac{z^l}{l} \sum_k \Big[\Big(\frac{1}{b_k}\Big)^l - \Big(\frac{1}{a_k}\Big)^l\Big].$$

Hence $\tilde{P}$ is in fact a polynomial of degree at most ρ. It follows that A is of finite order at most ρ. Since in this argument the roles of A and B can be exchanged, we conclude that the order of A actually equals ρ. The assertion of the lemma now follows from what was said above by Lindelöf's Theorem, see, e.g., [L, I. Lehrsatz 14, 15]. □

Let us remark that the assumption $\rho \notin \mathbb{Z}$ in the last part of Lemma 3.11 can not be dropped.

3.12. Proposition. *Let $W \in \mathcal{M}^{\text{ep}}_{<\infty}$. Then every entry w_{ij} is an entire function of growth at most order $\frac{1}{2}$, finite type. In particular, if $\mathfrak{P}$ is a semibounded dB-space such that $1 \in \operatorname{Assoc}\mathfrak{P} := \mathfrak{P} + z\mathfrak{P}$, then*

$$\sup_{F \in \operatorname{Assoc}\mathfrak{P}} \limsup_{r\to\infty} \frac{\log \max_{|z|=r} |F(z)|}{\sqrt{r}} < \infty \tag{3.14}$$

Proof. Consider the matrix $\hat{W} := \mathcal{T}_{2,0}(W) \in \mathcal{M}^{\text{sym}}_{<\infty}$. Since at most one of $\mathfrak{K}_-(\hat{W})$ and $\mathfrak{K}_+(\hat{W})$ is degenerated, we conclude that $1 \in \operatorname{Assoc}\mathfrak{P}(E)$ where E either is $\hat{w}_{11} - i\hat{w}_{12}$ or $\hat{w}_{22} + i\hat{w}_{21}$. In any case we conclude from [KWW2, Theorem 3.10] that E and, hence, one entry of $\hat{W}$ is of exponential type. Lemma 3.11 now implies that all entries of $\hat{W}$ are of exponential type. Therefore the entries of W are of growth at most order $\frac{1}{2}$, finite type.

Let $\mathfrak{P}$ be a semibounded dB-space and choose an inner product $(.,.)$ on $\mathfrak{P}$ such that $(\mathfrak{P}, (.,.)) = \mathfrak{P}(E)$ is a semibounded dB-Pontryagin space, this is possible by [KWW4]. Since $\mathfrak{P}$ and $\mathfrak{P}(E)$ coincide as sets, we have $\operatorname{Assoc}\mathfrak{P} = \operatorname{Assoc}\mathfrak{P}(E)$. If $1 \in \operatorname{Assoc}\mathfrak{P}$ hence also $1 \in \operatorname{Assoc}\mathfrak{P}(E)$, and we conclude that there exists a matrix $W \in \mathcal{M}^{\text{ep}}_{<\infty}$ such that $(0,1)W = (-B, A)$. By what was proved in the first paragraph the function E is of growth at most order $\frac{1}{2}$, finite type. The relation (3.14) now follows from [KW5, Theorem 3.4]. □

In order to give the second promised structure result on dB-spaces, we need to recall a construction introduced in [KWW4]: If $\mathfrak{P}$ is a symmetric dB-space, then define

$$\mathfrak{P}_+ := \{F \in \mathcal{O}(\mathbb{C}) : F(z^2) \in \mathfrak{P}\}.$$

If $\mathfrak{P}_+$ is endowed with a topology and inner product so that the map $F(z) \mapsto F(z^2)$ becomes an isometric homeomorphism, this space is a semibounded dB-space. The main result of [KWW4] states that every semibounded dB-space can be obtained in this way and determines the kernel of the assignment $\Upsilon : \mathfrak{P} \mapsto \mathfrak{P}_+$. Moreover, if $\mathfrak{P}$ and hence also $\mathfrak{P}_+$ is a dB-Pontryagin space, the action of the assignment Υ is explicitly determined in terms of the respective generating Hermite-Biehler functions.

This construction on the level of dB-spaces is the exact analogue of the transformations $\mathcal{T}_{\surd}$, $\mathcal{T}_{2,\gamma}$ on the level of matrix functions. This follows by comparing the definition of $\mathcal{T}_{\surd}$ and $\mathcal{T}_{2,\gamma}$ with [KWW4, Theorem 4.5].

3.13. Lemma. *Let $W \in \mathcal{M}^{\text{sym}}_{<\infty}$. Then $E_W \in \mathcal{HB}^{\text{sym}}$, $E_{\mathcal{T}_{\surd}(W)} \in \mathcal{HB}^{\text{ep}}$, and*

$$\mathfrak{P}(E_W)_+ = \mathfrak{P}(E_{\mathcal{T}_{\surd}(W)}).$$

Conversely, let $W \in \mathcal{M}^{\mathrm{ep}}_{<\infty}$ *and* $\gamma \in \mathbb{R}$. *Then* $E_W \in \mathcal{H}B^{\mathrm{ep}}$, $E_{\mathcal{T}_{2,\gamma}(W)} \in \mathcal{H}B^{\mathrm{sym}}$, *and*

$$\mathfrak{P}(E_{\mathcal{T}_{2,\gamma}(W)})_+ = \mathfrak{P}(E_W).$$

Each two spaces $\mathfrak{P}(E_{\mathcal{T}_{2,\gamma}(W)})$ *are not isometrically equal and every dB-Pontryagin space* $\mathfrak{P}$ *with* $\mathfrak{P}_+ = \mathfrak{P}(E_W)$ *is of this form.*

It was shown in [KWW4, Proposition 2.6] that every even function $F \in \operatorname{Assoc}\mathfrak{P}$ can be obtained as $F(z) = G(z^2)$ with $G \in \operatorname{Assoc}\mathfrak{P}_+$. In particular, if $1 \in \operatorname{Assoc}\mathfrak{P}$, then also $1 \in \operatorname{Assoc}\mathfrak{P}_+$. The converse does not hold in general.

3.14. Proposition. *Let* $E = A - iB \in \mathcal{H}B^{\mathrm{sym}}_{<\infty}$ *and put* $E_+ := A_+ - iB_+$ *with*

$$A_+(z^2) = A(z), \; B_+(z^2) = zB(z),$$

so that $\mathfrak{P}(E)_+ = \mathfrak{P}(E_+)$. *Assume that* $1 \in \operatorname{Assoc}\mathfrak{P}(E)_+$ *and let* $W_+ \in \mathcal{M}^{\mathrm{ep}}_{<\infty}$ *be such that* $\mathfrak{K}_-(W_+) = \mathfrak{K}(W_+)$ *and* $(0,1)W_+ = (-B_+, A_+)$. *Then* $1 \in \operatorname{Assoc}\mathfrak{P}(E)$ *if and only if* $m(W_+) \in \mathbb{R}$. *In this case there exists* $W \in \mathcal{M}^{\mathrm{sym}}_{<\infty}$, $\mathfrak{K}_-(W) = \mathfrak{K}(W)$, *such that* $(0,1)W = (-B, A)$.

Proof. By its definition the function B_+ has only finitely many zeros in $\mathbb{C} \setminus [0,\infty)$. Hence $W_+ \in \mathcal{M}^{\mathrm{ep}}_{<\infty}$ and thus $\hat{W} := \mathcal{T}_{2,0}(W_+) \in \mathcal{M}^{\mathrm{sym}}_{<\infty}$. By the definition of $\mathcal{T}_{2,0}$ the matrix $\hat{W}$ satisfies $(0,1)\hat{W} = (-B, A)$. We have $1 \in \operatorname{Assoc}\mathfrak{P}(E)$ if and only if $\mathfrak{K}_-(\hat{W})^\circ = \{0\}$, cf. [KW1, Proposition 10.3], [KW2, Lemma 5.11]. This, however, is in view of Proposition 3.9 equivalent to $m(W_+) \in \mathbb{R}$. □

4. Chains of matrix functions

In this section we investigate chains of matrix functions and introduce the appropriate analogues of the notion of symmetry and semiboundedness on the level of chains of matrices.

Let us recall the notion of a maximal chain of matrices as introduced in [KW3]. For a matrix $W \in \mathcal{M}$ denote by $\mathfrak{t}(W)$ the *trace function* $\mathfrak{t}(W) := \operatorname{tr}(W'(0)J)$.

4.1. Definition. A mapping $\omega : \mathcal{I} \to \mathcal{M}_{<\infty}$ is called a *maximal chain of matrices* if the following axioms are satisfied:

- **(W1)** The set $\mathcal{I}$ equals $(0, M)$, $0 < M < \infty$, with possible exception of finitely many points.
- **(W2)** The function ω is not constant on any interval contained in $\mathcal{I}$.
- **(W3)** For all $s, t \in \mathcal{I}$, $s \leq t$, we have $\omega(s)^{-1}\omega(t) \in \mathcal{M}_{<\infty}$ and
$$\operatorname{ind}_- \omega(t) = \operatorname{ind}_- \omega(s) + \operatorname{ind}_- \omega(s)^{-1}\omega(t).$$
- **(W4)** If $t \in \mathcal{I}$ and for some $W \in \mathcal{M}_{<\infty}$, $W \neq I$, we have $W^{-1}\omega(t) \in \mathcal{M}_{<\infty}$ and $\operatorname{ind}_- \omega(t) = \operatorname{ind}_- W + \operatorname{ind}_- W^{-1}\omega(t)$, then there exists a number $s \in \mathcal{I}$ such that $W = \omega(s)$.
- **(W5)** We have $\lim_{t \nearrow M} \mathfrak{t}(\omega(t)) = +\infty$. If $\mathcal{I}$ is not connected, there exist numbers $s < t$, both contained in the last connected component $\mathcal{I}_\infty$ of $\mathcal{I}$ (that is $\sup \mathcal{I}_\infty = M$), such that $\omega(s)^{-1}\omega(t)$ is not a linear polynomial.

It is proved in [KW3, Lemma 3.5] that the function $\operatorname{ind}_-\omega(t)$ is constant on each connected component of $\mathcal{I}$ and takes different values on different components. Moreover, by (W3), it is nondecreasing. In particular, it is bounded and attains its maximum on $\mathcal{I}_\infty$. Let us define $\operatorname{ind}_-\omega := \max_{t\in\mathcal{I}} \operatorname{ind}_-\omega(t)$. The set of all maximal chains ω with $\operatorname{ind}_-\omega = \kappa$ will be denoted by $\mathfrak{M}_\kappa$. Moreover,

$$\mathfrak{M}_{\leq\kappa} := \bigcup_{\nu\leq\kappa} \mathfrak{M}_\nu,\ \mathfrak{M}_{<\infty} := \bigcup_{\nu\in\mathbb{N}\cup\{0\}} \mathfrak{M}_\nu\,.$$

It is already seen from the axiom (W5) that points s,t where the transfer matrix $\omega(s)^{-1}\omega(t)$ is a linear polynomial play a special role. This is formalized by the notion of indivisible intervals. Let $\omega : \mathcal{I} \to \mathcal{M}_{<\infty}$ be a maximal chain of matrices. An interval $(s,t) \subseteq \mathcal{I}$ is called *indivisible of type* $\phi \in [0,\pi)$ if for all $s',t' \in (s,t)$

$$\omega(s')^{-1}\omega(t') = W_{(l(s',t'),\phi)}$$

The number $L := \sup\{l(s',t') : s' \leq t', s',t' \in (s,t)\}$ is called the *length* of the indivisible interval (s,t).

If (s_1,t_1) and (s_2,t_2) are indivisible intervals of types ϕ_1 and ϕ_2, respectively, which have nonempty intersection, then $\phi_1 = \phi_2$ and $(\min\{s_1,s_2\}, \max\{t_1,t_2\})$ is again indivisible of the same type. Hence every indivisible interval is contained in a maximal indivisible interval.

It can happen that for some $s,t \in \mathcal{I}$, $s \leq t$, we have $\omega(s)^{-1}\omega(t) = W_{(l,\phi)}$ for some $l < 0$. Although in this case $(s,t) \not\subseteq \mathcal{I}$ we shall speak of an *indivisible interval of negative length.*

Chains which can be obtained out of each other by a change of variable will share their important properties. More precisely: Let $\mathcal{J}_1, \mathcal{J}_2$ be open subsets of $\mathbb{R}$ and let $\omega_i : \mathcal{J}_i \to \mathcal{M}_{<\infty}$ be functions. Then we say that ω_2 is a *reparameterization* of ω_1 if there exists an increasing and bijective map $\alpha : \mathcal{J}_2 \to \mathcal{J}_1$ such that $\omega_2 = \omega_1 \circ \alpha$. In this case we write $\omega_2 \sim \omega_1$. It is obvious that $\sim$ is an equivalence relation.

A central role in the theory of maximal chains of matrices is played by the Weyl coefficient associated to a maximal chain. It is proved in [KW2] that for all functions $\tau \in \mathcal{N}_0$ the limit

$$q_\infty(\omega)(z) := \lim_{t\nearrow\sup\mathcal{J}} (\omega(t) \star \tau)(z)$$

exists locally uniformly on compact subsets of $\mathbb{C}\setminus\mathbb{R}$ with respect to the chordal metric and does not depend on the particular choice of τ. It is called the *Weyl coefficient* of the chain ω. The main result of [KW2] states that the set $\mathfrak{M}_\kappa/\sim$ bijectively corresponds to $\mathcal{N}_\kappa$ via

$$\omega/\sim\ \longmapsto\ q_\infty(\omega)\,.$$

The elements of maximal chains can be characterized by means of decompositions of the Weyl coefficient. Recall the following fact from [KW3]: Let $\omega : \mathcal{I} \to \mathcal{M}_{<\infty}$ be a maximal chain and let $W \in \mathcal{M}_{<\infty}$. Then $W \in \omega(\mathcal{I})$ if and only if $\operatorname{ind}_- W \leq \operatorname{ind}_-\omega$ and there exists a function $\tau \in \mathcal{N}_{\operatorname{ind}_-\omega - \operatorname{ind}_- W}$ such that $q_\infty(\omega) = W \star \tau$.

Let a function $\upsilon : \mathcal{J} \to \mathcal{M}_{<\infty}$ be given. In the next lemma we give conditions, adapted to our needs, under which υ can be extended to a maximal chain. This result is an immediate consequence of [KW3, Lemma 3.7] and [KW2, Lemma 8.5] with its proof.

4.2. Lemma. *Let $\upsilon : \mathcal{J} \to \mathcal{M}_{<\infty}$ be given and assume that υ satisfies (W3) and*

(C1) *We have $\kappa_m := \sup_{t\in\mathcal{J}} \operatorname{ind}_- \upsilon(t) < \infty$.*

(C2) *The following two implications hold true.*

(a) *If $\limsup_{t\nearrow\sup\mathcal{J}} \mathfrak{t}(\omega(t)) < +\infty$, then $\lim_{t\nearrow\sup\mathcal{J}} \omega(t) \in \mathcal{M}_\kappa$.*

(b) *If $\limsup_{t\nearrow\sup\mathcal{J}} \mathfrak{t}(\omega(t)) = +\infty$ and if there is a number $\phi \in [0,\pi)$ such that*

$$\omega(s)^{-1}\omega(t) = W_{(l(s,t),\phi)}, \ s,t \in \omega^{-1}\big(\mathcal{M}_{\kappa_m}\big),$$

then for one (and hence for all) $t \in \omega^{-1}\big(\mathcal{M}_{\kappa_m}\big)$

$$\omega(t) \star \cot\phi \in \mathcal{N}_{\kappa_m}.$$

Then there exists a maximal chain $\omega \in \mathfrak{M}_{\kappa_m}$ and a nondecreasing function $\lambda : \mathcal{J} \to \mathcal{I}$ such that $\upsilon = \omega \circ \lambda$. If $\limsup_{t\nearrow\sup\mathcal{J}} \mathfrak{t}(\omega(t)) = +\infty$, the chain ω is unique and we have

$$q_\infty(\omega) = \lim_{t\nearrow\sup\mathcal{J}} \big[\upsilon(t) \star \tau\big], \ \tau \in \mathcal{N}_0.$$

Otherwise, the set of all extensions in $\mathfrak{M}_{\kappa_m}$ is parameterized by the set of all functions $\tau \in \mathcal{N}_0$ with the property that

$$\operatorname{ind}_- \Big[\lim_{t\nearrow\sup\mathcal{J}} \upsilon(t)\Big] \star \tau = \operatorname{ind}_- \lim_{t\nearrow\sup\mathcal{J}} \upsilon(t).$$

This correspondence is established via the relation

$$q_\infty(\omega) = \Big[\lim_{t\nearrow\sup\mathcal{J}} \upsilon(t)\Big] \star \tau.$$

For the purposes of the present paper two particular kinds of chains of matrices are of interest.

4.3. Definition. Let $\omega \in \mathfrak{M}_\kappa$. We write $\omega \in \mathfrak{M}_\kappa^{\mathrm{sym}}$ if $\omega(t) \in \mathcal{M}_{<\infty}^{\mathrm{sym}}$ for all $t \in \mathcal{I}$. Moreover, we write $\omega \in \mathfrak{M}_\kappa^{\mathrm{ep}}$ if $\omega(t) \in \mathcal{M}_{<\infty}^{\mathrm{ep}}$ for all $t \in \mathcal{I}$ and if the number of zeros which an entry of $\omega(t)$ possesses in $\mathbb{C} \setminus [0,\infty)$ is bounded independently of $t \in \mathcal{I}$.

The following inverse result gives a connection between the classes $\mathfrak{M}_{<\infty}^{\mathrm{sym}}$ and $\mathcal{N}_{<\infty}^{\mathrm{sym}}$. The corresponding result for the class $\mathfrak{M}_{<\infty}^{\mathrm{ep}}$ will be seen later, cf. Proposition 5.6.

4.4. Proposition. *Let ω be a maximal chain of matrices. Then $\omega \in \mathfrak{M}_{<\infty}^{\mathrm{sym}}$ if and only if $q_\infty(\omega) \in \mathcal{N}_{<\infty}^{\mathrm{sym}}$.*

Proof. Assume that $\omega \in \mathfrak{M}_{<\infty}^{\mathrm{sym}}$. Then for all $t \in \mathcal{I}$ the function $\omega(t) \star \infty$ is odd. Since

$$\lim_{t\to\sup\mathcal{I}} \big(\omega(t) \star \infty\big)(z) = q_\infty(\omega)$$

we conclude that also $q_\infty(\omega)$ belongs to $\mathcal{N}_{<\infty}^{\mathrm{sym}}$.

Assume conversely that $q_\infty(\omega)$ is odd and let $t \in \mathcal{I}$ be given. Then there exists a function $\tau \in \mathcal{N}_{<\infty}$, $\operatorname{ind}_- \tau = \operatorname{ind}_- \omega - \operatorname{ind}_- \omega(t)$, such that

$$q_\infty(\omega)(z) = \big(\omega(t) \star \tau\big)(z) = \frac{\omega(t)_{11}(z)\tau(z) + \omega(t)_{12}(z)}{\omega(t)_{21}(z)\tau(z) + \omega(t)_{22}(z)}\,.$$

It follows that

$$\begin{aligned} q_\infty(\omega)(z) = -q_\infty(\omega)(-z) &= \frac{\omega(t)_{11}(-z)[-\tau(-z)] - \omega(t)_{12}(-z)}{-\omega(t)_{21}(-z)[-\tau(-z)] + \omega(t)_{22}(-z)} \\ &= \big(-JV\omega(t)^{-1}VJ\big)(-z) \star [-\tau(-z)]\,. \end{aligned}$$

Hereby the constant matrices J and V are defined as in Lemma 2.3. With $\omega(t)$ and τ we also have $(-JV\omega(t)^{-1}VJ)(-z) \in \mathcal{M}_{<\infty}$ and $-\tau(-z) \in \mathcal{N}_{<\infty}$, respectively. In fact

$$\operatorname{ind}_-\big(-JV\omega(t)^{-1}VJ\big)(-z) = \operatorname{ind}_- \omega(t)(z),\ \operatorname{ind}_-\big(-\tau(-z)\big) = \operatorname{ind}_- \tau(z)\,.$$

Hence

$$\begin{aligned} \operatorname{ind}_- q_\infty(\omega)(z) &= \operatorname{ind}_- \omega(t) + \operatorname{ind}_- \tau \\ &= \operatorname{ind}_-\big(-JV\omega(t)^{-1}VJ\big)(-z) + \operatorname{ind}_-\big(-\tau(-z)\big)\,. \end{aligned}$$

We conclude that both matrices, $\omega(t)$ and $(-JV\omega(t)^{-1}VJ)(-z)$, are members of the maximal chain of matrices having $q_\infty(\omega)$ as its Weyl coefficient. Since $\operatorname{ind}_-(-JV\omega(t)^{-1}VJ)(-z) = \operatorname{ind}_- \omega(t)$ and $\mathfrak{t}((-JV\omega(t)^{-1}VJ)(-z)) = \mathfrak{t}(\omega(t))$ we conclude that $(-JV\omega(t)^{-1}VJ)(-z) = \omega(t)$, cf. [KW3, Lemma 3.5]. This means that $\omega(t) \in \mathcal{M}^{\text{sym}}_{<\infty}$. □

The following technical condition on a maximal chain ω will appear frequently:

($\mathbf{K}_-$) For all $t \in \mathcal{I}$ we have $\mathfrak{K}_-(\omega(t)) = \mathfrak{K}(\omega(t))$.

Recall from [KW2] that ω satisfies (K_-) if and only if for one $t \in \mathcal{I}$ the equality $\mathfrak{K}_-(\omega(t)) = \mathfrak{K}(\omega(t))$ holds. Moreover, the inverse result [KW2, Theorem 5.7] shows that ω satisfies (K_-) if and only if $\lim_{y\to+\infty} y^{-1}q_\infty(\omega)(iy) = 0$.

We need two general constructions which can be made with chains of matrices. The first one formalizes the intuitive idea of *linking of chains*, compare the discussion after [KW2, Theorem 7.1]. Let υ_1 and υ_2 be functions into $\mathcal{M}_{<\infty}$ defined on open subsets $\mathcal{J}_1, \mathcal{J}_2$ of $\mathbb{R}$. If both, $\mathcal{J}_1$ and $\mathcal{J}_2$ are nonempty we define a function $\upsilon_1 \uplus \upsilon_2$ as follows: Choose increasing bijections α_i of $\mathcal{J}_i$ onto open sets with the property that

$$\inf \alpha_i(\mathcal{J}_i) = i-1,\ \sup \alpha_i(\mathcal{J}_i) = i,\ i = 1,2\,,$$

and define $\upsilon_1 \uplus \upsilon_2 : \alpha_1(\mathcal{J}_1) \cup \alpha_2(\mathcal{J}_2) \to \mathcal{M}_{<\infty}$ by

$$\upsilon_1 \uplus \upsilon_2(t) := \begin{cases} (\upsilon_1 \circ \alpha_1^{-1})(t) & ,\ t \in \alpha_1(\mathcal{J}_1) \\ (\upsilon_2 \circ \alpha_2^{-1})(t) & ,\ t \in \alpha_2(\mathcal{J}_2) \end{cases}$$

Note that this definition is independent of the choice of α_1, α_2 if we identify functions which are equal up to reparameterization. For the function ε with empty domain we set

$$\upsilon \uplus \varepsilon = \varepsilon \uplus \upsilon = \upsilon .$$

Up to reparameterization the operation $\uplus$ is associative.

The second construction is simply *extension by continuity*: Assume that the function υ is defined on a set $\mathcal{J}$ of the form $(a,b) \setminus \{x_1, \dots, x_n\}$. Let L be the set of all those points x_i such that the limit $\lim_{x \to x_i} \upsilon(x)$ exists. Then we can define $\mathsf{C}\upsilon : \mathcal{J} \cup L \to \mathcal{M}_{<\infty}$ by

$$\mathsf{C}\upsilon(t) := \begin{cases} \upsilon(t) & , \; t \in \mathcal{I} \\ \lim_{x \to x_i} \upsilon(x) & , \; t = x_i \in L \end{cases}$$

5. Transformation of matrix chains

The transformation $\mathcal{T}_{\surd}$ can be applied pointwise to a maximal chain of matrices. The outcome will almost be a maximal chain.

5.1. The square root transformation

Consider a chain $\omega : \mathcal{I} \to \mathcal{M}_{<\infty}$ in $\mathfrak{M}_\kappa^{\mathrm{sym}}$. Write the index set $\mathcal{I}$ as

$$\mathcal{I} = (0, \sigma_1) \cup (\sigma_1, \sigma_2) \cup \cdots \cup (\sigma_n, M) ,$$

and put $\sigma_0 := 0, \sigma_{n+1} := M$. To each point σ_i, $i = 1, \dots, n+1$, we associate a function ς_i according to the table on top of the next page.

Define a function $\omega_{\surd}$ as

$$\omega_{\surd} := \mathsf{C}\big(\mathcal{T}_{\surd} \circ \omega|_{(0,\sigma_1)} \uplus \varsigma_1 \uplus \mathcal{T}_{\surd} \circ \omega|_{(\sigma_1,\sigma_2)} \uplus \varsigma_2 \uplus \cdots \uplus \varsigma_n \uplus \mathcal{T}_{\surd} \circ \omega|_{(\sigma_n,M)} \uplus \varsigma_{n+1}\big)$$

5.1. Theorem. *Let $\omega \in \mathfrak{M}_\kappa^{\mathrm{sym}}$ and assume that ω satisfies (K_-). Let ϖ be the maximal chain with $zq_\infty(\varpi)(z^2) = q_\infty(\omega)(z)$. Then there exist functions $\lambda : \operatorname{dom} \varpi \to \operatorname{dom} \omega_{\surd}$, $\mu : \operatorname{dom} \omega_{\surd} \to \operatorname{dom} \varpi$ such that*

$$\varpi = \omega_{\surd} \circ \lambda, \;\; \omega_{\surd} = \varpi \circ \mu .$$

The proof of this theorem will be carried out in four steps. Before we come to the first step, we need to provide two lemmata. The first of which is implicitly contained in [dB1], the second one follows from the considerations in [KW2]. We shall however provide complete proofs.

5.2. Lemma. *Let $\omega = (W_t)_{t \in \mathcal{I}} \in \mathfrak{M}_{<\infty}$ and let $\mathcal{I}_1$ be a connected component of $\mathcal{I}$. Assume that for all $s, t \in \mathcal{I}_1$ the transfer matrix $W_{st} = \omega(s)^{-1}\omega(t)$ is a polynomial and that*

$$n := \sup_{s,t \in \mathcal{I}_1} \deg W_{st} < \infty .$$

Then

$$\mathcal{I}_1 = (m_0, m_1] \cup [m_1, m_2] \cup \cdots \cup [m_{n-1}, m_n) ,$$

$\lim_{t\nearrow\sigma_i} \mathcal{T}_{\surd}(\omega(t))$	$\lim_{t\searrow\sigma_i} \mathcal{T}_{\surd}(\omega(t))$	Definition of ς_i $i=1,\dots,n:$ $\alpha := \operatorname{ind}_- \lim_{t\searrow\sigma_i} \mathcal{T}_{\surd}(\omega(t)) - \operatorname{ind}_- \lim_{t\nearrow\sigma_i} \mathcal{T}_{\surd}(\omega(t))$	
$\exists$	$\exists$	$\alpha=0$	$l \mapsto \lim_{t\nearrow\sigma_i} \mathcal{T}_{\surd}(\omega(t)) \cdot W_{(l,0)}\,, \quad l\in(0,L)$ $L := \mathfrak{t}(\lim_{t\searrow\sigma_i} \mathcal{T}_{\surd}(\omega(t))) - \mathfrak{t}(\lim_{t\nearrow\sigma_i} \mathcal{T}_{\surd}(\omega(t)))$
		$\alpha\neq 0$	$\left[\begin{array}{c} l \mapsto \lim\limits_{t\nearrow\sigma_i} \mathcal{T}_{\surd}(\omega(t)) W_{(l,0)} \\ l\in(0,\infty) \end{array} \right] \uplus \left[\begin{array}{c} l \mapsto \lim\limits_{t\searrow\sigma_i} \mathcal{T}_{\surd}(\omega(t)) W_{(l,0)} \\ l\in(-\infty,0) \end{array} \right]$
$\nexists$	$\exists$	$l \mapsto \lim_{t\searrow\sigma_i} \mathcal{T}_{\surd}(\omega(t)) \cdot W_{(l,0)}\,, \quad l\in(-\infty,0)$	
$\exists$	$\nexists$	$l \mapsto \lim_{t\nearrow\sigma_i} \mathcal{T}_{\surd}(\omega(t)) \cdot W_{(l,0)}\,, \quad l\in(0,\infty)$	
$\nexists$	$\nexists$	ε	

$i=n+1:$

$\exists$		$l \mapsto \lim_{t\nearrow\sigma_i} \mathcal{T}_{\surd}(\omega(t)) \cdot W_{(l,0)}\,, \quad l\in(0,\infty)$
$\nexists$		ε

where

$$\inf \mathcal{I}_1 = m_0 < m_1 < \cdots < m_{n-1} < m_n = \sup \mathcal{I}_1\,,$$

and where the intervals (m_{i-1}, m_i) *are indivisible in* ω *of certain types* $\phi_i \in [0,\pi)$ *with* $\phi_i \neq \phi_{i+1}$.

Proof. Let us start with the following remark: If $E \in \mathcal{H}B_{<\infty}$ is a polynomial, then the chain of dB-subspaces of $\mathfrak{P}(E)$ is given by

$$\{0\} \subsetneq \operatorname{span}\{1\} \subsetneq \operatorname{span}\{1,z\} \subsetneq \cdots \subsetneq \operatorname{span}\{1,z,\dots,z^{\deg E-1}\} = \mathfrak{P}(E)\,.$$

If $\operatorname{ind}_- \mathfrak{P}(E) = 0$, every dB-subspace is also positive definite and in particular non-degenerated.

It follows from [dB1] that a polynomial matrix $W \in \mathcal{M}_0$ can be factorized uniquely as

$$W = W_{(l_1,\phi_1)} \cdot \cdots \cdot W_{(l_n,\phi_n)}$$

with $n = \deg W$, $l_i > 0$, $\phi_i \in [0,\pi)$, $\phi_i \neq \phi_{i+1}$.

Let us come to the proof of the present assertion. Choose $s_0, t_0 \in \mathcal{I}_1$, $s_0 < t_0$, such that $\deg W_{s_0 t_0} = n$. Since $\operatorname{ind}_- W_{s_0 t_0} = 0$, we can factorize $W_{s_0 t_0}$ as

$$W_{s_0 t_0} = W_{(l_1,\phi_1)} \cdot \cdots \cdot W_{(l_n,\phi_n)}\,.$$

If $t \in \mathcal{I}_1$, $t > t_0$, we have

$$W_{s_0 t} = W_{s_0 t_0} W_{t_0 t} = W_{(l_1,\phi_1)} \cdot \cdots \cdot W_{(l_n,\phi_n)} W_{t_0 t}\,.$$

On the other hand we have the factorization

$$W_{s_0t} = W_{(l'_1,\phi'_1)} \cdot \cdots \cdot W_{(l'_k,\phi'_k)}\,.$$

By uniqueness of the factorization we obtain $k = n$, $l_i = l'_i$, $\phi_i = \phi'_i$ for $i = 1,\dots,n-1$ and

$$W_{(l'_n,\phi'_n)} = W_{(l_n,\phi_n)} W_{t_0t}\,.$$

This implies $\phi_n = \phi'_n$ and $W_{t_0t} = W_{(l,\phi_n)}$ for some $l > 0$.

The same argument shows that for all $s \in \mathcal{I}_1$, $s < s_0$, the transfer matrix W_{ss_0} is of the form $W_{(l,\phi_1)}$.

If we choose $m_i \in \mathcal{I}_1$, $i = 1,\dots,n-1$, such that

$$W_{m_i} = W_{s_0} \prod_{j=1}^{i} W_{(l_j,\phi_j)}$$

we obtain the desired result. □

Note that in the situation of the previous lemma the types ϕ_i in the assertion are exactly the types occurring in the factorization of any transfer matrix of maximal degree.

5.3. Lemma. *If $W \in \mathcal{M}_{<\infty}$ and $W \star \infty = \infty$, then*

$$W = \begin{pmatrix} 1 & p(z) \\ 0 & 1 \end{pmatrix} \tag{5.1}$$

for some polynomial p. The matrix (5.1) can not be decomposed as W_1W_2 with $W_1, W_2 \in \mathcal{M}_{<\infty}$ and $\operatorname{ind}_- W = \operatorname{ind}_- W_1 + \operatorname{ind}_- W_2$ *differently than in the form $W = [WW_{(-l,0)}]W_{(l,0)}$ or $W = W_{(l,0)}[W_{(-l,0)}W]$.*

Proof. The assumption $W \star \infty = \infty$ just means that $W_{21} \equiv 0$. As $\det W = 1$, the functions W_{11} and W_{22} are zero-free and hence equal to e^{v_1} and e^{v_2}, respectively. Thereby $v_i^\# = v_i$ and $v_i(0) = 0$. Every entry of W is of finite exponential type, and therefore $v_1(z) = az$ (and thus $v_2(z) = -az$) for some $a \in \mathbb{R}$. Since every entry of W is of bounded type in $\mathbb{C}^+$, it follows that $a = 0$.

Since $W \star 0 \in \mathcal{N}_{<\infty}$, the function W_{12} can have only finitely many zeros. The same argument as above shows that W_{12} must be a polynomial.

If p is linear, the assertion is clear. Assume that the degree of p is at least 2. Then $\operatorname{ind}_- W > 0$. Consider the chain $(W_t)_{t\in\mathcal{J}}$ which goes downwards from W as constructed in [KW2, Theorem 7.1]. Its domain is of the form $(c_-,\xi_1)\cup(\xi_1,\xi_2)\cup\cdots\cup(\xi_m,0]$. The interval (c_-,ξ_1) must be indivisible of type 0 and infinite length since $(1,0)^T$ belongs to $\mathfrak{K}(W)$ and is neutral. Also $(\xi_m,0]$ must be indivisible of type 0 and infinite length since $\operatorname{ind}_- W \star \infty < \operatorname{ind}_- W$. Since $\lim_{t\nearrow\xi_1} W_t \circ = \lim_{t\searrow\xi_n} W_t \circ$ it follows from the results of [KW3, §5] on intermediate Weyl coefficients that $\xi_1 = \xi_n$. □

In the proof of Theorem 5.1 we mainly deal with the function $\mathcal{T}_{\checkmark} \circ \omega : \mathcal{I} \to \mathcal{M}_{<\infty}$.

Step 1: The function $\mathcal{T}_{\surd} \circ \omega$ satisfies (C1). This follows immediately from:

5.4. Lemma. *Let $\omega \in \mathfrak{M}_{<\infty}^{\text{sym}}$ and assume that ω satisfies (K_-). Then* $\operatorname{ind}_-(\mathcal{T}_{\surd}\circ\omega)(t)$ *is constant on each connected component of $\mathcal{I}$.*

Proof. Put $E_t := E_{\omega(t)}$, $t \in \mathcal{I}$. If $E_{t,+}$ and $E_{t,-}$ are defined by

$$B_{t,+}(z^2) := zB_t(z),\ A_{t,+}(z^2) := A_t(z),$$

$$B_{t,-}(z^2) := \frac{B_t(z)}{z},\ A_{t,-}(z^2) := A_t(z),$$

then, cf. Lemma 3.13, [KWW4, Proposition 4.9],

$$\mathfrak{P}(E_t)_e \cong \mathfrak{P}(E_{t,+}) \cong \mathfrak{K}((\mathcal{T}_{\surd} \circ \omega)(t)),\ \mathfrak{P}(E_t)_o \cong \mathfrak{P}(E_{t,-}).$$

In particular, we get

$$\operatorname{ind}_- \mathfrak{P}(E_t) = \operatorname{ind}_- \mathfrak{P}(E_{t,+}) + \operatorname{ind}_- \mathfrak{P}(E_{t,-}).$$

Recall that $\operatorname{ind}_- \mathfrak{P}(E_t)$ is constant on $\mathcal{I}_1$, say $\operatorname{ind}_- \mathfrak{P}(E_t) = \kappa_1$, $t \in \mathcal{I}_1$. We shall show that the set

$$M_\nu := \left\{t \in \mathcal{I}_1 : \operatorname{ind}_-(\mathcal{T}_{\surd} \circ \omega)(t) = \nu\right\}$$

is closed in $\mathcal{I}_1$. Let $(t_n)_{n\in\mathbb{N}}$, $t_n \in M_\nu$, and assume that $t_n \to t_0 \in \mathcal{I}_1$. Then

$$\operatorname{ind}_- \mathfrak{P}(E_{t_n,+}) = \operatorname{ind}_-(\mathcal{T}_{\surd} \circ \omega)(t) = \nu,\ n \in \mathbb{N}$$

and hence

$$\operatorname{ind}_- \mathfrak{P}(E_{t_n,-}) = \kappa_1 - \nu,\ n \in \mathbb{N}.$$

Recall that each set $\mathcal{H}B_{\leq\kappa}$ is closed with respect to locally uniform convergence. Since $\lim_{n\to\infty} \omega(t_n) = \omega(t_0)$ locally uniformly, it follows that also $\lim_{n\to\infty} E_{t_n,\pm} = E_{t_0,\pm}$. Therefore

$$\operatorname{ind}_- \mathfrak{P}(E_{t_0,+}) \leq \nu,\ \operatorname{ind}_- \mathfrak{P}(E_{t_0,-}) \leq \kappa_1 - \nu. \tag{5.2}$$

However, we must have

$$\operatorname{ind}_- \mathfrak{P}(E_{t_0,+}) + \operatorname{ind}_- \mathfrak{P}(E_{t_0,-}) = \kappa_1.$$

Hence in both inequalities (5.2) equality must hold, and we conclude that $t_0 \in M_\nu$.

Since $\mathcal{I}_1 = \bigcup_{\nu=0}^{\kappa_1} M_\nu$ and $\mathcal{I}_1$ is connected, it follows that all but one of the sets M_ν is empty, i.e., that $\operatorname{ind}_-(\mathcal{T}_{\surd} \circ \omega)(t)$ is constant on $\mathcal{I}_1$. □

Step 2: We show that $\mathcal{T}_{\surd} \circ \omega$ satisfies (W3).

Proof of Step 2. Let $t, s \in \mathcal{I}$, $t \leq s$, be given. Put $W_t := \omega(t)$, $\hat{W}_t := (\mathcal{T}_{\surd}\circ\omega)(t)$.

Case 1: Assume first that t is not contained in the interior of an indivisible interval of the chain ω. Then $\mathfrak{P}(E_{W_t}) \subseteq \mathfrak{P}(E_{W_s})$ isometrically, and hence also $\mathfrak{P}(E_{W_t})_e \subseteq \mathfrak{P}(E_{W_s})_e$. It follows that

$$\mathfrak{P}(E_{\hat{W}_t}) = \mathfrak{P}(E_{W_t})_+ \subseteq \mathfrak{P}(E_{W_s})_+ = \mathfrak{P}(E_{\hat{W}_s})$$

isometrically. By [KW1, Theorem 12.2] there exists a matrix

$$\hat{W}_{ts} \in \mathcal{M}_{<\infty},\ \operatorname{ind}_- \hat{W}_{ts} = \operatorname{ind}_- \mathfrak{P}(E_{\hat{W}_s}) - \operatorname{ind}_- \mathfrak{P}(E_{\hat{W}_t})$$

such that

$$(-B_{\hat{W}_s}, A_{\hat{W}_s}) = (-B_{\hat{W}_t}, A_{\hat{W}_t})\,\hat{W}_{ts}\,.$$

From our assumption $\mathfrak{K}_-(W_t) = \mathfrak{K}(W_t)$, [KW1, Corollary 10.4] and [KWW2, Proposition 2.6] we know that $1 \in \operatorname{Assoc}\mathfrak{P}(E_{\hat{W}_t}), \operatorname{Assoc}\mathfrak{P}(E_{\hat{W}_s})$. The matrices $\hat{W}_t$ and $\hat{W}_s$ are the unique matrices belonging to $\mathcal{M}_{\operatorname{ind}_-\mathfrak{P}(E_{\hat{W}_t})}$ and $\mathcal{M}_{\operatorname{ind}_-\mathfrak{P}(E_{\hat{W}_s})}$, respectively, with (cf. [KW1, Corollary 10.4])

$$\mathfrak{K}_-(\hat{W}_t) = \mathfrak{K}(\hat{W}_t),\ \mathfrak{K}_-(\hat{W}_s) = \mathfrak{K}(\hat{W}_s)\,, \tag{5.3}$$

and

$$(-B_{\hat{W}_t}, A_{\hat{W}_t}) = (0,1)\hat{W}_t,\ (-B_{\hat{W}_s}, A_{\hat{W}_s}) = (0,1)\hat{W}_s\,.$$

It follows that (cf. proof of [KW2, Theorem 7.1])

$$\hat{W}_s = \hat{W}_t\,\hat{W}_{ts}\,.$$

By (5.3) we have $\operatorname{ind}_-\mathfrak{P}(E_{\hat{W}_t}) = \operatorname{ind}_-\hat{W}_t$ and $\operatorname{ind}_-\mathfrak{P}(E_{\hat{W}_s}) = \operatorname{ind}_-\hat{W}_s$, respectively, i.e.,

$$\operatorname{ind}_-\hat{W}_{ts} = \operatorname{ind}_-\hat{W}_s - \operatorname{ind}_- W_t\,,$$

and we have proved the requirement of (W3) in the considered case.

Case 2: Next assume that t belongs to the interior of a maximal indivisible interval (t_-, t_+) of type 0. Then, by [KW3, Proposition 3.16], at least one of t_- and t_+ belongs to $\mathcal{I}$, say $t_- \in \mathcal{I}$. By Proposition 3.6 we have $\hat{W}_t = \hat{W}_{t_-}$ and hence the assertion follows from what we already proved.

Case 3: Finally consider the case that t belongs to a maximal indivisible interval (t_-, t_+) of type $\frac{\pi}{2}$ (regardless whether this indivisible interval is of positive, negative or infinite length). We divide the proof of this case into three subcases.

$s \leq t_+$: In this case we have $W_s = W_t W_{(l,\frac{\pi}{2})}$ for some $l \in \mathbb{R}\setminus\{0\}$. It follows from Proposition 3.6 that $\hat{W}_s = \hat{W}_t W_{(l',\phi)}$, i.e., $\hat{W}_{ts} = W_{(l',\phi)}$. Thereby $\phi \in (0,\pi)$ and $\operatorname{sgn} l' = \operatorname{sgn} l$.

Consider the space $\mathfrak{P}(E_{W_s})$. Since s is the right endpoint of an indivisible interval of type $\frac{\pi}{2}$ we have $A_{W_s} \in \mathfrak{P}(E_{W_s})$. It follows from the fact $\overline{\operatorname{dom} S} = \operatorname{span}\{A\}^\perp$ that $\mathfrak{P}(E_{W_s})_o \subseteq \overline{\operatorname{dom} S}$. The spaces $\mathfrak{P}(E_{W_s})$ and $\mathfrak{P}(E_{W_t})$ are equal as sets but not isometrically. However, on the subspace $\overline{\operatorname{dom} S}$ the inner products of $\mathfrak{P}(E_{W_t})$ and $\mathfrak{P}(E_{W_s})$ coincide. Hence $\mathfrak{P}(E_{W_t})_o = \mathfrak{P}(E_{W_s})_o$ isometrically. We have

$$\operatorname{ind}_-\mathfrak{P}(E_{W_s})_e + \operatorname{ind}_-\mathfrak{P}(E_{W_s})_o = \operatorname{ind}_-\mathfrak{P}(E_{W_s})$$

$$= \operatorname{ind}_-\mathfrak{P}(E_{W_t}) + \begin{cases} 0 & , l > 0 \\ 1 & , l < 0 \end{cases} = \operatorname{ind}_-\mathfrak{P}(E_{W_t})_e + \operatorname{ind}_-\mathfrak{P}(E_{W_t})_o + \begin{cases} 0 & , l > 0 \\ 1 & , l < 0 \end{cases}.$$

Since $\operatorname{sgn} l' = \operatorname{sgn} l$, $\operatorname{ind}_-\mathfrak{P}(E_{W_t})_o = \operatorname{ind}_-\mathfrak{P}(E_{W_s})_o$ and $\mathfrak{P}(E_{\hat{W}_t}) \cong \mathfrak{P}(E_{W_t})_e$, $\mathfrak{P}(E_{\hat{W}_s}) \cong \mathfrak{P}(E_{W_s})_e$, it follows that

$$\operatorname{ind}_-\mathfrak{P}(E_{\hat{W}_s}) = \operatorname{ind}_-\mathfrak{P}(E_{\hat{W}_s}) + \operatorname{ind}_-\hat{W}_{ts}.$$

$s > t_+, t_+ \in \mathcal{I}$: We decompose $\hat{W}_s$ as $\hat{W}_s = \hat{W}_{t_+}\hat{W}_{t_+s}$. Since t_+ is not contained in the interior of an indivisible interval, it follows from what we already proved that $\operatorname{ind}_- \hat{W}_s = \operatorname{ind}_- \hat{W}_{t_+} + \operatorname{ind}_- \hat{W}_{t_+s}$. Moreover, we decompose $\hat{W}_{t_+} = \hat{W}_t\hat{W}_{tt_+}$. By the above treated case also in this relation negative indices add up. Since $\operatorname{ind}_-(\hat{W}_{tt_+}\hat{W}_{t_+s}) \leq \operatorname{ind}_- \hat{W}_{tt_+} + \operatorname{ind}_- \hat{W}_{t_+s}$, it follows that also in the factorization $\hat{W}_s = \hat{W}_t(\hat{W}_{tt_+}\hat{W}_{t_+s})$ negative indices add up.

$s > t_+, t_+ \notin \mathcal{I}$: In this case we must have $t_- \in \mathcal{I}$. To shorten notation put $E_t := E_{W_t}$ and let $\hat{E}_t$, A_t, etc. be defined correspondingly.

The interval $[t_-, t]$ is in ω indivisible of type $\frac{\pi}{2}$ and positive length $l > 0$. Hence $A_{t_-} = A_t \in \mathfrak{P}(E_t)$. Since $t_+ \notin \mathcal{I}$, we have $[A_t, A_t]_{\mathfrak{P}(E_s)} = 0$. It follows that

$$\mathfrak{P}(E_t)_e \supsetneq \mathfrak{P}(E_{t_-})_e$$

with codimension 1 and hence also $\mathfrak{P}(\hat{E}_t) \supsetneq \mathfrak{P}(\hat{E}_{t_-})$ with codimension 1. The set $\mathfrak{P}(\hat{E}_t)$ endowed with the inner product inherited from $\mathfrak{P}(\hat{E}_s)$ is a degenerated dB-subspace $\mathfrak{P}$ of $\mathfrak{P}(\hat{E}_s)$, and we have

$$\mathfrak{P} = \mathfrak{P}(\hat{E}_{t_-})[\dot{+}] \operatorname{span}\left\{A_{t_-}(\sqrt{z})\right\}, \qquad \mathfrak{P}^\circ = \operatorname{span}\left\{A_{t_-}(\sqrt{z})\right\}.$$

Consider the space $\mathfrak{K}(\hat{W}_{t_-s})$.

By [KW2, Lemma 7.6] it contains a constant $(\cos\psi, \sin\psi)^T$. This constant has the property that the space

$$\mathfrak{P}(\hat{E}_{t_-})[\dot{+}] \operatorname{span}\left\{ -\hat{B}_{t_-}\cos\psi + \hat{A}_{t_-}\sin\psi\right\}$$

where

$$\left[-\hat{B}_{t_-}\cos\psi + \hat{A}_{t_-}\sin\psi, -\hat{B}_{t_-}\cos\psi + \hat{A}_{t_-}\sin\psi\right] = \left[\begin{pmatrix}\cos\psi\\ \sin\psi\end{pmatrix}, \begin{pmatrix}\cos\psi\\ \sin\psi\end{pmatrix}\right]_{\mathfrak{K}(\hat{W}_{t_-s})}$$

is a dB-subspace of $\mathfrak{P}(\hat{E}_s)$. It follows that

$$\mathfrak{P}(\hat{E}_{t_-})[\dot{+}] \operatorname{span}\left\{ -\hat{B}_{t_-}\cos\psi + \hat{A}_{t_-}\sin\psi\right\} = \mathfrak{P},$$

and hence that

$$\operatorname{span}\left\{ -\hat{B}_{t_-}\cos\psi + \hat{A}_{t_-}\sin\psi\right\} = \operatorname{span}\left\{A_{t_-}(\sqrt{z})\right\},$$

i.e.,

$$A_{t_-}(\sqrt{z}) = \lambda\big(-\hat{B}_{t_-}(z)\cos\psi + \hat{A}_{t_-}(z)\sin\psi\big) \tag{5.4}$$

for some $\lambda \in \mathbb{C}$, and that

$$\left[\begin{pmatrix}\cos\psi\\ \sin\psi\end{pmatrix}, \begin{pmatrix}\cos\psi\\ \sin\psi\end{pmatrix}\right]_{\mathfrak{K}(\hat{W}_{t_-s})} = 0.$$

From the definition of the transformation $\mathcal{T}_{\sqrt{}}$ we obtain

$$\hat{B}_{t_-}(z^2) = zB_{t_-}(z),\ \hat{A}_{t_-}(z^2) = A_{t_-}(z) + w'_{t_-,12}(0)zB_{t_-}(z).$$

It follows that

$$A_{t_-}(\sqrt{z}) = \hat{A}_{t_-}(z) - w'_{t_-,12}(0)\hat{B}_{t_-}(z).$$

Comparing this with (5.4) we obtain $\psi = \operatorname{Arccot} w'_{t_-,12}(0)$.

Proposition 3.6 shows that $\hat{W}_{t\_t} = W_{(l',\psi)}$ with some $l' > 0$. Hence the constant $(\cos\psi, \sin\psi)^T$ belongs to $\mathfrak{K}(\hat{W}_{t\_s})$ as well as to $\mathfrak{K}(\hat{W}_{t\_t}^{-1})$. In the first space it is neutral, in the second (one-dimensional) space it is negative. It follows (cf. [ADRS]) that

$$\operatorname{ind}_- \hat{W}_{t\_t}^{-1}\hat{W}_{t\_s} = \operatorname{ind}_- \hat{W}_{t\_s}\,.$$

We have $\operatorname{ind}_- \hat{W}_t = \operatorname{ind}_- \hat{W}_{t\_}$ and hence $(\hat{W}_{ts} = W_t^{-1}W_s = W_{t\_t}^{-1}W_{t\_s})$

$$\operatorname{ind}_- \hat{W}_s = \operatorname{ind}_- \hat{W}_{t\_} + \operatorname{ind}_- \hat{W}_{t\_s} = \operatorname{ind}_- \hat{W}_t + \operatorname{ind}_- \hat{W}_{ts}\,.$$

We have seen that in any case in the relation $W_s = W_t W_{ts}$ negative indices add. □

Step 3: We show that $\mathcal{T}_{\surd} \circ \omega$ satisfies (C2).

Proof of Step 3. Also this proof is divided into several cases. We use again the notation $\omega(t) = W_t$ and $(\mathcal{T}_{\surd} \circ \omega)(t) = \hat{W}_t$.

Case 1: ω ends with an indivisible interval $(m, \sup\mathcal{I})$ of type 0.

Then $\hat{W}_t = \hat{W}_m$, $t \in (m, \sup\mathcal{I})$, and hence

$$\limsup_{t\nearrow\sup\mathcal{I}} \mathfrak{t}(\hat{W}_t) = \mathfrak{t}(\hat{W}_m) < +\infty$$

and

$$\lim_{t\nearrow\sup\mathcal{I}} \hat{W}_t = \hat{W}_m\,.$$

We are therefore in case (a) of (C2). By the already proved property (W3) we have

$$\operatorname{ind}_- \hat{W}_t \le \operatorname{ind}_- \hat{W}_m,\ t \in \mathcal{I}, t \le m\,.$$

For $t \in (m, \sup\mathcal{I})$ trivially $\operatorname{ind}_- \hat{W}_t = \operatorname{ind}_- \hat{W}_m$. Hence the implication (a) of (C2) holds true.

Case 2: ω ends neither with an indivisible interval of type zero nor is the last component $\mathcal{I}_\infty$ of ω of the form $(m_-, m] \cup [m, \sup\mathcal{I})$ with an indivisible interval (m_-, m) of type 0 and an indivisible interval $(m, \sup\mathcal{I})$ of type $\frac{\pi}{2}$.

We claim that there exist $s, t \in \hat{\mathcal{I}}_\infty := \{u \in \mathcal{I} : \operatorname{ind}_- \hat{W}_u = \max_{v\in\mathcal{I}} \operatorname{ind}_- \hat{W}_v\}$ such that $\hat{W}_{st}$ $(:= \hat{W}_s^{-1}\hat{W}_t)$ is not a linear polynomial. This excludes the occurrence of case (b) of (C2). In case (a) we appeal to [KW3, Lemma 3.4] to obtain the validity of the desired implication.

In order to establish our claim assume on the contrary that for some $\phi \in [0, \pi)$ we have $\hat{W}_{st} = W_{(l(s,t),\phi)}$ whenever $s, t \in \hat{\mathcal{I}}_\infty$. By Lemma 5.4 this relation holds especially for $s, t \in \mathcal{I}_\infty = \{u \in \mathcal{I} : \operatorname{ind}_- W_u = \max_{v\in\mathcal{I}} \operatorname{ind}_- W_v\}$.

$\phi = 0$: Choose $s, t \in \mathcal{I}_\infty$, $s < t$, such that W_{st} is not a linear polynomial. By Remark 3.7 we have for some $\alpha \in \mathbb{R}$

$$W_{st} = \begin{pmatrix} 1 & \alpha z + l(s,t)z^3 \\ 0 & 1 \end{pmatrix}.$$

Since W_{st} is not a linear polynomial we must have $l(s,t) \neq 0$. Thus $\operatorname{ind}_- W_{st} > 0$, a contradiction.

$\phi \neq 0$: We conclude from Remark 3.7 that for all $s,t \in \mathcal{I}_\infty$, $s<t$,

$$W_{st} = W_{(l_1,0)} W_{(l_2,\frac{\pi}{2})} W_{(l_3,0)}$$

with some $l_i \geq 0$. By Lemma 5.2 we must be in one of the following situations:

$$\mathcal{I}_\infty = \begin{cases} (m_0,m_1] \cup [m_1,m_2] \cup [m_2,m_3) & \text{of types } 0, \frac{\pi}{2}, 0 \\ (m_0,m_1] \cup [m_1,m_2) & \text{of types } \frac{\pi}{2}, 0 \\ (m_0,m_1] \cup [m_1,m_2) & \text{of types } 0, \frac{\pi}{2} \end{cases}$$

Again a contradiction since these are exactly those cases we do not consider in the present step of the proof.

Case 3: The last component $\mathcal{I}_\infty$ of ω is equal to $(m_0,m_1] \cup [m_1,m_2)$ with an indivisible interval $(m_0,m_1]$ of type 0 and an indivisible interval $[m_1,m_2)$ of type $\frac{\pi}{2}$.

In this case the interval (m_0,m_2) is indivisible in $(\hat{W}_t)_{t\in\mathcal{I}}$ of some type $\psi \in (0,\pi)$. It follows that $\lim_{t\nearrow\sup\mathcal{I}} \mathfrak{t}(\hat{W}_t) = +\infty$, which rules out the occurrence of case (a) in (C2).

Fix $t_0 \in (m_1,m_2)$. Since $[m_1,t_0]$ is indivisible of type $\frac{\pi}{2}$ in $(W_t)_{t\in\mathcal{I}}$ and has positive length, we obtain that

$$A_{m_1} \in \mathfrak{P}(E_{t_0}),\ [A_{m_1}, A_{m_1}]_{\mathfrak{P}(E_{t_0})} > 0\,.$$

As we saw in the proof of (W3), the linear combination $\hat{S}_\psi$ of $\hat{A}_{t_0}$ and $\hat{B}_{t_0}$ which belongs to the space $\mathfrak{P}(\hat{E}_{t_0})$ is linearly dependent with $A_{m_1}(\sqrt{z})$ and hence

$$[\hat{S}_\psi, \hat{S}_\psi]_{\mathfrak{P}(\hat{E}_{t_0})} > 0\,.$$

It follows from [KW2, Lemma 5.12] that

$$\operatorname{ind}_- \hat{W}_{t_0} \star \cot\psi = \operatorname{ind}_- \hat{W}_{t_0}\,.$$

We conclude that the implication (b) of (C2) holds true. □

We have established that $\mathcal{T}_{\surd} \circ \omega$ can be extended to a maximal chain ϖ by means of Lemma 4.2. If $\lim_{t\nearrow M} \mathfrak{t}((\mathcal{T}_{\surd}\circ\omega)(t)) = +\infty$, this extension is unique and satisfies

$$q_\infty(\varpi) = \lim_{t\nearrow M} (\mathcal{T}_{\surd} \circ \omega(t)) \star \infty\,.$$

By (3.4) this yields

$$z q_\infty(\varpi)(z^2) = q_\infty(\omega)(z)\,. \tag{5.5}$$

If $\lim_{t\nearrow M} \mathfrak{t}((\mathcal{T}_{\surd}\circ\omega)(t)) < +\infty$, so that the extension of $\mathcal{T}_{\surd}\circ\omega$ is not unique, we should choose the parameter $\tau = \infty$ in Lemma 4.2 in order to achieve the relation (5.5). The fact that this choice is permitted needs justification. However, we saw in the proof of Step 3 that there exist $s,t \in \mathcal{I}_\infty$ such that $\hat{W}_{st}$ is not of the form $W_{(l,0)}$. As indicated in the proof of [KW2, Lemma 8.5] this implies that $\operatorname{ind}_-[\lim_{t\nearrow M} \mathcal{T}_{\surd}\circ\omega(t)] \star \infty = \operatorname{ind}_- \lim_{t\nearrow M} \mathcal{T}_{\surd}\circ\omega(t)$.

Let ϖ be the maximal chain with $zq_\infty(\varpi)(z^2) = q_\infty(\omega)(z)$. Due to the previous steps there exists a nondecreasing function $\hat{\mu} : \mathcal{I} \to \operatorname{dom}\varpi$ such that $\mathcal{T}_{\surd}\circ\omega = \varpi\circ\hat{\mu}$.

Step 4: There exists a surjective function $\mu : \operatorname{dom}\omega_{\surd} \to \operatorname{dom}\varpi$ such that $\omega_{\surd} = \varpi \circ \mu$.

Proof of Step 4. Write $\mathcal{I} = (\sigma_0, \sigma_1) \cup \cdots \cup (\sigma_n, \sigma_{n+1})$. By Lemma 5.4 for each i the set $\hat{\mu}((\sigma_i, \sigma_{i+1}))$ is contained in one connected component of $\operatorname{dom}\varpi$. As it is seen from (3.3), the function $(\mathfrak{t} \circ \varpi)(\hat{\mu}(t))$ depends continuously on $t \in (\sigma_i, \sigma_{i+1})$. Thus $\hat{\mu}$ is continuous, and it follows that $\hat{\mu}((\sigma_i, \sigma_{i+1}))$ is an interval. Put

$$\xi_{i,-} := \lim_{t \to \sigma_i -} \hat{\mu}(t), \ \xi_{i,+} := \lim_{t \to \sigma_i +} \hat{\mu}(t) .$$

Since $\mathcal{T}_{\surd}(W)$ depends continuously on W and $\mathcal{T}_{\surd}(I) = I$, we have $\xi_{0,+} = 0$.

We have $\xi_{n+1,-} = \sigma_{n+1}$ if and only if $\lim_{t \nearrow \sigma_{n+1}} \mathfrak{t}((\mathcal{T}_{\surd} \circ \omega)(t)) = +\infty$. Otherwise, by (5.5) and the definition of ϖ, we have

$$q_\infty(\varpi) = \lim_{t \nearrow \sigma_{n+1}} (\mathcal{T}_{\surd} \circ \omega)(t) \star \infty = \varpi(\xi_{n+1,-}) \star \infty .$$

This implies that the interval $(\xi_{n+1,-}, \sup\operatorname{dom}\varpi)$ is indivisible of type 0 in ϖ.

Consider a point σ_i, $i \in \{1, \dots, n\}$. The essential observation for the following proof is that, by (3.4) and the existence of intermediate Weyl-coefficients (see [KW3, §5]), the function $(\mathcal{T}_{\surd} \circ \omega) \star \infty$ has a continuous extension to $[\sigma_0, \sigma_{n+1}]$.

We divide cases similar as in the definition of ς_i.

$\xi_{i,-}, \xi_{i,+} \notin \operatorname{dom}\varpi$: Denote by $q_{\xi_{i,\pm}}$ the intermediate Weyl coefficient of ϖ at the singularity $\xi_{i,\pm}$. We have

$$q_{\xi_{i,-}} = \lim_{s \to \xi_{i,-} -} \varpi(s) \star \infty = \lim_{t \to \sigma_i -} (\mathcal{T}_{\surd} \circ \omega)(t) \star \infty = \lim_{t \to \sigma_i +} (\mathcal{T}_{\surd} \circ \omega)(t) \star \infty =$$
$$= \lim_{s \to \xi_{i,+} +} \varpi(s) \star \infty = q_{\xi_{i,+}} ,$$

and hence $\xi_{i,-} = \xi_{i,+}$.

$\xi_{i,-} \in \operatorname{dom}\varpi, \xi_{i,+} \notin \operatorname{dom}\varpi$: We have

$$q_{\xi_{i,+}} = \lim_{t \to \sigma_i +} (\mathcal{T}_{\surd} \circ \omega)(t) \star \infty = \lim_{t \to \sigma_i -} (\mathcal{T}_{\surd} \circ \omega)(t) \star \infty = \varpi(\xi_{i,-}) \star \infty .$$

Since $q_{\xi_{i,+}}$ is the Weyl-coefficient of the maximal chain $\varpi|_{\operatorname{dom}\varpi \cap (0, \xi_{i,+})}$, it follows that $(\xi_{i,-}, \xi_{i,+})$ is indivisible of type 0 and infinite length.

$\xi_{i,-} \notin \operatorname{dom}\varpi, \xi_{i,+} \in \operatorname{dom}\varpi$: The same argument as above yields $q_{\xi_{i,-}} = \varpi(\xi_{i,+}) \star \infty$ and hence that $\operatorname{ind}_- \varpi(\xi_{i,+}) \star \infty < \operatorname{ind}_- \varpi(\xi_{i,+})$. This implies that the interval $(\xi_{i,-}, \xi_{i,+})$ is indivisible of type 0 and infinite length.

$\xi_{i,-} \in \operatorname{dom}\varpi, \xi_{i,+} \in \operatorname{dom}\varpi$: In this case we find $\varpi(\xi_{i,-}) \star \infty = \varpi(\xi_{i,+}) \star \infty$ and hence, by Lemma 5.3,

$$\varpi(\xi_{i,-})^{-1} \varpi(\xi_{i,+}) = \begin{pmatrix} 1 & p(z) \\ 0 & 1 \end{pmatrix}$$

If $\operatorname{ind}_- \varpi(\xi_{i,-}) = \operatorname{ind}_- \varpi(\xi_{i,+})$, the interval $(\xi_{i,-}, \xi_{i,+})$ is indivisible of type 0 and positive length

$$\mathfrak{t}(\lim_{t \searrow \sigma_i} \mathcal{T}_{\surd}(\omega(t))) - \mathfrak{t}(\lim_{t \nearrow \sigma_i} \mathcal{T}_{\surd}(\omega(t))) .$$

If $\operatorname{ind}_- \varpi(\xi_{i,-}) < \operatorname{ind}_- \varpi(\xi_{i,+})$, there exists a singularity ξ of ϖ in the interval $(\xi_{i,-}, \xi_{i,+})$. By Lemma 5.3 the intervals $(\xi_{i,-}, \xi)$ and $(\xi, \xi_{i,+})$ are indivisible of type 0 and infinite length.

The case $\operatorname{ind}_- \varpi(\xi_{i,-}) > \operatorname{ind}_- \varpi(\xi_{i,+})$ can not occur.

We saw that in any case $\varpi|_{(\xi_{i,-},\xi_{i,+})} \sim \varsigma_i$. The required function μ is now defined in the obvious way. □

We have constructed the function μ required in Theorem 5.1. To complete the proof of Theorem 5.1 choose for λ a right inverse of μ. ✌

5.5. Corollary. *Let $\omega \in \mathfrak{M}_\kappa^{\mathrm{sym}}$ and assume that ω satisfies (K$_-$). Then the maximal chain ϖ with $zq_\infty(\varpi)(z^2) = q_\infty(\omega)(z)$ belongs to $\mathfrak{M}_{\leq\kappa}^{\mathrm{ep}}$ and also satisfies (K$_-$).*

Proof. We know from Theorem 5.1 that $\varpi = \omega_{\surd} \circ \lambda$. By (3.4) the number of zeros of $\mathcal{T}_{\surd}(\omega(t))_{21}$ in $\mathbb{C} \setminus [0,\infty)$ equals the number of non-real zeros of $\omega(t)_{21}$ and is therefore, by Corollary 2.10, bounded uniformly by $\operatorname{ind}_- \omega$. On indivisible intervals of type 0 the entry $\varpi(s)_{21}$ is constant. Hence the number of zeros of $\varpi(s)_{21}$ in $\mathbb{C} \setminus [0,\infty)$ is bounded independently of s. By Corollary 2.10, (ii), the maximum number of zeros that any entry of $\varpi(s)$ can have in $\mathbb{C}\setminus[0,\infty)$ is bounded independently of s.

The fact that ϖ satisfies (K$_-$) readily follows from the relation of Weyl coefficients and [KW2, Theorem 5.7]. □

We deduce an inverse result for the class $\mathfrak{M}_{<\infty}^{\mathrm{ep}}$, the analogue to Proposition 4.4.

5.6. Proposition. *Let ϖ be a maximal chain. Then $\varpi \in \mathfrak{M}_{<\infty}^{\mathrm{ep}}$ if and only if $q_\infty(\varpi) \in \mathcal{N}_{<\infty}^{\mathrm{ep}}$.*

Proof. Assume first that $\varpi \in \mathfrak{M}_{<\infty}^{\mathrm{ep}}$. For every $t \in \operatorname{dom}\varpi$ the function $\varpi(t) \star \infty$ belongs to $\mathcal{N}_{\leq \operatorname{ind}_- \varpi}^{\mathrm{ep}}$. Moreover, the number of poles $\gamma(\varpi(t) \star \infty)$ of $\varpi(t) \star \infty$ which are located in $\mathbb{C} \setminus [0,\infty)$ is equal to the number of zeros of $\varpi(t)_{21}$. Hence it is bounded independently of $t \in \operatorname{dom}\varpi$. We have $\lim_{t\to\sup\operatorname{dom}\varpi} \varpi(t)\star\infty = q_\infty(\varpi)$ and hence [KWW2, Proposition 4.10] implies that $q_\infty(\varpi) \in \mathcal{N}_{<\infty}^{\mathrm{ep}}$.

Conversely, let ϖ be given such that $q_\infty(\varpi) \in \mathcal{N}_{<\infty}^{\mathrm{ep}}$ and assume first that additionally $\lim_{x\to-\infty} q_\infty(\varpi)(x) = 0$. By [KWW2, Theorem 4.1] the function $zq_\infty(\varpi)(z^2)$ belongs to $\mathcal{N}_{<\infty}^{\mathrm{sym}}$. Let $\omega \in \mathfrak{M}_{<\infty}^{\mathrm{sym}}$ be the maximal chain with $q_\infty(\omega) = zq_\infty(\omega)(z^2)$. Since

$$\lim_{y\to\infty} \frac{1}{iy} q_\infty(\omega)(iy) = \lim_{y\to\infty} q_\infty(\varpi)(-y^2) = 0\,,$$

we know that ω satisfies (K$_-$). Corollary 5.5 implies that $\varpi \in \mathfrak{M}_{<\infty}^{\mathrm{ep}}$.

The general case is reduced to the already proved particular instance by a possible application of one of the transforms $\mathcal{T}_J$ or $\mathcal{T}_\alpha$ of [KW2, Section 10]. □

5.7. *Remark.* Note that Proposition 5.6 could most likely also be deduced without help of the transformation $\mathcal{T}_{\surd}$ by employing the the theory of isometric embeddings of dB-spaces into 'L^2-spaces' induced by distributions associated to integral representations of generalized Nevanlinna functions. For this not yet fully developed theory see [KW2, §4,§6], [KWW2, §5].

5.8. Corollary. *Let $\varpi \in \mathfrak{M}^{\mathrm{ep}}_{<\infty}$ and assume that $\lim_{x\to-\infty} q_\infty(\varpi)(x) = 0$. Then there exist only finitely many indivisible intervals of type 0 in ϖ. In fact, the number of such intervals is bounded by $2\,\mathrm{ind}_-\varpi + 1$.*

Proof. We know that $\varpi = \omega_{\surd} \circ \lambda$ where $\omega : \mathcal{I} \to \mathcal{M}_{<\infty}$, $\mathcal{I} = (\sigma_0, \sigma_1) \cup \dots \cup (\sigma_n, \sigma_{n+1})$, is the maximal chain with $q_\infty(\omega)(z) = z q_\infty(\varpi)(z^2)$. By Remark 3.7 the parts $(\mathcal{T}_{\surd} \circ \omega)|_{(\sigma_i,\sigma_{i+1})}$ of $\omega_{\surd}$ can not contain indivisible intervals of type 0. Thus ϖ contains at most $2n + 1$ indivisible intervals of type 0. □

5.2. The inverse transformation

We employ the rather detailed discussion of Theorem 5.1 to define and investigate the inverse of the square root transformation.

Let $\varpi \in \mathfrak{M}^{\mathrm{ep}}_{<\infty}$. Then by [KWW2, Proposition 4.9] a function $\Lambda : \mathrm{dom}\,\varpi \to \mathbb{R} \cup \{\pm\infty\}$ is well defined by

$$\Lambda(t) := -\lim_{x\to-\infty} \frac{\varpi(t)_{22}(x)}{\varpi(t)_{21}(x)}\,.$$

The following lemma will play a central role.

5.9. Lemma. *Let $\varpi \in \mathfrak{M}^{\mathrm{ep}}_{<\infty}$ and assume that $\lim_{x\to\infty} q_\infty(\varpi)(x) = 0$. Then there exist pairwise disjoint open intervals (a_k, b_k), $k = 0, \dots, n$, with $a_k < b_k \le a_{k+1}$, such that*

$$\mathrm{dom}\,\varpi \setminus \bigcup \big\{[t_-, t_+] : (t_-, t_+) \textit{ maximal indivisible of type } 0 \textit{ in } \varpi\big\}$$

$$= \bigcup_{k=0}^{n} (a_k, b_k) \cup \big\{a_k : a_k \in \mathrm{dom}\,\varpi, a_k = b_{k-1}\big\}$$

and that the function Λ has the following properties:

(i) *$\Lambda|_{(a_k,b_k)}$ is finite, nondecreasing and continuous from the left.*

(ii) *If $\alpha \in \{a_k : a_k \in \mathrm{dom}\,\varpi, a_k = b_{k-1}\}$, then either $\lim_{t\to\alpha\pm} \Lambda(t) \in \mathbb{R}$ and $\lim_{t\to\alpha+} \Lambda(t) - \lim_{t\to\alpha-} \Lambda(t) < 0$, or $\limsup_{t\to\alpha} |\Lambda(t)| = \infty$.*

The intervals (a_k, b_k) are uniquely determined by these properties.

For each k the function $\tau_k(t) := \Lambda(t) - \varpi(t)'_{21}(0)$, $t \in (a_k, b_k)$, is strictly increasing and continuous from the left.

The proof of this lemma will also show how the inverse of the square root transformation acts. Let us describe this action precisely.

Let $\varpi \in \mathfrak{M}^{\mathrm{ep}}_{<\infty}$, assume that $\lim_{x\to\infty} q_\infty(\varpi)(x) = 0$, and let (a_k, b_k) be the intervals from the above lemma. For each $k \in \{1, \dots, n\}$ define $\varpi_k : \mathbb{R} \to \mathcal{M}_{<\infty}$ as follows: If $t \in \mathrm{ran}\,\tau_k$, put

$$\varpi_k(t) := \big(\mathcal{T}_{2,\Lambda(t)} \circ \varpi\big)(\tau_k^{-1}(t))\,.$$

Since τ_k is strictly increasing and continuous from the left, we have

$$\big(\inf \operatorname{ran} \tau_k, \sup \operatorname{ran} \tau_k\big) = \operatorname{ran} \tau_k \dot\cup \bigcup_l (\alpha_l, \beta_l] ,$$

with at most countably many pairwise disjoint intervals $(\alpha_l, \beta_l]$. If $t \in (\alpha_l, \beta_l]$ define

$$\varpi_k(t) := \begin{cases} \big[\lim_{\substack{t \nearrow \alpha_l \\ t \in \operatorname{ran} \tau_k}} \varpi_k(t)\big] \cdot W_{(t-\alpha_l,0)} & , t \in (\alpha_l, \beta_l] \\ \big[\lim_{\substack{t \nearrow \sup \operatorname{ran} \tau_k \\ t \in \operatorname{ran} \tau_k}} \varpi_k(t)\big] \cdot W_{(t-\sup \operatorname{ran} \tau_k,0)} & , t \in (\sup \operatorname{ran} \tau_k, \infty) \\ \big[\lim_{\substack{t \searrow \inf \operatorname{ran} \tau_k \\ t \in \operatorname{ran} \tau_k}} \varpi_k(t)\big] \cdot W_{(t-\inf \operatorname{ran} \tau_k,0)} & , t \in (-\infty, t, \inf \operatorname{ran} \tau_k) \end{cases}$$

For $k = 0$ we define a function $\varpi_0 : (0, \infty) \to \mathcal{M}_{<\infty}$ in exactly the same manner.

5.10. Theorem. *Let* $\varpi \in \mathfrak{M}^{\text{ep}}_{<\infty}$ *and assume that* $\lim_{x\to\infty} q_\infty(\varpi)(x) = 0$. *Then*

$$\omega := \varpi_0 \uplus \varpi_1 \uplus \cdots \uplus \varpi_n \in \mathfrak{M}^{\text{sym}}_{<\infty} ,$$

and

$$q_\infty(\omega)(z) = z q_\infty(\varpi)(z^2) .$$

Proof of Lemma 5.9 *and Theorem* 5.10. We use the same notation as in the first paragraph of the proof of Step 4. Consider the continuous and nondecreasing function $\hat\mu : \operatorname{dom} \omega \to \operatorname{dom} \varpi$. Put

$$\mathcal{L} := \operatorname{dom} \omega \setminus \bigcup \big\{(t_-, t_+] : (t_-, t_+) \text{ maximal indivisible of type } 0\big\} .$$

Then, by Lemma 3.4, $\hat\mu|_{\mathcal{L}}$ is injective. The set $\mathcal{L}$ has the property that it is closed in $\operatorname{dom} \omega$ with respect to monotonically increasing limits. We show that the function $(\hat\mu|_{\mathcal{L}})^{-1}$ is continuous from the left: Assume that $s_n \nearrow s$, $s_n, s \in \hat\mu(\mathcal{L})$, and put $t_n := (\hat\mu|_{\mathcal{L}})^{-1}, t := (\hat\mu|_{\mathcal{L}})^{-1}$. Then $(t_n)_{n\in\mathbb{N}}$ is increasing and bounded above by t. Hence $\lim_{n\to\infty} t_n =: t_0$ exists and belongs to $\operatorname{dom} \omega$. Therefore it belongs to $\mathcal{L}$. We have

$$\hat\mu(t_0) = \lim_{n\to\infty} \hat\mu(t_n) = s = \hat\mu(t) ,$$

and therefore $t_0 = t$.

For $k = 1, \dots, n$ we define numbers $a_k := \xi_{k,+}$ and $b_k := \xi_{k+1,-}$. Then

$$(a_k, b_k) \subseteq \hat\mu\big((\sigma_k, \sigma_{k+1})\big) \subseteq [a_k, b_k] .$$

By Remark 3.7 (a_k, b_k) does not contain any indivisible interval of type 0. Moreover, by Step 4, every interval in $\operatorname{dom} \omega$ which has empty intersection with $\bigcup_{k=0}^n (a_k, b_k)$ is indivisible of type 0. It follows that

$$\operatorname{dom} \varpi = \bigcup_{k=0}^n (a_k, b_k) \dot\cup \big\{a_k : a_k \in \operatorname{dom} \varpi, a_k = b_{k-1}\big\} \dot\cup$$

$$\dot\cup \bigcup \big\{[t_-, t_+] : (t_-, t_+) \text{ maximal indivisible of type } 0\big\} ,$$

in particular $\bigcup_{k=0}^n (a_k, b_k) \subseteq \mathcal{L}$.

Let $t \in \mathcal{L}$. Then t is not the right endpoint of an indivisible interval of type 0 and hence $\omega(t)_{21} \notin \mathfrak{P}(E_{\omega(t)})$. Hence

$$\lim_{y\to+\infty} \frac{1}{iy}\frac{\omega(t)_{22}(iy)}{\omega(t)_{21}(iy)} = 0\,,$$

and it follows from (3.4) that

$$\omega(t)_{12}'(0) = (\Lambda \circ \hat{\mu})(t),\ t \in \mathcal{L}\,.$$

It readily follows that Λ is real and nondecreasing on (a_k, b_k). Moreover, since $\omega(t)_{12}'(0)$ depends continuously on t, it follows from

$$\Lambda(t) = \omega\big((\hat{\mu}|_{\mathcal{L}})^{-1}\big)_{12}'(0),\ t \in (a_k, b_k)\,,$$

that Λ is continuous from the left. By Lemma 3.3 we also obtain

$$\big(\mathcal{T}_{2,\Lambda(t)} \circ \varpi\big)(t) = \big(\omega \circ (\hat{\mu}|_{\mathcal{L}}^{-1})\big)(t),\ t \in (a_k, b_k)\,.$$

Since for all $t \in \operatorname{dom}\omega$ the relation $\omega(t)_{21}'(0) = \varpi(t)_{21}'(0)$ holds, we see that

$$\tau_k(t) = \big(\mathfrak{t} \circ \omega \circ (\hat{\mu}|_{\mathcal{L}}^{-1})\big)(t),\ t \in (a_k, b_k)\,.$$

Thus τ_k is nondecreasing, injective and continuous from the left.

We conclude from the above discussion that, what is missing from $\varpi_k|_{\operatorname{ran}\tau_k}$ to all of ω are just indivisible intervals of type 0. However, the definition of ϖ_k on $\mathbb{R} \setminus \operatorname{ran}\tau_k$ just fills in indivisible intervals of type 0. Note that

$$\mathfrak{t}\big[\big(\mathcal{T}_{2,\Lambda(t)} \circ \varpi\big)(\tau_k^{-1}(t))\big] = t,\ t \in \operatorname{ran}\tau_k\,.$$

Thus the filled in indivisible intervals have the proper length. □

5.11. *Remark.* We know from [KWW2, Corollary 3.4, Proposition 4.7] that, if $q \in \mathcal{N}_{<\infty}$, then $zq(z) \in \mathcal{N}_{<\infty}$ if and only if $q \in \mathcal{N}_{<\infty}^{\mathrm{ep}}$. Under an additional regularity condition on q, the previous results lead to an explicit method to construct the maximal chain with Weyl-coefficient $zq(z)$ out of the one with Weyl-coefficient q.

To see this recall from [KW2, Lemma 10.1] that, if $\omega \in \mathfrak{M}_\kappa$ and if we define $\upsilon(t) := -J\omega(t)J$, then $\upsilon \in \mathfrak{M}_\kappa$ and the respective Weyl-coefficients are related by $q_\infty(\upsilon) = -q_\infty(\omega)^{-1}$. Denote this transformation by $\mathcal{T}_J$, let $\mathcal{T}_{\surd}$ be the square root transformation and $\mathcal{T}_2$ its inverse as studied above. Then for each $q \in \mathcal{N}_{<\infty}^{\mathrm{ep}}$ we have

$$q(z) \overset{\mathcal{T}_2}{\rightsquigarrow} zq(z^2) \overset{\mathcal{T}_J}{\rightsquigarrow} \frac{-1}{zq(z^2)} \overset{\mathcal{T}_{\surd}}{\rightsquigarrow} \frac{-1}{zq(z)} \overset{\mathcal{T}_J}{\rightsquigarrow} zq(z)$$

In order to justify the application of Theorem 5.1 and Theorem 5.10, we have to assume that

$$\lim_{x\to-\infty} q(x) = \lim_{x\to-\infty} \frac{1}{xq(x)} = 0\,.$$

This just means that the chain ω whose Weyl-coefficient is $zq(z^2)$ does not start with an indivisible interval.

The same procedure can be applied in order to construct the chain with Weyl-coefficient $z^{-1}q(z)$ out of the chain corresponding to $q(z) \in \mathcal{N}_{<\infty}^{\mathrm{ep}}$. This is exactly what will be needed in the discussion of generalized strings. In fact, under the assumption that

$$\lim_{x\to-\infty}\frac{1}{q(x)}=\lim_{x\to-\infty}\frac{q(x)}{x}=0\,,$$

we have

$$q(z)\overset{\mathcal{T}_J}{\rightsquigarrow}\frac{-1}{q(z)}\overset{\mathcal{T}_2}{\rightsquigarrow}\frac{-z}{q(z^2)}\overset{\mathcal{T}_J}{\rightsquigarrow}\frac{q(z^2)}{z}\overset{\mathcal{T}_{\surd}}{\rightsquigarrow}\frac{q(z)}{z}$$

In both cases the essential part of the sequence is $\mathcal{T}_{\surd}\circ\mathcal{T}_J\circ\mathcal{T}_2$.

6. Evolution of singularities

In the following discussion we recall some facts and notions on singularities of maximal chains from [KW3]. If $\omega\in\mathfrak{M}_{<\infty}$, $\operatorname{dom}\omega=(0,\sigma_1)\cup(\sigma_1,\sigma_2)\cup\cdots\cup(\sigma_n,M)$, then the numbers σ_i are called the *singularities of the chain* ω. One reason is that $\lim_{t\to\sigma_i}|\mathfrak{t}(\omega(t))|=\infty$, another one is that the σ_i are exactly the points of increase of $\operatorname{ind}_-\omega(t)$. A first characteristic value attached to a singularity is

$$\kappa(\sigma_i):=\lim_{t\searrow\sigma_i}\operatorname{ind}_-\omega(t)-\lim_{t\nearrow\sigma_i}\operatorname{ind}_-\omega(t)\in\mathbb{N}\,.$$

Another characteristic of a singularity is whether or not there is an indivisible interval to the left or to the right of it. Put

$$\sigma_i^+:=\sup\big(\{t\in\mathcal{I}:\,(\sigma_i,t)\text{ indivisible}\}\cup\{\sigma_i\}\big),$$

$$\sigma_i^-:=\inf\big(\{t\in\mathcal{I}:\,(t,\sigma_i)\text{ indivisible}\}\cup\{\sigma_i\}\big)\,.$$

We call σ_i of *polynomial type*, if $\sigma_i^-<\sigma_i<\sigma_i^+$. Moreover, σ_i is called *left dense*, *right dense* or *dense*, if $\sigma_i^-=\sigma_i<\sigma_i^+$, $\sigma_i^-<\sigma_i=\sigma_i^+$ or $\sigma_i^-=\sigma_i=\sigma_i^+$, respectively.

A deeper insight in the structure of a singularity is obtained by considering the chain of dB-spaces associated with the chain ω. If ω satisfies (K_-), then $\mathfrak{K}(\omega(t))\cong\mathfrak{K}_-(\omega(t))\cong\mathfrak{P}(E_{\omega(t)})$. Moreover, if $s,t\in\operatorname{dom}\omega$, $s\leq t$, then $\mathfrak{P}(E_{\omega(s)})\subseteq\mathfrak{P}(E_{\omega(t)})$ and this inclusion is isometric unless for some $\epsilon>0$ the interval $(s-\epsilon,t)$ is indivisible. Conversely, if $t\in\mathcal{I}$, then every non-degenerated dB-subspace of $\mathfrak{P}(E_{\omega(t)})$ is of the form $\mathfrak{P}(E_{\omega(s)})$. It is an important observation that the singularities of ω correspond to the degenerated dB-spaces in this chain. In fact, if we put

$$\mathfrak{P}_{\sigma_i^-}:=\operatorname{cls}\big\{\mathfrak{P}(E_{\omega(t)}):\,t<\sigma_i^-\big\},\ \mathfrak{P}_{\sigma_i^+}:=\bigcap\big\{\mathfrak{P}(E_{\omega(t)}):\,t>\sigma_i^+\big\}\,,$$

then every dB-space $\mathfrak{P}_{\sigma_i^-}\subsetneq\mathfrak{P}\subsetneq\mathfrak{P}_{\sigma_i^+}$ is degenerated. Conversely, unless (σ_i^-,σ_i^+) is indivisible with negative length, there exists a degenerated dB-space $\mathfrak{P}_{\sigma_i^-}\subseteq\mathfrak{P}\subseteq\mathfrak{P}_{\sigma_i^+}$. More exactly: Put $\delta:=\dim\mathfrak{P}_{\sigma_i^+}/\mathfrak{P}_{\sigma_i^-}\in\mathbb{N}\cup\{0\}$, and $\delta_-:=\dim\mathfrak{P}^\circ_{\sigma_i^-}$, $\delta_+:=\dim\mathfrak{P}^\circ_{\sigma_i^+}$. If $\delta>1$, then there exist degenerated dB-spaces $\mathfrak{P}_1,\ldots,\mathfrak{P}_{\delta-1}$ with

$$\mathfrak{P}_{\sigma_i^-}\subsetneq\mathfrak{P}_1\subsetneq\mathfrak{P}_2\subsetneq\cdots\subsetneq\mathfrak{P}_{\delta-1}\subsetneq\mathfrak{P}_{\sigma_i^+}\,.$$

It was proved in [KW3] that the isotropic parts of the members of a chain of subsequent degenerated dB-spaces show a very particular behavior. If $\mathfrak{Q}_1, \dots, \mathfrak{Q}_n$ are degenerated dB-spaces, $\mathfrak{Q}_i \subsetneq \mathfrak{Q}_{i+1}$ with codimension 1, then there exists an index $i_{max} \in \{1, \dots, n\}$ such that

$$\mathfrak{Q}_1^\circ \subsetneq \mathfrak{Q}_2^\circ \subsetneq \cdots \subsetneq \mathfrak{Q}_{i_{max}}^\circ \supseteq \mathfrak{Q}_{i_{max}+1}^\circ \supsetneq \cdots \supsetneq \mathfrak{Q}_n^\circ ,$$

where in each inclusion the codimension is at most 1 and only in the middle inclusion equality can hold.

The singularity σ_i is of polynomial type, left dense, right dense or dense, if and only if $\delta_- = 0 = \delta_+$, $\delta_- > 0 = \delta_+$, $\delta_- = 0 < \delta_+$ or $\delta_- > 0 < \delta_+$, respectively. We can have $\delta = 0$ only if σ_i is dense. The case $\delta = 1$, $\delta_- = \delta_+ = 0$, just means that (σ_i^-, σ_i^+) is indivisible with negative length.

We will visualize this inner structure of a singularity in the following way: For example

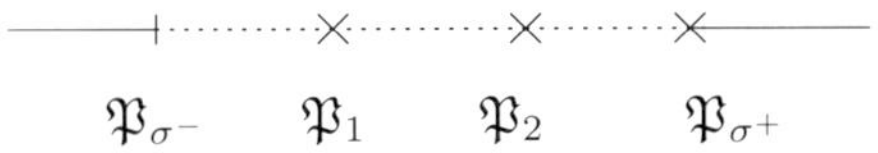

should describe a singularity which is right dense with $\delta = 3$.

It is our aim in this section to describe the evolution of singularities when performing the transformation $\mathcal{T}_{\surd} \circ \mathcal{T}_J \circ \mathcal{T}_2$. In view of our needs in the investigation of generalized strings we will content ourselves with a sound discussion of the case that this transformation is applied to a chain ϖ with $\operatorname{ind}_- \varpi = 0$.

In the first step we deal with the transformation $\mathcal{T}_2$. To this end let $\varpi \in \mathfrak{M}_0^{\mathrm{ep}}$ be given, assume that $\lim_{x \to -\infty} q_\infty(\varpi)(x) = 0$, and let $\omega \in \mathfrak{M}_{<\infty}^{\mathrm{sym}}$ be such that $q_\infty(\omega)(z) = z q_\infty(\varpi)(z^2)$.

We have to investigate the structure of the chain of dB-spaces arising from $\mathfrak{P}(E_{\omega(t)})$. By Lemma 2.13 all these spaces $\mathfrak{P}$ are symmetric. Moreover, since $\operatorname{ind}_- \varpi = 0$, we know that always $\mathfrak{P}_+$ is a Hilbert space, see Lemma 3.13. In particular, by [KWW4, Lemma 2.4] for every dB-space $\mathfrak{P}$ in this chain we have $\dim \mathfrak{P}^\circ \leq 1$. By the structure theory of degenerated dB-spaces developed in [KW3] this knowledge already has a big influence on the kind of singularities that may appear.

6.1. Proposition. *Let* $\varpi \in \mathfrak{M}_0^{\mathrm{ep}}$ *and assume that* $\lim_{x \to -\infty} q_\infty(\varpi)(x) = 0$. *Moreover, let* $\omega \in \mathfrak{M}_{<\infty}^{\mathrm{sym}}$ *be such that* $q_\infty(\omega)(z) = z q_\infty(\varpi)(z^2)$ *and write* $\operatorname{dom} \omega = (\sigma_0, \sigma_1) \cup \cdots \cup (\sigma_n, \sigma_{n+1})$. *We have for every* $k \in \{1, \dots, n\}$

$$\kappa(\sigma_k) = 1, \ \delta(\sigma_k) \leq 3, \ \delta_\pm(\sigma_k) \leq 1 .$$

Let a_k, b_k, $k = 0, \dots, n$, *and* Λ *be as in Lemma 5.9. According to the following table the structure of* σ_k *can be read off the behavior of* Λ *at* a_k *and* b_{k-1} *and the fact whether or not there is an indivisible interval between* b_{k-1} *and* a_k. *Thereby we set*

$$l := \mathfrak{t}(\varpi(a_k)) - \mathfrak{t}(\varpi(b_{k-1})) ,$$

$$\Lambda(b_{k-1}-) := \lim_{t \to b_{k-1}-} \Lambda(t), \ \Lambda(a_k+) := \lim_{t \to a_k+} \Lambda(t) ,$$

and, in case both of these limits are finite, $m := \Lambda(a_k+) - \Lambda(b_{k-1}-)$.

$\Lambda(b_{k-1}-)$	$\Lambda(a_k+)$	$b_{k-1}=a_k$	$b_{k-1}<a_k$	Structure of σ_k	
$\in\mathbb{R}$	$\in\mathbb{R}$	✓ $m<0$		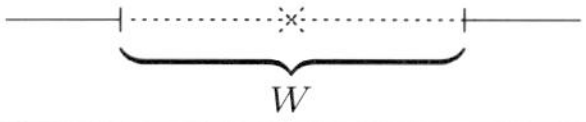	$W=W_{(m,0)}$
$\in\mathbb{R}$	$\in\mathbb{R}$		✓	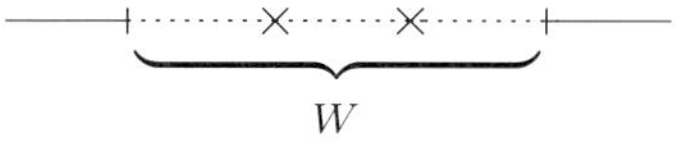	$W=\begin{pmatrix}1 & mz+lz^3\\ 0 & 1\end{pmatrix}$
$+\infty$	$\in\mathbb{R}$	✓		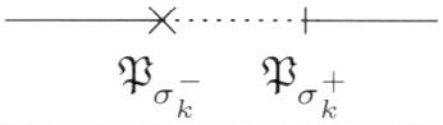	$\mathfrak{P}_{\sigma_k^-,e}=\mathfrak{P}_{\sigma_k^+,e}$
$+\infty$	$\in\mathbb{R}$		✓	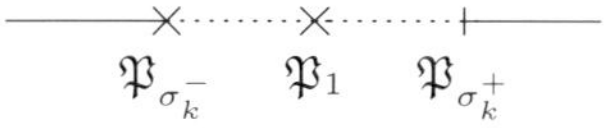	$\mathfrak{P}_{\sigma_k^-,e}\neq\mathfrak{P}_{1,e}=\mathfrak{P}_{\sigma_k^+,e}$
$\in\mathbb{R}$	$-\infty$	✓		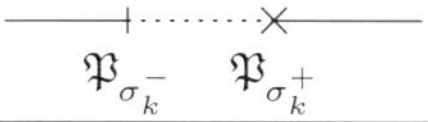	$\mathfrak{P}_{\sigma_k^-,e}=\mathfrak{P}_{\sigma_k^+,e}$
$\in\mathbb{R}$	$-\infty$		✓	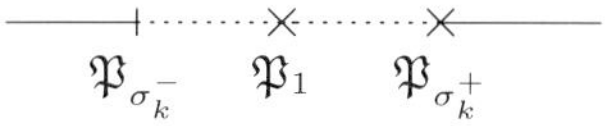	$\mathfrak{P}_{\sigma_k^-,e}=\mathfrak{P}_{1,e}\neq\mathfrak{P}_{\sigma_k^+,e}$
$+\infty$	$-\infty$	✓			
$+\infty$	$-\infty$		✓	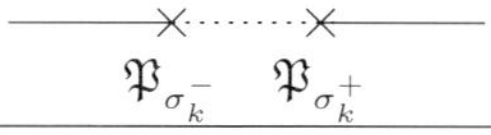	$\mathfrak{P}_{\sigma_k^-,e}\neq\mathfrak{P}_{\sigma_k^+,e}$

Proof. As we already noted for every dB-space $\mathfrak{P}$ in the chain arising from $\mathfrak{P}(E_{\omega(t)})$, we must have $\dim\mathfrak{P}^\circ\leq 1$. From this it is immediate that $\delta_\pm(\sigma_k)\leq 1$. The fact that $\delta(\sigma_k)\leq 3$ is clear from the structure of the isotropic parts of a chain of subsequent degenerated dB-spaces.

The proof of the remaining assertions of the present proposition is based on the following observations:

(i) The limit $\Lambda(a_k+)$ is finite if and only if $\mathfrak{P}_{\sigma_k^+}$ is non-degenerated. Similarly, the limit $\Lambda(b_{k-1}-)$ is finite if and only if $\mathfrak{P}_{\sigma_k^-}$ is non-degenerated.

(ii) We have $\mathfrak{P}_{\sigma_k^+,+}=\mathfrak{P}(E_{\varpi(a_k)})$ and $\mathfrak{P}_{\sigma_k^-,+}=\mathfrak{P}(E_{\varpi(b_{k-1})})$.

(iii) If $\mathfrak{P}_1\subsetneq\mathfrak{P}_2$ are two degenerated symmetric dB-spaces such that $\mathfrak{P}_{1,+}$ and $\mathfrak{P}_{2,+}$ are non-degenerated, then $\mathfrak{P}_{1,+}\subsetneq\mathfrak{P}_{2,+}$.

(iv) If $b_{k-1}=a_k$, then $\delta(\sigma_k)\leq 1$.

(v) There exist at most two degenerated dB-spaces $\mathfrak{P}$ with $\mathfrak{P}_{\sigma_k^-}\subseteq\mathfrak{P}\subseteq\mathfrak{P}_{\sigma_k^+}$.

ad (i): We have $\Lambda(a_k+)>-\infty$ if and only if $\lim_{t\searrow a_k}\tau_k(t)>-\infty$. This implies that $\zeta_+:=\lim_{t\searrow a_k}(\hat{\mu}|_{\mathcal{L}})^{-1}>\sigma_k$. Since in this case (σ_k,ζ_+) is indivisible, we have $\mathfrak{P}_{\sigma_k^+}=\mathfrak{P}(E_{\omega(\zeta_+)})$. The same argument applies to $\lambda(b_{k-1}-)$.

ad (ii): We have

$$\mathfrak{P}_{\sigma_k^+,e} = \Big(\bigcap_{t>\sigma_k^+} \mathfrak{P}(E_{\omega(t)})\Big)_e = \bigcap_{t>\sigma_k^+} \mathfrak{P}(E_{\omega(t)})_e\,.$$

However,

$$\mathfrak{P}(E_{\omega(t)})_e \cong \mathfrak{P}(E_{\omega(t)})_+ = \mathfrak{P}(E_{\varpi(\hat\mu(t))})\,,$$

and, since $\hat\mu(\sigma_k^+) = a_k$,

$$\bigcap_{t>\sigma_k^+} \mathfrak{P}(E_{\varpi(\hat\mu(t))}) = \mathfrak{P}(E_{\varpi(a_k)})\,.$$

The same argument applies to $\mathfrak{P}(E_{\varpi(b_{k-1})})$.

ad (iii): Assume that $\mathfrak{P}_{1,+} = \mathfrak{P}_{2,+} =: \mathfrak{P}$. By [KWW4, Proposition 4.7] there exists exactly one degenerated dB-space $\mathfrak{Q}$ with $\mathfrak{Q}_+ = \mathfrak{P}$, a contradiction.

ad (iv): If $b_{k-1} = a_k$, then, by (ii), $\mathfrak{P}_{\sigma_k^+,+} = \mathfrak{P}_{\sigma_k^-,+}$. By [KWW4, Theorem 3.11] this implies that $\dim \mathfrak{P}_{\sigma_k^+}/\mathfrak{P}_{\sigma_k^-} \le 1$.

ad (v): Since $\dim \mathfrak{P}^\circ \le 1$, this follows from the structure of a chain of subsequent degenerated dB-spaces.

We will now go through the cases listed in the above table.

$\Lambda(b_{k-1}-), \Lambda(a_k+) \in \mathbb{R},\ b_{k-1} = a_k$: We know that $\delta_- = \delta_+ = 0$ and that $\mathfrak{P}_{\sigma_k^-,+} = \mathfrak{P}_{\sigma_k^+,+}$ It follows from Lemma 3.4 that $\omega(\sigma_k^-)^{-1}\omega(\sigma_k^+) = W_{(\alpha,0)}$ for some α. However, in the present case we have

$$\alpha = \lim_{t\searrow a_k} \tau_k(t) - \lim_{t\nearrow b_{k-1}} \tau_{k-1}(t) = \Lambda(a_k+) - \Lambda(b_{k-1}-) = m\,.$$

$\Lambda(b_{k-1}-), \Lambda(a_k+) \in \mathbb{R},\ b_{k-1} < a_k$: Again we know that $\delta_- = \delta_+ = 0$. We have $\mathcal{T}_{\surd}(\omega(\sigma_k^-))^{-1}\mathcal{T}_{\surd}(\omega(\sigma_k^+)) = W_{(l,0)}$. Since $\omega(\sigma_k^-) = \mathcal{T}_{2,\Lambda(b_{k-1}-)}(\varpi(b_{k-1}))$ and $\omega(\sigma_k^+) = \mathcal{T}_{2,\Lambda(a_k+)}(\varpi(a_k))$, the assertion follows from Remark 3.7.

$\Lambda(b_{k-1}-) = +\infty, \Lambda(a_k+) \in \mathbb{R},\ b_{k-1} = a_k$: It follows from (iv) that $\delta \le 1$. Moreover, we know from (i) that $\delta_- = 1$ and $\delta_+ = 0$, which implies that $\delta \ne 0$.

$\Lambda(b_{k-1}-) = +\infty, \Lambda(a_k+) \in \mathbb{R},\ b_{k-1} < a_k$: We have $\delta_- = 1$, $\delta_+ = 0$, and know that $\mathfrak{P}_{\sigma_k^-,+} \ne \mathfrak{P}_{\sigma_k^+,e}$. Since (σ_k, σ_k^+) is indivisible of type 0, it follows that $\omega(\sigma_k^+)_{21} \in \mathfrak{P}(E_{\omega(\sigma_k^+)})$, and hence that $\mathfrak{P}_{\sigma_k^+,e} \subseteq \overline{\operatorname{dom} \mathcal{S}_{\mathfrak{P}_{\sigma_k^+}}}$. This space is a dB-subspace with codimension 1 in $\mathfrak{P}_{\sigma_k^+}$ and is degenerated since $\delta_- > 0$. However, it contains the same even functions than $\mathfrak{P}_{\sigma_k^+}$, and thus can not be equal to $\mathfrak{P}_{\sigma_k^-}$. By (v) it must be the only space which lies strictly between $\mathfrak{P}_{\sigma_k^+}$ and $\mathfrak{P}_{\sigma_k^-}$.

$\Lambda(b_{k-1}-) \in \mathbb{R}, \Lambda(a_k+) = -\infty,\ b_{k-1} = a_k$: It follows from (iv) that $\delta \le 1$. We know, moreover, from (i) that $\delta_- = 0$ and $\delta_+ = 1$, which also implies that $\delta \ne 0$.

$\Lambda(b_{k-1}-) \in \mathbb{R}, \Lambda(a_k+) = -\infty,\ b_{k-1} < a_k$: We have $\delta_- = 0$, $\delta_+ = 1$, and $\mathfrak{P}_{\sigma_k^+,+} \ne \mathfrak{P}_{\sigma_k^-,+}$. Let $\mathfrak{P}_1$ be the smallest degenerated dB-space which contains $\mathfrak{P}_{\sigma_k^-}$. Since

(σ_k^-, σ_k) is indivisible of type 0, we have $\mathfrak{P}_1 = \operatorname{span}(\mathfrak{P}_{\sigma_k,-} \cup \{\omega(\sigma_k^-)_{21}\})$, and therefore $\mathfrak{P}_{1,e} = \mathfrak{P}_{\sigma_k^-,e}$. This rules out the possibility that $\delta = 1$.

$\Lambda(b_{k-1}-) = +\infty, \Lambda(a_k+) = -\infty,\ b_{k-1} = a_k$: We have $\delta_- = \delta_+ = 1$ and $\mathfrak{P}_{\sigma_k^-,+} = \mathfrak{P}_{\sigma_k^+,+}$. Thus (iii) implies that $\mathfrak{P}_{\sigma_k^-} = \mathfrak{P}_{\sigma_k^+}$.

$\Lambda(b_{k-1}-) = +\infty, \Lambda(a_k+) = -\infty,\ b_{k-1} < a_k$: We have $\delta_- = \delta_+ = 1$ and $\mathfrak{P}_{\sigma_k^-,+} \neq \mathfrak{P}_{\sigma_k^+,+}$. Thus $\delta > 0$, and by (v) it follows that $\delta = 1$.

It remains to show that $\kappa(\sigma_k) = 1$. To this end recall that $\operatorname{ind}_- \mathfrak{P}_{\sigma_k^-} = \max_{t<\sigma_k} \operatorname{ind}_- \omega(t)$ and $\operatorname{ind}_- \mathfrak{P}_{\sigma_k^+} + \delta_+ = \min_{t>\sigma_k} \operatorname{ind}_- \omega(t)$, cf. [KW3].

Consider the case that $\delta_- = \delta_+ = 0$. If $\delta = 1$ the assertion is clear. If $\delta = 3$ it follows since $l > 0$. Assume that not both of $\delta_\pm$ are equal to 0. If $\delta = 1$, we have $\operatorname{ind}_- \mathfrak{P}_{\sigma_k^+} = \operatorname{ind}_- \mathfrak{P}_{\sigma_k^-}$, and again $\kappa(\sigma_k) = 1$. Consider the fourth of the cases in the table. Then $\operatorname{ind}_- \mathfrak{P}_{\sigma_k^-} = \operatorname{ind}_- \mathfrak{P}_1$ since $\mathfrak{P}_{\sigma_k^-,e}$ and $\mathfrak{P}_{1,e}$ have the same negative index. The increase of negative squares from $\mathfrak{P}_1$ to $\mathfrak{P}_{\sigma_k^+}$ is then 1. In the last remaining case, the same argument shows that $\operatorname{ind}_- \mathfrak{P}_1 = \operatorname{ind}_- \mathfrak{P}_{\sigma_k^+}$, and thus that $\kappa(\sigma_k) = 1$. □

Next we deal with $\mathcal{T}_J$. It is a consequence of Lemma 2.14 that an application of this transformation does not change the structure of singularities.

6.2. Corollary. *Let $\omega \in \mathfrak{M}_{<\infty}$ and assume that $\mathfrak{K}_-(\omega(t)) = \mathfrak{K}(\omega(t))$ and $\mathfrak{K}_+(\omega(t)) = \mathfrak{K}(\omega(t))$. Moreover, let $\upsilon := \mathcal{T}_J(\omega)$. Let π_+, π_- be the projections of $\mathfrak{K}(\omega(t))$ onto the first and second component, respectively, and put $\phi := \pi_+ \circ \pi_-^{-1}$. Then ϕ induces an order preserving bijection of the chain of all dB-subspaces of $\mathfrak{P}(E_{\omega(t)})$ onto the chain of all dB-subspaces of $\mathfrak{P}(E_{\upsilon(t)})$. Thereby for all dB-subspaces $\mathfrak{Q}$ of $\mathfrak{P}(E_{\omega(t)})$*

$$\operatorname{ind}_- \phi(\mathfrak{Q}) = \operatorname{ind}_- \mathfrak{Q},\ \dim \phi(\mathfrak{Q})^\circ = \dim \mathfrak{Q}^\circ .$$

If $\omega \in \mathfrak{M}_{<\infty}^{\mathrm{sym}}$, then also υ is symmetric, and we have

$$\phi(\mathfrak{Q}_e) = \phi(\mathfrak{Q})_o,\ \phi(\mathfrak{Q}_o) = \phi(\mathfrak{Q})_e . \tag{6.1}$$

Proof. With the notation of Lemma 2.14 we have $E_{\upsilon(t)} = \tilde{E}_{\omega(t)}$. Hence the first assertion is an immediate corollary of Lemma 2.14. It remains to investigate the symmetric situation. However, by Lemma 2.7, $\phi = \pi_+ \circ \pi_-^{-1}$ maps even to odd functions and odd to even functions. Since ϕ^{-1} has the same property, the relation (6.1) follows. □

We can now deduce which singularities are created by an application of $\mathcal{T}_{\surd} \circ \mathcal{T}_J \circ \mathcal{T}_2$.

6.3. Proposition. *Let $\varpi \in \mathfrak{M}_0^{\mathrm{ep}}$ and assume that*

$$\lim_{x\to-\infty} q_\infty(\varpi)(x) = \lim_{x\to-\infty} \frac{1}{q_\infty(\varpi)(x)} = 0 .$$

Moreover, let $\upsilon \in \mathfrak{M}^{\mathrm{ep}}_{<\infty}$ *be such that* $q_\infty(\upsilon)(z) = -(zq_\infty(\varpi)(z))^{-1}$. *Let* a_k, b_k, $k = 0, \dots, n$, *and* Λ *be as in Lemma* 5.9. *Then* υ *has exactly* n *singularities, say* $\gamma_1 < \cdots < \gamma_n$. *We have for every* $k \in \{1, \dots, n\}$

$$\kappa(\gamma_k) = 1,\ \delta(\gamma_k) \le 2,\ \delta_\pm(\gamma_k) \le 1\,.$$

The structure of γ_k *can be read off the behavior of* Λ *at* a_k *and* b_{k-1} *and the fact whether or not there is an indivisible interval between* b_{k-1} *and* a_k. *Let* $l, \Lambda(b_{k-1}-), \Lambda(a_k+)$ *and* m *be as in Proposition* 6.1. *Moreover, put*

$$\alpha := \varpi(b_{k-1})'_{21}(0),\ p(z) := -(mz + lz^2),\ m' := m(1+\alpha^2),\ \phi := -\operatorname{Arccot}\alpha\,.$$

$\Lambda(b_{k-1}-)$	$\Lambda(a_k+)$	Structure of σ_k		
$\in\mathbb{R}$	$\in\mathbb{R}$	$b_{k-1} = a_k$	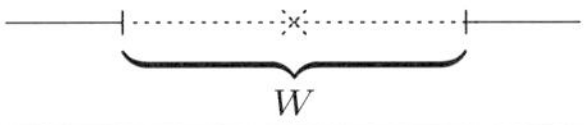	$W = W_{(m',\phi)}$
		$b_{k-1} < a_k$	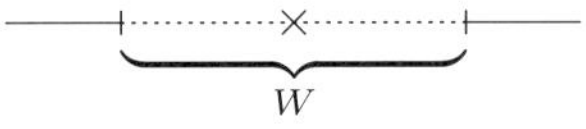	$W=\begin{pmatrix} 1-\alpha p(z) & -\alpha^2 p(z) \\ p(z) & 1+\alpha p(z)\end{pmatrix}$
$+\infty$	$\in\mathbb{R}$			
$\in\mathbb{R}$	$-\infty$			
$+\infty$	$-\infty$			

Proof. Let $\omega := \mathcal{T}_2(\varpi)$ and $\tilde\omega := \mathcal{T}_J(\omega)$, so that $\upsilon = \mathcal{T}_{\sqrt{\ }}(\tilde\omega)$. By Theorem 5.1 and Corollary 6.2 singularities of υ can only occur at singularities of ω. We shall go through the different possibilities of singularities of ω and show that each of them gives rise to a singularity of υ with the asserted structure. We will use the same notation as in the previous proofs.

The first two cases require explicit computation.

$\Lambda(b_{k-1}-) \in \mathbb{R}, \Lambda(a_k+) \in \mathbb{R},\ b_{k-1} = a_k$: Then we have $\omega(\sigma_k^-)^{-1}\omega(\sigma_k^+) = W_{(m,0)}$, and thus $\tilde\omega(\sigma_k^-)^{-1}\tilde\omega(\sigma_k^+) = W_{(m,\frac{\pi}{2})}$. Moreover, $\tilde\omega(\sigma_k^-)'_{21}(0) = -\omega(b_{k-1})'_{21}(0) = -\varpi(b_{k-1})'_{21}(0) = -\alpha$. We obtain from Proposition 3.6 that

$$\upsilon(\hat\mu(\sigma_k^-))^{-1}\upsilon(\hat\mu(\sigma_k^+)) = W_{(m',\phi)}\,.$$

$\Lambda(b_{k-1}-) \in \mathbb{R}, \Lambda(a_k+) \in \mathbb{R},\ b_{k-1} < a_k$: In this case

$$\tilde\omega(\sigma_k^-)^{-1}\tilde\omega(\sigma_k^+) = \begin{pmatrix} 1 & 0 \\ -p(z) & 1\end{pmatrix}$$

and $\tilde\omega(\sigma_k^-)'_{12}(0) = \tilde\omega(\sigma_k^+)'_{12}(0) = -\alpha$. By (3.2) it follows that

$$\upsilon(\hat\mu(\sigma_k^-))^{-1}\upsilon(\hat\mu(\sigma_k^+)) = \begin{pmatrix} 1-\alpha p(z) & -\alpha^2 p(z) \\ p(z) & 1+\alpha p(z)\end{pmatrix}$$

For the investigation of the remaining cases note that, by Corollary 6.2, we just have to inspect the behavior of the odd parts of the dB-spaces arising from the chain ω.

$\Lambda(b_{k-1}-) = +\infty, \Lambda(a_k+) \in \mathbb{R},\ b_{k-1} = a_k$: By Proposition 6.1 $\mathfrak{P}_{\sigma_k^-,o}$ is degenerated, $\mathfrak{P}_{\sigma_k^+,o}$ is non-degenerated and $\dim \mathfrak{P}_{\sigma_k^+,o}/\mathfrak{P}_{\sigma_k^-,o} = 1$. Hence we have a singularity with $\delta_- = 1$, $\delta_+ = 0$ and $\delta = 1$.

$\Lambda(b_{k-1}-) = +\infty, \Lambda(a_k+) \in \mathbb{R},\ b_{k-1} < a_k$: We have $\mathfrak{P}_{\sigma_k^-,o} = \mathfrak{P}_{1,o}$ and this space is degenerated. The space $\mathfrak{P}_{\sigma_k^+,o}$ is non-degenerated. Thus we also have a singularity with $\delta_- = 1$, $\delta_+ = 0$ and $\delta = 1$.

The remaining cases are treated exactly in the same manner. This knowledge on the structure of the singularities implies that in any case $\delta \leq 2$, $\delta_\pm \leq 1$ and, cf. [KW3, Corollary 2.12] and its proof, that the increase of the negative index is equal to 1. □

7. On generalized strings

Recall [KaK1] that a *string* $S[L,m]$ is given by its length L, $0 \leq L \leq \infty$, and a nonnegative and nondecreasing function m defined on $[0,L)$ which may be chosen to be left-continuous. This concept can be generalized as follows: Let m be a function defined on $[0,L)$ which is nondecreasing and left-continuous except a finite number of points from $[0,L)$, and let D be a nondecreasing and left-continuous step function defined on $[0,L)$ which is constant except a finite number of growth points. Note that D corresponds to the so-called dipoles in [KL2].

Some point $x_e \in [0,L)$ is called a *dipole* if $D(x_e+) - D(x_e) > 0$, a *negative jump*, if $m(x_e+) - m(x_e) < 0$, and a *singularity*, if $m(x) \to +\infty$ for $x \nearrow x_e$, or $m(x) \to -\infty$ for $x \searrow x_e$, and $\int_{(x_e-\epsilon, x_e+\epsilon)} m(t)^2 dt < \infty$. The point $x_e \in [0,L)$ is called *critical* if it is a dipole, a singularity or a negative jump. The point 0 is critical if it is a dipole or if $-\infty \leq m(0+) < 0$, the point L is never a critical point. Observe that at a critical point can be both a dipole and a singularity or a negative jump. The triple $S[L,m,D]$ is called *generalized string*. The relation

$$f'(x) + z\int_{[0,x]} f(x)dm(x) + z^2 \int_{[0,x]} f(x)dD(x) = 0, \quad f'(0-) = 0. \tag{7.1}$$

is the differential equation of a generalized string. Of course, this equation requires some explanation if $S[L,m,D]$ has singularities among its critical points. The appropriate interpretation is given by means of canonical systems, and for the sake of completeness we continue to recall some basic facts about these systems.

Let H be a real, symmetric and non-negative 2×2–matrix function on the interval $[0,l_H)$:

$$H(x) = \begin{pmatrix} h_1(x) & h_3(x) \\ h_3(x) & h_2(x) \end{pmatrix}, \quad x \in [0,l_H),$$

with locally integrable functions h_1, h_2 and h_3. A *canonical system* is a boundary value problem of the form

$$Jf'(x) = -zH(x)f(x), \; x \in [0, l_H), \quad f_1(0) = 0, \tag{7.2}$$

with $f(x) = (f_1(x) \; f_2(x))^T$, $J = \begin{pmatrix} 0 & -1 \\ 1 & 0 \end{pmatrix}$, and a complex parameter z. Here the differential equation in (7.2) is considered to hold almost everywhere on $[0, l_H)$. Weyl's limit point case prevails at the point l_H for the canonical system (7.2) if and only if

$$\int_0^{l_H} \operatorname{tr} H(x)dx = \infty, \tag{7.3}$$

and from now on we assume that for each Hamiltonian H the relation (7.3) holds. The *fundamental matrix function*

$$W(x, z) = \begin{pmatrix} w_{11}(x, z) & w_{12}(x, z) \\ w_{21}(x, z) & w_{22}(x, z) \end{pmatrix}$$

of a canonical system (7.2) with Hamiltonian H is the unique solution of the integral equation

$$W(x, z)J - J = z \int_0^x W(s, z)H(s)ds. \tag{7.4}$$

Note that $W(0, z) = I$, and that $x \to W(x, z)$, $0 \le x < l_H$ is a maximal chain of matrix functions belonging to $\mathfrak{M}_0$.

For each $\omega \in \widetilde{\mathcal{N}_0}$ and $z \in \mathbb{C}^+$ the limit

$$Q(z) := \lim_{x \to l_H} \frac{w_{11}(x, z)\omega(z) + w_{12}(x, z)}{w_{21}(x, z)\omega(z) + w_{22}(x, z)} \tag{7.5}$$

exists, is independent of ω, and, as a function of z, belongs to the set of Nevanlinna functions $\mathcal{N}_0$, cf. [dB2]. The function Q is called the *Titchmarsh-Weyl coefficient* of the canonical system (7.2) or of the Hamiltonian H. Note that $W(\cdot, z)$ is a maximal matrix chain, and that Q coincides with its Weyl coefficient.

Let $\xi_\phi := (\cos\phi, \sin\phi)^T$ for some $\phi \in [0, \pi)$. The open interval $I_\phi \subset [0, l_H)$ is called H-indivisible of type ϕ if the relation

$$\xi_\phi^T JH = 0, \text{ a.e. on } I_\phi, \tag{7.6}$$

holds, see [Ka], [dB1]. This notion is the same as introduced for maximal chains in Section 4. In particular, $\det H = 0$ a.e. on I_ϕ. If (x_1, x_2) is a H-indivisible of type ϕ and length l, the fundamental matrix W satisfies the relation $W(x_2, z) = W(x_1, z)W_{(l,\phi)}(z)$, where the factor $W_{(l,\phi)}(z)$ is defined in relation (2.2), that is, H-indivisible intervals are also indivisible.

A Hamiltonian H is called *trace normed* if $h_1 + h_2 = 1$ a.e. on $[0, \infty)$. For the class of trace normed Hamiltonians a basic inverse result in [dB1] can be formulated as follows (see [W1]): Each function $Q \in \mathcal{N}_0$ is the Titchmarsh-Weyl coefficient of a canonical system with a trace normed Hamiltonian H on $[0,\infty)$. This correspondence is bijective if two Hamiltonians which coincide almost everywhere are identified.

Let Q_H denote the Titchmarsh-Weyl coefficient corresponding to some Hamiltonian H, and let

$$\widehat{H} = JHJ^T. \tag{7.7}$$

Then

$$\widehat{W}(x,z) = JW(x,z)J^T \tag{7.8}$$

is the fundamental matrix corresponding to $\widehat{H}$, and the relation (7.5) implies that

$$Q_{\widehat{H}}(z) = -(Q_H(z))^{-1} \tag{7.9}$$

If H is of diagonal form, that is $H = \operatorname{diag}(h_1, h_2)$, then $Q_H \in \mathcal{N}_0^{\mathrm{sym}}$. The following proposition will be of use in what follows.

For the following we need the fact that any $Q \in \mathcal{N}_0 \setminus \{\infty\}$ admits a unique integral representation:

$$a + bz + \int_{\mathbb{R}} \left(\frac{1}{t-z} - \frac{t}{1+t^2} \right) d\sigma(t)\,, \tag{7.10}$$

where $a, b \in \mathbb{R}$, $b \geq 0$, σ is a nonnegative Borel measure on $\mathbb{R}$ with $\int_{\mathbb{R}} \frac{1}{1+t^2}\, d\sigma(t) < \infty$. σ is called the spectral measure of $Q(z)$.

7.1. Proposition. *Let $Q \neq \infty$ be some Nevanlinna function with a semibounded spectral measure,* $\operatorname{supp} \sigma \subset [c,\infty)$, $c \in \mathbb{R}$. *Let H be some Hamiltonian corresponding to Q with left-continuous components h_i, $1 \leq i \leq 3$ and* $\det H = 0$. *If W denotes the corresponding fundamental matrix, the relations*

$$\frac{h_3(x)}{h_2(x)} = \lim_{z\to-\infty} -\frac{w_{12}(x,z)}{w_{11}(x,z)} = \lim_{z\to-\infty} -\frac{w_{22}(x,z)}{w_{21}(x,z)}, \qquad x>0,\ h_2(x)>0, \tag{7.11}$$

$$\frac{h_3(x)}{h_1(x)} = \lim_{z\to-\infty} -\frac{w_{11}(x,z)}{w_{12}(x,z)} = \lim_{z\to-\infty} -\frac{w_{21}(x,z)}{w_{22}(x,z)}, \qquad x>0,\ h_1(x)>0, \tag{7.12}$$

hold.

Proof. The existence of a Hamiltonian H corresponding to Q with the required properties was shown in [W3], Theorem 3.1. We continue in two steps

Step 1: First we assume $\operatorname{supp} \sigma \subseteq (0,\infty)$ and $b = \lim_{y\to\infty} Q(iy)/iy = 0$. According to Corollary 3.2 of [W3], the function

$$v(x) = \frac{h_3(x)}{h_2(x)}, \qquad x \in (0, l_H)$$

is nondecreasing with and left-continuous. A rescaling allows to assume without loss of generality that the Hamiltonian H is of the form

$$H(x) = \begin{pmatrix} v(x)^2 & v(x) \\ v(x) & 1 \end{pmatrix}. \tag{7.13}$$

We are going to show that

$$v(x) = \lim_{z\to-\infty} -\frac{w_{22}(x,z)}{w_{21}(x,z)}, \qquad x > 0. \tag{7.14}$$

If $v(x) = v(x_0)$ for some $x_0 < x$, the interval (x_0, x) is H-indivisible with constant v. It follows with $l = x - x_0$ that

$$\begin{pmatrix} w_{21}(x,z) \\ w_{22}(x,z) \end{pmatrix} = \begin{pmatrix} 1 - zlv & -zl \\ zlv^2 & 1 + zlv \end{pmatrix} \begin{pmatrix} w_{21}(x_0,z) \\ w_{22}(x_0,z) \end{pmatrix},$$

which leads to the relation

$$-\frac{w_{22}(x,z)}{w_{21}(x,z)} = v + \frac{1}{zl + N(z)}, \quad N(z) = -\left(\frac{w_{22}(x_0,z)}{w_{21}(x_0,z)} + v\right)^{-1}. \tag{7.15}$$

As $w_{22}(x_0,z)/w_{21}(x_0,z)$ is a Nevanlinna function, also $N(z)$ is a Nevanlinna function, and the relation (7.14) follows from the relation (7.15). Now we assume that $v(x) > v(t)$ for all $0 < t < x$. The relation (7.4) implies that

$$\begin{pmatrix} w_{21}'(x,z) \\ w_{22}'(x,z) \end{pmatrix} = -z(v(x)w_{21}(x,z) + w_{22}(x,z)) \begin{pmatrix} 1 \\ -v(x) \end{pmatrix}. \tag{7.16}$$

It follows that $-w_{22}'(x,z) = v(x)w_{21}'(x,z)$, which implies that

$$w_{22}(x,z) = 1 - v(x)w_{21}(x,z) + \int_0^x w_{21}(t,z)dv(t). \tag{7.17}$$

We will show that

$$\lim_{z\to-\infty} w_{21}(x,z) = \infty \tag{7.18}$$

and

$$\lim_{z\to-\infty} \frac{w_{21}(t,z)}{w_{21}(x,z)} = 0, \quad 0 \le t < x, \tag{7.19}$$

then the relation (7.17) implies the relation (7.14). Because of

$$\begin{aligned} w_{21}'(x,z) &= -z(v(x)w_{21}(x,z) + w_{22}(x,z)) \\ &= -zv(x)w_{21}(x,z) - z + z\int_0^x v(t)w_{21}'(t,z)dt, \end{aligned}$$

one finds with integration by parts that

$$w_{21}'(x,z) = -z\left(1 + \int_0^x w_{21}(t,z)dv(t)\right), \tag{7.20}$$

and it follows that

$$w_{21}(x,z) = -zx - z\int_0^x (x-u)w_{21}(u,z)dv(u). \tag{7.21}$$

Let $z < 0$. The relation (7.20) implies that the function $w_{21}'(\cdot,z)$ is positive in a neighborhood of 0. Then the relation (7.21) implies that $w_{21}(\cdot,z)$ is positive and nondecreasing, and the relations (7.18) and

$$w_{21}(x,z) \geq -zw_{21}(t,z)\int_t^x (x-u)dv(u) \tag{7.22}$$

follow. As $\int_t^x (x-u)dv(u) > 0$, the relation (7.22) implies the relation (7.19).

Step 2: Now we assume that $\operatorname{supp}\sigma \subset (c,\infty)$ for some $c < 0$. Recall from [W2] that then the Hamiltonian $\tilde{H}$ defined by

$$\tilde{H}(x) = W(x,-c)H(x)W(x,-c)^T \tag{7.23}$$

corresponds to the Titchmarsh-Weyl coefficient $\tilde{Q}(z)$ given by

$$\tilde{Q}(z) = Q(z-c),$$

and the fundamental matrix

$$\tilde{W}(x,z) = W(x,z-c)W(x,-c)^{-1}.$$

In particular, $\operatorname{supp}\tilde{\sigma} \subset (0,\infty)$, and it follows that there exists a nondecreasing function $\tilde{v}$ such that the Hamiltonian $\tilde{H}$ is of the form (7.13). Moreover, the relation (7.14) is satisfied. It follows that

$$\frac{h_3(x)}{h_2(x)} = \frac{w_{22}(x,-c)\tilde{v}(x) - w_{12}(x,-c)}{-w_{21}(x,-c)\tilde{v}(x) + w_{11}(x,-c)},$$

and that

$$\begin{pmatrix} w_{21}(x,z-c) \\ w_{22}(x,z-c) \end{pmatrix} = \begin{pmatrix} \tilde{w}_{21}(x,z)w_{11}(x,-c) + \tilde{w}_{22}(x,z)w_{21}(x,-c) \\ \tilde{w}_{21}(x,z)w_{12}(x,-c) + \tilde{w}_{22}(x,z)w_{22}(x,-c) \end{pmatrix}.$$

Together with (7.14) the last relation implies

$$\lim_{z\to-\infty} -\frac{w_{22}(x,z)}{w_{21}(x,z)} = \frac{h_3(x)}{h_2(x)}.$$

The second relation in (7.11) follows from

$$-\frac{w_{12}(x,z)}{w_{11}(x,z)} + \frac{w_{22}(x,z)}{w_{21}(x,z)} = \frac{1}{w_{11}(x,z)w_{21}(x,z)} \to 0 \qquad (z \to -\infty),$$

where the relation $\det W(x,z) = 1$ has been used. The relations (7.12) can be shown in a similar way. If $\lim_{y\to\infty} Q(iy)/iy = b > 0$, the Hamiltonian is of the form $\operatorname{diag}(1,0)$ on the interval $(0,b)$, and it is easy to see that the relation (7.12) for $x \in (0,b)$ holds. □

If a generalized string $S[L,m,D]$ is given, a Hamiltonian H_0 of a canonical system can be constructed as follows: Define a new scale by

$$x(t) = t + \int_{[0,t)} dD(u),\, L_0 = L + \int_{[0,L)} dD(u). \tag{7.24}$$

Let $x(t+) = x(t) + D(t) - D(t-)$. If t_e is a dipole of $S[L, m, D]$, the interval $(x(t_e), x(t_e+))$ is assumed to be maximal H_0-indivisible of type $\pi/2$, that is, $H_0 = \operatorname{diag}(0,1)$ on $(x(t_e), x(t_e+))$. Define

$$m_0(x(t)) = m(t), \quad H_0(x(t)) = \begin{pmatrix} 1 & -m_0(x(t)) \\ -m_0(x(t)) & m_0(x(t))^2 \end{pmatrix}, \tag{7.25}$$

and put $H_0(x(t_e)) = \operatorname{diag}(0,1)$ if the point t_e is a singularity of $S[L, m, D]$. If $L + \int_{[0,L)} m(t)^2 dt < \infty$, it is assumed that $H_0 = \operatorname{diag}(0,1)$ on (L_0, ∞). Summing up, there exists a (possibly empty) finite sequence of n maximal H_0-indivisible intervals $\mathfrak{D}_k$ of type $\pi/2$ such that $\mathfrak{D}_k < \mathfrak{D}_{k+1}$, and with $\mathfrak{D} = \bigcup_{k=1}^n \mathfrak{D}_k$ and $\mathfrak{I} := [0, L_0) \setminus \mathfrak{D}$ the Hamiltonian H_0 is given as follows.

$$H_0(x) = \begin{cases} \begin{pmatrix} 1 & -m_0(x) \\ -m_0(x) & m_0(x)^2 \end{pmatrix} & \text{if } x \in \mathfrak{I}, \\ \begin{pmatrix} 0 & 0 \\ 0 & 1 \end{pmatrix} & \text{if } x \in \mathfrak{D}. \end{cases} \tag{7.26}$$

Moreover, the construction of H_0 implies that the limit point case prevails. Conversely, if a Hamiltonian H_0 of the form (7.26) is given, the corresponding generalized string $S[L, m, D]$ can be recovered as follows: For $x \in \mathfrak{I}$, let

$$t(x) := \int_0^x \chi_{\{h_1 \neq 0\}}(u) du, \; L := t(L_0), \tag{7.27}$$

and

$$m(t) := m_0(x), \; D(t) := \int_0^t \chi_{\{h_1 = 0\}}(u) du. \tag{7.28}$$

If $Q \in \mathcal{N}_0$ with spectral measure σ has the property that $\operatorname{supp} \sigma \cap (-\infty, 0]$ consists of finitely many isolated points, there exists a Hamiltonian H of the form (7.26) such that Q is its Titchmarsh-Weyl coefficient, see [W3].

Let f be the solution of the relation (7.2) with Hamiltonian H_0. For $t(x)$ given by the relation (7.27), the function

$$f(t) := f_2(x) \tag{7.29}$$

is the solution of the relation (7.1). To justify the relation (7.29), note that if 0 is a dipole or a singularity the relations $Jf' = -zHf$, $f_1(0) = 0$, and $H_0(0) = \operatorname{diag}(0,1)$ imply that $f_2'(0) = 0$. Otherwise, the relation $f'(0) = f_2'(0) = zm(0)f(0)$ holds and matches with the equation (7.1). On intervals where m_0 is defined and bounded, the relation $Jf' = -zHf$ implies that f_2 satisfies the differential equation

$$df_2' = -zf_2 dm_0. \tag{7.30}$$

Now consider a maximal H-indivisible interval $\mathfrak{D}_k = (a, b)$ of type $\pi/2$. Then

$$f_2'(x) = 0, \; f_1'(x) = -zf_2(x) \text{ if } x \in (a, b),$$

which yields

$$f_2(a) = f_2(b),\ f_1(b) - f_1(a) = -z(b-a)f_2(a). \tag{7.31}$$

The relation $f_2'(x) = zf_1(x) - zm_0(x)f_2(x)$ for $x = a$ and $x = b$ and the relations (7.31) yield

$$f_2'(b) - f_2'(a) + z(m_0(b) - m_0(a))f_2(a) + z^2(b-a)f_2(a) = 0. \tag{7.32}$$

In particular, as f_2 is constant on (a, b), the function f is well defined by the relation (7.29).

Recall [LW] that the solution f of (7.1) can also be characterized in terms of m. Namely, if x_e is singularity of m, for a solution f of (7.1) on $[0, x_e)$ with $f'(0-) = 0$ the limits $f(x_e-)$ and $\int\limits_0^{x_e-} m(t)f'(t)dt$ exist. Moreover, if $x > x_e$, there is exactly one solution f on (x_e, x) of (7.1) such that $f(x_e-) = f(x_e+)$ and $\int\limits_{x_e-}^{x} m(t)f'(t)dt$ is finite. This solution f coincides with the function defined in (7.29).

Let Q_0 be the Titchmarsh-Weyl coefficient of the canonical system with the Hamiltonian (7.26). The function

$$Q_S(z) := z^{-1}Q_0(z) \tag{7.33}$$

is called the *principal Titchmarsh-Weyl coefficient of the generalized string* $S[L, m, D]$, see [LW]. Let $\mathcal{N}_\kappa^+$ be the set of all functions $q \in \mathcal{N}_\kappa$ such that $zq(z) \in \mathcal{N}_0$ is a Nevanlinna function. The basic inverse result from [LW] can be formulated as follows:

If $S[L, m, D]$ is a generalized string with κ critical points, then its principal Titchmarsh-Weyl coefficient Q_S belongs to the class $\mathcal{N}_\kappa^+$. Conversely, each function $Q \in \mathcal{N}_\kappa^+$ is the principal Titchmarsh-Weyl coefficient of a generalized string with κ critical points, which is uniquely determined by Q.

Let a generalized string $S[L, m, D]$ with principal Titchmarsh-Weyl coefficient $Q_S \in \mathcal{N}_\kappa^+$ be given. Then Q_S is also the Weyl coefficient of some matrix chain υ, and now the problem arises how the singularities of ϖ which are characterized in Proposition 6.3 can be described in terms of the generalized string $S[L, m, D]$.

As $Q_0 \in \mathcal{N}_0^{\mathrm{ep}}$ implies in particular that the corresponding spectral measure σ is semibounded, Proposition 7.1 can be applied to the Hamiltonian H_0 of (7.26):

7.2. Corollary. *Let W_0 be the fundamental matrix of the canonical system with Hamiltonian H_0 from the relation* (7.26). *Then*

$$m_0(x) = \lim_{z\to-\infty} \frac{w_{12}^0(x,z)}{w_{11}^0(x,z)}, \qquad x \in \mathfrak{I}. \tag{7.34}$$

Moreover,

$$\lim_{z\to-\infty} \frac{w_{12}^0(x,z)}{w_{11}^0(x,z)} = \infty, \qquad x \in \mathfrak{D}.$$

7.3. Theorem. *Let $Q_S \in \mathcal{N}_\kappa^+$ be the principal Titchmarsh-Weyl coefficient of some generalized string $S[L, m, D]$, and assume that υ is the maximal chain with Weyl coefficient Q_S. Then the function m_0 from the relation* (7.25) *is equal to the function Λ from Lemma* 5.9*, and the maximal chain υ has κ singularities, which correspond to the critical points of $S[L, m, D]$. The five possible cases concerning the structure of a singularity of the chain υ which are described in Proposition* 6.3 *correspond to a negative jump in case* 1*, a dipole in case* 2*, and to a singularity in the last* 3 *cases.*

Proof. Let ϖ be the maximal chain with Weyl coefficient

$$q_\infty(\varpi)(z) = -\frac{1}{zQ_S(z)}.$$

Then $q_\infty(\varpi) \in \mathcal{N}_0$, and the representation formulas for $\mathcal{N}_\kappa^+$ functions from [KL1] imply that $\lim_{z\to\infty} q_\infty(\varpi)(z) = 0$. One finds from the relations (7.9) and (7.33) that the the matrix chain $\widehat{W}_0$ is equal to the chain ϖ from Proposition 6.3, and the relations (7.34) and (7.8) imply that m_0 is equal to the function Λ from Lemma 5.9. Note that the intervals (a_k, b_k) from Lemma 5.9 are the maximal intervals of $\mathfrak{I}$ which contain no critical point. □

If W_0^D denotes the factor in the chain W_0 which corresponds to a maximal H_0-indivisible interval (x_1, x_2) of type $\pi/2$ and length d, that is, $W_0(x_2, z) = W_0(x_1, z)W_0^D(z)$, then

$$W_0^D(z) = \begin{pmatrix} 1 & 0 \\ -zd & 1 \end{pmatrix}. \tag{7.35}$$

Roughly speaking it can be said that W_0^D corresponds to a dipole interval of length d. If $\Delta m_0 = m_0(x_2+) - m_0(x_1)$, the corresponding factor in the maximal chain with Titchmarsh-Weyl coefficient $Q_d(z) = zQ_S(z^2)$ is equal to

$$W_d^{D,\Delta}(z) = \begin{pmatrix} 1 & 0 \\ -z\Delta m_0 - z^3 d & 1 \end{pmatrix}, \tag{7.36}$$

and factor in the chain with Titchmarsh-Weyl coefficient $Q_S(z)$ is equal to

$$W_s^{D,\Delta}(z) = I - (z^2 d + z\Delta m_0)(t(x_1), 1)^T (t(x_1), 1)J. \tag{7.37}$$

Note that both matrix functions generate 1 negative square if $d > 0$ or if $d = 0$ and $\Delta m_0 < 0$.

References

[ADRS] D. Alpay, A. Dijksma, J. Rovnyak, H. de Snoo: *Schur functions, operator colligations, and reproducing kernel Pontryagin spaces,* Oper. Theory Adv. Appl. **96**, Birkhäuser Verlag, Basel 1997.

[dB1] L. de Branges: *Hilbert spaces of entire functions,* Prentice-Hall, London 1968.

[dB2] L. de Branges: *Some Hilbert spaces of entire functions II,* Trans. Amer. Math. Soc. **99** (1961), 118–152.

[Br] P. Bruinsma: *Interpolation problems for Schur and Nevanlinna pairs,* Doctoral Dissertation, University of Groningen 1991.

[DLLS] A. Dijksma, H. Langer, A. Luger, Yu. Shondin: *A factorization result for generalized Nevanlinna functions of the class $\mathcal{N}_\kappa$,* Integral Equations Operator Theory **36** (2000), 121–125.

[Ka] I.S. Kac: *Linear relations, generated by a canonical differential equation on an interval with a regular endpoint, and expansibility in eigenfunctions (Russian),* Deposited in Ukr NIINTI, No. **1453**, 1984. (VINITI Deponirovannye Nauchnye Raboty, No. **1** (195), b.o. 720, 1985).

[KaK1] I.S. Kac and M.G. Krein: *On the spectral functions of the string,* Supplement II to the Russian edition of F.V. Atkinson, *Discrete and continuous boundary problems.* Mir, Moscow, 1968 (Russian) (English translation: Amer. Math. Soc. Transl. (2) **103** (1974), 19–102).

[KaK2] I.S. Kac and M.G. Krein: *R-functions–analytic functions mapping the upper halfplane into itself,* Supplement II to the Russian edition of F.V. Atkinson, *Discrete and continuous boundary problems.* Mir, Moscow, 1968 (Russian) (English translation: Amer. Math. Soc. Transl. (2) **103** (1974), 1–18.

[KWW1] M. Kaltenbäck, H. Winkler, H. Woracek: *Almost Pontryagin spaces,* Operator Theory Adv.Appl, to appear.

[KWW2] M. Kaltenbäck, H. Winkler, H. Woracek: *Generalized Nevanlinna functions with essentially positive spectrum,* J.Oper.Theory, to appear.

[KWW3] M. Kaltenbäck, H. Winkler, H. Woracek: *Strings, dual strings, and related canonical systems,* submitted.

[KWW4] M. Kaltenbäck, H. Winkler, H. Woracek: *De Branges spaces of entire functions symmetric about the origin,* submitted.

[KW1] M. Kaltenbäck, H. Woracek: *Pontryagin spaces of entire functions I,* Integral Equations Operator Theory **33** (1999), 34–97.

[KW2] M. Kaltenbäck, H. Woracek: *Pontryagin spaces of entire functions II,* Integral Equations Operator Theory **33** (1999), 305–380.

[KW3] M. Kaltenbäck, H. Woracek: *Pontryagin spaces of entire functions III,* Acta Sci.Math. (Szeged) **69** (2003), 241–310.

[KW4] M. Kaltenbäck, H. Woracek: *Pontryagin spaces of entire functions IV,* in preparation.

[KW5] M. Kaltenbäck, H. Woracek: *De Branges spaces of exponential type: General theory of growth,* Acta Sci.Math. (Szeged), to appear.

[KL1] M.G. Krein, H. Langer: *Über einige Fortsetzungsprobleme, die eng mit der Theorie hermitescher Operatoren im Raume Π_κ zusammenhängen. I. Einige Funktionenklassen und ihre Darstellungen,* Math. Nachr. **77** (1977), 187–236.

[KL2] M.G. Krein, H. Langer: *On some extension problems which are closely connected with the theory of hermitian operators in a space Π_κ, III. Indefinite analogues of the Hamburger and Stieltjes moment problems,* Part (1): Beiträge zur Analysis **14** (1979), 25–40. Part (2): Beiträge zur Analysis **15** (1981), 27–45.

[LW] H. Langer, H. Winkler: *Direct and inverse spectral problems for generalized strings,* Integral Equations Operator Theory **30** (1998), 409–431.

[L] B. Levin: *Nullstellenverteilung ganzer Funktionen,* Akademie Verlag, Berlin 1962.

[W1] H. Winkler: *The inverse spectral problem for canonical systems,* Integral Equations Operator Theory, **22** (1995), 360–374.

[W2] H. Winkler: *On transformations of canonical systems,* Operator Theory Adv. Appl. **80** (1995), 276–288.

[W3] H. Winkler: *Canonical systems with a semibounded spectrum,* Operator Theory Adv.Appl. **106** (1998), 397–417.

Michael Kaltenbäck
Institut für Analysis und Scientific Computing
Technische Universität Wien
Wiedner Hauptstr. 8–10/101
A–1040 Wien
Austria
e-mail: michael.kaltenbaeck@tuwien.ac.at

Henrik Winkler
Institute der Wiskunde en Informatica
Rijksuniversiteit Groningen
Postbus 800
9700 AV Groningen
The Netherlands
e-mail: winkler@math.rug.nl

Harald Woracek
Institut für Analysis und Scientific Computing
Technische Universität Wien
Wiedner Hauptstr. 8–10/101
A–1040 Wien
Austria
e-mail: harald.woracek@tuwien.ac.at

Operator Theory:
Advances and Applications, Vol. 163, 249–256

Orthogonal Polynomials on the Unit Circle with Respect to a Rational Weight Function

Levon V. Mikaelyan

Abstract. We use the analogue of the Christoffel's formula for orthogonal polynomials on the unit circle introduced in [5] to construct a system of orthogonal polynomials on the unit circle with respect to weights of the type $\left|\frac{p(z)}{g(z)}\right|^2$, where $p(z)$ and $g(z)$ are arbitrary polynomials. Exact formulas are established for Toeplitz determinants of these weights.

Mathematics Subject Classification (2000). Primary 42C05; Secondary 47B35.

Keywords. Orthogonal polynomials, Christoffel's formula, Toeplitz determinants .

For a given non-negative polynomial $\varrho(x)$ on (a,b) the well-known formula of Christoffel ([6]) represents the polynomials orthogonal on the interval (a,b) with respect to the measure $\varrho(x)d\alpha(x)$ through the orthogonal polynomials with respect to the measure $d\alpha(x)$.

In [5] (see also [2], [3]) the analogue of the above-mentioned formula for polynomials orthogonal on the unit circle was obtained. The objective of this note is to calculate the normalizing constants contained in this formula (Theorem 1 below) and give some applications of this result and Theorem 1.

We need some definitions and notation in order to formulate the analogue of the Christoffel's formula.

For a non-decreasing function $\mu(t)$ on $[-\pi,\pi]$ we denote by $d\mu(t)$ the measure generated by $\mu(t)$ on $[-\pi,\pi]$. This measure will be called of *infinite type* if the range of $\mu(t)$ consists of an infinite number of different values.

For a polynomial $P_k(z) = a_0 + a_1 z + \cdots + a_k z^k$ of degree k we introduce the following notation

$$\dot{P}_k(z) = \overline{a}_0 z^k + \overline{a}_1 z^{k-1} + \cdots + \overline{a}_k,$$

and we take for every number $\alpha \neq 0$, $\quad \dot{\alpha} = 1/\overline{\alpha}$.

The author supported in part by the grant NFSAT MA 070-02 / CRDF 12011.

It is well known (see [4]) that every non-negative trigonometrical polynomial $Q(e^{it}),\ t \in [-\pi, \pi]$, can be represented in the form

$$Q(z) = \left| a \prod_{j=1}^{n} (z - \alpha_j) \right|^2 = |a|^2 \frac{\prod_{j=1}^{n} (z - \alpha_j)(1 - z\overline{\alpha}_j)}{z^n}, \tag{1}$$

where $|\alpha_j| \neq 0$, and, from now on, let us assume that the variables t and z are connected via the formula $z = e^{it}$.

Under these agreements the following theorem holds (see [5]):

Theorem 1. *Let $d\mu(t)$ be a measure of infinite type on $[-\pi, \pi]$; $Q(z)$ be a trigonometrical polynomial of the form (1), where $\alpha_1, \alpha_2, \ldots, \alpha_n$ are pairwise distinct points and neither of them lies on the unit circle.*

If

$$d\tilde{\mu}(t) = Q(z) d\mu(t), \tag{2}$$

and $\{\varphi_k(z)\}_{k=0}^{\infty}$ and $\{\psi_k(z)\}_{k=0}^{\infty}$ are systems of orthogonal polynomials on the unit circle with respect to the measures $d\mu(t)$ and $d\tilde{\mu}(t)$ respectively, then for all $k = 0, 1, \ldots$

$$\psi_k(z) = \frac{c_k z^n}{|a|^2 \prod_{j=1}^{n} (z - \alpha_j)(1 - z\overline{\alpha}_j)} \times$$

$$\begin{vmatrix} \varphi_k(z) & \varphi_{k+1}(z) & \ldots & \varphi_{k+n}(z) & z^{-1}\dot{\varphi}_{k+1}(z) & \ldots & z^{-n}\dot{\varphi}_{k+n}(z) \\ \varphi_k(\alpha_1) & \varphi_{k+1}(\alpha_1) & \ldots & \varphi_{k+n}(\alpha_1) & \alpha_1^{-1}\dot{\varphi}_{k+1}(\alpha_1) & \ldots & \alpha_1^{-n}\dot{\varphi}_{k+n}(\alpha_1) \\ \varphi_k(\dot{\alpha}_1) & \varphi_{k+1}(\dot{\alpha}_1) & \ldots & \varphi_{k+n}(\dot{\alpha}_1) & \dot{\alpha}_1^{-1}\dot{\varphi}_{k+1}(\dot{\alpha}_1) & \ldots & \dot{\alpha}_1^{-n}\dot{\varphi}_{k+n}(\dot{\alpha}_1) \\ \vdots & \vdots & \ddots & \vdots & \vdots & \ddots & \vdots \\ \varphi_k(\alpha_n) & \varphi_{k+1}(\alpha_n) & \ldots & \varphi_{k+n}(\alpha_n) & \alpha_n^{-1}\dot{\varphi}_{k+1}(\alpha_n) & \ldots & \alpha_n^{-n}\dot{\varphi}_{k+n}(\alpha_n) \\ \varphi_k(\dot{\alpha}_n) & \varphi_{k+1}(\dot{\alpha}_n) & \ldots & \varphi_{k+n}(\dot{\alpha}_n) & \dot{\alpha}_n^{-1}\dot{\varphi}_{k+1}(\dot{\alpha}_n) & \ldots & \dot{\alpha}_1^{-n}\dot{\varphi}_{k+n}(\dot{\alpha}_n) \end{vmatrix}, \tag{3}$$

where $c_k \neq 0$ are normalizing constants.

Taking into account that the paper [5] is in Russian and is not easily accessible, for completeness reasons, we bring the proof of this theorem.

Proof. The determinant of the right-hand expression in (3) multiplied by z^n is a polynomial of degree not greater than $2n + k$ and vanishes for $z = \alpha_1, \alpha_2, \ldots, \alpha_n$ and $\dot{\alpha}_1, \dot{\alpha}_2, \ldots, \dot{\alpha}_n$; hence it is divisible by $\prod_{j=1}^{n}(z - \alpha_j)(1 - z\overline{\alpha}_j)$, so $\psi_k(z)$ is a polynomial of the degree not greater than k. Therefore, for the proof of the theorem it suffices to show that $\psi_k(z)$ is a polynomial of the degree k and satisfies the equations

$$J_{k,j} := \int_{-\pi}^{\pi} \psi_k(z) z^{-j} d\tilde{\mu}(t) = 0 \tag{4}$$

for $j = 0, 1, \ldots, k - 1$.

To prove (4), we first notice that $\psi_k(z)$ can be represented in the following manner:

$$\psi_k(z) = \frac{c_k}{Q(z)} \Big(\sum_{p=0}^{n} \mu_{p,k} \varphi_{k+p}(z) + \sum_{q=1}^{n} \nu_{q,k} z^{-q} \dot{\varphi}_{k+q}(z) \Big), \tag{5}$$

where $\mu_{p,k}$ and $\nu_{q,k}$ are some constants. From (5) we have

$$J_{k,j} = c_k\Big(\sum_{p=0}^{n}\mu_{p,k}\int_{-\pi}^{\pi}\varphi_{k+p}(z)z^{-j}d\mu(t) + \sum_{q=1}^{n}\nu_{q,k}\int_{-\pi}^{\pi}z^{-q-j}\dot{\varphi}_{k+q}(z)d\mu(t)\Big).$$

Now, for $0 \le j \le k-1$,

$$\int_{-\pi}^{\pi}\varphi_{k+p}(z)z^{-j}d\mu(t) = 0 \qquad \text{and} \qquad \int_{-\pi}^{\pi}z^{-q-j}\dot{\varphi}_{k+q}(z)d\mu(t) = 0 \tag{6}$$

for all $p = 0, 1, \dots, n$ and $q = 1, 2, \dots, n$, since the system of polynomials $\{\varphi_k(z)\}_{k=0}^{\infty}$ is orthogonal with respect to $d\mu(t)$. Note that the equations of the second part of (6) remain true also for $j = k$.

Let us prove that $\psi_k(z)$ is a polynomial of the degree k. For the leading coefficient $\psi_k^{(k)}$ of $\psi_k(z)$ we have from (5)

$$\psi_k^{(k)} = \frac{c_k\mu_{n,k}\varphi_{k+n}^{(k+n)}}{|a|^2\prod_{j=1}^{n}(-\overline{\alpha}_j)}, \tag{7}$$

where $\varphi_{k+n}^{(k+n)} \neq 0$ is the leading coefficient of $\varphi_{k+n}(z)$; so it suffices to show that $\mu_{n,k}$ is different from zero. From (3) we obtain

$$\mu_{n,k} = (-1)^n m_n^k, \tag{8}$$

where

$$m_n^k = \begin{vmatrix} \varphi_k(\alpha_1) & \varphi_{k+1}(\alpha_1) & \dots & \varphi_{k+n-1}(\alpha_1) & \alpha_1^{-1}\dot{\varphi}_{k+1}(\alpha_1) & \dots & \alpha_1^{-n}\dot{\varphi}_{k+n}(\alpha_1) \\ \varphi_k(\dot{\alpha}_1) & \varphi_{k+1}(\dot{\alpha}_1) & \dots & \varphi_{k+n-1}(\dot{\alpha}_1) & \dot{\alpha}_1^{-1}\dot{\varphi}_{k+1}(\dot{\alpha}_1) & \dots & \dot{\alpha}_1^{-n}\dot{\varphi}_{k+n}(\dot{\alpha}_1) \\ \vdots & \vdots & \ddots & \vdots & \vdots & \ddots & \vdots \\ \varphi_k(\alpha_n) & \varphi_{k+1}(\alpha_n) & \dots & \varphi_{k+n-1}(\alpha_n) & \alpha_n^{-1}\dot{\varphi}_{k+1}(\alpha_n) & \dots & \alpha_n^{-n}\dot{\varphi}_{k+n}(\alpha_n) \\ \varphi_k(\dot{\alpha}_n) & \varphi_{k+1}(\dot{\alpha}_n) & \dots & \varphi_{k+n-1}(\dot{\alpha}_n) & \dot{\alpha}_n^{-1}\dot{\varphi}_{k+1}(\dot{\alpha}_n) & \dots & \dot{\alpha}_1^{-n}\dot{\varphi}_{k+n}(\dot{\alpha}_n) \end{vmatrix}. \tag{9}$$

Assuming the contrary, we can find constants $a_0, a_1, \dots, a_{n-1}$ and $b_1, b_2, \dots, b_n$, not all of them zero, such that

$$B(z) = \frac{1}{z^n}\Big(\sum_{p=0}^{n-1}a_p z^n\varphi_{k+p}(z) + \sum_{q=1}^{n}b_q z^{n-q}\dot{\varphi}_{k+q}(z)\Big)$$

vanishes for $z = \alpha_1, \alpha_2, \dots, \alpha_n$ and $\dot{\alpha_1}, \dot{\alpha_2}, \dots, \dot{\alpha_n}$. Hence $B(z)$ is of the form $Q(z)q(z)$, where $q(z)$ is a polynomial of the degree not greater than $k-1$. It is easy to see from (6) that

$$\int_{-\pi}^{\pi}B(z)\overline{u(z)}d\mu(t) = 0$$

for all polynomials $u(z)$ of degree not greater than $k-1$; so

$$\int_{-\pi}^{\pi}Q(z)|q(z)|^2d\mu(t) = 0,$$

which contradicts the condition of $d\mu$ being a measure of infinite type. $\square$

To calculate the normalizing constants c_k let us assume that the system of polynomials $\{\varphi_k(z)\}_{k=0}^{\infty}$ is orthonormal with respect to the measure $\frac{1}{2\pi}d\mu$ with the leading coefficient of $\varphi_k(z)$ being positive for all $k = 0, 1, \ldots$. Assume also that the same is true for $\{\psi_k(z)\}_{k=0}^{\infty}$ with respect to the measure $\frac{1}{2\pi}d\tilde{\mu}$. The systems $\{\varphi_k(z)\}_{k=0}^{\infty}$ and $\{\psi_k(z)\}_{k=0}^{\infty}$ are uniquely determined under this normalization (see [4]).

For this normalization we get

$$I_k := \frac{1}{2\pi}\int_{-\pi}^{\pi} \psi_k(z)\overline{\psi_k(z)}d\tilde{\mu}(t) = 1.$$

From (5) we conclude that

$$2\pi I_k = c_k\Big(\sum_{p=0}^{n}\mu_{p,k}\int_{-\pi}^{\pi}\varphi_{k+p}(z)\overline{\psi_k(z)}d\mu(t) + \sum_{q=1}^{n}\nu_{q,k}\int_{-\pi}^{\pi} z^{-q}\dot{\varphi}_{k+q}(z)\overline{\psi_k(z)}d\mu(t)\Big),$$

and in view of (6);

$$I_k = c_k\mu_{0,k}\frac{1}{2\pi}\int_{-\pi}^{\pi}\varphi_k(z)\overline{\psi_k(z)}d\mu(t). \tag{10}$$

For $\mu_{0,k}$, from (3) we have $\mu_{0,k} = m_0^k$, where

$$m_0^k = \begin{vmatrix} \varphi_{k+1}(\alpha_1) & \varphi_{k+2}(\alpha_1) & \ldots & \varphi_{k+n}(\alpha_1) & \alpha_1^{-1}\dot{\varphi}_{k+1}(\alpha_1) & \ldots & \alpha_1^{-n}\dot{\varphi}_{k+n}(\alpha_1) \\ \varphi_{k+1}(\dot{\alpha}_1) & \varphi_{k+2}(\dot{\alpha}_1) & \ldots & \varphi_{k+n}(\dot{\alpha}_1) & \dot{\alpha}_1^{-1}\dot{\varphi}_{k+1}(\dot{\alpha}_1) & \ldots & \dot{\alpha}_1^{-n}\dot{\varphi}_{k+n}(\dot{\alpha}_1) \\ \vdots & \vdots & \ddots & \vdots & \vdots & \ddots & \vdots \\ \varphi_{k+1}(\alpha_n) & \varphi_{k+2}(\alpha_n) & \ldots & \varphi_{k+n}(\alpha_n) & \alpha_n^{-1}\dot{\varphi}_{k+1}(\alpha_n) & \ldots & \alpha_n^{-n}\dot{\varphi}_{k+n}(\alpha_n) \\ \varphi_{k+1}(\dot{\alpha}_n) & \varphi_{k+2}(\dot{\alpha}_n) & \ldots & \varphi_{k+n}(\dot{\alpha}_n) & \dot{\alpha}_n^{-1}\dot{\varphi}_{k+1}(\dot{\alpha}_n) & \ldots & \dot{\alpha}_1^{-n}\dot{\varphi}_{k+n}(\dot{\alpha}_n) \end{vmatrix}. \tag{11}$$

Now, if we rearrange the polynomial $\psi_k(z)$ as a linear combination of the polynomials $\varphi_j(z)$, $j = 0, 1, \ldots, k$, i.e.,

$$\psi_k(z) = \sum_{j=0}^{k} d_j\varphi_j(z); \tag{12}$$

from (10) and (12) we obtain

$$c_k m_0^k \overline{d_k} = 1. \tag{13}$$

Let $\varphi_k^{(k)}$ and $\psi_k^{(k)}$ be the leading coefficients of $\varphi_k(z)$ and $\psi_k(z)$ respectively. From (7), (8) and (12) it is easy to conclude that

$$\psi_k^{(k)} = d_k\varphi_k^{(k)} = \frac{c_k m_n^k \varphi_{k+n}^{(k+n)}}{|a|^2\prod_{j=1}^{n}\overline{\alpha}_j}.$$

Finally, in view of (13) and the last relation, taking into account that $\varphi_k^{(k)}$, $\varphi_{k+n}^{(k+n)}$ and $\psi_k^{(k)}$ are positive; we obtain

$$c_k = |a|e^{-i\theta_k}\sqrt{\frac{\varphi_k^{(k)}\Pi_{j=1}^{n}\alpha_j}{\varphi_{k+n}^{(k+n)} m_0^k\overline{m}_n^k}}, \qquad \psi_k^{(k)} = \frac{1}{|a|}\sqrt{\frac{\varphi_k^{(k)}\varphi_{k+n}^{(k+n)} m_n^k}{m_0^k\Pi_{j=1}^{n}\overline{\alpha}_j}}, \tag{14}$$

where m_n^k and m_0^k are defined by (9) and (11) respectively, and

$$\theta_k = \arg \frac{m_n^k}{\prod_{j=1}^n \overline{\alpha}_j} .$$

Let us now give some applications of these results.

First we consider weights $h(z)$ on the unit circle, which are ratios of non-negative trigonometrical polynomials. It is clear that such weights admit the following representation:

$$h(z) = \left| \frac{\prod_{j=1}^n (z - \alpha_j)}{g(z)} \right|^2 \qquad (z = e^{it}), \tag{15}$$

where $g(z)$ is an ordinary polynomial having zeros only inside the unit circle and with a positive leading coefficient. Using the fact that, if $g(z)$ is a polynomial of degree m, then the orthonormal polynomials $\varphi_k(z)$ on the unit circle with respect to the measure

$$\frac{1}{2\pi |g(z)|^2} dt$$

(with respect to the weight $1/2\pi|g(z)|^2$) have the form (see [4])

$$\varphi_k(z) = z^{k-m} g(z), \qquad k = m, m+1, \dots ; \tag{16}$$

from Theorem 1 we immediately obtain

Theorem 2. *Let the weight $h(z)$ be representable in the form (15), where $g(z)$ is a polynomial of the degree m with positive leading coefficient having zeros inside the unit circle only, then the polynomials $\psi_k(z)$, $k = m, m+1, \dots$, orthogonal on the unit circle with respect to the weight $h(z)$, admit the following representation:*

$$\psi_k(z) = \frac{c_k z^n}{\prod_{j=1}^n (z - \alpha_j)(1 - z\overline{\alpha}_j)} \times$$

$$\begin{vmatrix} z^{k-m} g(z) & z^{k-m+1} g(z) & \dots & z^{k-m+n} g(z) & z^{-1} \dot{g}(z) & \dots & z^{-n} \dot{g}(z) \\ \alpha_1^{k-m} g(\alpha_1) & \alpha_1^{k-m+1} g(\alpha_1) & \dots & \alpha_1^{k-m+n} g(\alpha_1) & \alpha_1^{-1} \dot{g}(\alpha_1) & \dots & \alpha_1^{-n} \dot{g}(\alpha_1) \\ \dot{\alpha}_1^{k-m} g(\dot{\alpha}_1) & \dot{\alpha}_1^{k-m+1} g(\dot{\alpha}_1) & \dots & \dot{\alpha}_1^{k-m+n} g(\dot{\alpha}_1) & \dot{\alpha}_1^{-1} \dot{g}(\dot{\alpha}_1) & \dots & \dot{\alpha}_1^{-n} \dot{g}(\dot{\alpha}_1) \\ \vdots & \vdots & \ddots & \vdots & \vdots & \ddots & \vdots \\ \alpha_n^{k-m} g(\alpha_n) & \alpha_n^{k-m+1} g(\alpha_n) & \dots & \alpha_n^{k-m+n} g(\alpha_n) & \alpha_1^{-1} \dot{g}(\alpha_n) & \dots & \alpha_1^{-n} \dot{g}(\alpha_n) \\ \dot{\alpha}_n^{k-m} g(\dot{\alpha}_n) & \dot{\alpha}_n^{k-m+1} g(\dot{\alpha}_n) & \dots & \dot{\alpha}_n^{k-m+n} g(\dot{\alpha}_n) & \dot{\alpha}_1^{-1} \dot{g}(\dot{\alpha}_n) & \dots & \dot{\alpha}_1^{-n} \dot{g}(\dot{\alpha}_n) \end{vmatrix}, \tag{17}$$

where $c_k \neq 0$ are the normalizing constants.

The special case of this theorem, when $g(z) \equiv 1$, yields the following statement:

Corollary 3. *The system of the polynomials $\{\psi_k(z)\}_{k=0}^{\infty}$, orthogonal on the unit circle with respect to the weight*

$$Q(z) = \left| a \prod_{j=1}^{n} (z - \alpha_j) \right|^2,$$

has the form

$$\psi_k(z) = \frac{c_k}{Q(z)} \begin{vmatrix} z^k & z^{k+1} & \dots & z^{k+n} & z^{-1} & \dots & z^{-n} \\ \alpha_1^k & \alpha_1^{k+1} & \dots & \alpha_1^{k+n} & \alpha_1^{-1} & \dots & \alpha_1^{-n} \\ \dot{\alpha}_1^k & \dot{\alpha}_1^{k+1} & \dots & \dot{\alpha}_1^{k+n} & \dot{\alpha}_1^{-1} & \dots & \dot{\alpha}_1^{-n} \\ \vdots & \vdots & \ddots & \vdots & \vdots & \ddots & \vdots \\ \alpha_n^k & \alpha_n^{k+1} & \dots & \alpha_n^{k+n} & \alpha_n^{-1} & \dots & \alpha_n^{-n} \\ \dot{\alpha}_n^k & \dot{\alpha}_n^{k+1} & \dots & \dot{\alpha}_n^{k+n} & \dot{\alpha}_n^{-1} & \dots & \dot{\alpha}_n^{-n} \end{vmatrix}, \tag{18}$$

where $c_k \neq 0$ are the normalizing constants.

Now, let us show some applications of the second formula in (14). Assume that $d\mu(t)$ is a measure of infinite type on $[-\pi, \pi]$, the constants

$$\hat{\mu}_k = \frac{1}{2\pi} \int_{-\pi}^{\pi} e^{-ikt} d\mu(t)$$

are its Fourier-Stieltjes coefficients and $\{\varphi_k(z)\}_{k=0}^{\infty}$ is the system of polynomials orthonormal on the unit circle with respect the measure $\frac{1}{2\pi} d\mu(t)$. If $\varphi_k(z)$ are normalized in such a way that their leading coefficients $\varphi_k^{(k)}$ are positive, then, as it is known (see for instance [4]):

$$[\varphi_k^{(k)}]^2 = \frac{D_{k-1}[\mu]}{D_k[\mu]}, \quad k = 0, 1, \dots,$$

where $D_k[\mu] = det\|\hat{\mu}_{i-j}\|_{i,j=0}^{k}$ are the Toeplitz determinants associated with the measure $d\mu(t)$, $D_{-1}[\mu] = 1$. According to the last relation, from (14) we deduce the following:

Proposition 4. *If the measures $d\mu(t)$ and $d\tilde{\mu}(t)$ satisfy the conditions of Theorem 1, then for the Toeplitz determinants associated with these measures holds:*

$$\frac{D_{k-1}[\tilde{\mu}]}{D_k[\tilde{\mu}]} = \frac{m_n^k}{|a|^2 m_0^k \Pi_{j=1}^n \overline{\alpha}_j} \sqrt{\frac{D_{k-1}[\mu] D_{k+n-1}[\mu]}{D_k[\mu] D_{k+n}[\mu]}}, \; k = 0, 1, \dots,$$

where m_n^k and m_0^k are as in (9) and (11).

In particular, if we assume that

$$d\mu(t) = \frac{1}{|g(z)|^2} dt,$$

where $g(z)$ is a polynomial of the degree m with the leading coefficient $g_m > 0$, then from (16) we obtain $\varphi_k^{(k)} = g_m$ for $k = m, m+1, \dots$.

Hence for the Toeplitz determinants associated with the measure

$$d\tilde{\mu}(t) = \left| \frac{\prod_{j=1}^{n}(z-\alpha_j)}{g(z)} \right|^2 dt,$$

we get from (14) the following representation:

$$\frac{D_{k-1}[\tilde{\mu}]}{D_k[\tilde{\mu}]} = \frac{g_m^2 m_n^k}{m_0^k \Pi_{j=1}^n \overline{\alpha}_j}$$

for $k = m, m+1, \ldots$. Applying (9) and (11) to this special case, we obtain

$$m_0^k = \begin{vmatrix} \alpha_1^{k-m+1} g(\alpha_1) & \alpha_1^{k-m+2} g(\alpha_1) & \ldots & \alpha_1^{k-m+n} g(\alpha_1) & \alpha_1^{-1} \dot{g}(\alpha_1) & \ldots & \alpha_1^{-n} \dot{g}(\alpha_1) \\ \dot{\alpha}_1^{k-m+1} g(\dot{\alpha}_1) & \dot{\alpha}_1^{k-m+2} g(\dot{\alpha}_1) & \ldots & \dot{\alpha}_1^{k-m+n} g(\dot{\alpha}_1) & \dot{\alpha}_1^{-1} \dot{g}(\dot{\alpha}_1) & \ldots & \dot{\alpha}_1^{-n} \dot{g}(\dot{\alpha}_1) \\ \vdots & \vdots & \ddots & \vdots & \vdots & \ddots & \vdots \\ \alpha_n^{k-m+1} g(\alpha_n) & \alpha_n^{k-m+2} g(\alpha_n) & \ldots & \alpha_n^{k-m+n} g(\alpha_n) & \alpha_1^{-1} \dot{g}(\alpha_n) & \ldots & \alpha_1^{-n} \dot{g}(\alpha_n) \\ \dot{\alpha}_n^{k-m+1} g(\dot{\alpha}_n) & \dot{\alpha}_n^{k-m+2} g(\dot{\alpha}_n) & \ldots & \dot{\alpha}_n^{k-m+n} g(\dot{\alpha}_n) & \dot{\alpha}_1^{-1} \dot{g}(\dot{\alpha}_n) & \ldots & \dot{\alpha}_1^{-n} \dot{g}(\dot{\alpha}_n) \end{vmatrix}$$

and

$$m_n^k = \begin{vmatrix} \alpha_1^{k-m} g(\alpha_1) & \alpha_1^{k-m+1} g(\alpha_1) & \ldots & \alpha_1^{k-m+n-1} g(\alpha_1) & \alpha_1^{-1} \dot{g}(\alpha_1) & \ldots & \alpha_1^{-n} \dot{g}(\alpha_1) \\ \dot{\alpha}_1^{k-m} g(\dot{\alpha}_1) & \dot{\alpha}_1^{k-m+1} g(\dot{\alpha}_1) & \ldots & \dot{\alpha}_1^{k-m+n-1} g(\dot{\alpha}_1) & \dot{\alpha}_1^{-1} \dot{g}(\dot{\alpha}_1) & \ldots & \dot{\alpha}_1^{-n} \dot{g}(\dot{\alpha}_1) \\ \vdots & \vdots & \ddots & \vdots & \vdots & \ddots & \vdots \\ \alpha_n^{k-m} g(\alpha_n) & \alpha_n^{k-m+1} g(\alpha_n) & \ldots & \alpha_n^{k-m+n-1} g(\alpha_n) & \alpha_1^{-1} \dot{g}(\alpha_n) & \ldots & \alpha_1^{-n} \dot{g}(\alpha_n) \\ \dot{\alpha}_n^{k-m} g(\dot{\alpha}_n) & \dot{\alpha}_n^{k-m+1} g(\dot{\alpha}_n) & \ldots & \dot{\alpha}_n^{k-m+n-1} g(\dot{\alpha}_n) & \dot{\alpha}_1^{-1} \dot{g}(\dot{\alpha}_n) & \ldots & \dot{\alpha}_1^{-n} \dot{g}(\dot{\alpha}_n) \end{vmatrix}.$$

Remark 5. If the determinants in formulas (17) and (18) are expanded according to the Laplace's formula with respect to the first $(n+1)$ columns, then the multipliers in the corresponding sum will be the Vandermonde determinants, or will differ from Vandermonde determinants by a multiplier, which is easy to calculate. Consequently, another representation of orthogonal polynomials can be obtained from (17) and (18). In this way, using the last two equations one can obtain Day's formula for Toeplitz determinants of rational functions, given the condition that they are non-negative on the unit circle (see [1]).

Remark 6. In the case, if in the theorems above and the corollary some α-s have multiplicity greater than one, one should proceed as follows:

i) If $\alpha_j, (|\alpha_j| \neq 1)$ has multiplicity $m\,(m > 1)$, i.e., if there are m multipliers $(z-\alpha_j)(1-z\overline{\alpha}_j)$ in the representation of $Q(z)$, then the rows $2k_j + 2$, $2k_j + 4, \ldots, 2k_j + 2m - 2$ of the right-hand determinant of (3), (17), (18) must be replaced by derivatives of the orders $1, 2, \ldots, m-1$ of the first row at the point α_j, and the rows $2k_j + 3$, $2k_j + 5, \ldots, 2k_j + 2m - 1$ must be replaced by the same derivatives at the point $\dot{\alpha}_j$. Here $2k_j$ is the number of the first row, which contains α_j.

ii) If $|\alpha_j| = 1$, i.e., $\alpha_j = \dot{\alpha}_j$ and there are m multipliers

$$(z - \alpha_j)(1 - z\overline{\alpha}_j) = -\frac{(z - \alpha_j)^2}{\alpha_j}$$

in the representation of $Q(z)$, then the rows $2k_j$, $2k_j + 1$, $\dots$, $2k_j + 2m - 1$ must be replaced by derivatives of orders $0, 1, \dots, 2m - 1$ of the first row at the point α_j.

Remark 7. We would like to mention that one can use Theorem 1 for the construction of an algorithm for finding orthogonal polynomials on the unit circle with respect to the given rational weight.

Acknowledgements

I would like to express my gratitude to the organizers of the Colloquium on Operator Theory in Vienna and especially to Professor Heinz Langer, whose efforts made possible my participation. I also convey my thanks to the Referees for their valuable suggestions.

References

[1] A. Böttcher, B. Silbermann, *Analysis of Toeplitz Operators.* Springer-Verlag, Berlin, 1990.

[2] B.L. Golinskiĭ, *An analogue of the Christoffel formula for orthogonal polynomials on the unit circle and some applications.* Izv. Vyssh. Uchebn. Zaved. Mat. **1**:(2) (1958), 33–42.

[3] E. Godoy, F. Marcellan, *An analog of the Christoffel formula for polynomial modification of a measure on the unit circle.* Boll. Un. Mat. Ital. A (7) **5** (1991), 1–12.

[4] U. Grenander, G. Szegő, *Toeplitz forms and their applications.* Univ. Calif. Press, Berkeley, Los Angeles, 1958; Russ. transl.: Izdat. Inostr. Lit., Moscow, 1961.

[5] L. Mikaelyan, *The analogue of Christoffel formula for orthogonal polynomials on unit circle.* Dokl. Akad. Nauk Arm. SSR, **67**:5 (1978), 257–263.

[6] G. Szegő, *Orthogonal Polynomials.* Volume 23 of AMS Colloquium Publications, Providence, R.I., 1959; Russ. transl.: Izdat. Fiz.-Mat. Lit., Moscow 1962.

Levon V. Mikaelyan
Dept. of Applied Mathematics
Yerevan State University
A. Manoukian 1
375025 Yerevan
Armenia
e-mail: mikaelyanl@ysu.am

Operator Theory:
Advances and Applications, Vol. 163, 257–274

Bi-dimensional Moment Problems and Regular Dilations

Dan Popovici

To Professor Heinz Langer

Abstract. Given a map f on $\mathbb{Z}_+^2 \times X$ (X is just a non-empty set) into a Hilbert space $\mathcal{H}$ we provide necessary and sufficient conditions in order to ensure the existence of a commuting pair (S, T) of contractions on $\mathcal{H}$ having regular dilation such that

$$S^m T^n f(0,0,x) = f(m,n,x), \quad (m,n) \in \mathbb{Z}_+^2, \ x \in X.$$

Such moment problems are strongly related to the theory of harmonizable and stationary processes. Isometric or unitary solutions are also characterized in terms of the initial data.

Mathematics Subject Classification (2000). Primary 47A57, 47A20, 60G10; Secondary 43A35, 47A45, 47B15.

Keywords. Moment problem, regular dilation, unitary extension, spectral measure, positive definite function, harmonizable process, stationary process.

1. Introduction

One of the most important directions in the theory of non-selfadjoint operators was opened by the theorem of Sz.-Nagy [20] on the existence of a unitary dilation for every Hilbert space contraction. Among numerous applications of this last theorem we want to mention invariant subspace theory, interpolation theory or prediction theory for stochastic processes ([22], [4], [7]).

More recently, several authors have contemplated the idea to extend known results for single operators to systems of (at least two) commuting operators on the same Hilbert space. It is expected that such generalizations would provide a larger set of applications. The problem of finding a unitary (or isometric) dilation of a

This work was supported by the EU Research Training Network "Analysis and Operators" with contract no. HPRN-CT-2000-00116.

system of several commuting contractions was proposed and solved by Sz.-Nagy under the additional assumption that the given system is doubly commuting ([21]). A special kind of a joint dilation for n commuting contractions, called regular and based on the fact that a certain operator function is positive definite on $\mathbb{Z}^n$, was introduced and developed by Brehmer in [1]. We should mention here the work of Sz.-Nagy [19] and Halperin [5] which brought more light on the subject. Regular dilations have been recently studied, for example, in [2], [3], [23], [8], [9], [11].

It is our aim in the present paper to give detailed solutions to the following problem:

Problem 1.1. *Provide necessary and sufficient conditions on a function $f : \mathbb{Z}_+^2 \times X \to \mathcal{H}$ (X is a non-empty set and $\mathcal{H}$ a Hilbert space) in order to ensure the existence of a commuting pair (S, T) of contractions on $\mathcal{H}$ having regular dilation and such that*

$$f(m, n, x) = S^m T^n f(0, 0, x), \quad (m, n) \in \mathbb{Z}_+^2, \ x \in X.$$

Similar one-dimensional problems have been proposed and completely solved in [14], [15], [16], [17], [18] or [12]. For the multi-dimensional case we refer to [13] and [6].

From the applications point of view we should mention the connection of such problems with the theory of harmonizable stochastic processes[1]:

An $\mathcal{H}$-valued discrete process $\{f(n)\}_n$ on $\mathbb{Z}$ is weakly harmonizable if and only if it has a stationary dilation. In other words, there exist a Hilbert space $\mathcal{K}$ containing $\mathcal{H}$ (as a closed subspace) and a $\mathcal{K}$-valued discrete stationary process $\{g(n)\}_n$ such that $f(n) = P_{\mathcal{H}}^{\mathcal{K}} g(n)$, $n \in \mathbb{Z}$. In particular, if an operator function Φ is positive definite on $\mathbb{Z}$ with $\Phi(0) = I_{\mathcal{H}}$ then the process $\{\Phi(n)h\}_n$ is weakly harmonizable, for every $h \in \mathcal{H}$. In addition, Φ has the form

$$\Phi(n) = \begin{cases} T^n, & n \geq 0 \\ T^{*|n|}, & n < 0 \end{cases},$$

for a certain contraction T on $\mathcal{H}$ ([21]).

Passing to the bi-dimensional case, we consider $\mathcal{H}$-valued two-time parameter discrete weakly harmonizable processes on $\mathbb{Z}^2$ of the form $\{\Phi(m, n)h\}_{(m,n)}$ ($h \in \mathcal{H}$ is fixed), where Φ is a given positive definite function on $\mathbb{Z}^2$. In particular, we assume that Φ has the form

$$\Phi_{S,T}(m, n) = \begin{cases} S^m T^n, & m, n \geq 0 \\ S^{*|m|} T^n, & m < 0, n \geq 0 \\ T^{*|n|} S^m, & m \geq 0, n < 0 \\ S^{*|m|} T^{*|n|}, & m, n < 0 \end{cases}, \tag{1.1}$$

defined in terms of a certain pair (S, T) of commuting contractions on $\mathcal{H}$. $\Phi_{S,T}$ is positive definite on $\mathbb{Z}^2$ if and only if (S, T) has a regular unitary dilation (U, Z). In this case $\{U^m Z^n h\}_{(m,n) \in \mathbb{Z}^2}$ is a stationary dilation of $\{\Phi_{S,T}(m, n)h\}_{(m,n) \in \mathbb{Z}^2}$.

[1]for the theory of harmonizable processes we refer to [7]

We choose to characterize two-time parameter weakly harmonizable processes of the form $\{\Phi_{S,T}(m,n)h\}_{(m,n)}$ defined on $\mathbb{Z}_+^2$, rather than on $\mathbb{Z}^2$. In fact, we consider a family $\{\{f(m,n,x)\}_{(m,n)}\}$ (indexed by X) of two-time parameter processes and provide necessary and sufficient conditions in order to ensure an expression as above, but defined in terms of the same (for all processes in the family) pair (S,T) of commuting contractions having regular dilation.

We also characterize the cases when (S,T) is a pair of commuting isometries (i.e., the process is stationary[2]) or unitary operators on $\mathcal{H}$. Various kind of applications conclude the paper.

2. Notation and preliminaries

Let $\mathcal{H}$ and $\mathcal{K}$ be complex Hilbert spaces. Denote by $\mathcal{L}(\mathcal{H},\mathcal{K})$ ($\mathcal{L}(\mathcal{H})$ if $\mathcal{H}=\mathcal{K}$) the Banach space of all linear and bounded operators between $\mathcal{H}$ and $\mathcal{K}$. If $\{\mathcal{H}_\alpha\}_{\alpha\in I}$ is a family of (closed) subspaces in $\mathcal{H}$ then its closed linear span is expressed by $\bigvee_{\alpha\in I}\mathcal{H}_\alpha$.

A function $\Phi:\mathbb{Z}^2\to\mathcal{L}(\mathcal{H})$ is said to be *positive definite* if

$$\sum_{(m,n),(p,q)\in\mathbb{Z}^2}\langle\Phi(m-p,n-q)h_{m,n};h_{p,q}\rangle\geq 0,$$

for every sequence $\{h_{m,n}\}\subset\mathcal{H}$ with finite support.

A *unitary dilation* of a commuting pair (S,T) of contractions on $\mathcal{H}$ is a commuting pair (U,Z) of unitary operators on a Hilbert space $\mathcal{K}$ containing $\mathcal{H}$ (as a closed subspace) such that

$$S^mT^nh=P_{\mathcal{H}}U^mZ^nh,\quad (m,n)\in\mathbb{Z}_+^2,\ h\in\mathcal{H}$$

($P_{\mathcal{H}}$ is the orthogonal projection of $\mathcal{K}$ onto $\mathcal{H}$). Such a dilation is said to be *minimal* if

$$\mathcal{K}=\bigvee_{(m,n)\in\mathbb{Z}^2}U^mZ^n\mathcal{H}.$$

A unitary dilation (U,Z) on $\mathcal{K}$ for a commuting pair (S,T) of contractions on $\mathcal{H}$ is said to be *regular* if, in addition,

$$S^{*m}T^nh=P_{\mathcal{H}}U^{*m}Z^nh,\quad (m,n)\in\mathbb{Z}_+^2,\ h\in\mathcal{H}.$$

It was stated by Brehmer ([1], [19], [5]) that a pair (S,T) of commuting contractions has regular unitary dilation if and only if

$$I_{\mathcal{H}}-S^*S-T^*T+S^*T^*ST\geq 0.$$

[2]A process $\{f(m,n)\}_{(m,n)\in\mathbb{Z}_+^2}\subset\mathcal{H}$ is supposed to be *stationary* if $\langle f(m+1,n),f(p+1,q)\rangle=\langle f(m,n+1),f(p,q+1)\rangle=\langle f(m,n),f(p,q)\rangle$, $(m,n),(p,q)\in\mathbb{Z}_+^2$.

If (S,T) has a regular unitary dilation (U,Z) on $\mathcal{K}$ then, by Stone's theorem, there exists an $\mathcal{L}(\mathcal{K})$-valued spectral measure E on the bitorus $\mathbb{T}^2$ such that

$$U^m Z^n = \int_{\mathbb{T}^2} \bar{\lambda}^m \bar{\mu}^n dE(\lambda,\mu), \quad (m,n) \in \mathbb{Z}^2. \tag{2.1}$$

Setting, for any Borel subset σ of $\mathbb{T}^2$,

$$\omega(\sigma) = P_{\mathcal{H}} E(\sigma)|_{\mathcal{H}}, \tag{2.2}$$

we obtain an $\mathcal{L}(\mathcal{H})$-valued semispectral measure on $\mathbb{T}^2$, uniquely determined by (S,T) such that

$$\Phi_{S,T}(m,n) = \int_{\mathbb{T}^2} \bar{\lambda}^m \bar{\mu}^n d\omega(\lambda,\mu), \quad (m,n) \in \mathbb{Z}^2. \tag{2.3}$$

Conversely, suppose that (2.3) holds for an $\mathcal{L}(\mathcal{H})$-valued semispectral measure on $\mathbb{T}^2$. Then, by Naimark's theorem, there exists a larger Hilbert space $\mathcal{K}$ and an $\mathcal{L}(\mathcal{K})$-valued spectral measure E on $\mathbb{T}^2$ such that (2.2) holds. The pair (U,Z) defined on $\mathcal{K}$ by

$$U = \int_{\mathbb{T}^2} \bar{\lambda} dE(\lambda,\mu) \text{ and } Z = \int_{\mathbb{T}^2} \bar{\mu} dE(\lambda,\mu)$$

is clearly a regular unitary dilation of (S,T) and satisfies (2.1).

Consequently, the following conditions are equivalent:

(i) (S,T) has a regular unitary dilation;

(ii) There exists an $\mathcal{L}(\mathcal{H})$-valued semispectral measure ω on $\mathbb{T}^2$ such that (2.3) holds;

(iii) $\Phi_{S,T}$ is positive definite.

In other words our problem for a function $f : \mathbb{Z}_+^2 \times X \to \mathcal{H}$ can be reformulated as follows:

Find necessary and sufficient conditions in order to ensure the existence of an $\mathcal{L}(\mathcal{H})$-valued semispectral measure ω on $\mathbb{T}^2$ such that

$$f(m,n,x) = \int_{\mathbb{T}^2} \bar{\lambda}^m \bar{\mu}^n d\omega(\lambda,\mu) f(0,0,x), \quad (m,n) \in \mathbb{Z}_+^2, \ x \in X$$

and

$$\Phi_{\int_{\mathbb{T}^2} \bar{\lambda} d\omega(\lambda,\mu), \int_{\mathbb{T}^2} \bar{\mu} d\omega(\lambda,\mu)}(m,n) = \int_{\mathbb{T}^2} \bar{\lambda}^m \bar{\mu}^n d\omega(\lambda,\mu), \quad (m,n) \in \mathbb{Z}^2.$$

A (regular) unitary dilation for a pair (V,W) of commuting isometries on $\mathcal{H}$ is, in fact, a unitary extension. If (U_1,Z_1) on $\mathcal{K}_1$ and (U_2,Z_2) on $\mathcal{K}_2$ are minimal unitary extensions for (V,W) then there exists a unique (up to an isomorphism) unitary operator $\varphi : \mathcal{K}_1 \to \mathcal{K}_2$ which leaves $\mathcal{H}$ invariant ($\varphi|_{\mathcal{H}} = I_{\mathcal{H}}$) and intertwines U_1 and U_2 ($\varphi U_1 = U_2 \varphi$), respectively Z_1 and Z_2 ($\varphi Z_1 = Z_2 \varphi$). According to

[10], a model for the minimal unitary extension of (V,W) is given on $\mathcal{K} = \mathcal{H} \oplus H^2_{\ker(VW)^*}(\mathbb{T})$ by the formulas

$$U(h, f(z)) := \Big(Vh + W^*f(0), V(I - WW^*)f(z) + W^*\frac{f(z) - f(0)}{z}\Big) \tag{2.4}$$

and

$$Z(h, f(z)) := \Big(Wh + V^*f(0), W(I - VV^*)f(z) + V^*\frac{f(z) - f(0)}{z}\Big), \tag{2.5}$$
$$h \in \mathcal{H},\ f \in H^2_{\ker(VW)^*}(\mathbb{T}).$$

3. General solutions

Let X be a non-empty set, $\mathcal{H}$ a complex Hilbert space and $f : \mathbb{Z}^2_+ \times X \to \mathcal{H}$. The first result is a generalization of the one in [13]:

Theorem 3.1. *The following conditions are equivalent:*

(i) *Problem* 1.1 *has a solution;*

(ii)
$$\Big\| \sum_{(m,n),(p,q),x} c^x_{(m,n),(p,q)} f(m+p, n+q, x)\Big\|^2$$
$$\leq \sum_{\substack{(m,n),(m',n'),x\\(p,q),(p',q'),y}} c^x_{(m,n),(m',n')}\overline{c^y_{(p,q),(p',q')}}\langle f((m-p)^+ + m', (n-q)^+ + n', x),$$
$$f((m-p)^- + p', (n-q)^- + q', y)\rangle,$$
for every finite family $\{c^x_{(m,n),(m',n')}\}$ *of complex numbers;*

(iii)
$$\sum_{\substack{(m,n),(m',n'),x\\(p,q),(p',q'),y}} c^x_{(m,n),(m',n')}\overline{c^y_{(p,q),(p',q')}}\langle f((m-p)^+ + m', (n-q)^+ + n', x),$$
$$f((m-p)^- + p', (n-q)^- + q', y)\rangle \geq 0,$$
for every finite family $\{c^x_{(m,n),(m',n')}\}$ *of complex numbers;*

(iv)
$$\Big\| \sum_{(m,n),x} c^x_{m,n} f(m+1, n, x)\Big\| \leq \Big\| \sum_{(m,n),x} c^x_{m,n} f(m, n, x)\Big\|,$$
$$\Big\| \sum_{(m,n),x} c^x_{m,n} f(m, n+1, x)\Big\| \leq \Big\| \sum_{(m,n),x} c^x_{m,n} f(m, n, x)\Big\|$$
and
$$\Big\| \sum_{(m,n),x} c^x_{m,n} f(m+1, n, x)\Big\|^2 + \Big\| \sum_{(m,n),x} c^x_{m,n} f(m, n+1, x)\Big\|^2$$
$$\leq \Big\| \sum_{(m,n),x} c^x_{m,n} f(m, n, x)\Big\|^2 + \Big\| \sum_{(m,n),x} c^x_{m,n} f(m+1, n+1, x)\Big\|^2,$$
for every finite family $\{c^x_{m,n}\}$ *of complex numbers.*

Proof. (i) $\Rightarrow$ (ii) Suppose that Problem 1.1 has a solution (S,T) (on $\mathcal{H}$) having a regular unitary dilation (U,Z) (on $\mathcal{K}$). Then, for every finite family $\{c^x_{(m,n),(p,q)}\}$ of complex numbers,

$$\begin{aligned}
&\|\sum_{(m,n),(p,q),x} c^x_{(m,n),(p,q)} S^{m+p}T^{n+q}f(0,0,x)\|^2\\
&= \|\sum_{(m,n),(p,q),x} c^x_{(m,n),(p,q)} P_{\mathcal{H}}U^mZ^nS^pT^qf(0,0,x)\|^2\\
&\leq \langle \sum_{(m,n),(m',n'),x} c^x_{(m,n),(m',n')} U^mZ^nS^{m'}T^{n'}f(0,0,x),\\
&\qquad\qquad \sum_{(p,q),(p',q'),y} c^y_{(p,q),(p',q')} U^pZ^qS^{p'}T^{q'}f(0,0,y)\rangle\\
&= \sum_{\substack{(m,n),(m',n'),x\\(p,q),(p',q'),y}} c^x_{(m,n),(m',n')}\overline{c^y_{(p,q),(p',q')}}\\
&\qquad\qquad \langle U^{m-p}Z^{n-q}S^{m'}T^{n'}f(0,0,x), S^{p'}T^{q'}f(0,0,y)\rangle\\
&= \sum_{\substack{(m,n),(m',n'),x\\(p,q),(p',q'),y}} c^x_{(m,n),(m',n')}\overline{c^y_{(p,q),(p',q')}}\langle S^{(m-p)^+}T^{(n-q)^+}S^{m'}T^{n'}f(0,0,x),\\
&\qquad\qquad S^{(m-p)^-}T^{(n-q)^-}S^{p'}T^{q'}f(0,0,y)\rangle.
\end{aligned}$$

(ii) $\Rightarrow$ (iii) is obvious.

(iii) $\Rightarrow$ (iv) Fix a finite family $\{c^x_{m,n}\}$ of complex numbers. The inequalities of (iv) can be obtained if the family $\{c^x_{(m,n),(m',n')}\}$ in (iii) take, respectively, one of the following forms:

(a)
$$c^x_{(m,n),(m',n')} = \begin{cases} -c^x_{m'-1,n'}, & \text{if } m=n=0;\ m'\geq 1\\ c^x_{m',n'}, & \text{if } m=1, n=0\\ 0, & \text{for the other cases}\end{cases};$$

(b)
$$c^x_{(m,n),(m',n')} = \begin{cases} -c^x_{m',n'-1}, & \text{if } m=n=0;\ n'\geq 1\\ c^x_{m',n'}, & \text{if } m=0, n=1\\ 0, & \text{for the other cases}\end{cases};$$

(c)
$$c^x_{(m,n),(m',n')} = \begin{cases} c^x_{m'-1,n'-1}, & \text{if } m=n=0;\ m',n'\geq 1\\ -c^x_{m'-1,n'}, & \text{if } m=0, n=1;\ m'\geq 1\\ -c^x_{m',n'-1}, & \text{if } m=1, n=0;\ n'\geq 1\\ c^x_{m',n'}, & \text{if } m=n=1\\ 0, & \text{for the other cases}\end{cases},$$

$$((m,n),(m',n')\in\mathbb{Z}^2_+,\ x\in X).$$

Observe that the sum in (iii) reduces to the terms for which $m,n,p,q \in \{0,1\}$. In view of the formulas

$$(a-b)^+ = \begin{cases} 0, & a=0 \\ 1-b, & a=1 \end{cases}$$

and

$$(a-b)^- = \begin{cases} b, & a=0 \\ 0, & a=1 \end{cases} \qquad (b \in \{0,1\})$$

we only have to change some summation variables to get the conclusion.

(iv) $\Rightarrow$ (i) Let

$$\mathcal{H}^f := \bigvee_{(m,n),x} f(m,n,x).$$

By (iv), operators

$$\mathcal{H}^f \ni \sum_{(m,n),x} c^x_{m,n} f(m,n,x) \overset{S_0}{\mapsto} \sum_{(m,n),x} c^x_{m,n} f(m+1,n,x) \in \mathcal{H}^f$$

and

$$\mathcal{H}^f \ni \sum_{(m,n),x} c^x_{m,n} f(m,n,x) \overset{T_0}{\mapsto} \sum_{(m,n),x} c^x_{m,n} f(m,n+1,x) \in \mathcal{H}^f$$

are well defined contractions on $\mathcal{H}^f$. They commute since

$$\begin{aligned} S_0T_0f(m,n,x) &= f(m+1,n+1,x) \\ &= T_0S_0f(m,n,x), \qquad (m,n) \in \mathbb{Z}^2_+, x \in X. \end{aligned}$$

It is easily shown (inductively) that formulas

$$f(m,n,x) = S_0^m T_0^n f(0,0,x), \quad (m,n) \in \mathbb{Z}^2_+, x \in X$$

hold true. The last inequality of (iv) becomes

$$\|S_0h\|^2 + \|T_0h\|^2 \le \|h\|^2 + \|S_0T_0h\|^2,$$

for every h in the linear space generated by $\{f(m,n,x) \mid (m,n) \in \mathbb{Z}^2_+,\ x \in X\}$, so for every $h \in \mathcal{H}^f$. Deduce that

$$I_{\mathcal{H}^f} - S_0^*S_0 - T_0^*T_0 + S_0^*T_0^*S_0T_0 \ge 0,$$

and the pair (S_0,T_0) has regular unitary dilation. It can be extended (by 0 on $\mathcal{H} \ominus \mathcal{H}^f$) to a commuting pair (S,T) of contractions on $\mathcal{H}$ having regular unitary dilation. Obviously,

$$f(m,n,x) = S^mT^nf(0,0,x), \quad (m,n) \in \mathbb{Z}^2_+,\ x \in X,$$

as required. □

Remark 3.2.

- There is another method, similar to the one used in [13] (cf. [14]) to build a solution to Problem 1.1 starting from the inequality (ii) of Theorem 3.1, but without using (iv). On the linear space $\mathcal{F}_0$ of all finite families $\{c^x_{(m,n),(p,q)}\}$ of complex numbers we define (in view of (ii)) the semi-inner product

$$\begin{aligned}&\langle\{c^x_{(m,n),(m',n')}\},\{d^y_{(p,q),(p',q')}\}\rangle \qquad (3.1)\\ &:= \sum_{\substack{(m,n),(m',n'),x\\(p,q),(p',q'),y}} c^x_{(m,n),(m',n')}\overline{d^y_{(p,q),(p',q')}}\langle f((m-p)^+ + m',(n-q)^+ + n',x),\\ &\qquad f((m-p)^- + p',(n-q)^- + q',y)\rangle,\\ &\qquad\qquad \{c^x_{(m,n),(m',n')}\},\{d^y_{(p,q),(p',q')}\}\in\mathcal{F}_0.\end{aligned}$$

  By factorization and completion $\mathcal{F}_0$ becomes a Hilbert space $\mathcal{F}$. The map

$$\mathcal{F}\ni\{c^x_{(m,n),(p,q)}\}\overset{\varphi}{\mapsto}\sum_{(m,n),(p,q),x} c^x_{(m,n),(p,q)}f(m+p,n+q,x)\in\mathcal{H}$$

  is a well-defined contraction (consequence of (ii)). Operators

$$\mathcal{F}\ni\{c^x_{(m,n),(p,q)}\}\overset{A}{\mapsto}\{c^x_{(m+1,n),(p,q)}\}\in\mathcal{F} \qquad (3.2)$$

  and

$$\mathcal{F}\ni\{c^x_{(m,n),(p,q)}\}\overset{B}{\mapsto}\{c^x_{(m,n+1),(p,q)}\}\in\mathcal{F} \qquad (3.3)$$

  have isometric adjoints and the pair $(\varphi A^*\varphi^*,\varphi B^*\varphi^*)$ is a solution to Problem 1.1.
- A solution to Problem 1.1 is unique if and only if $\mathcal{H}=\mathcal{H}^f$.
- If $f:\mathbb{Z}^2_+\times X\to\mathcal{H}$ satisfies one of the equivalent conditions of Theorem 3.1 then it can be extended to a function $F:\mathbb{Z}^2\times X\to\mathcal{H}$ such that

$$\begin{aligned}&\|\sum_{(m,n),x} c^x_{m,n}F(m,n,x)\|^2\\ &\qquad\le\sum_{(m,n),(p,q),x,y} c^x_{m,n}\overline{c^y_{p,q}}\langle F(m-p,n-q,x),f(0,0,y)\rangle,\end{aligned}$$

  for every finite family $\{c^x_{m,n}\}_{(m,n)\in\mathbb{Z}^2,x\in X}$ of complex numbers. More precisely, let

$$F(m,n,x):=\Phi_{S,T}(m,n)f(0,0,x),\quad (m,n)\in\mathbb{Z}^2,\ x\in X$$

  be such a function defined in terms of a solution (S,T) of Problem 1.1. Then, for every finite family $\{c^x_{m,n}\}$ of complex numbers and any regular unitary dilation (U,Z) of (S,T),

$$
\begin{aligned}
&\|\sum_{(m,n),x} c^x_{m,n}\Phi_{S,T}(m,n)f(0,0,x)\|^2\\
&\leq \|\sum_{(m,n),x} c^x_{m,n}U^mZ^nf(0,0,x)\|^2\\
&= \langle \sum_{(m,n),x} c^x_{m,n}U^mZ^nf(0,0,x), \sum_{(p,q),y} c^y_{p,q}U^pZ^qf(0,0,y)\rangle\\
&= \sum_{(m,n),(p,q),x,y} c^x_{m,n}\overline{c^y_{p,q}}\langle U^{m-p}Z^{n-q}f(0,0,x), f(0,0,y)\rangle\\
&= \sum_{(m,n),(p,q),x,y} c^x_{m,n}\overline{c^y_{p,q}}\langle \Phi_{S,T}(m-p,n-q)f(0,0,x), f(0,0,y)\rangle.
\end{aligned}
$$

The first application generalizes the one in [14]:

Corollary 3.3. *Let $f:\mathbb{Z}_+\times X\to\mathcal{H}$. The following conditions are equivalent:*

(i) *There exists a contraction S on $\mathcal{H}$ such that*

$$f(m,x)=S^mf(0,x),\quad m\geq 0,\ x\in X;$$

(ii)
$$
\begin{aligned}
&\|\sum_{(m,n),x} c^x_{m,n}f(m+n,x)\|^2\\
&\leq \sum_{(m,n),(p,q),x,y} c^x_{m,n}\overline{c^y_{p,q}}\langle f((m-p)^+ +n,x), f((m-p)^- +q,y)\rangle,
\end{aligned}
$$

for every finite family $\{c^x_{m,n}\}$ of complex numbers;

(iii)
$$\sum_{(m,n),(p,q),x,y} c^x_{m,n}\overline{c^y_{p,q}}\langle f((m-p)^+ +n,x), f((m-p)^- +q,y)\rangle\geq 0,$$

for every finite family $\{c^x_{m,n}\}$ of complex numbers;

(iv)
$$\|\sum_{m,x} c^x_m f(m+1,x)\|\leq \|\sum_{m,x} c^x_m f(m,x)\|,$$

for every finite family $\{c^x_{m,n}\}$ of complex numbers.

Proof. It is easy to observe that a solution to Problem 1.1 (if it exists) for the map

$$\mathbb{Z}^2_+\times X\ni (m,n,x)\mapsto f(m,x)\in\mathcal{H}$$

can be taken of the form $(S,I_{\mathcal{H}})$ such that S is a contraction on $\mathcal{H}$. Finally, the link between two finite families $\{c^x_{m,n}\}$ and $\{c^x_{(m,n),(p,q)}\}$ of complex numbers is given by

$$c^x_{m,n}=\sum_{p,q} c^x_{(m,n),(p,q)}.$$

□

For a given double sequence of operators $\{A_{m,n}\}_{(m,n)\in\mathbb{Z}_+^2} \subset \mathcal{L}(\mathcal{K},\mathcal{H})$, if we re-write the equivalent conditions of Theorem 3.1 for the function

$$\mathbb{Z}_+^2 \times \mathcal{K} \ni (m,n,k) \overset{f}{\mapsto} A_{m,n}(k) \in \mathcal{H}$$

and replace any finite family $\{c^x_{(m,n),(p,q)}\}$ of complex numbers with the finite double sequence $\{k_{(m,n),(p,q)}\} \subset \mathcal{K}$ given by

$$k_{(m,n),(p,q)} := \sum_{k\in\mathcal{K}} c^k_{(m,n),(p,q)} k, \quad (m,n),(p,q) \in \mathbb{Z}_+^2,$$

then we obtain an improvement and an extension of [15]:

Corollary 3.4. *The following conditions are equivalent:*

(i) *There exists a commuting pair (S,T) of contractions on $\mathcal{H}$ having regular unitary dilation and such that*

$$A_{m,n} = S^m T^n A_{0,0}, \quad (m,n) \in \mathbb{Z}_+^2;$$

(ii)
$$\begin{aligned}\Big\| \sum_{(m,n),(p,q)} & A_{m+p,n+q}(k_{(m,n),(p,q)})\Big\|^2 \\ &\leq \sum_{\substack{(m,n),(m',n')\\(p,q),(p',q')}} \langle A_{(m-p)^+ +m',(n-q)^+ +n'}(k_{(m,n),(m',n')}), \\ &\qquad\qquad A_{(m-p)^- +p',(n-q)^- +q'}(k_{(p,q),(p',q')})\rangle,\end{aligned}$$

for every finite sequence $\{k_{(m,n),(m',n')}\}$ of vectors in $\mathcal{K}$;

(iii)
$$\begin{aligned}\sum_{\substack{(m,n),(m',n')\\(p,q),(p',q')}} & \langle A_{(m-p)^+ +m',(n-q)^+ +n'}(k_{(m,n),(m',n')}), \\ &\qquad A_{(m-p)^- +p',(n-q)^- +q'}(k_{(p,q),(p',q')})\rangle \geq 0,\end{aligned}$$

for every finite sequence $\{k_{(m,n),(m',n')}\}$ of vectors in $\mathcal{K}$;

(iv)
$$\Big\|\sum_{m,n} A_{m+1,n}(k_{m,n})\Big\| \leq \Big\|\sum_{m,n} A_{m,n}(k_{m,n})\Big\|,$$
$$\Big\|\sum_{m,n} A_{m,n+1}(k_{m,n})\Big\| \leq \Big\|\sum_{m,n} A_{m,n}(k_{m,n})\Big\|$$

and

$$\begin{aligned}\Big\|\sum_{m,n} & A_{m+1,n}(k_{m,n})\Big\|^2 + \Big\|\sum_{m,n} A_{m,n+1}(k_{m,n})\Big\|^2 \\ &\leq \Big\|\sum_{m,n} A_{m,n}(k_{m,n})\Big\|^2 + \Big\|\sum_{m,n} A_{m+1,n+1}(k_{m,n})\Big\|^2,\end{aligned}$$

for every finite sequence $\{k_{m,n}\}$ of vectors in $\mathcal{K}$.

A similar use of Corollary 3.3 provides:

Corollary 3.5. *([12]) Let $\{A_m\}_{m\geq 0}\subset \mathcal{L}(\mathcal{K},\mathcal{H})$. The following conditions are equivalent:*

(i) *There exists a contraction S on $\mathcal{H}$ such that*

$$A_m = S^m A_0, \quad m\geq 0;$$

(ii) $$\|\sum_{m,n} A_{m+n}(k_{m,n})\|^2 \leq \sum_{m,n,p,q} \langle A_{(m-p)^+ +n}(k_{m,n}), A_{(m-p)^- +q}(k_{p,q})\rangle,$$

for every finite sequence $\{k_{m,n}\}$ of vectors in $\mathcal{K}$;

(iii) $$\sum_{m,n,p,q} \langle A_{(m-p)^+ +n}(k_{m,n}), A_{(m-p)^- +q}(k_{p,q})\rangle \geq 0,$$

for every finite sequence $\{k_{m,n}\}$ of vectors in $\mathcal{K}$;

(iv) $$\|\sum_{m} A_{m+1}(k_m)\| \leq \|\sum_{m} A_m(k_m)\|,$$

for every finite sequence $\{k_m\}$ of vectors in $\mathcal{K}$.

The following corollary may also have some independent interest:

Corollary 3.6. *Let P,Q be linear and bounded operators on $\mathcal{H}$. Then (P,Q) is a commuting pair of orthogonal projections if and only if*

$$\|P(a+b)+PQ(c+d)\| \leq \|a+Pb+Qc+PQd\|$$

and

$$\|Q(a+b)+PQ(c+d)\| \leq \|a+Qb+Pc+PQd\|,$$

for every $a,b,c,d\in\mathcal{H}$.

Proof. We can use Corollary 3.4 for the sequence $\{A_{m,n}\}\subset\mathcal{L}(\mathcal{H})$ given by

$$A_{m,n} = \begin{cases} I_{\mathcal{H}}, & \text{if } m=n=0\\ P, & \text{if } m\neq 0, n=0\\ Q, & \text{if } m=0, n\neq 0\\ PQ, & \text{if } m,n\neq 0\end{cases}.$$

The assumption in Corollary 3.4 (i) take the following form: there exists a commuting pair (S,T) of contractions on $\mathcal{H}$ having regular unitary dilation and such that

$$S^m = P\ (m>0), \quad T^n = Q\ (n>0), \quad S^mT^n = PQ\ (m,n>0).$$

We obtain that $S=P$ and $T=Q$ are both idempotent contractions on $\mathcal{H}$, hence orthogonal projections. We also observe that a pair of commuting orthogonal projections has regular unitary dilation (being double commuting).

The conclusion follows if we take, for any finite sequence $\{h_{m,n}\}\subset\mathcal{H}$, $a=h_{0,0}$, $b=\sum_{m\geq 1}h_{m,0}$, $c=\sum_{n\geq 1}h_{0,n}$ and $d=\sum_{m,n\geq 1}h_{m,n}$. □

4. Isometric solutions

In order to obtain a commuting pair of isometries (briefly bi-isometry) as solution to Problem 1.1 it is necessary and sufficient that equality holds in Theorem 3.1 (ii). More precisely, we have:

Theorem 4.1. *Let f be as above. The following conditions are equivalent:*

(i) *There exists a bi-isometry (V,W) on $\mathcal{H}$ as a solution to Problem 1.1;*

(ii) $$\|\sum_{(m,n),(p,q),x} c^x_{(m,n),(p,q)} f(m+p,n+q,x)\|^2$$
$$= \sum_{\substack{(m,n),(m',n'),x\\(p,q),(p',q'),y}} c^x_{(m,n),(m',n')}\overline{c^y_{(p,q),(p',q')}}\langle f((m-p)^+ + m',(n-q)^+ + n',x),$$
$$f((m-p)^- + p',(n-q)^- + q',y)\rangle,$$

for every finite family $\{c^x_{(m,n),(m',n')}\}$ of complex numbers;

(iii) $$\langle f(m+1,n,x), f(p+1,q,y)\rangle = \langle f(m,n+1,x), f(p,q+1,y)\rangle$$
$$= \langle f(m,n,x), f(p,q,y)\rangle,$$
$$(m,n),(p,q)\in\mathbb{Z}^2_+,\ x,y\in X.$$

Proof. (i) $\Rightarrow$ (ii) We proceed as in the corresponding part in the proof of Theorem 3.1. We just have to observe that, for every commuting pair (V,W) of isometries on $\mathcal{H}$,

$$\langle V^mW^nh, V^pW^qh\rangle = \langle V^{(m-p)^+}W^nh, V^{(m-p)^-}W^qh\rangle$$
$$= \langle V^{(m-p)^+}W^{(n-q)^+}h, V^{(m-p)^-}W^{(n-q)^-}h\rangle,$$
$$(m,n),(p,q)\in\mathbb{Z}^2_+,\ h\in\mathcal{H}.$$

(ii) $\Rightarrow$ (iii) Fix a finite family $\{c^x_{m,n}\}$ of complex numbers and apply (ii) to the (finite) family $\{c^x_{(m,n),(m',n')}\}$ given by

$$c^x_{(m,n),(m',n')} = \begin{cases} -c^x_{m'-1,n'}, & \text{if } m=n=0;\ m'\geq 1\\ c^x_{m',n'}, & \text{if } m=1, n=0\\ 0, & \text{for the other cases}\end{cases},$$

for $(m,n),(m',n')\in\mathbb{Z}^2_+$ and $x\in X$.

The left-hand side of (ii) becomes

$$\|\sum_{(p,q),x} c^x_{(0,0),(p,q)} f(p,q,x) + \sum_{(p,q),x} c^x_{(1,0),(p,q)} f(p+1,q,x)\|^2$$
$$= \| -\sum_{p\geq 1,q\geq 0,x} c^x_{p-1,q} f(p,q,x) + \sum_{(p,q),x} c^x_{p,q} f(p+1,q,x)\|^2 = 0.$$

Similarly, the corresponding right-hand side can be re-written as

$$\sum_{\substack{(m,n),(p,q)\in\{(0,0),(1,0)\}\\(m',n'),(p',q'),x,y}} c^x_{(m,n),(m',n')}\overline{c^y_{(p,q),(p',q')}}$$

$$\langle f((m-p)^+ + m', (n-q)^+ + n', x),$$
$$f((m-p)^- + p', (n-q)^- + q', y)\rangle$$

$$= \sum_{\substack{m',p'\geq 1\\n',q'\geq 0}} c^x_{m'-1,n'}\overline{c^y_{p'-1,q'}}\langle f(m',n',x), f(p',q',y)\rangle$$

$$- \sum_{\substack{m'\geq 1\\n',p',q'\geq 0}} c^x_{m'-1,n'}\overline{c^y_{p',q'}}\langle f(m',n',x), f(p'+1,q',y)\rangle$$

$$- \sum_{\substack{p'\geq 1\\m',n',q'\geq 0}} c^x_{m',n'}\overline{c^y_{p'-1,q'}}\langle f(m'+1,n',x), f(p',q',y)\rangle$$

$$+ \sum_{m',n',p',q'\geq 0} c^x_{m',n'}\overline{c^y_{p',q'}}\langle f(m',n',x), f(p',q',y)\rangle$$

$$= \|\sum_{(m,n),x} c^x_{m,n} f(m,n,x)\|^2 - \|\sum_{(m,n),x} c^x_{m,n} f(m+1,n,x)\|^2.$$

We deduce that

$$\|\sum_{(m,n),x} c^x_{m,n} f(m,n,x)\| = \|\sum_{(m,n),x} c^x_{m,n} f(m+1,n,x)\|,$$

for every finite family $\{c^x_{m,n}\}$ of complex numbers. Equivalently,

$$\langle f(m+1,n,x), f(p+1,q,y)\rangle = \langle f(m,n,x), f(p,q,y)\rangle$$

and, analogously,

$$\langle f(m,n+1,x), f(p,q+1,y)\rangle = \langle f(m,n,x), f(p,q,y)\rangle$$

for every $(m,n),(p,q)\in\mathbb{Z}^2_+$ and $x,y\in X$.

(iii) $\Rightarrow$ (i) By

$$\mathcal{H}^f \ni \sum_{(m,n),x} c^x_{m,n} f(m,n,x) \overset{V_0}{\mapsto} \sum_{(m,n),x} c^x_{m,n} f(m+1,n,x) \in \mathcal{H}^f \tag{4.1}$$

and

$$\mathcal{H}^f \ni \sum_{(m,n),x} c^x_{m,n} f(m,n,x) \overset{W_0}{\mapsto} \sum_{(m,n),x} c^x_{m,n} f(m,n+1,x) \in \mathcal{H}^f \tag{4.2}$$

we define (in view of (iii)) a commuting isometric pair (V_0, W_0) on $\mathcal{H}^f$ which satisfies

$$f(m,n,x) = V_0^m W_0^n f(0,0,x), \quad (m,n)\in\mathbb{Z}^2_+,\ x\in X.$$

It can be extended to a commuting isometric pair $(V_0\oplus I_{\mathcal{H}\ominus\mathcal{H}^f}, W_0\oplus I_{\mathcal{H}\ominus\mathcal{H}^f})$ on $\mathcal{H}$ with the same property. $\square$

Remark 4.2.

- The method used in Remark 3.2 for an alternate solution to Theorem 3.1 can be modified correspondingly to obtain another proof for the implication $(ii) \Rightarrow (i)$ of Theorem 4.1. Following the same notations we obtain a Hilbert space $\mathcal{F}$ (with the inner product defined by (3.1)) and operators A and B (introduced in (3.2) and (3.3)) having isometric adjoints. The map

$$\mathcal{F} \ni \{c^x_{(m,n),(p,q)}\} \overset{\varphi}{\mapsto} \sum_{(m,n),(p,q),x} c^x_{(m,n),(p,q)} f(m+p, n+q, x) \in \mathcal{H}^f$$

is unitary (being isometric-by (ii)-and surjective). The commuting isometric pair $(\varphi A^*\varphi^*, \varphi B^*\varphi^*)$ is a solution to our problem on $\mathcal{H}^f$ and can be obviously extended to a solution on $\mathcal{H}$.
- A pair (V, W) satisfying Theorem 4.1 (i) is uniquely determined if and only if $\mathcal{H} = \mathcal{H}^f$.

Particular choices for the map f will lead to different applications, corresponding to the ones in the previous section:

Corollary 4.3. *Let $f : \mathbb{Z}_+ \times X \to \mathcal{H}$. The following conditions are equivalent:*

(i) *There exists an isometric operator V on $\mathcal{H}$ such that*

$$f(m, x) = V^m f(0, x), \quad m \geq 0,\ x \in X;$$

(ii)
$$\|\sum_{(m,n),x} c^x_{m,n} f(m+n, x)\|^2 = \sum_{(m,n),(p,q),x,y} c^x_{m,n}\overline{c^y_{p,q}}\langle f((m-p)^+ + n, x), f((m-p)^- + q, y)\rangle,$$

for every finite family $\{c^x_{m,n}\}$ of complex numbers;

(iii) $\langle f(m+1, x), f(n+1, y)\rangle = \langle f(m, x), f(n, y)\rangle, \quad m, n \geq 0,\ x, y \in X.$

Corollary 4.4. *Let $\{A_{m,n}\}_{(m,n)\in\mathbb{Z}^2_+}$ be a sequence of operators between $\mathcal{K}$ and $\mathcal{H}$. The following conditions are equivalent:*

(i) *There exists a commuting pair (V, W) of isometries on $\mathcal{H}$ such that*

$$A_{m,n} = V^m W^n A_{0,0}, \quad (m, n) \in \mathbb{Z}^2_+;$$

(ii)
$$\|\sum_{(m,n),(p,q)} A_{m+p,n+q}(k_{(m,n),(p,q)})\|^2 = \sum_{\substack{(m,n),(m',n')\\(p,q),(p',q')}} \langle A_{(m-p)^+ + m',(n-q)^+ + n'}(k_{(m,n),(m',n')}), A_{(m-p)^- + p',(n-q)^- + q'}(k_{(p,q),(p',q')})\rangle,$$

for every finite sequence $\{k_{(m,n),(m',n')}\}$ of vectors in $\mathcal{K}$;

(iii) $A^*_{m+1,n}A_{p+1,q} = A^*_{m,n+1}A_{p,q+1} = A^*_{m,n}A_{p,q}, \quad (m, n), (p, q) \in \mathbb{Z}^2_+.$

Corollary 4.5. *([12]) Let $\{A_m\}_{m\geq 0}$ be a sequence of operators between $\mathcal{K}$ and $\mathcal{H}$. The following conditions are equivalent:*

(i) *There exists an isometry V on $\mathcal{H}$ such that*

$$A_m = V^m A_0, \quad m \geq 0;$$

(ii) $$\|\sum_{(m,n),(p,q)} A_{m+n}(k_{m,n})\|^2 = \sum_{m,n,p,q} \langle A_{(m-p)^+ + n}(k_{m,n}), A_{(m-p)^- + q}(k_{p,q})\rangle,$$

for every finite sequence $\{k_{m,n}\}$ of vectors in $\mathcal{K}$;

(iii) $$A^*_{m+1} A_{n+1} = A^*_m A_n, \quad m, n \geq 0.$$

5. Unitary solutions

Let $f : \mathbb{Z}^2_+ \times X \to \mathcal{H}$, $\mathcal{H}^f_{p,q} := \bigvee_{m\geq p, n\geq q, x\in X} f(m,n,x)$, $(p,q) \in \mathbb{Z}^2_+$ and $\mathcal{H}^f := \mathcal{H}^f_{0,0}$.

In order to obtain a commuting pair of unitary operators as a solution to Problem 1.1 attached to f we proceed as follows:

- find a bi-isometric solution (according to Theorem 4.1);
- extend its restriction to $\mathcal{H}^f$ to a commuting pair of unitary operators on $\mathcal{H}$.

Unfortunately, even if f satisfies the equivalent conditions of Theorem 4.1, it is not always possible to perform this last step:

Remark 5.1. Suppose that there exists a commuting pair (U, Z) of unitary operators on $\mathcal{H}$ such that

$$f(m,n,x) = U^m Z^n f(0,0,x), \quad (m,n) \in \mathbb{Z}^2_+, \ x \in X. \tag{5.1}$$

Then $(U|_{\mathcal{H}^f}, Z|_{\mathcal{H}^f})$ is a commuting pair of isometric operators on $\mathcal{H}^f$. If $(\tilde{U}, \tilde{Z})$ is its minimal unitary extension (given by (2.4) and (2.5)) on the Hilbert space $\tilde{\mathcal{H}} = \mathcal{H}^f \oplus H^2_{\ker(UZ|_{\mathcal{H}^f})^*}(\mathbb{T})$ then the (Hilbert) dimension of $\tilde{\mathcal{H}}$ (which clearly depends only on f and not on the particular choice of (U, Z)) is not greater than the dimension of $\mathcal{H}$. Consequently, if, for example, $\mathcal{H}^f = \mathcal{H}$ and one of the operators V_0 and W_0 defined by (4.1) and (4.2) is not unitary then our problem cannot have a bi-unitary solution.

The following theorem is an improvement of a recent result in [6]:

Theorem 5.2. *The following conditions are equivalent:*

(i) *There exists a commuting pair (U, Z) of unitary operators on $\mathcal{H}$ such that* (5.1) *holds true;*

(ii) (a) *f verifies one of the equivalent conditions of Theorem* 4.1*;*
(b) $\aleph_0 \dim(\mathcal{H}^f \ominus \mathcal{H}^f_{1,1}) \leq \dim(\mathcal{H} \ominus \mathcal{H}^f)$.

Proof. Since the restriction of a commuting pair of unitary operators on $\mathcal{H}$ to one of its invariant subspaces (namely $\mathcal{H}^f$) is a commuting pair of isometries, we can always suppose that the equivalent conditions of Theorem 4.1 hold.

In other words, the problem reduces to the following: *find necessary and sufficient conditions on f such that the pair (V_0, W_0) given on $\mathcal{H}^f$ by (4.1) and (4.2) extends to a commuting pair of unitary operators on $\mathcal{H}$.*

Let $(\tilde{U}, \tilde{Z})$ be the minimal unitary extension of (V_0, W_0) given on $\tilde{\mathcal{H}} = \mathcal{H}^f \oplus H^2_{\mathcal{H}^f \ominus \mathcal{H}^f_{1,1}}(\mathbb{T})$ by (2.4) and (2.5) (remark that $\ker(V_0 W_0)^* = \mathcal{H}^f \ominus V_0 W_0 \mathcal{H}^f = \mathcal{H}^f \ominus \mathcal{H}^f_{1,1}$).

If $\aleph_0 \dim(\mathcal{H}^f \ominus \mathcal{H}^f_{1,1}) \leq \dim(\mathcal{H} \ominus \mathcal{H}^f)$ then there exists an isometric operator $Y : \tilde{\mathcal{H}} \ominus \mathcal{H}^f \to \mathcal{H} \ominus \mathcal{H}^f$. Deduce that $I_{\mathcal{H}^f} \oplus Y$ is a unitary operator between $\tilde{\mathcal{H}}$ and $\mathcal{H}' := \mathcal{H}^f \oplus Y(\tilde{\mathcal{H}} \ominus \mathcal{H}^f)$. Consequently, $((I_{\mathcal{H}^f} \oplus Y)\tilde{U}(I_{\mathcal{H}^f} \oplus Y)^*, (I_{\mathcal{H}^f} \oplus Y)\tilde{Z}(I_{\mathcal{H}^f} \oplus Y)^*)$ is a commuting pair of unitary operators acting on $\mathcal{H}'$ ($\subset \mathcal{H}$) that extends (V_0, W_0). Take

$$U = I_{\mathcal{H} \ominus \mathcal{H}'} \oplus (I_{\mathcal{H}^f} \oplus Y)\tilde{U}(I_{\mathcal{H}^f} \oplus Y)^* \text{ and } Z = I_{\mathcal{H} \ominus \mathcal{H}'} \oplus (I_{\mathcal{H}^f} \oplus Y)\tilde{Z}(I_{\mathcal{H}^f} \oplus Y)^*$$

and observe that (U, Z) is a commuting pair of unitary operators on $\mathcal{H}$ which verifies (5.1).

Conversely, suppose that (V_0, W_0) can be extended to a commuting pair (U, Z) of unitary operators on $\mathcal{H}$. Then (U, Z) contains a minimal unitary extension $(U_0 := U|_{\mathcal{H}_0}, Z_0 := Z|_{\mathcal{H}_0})$ of (V_0, W_0) on $\mathcal{H}_0 := \bigvee_{(m,n)\in\mathbb{Z}^2} U^m Z^n \mathcal{H}^f$. Moreover,

$$\aleph_0 \dim(\mathcal{H}^f \ominus \mathcal{H}^f_{1,1}) = \dim(\mathcal{H}_0 \ominus \mathcal{H}^f) \leq \dim(\mathcal{H} \ominus \mathcal{H}^f).$$

□

We conclude the paper with some remarks:

Remark 5.3.

- Suppose that f satisfies one of the equivalent conditions of Theorem 4.1 and $\dim \mathcal{H}^f \leq \aleph_0$. If $\mathcal{H}^f_{1,1} = \mathcal{H}^f$ then

  $$\dim(\mathcal{H} \ominus \mathcal{H}^f) \geq 0 = \aleph_0 \dim(\mathcal{H}^f \ominus \mathcal{H}^f_{1,1})$$

  and our problem has a unitary solution. If $\mathcal{H}^f_{1,1} \neq \mathcal{H}^f$ we obtain the same conclusion since, in this case,

  $$\aleph_0 \dim(\mathcal{H}^f \ominus \mathcal{H}^f_{1,1}) = \aleph_0 \leq \dim \mathcal{H} = \dim(\mathcal{H} \ominus \mathcal{H}^f).$$

- Jabłoński and Stochel showed in [6] that it is possible to produce unitary solutions to our problem if $|X| = 1$, f verifies the conditions of Theorem 4.1 (iii) and $\dim(\mathcal{H} \ominus \mathcal{H}^f) \geq \dim \mathcal{H}^f$. This is, however, a consequence of Theorem 5.2: as mentioned above we can consider the case $\dim \mathcal{H}^f > \aleph_0$. Then

  $$\dim(\mathcal{H} \ominus \mathcal{H}^f) \geq \dim \mathcal{H}^f \geq \aleph_0 \dim(\mathcal{H}^f \ominus \mathcal{H}^f_{1,1}).$$

- If it exists, a pair (U, Z) of commuting unitary operators on $\mathcal{H}$ such that (5.1) holds is uniquely determined by f if and only if
$$\mathcal{H} = \mathcal{H}^f = \mathcal{H}^f_{1,1}.$$
Suppose that $\mathcal{H}^f = \mathcal{H}^f_{1,1}$. Then operators V_0 and W_0 defined on $\mathcal{H}^f$ by (4.1) and (4.2) are unitary and (V_0, W_0) has a unique unitary extension to $\mathcal{H}$ if and only if $\mathcal{H} = \mathcal{H}^f$.
If the pair (U, Z) is unique then we can freely suppose that this unitary extension of (V_0, W_0) is minimal. Let φ be any unitary operator on $\mathcal{H} \ominus \mathcal{H}^f$. Then $((I_{\mathcal{H}^f} \oplus \varphi)U(I_{\mathcal{H}^f} \oplus \varphi)^*, (I_{\mathcal{H}^f} \oplus \varphi)Z(I_{\mathcal{H}^f} \oplus \varphi)^*)$ is a minimal unitary extension of (V_0, W_0). By uniqueness, we conclude that (U, Z) commutes with any operator of the form $I_{\mathcal{H}^f} \oplus A$, A being linear and bounded on $\mathcal{H} \ominus \mathcal{H}^f$. In particular,
$$U(I_{\mathcal{H}^f} \oplus (x \otimes x))U^*(x) = (I_{\mathcal{H}^f} \oplus (x \otimes x))(x)$$
or, equivalently,
$$\langle x, Ux\rangle Ux = x,$$
for any unit vector $x \in \mathcal{H} \ominus \mathcal{H}^f$ (if $\mathcal{H} = \mathcal{H}^f$ then $V_0 = U$ and $W_0 = Z$ are unitary and $\mathcal{H}^f_{1,1} = V_0 W_0 \mathcal{H}^f = \mathcal{H}^f$; $x \otimes x$ is defined on $\mathcal{H} \ominus \mathcal{H}^f$ by $x \otimes x(y) := \langle y, x\rangle x$). Hence, $U(\mathcal{H} \ominus \mathcal{H}^f) = \mathcal{H} \ominus \mathcal{H}^f$ and, analogously, $Z(\mathcal{H} \ominus \mathcal{H}^f) = \mathcal{H} \ominus \mathcal{H}^f$. We obtain that $\mathcal{H}^f$ is reducing for (U, Z) and, consequently $(V_0 = U|_{\mathcal{H}^f}, W_0 = Z|_{\mathcal{H}^f})$ should be a unitary pair. But then $\mathcal{H}^f_{1,1} = \mathcal{H}^f$.
- The conclusion of Theorem 5.2 holds for any particular choice of f: the same kind of solutions can be obtained for the moment problems proposed in Corollaries 4.3-4.5.

Acknowledgement

The work on this paper was partly done during the author's postdoctoral visit at the Department of Analysis and Scientific Computing, Technical University of Vienna. The author would like to express his appreciation to Professor Heinz Langer for his kind support. Very helpful remarks of the referees are also acknowledged with gratitude.

References

[1] S. Brehmer, *Über vertauschbare Kontraktionen des Hilbertschen Raumes.* Acta Sci. Math. **22** (1961), 106–111.

[2] R.E. Curto, F.H. Vasilescu, *Standard operator models in the polydisc.* Indiana Univ. Math. J. **42** (1993), 791–810.

[3] R.E. Curto, F.H. Vasilescu, *Standard operator models in the polydisc II.* Indiana Univ. Math. J. **44** (1995), 727–746.

[4] C. Foiaş, A.E. Frazho, I. Gohberg, M.A. Kaashoek, *Metric Constrained Interpolation, Commutant Lifting and Systems.* Operator Theory: Advances and Applications 100, Birkhäuser-Verlag, Basel, 1998.

[5] I. Halperin, *Sz.-Nagy–Brehmer dilations.* Acta Sci. Math. **23** (1962), 279–289.

[6] Z.J. Jabłoński, J. Stochel, *Subnormality and operator multidimensional moment problems.* J. London Math. Soc. **71** (2005), 438–466.

[7] Y. Kakihara, *Multidimensional second order stochastic processes.* Series on Multivariate Analysis 2, World Scientific Publ. Co., Inc., River Edge, NJ, 1997.

[8] A. Olofsson, *Operator valued n-harmonic measure in the polydisc.* Studia Math. **163** (2004), pp. 203–216.

[9] D. Popovici, *Matrix representations for regular dilations.* Preprint (2004).

[10] D. Popovici, *On the unitary extension for a system of commuting isometries.* Preprint (2004).

[11] D. Popovici, *Regular dilations on Kreĭn spaces.* Preprint (2004).

[12] D. Popovici, Z. Sebestyén, *Positive definite functions and Sebestyén's operator moment problem.* Glasgow Math. J. **47** (2005), 471–488.

[13] D. Popovici, Z. Sebestyén, *Sebestyén moment problem: The multidimensional case.* Proc. Amer. Math. Soc. **132** (2004), 1029–1035.

[14] Z. Sebestyén, *Moment theorems for operators on Hilbert space.* Acta Sci. Math. **44** (1982), 165–171.

[15] Z. Sebestyén, *Moment theorems for operators on Hilbert space II.* Acta Sci. Math. **47** (1984), 101–106.

[16] Z. Sebestyén, *A moment theorem for contractions on Hilbert spaces.* Acta Math. Hung. **47** (1986), 391–393.

[17] Z. Sebestyén, *Moment problem for dilatable semigroups of operators.* Acta Sci. Math. **45** (1983), 365–376.

[18] Z. Sebestyén, *Moment type theorems for $*$-semigroups of operators on Hilbert space I.* Ann. Univ. Sci. Budapest, Eötvös Sect. Math. **26** (1983), 213–218.

[19] B. Sz.-Nagy, *Bemerkungen zur vorstehenden Arbeit des Herrn S. Brehmer.* Acta Sci. Math. **22** (1961), 112–114.

[20] B. Sz.-Nagy, *Sur les contractions de l'espace de Hilbert.* Acta Sci. Math. **15** (1953), 87–92.

[21] B. Sz.-Nagy, *Transformations de l'espace de Hilbert, fonctions de type positif sur un groupe.* Acta Sci. Math. **15** (1954), 104–114.

[22] B. Sz.-Nagy, C. Foiaş, *Harmonic analysis of operators on Hilbert space.* North-Holland, Amsterdam, 1970.

[23] D. Timotin, *Regular dilations and models for multi-contractions.* Indiana Univ. Math. J. **47** (1998), 671–684.

Dan Popovici
Department of Mathematics and Computer Science
University of the West Timişoara
B-dul Vasile Pârvan 4
RO-300223 Timişoara
Romania
e-mail: popovici@math.uvt.ro

Operator Theory:
Advances and Applications, Vol. 163, 275–298

Isospectral Vibrating Systems, Part 2: Structure Preserving Transformations

Uwe Prells and Peter Lancaster

Abstract. The study of inverse problems for $n \times n$-systems of the form $L(\lambda) := M\lambda^2 + D\lambda + K$ is continued. In this paper it is assumed that one vibrating system is specified and the objective is to generate isospectral families of systems, i.e., systems which reproduce precisely the eigenvalues of the given system together with their multiplicities. Two central ideas are developed and used, namely, standard triples of matrices, and structure preserving transformations.

Mathematics Subject Classification (2000). Primary 74A15; Secondary 15A29.

Keywords. Vibrating systems, inverse problems, damping.

1. Introduction

A broad spectrum of physical problems concerning vibrations is covered by classical models of second order constant-coefficient differential equations. In many cases it is natural to refer to the three coefficient matrices as the mass (M), damping (D), and stiffness matrices (K) and we will use this terminology. However, by admitting the possibility of complex coefficients (not just real matrices), the number of physical problems covered is enlarged (admitting gyroscopic effects, for example) and, not only this, it turns out that a broader mathematical analysis is obtained which casts some light on those problems in which M, D, K are real. This leads to our first definition:

Definition 1. A *(vibrating) system* is a triple of $n \times n$ complex matrices $\{M, D, K\}$ for which M is nonsingular.

When convenient, the quadratic matrix function $L(\lambda) := \lambda^2 M + \lambda D + K$ may also be described as a vibrating system.

It is well known that the solutions of the corresponding differential equations can be described entirely in terms of the solutions of the *algebraic* eigenvalue problem: find those $\lambda \in \mathbb{C}$ and non-zero $x \in \mathbb{C}^n$ for which

$$L(\lambda)x = (\lambda^2 M + \lambda D + K)x = 0.$$

In Part 1 of this work (reference [6]) it was shown how, in some cases, the three coefficient matrices can be recovered if (well-defined) complete information is given on the eigenvalue and eigenvector structures (known as the spectral data). Here, the objective is to start with one system, say (M_0, D_0, K_0), and, implicitly, a complete set of corresponding spectral data, and to show how to generate *isospectral* vibrating systems, i.e., systems (M, D, K) which share the same set of eigenvalues (including their multiplicity structures). And this is to be done without explicit reference to eigenvalues and eigenvectors. There is an interesting contrast with the approach taken by Chu et al. [2], where *partial* eigenstructure is prescribed in a scheme to generate monic real symmetric systems.

Our objective will be achieved using two parallel (but closely linked) notions. The first is the tool of "standard triples" of matrices for $L(\lambda)$ introduced and developed in great detail in [4] and earlier works cited there (see Section 3). The second is the idea of "structure preserving transformations" developed in this context in [3] (and introduced here in Section 2). Computational procedures for solution of the problem for general systems (with no symmetries) are contained in Section 5. Sections 6, 7, and 8 are concerned with the important (and more complex) problems in which M, D, K are required to be Hermitian (or real symmetric). It is of interest to ask when there is an isospectral system of simplest possible form – in which all three coefficients are diagonal. This is the topic of Section 9. We conclude this introduction with a discussion of the basic idea of linearization.

It is very well known that linear second order systems can be transformed to first order in an elementary way; generally known as "linearization". Those linearizations associated with vibrating systems are our present concern. In the first two sections their properties under transformations of three types are reviewed, namely, "equivalence", "strict equivalence", and "similarity" in decreasing order of generality. There have been several recent publications concerning equivalence and similarity transformations of vibrating systems. The authors' formulation of these ideas together with new results appear in Sections 2 and 4 (especially Theorem 7 and its Corollary).

Definition 2. Let $L(\lambda)$ be a vibrating system. A $2n \times 2n$ matrix pencil $\lambda X - Y$ is a *linearization* of $L(\lambda)$ if

$$\begin{bmatrix} L(\lambda) & 0 \\ 0 & I_n \end{bmatrix} = E(\lambda)(\lambda X - Y)F(\lambda) \tag{1}$$

for some matrix polynomials $E(\lambda)$, $F(\lambda)$ with constant non-zero determinants.

Notice that it follows from this definition that the matrix X of a linearization is necessarily nonsingular. (This follows from the assumption that M is nonsingular.) The transformation of $\lambda X - Y$ on the right-hand side of (1) is known as an *equivalence transformation.* The "equivalence class" of all pencils equivalent to

$\lambda X - Y$ in this sense determines the set of all linearizations of $L(\lambda)$. A vital property of equivalence transformations is that they preserve the set of eigenvalues and their multiplicity structures (i.e., they have a common Jordan structure). Thus a linearization of a vibrating system shares its "spectrum" (its eigenvalues – with their multiplicities); it is *isospectral.*

Define the right and left companion matrices, C_R and C_L, of a vibrating system by:

$$C_R = \begin{bmatrix} 0 & I \\ -M^{-1}K & -M^{-1}D \end{bmatrix}, \qquad C_L = \begin{bmatrix} 0 & -KM^{-1} \\ I & -DM^{-1} \end{bmatrix}. \tag{2}$$

The first lemma is very well known:

Lemma 1. *The following pencils are linearizations of the vibrating system $L(\lambda)$:*

1. $\lambda A - B$ *with*

$$A := \begin{bmatrix} D & M \\ M & 0 \end{bmatrix}, \quad B := \begin{bmatrix} -K & 0 \\ 0 & M \end{bmatrix}, \tag{3}$$

2. $$\lambda I_{2n} - A^{-1}B = \lambda I_{2n} - C_R,$$

3. $$\lambda I_{2n} - BA^{-1} = \lambda I_{2n} - C_L.$$

2. Structure preserving transformations

If E and F are nonsingular $2n$-by-$2n$ matrices in $\mathbb{C}^{2n\times 2n}$, then the pencils $\lambda A - B$ and $E(\lambda A - B)F$ are said to *strictly equivalent.* Clearly, a strict equivalence is also an equivalence in the sense used in Definition 2 *et seq.*, and so strict equivalence also preserves the spectrum. The set of all pencils strictly equivalent to $\lambda A - B$ in this sense form an equivalence class. (The relation of "strict equivalence" is reflexive, symmetric and transitive.)

Now consider the linearization $\lambda A - B$ of (3) and let $\lambda A' - B'$ be a strictly equivalent pencil, i.e., there exist E and F such that $A' = EAF$, $B' = EBF$. The question is, when does $\lambda A' - B'$ define a vibrating system? The answer is, when A', B' have the defining properties of A and B, namely, when there exist M', D', K' with M' nonsingular such that

$$A' = \begin{bmatrix} D' & M' \\ M' & 0 \end{bmatrix}, \quad B' = \begin{bmatrix} -K' & 0 \\ 0 & M' \end{bmatrix}. \tag{4}$$

Definition 3. (cf. Garvey et al. [3].) Let $\lambda A - B$ be defined from a vibrating system as in (3). Let $\lambda A' - B'$ be obtained from $\lambda A - B$ by a strict equivalence. Then this strict equivalence is said to be *structure preserving* if and only if A', B' have the form (4) and M' is nonsingular.

For brevity, a structure preserving strict equivalence of this kind is called an SPE. When the linearizations of two vibrating systems are connected in this way they will be said to be *related by an SPE.* Observe that strict equivalence transformations of $L(\lambda)$ itself, say $L(\lambda) \mapsto SL(\lambda)T$ where S and T are nonsingular, are included in this definition. Such a transformation corresponds to an SPE with

$$E = \begin{bmatrix} S & 0 \\ 0 & S \end{bmatrix}, \quad F = \begin{bmatrix} T & 0 \\ 0 & T \end{bmatrix}.$$

Definition 4. (Lancaster and Prells [8].) Let C_R be the right companion matrix (2) of $L(\lambda)$. The nonsingular $2n$-by-$2n$ matrix S_R is called a (*right*) *structure preserving similarity* (SPS) if $S_R^{-1}C_RS_R = C'_R$ is a (right) companion matrix.

(Clearly, a similar definition can be made in terms of the left companion matrix.) When the companion matrices of two vibrating systems are connected in this way they will be said to be *related by an SPS*. The close relationship between SPE and SPS transformations is the subject of the easily proved result:

Theorem 2. *Two vibrating systems are related by an SPE if and only if they are related by an SPS.*

Proof. Consider a vibrating system $L_0(\lambda) = \lambda^2 M_0 + \lambda D_0 + K_0$. Let

$$A_0 := \begin{bmatrix} D_0 & M_0 \\ M_0 & 0 \end{bmatrix}, \quad B_0 := \begin{bmatrix} -K_0 & 0 \\ 0 & M_0 \end{bmatrix}, \tag{5}$$

and let $C_R^{(0)}$ be the corresponding right companion matrix. Consider the SPE $\lambda A - B = E(\lambda A_0 - B_0)F$. Since the structure is preserved A and B have the form of equations (3).This implies that A is nonsingular and

$$A^{-1}B = (EA_0F)^{-1}(EB_0F) = F^{-1}A_0^{-1}E^{-1}EB_0F = F^{-1}(A_0^{-1}B_0)F,$$

i.e., $C_R = F^{-1}C_R^{(0)}F$, so that the companion matrices C_R and $C_R^{(0)}$ are similar. Thus, the systems are related by an SPS.

Conversely, let $C_R = S^{-1}C_R^{(0)}S$ be an SPS (with the definitions above for C_R, $C_R^{(0)}$, A, B, A_0, B_0). Then

$$\begin{aligned} \lambda A - B &= A(\lambda I - C_R) = A(\lambda I - S^{-1}C_R^{(0)}S) = (AS^{-1})(\lambda I - C_R^{(0)})S \\ &= (AS^{-1}A_0^{-1})(\lambda A_0 - B_0)S, \end{aligned}$$

and this is an SPE. □

Corollary 3. *Let A, B be defined by a vibrating system as in* (3) *and let $E(\lambda A - B)F = \lambda A_0 - B_0$ be an SPE, then E defines an SPS for C_L, ($C_L = EC_L^{(0)}E^{-1}$) and F defines an SPS for C_R, ($C_R = F^{-1}C_R^{(0)}F$).*

Proof. It has been shown in the above argument that, given the SPE defined by matrices E and F, the matrix F defines an SPS for C_R. In a similar way it follows that E defines an SPS for C_L. □

These results suggest that one might equally well work with SPS transformations as with SPE's (which require more parameters). Indeed, if one is interested only in *monic* systems (with $M = I$), SPS transformations tell the whole story.

Otherwise, SPS transformations tell us about the products $M^{-1}D$ and $M^{-1}K$ and it remains to specify M itself, and hence D and K. In contrast, SPE's are defined in terms of the explicit coefficients M, D, and K. This point becomes more significant when working with systems for which the three coefficients are Hermitian (or real symmetric) in which case the corresponding symmetry is reflected explicitly in A and B of (3).

3. Standard pairs and triples

The following notions of "standard pairs and triples" play an important unifying role in the discussion of linearizations (see [4] and earlier references given there). Our definitions differ in some respects from earlier sources. They are consistent with [6], and are formulated so that they are better suited to inverse problems – they do not refer explicitly to the matrix coefficients of the vibrating system.

Definition 5. A pair of matrices $U \in \mathbb{C}^{n\times 2n}$ and $T \in \mathbb{C}^{2n\times 2n}$ form a *standard pair* for a vibrating system if

(a) the dimension of each eigenspace of T does not exceed n, and

(b) the $2n \times 2n$ matrix $\begin{bmatrix} U \\ UT \end{bmatrix}$ is nonsingular.

Note that the technical condition (a) is vital if T is to relate to a vibrating system (see Theorem 1.7 of [4]).

Definition 6. Three matrices $U \in \mathbb{C}^{n\times 2n}$ and $T \in \mathbb{C}^{2n\times 2n}$ and $V \in \mathbb{C}^{2n\times n}$ form a *standard triple* if (U, T) is a standard pair, $UV = 0$ and the $n \times n$ matrix UTV is nonsingular.

Notice that, if (U, T, V) is a standard triple, then

$$\begin{bmatrix} U \\ UT \end{bmatrix} V = \begin{bmatrix} 0 \\ UTV \end{bmatrix}.$$

Since UTV is nonsingular (by definition) we may define $UTV = M^{-1}$ so that the equation

$$\begin{bmatrix} U \\ UT \end{bmatrix} V = \begin{bmatrix} 0 \\ M^{-1} \end{bmatrix} \tag{6}$$

is satisfied. In other words, given only a standard pair, the third member of a triple can be obtained by first assigning a nonsingular M and then solving equation (6) for V.

Important examples of standard triples associated with vibrating systems involve the companion matrices of (2) and are:

$$U_1 = \begin{bmatrix} I_n & 0 \end{bmatrix}, \quad T_1 = C_R = \begin{bmatrix} 0 & I_n \\ -M^{-1}K & -M^{-1}D \end{bmatrix}, \quad V_1 = \begin{bmatrix} 0 \\ M^{-1} \end{bmatrix}, \tag{7}$$

and

$$U_2 = \begin{bmatrix} 0 & M^{-1} \end{bmatrix}, \quad T_2 = C_L = \begin{bmatrix} 0 & -KM^{-1} \\ I_n & -DM^{-1} \end{bmatrix}, \quad V_2 = \begin{bmatrix} I_n \\ 0 \end{bmatrix}, \tag{8}$$

(see [4] for details). It is easy to see that, if (U, T, V) is a standard triple then the three matrices obtained by a (generalized) similarity,

$$U_1 = UP^{-1}, \quad T_1 = PTP^{-1}, \quad V_1 = PV, \tag{9}$$

also form a standard triple. For example, the similarity from (U_1, T_1, V_1) to (U_2, T_2, V_2) in the examples of (7) and (8) is determined by the matrix A of equation (3), i.e., in this case $P = A = \begin{bmatrix} D & M \\ M & 0 \end{bmatrix}$.

Another important case is obtained by choosing P in (9) to be a matrix for which PTP^{-1} is in Jordan canonical form. The resulting (very special) standard triple is labelled a *Jordan triple* (see [4] and [6]).

Now let (U, T, V) be a standard triple for a vibrating system and consider the $n \times n$ matrices $\Gamma_0, \Gamma_1, \Gamma_2, \ldots$ defined by

$$\Gamma_j = UT^jV, \qquad j = 0, 1, 2, \ldots. \tag{10}$$

They are known as the *moments* of the triple. Observe that, by Definition 6, $\Gamma_0 = 0$ and Γ_1 is nonsingular. (Also, if it is known that T is nonsingular then moments can be defined for negative integers j.)

It is an important fact that *any transformation to a similar triple does not affect the moments*:

$$\Gamma_j = UT^jV = (UP^{-1})(PT^jP^{-1})(PV) = (UP^{-1})(PTP^{-1})^j(PV).$$

Thus, for a vibrating system the moments are the same whether they are found from the triples of (7), (8), or from a Jordan triple. This invariance property suggests that the moments reflect intrinsic properties of the vibrating system. Indeed, it has been shown in Theorem 2 of [6] that the moments generate a unique corresponding vibrating system through the recursive equations

$$M = \Gamma_1^{-1}, \quad D = -M\Gamma_2M, \quad K = -M(\Gamma_3)M + D\Gamma_1D. \tag{11}$$

Furthermore, for any standard pair, (U, T),

$$MUT^2 + DUT + KU = 0. \tag{12}$$

Theorem 4. *Let (U, T, V) be any standard triple and define moments by* (10). *Then equations* (11) *generate a unique corresponding vibrating system.*

Furthermore, any two standard triples for the same system are similar (*in the sense of equations* (9)).

Proof. The first statement is obtained by adapting the proof of Theorem 2 of [6] to standard triples, rather than Jordan triples. The second follows from Theorem 1.25 of [4], for example. □

Naturally, the moments generated by a standard triple may now be described without ambiguity as *the moments of a vibrating system.*

Corollary 5. *If (U, T, V) is a standard triple for a vibrating system $L(\lambda)$, then $\lambda I - T$ is a linearization for $L(\lambda)$.*

Proof. Note that the standard triple (7) has C_R as the main matrix and (Lemma 1) $\lambda I - C_R$ is a linearization. Also, if (U, T, V) is any standard triple, then the second statement of the theorem implies that C_R and T are similar. Hence, $\lambda I - C_R = P(\lambda I - T)P^{-1}$ and it follows from Definition 2 that $\lambda I - T$ is a linearization. □

It is useful to note that the transforming matrix P used in the last step of this proof can be written explicitly in the form $P = \begin{bmatrix} U \\ UT \end{bmatrix}$. To see this, first use (12) to verify that $C_R P = PT$, and hence $C_R = PTP^{-1}$.

The following technical lemma concerning the moments will assist in the proof of a theorem showing how SPE (and SPS) transformations can be obtained from standard triples.

Lemma 6. (a)
$$\begin{bmatrix} \Gamma_0 & \Gamma_1 \\ \Gamma_1 & \Gamma_2 \end{bmatrix} = A^{-1}.$$

(b)
$$\begin{bmatrix} \Gamma_1 & \Gamma_2 \\ \Gamma_2 & \Gamma_3 \end{bmatrix} = A^{-1} \begin{bmatrix} -K & 0 \\ 0 & M \end{bmatrix} A^{-1}.$$

Proof. (a) Using equations (11) we have

$$A^{-1} = \begin{bmatrix} D & M \\ M & 0 \end{bmatrix}^{-1} = \begin{bmatrix} 0 & M^{-1} \\ M^{-1} & -M^{-1}DM^{-1} \end{bmatrix} = \begin{bmatrix} \Gamma_0 & \Gamma_1 \\ \Gamma_1 & \Gamma_2 \end{bmatrix}.$$

(b) Using equations (11) again,

$$\begin{bmatrix} \Gamma_1 & \Gamma_2 \\ \Gamma_2 & \Gamma_3 \end{bmatrix} = \begin{bmatrix} M^{-1} & -M^{-1}DM^{-1} \\ -M^{-1}DM^{-1} & M^{-1}(DM^{-1}D - K)M^{-1} \end{bmatrix}.$$

Multiplying on left and right with A gives

$$A \begin{bmatrix} \Gamma_1 & \Gamma_2 \\ \Gamma_2 & \Gamma_3 \end{bmatrix} A = \begin{bmatrix} 0 & -KM^{-1} \\ I & -DM^{-1} \end{bmatrix} \begin{bmatrix} D & M \\ M & 0 \end{bmatrix} = \begin{bmatrix} -K & 0 \\ 0 & M \end{bmatrix},$$

from which (b) follows. □

4. Constructing SPE and SPS transformations

The next theorem is the main result of this paper. First recall the standard triple (7) for a vibrating system. To study isospectral systems it is convenient to write associated standard triples in the form (X, C_R, Y) where X and Y are still to be determined. The second matrix, C_R, is then common to all of these systems, so they must be isospectral. Furthermore, C_R automatically satisfies the condition (a) of Definition 5.

Now it will also be shown how SPE and SPS transformations can be constructed from such a set of standard triples. In the process, a constructive method for finding isospectral families of systems will appear.

Theorem 7. *Let a vibrating system $L_0(\lambda)$ be given and matrices A_0 and B_0 be formed as in* (5)*, and write $C_0 = A_0^{-1}B_0$. Consider any standard triple of the form $(X,\, C_0,\, Y)$.*

Then the $2n \times 2n$ matrices $\begin{bmatrix} A_0Y & B_0Y \end{bmatrix}$ and $\begin{bmatrix} X \\ XC_0 \end{bmatrix}$ are nonsingular and

$$E := \begin{bmatrix} Y & C_0Y \end{bmatrix}^{-1} A_0^{-1}, \quad F := \begin{bmatrix} X \\ XC_0 \end{bmatrix}^{-1} \tag{13}$$

determine an SPE of the given system.

Furthermore, if the moments of the transformed system $L(\lambda)$ are written $\Gamma_j = XC_0^jY$, then the transformed coefficients are

$$M = (\Gamma_1)^{-1}, \quad D = -M\Gamma_2M, \quad K = -M\Gamma_3M + D\Gamma_1D. \tag{14}$$

Conversely, every system $L(\lambda)$ isospectral with $L_0(\lambda)$ can be obtained from some choice of X and Y in a standard triple $(X,\, C_0,\, Y)$

Before going into the proof of this theorem, the importance of two properties of the standard triple $(X,\, C_0,\, Y)$ are emphasized. Namely, that

$$XY = 0, \quad \text{and} \quad \det(XC_0Y) \neq 0. \tag{15}$$

Keep in mind that A_0 and B_0 (and hence C_0) are prescribed by the coefficients M_0, D_0, K_0 of a given vibrating system. The design of isospectral systems depends only on these two conditions and, generally, they can be satisfied in infinitely many ways. The first can be interpreted as a biorthogonality condition imposed on the rows of X and columns of Y, and the second as a non-degeneracy condition which, generically, will be satisfied once the first condition is resolved. This interpretation is particularly useful in the discussion of systems with Hermitian (or real symmetric) coefficients. Conditions (15) are also natural generalizations of corresponding conditions used in Part I of this work, [6]. There, the analysis is confined to the case of *Jordan triples*, a special case of the present formulation. In Part I the data consists of suitable sets of eigenvalue and eigenvector data. Here, the data consists of the coefficients M_0, D_0, K_0 of a given system and the spectral information remains implicit.

Proof of Theorem 7. Consider the following product and use (15) to obtain

$$\begin{bmatrix} X \\ XC_0 \end{bmatrix} \begin{bmatrix} Y & C_0Y \end{bmatrix} = \begin{bmatrix} 0 & XC_0Y \\ XC_0Y & XC_0^2Y \end{bmatrix}.$$

But we also have $\det(XC_0Y) \neq 0$ and it follows that the product is nonsingular. So each factor on the left is nonsingular and E, F can be defined as in (13). So we may define the strict equivalence

$$\lambda A - B = E(\lambda A_0 - B_0)F. \tag{16}$$

It is to be shown that this is an SPE.

Since $A = EA_0F$ it follows that A is nonsingular and

$$\begin{aligned} A &= EA_0F = \begin{bmatrix} Y & C_0Y \end{bmatrix}^{-1} \begin{bmatrix} X \\ XC_0 \end{bmatrix}^{-1}, \\ &= \begin{bmatrix} XY & XC_0Y \\ XC_0Y & XC_0^2Y \end{bmatrix}^{-1} = \begin{bmatrix} \Gamma_0 & \Gamma_1 \\ \Gamma_1 & \Gamma_2 \end{bmatrix}^{-1}. \end{aligned}$$

Now use Part (a) of Lemma 6 to obtain

$$A = \begin{bmatrix} D & M \\ M & 0 \end{bmatrix},$$

and the structure of A_0 is preserved.

For the transformation of B_0, observe that

$$B_0 = E^{-1}BF^{-1} = \begin{bmatrix} A_0Y & B_0Y \end{bmatrix} B \begin{bmatrix} X \\ XC_0 \end{bmatrix}.$$

Multiply on the left and on the right by

$$\begin{bmatrix} X \\ XC_0 \end{bmatrix} A_0^{-1} \quad \text{and} \quad \begin{bmatrix} Y & C_0Y \end{bmatrix},$$

respectively, to obtain

$$\begin{bmatrix} \Gamma_1 & \Gamma_2 \\ \Gamma_2 & \Gamma_3 \end{bmatrix} = \begin{bmatrix} \Gamma_0 & \Gamma_1 \\ \Gamma_1 & \Gamma_2 \end{bmatrix} B \begin{bmatrix} \Gamma_0 & \Gamma_1 \\ \Gamma_1 & \Gamma_2 \end{bmatrix}.$$

Using Part (a) of Lemma 6 it follows that

$$B = A \begin{bmatrix} \Gamma_1 & \Gamma_2 \\ \Gamma_2 & \Gamma_3 \end{bmatrix} A.$$

Finally, Part (b) of Lemma 6 yields

$$B = \begin{bmatrix} -K & 0 \\ 0 & M \end{bmatrix}.$$

Thus, the structure of B_0 is also preserved and we have an SPE.

The formulae (13) are just a repetition of (11).

For the converse, if L and L_0 are isospectral systems they share a common Jordan form J, and so L has a Jordan triple $(\hat{X}, J, \hat{Y})$, say. But J is also a Jordan form for C_0, say $C_0 = TJT^{-1}$, and the triple $(\hat{X}T^{-1}, TJT^{-1}, T\hat{Y})$ obtained by similarity is also a standard triple for L. Since $TJT^{-1} = C_0$, the result is obtained if we define $X = \hat{X}T^{-1}$ and $Y = T\hat{Y}$. □

Notice also that an application of Corollary 3 gives:

Corollary 8. *With E and F defined as in Theorem 7, E defines an SPS for C_L, ($C_L = EC_L^{(0)}E^{-1}$) and F defines an SPS for C_R, ($C_R = F^{-1}C_R^{(0)}F$).*

Example 1. Consider the vibrating system defined by

$$M_0 = \begin{bmatrix} 1 & 0 \\ 0 & 1 \end{bmatrix}, \quad D_0 = \begin{bmatrix} 1 & 1 \\ 1 & 1 \end{bmatrix}, \quad K_0 = \begin{bmatrix} 1 & 0 \\ 0 & 2 \end{bmatrix},$$

with (truncated) eigenvalues

$$-0.9567 \pm 0.6412i, \quad \text{and} \quad -0.0433 \pm 1.2272i.$$

(Note that, although D is singular, the damping is pervasive.) Making the (almost arbitrary) choice

$$X = \begin{bmatrix} 1 & 2 & 0 & -3 \\ 0 & 1 & 1 & 2 \end{bmatrix},$$

it is found that $\begin{bmatrix} X \\ XC_0 \end{bmatrix}$ is nonsingular. Since C_0 is a companion matrix, condition (a) of Definition 5 is satisfied and it follows that (X, C_0) form a standard pair.

Assign the mass matrix $M = \begin{bmatrix} 1 & 0 \\ 0 & 1 \end{bmatrix}$, and compute the third member of the standard triple (cf. equation (6)):

$$Y = \frac{1}{5}\begin{bmatrix} -5 & 0 \\ 1 & 3 \\ 1 & -7 \\ -1 & 2 \end{bmatrix}.$$

Now moments Γ_1, Γ_2 and Γ_3 can be computed (as in equation (10)), and the new (exact) coefficients come from (14):

$$M = \begin{bmatrix} 1 & 0 \\ 0 & 1 \end{bmatrix}, \quad D = \begin{bmatrix} -0.8 & -5.4 \\ 1.6 & 2.8 \end{bmatrix}, \quad K = \begin{bmatrix} -1.4 & -4.0 \\ -0.2 & -2 \end{bmatrix}.$$

Calculations confirm that the spectrum is unchanged.

Alternatively, after finding Y as above, E and F could be computed from equations (13) and then the coefficients of the system are obtained from $A = EA_0F$, $B = EB_0F$.

5. Algorithms

The computational procedures for finding isospectral systems resulting from Theorem 7 are summarized in this section. Some simplifications arise if one is interested only in monic systems. The reader will easily make these adjustments when required.

Algorithm 1. The SPE method.

Data: Coefficients M_0, D_0, K_0 of a vibrating system and corresponding matrices

$$A_0 := \begin{bmatrix} D_0 & M_0 \\ M_0 & 0 \end{bmatrix}, \quad B_0 := \begin{bmatrix} -K_0 & 0 \\ 0 & M_0 \end{bmatrix}.$$

Step 1. Compute the matrix C_R of equation (2).

Step 2. Choose an $n \times 2n$ matrix X for which $\begin{bmatrix} X \\ XC_R \end{bmatrix}$ is nonsingular.

Step 3. Choose a new mass matrix M and solve for Y:

$$\begin{bmatrix} X \\ XC_R \end{bmatrix} Y = \begin{bmatrix} 0 \\ M^{-1} \end{bmatrix}.$$

Step 4. Compute

$$E := \begin{bmatrix} Y & C_R Y \end{bmatrix}^{-1} A_0^{-1}, \quad F := \begin{bmatrix} X \\ XC_R \end{bmatrix}^{-1}$$

Step 5. Compute $A = EA_0F$, and EB_0F and read off the new coefficients D and K.

End

Algorithm 2. The moment method.

Data: Coefficients M_0, D_0, K_0 of a vibrating system.

Steps 1, 2, and 3: As steps 1, 2, and 3 of Algorithm 1.

Step 4. Compute the moments $\Gamma_1 = XC_RY$, $\Gamma_2 = XC_R^2Y$, $\Gamma_3 = XC_R^3Y$.

Step 5. Compute the new coefficients

$$M = \Gamma_1^{-1}, \quad D = -M\Gamma_2M, \quad K = -M\Gamma_3M + D\Gamma_1D.$$

End

Of course, these summaries cannot be described as efficient general purpose algorithms. At a superficial level, Algorithm 1 may seem more direct. On the other hand, Algorithm 2 may gain in efficiency because it is formulated in terms of matrices of size n rather than $2n$, as in the first algorithm.

6. Systems with symmetries

In this section the study of systems isospectral with a given system (M_0, D_0, K_0) is continued, but now all of these matrices are Hermitian (in particular, they may be real and symmetric). Isospectral systems are to be generated which preserve the symmetry of (M_0, D_0, K_0).

Here, an important role is played by the nonsingular Hermitian matrix

$$A_0 = \begin{bmatrix} D_0 & M_0 \\ M_0 & 0 \end{bmatrix}. \tag{17}$$

Continuing the practice of Section 4, C_0 will denote the right companion matrix associated with the given system.

Matrix A_0 is always indefinite with inertia $\{n, n, 0\}$, and it has the important property that it symmetrises C_0 in the sense that $A_0C_0 = C_0^*A_0 = (A_0C_0)^*$. (The * denotes the complex-conjugate transposed matrix.) The matrix C_0 is said to be *self-adjoint in the indefinite inner product defined by* A_0, see [5]. Indeed, it is easy to prove the more general statement:

Lemma 9. *The matrix A_0 of (17) has the property that, for $j = 0, 1, 2, \ldots$,*

$$A_0 C_0^j = (A_0 C_0^j)^* = (C_0^j)^* A_0.$$

Let (X, C_0, Y) be a standard triple generated as in the algorithms of Section 5, for example. It will now be shown that if X and Y satisfy a geometric constraint, then the system generated will be self-adjoint (in the sense that the coefficients are Hermitian). But first note the following lemma:

Lemma 10. *Let (X, T, Y) be a standard triple for a vibrating system (M, D, K). Then M, D and K are Hermitian if and only if the moments Γ_1, Γ_2, and Γ_3 are Hermitian.*

Proof. The proof is an easy verification using the formulae of (11) (or see [6]). □

Proposition 11. *Let (X, C_0, Y) be a standard triple. Then the corresponding vibrating system has Hermitian coefficients if $Y = A_0^{-1} X^*$.*

Proof. If $Y = A_0^{-1} X^*$ then $X = Y^* A_0$ and

$$\Gamma_j = X C_0^j Y = Y^* (A_0 C_0^j) Y.$$

But, by Lemma 9, $A_0 C_0^j = (A_0 C_0^j)^*$ and it follows that Γ_j is Hermitian for each j. Then it follows from Lemma 10 that the corresponding system has Hermitian coefficients. □

Now the condition $X = Y^* A_0$ is reformulated in terms of X alone, and the geometric interpretation will be clarified. Since $X^* = A_0 Y$, the definition of Y in terms of X and C_0 (from Section 3) gives

$$X^* = A_0 \begin{bmatrix} X \\ X C_0 \end{bmatrix}^{-1} \begin{bmatrix} 0 \\ M^{-1} \end{bmatrix},$$

whence

$$\begin{bmatrix} X \\ X C_0 \end{bmatrix} A_0^{-1} X^* = \begin{bmatrix} 0 \\ M^{-1} \end{bmatrix}.$$

Thus, Hermitian coefficients are generated provided the two following conditions are satisfied:

$$X A_0^{-1} X^* = 0, \tag{18}$$

$$X (C_0 A_0^{-1}) X^* = M^{-1}. \tag{19}$$

These equations say that, *first, the subspace ImX^* must be isotropic with respect to the known indefinite matrix A_0^{-1} and, at the same time, (if M is to be positive definite) this subspace must be positive with respect to the Hermitian matrix $C_0 A_0^{-1}$.*

(Here, it is simply convenient to suppose that the ubiquitous condition $M > 0$ applies.) Naturally, equations (18) and (19) are the symmetrised form of the fundamental equations (15). They are also the analogues of very similar conditions formulated in [6] in terms of canonical forms for A_0 and C_0.

A procedure for generating Hermitian (or real and symmetric) isospectral systems can now be formulated as follows:

Algorithm 3. Hermitian, or real symmetric systems.

Step 1. From the given Hermitian (or real symmetric) system $\{M_0,\ D_0,\ K_0\}$ with $M_0 > 0$ formulate

$$C_0 = \begin{bmatrix} 0 & I \\ -M_0^{-1}K_0 & -M_0^{-1}D_0 \end{bmatrix}$$

and

$$A_0^{-1} = \begin{bmatrix} 0 & M_0^{-1} \\ M_0^{-1} & -M_0^{-1}D_0M_0^{-1} \end{bmatrix}.$$

Step 2. Compute an n-dimensional subspace $\mathbb{S}$ which is A_0^{-1}-isotropic and is also $C_0A_0^{-1}$-positive.

Step 3. Form the columns of matrix X^* from basis vectors for $\mathbb{S}$ and compute $Y = A_0^{-1}X^*$.

Step 4. Compute the moments

$$\Gamma_1 = XC_0Y, \qquad \Gamma_2 = XC_0^2Y, \qquad \Gamma_3 = XC_0^3Y.$$

Step 5. Compute the new coefficients

$$M = \Gamma_1^{-1}, \qquad D = -M\Gamma_2M, \qquad K = -M\Gamma_3M + D\Gamma_1D.$$

End

From the computational point of view, the most serious challenge here is to complete Step 2. To the authors' knowledge algorithms for this step are not generally available. The first essential is a general-purpose algorithm for finding a family of n-dimensional subspaces which are isotropic with respect to a given $2n \times 2n$ matrix with inertia $\{n, n, 0\}$. Then it is necessary to scan these subspaces to find one (or more) satisfying the positivity condition of Step 2. If the context is that of real and symmetric systems then C_0 and A_0 are real, of course, and the algorithms may be in real arithmetic.

Analysis of the relevant algorithms may be assisted by using the spectral decompositions:

$$C_0 = \begin{bmatrix} X \\ XJ \end{bmatrix} J \begin{bmatrix} X \\ XJ \end{bmatrix}^{-1}, \qquad A_0^{-1} = \begin{bmatrix} X \\ XJ \end{bmatrix} P \begin{bmatrix} X \\ XJ \end{bmatrix}^{*}.$$

Definitions of J and P can be found in [6] and the complete theory in [5], for example.

Example 2. It is shown here that, by defining $\mathbb{S} = \text{span}\{e_1, e_2, \ldots, e_n\}$ at Step 2 of Algorithm 3, the original problem can be reproduced. First, it is easily seen that this subspace is A_0^{-1}-isotropic and $C_0A_0^{-1}$-positive. Indeed, it is found that $X = \begin{bmatrix} I & 0 \end{bmatrix}$ and $X(C_0A_0^{-1})X^* = M_0^{-1}$ so that (from (19)), $M = M_0$.

Step 3 gives $Y = \begin{bmatrix} 0 \\ M_0^{-1} \end{bmatrix}$.

At Step 4 it is found that

$$\Gamma_1^{(1)} = M_0^{-1}, \quad \Gamma_2 = -M_0^{-1} D_0 M_0^{-1}, \quad \Gamma_3 = \left(-M_0^{-1} K_0 + (M_0^{-1} D_0)^2\right) M_0^{-1}$$

and then that $M = M_0,\ D = D_0,\ K = K_0$.

Example 3. Data:

$$M_0 = \begin{bmatrix} 1 & 0 \\ 0 & 1 \end{bmatrix}, \quad D_0 = \begin{bmatrix} 1 & 0 \\ 0 & 2 \end{bmatrix}, \quad K_0 = \begin{bmatrix} 4 & 3 \\ 3 & 6 \end{bmatrix}.$$

The truncated eigenvalues are found to be (in truncated form), $-0.6959 \pm 1.1943i$, and $-0.8041 \pm 2.6841i$. Step 1 yields

$$C_0 = \begin{bmatrix} 0 & 0 & 1 & 0 \\ 0 & 0 & 0 & 1 \\ -4 & -3 & -1 & 0 \\ -3 & -6 & 0 & -2 \end{bmatrix}, \quad A_0^{-1} = \begin{bmatrix} 0 & 0 & 1 & 0 \\ 0 & 0 & 0 & 1 \\ 1 & 0 & -1 & 0 \\ 0 & 1 & 0 & -2 \end{bmatrix}.$$

For Step 2 it is found that, for example, the image (or range) of the (truncated) matrix

$$X^* = \begin{bmatrix} 0.2923 & -1.3743 \\ -1.6097 & -0.1742 \\ -0.4729 & -0.1959 \\ -0.1742 & 0.4204 \end{bmatrix}$$

is A_0^{-1}-isotropic and $C_0 A_0^{-1}$-positive. Then step 3 yields

$$Y = \begin{bmatrix} -0.4729 & -0.1959 \\ -0.1742 & 0.4204 \\ 0.7651 & -1.1784 \\ -1.2613 & -1.0151 \end{bmatrix}.$$

The calculations of Steps 3 and 4 produce

$$M = \begin{bmatrix} 11.9099 & -6.6010 \\ -6.6010 & 4.2471 \end{bmatrix}, \quad D = \begin{bmatrix} 46.9830 & 14.7497 \\ 14.7497 & -31.3384 \end{bmatrix},$$

$$K = \begin{bmatrix} 47.5991 & 56.5223 \\ 56.5223 & 69.3272 \end{bmatrix}.$$

Then it can be verified that, indeed, this system and the system M_0, D_0, K_0 are isospectral.

7. Monic Hermitian systems with $D \geq 0$

Hermitian systems arise in many practical problems and, frequently, the conditions $M > 0$ and $D \geq 0$ hold. In this case, the system is easily reduced to *monic* form, i.e., with $M = I$. In this section it is shown that, under these conditions, a direct attack can be made on the problem (of generating isospectral systems) using the methods of the preceding section. Note first of all that, in the monic case,

$$A_0^{-1} = \begin{bmatrix} 0 & I \\ I & -D \end{bmatrix}, \quad C_0 = \begin{bmatrix} 0 & I \\ -K & -D \end{bmatrix}. \tag{20}$$

First, let us examine the possibility of generating an A_0^{-1}-isotropic subspace of the form $\mathrm{Im} X^*$ with $X = \begin{bmatrix} I & Z \end{bmatrix}$ for some $n \times n$ matrix Z. It is easily seen that such a subspace is A_0^{-1}-isotropic if and only if

$$0 = Z + Z^* - ZDZ^*. \tag{21}$$

Suppose that D has rank h and consider a factorization $D = WW^*$ where W is an $n \times h$ matrix of full rank.

Solution sets for (21) are easily formulated in the two extreme cases $h = 0$ and $h = n$:

- When $h = 0$ then $D = 0$ and every skew-symmetric matrix Z is a solution of (21).
- When $h = n$ it is easily verified that every matrix of the form

  $$Z = (W^{-1})^*(I - U)W^{-1}, \tag{22}$$

  where U is a unitary matrix, is a solution of equation (21).

The next proposition gives a generalization of these families of solutions to the general case $0 \leq h \leq n$.

First, for any $n \times h$ matrix W of rank h define a *unitary completion* of W to be an $n \times (n - h)$ matrix $\widehat{W}$ for which $W^*\widehat{W} = 0$ and $\widehat{W}^*\widehat{W} = I_{n-h}$. Then it can be shown that the matrix

$$E := \begin{bmatrix} W & \widehat{W} \end{bmatrix}$$

is nonsingular.

Proposition 12. *Given the full-rank factorization $D = WW^*$, let $\widehat{W}$ be a unitary completion of W, and construct the nonsingular matrix E as above. Then for any unitary matrix U of size h, any skew-Hermitian matrix S of size $n - h$, and an arbitrary matrix N of size $(n - h) \times h$, the matrix*

$$Z = (E^{-1})^* \begin{bmatrix} I_h - U & UN^* \\ -N & \frac{1}{2}N^*N - S \end{bmatrix} E^{-1} \tag{23}$$

is a solution of (21).

Proof. The proof is by verification. Substitute for Z from (23) in (21) and simplify. □

It can be shown that the number of free (complex) parameters in the representation (23) is $\frac{1}{2}n(n - 1)$, and that this is *independent of h*. The parameters will be further constrained on applying the second condition, namely, that the subspace $\mathbb{S}$ =Im(X^*) is $C_0A_0^{-1}$ positive. Note that, in the monic case,

$$C_0A_0^{-1} = \begin{bmatrix} I & -D \\ -D & D^2 - K \end{bmatrix},$$

so, with $X = \begin{bmatrix} I & Z \end{bmatrix}$, the positivity condition becomes

$$I - ZD - DZ^* + Z(D^2 - K)Z^* > 0.$$

Example 4. Consider the monic Hermitian system given by $D_0 = \begin{bmatrix} 2 & i \\ -i & 2 \end{bmatrix}$ and $K_0 = \text{diag}(1, -1)$. The eigenvalues (rounded) are $\lambda_1 = -3.0523, \lambda_2 = 0.4476$ and $\lambda_3 = \overline{\lambda_4} = -0.6977 + 0.4952i$.

Since D_0 is positive definite there is a solution of (21) of the form $Z = (W^{-1})^*(I_2 - U)W^{-1}$ where U is unitary. From the singular value decomposition, $D = W_0\Sigma^2W_0^*$, it is found that $D = WW^*$ with

$$W = W_0\Sigma = \begin{bmatrix} i/\sqrt(2) & i\sqrt{3/2} \\ -1/\sqrt{2} & \sqrt{3/2} \end{bmatrix}$$

and the choice

$$U = \begin{bmatrix} \cos(2\pi/3) & -\sin(2\pi/3) \\ \sin(2\pi/3) & \cos(2\pi/3) \end{bmatrix}$$

yields

$$Z = (W^{-1})^*(I_2 - U)W^{-1} = \begin{bmatrix} 1 & 0 \\ i & 1 \end{bmatrix}.$$

It is now a matter of computation to verify that, for $X = [I_2\ Z]$, we have

$$XPX^* = I_2 - ZD_0 - D_0X^* + Z(D_0^2 - K_0)Z^* = I_2$$

and hence ImX^* is $C_0A_0^{-1}$-positive. After first computing Γ_2 and Γ_3 in Step 4 of Algorithm 3, new monic system matrices are obtained:

$$D = \text{diag}(4, 0), \quad K = \begin{bmatrix} 3 & i \\ -i & 0 \end{bmatrix}.$$

This monic system has, indeed, the same eigenvalues as the initial system.

In this section we have focused on the case of standard triples (X, C_0, Y) for which $X = [I_n\ Z]$, i.e., for which the first n-by-n block of X is nonsingular. In the next section the case where the *second* n-by-n block is nonsingular will lead to another family of monic isospectral systems.

8. Isospectral families of Hermitian, or real symmetric systems

The notations of Section 7 are maintained here. Once again, as the context is Hermitian systems with positive definite leading coefficient, there is no significant loss in assuming that the system is already reduced to monic form.

In generating a parametrized set of isospectral systems, keep in mind that, from (18) and (19), such a system corresponds to an n-dimensional subspace of $\mathbb{C}^{2n}$, namely $\operatorname{Im} X^*$. Decompose

$$X = \left[\begin{array}{cc} X_1 & X_2 \end{array}\right], \quad \text{where} \quad X_1, X_2 \in \mathbb{C}^{n\times n}.$$

Using C_0 and A_0^{-1} for the monic system, the isotropy condition (18) takes the form

$$X_1 X_2^* + X_2 X_1^* = X_2^* D_0 X_2. \tag{24}$$

Now a general complex matrix C can be written in the form $C = A + iB$ where $A = \frac{1}{2}(C + C^*)$ and $B = \frac{1}{2i}(C - C^*)$ are Hermitian. Using this, it follows from the last equation that

$$X_1 X_2^* = \frac{1}{2}(X_2^* D_0 X_2) + S \tag{25}$$

where S is an arbitrary skew-Hermitian matrix.

In applying this technique, it is important to recognize that it is the n-dimensional subspace $\operatorname{Im}(X^*)$ which entirely determines the new system (isospectral with the (A_0, C_0) system). Furthermore, $\operatorname{Im}(X^*) = \operatorname{Im}(X^* A)$ for any nonsingular $n \times n$ matrix A. Using this idea, if we confine attention to subspaces for which X_2 is nonsingular then we can, without further loss of generality, set $X_2 = I$. However, it is convenient to introduce another real parameter $\epsilon \neq 0$ and set $X_2 = \epsilon I$. In this case, using (25) (and absorbing a real parameter into S), we may take $X_1 = \frac{1}{2}\epsilon D_0 + S$ and

$$X^* = \left[\begin{array}{c} \frac{1}{2}\epsilon D_0 + S \\ \epsilon I \end{array}\right]. \tag{26}$$

Now reconsider Algorithm 3, and notice first that in Step 1 we can set $M_0 = I$.

For Step 2 a parametrized set of subspaces is to be defined with the necessary isotropic and positivity properties. Let S be an arbitrary skew-Hermitian matrix ($n \geq 2$), ϵ be a real parameter, and define X by (26). Then

$$XA_0^{-1}X^* = \left[\begin{array}{cc} \frac{1}{2}\epsilon D_0 - S & \epsilon I \end{array}\right] \left[\begin{array}{cc} 0 & I \\ I & -D_0 \end{array}\right] \left[\begin{array}{c} \frac{1}{2}\epsilon D_0 + S \\ \epsilon I \end{array}\right] = 0$$

and the isotropic property is verified independently of ϵ and S.

Then, with a little computation,

$$\begin{aligned} X(C_0 A_0^{-1})X^* &= \left[\begin{array}{cc} \frac{1}{2}\epsilon D_0 - S & \epsilon I \end{array}\right] \left[\begin{array}{cc} I & -D_0 \\ -D_0 & -K_0 + D_0^2 \end{array}\right] \left[\begin{array}{c} \frac{1}{2}\epsilon D_0 + S \\ \epsilon I \end{array}\right] \\ &= (\frac{1}{2}\epsilon D_0 + S)(\frac{1}{2}\epsilon D_0 + S)^* - \epsilon^2 K_0, \end{aligned}$$

and it is apparent from this last expression that, provided only that ϵ and S are chosen so that $\frac{1}{2}\epsilon D_0 + S$ is nonsingular, $X(C_0A_0^{-1})X^* > 0$ for all sufficiently small ϵ.

Then it is found in Step 3 (with $M = I$) that

$$Y = \begin{bmatrix} 0 & I \\ I & -D_0 \end{bmatrix} \begin{bmatrix} \frac{1}{2}\epsilon D_0 + S \\ \epsilon I \end{bmatrix} = \begin{bmatrix} \epsilon I \\ -\frac{1}{2}\epsilon D_0 + S \end{bmatrix}.$$

Now all the necessary information is generated to complete Steps 4 and 5, and hence a family of isospectral systems determined by $\epsilon \in \mathbb{R}$ and skew-Hermitian matrices S.

It is clear that, if the given system is real and symmetric, then *real and symmetric isospectral systems are determined by choosing a real skew-symmetric matrix S*. The following examples are of this kind.

Example 5. (Data are taken from Example 7 of [8].) Let $M_0 = I_4$,

$$D_0 = \begin{bmatrix} 1 & 0 & 0 & 0 \\ 0 & 0.5 & 0 & 0 \\ 0 & 0 & 1 & 0 \\ 0 & 0 & 0 & 0.5 \end{bmatrix}, \quad K_0 = \begin{bmatrix} 5 & -2 & 0 & 0 \\ -2 & 4 & -2 & 0 \\ 0 & -2 & 5 & -3 \\ 0 & 0 & -3 & 3 \end{bmatrix}.$$

The spectrum of this system is

$$-0.3951 \pm 2.8146i, \quad -0.4247 \pm 2.3936i, \quad -0.3302 \pm 1.5813i, \quad -0.3500 \pm 0.4081i.$$

Assign the arbitrary skew-symmetric matrix

$$S = \begin{bmatrix} 0 & 1 & 1 & 1 \\ -1 & 0 & 1 & 1 \\ -1 & -1 & 0 & 1 \\ -1 & -1 & -1 & 0 \end{bmatrix},$$

and it is found experimentally that the positivity condition is satisfied for $0 \le \epsilon \le \frac{1}{8}$.

For the purpose of comparison, this technique has been applied to generate isospectral systems at $\epsilon = 1/16$ and at $\epsilon = 1/8$. The results are then scaled to produce monic systems. First compare the damping matrices $D(\epsilon)$:

$$D(1/16) = \begin{bmatrix} .8210 & -.0985 & .2144 & .0484 \\ & 1.0410 & -.1285 & -.0760 \\ & & .5334 & .2769 \\ & & & .6046 \end{bmatrix},$$

$$D(1/8) = \begin{bmatrix} 1.440 & -.6155 & .8599 & .0396 \\ & 1.5039 & -.4797 & -.4254 \\ & & .3818 & 1.1442 \\ & & & -.3297 \end{bmatrix},$$

Note that the positive definite property of D_0 is maintained at $\epsilon = 1/16$ and lost at $\epsilon = 1/8$.

The stiffness matrices are:

$$K(1/16) = \begin{bmatrix} 3.7321 & -1.7603 & -.5248 & 3.6499 \\ & 5.0111 & -1.2465 & -.2472 \\ & & 3.3212 & -.5021 \\ & & & 5.0464 \end{bmatrix},$$

$$K(1/8) = \begin{bmatrix} 5.8149 & -2.9548 & -.8040 & 6.0288 \\ & 5.3074 & -.1764 & -2.2789 \\ & & 2.3600 & -.2299 \\ & & & 7.4087 \end{bmatrix}.$$

The technique developed and illustrated here can include Example 3 as a limiting case (i.e., in the limit as $\epsilon \to 0$) and, consequently, determines smooth isospectral continuations of the given system. However, some care must be taken if n is odd, for then the perturbing skew-symmetric matrix S is necessarily singular and (see (26)) X does not have full rank at $\epsilon = 0$.

Now Example 1 is revisited to construct real and symmetric isospectral systems.

Example 6. As in Example 1 take

$$M_0 = \begin{bmatrix} 1 & 0 \\ 0 & 1 \end{bmatrix}, \quad D_0 = \begin{bmatrix} 1 & 1 \\ 1 & 1 \end{bmatrix}, \quad K_0 = \begin{bmatrix} 1 & 0 \\ 0 & 2 \end{bmatrix},$$

with (truncated) eigenvalues

$$-0.9567 \pm 0.6412i, \quad \text{and} \quad -0.0433 \pm 1.2272i.$$

Since $n = 2$, we take $S = \begin{bmatrix} 0 & 1 \\ -1 & 0 \end{bmatrix}$ and, in effect, there is only one free parameter, namely ϵ. It is easily seen that the positivity condition is satisfied if $-0.6 \le \epsilon \le 0.5$.

The damping and stiffness matrices are tabulated below for $\epsilon = -0.1$ and $\epsilon = 0.1$. Observe that the singular damping matrix D_0 is perturbed to a positive definite matrix by a negative shift of the parameter, but not by the positive shift.

$$D(-0.1) = \begin{bmatrix} .8988 & -.9235 \\ -.9235 & 1.1012 \end{bmatrix} \qquad K(-0.1) = \begin{bmatrix} 1.6477 & .0504 \\ .0504 & 1.2153 \end{bmatrix}$$

$$D(0.1) = \begin{bmatrix} 1.1013 & -1.1276 \\ -1.1276 & .8987 \end{bmatrix} \qquad K(0.1) = \begin{bmatrix} 2.4715 & -.0503 \\ -.0503 & .8103 \end{bmatrix}$$

9. Systems in reduced form

A wide class of systems of practical interest are intimately related to a diagonal system, or to a system which is "close to diagonal". They will be investigated in this section.

Observe first that, for each companion matrix appearing in a standard triple there is a class of isospectral systems. If a relation $\leftrightarrow$ is defined on $n \times n$ systems by saying $L_1(\lambda) \leftrightarrow L_2(\lambda)$ when they are isospectral, then $\leftrightarrow$ determines an *equivalence relation*, and it is natural to ask for canonical systems in each equivalence class. The fact that this is an equivalence relation is most easily verified by noting that all companion matrices associated with one of these classes have the same Jordan form (and this is the true canonical representation).

Definition 7 A system $L(\lambda)$ is said to be *reduced* or in *reduced form* if

$$L(\lambda) = (I\lambda - J_1)(I\lambda - J_2) = \lambda^2 - \lambda(J_1 + J_2) + J_1 J_2 \tag{27}$$

where J_1 and J_2 are in Jordan canonical form.

(Definition 7 includes a similar notion of "canonical systems" introduced in [8] under more restrictive hypotheses. There is a thorough treatment of the conditions under which $L(\lambda)$ in (27) is self-adjoint in [7].)

Not all equivalence classes contain such a reduced form. For example, a system with $n = 3$ and three distinct double eigenvalues each with an associated block of size two has no reduced form in its equivalence class (such an example is easily constructed). For this reason, the word "reduced" is used here, rather than "canonical". If J_1 and J_2 are consistent with a real, or a Hermitian system then, of course, there is interest in isospectral families and reduced forms which preserve these properties.

In many cases, when there is a reduced form (27) the class has an associated Jordan matrix

$$J = \begin{bmatrix} J_1 & 0 \\ 0 & J_2 \end{bmatrix}, \tag{28}$$

but this is not always the case. For example, when $n = 1$ the system $L(\lambda) = \lambda^2$ is already in reduced form (with $J_1 = J_2 = 0$), but

$$J = \begin{bmatrix} 0 & 1 \\ 0 & 0 \end{bmatrix} \neq \begin{bmatrix} J_1 & 0 \\ 0 & J_2 \end{bmatrix}.$$

To put this another way, it is **not** the case that a structure like (28) for J implies that there is a reduced form like (27) in the associated equivalence class. For example, if $n = 1$ and $J_1 = J_2 = \lambda_0$, then

$$(\lambda I - J_1)(\lambda I - J_2) = (\lambda - \lambda_0)^2,$$

but the Jordan form of this system is $\begin{bmatrix} \lambda_0 & 1 \\ 0 & \lambda_0 \end{bmatrix}$ and not $\begin{bmatrix} \lambda_0 & 0 \\ 0 & \lambda_0 \end{bmatrix}$.

Sufficient conditions for a Jordan form with the structure of (28) to have a reduced form in its equivalence class of systems are contained in:

Proposition 13. *An isospectral equivalence class contains a reduced system if the class has a corresponding Jordan matrix J of the form* (28) *where J_1 and J_2 are $n \times n$ Jordan forms and either* (a) *J_1 and J_2 have no common eigenvalues, or* (b) *J_1 and J_2 commute and* $\det(J_1 - J_2) \neq 0$.

Proof. Case (a) follows because the first factor in (27) is nonsingular at each eigenvalue of the second, and *vice versa.* This ensures that chains of right (left) eigenvectors for J_2 (resp. J_1) are inherited by $L(\lambda)$.

For case (b) observe that the determinantal condition implies that the matrix

$$X = \begin{bmatrix} I & I \\ J_1 & J_2 \end{bmatrix}$$

is nonsingular. Then (with L as in (27))

$$C_R X = \begin{bmatrix} J_1 & J_2 \\ J_1^2 + (J_2 J_1 - J_1 J_2) & J_2^2 \end{bmatrix} \quad \text{and} \quad XJ = \begin{bmatrix} J_1 & J_2 \\ J_1^2 & J_2^2 \end{bmatrix}.$$

Since J_1 and J_2 commute it follows that $C_R X = XJ$ (so that $C_R = XJX^{-1}$). Thus, J is characteristic of this equivalence class. □

Notice that in case (b) J_1 and J_2 may have common eigenvalues.

Definition 8. A system is said to be *regular* if the following properties hold:

1. The non-real eigenvalues (if any) arise in conjugate pairs with the same multiplicities. (Denote the corresponding Jordan matrices by J_c and $\overline{J_c}$.)
2. The real eigenvalues, if any, (say $2r$ in number) can be divided into two subsets of size r with corresponding $r \times r$ Jordan matrices J_s and J_t.
3. $\det(J_s - J_t) \neq 0$.
4. J_s and J_t commute.

Property 1 is enjoyed by two important classes of systems, namely, those with real coefficient matrices and those with Hermitian coefficients (and this is the main reason for introducing this definition). Of course, Properties 2, 3, and 4 are void if there are no real eigenvalues. Property 2 excludes some systems; for example, those with exactly one real eigenvalue which is not semisimple and has algebraic multiplicity two. Notice that there are *no constraints* on the multiplicities of the non-real eigenvalues (other than the fact that their algebraic multiplicity cannot exceed $n - r$).

Given Properties 1-4, it is possible to construct an associated $2n \times 2n$ Jordan matrix as in (28) with

$$J_1 = \begin{bmatrix} J_c & 0 \\ 0 & J_s \end{bmatrix}, \quad J_2 = \begin{bmatrix} \overline{J_c} & 0 \\ 0 & J_t \end{bmatrix}.$$

Now form the reduced system $L(\lambda)$ of (27) and it is easily seen that either or both of the conditions (a) and (b) of Theorem 10 hold. Then it is easily verified that:

Proposition 14. *The isospectral equivalence class of a regular system contains a real reduced system:*

$$D = -(J_1 + J_2) = -\begin{bmatrix} J_c + \overline{J_c} & 0 \\ 0 & J_s + J_t \end{bmatrix}, \quad K = J_1 J_2 = \begin{bmatrix} J_c \overline{J_c} & 0 \\ 0 & J_s J_t \end{bmatrix}.$$

If in addition, all eigenvalues are semisimple, then there is a real diagonal reduced system.

One might ask whether a reduced system will be included in the isospectral families of Sections 7 and 8. As the parametrized systems are all Hermitian it is, of course, necessary that the reduced form be symmetric so, as in Proposition 12, the answer can be "yes" if all real eigenvalues are semisimple or if there are no real eigenvalues, as in:

Example 7. The systems generated in Example 6 include a reduced system – up to a congruence transformation. If we choose (truncated) $\epsilon = -0.1728$ then we find

$$X^* = \begin{bmatrix} \frac{\epsilon}{2} D_0 + S \\ \epsilon\, I_2 \end{bmatrix} = \begin{bmatrix} -0.0864 & 0.9136 \\ -1.0864 & -0.0864 \\ -0.1728 & 0 \\ 0 & -0.1728 \end{bmatrix}$$

and

$$V_1 = \begin{bmatrix} \epsilon\, I_2 \\ S - \frac{\epsilon}{2} D_0 \end{bmatrix} = \begin{bmatrix} -0.1728 & 0 \\ 0 & -0.1728 \\ 0.0864 & 1.0864 \\ -0.9136 & 0.0864 \end{bmatrix}$$

from which we may compute

$$\Gamma_1 = U_1 T V_1 = \begin{bmatrix} 0.8123 & 0.0149 \\ 0.0149 & 1.1280 \end{bmatrix}, \qquad \Gamma_2 = U_1 T^2 V_1 = \begin{bmatrix} -0.6545 & 0.8422 \\ 0.8422 & -1.3157 \end{bmatrix},$$

$$\Gamma_3 = U_1 T^3 V_1 = \begin{bmatrix} 0.0075 & -1.7931 \\ -1.7931 & 0.8645 \end{bmatrix},$$

and hence we have

$$M = \begin{bmatrix} 1.2314 & -0.0163 \\ -0.0163 & 0.8867 \end{bmatrix}, \qquad D = \begin{bmatrix} 1.0265 & -0.9519 \\ -0.9519 & 1.0591 \end{bmatrix},$$

$$K = \begin{bmatrix} 1.7653 & 0.0699 \\ 0.0699 & 1.2395 \end{bmatrix}.$$

It can now be verified that $KM^{-1}D = DM^{-1}K$ and hence the three matrices M, D, K can be diagonalized by a congruence transformation [1]. Define a matrix of eigenvectors of the undamped system:

$$X_0 := \begin{bmatrix} -0.5655 & -0.7018 \\ 0.8165 & -0.6792 \end{bmatrix}.$$

Because of the commutativity assumption above, this matrix will also diagonalize D. Indeed, $X_0^T M X_0 = I_2$, $X_0^T D X_0 = \mathrm{diag}(1.9134, 0.0866) = -\Lambda - \Lambda^*$ and

$$X_0^T K X_0 = \mathrm{diag}(1.3264, 1.5079) = \Lambda\Lambda^*$$

where $\Lambda := \mathrm{diag}(-0.9567 + 0.6412i, -0.0433 + 1.2272i)$.

Instead of this indirect derivation of the reduced system it is also possible to obtain the same result by using $U_d^* := U_1^*(X_0^T)^{-1}$ which is A_0^{-1}-isotropic and leaves $M_0 = I_2$ invariant, i.e., $U_d^* C_0 A_0^{-1} U_d = I_2$.

10. Conclusions

The spectral theory of vibrating systems has been reviewed and and re-examined from the point of view of inverse spectral problems: i.e., the construction of systems with spectral characteristics defined implicitly via a given system. In Part 1 of this work (see [8]) the spectral characteristics were prescribed explicitly in the form of complete sets of eigenvalue and eigenvector data. If no symmetry properties are required of the systems generated, the problem has a relatively easy solution summarized in the five-step procedures summarized in Section 5.

If symmetries are imposed on the coefficients of all systems considered, then the situation is more involved. However, an explicit construction is given in Section 6 for the determination of isospectral families of symmetric systems (whether complex Hermitian or real symmetric). If efficient general-purpose algorithms are to be created, then an efficient procedure is required for the determination of n-dimensional subspaces of a $2n$-dimensional space which are neutral with respect to a known real symmetric indefinite matrix (see Step 2 of Algorithm 3 in Section 6). The construction of symmetric systems with positivity conditions imposed on the coefficient matrices remains an essentially open problem, although methods developed in [8] should cast some light on this problem. Special cases of problems with positivity constraints have been discussed in Sections 7 and 8.

Finally, in Section 9 it has been clarified under what circumstances an isospectral family may contain a diagonal (or "close to diagonal") system.

Acknowledgements

The first named author gratefully acknowledges Prof. Seamus D. Garvey for many stimulating discussions and the support of the EPSRC in this work through the research grant GR/S31679/01 on "Model Compaction and Model Reduction Methods for Large-Scale Dynamic Systems in Engineering". The second named author gratefully acknowledges support from Prof. N. J. Higham of the University of Manchester and from the Natural Sciences and Engineering Research Council of Canada.

References

[1] T.K. Caughey and M.E. O'Kelly, *Classical normal modes in damped linear system*, Journal of Applied Mechanics, Transaction of the ASME, **32**, 1965, 583–588.

[2] M.T. Chu, Y.-C. Kuo and W.-W. Lin, *On inverse quadratic eigenvalue problems with partially prescribed eigenstructure* SIAM J. Matrix Anal. Appl., **25**, 2004, 995–1020.

[3] S.G. Garvey, M.I. Friswell and U. Prells, *Co-ordinate Transformations for Second Order Systems, Part I: General Transformations*, J. Sound & Vibration, **258**(5), 2002, 885–909.

[4] I. Gohberg, P. Lancaster and L. Rodman, *Matrix Polynomials*, Academic Press, New York, 1982.

[5] I. Gohberg, P. Lancaster and L. Rodman, *Matrices and Indefinite Scalar Products*, Birkhäuser, Basel, 1983.

[6] P. Lancaster, *Isospectral vibrating systems. Part 1: The spectral method*, Linear Algebra Appl., **409**, 2005, 51–69.

[7] P. Lancaster and J. Maroulas, *Inverse eigenvalue problems for damped vibrating systems*, J. Math. Anal. Appl. **123**, 1987, 238–261.

[8] P. Lancaster and U. Prells, *Inverse problems for vibrating systems*, J. Sound & Vibration, **283**, 2005, 891–914.

[9] P. Lancaster and M. Tismenetsky, *The Theory of Matrices Second Edition*, Academic Press, Orlando, 1985.

Uwe Prells
School of Mechanical, Materials,
Manufacturing Engineering & Management
University of Nottingham
Nottingham NG7 2RD
United Kingdom
e-mail: `prells@penmaen.demon.co.uk`

Peter Lancaster
Department of Mathematics and Statistics
University of Calgary
Calgary AB T2N 1N4
Canada
e-mail: `lancaste@ucalgary.ca`

Operator Theory:
Advances and Applications, Vol. 163, 299–335

A Functional Description for the Commutative WJ^*-algebras of the D^+_κ-class

Vladimir Strauss

Dedicated to Professor Heinz Langer

Abstract. We consider the action in Krein spaces of weakly closed J-symmetric operator algebras with identity possessing an invariant maximal non-negative subspace, presented as a direct sum of a finite-dimensional neutral subspace and a uniformly positive subspace. A relation between these algebras (that in addition are assumed to be commutative) and function spaces of the type $L^\infty_\sigma \cap L^2_\nu$ is established.

Mathematics Subject Classification (2000). Primary 46C20, 47B50; Secondary 47B40, 47A60.

Keywords. Indefinite metric, operator algebras, model representation, functional calculus.

Introduction

This work has a direct connection with the paper [40]. It is assumed the reader is familiar with the elements of Krein space geometry and operator theory (see [9], [2], [16], [26]). In this paper the terminology introduced in [3] will be used.

One of the purposes of this paper is the study of relations between commutative WJ^*-algebras (i.e., weakly closed J-symmetric operator algebras with identity) in separable Krein spaces and certain function spaces. It is assumed that a WJ^*-algebra has a maximal non-negative invariant subspace, presented as a direct sum of a finite-dimensional neutral subspace and a uniformly positive subspace. As is known ([6], [39]) this algebra generates a spectral function E_λ with a peculiar spectral set Λ that provides a resolution of spectral type for the operators in the algebra. In particular to every operator A one can associate a scalar function

This work was completed with the support of project CONICIT (Venezuela) No 97000668.

$f_A(\lambda)$ (the portrait of A) such that

$$AE(\Delta) = \int_{\Delta} f_A(\lambda) dE_\lambda,$$

where Δ runs over the set of all closed intervals of the real line disjoint from Λ. The main results (Theorems 5.1, 5.4 and Corollary 5.5) show that the set of all portraits corresponding to operators in the algebra contains, as a principal part, a space $L^\infty_\sigma \cap L^2_\nu$. Here the space L^∞_σ is generated by a bounded measure while the space L^2_ν is generated by an unbounded one. Points at which the unbounded measure has an infinite rate of growth belong to the set of peculiarities Λ mentioned above but, generally speaking, do not exhaust the set.

Section 1 contains mainly definitions and well-known results used in the course of the paper. In particular, in Subsection 1.2 some function spaces are described and Subsection 1.4 deals with a model representation of a resolution of the identity that is simultaneously J-orthogonal and similar to an orthogonal resolution of the identity. In Section 2 a notion of unbounded elements conformed with a resolution of the identity is introduced and studied. Roughly speaking a resolution defines a correspondence between a Banach space and a (vector-valued) function space. From this point of view unbound elements correspond to measurable functions that do not belong to the function space. Section 3 represents a short description of results from [40] that are used in Sections 4 and 5. In Section 4 a functional description is given for an algebra with a spectral resolution having only one peculiarity, while Section 5 deals with the general case. In Subsection 4.3 a functional calculus for J-self-adjoint operators is constructed. Historical and bibliographical remarks are situated in Section 6.

1. Preliminaries

1.1. Krein spaces

Let $\mathcal{H}$ be a Krein space with an indefinite sesquilinear form $[\cdot,\cdot]$, let $\mathcal{H} = \mathcal{H}_+[\dot{+}]\mathcal{H}_-$ be its canonical decomposition, let P_+ and P_- be canonical projections: $\mathcal{H}_+ = P_+\mathcal{H}$, $\mathcal{H}_- = P_-\mathcal{H}$, let $J = P_+ - P_-$ be a canonical symmetry, and let $(\cdot,\cdot) = [J\cdot,\cdot]$ be a canonical scalar product. Note that any one of these canonical objects uniquely determines the others. Everywhere below we fix on $\mathcal{H}$ a unique form $[x,y] = (Jx,y)$. At the same time let us note that in the question we consider, a concrete choice of Hilbert scalar product is not really essential. One needs only to fix the topology (defined by the above-mentioned scalar product) and the structure of J.

Below non-negative (especially maximal non-negative) subspaces will play an important role (see [22] for references). The set of all maximal non-negative subspaces of the Krein space $\mathfrak{H}$ is denoted $\mathfrak{M}^+(\mathcal{H})$.

A subspace $\mathcal{L}$ is called *pseudo-regular* ([15]) if it can be presented in the form

$$\mathcal{L} = \hat{\mathcal{L}} \dot{+} \mathcal{L}_1, \tag{1.1}$$

where $\hat{\mathcal{L}}$ is a regular subspace and $\mathcal{L}_1$ is a neutral subspace (i.e., $\mathcal{L}_1$ is an isotropic part of $\mathcal{L}$).

Proposition 1.1. ([4], [5]) *Let:*

- $\mathfrak{L}_+$ *be a pseudo-regular subspace belonging to* $\in \mathfrak{M}^+(\mathfrak{H})$*;*
- $\mathfrak{L}_0$ *be the isotropic subspace of* $\mathfrak{L}_+$*;*
- $(\cdot,\cdot)'$ *be a scalar product on* $\mathfrak{L}_0$*, such that the norm* $\sqrt{(x,x)'}$ *is equivalent to the original one;*
- $\mathfrak{L}_- = \mathfrak{L}_+^{[\perp]}$*;*

and let

$$\mathfrak{L}_+ = \hat{\mathfrak{L}}_+ \dot{+} \mathfrak{L}_0, \; \mathfrak{L}_- = \hat{\mathfrak{L}}_- \dot{+} \mathfrak{L}_0, \tag{1.2}$$

where $\hat{\mathfrak{L}}_+$ *and* $\hat{\mathfrak{L}}_-$ *are uniformly definite subspaces. Then one can define on* $\mathfrak{H}$ *a canonical scalar product* $(\cdot,\cdot)$ *such that:*

$$\left.\begin{array}{llcl} \text{a)} & \text{on } \mathfrak{L}_0 & : & (\cdot,\cdot) \equiv (\cdot,\cdot)' \\ \text{b)} & \mathfrak{L}_0 \perp \hat{\mathfrak{L}}_+ & , & \mathfrak{L}_0 \perp \hat{\mathfrak{L}}_- \\ \text{c)} & \text{on } \hat{\mathfrak{L}}_+ & : & (\cdot,\cdot) = [\cdot,\cdot] \\ \text{d)} & \text{on } \hat{\mathfrak{L}}_- & : & (\cdot,\cdot) = -[\cdot,\cdot] \end{array}\right\} \tag{1.3}$$

Definition 1.2. If a canonical scalar product of a Krein space $\mathfrak{H}$ has the properties (1.3), it is said to be *compatible* with Decomposition (1.2) and the choice of the scalar product $(\cdot,\cdot)'$ on $\mathfrak{L}_1$.

Define a special case of pseudo-regular subspaces: a non-negative (non-positive) subspace $\mathcal{L}$ is called *a subspace of the class* h^+ (h^-) if it is pseudo-regular and $\dim \mathcal{L}_1 < \infty$ for $\mathcal{L}_1$ as in (1.1). In Pontryagin spaces every subspace is pseudo-regular and every semi-definite subspace belongs to class h^+ or h^-.

Here the term "operator" means "bounded linear operator". By the symbol $B^\#$ we denote the operator J-adjoint (J-a.) to an operator B. For an operator A the symbol $\sigma(A)$ denotes its *spectrum* treated in the same way as in [13] or [3].

If an operator family $\mathfrak{Y}$ is such that the condition $A \in \mathfrak{Y}$ implies $A^\# \in \mathfrak{Y}$, then this family is said to be *J-symmetric.* An operator algebra $\mathfrak{A}$ is said to be WJ^*-algebra if it is closed in the weak operator topology, J-symmetric and contains the identity I. The symbol $\operatorname{Alg}\mathfrak{Y}$ means the minimal WJ^*-algebra which contains $\mathfrak{Y}$.

One of the most important directions in the development of the operator theory is connected to the existence of invariant maximal semi-definite subspaces for certain operator sets and the study of the properties of the operators in such sets. A subspace $\mathcal{L}$ is said to be *A-invariant* (*$\mathfrak{Y}$-invariant*) if it is invariant with respect to the operator A (operator family $\mathfrak{Y}$).

Definition 1.3. A J-symmetric operator family $\mathfrak{Y}$ belongs to the class D_κ^+ if there is a subspace $\mathcal{L}_+$ in $\mathcal{H}$, such that

- $\mathcal{L}_+$ is $\mathfrak{Y}$-invariant,
- $\mathcal{L}_+ \in \mathfrak{M}^+(\mathcal{H}) \cap h^+$,
- $\dim(\mathcal{L}_+ \cap \mathcal{L}_+^{[\perp]}) = \kappa$.

Remark 1.4. If a J-symmetric family $\mathfrak{Y} \in D_\kappa^+$ and $\mathcal{L}_+$ is a $\mathfrak{Y}$-invariant subspace corresponding to Definition 1.3, then the pseudo-regular subspace $\mathcal{L}_+^{[\perp]}$ is $\mathfrak{Y}$-invariant too.

1.2. Some function spaces

Assume that $\sigma(t)$ is a non-decreasing function defined on the segment $[-1;1]$, continuous in the points -1; 0; 1, continuous (at least) from the left in all other points of the segment and having an infinite number of growth points, where zero is one of these points. The mentioned function generates on $[-1;1]$ the Lebesgue-Stieltjes measure μ_σ and spaces (L^2_σ, L^∞_σ, etc.) of complex-valued functions. At the same time we shall consider also some spaces of vector-valued functions so from time to time we shall note after a symbol of a space a symbol of a range for the functions forming this space, for instance, $L^2_\sigma(\mathbb{C})$. Next, let us consider a slightly different construction. Let $G(t)$ be a μ_σ-measurable function defined a.e. on $[-1;1]$ and such that

- a.e. $G(t) \geq 1$,
- $\int_{-1}^{-\tau} G(t)d\sigma(t) < \infty$, $\int_\tau^1 G(t)d\sigma(t) < \infty$ for every $\tau \in (0;1]$,
- $\int_{-1}^1 G(t)d\sigma(t) = \infty$.

Set

$$\left.\begin{array}{rcl} \nu(\tau) & = & \begin{cases} \int_{-1}^{\tau} G(t)d\sigma(t), & \text{if } \tau \in [-1;0); \\ -\int_{\tau}^{1} G(t)d\sigma(t), & \text{if } \tau \in (0;1]. \end{cases} \\ \eta(\tau) & = & \int_{-1}^{\tau}(1/G(t))d\sigma(t) \text{ for } \tau \in [-1;1]. \end{array}\right\} \tag{1.4}$$

The function $\nu(t)$ is non-decreasing in both segments $[-1;0)$ and $(0;1]$ but it is unbounded in neighborhoods of zero. Define for it a corresponding function space. Let $f(t)$ and $g(t)$ be arbitrary functions continuous in $[-1;1]$ and vanishing in some neighborhoods (different in the general case for $f(t)$ and $g(t)$) of zero. Then the integral $\int_{-1}^1 f(t)\overline{g(t)}d\nu(t)$ is well defined and generates a structure of pre-Hilbert space on the set of all such functions. The completion of the space will be denote L^2_ν (or $L^2_\nu(\mathbb{C})$). In a similar way one can define the space L^1_ν. At the same time the function $\eta(t)$ is non-decreasing on the whole interval $[-1;1]$, hence $\eta(t)$ defines on this interval the ordinary Lebesgue-Stieltjes measure μ_η that is absolutely continuous with respect to μ_σ. Thus, the space L^2_η and others are defined as usual.

Note that due to (1.4) the spaces L^∞_σ and L^2_ν, as well as the spaces L^1_σ and L^2_η, form compatible pairs or Banach pairs (for details see [8] or [24]). Thus, the spaces $L^1_\sigma + L^2_\eta$ and $L^\infty_\sigma \cap L^2_\nu$ are well defined. In particular, the standard norm on $L^1_\sigma + L^2_\eta$ is given by the formula

$$\|f\| = \inf_{f_1+f_2=f}\{\|f_1\|_{L^1_\sigma} + \|f_2\|_{L^2_\eta}\}.$$

The space $L^\infty_\sigma \cap L^2_\nu$ can be considered as adjoint to the space $L^1_\sigma + L^2_\eta$ if the duality between these space is given by the formula $\langle f(t), g(t)\rangle = \int_{-1}^{1} f(t)\overline{g(t)}d\sigma(t)$, where $f(t) \in L^1_\sigma + L^2_\eta$ and $g(t) \in L^\infty_\sigma \cap L^2_\nu$.

Let us pass to some notations relating to direct integrals of Hilbert spaces and corresponding model descriptions of self-adjoint operators (see [33], §41; [10], Chapter 7; [11], Chapter 4.4; [34], Chapter VII). Let $\mathcal{E}$ be some separable Hilbert space ($\mathcal{E}$ can be finite-dimensional), let $\sigma(t)$ be as above. Consider a mapping $t \mapsto \mathcal{E}_t$, $t \in [-1; 1]$, where $\mathcal{E}_t \subset \mathcal{E}$, $\dim(\mathcal{E}_t)$ is a μ_σ-measurable (but not necessarily finite a.e.) function, and if $\dim(\mathcal{E}_{t_1}) = \dim(\mathcal{E}_{t_2})$, then $\mathcal{E}_{t_1} = \mathcal{E}_{t_2}$. Denote by $M_{\vec{\sigma}}(\mathcal{E})$ the space of the vector-valued functions $f(t)\colon\, t \mapsto \mathcal{E}_t$ μ_σ-measurable in the weak sense, defined a.e. and finite a.e. on the segment $[-1; 1]$. Next, the symbol $L^2_{\vec{\sigma}}(\mathcal{E})$ means here a Hilbert space of functions $f(t) \in M_{\vec{\sigma}}(\mathcal{E})$, such that $\int_{-1}^{1} \|f(t)\|^2_{\mathcal{E}} d\sigma(t) < \infty$

Introduce some notation related with multiplication operators by scalar function. Everywhere below we assume a scalar function $\varphi(t)$ to be defined a.e. on $[-1; 1]$, μ_σ-measurable and a.e. bounded. For $f(t) \in M_{\vec{\sigma}}(\mathcal{E})$ set

$$(\Phi f)(t) = \varphi(t) f(t). \tag{1.5}$$

It is clear that $(\Phi f)(t) \in M_{\vec{\sigma}}(\mathcal{E})$, so equality (1.5) defines on $M_{\vec{\sigma}}(\mathcal{E})$ the continuous operator Φ (= the multiplication operator by the function $\varphi(t)$). If $\varphi(t)$ satisfies some additional conditions one can consider the operator Φ as acting simultaneously on different spaces. If, for instance, $\varphi(t)$ is continuous then the operator Φ is well defined on every space $M_\sigma(\mathcal{E})$ independently of $\vec{\sigma}(t)$ and $\mathcal{E}$. If $\varphi(t) \in L^\infty_\sigma(\mathbb{C})$ then $L^2_{\vec{\sigma}}(\mathcal{E})$ can also be taken as a domain of Φ. So, if it is necessary, we'll mention simultaneously the operator Φ and its domain using the notation $\{\Phi, \mathfrak{D}(\Phi)\}$, say, $\{\Phi, L^2_{\vec{\sigma}}(\mathcal{E})\}$.

1.3. Spectral functions with peculiarities

Spectral resolution for different operator classes is one of the important problems in the operator theory.

Let $\Lambda = \{\lambda_k\}_1^n$ be a finite set of real numbers and let $\mathfrak{R}_\Lambda$ be the family $\{X\}$ of all Borel subsets of $\mathbb{R}$ such that $\partial X \cap \Lambda = \emptyset$, where ∂X is the boundary of X in $\mathbb{R}$. Let $E\colon\, X \mapsto E(X)$ be a countably additive (with respect to weak topology) function, that maps $\mathfrak{R}_\Lambda$ to a commutative algebra of projections in a Hilbert space $\mathcal{H}$, $E(\mathbb{R}) = I$. $E(X)$ is called *a spectral function* (*on* $\mathbb{R}$) *with the peculiar spectral set* Λ, the mention of Λ can be omitted. The symbol $\mathrm{Supp}(E)$ means the minimal closed subset $S \subset \mathbb{R}$, such that $E(X) = 0$ for every X: $X \subset \mathbb{R}\backslash S$ and $X \in \mathfrak{R}_\Lambda$. Besides the symbol E we shall use also the symbol E_λ, $\lambda \in \mathbb{R}$ as notation for a spectral function, where $E_\lambda = E((-\infty, \lambda))$. Note that the notion of peculiar set has no any direct connection with the behavior of the spectral function and it means only that some points on $\mathbb{R}$ are distinguished. See below Definition 1.8 for some explanations.

A spectral function E that acts in a Krein space, is said to be *J-orthogonal* (*J*-orth.sp.f.) if $E(X)$ is a *J*-ortho-projection for every $X \in \mathfrak{R}_\Lambda$.

Let us recall the definition of a *scalar spectral operator* with real spectrum ([14]). An operator A acting in a Hilbert space is said to be a scalar spectral operator if there exists a spectral function E with empty peculiar spectral set Λ, such that for every $X \in \mathfrak{R}_\Lambda$: $E(X)A = AE(X)$, $\sigma(A|_{E(X)\mathcal{H}}) \subset \bar{X}$ and $AE(X) = \int_X \xi E(d\xi)$ in the weak sense.

Now let E be a spectral function with peculiar spectral set Λ. A scalar function $f(\xi)$ is said to be *defined almost everywhere* (with respect to E), *to have finite value almost everywhere*, etc., if the corresponding property holds almost everywhere in the weak sense on an arbitrary set $X \in \mathfrak{R}_\Lambda$, $X \cap \Lambda = \emptyset$. We shall assume that the function $f(\xi)$ is not defined at Λ. The following theorem was announced in [39] and proved in [6].

Theorem 1.5. *Let $\mathfrak{Y} \in D_\kappa^+$ be a commutative family of J-s.a. operators with real spectra. Then there exists a J-orth.sp.f. E with a finite peculiar spectral set Λ (Λ may be the empty set), such that the following conditions hold*

$$\left.\begin{array}{ll}
\text{a)} & E_\lambda \in \operatorname{Alg}\mathfrak{Y} \textit{ for all } \lambda \in \mathbb{R}\backslash\Lambda; \\
\text{b)} & \textit{there is a non-negative subspace } \mathcal{L}_+, \textit{ corresponding to Definition 1.3, for which the descomposition } E(\Delta)\mathcal{H} = E(\Delta)\mathcal{L}_+[\dot{+}]E(\Delta)\mathcal{L}_- \textit{ holds, } \Delta \textit{ being any closed segment } \Delta \subset \mathbb{R} \textit{ satisfying } \Delta \in \mathfrak{R}_\Lambda \textit{ and } \Delta \cap \Lambda = \emptyset; \\
\text{c)} & \textit{for every operator } A \in \mathfrak{Y}, \textit{ there exists a defined almost everywhere and (uniformly) bounded function } \phi(\lambda) \textit{ such that for every interval } \Delta \subset \mathbb{R},\ \Delta \in \mathfrak{R}_\Lambda,\ \Delta \cap \Lambda = \emptyset, \textit{ the descomposition } AE(\Delta) = \int_\Delta \phi(\lambda)E(d\lambda) \textit{ is valid;} \\
\text{d)} & \textit{the subspace } \widetilde{\mathcal{H}} = \operatorname*{CLin}_{\Delta \in \mathfrak{R}_\Lambda,\ \Delta \cap \Lambda = \emptyset} \{E(\Delta)\mathcal{H}\} \textit{ is pseudo-regular and its isotropic part has finite dimension;} \\
\text{e)} & \textit{for every point } \lambda_0 \in \Lambda \textit{ the corresponding subspace } \mathcal{H}_{\lambda_0} = \bigcap_{\lambda_0 \in \Delta} E(\Delta)\mathcal{H} \textit{ is a subspace of eigenvectors and root vectors associated with an eigenvalue, or is a part of such a subspace for every operator } A \in \mathfrak{Y}; \\
\text{f)} & \textit{if } \lambda_0 \in \Lambda, \textit{ then either } \limsup_{\lambda \to \lambda_0} \|E_\lambda\| = \infty \textit{ or at least for one } A \in \mathfrak{Y} \textit{ the operator } A|_{\mathcal{H}_{\lambda_0}} \textit{ is not a spectral operator of scalar type.}
\end{array}\right\} \quad (1.6)$$

A spectral function E with a peculiar spectral set Λ satisfying Conditions (1.6) are called *an eigen spectral function* (e.s.f.) of the operator family $\mathfrak{Y}$.

Note that the restriction that all operators from $\mathfrak{Y}$ have real spectra is not very strong due to the following proposition.

Proposition 1.6. *If a commutative family $\mathfrak{Y}$ of J-s.a. operators belongs to D_κ^+ and $\sigma(A_0)\backslash\mathbb{R} \neq \emptyset$ at least for one $A_0 \in \mathfrak{Y}$, then the subspaces $\mathcal{H} = \mathcal{H}'[\dot{+}]\mathcal{H}''$, where $\mathcal{H}'$ and $\mathcal{H}''$ are $\mathfrak{Y}$-invariant, $\sigma(A|_{\mathcal{H}'}) \subset \mathbb{R}$ for all $A \in \mathfrak{Y}$, the subspace $\mathcal{H}''$ has finite dimension and the family $\mathfrak{Y}|_{\mathcal{H}'}$ belongs to the class $D_{\kappa'}^+$ with $\kappa' < \kappa$ (it is possible that $\kappa' = 0$).*

If a commutative family $\mathfrak{Y} \in D_\kappa^+$ of J-s.a. operators is such that $\sigma(A_0)\backslash\mathbb{R} \neq \emptyset$ at least for one $A_0 \in \mathfrak{Y}$, then, generally speaking, the subspace $\mathcal{H}'$ from Proposition 1.6 is not uniquely determined, but there exists a subspace $\mathcal{H}'_{max}$ that is maximal in the following sense: $\mathcal{H}'_{max}$ satisfies the conditions of Proposition 1.6 and for every subspace $\mathcal{H}'$ satisfying the same conditions $\mathcal{H}' \subset \mathcal{H}'_{max}$. For the operator family in question we, first, find an e.s.f. E_λ for the operator family $\mathfrak{Y}|_{\mathcal{H}'_{max}}$, and, second, set $E(\mathbb{C}\backslash\mathbb{R})\colon = I - E(\mathbb{R})$. The constructed J-orthogonal spectral function is said (as before) to be an *e.s.f.* of $\mathfrak{Y}$.

Definition 1.7. Let E_λ be an e.s.f. of an operator family $\mathfrak{Y}$ and let an operator $A \in \mathfrak{Y}$ and a function $\phi(\lambda)$ be connected by the system of equalities from (1.6c). Then the function $\phi(\lambda)$ is said to be *the portrait* of the operator A and the operator A is said to be *the original* of $\phi(\lambda)$ in $\mathfrak{Y}$ (with respect to E_λ).

Definition 1.8. Let a spectral function E with a peculiar spectral set Λ be an e.s.f. of $\mathfrak{Y}$. If $\lambda \in \Lambda$ then λ will be called a peculiarity (of $\mathfrak{Y}$). Let λ be a peculiarity. Fix a set $X \in \mathfrak{R}_\Lambda$: $X \cap \Lambda = \{\lambda\}$. The peculiarity λ is called *regular* if the operator family $\{E(X \cap Y)\}_{Y \in \mathfrak{R}_\Lambda}$ is bounded, otherwise it is called *singular.*

Note that the notion of regular and singular peculiarities is correctly defined since the boundedness of the family $\{E(X \cap Y)\}_{Y \in \mathfrak{R}_\Lambda}$ does not depend on X.

1.4. On a model representation for J-orthogonal spectral functions without peculiarities

Let us introduce an analog of $L^2_{\vec{\sigma}}(\mathcal{E})$ that can be used for a model representation of Krein spaces. Assume that the scalar functions $\sigma_+(t)$ and $\sigma_-(t)$ are such that

$$\sigma_+(t) = \int_{-1}^{t} \rho_+(\lambda)d\sigma(\lambda), \quad \sigma_-(t) = \int_{-1}^{t} \rho_-(\lambda)d\sigma(\lambda), \quad \rho_+^2(\lambda) = \rho_+(\lambda),$$

$$\rho_-^2(\lambda) = \rho_-(\lambda), \quad \sigma(t) = \int_{-1}^{t} \Big(\rho_+(\lambda) + \rho_-(\lambda) - \rho_+(\lambda)\rho_-(\lambda)\Big)d\sigma(\lambda),$$

where $\sigma(\lambda)$ is the same as in the previous subsection, and set

$$\mathcal{J}\text{-}L^2_{\vec{\sigma}}(\mathcal{E}) := L^2_{\vec{\sigma}_+}(\mathcal{E}_+) \oplus L^2_{\vec{\sigma}_-}(\mathcal{E}_-), \quad [f(t), g(t)]\colon = (f_+(t), g_+(t)) - (f_-(t), g_-(t)),$$

where $f(t) = f_+(t),) + (f_-(t),\ g(t) = g_+(t) + g_-(t),\ f_+(t), g_+(t)) \in L^2_{\vec{\sigma}_+}(\mathcal{E}_+)$, $f_-(t), g_-(t) \in L^2_{\vec{\sigma}_-}(\mathcal{E}_-)$. The space $\mathcal{J}\text{-}L^2_{\vec{\sigma}}(\mathcal{E})$ is said to be *a standard Krein space.* As a slight abuse of the previous notation put also $M_{\vec{\sigma}}(\mathcal{E})\colon = M_{\vec{\sigma}_+}(\mathcal{E}_+) \oplus M_{\vec{\sigma}_-}(\mathcal{E}_-)$.

A standard Krein space will be used for a model representation of J-orth.sp.f. E_λ without peculiarities. For simplicity, everywhere below we will assume that

$$E_{-1} = 0, \quad E_{+1} = I, \quad E_{-1} = E_{-1+0}. \tag{1.7}$$

Definition 1.9. Let E_λ be a J-orth.sp.f. and let its set of peculiarities be empty. A space $\mathcal{J}\text{-}L^2_{\vec{\sigma}}(\mathcal{E})$ is said to be a *model space* for E_λ if for some canonical scalar product on $\mathfrak{H}$ there is an isometric J-isometric operator $W\colon\ \mathcal{J}\text{-}L^2_{\vec{\sigma}}(\mathcal{E}) \mapsto \mathfrak{H}$, such that for every $\lambda \in [-1;1]$

$$E_\lambda = W X_\lambda W^{-1},$$

where $X_\lambda = \{X_\lambda,\ \mathcal{J}\text{-}L^2_{\vec{\sigma}}(\mathcal{E})\}$. The operator W is said to be an *operator of similarity.*

Proposition 1.10. *Every J-orth.sp.f. E_λ with empty set of peculiarities has a model space $\mathcal{J}\text{-}L^2_{\vec{\sigma}}(\mathcal{E})$.*

2. Unbounded elements in Banach spaces

2.1. General case

For future reference note the following simple algebraic fact.

Proposition 2.1. *Let $\mathfrak{L}$ be a vector space, let $f_1, f_2, \dots, f_m, f$ be linear functionals on $\mathfrak{L}$. Then $f \in \operatorname{Lin}\{f_j\}_1^m$ if and only if $\bigcap_{j=1}^{m} \operatorname{Ker} f_j \subset \operatorname{Ker} f$.*

Now we turn to the notion of unbounded elements.

Assume that $\mathcal{B}$ is a Banach space, P_t is a resolution of the identity (= a spectral function with the empty set of peculiar points) defined on the segment $[-1;1]$, continuous in zero (with respect to the weak topology) and

$$\left.\begin{array}{ll} \text{a)} & P_{-1} = 0,\ P_1 = I; \\ \text{b)} & \text{for every } t \in [-1;1] \text{ there exist the unilateral limits} \\ & \operatorname{w-lim}\limits_{\mu\to t-0} P_\mu \text{ and } \operatorname{w-lim}\limits_{\mu\to t+0} P_\mu, \text{ where for definiteness } P_{t-0} = \\ & P_t; \end{array}\right\} \tag{2.1}$$

Set $P_{\lambda,\mu} = I + P_\lambda - P_{\mu+0}$, where $\lambda \in [-1;0)$, $\mu \in (0;1]$.

Next, let $x_{\lambda,\mu}$ be a mapping of the numerical set $[-1;0) \times (0;1]$ into $\mathcal{B}$ ($\lambda \in [-1;0)$, $\mu \in (0;1]$). The function $x_{\lambda,\mu}$ is said to be *conformed* with P_t if the following condition is fulfilled: for every $\lambda, \alpha \in [-1;0)$, $\mu, \beta \in (0;1]$ the equality $P_{\lambda,\mu} x_{\alpha,\beta} = x_{\gamma,\delta}$ holds, where $\gamma = \min\{\lambda,\alpha\}$, $\delta = \max\{\mu,\beta\}$.

Assume now that the space $\mathcal{B}$ and the resolution of the identity P_t have additionally the following property

$$\left.\begin{array}{l} \text{if } \sup\limits_{\substack{\lambda\in[-1;0)\\ \mu\in(0;1]}} \{\|x_{\lambda,\mu}\|\} < \infty \text{ then there is an element } x \in \mathcal{B} \\ \text{such that for every } \lambda \in [-1;0),\ \mu \in (0;1] \text{ the equality} \\ x_{\lambda,\mu} = P_{\lambda,\mu} x \text{ holds.} \end{array}\right\} \tag{2.2}$$

It is clear that the element x from (2.2) is uniquely defined by $x_{\lambda,\mu}$ and can be found by the formula $x = \operatorname{w-lim}\limits_{\substack{\lambda\to-0\\ \mu\to+0}} x_{\lambda,\mu}$.

Definition 2.2. A function $x_{\lambda,\mu}$ conformed with P_λ is said to be *an unbounded element conformed with* P_λ (or, if it cannot produce a misunderstanding, an unbounded element) if $\sup_{\substack{\lambda\in[-1;0)\\ \mu\in(0;1]}} \{\|x_{\lambda,\mu}\|\} = \infty$.

Note that unbounded elements conformed with P_t exist if and only if zero is a point of growth for P_t, i.e., $P_{+\epsilon} - P_{-\epsilon} \neq 0$ for every $\epsilon > 0$. Everywhere below in this Section the mentioned condition for P_t is assumed to be fulfilled.

For brevity everywhere below we denote by symbols $\widetilde{x}$, $\widetilde{y}$, etc. unbounded elements. For $\lambda \in [-1;0)$, $\mu \in (0;1]$ we set $x_{\lambda,\mu} := P_{\lambda,\mu}\widetilde{x}$.

Definition 2.3. Unbounded elements $\widetilde{x}_1, \widetilde{x}_2, \dots, \widetilde{x}_k$, conformed with a (common) resolution of the identity P_λ are said to be *linearly independent modulo* $\mathcal{B}$ (*or* Mod $\mathcal{B}$) if every non-trivial linear combination of them is an unbounded element from $\mathcal{B}$. In a similar way two unbounded elements $\widetilde{x}$ and $\widetilde{y}$ is said to be *equal modulo* $\mathcal{B}$ ($\widetilde{x} = \widetilde{y}\,|\,\mathrm{Mod}\ \mathcal{B}$) if $\widetilde{x} - \widetilde{y} \in \mathcal{B}$.

Although values of functionals from $\mathcal{B}^*$ are not defined for unbounded elements from $\mathcal{B}$, at least for some functionals there is a natural way for including the unbounded elements in their domains. Let $\mathcal{B}_{\lambda,\mu} := P_{\lambda,\mu}\mathcal{B}$, where $\lambda \in [-1;0)$ and $\mu \in (0;1]$. The space $\mathcal{B}^*_{\lambda,\mu}$ can be identified with the space $\{(P_{\mu+0} - P_\lambda)\mathcal{B}\}^\perp$, i.e., we assume that functionals from $\mathcal{B}^*_{\lambda,\mu}$ are extended to the subspace $(P_{\mu+0} - P_\lambda)\mathcal{B}$ by zero. Although generally speaking the norms of functionals from $\mathcal{B}^*_{\lambda,\mu}$ are not preserved during this procedure, it is easy to re-define the norm on $\mathcal{B}$ and, consequently, also the norm on $\mathcal{B}^*$ without changing the norm topology in such a manner that the norm of every functional from $\mathcal{B}^*_{\lambda,\mu}$ coincides with the norm of its above-mentioned extension to all of $\mathcal{B}$. The linear span of the subspaces $\mathcal{B}^*_{\lambda,\mu}$ denote by the symbol $\mathfrak{M}$. If $f \in \mathcal{B}^*_{\lambda,\mu}$, then define $f\widetilde{x} := fP_{\lambda,\mu}\widetilde{x} = fx_{\lambda,\mu}$. Thus, for every functional $f \in \mathfrak{M}$ and an unbounded element $\widetilde{x}$ the value $f\widetilde{x}$ is well defined. In what follows $\{\widetilde{x}\}^\perp$ means the set of all functionals from $f \in \mathfrak{M}$ such that $f\widetilde{x} = 0$.

Lemma 2.4. *Let unbounded elements* $\widetilde{x}$, $\widetilde{x}_1$, $\widetilde{x}_2, \dots, \widetilde{x}_m$ *be linearly independent modulo* $\mathcal{B}$. *Then*

$$\sup_{f\in\bigcap_{j=1}^m\{\widetilde{x}_j\}^\perp,\ \|f\|=1} \{|f\widetilde{x}|\} = \infty.$$

Proof. Suppose the contrary and set $\mathfrak{M}' = \cap_{j=1}^m\{\widetilde{x}_j\}^\perp$. By hypothesis one has $\sup_{f\in\mathfrak{M}',\|f\|=1} \{|f\widetilde{x}|\} < \infty$, so by the theorem of Hahn-Banach there is an element $y \in \mathcal{B}^{**}$, such that $fy = f\widetilde{x}$ for every $f \in \mathfrak{M}'$. Then by virtue of Proposition 2.1 the element $P^{**}_{\lambda,\mu}(y-\widetilde{x})$ (considered as a functional on $\mathcal{B}^*_{\lambda,\mu}$) is a linear combination of the functionals $P_{\lambda,\mu}\widetilde{x}_1$, $P_{\lambda,\mu}\widetilde{x}_2, \dots, P_{\lambda,\mu}\widetilde{x}_m$ for all $\lambda \in [-1;0)$, $\mu \in (0;1]$. Thus, for all such λ and μ, $P^{**}_{\lambda,\mu}y \in \mathcal{B}_{\lambda,\mu}$. It is clear that $P^{**}_{\lambda,\mu}y$ is a bounded vector-valued function conformed with P_λ, so thanks to Condition (2.2) there is an element $z \in \mathcal{B}$, such that $P^{**}_{\lambda,\mu}y = P_{\lambda,\mu}z$ for all $\lambda \in [-1;0)$, $\mu \in (0;1]$. Finally note that for every element $f \in \mathfrak{M}'$ there exist λ and μ, such that $yf = P^{**}_{\lambda,\mu} yf$, hence $yf = fz$. Thus, there is $z \in \mathcal{B}$, such that $fz = f\widetilde{x}$ for every $f \in \mathfrak{M}'$. Then Proposition 2.1 implies that the element $z - \widetilde{x}$ is a linear combination of the elements $\widetilde{x}_1, \widetilde{x}_2, \dots, \widetilde{x}_m$ considered as functionals on $\mathfrak{M}'$. If $z = \widetilde{x} + \Sigma_{j=1}^m \beta_j \widetilde{x}_j$ on $\mathfrak{M}'$, then (recall that $\mathcal{B}^*_{\lambda,\mu} \subset \mathfrak{M}'$ for all $\lambda \in [-1;0)$, $\mu \in (0;1]$) $P_{\lambda,\mu} z = P_{\lambda,\mu}\widetilde{x} + \Sigma_{j=1}^m P_{\lambda,\mu}\beta_j \widetilde{x}_j$. Since $z \in \mathcal{B}$, then for $\lambda \to -0$ and $\mu \to +0$ the last equality implies the representation $z = \widetilde{x} + \Sigma_{j=1}^m \beta_j \widetilde{x}_j$ considered now in the space $\mathcal{B}$. This is a contradiction! □

Lemma 2.5. *Assume that $\mathcal{B}$ is a separable Banach space, $\widetilde{x}_1, \widetilde{x}_2, \dots, \widetilde{x}_k$ are a collection of unbounded elements conformed with P_λ and linearly independent modulo $\mathcal{B}$, and that $\{\mathfrak{C}_{\lambda,\mu}\}_{\lambda\in[-1;0),\,\mu\in(0;1]}$ is a family of vector subspaces of $\mathcal{B}^*$ possessing the following properties*

$$\left.\begin{array}{ll} \text{a)} & P^{**}_{\lambda,\mu}\mathfrak{C}_{\lambda,\mu} \subset \mathfrak{C}_{\lambda,\mu};\\ \text{b)} & \textit{if } 0 \in (\lambda_2,\mu_2) \subset (\lambda_1,\mu_1), \textit{ then } \mathfrak{C}_{\lambda_1,\mu_1} \subset \mathfrak{C}_{\lambda_2,\mu_2};\\ \text{c)} & \textit{for every functional } f \in \mathcal{B}^*_{\lambda,\mu} \textit{ there is a sequence } \{f_m\}_1^\infty \subset \\ & \mathfrak{C}_{\lambda,\mu} \textit{ with properties } \mathrm{w}^* - \lim_{m\to\infty} f_m = f,\ \lim_{m\to\infty}\|f_m\| = \|f\|,\\ & \textit{where } \mathrm{w}^* - \lim \textit{ is a limit considered in } w^*\textit{-topology.} \end{array}\right\} \tag{2.3}$$

Then for every functional $f \in \mathcal{B}^*$ and an arbitrary collection of numbers $\{\alpha_j\}_{j=1}^k$ there is a sequence $\{g_m\}_1^\infty \subset \cup_{\lambda,\mu}\mathfrak{C}_{\lambda,\mu}$, such that

$$\left.\begin{array}{ll} \text{a)} & \mathrm{w}^* - \lim_{m\to\infty} g_m = f;\\ \text{b)} & \lim_{m\to\infty} g_m \widetilde{x}_j = \alpha_j. \end{array}\right\}$$

Proof. First, we show that for every $f \in \mathcal{B}^*$ there is a sequence $\{u_m\}_1^\infty \subset \cup_{\lambda,\mu}\mathfrak{C}_{\lambda,\mu}$ converging to f in the w^*-topology. Assume (with no loss of generality) that $\|f\| = 1$. It is well known (see [21] Part V, §7, Th. 6) that a bounded closed ball is metrizable in the w^*-topology. Due to Conditions (2.3) and the equality $f = \mathrm{w}^* - \lim_{\lambda\to-0,\,\mu\to+0} P^*_{\lambda,\mu} f$ the functional f belongs to the closure (in the w^*-topology) of the intersection of $\mathfrak{M}$ and the unitary ball from $\mathcal{B}^*$. Since in the metrizable topology the sequential closure of a subset coincides with its standard closure, the existence of the desired sequence $\{u_m\}_1^\infty$ is now evident. Second, let $\{u_m\}_1^\infty$ be fixed. Denote $u_m\widetilde{x}_j := \alpha_m^{(j)}$, $j = 1, 2, \dots, k$. Lemma 2.4 implies that for every natural integer m and $j = 1, 2, \dots, k$ there is a functional $w_m^{(j)} \in \mathfrak{M}$, such that

$$|w_m^{(j)}\widetilde{x}_j| \ge 2^m(|\alpha_j - \alpha_m^{(j)}| + 1),\ w_m^{(j)}\widetilde{x}_l = 0 \text{ for } l \ne j,\ \|w_m^{(j)}\| = 1.$$

Set

$$v_m^{(j)} = \frac{(\alpha_j - \alpha_m^{(j)})}{w_m^{(j)}\widetilde{x}_j}\cdot w_m^{(j)}.$$

Then $\|v_m^{(j)}\| \le 1/2^m$. Third, Condition (2.3c) implies the existence of a functional $z_m^{(j)} \in \cup_{\lambda,\mu}\mathfrak{C}_{\lambda,\mu}$ such that $\|z_m^{(j)}\| < 1/2^{m-1}$, $|(z_m^{(j)} - v_m^{(j)})\widetilde{x}_l| < 1/2^m$, $l = 1, 2, \dots$. As a last step one can set $g_m = u_m + \Sigma_{j=1}^k z_m^{(j)}$. □

Remark 2.6. The proof of Lemma 2.5 shows that under Conditions (2.2), for any sequence $\{u_m\}_1^\infty \subset \cup_{\lambda,\mu}\mathfrak{C}_{\lambda,\mu}$ converging (in the w^*-topology) to an arbitrary functional $f \in \mathfrak{B}^*$ and for every collection of numbers $\{\alpha_j\}_1^k$, there is a "correcting" sequence $\{v_m\}_1^\infty \subset \cup_{\lambda,\mu}\mathfrak{C}_{\lambda,\mu}$, such that $\lim_{m\to\infty}\|v_m\| = 0$, $\lim_{m\to\infty}(u_m+v_m)\widetilde{x}_j = \alpha_j$, $j = 1, 2, \ldots, k$.

Now show that the concept of unbounded elements is applicable to the space $L^1_\sigma + L^2_\eta$, where the relation between $\sigma(t)$ and $\eta(t)$ is given by (1.4).

Proposition 2.7. *If $\mathcal{B} = L^1_\sigma + L^2_\eta$ and $P_t = \{\mathrm{X}_t, L^1_\sigma + L^2_\eta\}$, where X_t is the multiplication operator by the indicator of the set $(-\infty; t)$ (denoted $\chi_{(-\infty;t)}(\tau)$), then for $\mathcal{B}$ and P_t Conditions* (2.1) *and* (2.2) *are fulfilled.*

Proof. For the given P_t the fulfillment of Condition (2.1) is evident, so we shall concentrate on Condition (2.2) only. Consider on the segment $[-1; 1]$ a μ_σ-measurable, a.e. finite function $x(t)$ satisfying the condition

$$\sup_{\lambda\in[-1;0),\ \mu\in(0;1]} \{\|x_{\lambda,\mu}(t)\|_{L^1_\sigma+L^2_\nu}\} < \infty, \tag{2.4}$$

where

$$x_{\lambda,\mu}(t) = \begin{cases} x(t), & \text{if } t \in [-1; \lambda) \cup (\mu; 1]; \\ 0, & \text{if } t \in [\lambda, \mu]. \end{cases}$$

By the theorem of Banach-Steinhaus Condition (2.4) can be re-written in the following form: for every function $f(t) \in L^\infty_\sigma \cap L^2_\nu$

$$\sup_{\lambda\in[-1;0),\ \mu\in(0;1]} \{|\int_{-1}^{\lambda} x(t)f(t)d\sigma(t) + \int_{\mu}^{1} x(t)f(t)d\sigma(t)|\} < \infty. \tag{2.5}$$

We need to show that $x(t) \in L^1_\sigma + L^2_\eta$. Note that this is equivalent to $|x(t)| \in L^1_\sigma + L^2_\eta$, therefore we suppose (with no loss of generality) that $x(t) \geq 0$. Under this hypothesis set

$$x_1(t) = \begin{cases} x(t), & \text{if } x(t) < G(t); \\ G(t), & \text{if } x(t) \geq G(t); \end{cases} \qquad x_2(t) = x(t) - x_1(t).$$

Now we show that

$$x_1(t) \in L^2_\eta. \tag{2.6}$$

Suppose the contrary, i.e., $x_1(t) \notin L^2_\eta$. Then there is a decreasing sequence of positive numbers $\{\alpha_n\}_1^\infty$, $\lim_{n\to\infty}\alpha_n = 0$, such that

$$\left(\int_{-1}^{-\alpha_1} + \int_{\alpha_1}^{1}\right) x_1^2(t)d\eta(t) = \beta_1 \geq 1, \qquad \left(\int_{-\alpha_1}^{-\alpha_2} + \int_{\alpha_2}^{\alpha_1}\right) x_1^2(t)d\eta(t) = \beta_2 \geq 1,$$

$$\ldots \quad , \left(\int_{-\alpha_{n-1}}^{-\alpha_1} + \int_{\alpha_n}^{\alpha_{n-1}}\right) x_1^2(t)d\eta(t) = \beta_n \geq 1,$$

$$\ldots \quad .$$

For $t \in [-\alpha_{n-1}; -\alpha_n) \cup (\alpha_n; \alpha_{n-1}]$ set $f(t) = \dfrac{x_1(t)}{G(t) \cdot n \cdot \sqrt{\beta_n}}$, $n = 1, 2, \dots, \alpha_0 = 1$. Then

$$\int_{-1}^{1} |f(t)|^2 G(t) d\sigma(t) = \sum_{n=1}^{\infty} \frac{1}{n^2} < \infty, \quad 0 \le f(t) \le 1,$$

so $f(t) \in L_\sigma^\infty \cap L_\nu^2$ but $\int_{-1}^{1} x_1(t) f(t) d\sigma(t) = \sum_{n=1}^{\infty} \frac{\sqrt{\beta_n}}{n} = \infty$. The latter contradicts (2.5). Thus, (2.6) has been proved.

To complete the proof we need to demonstrate that $x_2(t) \in L_\sigma^1$. Suppose the contrary, i.e., $x_2(t) \notin L_\sigma^1$. Set

$$f(t) = \begin{cases} 0, & \text{if } x(t) \le G(t); \\ 1, & \text{if } x(t) > G(t). \end{cases}$$

By Condition (2.6) we have

$$\int_{-1}^{1} |f(t)|^2 G(t) d\sigma(t) \le \int_{-1}^{1} |x_1(t)|^2 d\eta(t) < \infty\ ,$$

therefore $f(t) \in L_\sigma^\infty \cap L_\nu^2$. Also

$$\lim_{\lambda \to -0,\ \mu \to +0} \int_{-1}^{1} x_{\lambda,\mu}(t) f(t) d\sigma(t) \ge \int_{-1}^{1} x_2(t) d\sigma(t) = \infty,$$

but the latter contradicts (2.5). □

Remark 2.8. The proof of Proposition 2.7 gives the following result for a μ_σ-measurable function $x(t)$ (Cf. [8], Theorem 5.2.1): Let

$$x_1(t) = \begin{cases} x(t), & \text{if } |x(t)| < G(t); \\ G(t) \cdot e^{i \arg x(t)}, & \text{if } |x(t)| \ge G(t); \end{cases} \qquad x_2(t) = x(t) - x_1(t).$$

Then $x(t) \in L_\sigma^1 + L_\eta^2$ if and only if simultaneously $x_1(t) \in L_\eta^2$ and $x_2(t) \in L_\sigma^1$.

2.2. Some remarks for the case of Hilbert spaces

Here and up to the end of the section $\mathcal{H}$ means a fixed separable Hilbert space and P_t means a fixed orthogonal spectral function.

For every bounded P_t-measurable function $\phi(t)$ define on $\mathcal{H}$ the following operator Φ:

$$\Phi x \colon = \int_{-1}^{1} \phi(t) dP_t x\ .$$

Using the previous notation, re-write the last formula as

$$(\Phi \widetilde{x})_{\lambda,\mu} = \int_{-1}^{\lambda} \phi(t) dP_t \widetilde{x} + \int_{\mu}^{1} \phi(t) dP_t \widetilde{x}. \tag{2.7}$$

Representation (2.7) allows the possibility of treating the operator Φ in a more general sense: unbounded elements conformed with P_t can be naturally included to the domain of Φ. Note that the image of an unbounded element can be both a bounded element and an unbounded element.

Until the end of the present section a non-decreasing function $\sigma(t)$, such that the μ_σ-measurability on $[-1;1]$ coincides with P_t-measurability, will be fixed. Note that the existence of $\sigma(t)$ follows from the separability of $\mathcal{H}$ (see [1] Section 76, Theorem 1).

We give an additional notation. Let $\{\widetilde{x}_j\}_1^k$ be a fixed family of unbounded elements conformed with P_t and linearly independent modulo $\mathcal{H}$. Both the unbounded elements and the ordinary vectors from $\mathcal{H}$ can be considered as functions defined on $[-1;0)\times(0;1]$ and taking values in $\mathcal{H}$ (see (2.2)). The linear span of vectors from $\mathcal{H}$ and unbounded elements from $\{\widetilde{x}_j\}_1^k$ consistently taken as functions on $[-1;0)\times(0;1]$ is denoted $\widetilde{\mathcal{H}}$. Additionally $\widetilde{\mathcal{H}}$ will be considered as a Hilbert space, where $\mathcal{H}$ is a subspace with the the same scalar product that was given on $\mathcal{H}$ initially and unbounded elements from $\{\widetilde{x}_j\}_1^k$ are mutually orthogonal and orthogonal to $\mathcal{H}$. The space $\widetilde{\mathcal{H}}$ is said to be a expansion of $\mathcal{H}$ (generated by $\{\widetilde{x}_j\}_1^k$).

Next, using the system $\{\widetilde{x}_j\}_1^k$ introduce the function $\nu(t)$:

$$\nu(t)=\begin{cases}\sigma(t)+\sum_{j=1}^k\|P_t\widetilde{x}_j\|^2, & \text{if } t\in[-1;0);\\ -\sigma(1)+\sigma(t)-\sum_{j=1}^k\|(P_t-I)\widetilde{x}_j\|^2, & \text{if } t\in(0;1].\end{cases} \tag{2.8}$$

The connection between P_t and $\sigma(t)$ implies that the function $\nu(t)$ introduced in (2.8) has Representation (1.4). In this case the function $G(t)$ from (1.4) can be calculated directly through $\nu(t)$ and $\sigma(t)$.

Proposition 2.9. ([40]) *Let $\{\widetilde{x}_j\}_1^k$ be a system of unbounded elements, forming together with $\mathcal{H}$ the space $\widetilde{\mathcal{H}}$, let $\phi(t)$ be a μ_σ-measurable function, let Φ be the operator defined by Formula* (2.7). *Then $\Phi\widetilde{\mathcal{H}}\subset\mathcal{H}$ if and only if $\phi(t)\in L_\sigma^\infty\cap L_\nu^2$.*

For future applications both the cases $\Phi\widetilde{\mathcal{H}}\subset\mathcal{H}$, and

$$\Phi\widetilde{\mathcal{H}}\subset\widetilde{\mathcal{H}} \tag{2.9}$$

are important.

Proposition 2.10. ([40]) *Let $\{\widetilde{x}_j\}_1^k$ be a system of unbounded elements, generating together with $\mathcal{H}$ the space $\widetilde{\mathcal{H}}$. Then there are no more than k^2 μ_σ-measurable functions $\phi(t)$ linearly independent modulo $L_\sigma^\infty\cap L_\nu^2$, such that Condition* (2.9) *is fulfilled.*

Remark 2.11. Let $\{\widetilde{x}_j\}_1^k$ be a system of unbounded elements, generating together with $\mathcal{H}$ the space $\widetilde{\mathcal{H}}$ and let $\mathcal{F}$ be a vector space of scalar μ_σ-measurable functions, such that (2.9) holds for every $\phi(t)\in\mathcal{F}$, and if $\phi(t)\in\mathcal{F}$ then $\bar{\phi}(t)\in\mathcal{F}$. If m is the maximal number of functions from $\mathcal{F}$ linearly independent modulo $L_\sigma^\infty\cap L_\nu^2$, then there are m *real*-valued functions from $\mathcal{F}$ linearly independent modulo $L_\sigma^\infty\cap L_\nu^2$.

Indeed, if the dimension of some real vector space is m then the dimension (as a complex space) of its complexification is also m.

3. A function model for a J-symmetric family of the class D_κ^+

Now pass to a function model of a J-symmetric family $\mathfrak{Y} \in D_\kappa^+$ with real spectrum and non-empty set of peculiarities. This model is defined with the help of an e.s.f. E_λ of $\mathfrak{Y}$. By virtue of Theorem 1.5, it is clear that the general situation can be reduced to the case of J-orth.sp.f. E_λ with a unique spectral peculiarity in zero. Furthermore, the case of a regular peculiarity is trivial because, under these conditions, all operators from $\mathfrak{Y}$ are spectral in the sense of Dunford and have a finite-dimensional nilpotent part. Thus, we can assume E_λ satisfies:

$$\left.\begin{array}{l} \text{a) } E_{-1} = E_{-1+0} = 0,\ E_{+1} = I; \\ \text{b) } \Lambda = \{0\}; \\ \text{c) } \sup\limits_{\lambda \in [-1;1]\setminus\{0\}} \{\|E_\lambda\|\} = \infty. \end{array}\right\} \tag{3.1}$$

Introduce some notation. Let

$$\left.\begin{array}{l} \mathcal{H}_1 = \widetilde{\mathcal{H}} \cap \widetilde{\mathcal{H}}^{[\perp]},\ \mathcal{H}_2 = \mathcal{H}_1^\perp \cap \widetilde{\mathcal{H}},\ \mathcal{H}_0 = J\mathcal{H}_1,\ P_j \text{ be an} \\ \text{orthoprojection (in the sense of Hilbert spaces) onto } \mathcal{H}_j, \\ j = 0, 1, 2,\ \widetilde{E}_\lambda := E_\lambda|_{\widetilde{\mathcal{H}}}. \end{array}\right\} \tag{3.2}$$

Note that by Condition (3.1c) the inequality $\mathcal{H}_1 \neq \{0\}$ holds.

In addition to (3.2) set

$$\widetilde{\mathcal{H}}^\uparrow = \mathcal{H}_0 \oplus \mathcal{H}_2,\ \widetilde{E}_\lambda = E_\lambda|_{\widetilde{\mathcal{H}}},\ \widetilde{E}_\lambda^\uparrow = (P_0 + P_2)E_\lambda|_{\widetilde{\mathcal{H}}^\uparrow}. \tag{3.3}$$

It is necessary to take in account that, generally speaking, the subspace $\mathcal{H}_2$ is indefinite. Since J-orth.sp.f. E_λ belongs to the class D_κ^+, there is an E_λ-invariant pair of J-orthogonal maximal semi-definite pseudo-regular subspaces $\mathfrak{L}_+$ and $\mathfrak{L}_-$ with finite-dimensional isotropic part, moreover by Condition (1.6b) we can assume that for every closed interval $\Delta \subset [-1;1]\setminus\{0\}$ the subspace $(E(\Delta)\mathcal{H}) \cap \mathfrak{L}_+$ is positive and the subspace $(E(\Delta)\mathcal{H}) \cap \mathfrak{L}_-$ is negative. Thanks to the last hypothesis the following subspaces are well defined

$$\widetilde{\mathcal{H}}_+ = \operatorname*{CLin}_{\Delta \subset [-1;1]\setminus\{0\}} \{E(\Delta)\mathfrak{L}_+\},\ \widetilde{\mathcal{H}}_- = \operatorname*{CLin}_{\Delta \subset [-1;1]\setminus\{0\}} \{E(\Delta)\mathfrak{L}_-\}. \tag{3.4}$$

Set

$$\mathcal{H}_2^+ = \mathcal{H}_2 \cap \widetilde{\mathcal{H}}_+, \quad \mathcal{H}_2^- = \mathcal{H}_2 \cap \widetilde{\mathcal{H}}_-, \tag{3.5}$$

and assume that a fundamental scalar product on $\mathcal{H}$ is also canonical for the subspace $\widetilde{\mathcal{H}} \oplus \mathcal{H}_0$ and on the last space is compatible (see Definition 1.2) with the given decompositions of the corresponding subspaces

$$\widetilde{\mathcal{H}}_+ = \mathcal{H}_1 \dot{+} \mathcal{H}_2^+ \text{ and } \widetilde{\mathcal{H}}_- = \mathcal{H}_1 \dot{+} \mathcal{H}_2^-. \tag{3.6}$$

Thus,

$$J|_{\mathcal{H}_0 \dot{+} \mathcal{H}_1 \dot{+} \mathcal{H}_2} = \begin{pmatrix} 0 & V^{-1} & 0 \\ V & 0 & 0 \\ 0 & 0 & J_2 \end{pmatrix}, \tag{3.7}$$

where the operator $V\colon \mathcal{H}_0 \mapsto \mathcal{H}_1$ is isometric, and J_2 is a canonical symmetry of the form $[\cdot,\cdot]$ on $\mathcal{H}_2$.

Let $\mathcal{J}\text{-}L^2_{\vec{\sigma}}(\mathcal{E})$ be a standard Krein space (see Section 1) and let $\{\widetilde{g}_j(t)\}_{j=1}^k \subset M_{\vec{\sigma}}(\mathcal{E})$ be a system of unbounded elements conformed with the operator-valued function X_τ and linearly independent modulo $\mathcal{J}\text{-}L^2_{\vec{\sigma}}(\mathcal{E})$. Denote by $\mathcal{J}\text{-}\widetilde{L}^2_{\vec{\sigma}}(\mathcal{E})$ the linear span generated by the space $\mathcal{J}\text{-}L^2_{\vec{\sigma}}(\mathcal{E})$ and the system $\{\widetilde{g}_j(t)\}_{j=1}^k$. Define on $\mathcal{J}\text{-}\widetilde{L}^2_{\vec{\sigma}}(\mathcal{E})$ structures of Hilbert and Krein spaces in the following way: on $\mathcal{J}\text{-}L^2_{\vec{\sigma}}(\mathcal{E})$ both structures coincide with the original structures, functions of the system $\{\widetilde{g}_j(t)\}_{j=1}^k$ are by definition positive (as elements of the Krein space), mutually orthogonal and J-orthogonal, normalized and J-normalized, orthogonal and J-orthogonal to $\mathcal{J}\text{-}L^2_{\vec{\sigma}}(\mathcal{E})$. The space $\mathcal{J}\text{-}\widetilde{L}^2_{\vec{\sigma}}(\mathcal{E})$ is said to be *the expansion of* $\mathcal{J}\text{-}L^2_{\vec{\sigma}}(\mathcal{E})$ *(generated by the collection* $\{\widetilde{g}_j(t)\}_{j=1}^k$*)*.

Theorem 3.1. *If a J-orth.sp.f. E_λ satisfies Condition* (3.1) *and a scalar product on $\mathcal{H}$ is compatible with* (3.6), *then there are, first, a subspace $\mathcal{J}\text{-}L^2_{\vec{\sigma}}(\mathcal{E})$ and a system of unbounded elements $\{\widetilde{g}_j(t)\}_{j=1}^k$ of this space forming together the space $\mathcal{J}\text{-}\widetilde{L}^2_{\vec{\sigma}}(\mathcal{E})$ and, second, an isometric J-isometric operator $W\colon \mathcal{J}\text{-}\widetilde{L}^2_{\vec{\sigma}}(\mathcal{E}) \mapsto \widetilde{\mathcal{H}}$, $WL^2_{\vec{\sigma}}(\mathcal{E}) = \mathcal{H}_2$, such that for every $\lambda \in [-1;1]$*

$$\widetilde{E}_\lambda = W \cdot X^{\#}_\lambda \cdot (W)^{-1}, \quad W^\uparrow = (I_2 \oplus V)W, \quad \widetilde{E}^\uparrow_\lambda = W^\uparrow \cdot X_\lambda \cdot (W^\uparrow)^{-1}, \qquad (3.8)$$

where $X_\lambda = \{X_\lambda,\ \mathcal{J}\text{-}\widetilde{L}^2_{\vec{\sigma}}(\mathcal{E})\}$, $k = \dim \mathcal{H}_0 = \dim \mathcal{H}_1$.

Definition 3.2. If for Decomposition (3.2), (3.5) a relation between a J-orth.sp.f. E_λ satisfying Condition (3.1) and a space $\mathcal{J}\text{-}\widetilde{L}^2_{\vec{\sigma}}(\mathcal{E})$ is given by Formulae (3.8), then $\mathcal{J}\text{-}\widetilde{L}^2_{\vec{\sigma}}(\mathcal{E})$ is said to be *a basic model space* for E_λ (compatible with (3.2), (3.3), (3.6)) and the operator W is said to be *an operator of similarity* corresponding to this space.

Remark 3.3. Let $\overset{*}{\mathcal{H}}$ be the factor-space de $\mathcal{H}$ generated by $\mathcal{H}_1$. Then $\overset{*}{\mathcal{H}}$ is a Krein space and $\widetilde{E}_\lambda$ induces on this space the J-orth.sp.f. $\overset{*}{E}_\lambda$. Note that the unique peculiarity of $\overset{*}{E}_\lambda$ is regular and can be removed, so, as a slight abuse of terminology, one can say that $\overset{*}{E}_\lambda$ has no peculiarities. It is clear that $W|_{\mathcal{J}\text{-}L^2_{\vec{\sigma}}(\mathcal{E})}$ in (3.8) exists if and only if there exists a J-isometric operator $\overset{*}{W}$: $\mathcal{J}\text{-}L^2_{\vec{\sigma}}(\mathcal{E}) \mapsto \overset{*}{\mathcal{H}}$, such that $\overset{*}{E}_\lambda = \overset{*}{W} X_\lambda \overset{*}{W}{}^{-1}$. Taking this reasoning into account one can say that an arbitrariness in the choice of a basic model space for J-orth.sp.f. E_λ satisfying (3.1) does not depend on the choice of P_2 but essentially depends on an arbitrariness in the choice of a basic model space for J-orth.sp.f. $\overset{*}{E}_\lambda$ with the empty set of peculiarities. This situation was considered in Subsection 1.4.

Up to this point we discussed model representations not for a family $\mathfrak{Y} \in D^+_\kappa$ but for a J-orth.sp.f. $E_\lambda \in D^+_\kappa$ with unique spectral peculiarity in zero. Now let $\mathfrak{Y} \in D^+_\kappa$ be a J-symmetric commutative operator family. Then its linear span contains a family (not uniquely defined) $\mathfrak{Y}_r$ of J-s.a. operators, such that the linear span of $\mathfrak{Y}_r$ coincides with the linear span of $\mathfrak{Y}$. Then (see Theorem 1.5) the family $\mathfrak{Y}_r$ has an e.s.f. E_λ. This spectral function is also said to be *an e.s.f.* of $\mathfrak{Y}$. Note (see Proposition 1.6 and accompanying comments) that the case $E(\mathbb{R}) \neq I$ is admissible.

Theorem 3.4. *Assume that $\mathfrak{Y} \in D_\kappa^+$ is a commutative J-symmetric family, whose e.s.f. E_λ satisfies Condition* (3.1)*, that a canonical scalar product on $\mathcal{H}$ is compatible with* (3.6)*, that $\mathcal{J}$-$\widetilde{L}^2_{\vec{\sigma}}(\mathcal{E})$ is a basic model space for E_λ, and W is a corresponding operator of similarity. Then for every operator $A \in \mathfrak{Y}$ there is a function $\varphi(t)$ such that*

$$\widetilde{A}^{\uparrow} = W^{\uparrow} \cdot \Phi \cdot (W^{\uparrow})^{-1}, \; \widetilde{A} = W \cdot \bar{\Phi}^{\#} \cdot W^{-1} \tag{3.9}$$

where $\widetilde{A}^{\uparrow} := (P_0 \oplus P_1)A|_{\widetilde{\mathcal{H}}^{\uparrow}}$, the space $\widetilde{\mathcal{H}}^{\uparrow}$ and the operator $W^{\uparrow}$ are defined via (3.3), (3.8), $\Phi = \{\Phi, \mathcal{J}\text{-}\widetilde{L}^2_{\vec{\sigma}}(\mathcal{E})\}$, *and $\bar{\Phi}$ is the multiplication operator by $\bar{\varphi}(t)$ acting in the space $\mathcal{J}$-$\widetilde{L}^2_{\vec{\sigma}}(\mathcal{E})$.*

Definition 3.5. If $\mathfrak{Y} \in D_\kappa^+$ is a commutative J-symmetric family and E_λ (satisfying Condition (3.1)) is its e.s.f., then a basic model space for E_λ is also said to be *a basic model space* for $\mathfrak{Y}$.

Remark 3.6. Assume that $\mathfrak{Y} \in D_\kappa^+$ is a commutative J-symmetric family, its e.s.f. E_λ satisfies Condition (3.1), $\mathcal{J}$-$\widetilde{L}^2_{\vec{\sigma}}(\mathcal{E})$ and W are, respectively, a basic model space and a corresponding operator of similarity for $\mathfrak{Y}$. If under these conditions an operator $A \in \mathfrak{Y}$ and a function $\varphi(t)$ are related by Formulae (3.9), then according to Definition 1.7 the function $\varphi(t)$ is a portrait of the operator A (with respect to E_λ), and the operator A is an original of $\varphi(t)$ in $\mathfrak{Y}$ (not unique in the general case). It is clear that the portrait of an operator does not depend on the choice of the basic model space $\mathcal{J}$-$\widetilde{L}^2_{\vec{\sigma}}(\mathcal{E})$ but does depend of the choice of E_λ and can be found via Representation (1.6c).

Remark 3.7. Theorem 3.4 raises a natural problem concerning a characterization of functions that can be portraits for operators from a given commutative J-symmetric operator family $\mathfrak{Y} \in D_\kappa^+$ with a fixed choice of J-orth.sp.f. E_λ. A partial answer to the problem is contained in Propositions 2.9 and 2.10. Indeed, under the present conditions the function $\nu(t)$ from (2.8) has the form

$$\nu(t) = \begin{cases} \displaystyle\int_{-1}^{t} G(t)d\sigma(t), \; t \in [-1; 0), \\ \\ -\displaystyle\int_{t}^{1} G(t)d\sigma(t), \; t \in (0; 1], \end{cases} \tag{3.10}$$

where $G(t) = 1 + \sum_{j=1}^{k} \|\widetilde{g}_j(t)\|^2_{\mathcal{E}}$.

4. Functional representation of commutative WJ^*-algebras of the class D_κ^+ with a unique singular peculiarity

4.1. Preliminary remarks

Below the symbol $\mathfrak{A}$ means a WJ^*-algebra belonging to the class D_κ^+. Let E_λ be its e.s.f. In this section we assume that all J-s.a. operators from $\mathfrak{A}$ have real spectra and that E_λ satisfies Conditions (3.1).

Let $\varphi(t)$ be a continuous function vanishing in a neighborhood of zero. Set

$$B_\varphi := \int_{-1}^{1} \varphi(t) dE_\lambda = \int_{\mathrm{Supp}(\varphi)} \varphi(t) dE_\lambda. \tag{4.1}$$

By Property (1.6a) $B_\varphi \in \mathfrak{A}$. Denote by $\mathfrak{A}_0$ the weak closure of the set of operators having the form (4.1), where $\varphi(t)$ runs through the set of all functions of the indicated type. Our immediate aim is to characterize the subalgebra $\mathfrak{A}_0$ in terms of the portraits of the operators from $\mathfrak{A}_0$. For this we shall use the notation from (3.2) and (3.5). First, for every $A \in \mathfrak{A}_0$

$$A|_{\mathfrak{H}_1} = 0, \quad A\mathfrak{H} \subset \widetilde{\mathfrak{H}}. \tag{4.2}$$

Indeed, for J-s.a. operator from $\mathfrak{A}_0$ the fulfillment of Condition (4.2) is evident, but the family $\mathfrak{A}_0$ is J-symmetric and linear, so the rest is straightforward.

Let J-$\widetilde{L}^2_{\vec{\sigma}}(\mathfrak{E})$ be a basic model space for $\mathfrak{A}$ generated by J-$L^2_{\vec{\sigma}}(\mathfrak{E})$ and a system of unbounded elements $\{\widetilde{g}_j(t)\}_{j=1}^k$ and let W be a corresponding operator of similarity. If a function $\psi(t)$ is the portrait of an operator $A \in \mathfrak{A}_0$, then by Proposition 2.9 and Remark 3.7

$$\psi(t) \in L^\infty_\sigma \cap L^2_\nu, \tag{4.3}$$

where $\nu(t)$ is defined by Formulae (3.10).

Simultaneously with the space $L^\infty_\sigma \cap L^2_\nu$ consider the space $L^1_\sigma + L^2_\eta$, where $\eta(t)$ is given by Formulae (1.4) and (3.10). Note that the space $L^1_\sigma + L^2_\eta$ has not only the direct relation with the space $L^\infty_\sigma \cap L^2_\nu$ but also with the space J-$\widetilde{L}^2_{\vec{\sigma}}(\mathfrak{E})$.

Proposition 4.1. *Let the spaces J-$\widetilde{L}^2_{\vec{\sigma}}(\mathfrak{E})$ and $L^1_\sigma + L^2_\eta$ be related by Conditions (1.4) and (3.10). Then for every function $\psi(t) \in L^1_\sigma + L^2_\eta$ there are vector-functions $g_{-1}(t), g_0(t), f_{-1}(t), f_0(t), \ldots, f_k(t) \in L^2_{\vec{\sigma}}(\mathfrak{E})$, such that*

$$\psi(t) = (g_{-1}(t), f_{-1}(t))_{\mathfrak{E}} + (g_0(t), f_0(t))_{\mathfrak{E}} + \sum_{j=1}^{k} (\widetilde{g}_j(t), f_j(t))_{\mathfrak{E}}.$$

Proof. Fix a function $g_0(t) \in L^2_{\vec{\sigma}}(\mathfrak{E})$, such that $\|g_0(t)\|_{\mathfrak{E}} = 1$, and a decomposition $\psi(t) = \psi_1(t) + \psi_2(t)$, where $\psi_1(t) \in L^1_\sigma$, $\psi_2(t) \in L^2_\eta$. Let

$$\theta(t) = \begin{cases} 0, & \text{if } \psi_1(t) = 0; \\ \bar{\psi}_1(t)/|\psi_1(t)|, & \text{if } \psi_1(t) \neq 0. \end{cases}$$

The desired system can be constructed as follows. Set $f_{-1}(t) = \theta(t)g_0(t)\sqrt{|\psi_1(t)|}$, $g_{-1}(t) = g_0(t)\sqrt{|\psi_1(t)|}$, $f_0(t) = (\bar{\psi}_2(t)/G(t))g_0(t)$, $f_j(t) = (\bar{\psi}_2(t)/G(t))\widetilde{g}_j(t)$, $j = 1, 2, \ldots$. □

Introduce an additional notation. Let (see (3.8))

$$e_j = W\widetilde{g}_j(t), \quad h_j = W^{\uparrow}\widetilde{g}_j(t) = V^{-1}e_j, \quad \iota_{lj}(t) = [\widetilde{g}_l(t), \widetilde{g}_j(t)], \tag{4.4}$$

$l, j = 1, 2, \ldots, k$. Then for the operator B_φ from (4.1)

$$(B_\varphi h_l, e_j) = \int_{-1}^{1} \varphi(t)\iota_{lj}(t)d\sigma(t). \tag{4.5}$$

It is clear that $\iota_{lj}(t) \in M_{\vec{\sigma}}(\mathbb{C})$. Note that functions from the collection $\{\iota_{lj}(t)\}_1^k$ can be both bounded and unbounded elements of the space $L^1_\sigma + L^2_\eta$.

Denote by $\mathfrak{S}(E_\lambda)$ the set of operators $\{S\}$ acting on $\mathfrak{H}$ and satisfying the following conditions

$$\left.\begin{array}{l} \text{a) } S = P_1 S P_0; \\ \text{b) if } \sum\limits_{l,j=1}^{k} \alpha_{lj}\iota_{lj}(t) \in L^1_\sigma + L^2_\eta, \text{ then } \sum\limits_{l,j=1}^{k} \alpha_{lj}(Sh_l, e_j) = 0. \end{array}\right\} \tag{4.6}$$

Conditions (4.6) imply that $\mathfrak{S}(E_\lambda)$ is a weakly closed vector set with nilpotent elements and for every $S_1, S_2 \in \mathfrak{S}(E_\lambda)$, $S_1 S_2 = 0$. One of the essential questions concerns the linear dimension of $\mathfrak{S}(E_\lambda)$. Conditions (4.6) give the following simple result.

Proposition 4.2. $\dim \mathfrak{S}(E_\lambda)$ *is equal to the maximal number of functions in the collection* $\{\iota_{lj}(t)\}_1^k$ *linearly independent modulo* $L^1_\sigma + L^2_\eta$.

The following example is from [40].

Example 4.3. Take the space $\mathcal{H}$ coinciding with $\mathbf{l}_2$, denote by $\{u_m\}_{m=1}^\infty$ the canonical basis of this space, i.e., $u_1 = (1, 0, 0, \ldots)$, $u_2 = (0, 1, 0, \ldots)$, …, define the canonical symmetry J by the equalities $Ju_0 = u_1$, $Ju_1 = u_0$; $Ju_{2m+1} = u_{2m+1}$, $Ju_{2m} = -u_{2m}$, $m = 1, 2, \ldots$; and the spectral function E_λ by the relations

a) $E_\lambda = 0$ for every $\lambda < 0$;

b) if $\lambda > 0$ then $(I - E_\lambda)u_0 = \sum\limits_{1/m \in [\lambda;1)} (u_{2(m-1)} - u_{2m-1})$, $(I - E_\lambda)u_1 = 0$, $(I - E_\lambda)u_{2(m-1)} = u_{2(m-1)} + u_1$, $(I - E_\lambda)u_{2m-1} = u_{2m-1} + u_1$ for $1/m \in [\lambda; 1)$ and $(I - E_\lambda)u_{2(m-1)} = (I - E_\lambda)u_{2m-1} = 0$ for $1/m \notin [\lambda; 1)$.

Here $k = 1$ and $\iota_{11}(t) = 0$, so for this case $\dim \mathfrak{S}(E_\lambda) = 0$, but for some subclasses of D^+_κ the inequality $\dim \mathfrak{S}(E_\lambda) > 0$ holds. Recall (see [3]) that a WJ^*-algebra $\mathfrak{A}$ belongs to the class $\mathbf{H}$ if there is at least one $\mathfrak{A}$-invariant subspace $L_+ \in \mathfrak{M}^+(\mathcal{H})$ and all $\mathfrak{A}$-invariant maximal non-negative subspaces belong to the class h^+. In [7] (Theorem 5.6) the following result was proved.

Theorem 4.4. *Assume that $\mathcal{H}$ is a separable space, that $A \in \mathbf{H}$ is a J-s.a. operator, and that E_λ is its e.s.f. with a set of peculiarities Λ. Then there exist a finite set $\widehat{\Lambda} \subset \mathbb{R}$, $\Lambda \subset \widehat{\Lambda}$ and E_λ-measurable sets $X_+, X_- \subset \mathbb{R}$, $X_+ \cap X_- = \emptyset$, $X_+ \cup X_- = \mathbb{R}\backslash\widehat{\Lambda}$, such that for every interval $\Delta \in \mathfrak{R}_\Lambda \cap \mathbb{R}$ each of the subspaces*

$$E(\Delta \cap X_+)\mathcal{H} \text{ and } E(\Delta \cap X_-)\mathcal{H} \tag{4.7}$$

either degenerates to $\{0\}$, or is definite (positive and, respectively, negative).

Note that this theorem can be used also for a WJ^*-algebra from $\mathbf{H}$ because in the statement of the theorem the condition $A \in \mathbf{H}$ can be replaced by the condition $E_\lambda \in \mathbf{H}$.

Proposition 4.5. *If $\mathfrak{A} \in \mathbf{H}$, then* $\dim \mathfrak{S}(E_\lambda) \geq 1$.

Proof. The set $\mathfrak{S}(E_\lambda)$ is determined by the behavior of E_λ in a neighborhood (no matter how small) of zero, so one can assume that (see Theorem 4.4) for some decomposition $[-1;1]/\{0\} = X_+ \cup X_-$ and every closed interval $\Delta \subset [-1;1]/\{0\}$ each of the subspaces $E(X_+ \cap \Delta)\mathfrak{H}$ and $E(X_- \cap \Delta)\mathfrak{H}$ is either trivial or definite. By this hypothesis $\|\widetilde{g}_j(t)\|^2 = \pm[\widetilde{g}_j(t), \widetilde{g}_j(t)]$ for a.e. $t \in [-1;1]/\{0\}$, $j = 1, 2, \dots, k$ with the choice of sign depending of the fulfillment of one of the conditions $t \in X_+$ and $t \in X_-$. Thus, a.e.

$$G(t) = 1 + \Big|\sum_{j=1}^{k} \iota_{jj}(t)\Big| \tag{4.8}$$

and thanks to Remark 2.8 one can conclude that the collection $\{\iota_{jj}(t)\}_1^k$ contains functions that do not belong to $L^1_\sigma + L^2_\eta$. The rest follows from Proposition 4.2. $\square$

It is clear that, generally speaking, $\dim \mathfrak{S}(E_\lambda) \leq k^2$. Let us show that this estimate is sharp.

Example 4.6. We shall not wholly describe $\mathfrak{H}$ and E_λ indicating only that $\mathfrak{H}$ is a Pontryagin space with $\kappa = 2$, $\dim \mathfrak{H}_1 = 2$ and adducing a basic model space for J-orth.sp.f. E_λ. So, let $L^2_{\widetilde{\sigma}}(\mathfrak{E})$ coincide with the space $L^2_t(\mathbb{C})$ of scalar functions with the standard Lebesgue measure and let $\widetilde{g}_1(t) = t^{-2}$, $\widetilde{g}_2(t) = |t|^{-3/2} + i|t|^{-5/4}$. Let us show that $\dim \mathfrak{S}(E_\lambda) = 4$. Here $G(t) = (1 + t^{-4} + |t|^{-3} + |t|^{-5/2})$, $\iota_{11}(t) = t^{-4}$, $\iota_{12}(t) = \bar{\iota}_{21}(t) = |t|^{-7/2} - i|t|^{-13/4}$, $\iota_{22}(t) = |t|^{-3} + |t|^{-5/2}$. The collection $\{\iota_{lj}(t)\}_1^k$ is linearly independent. In fact, let $\psi(t) = \sum_{l,j=1}^k \alpha_{lj}\iota_{lj}(t) \in L^1_\sigma + L^2_\eta$. With no loss of generality one can suppose that $\sum_{l,j=1}^k |\alpha_{lj}| < 1$. Then there is a neighborhood of zero where the inequality $|\psi(t)| < G(t)$ holds, so (see Proposition 2.7 and Remark 2.8) $\psi(t) \in L^2_\eta$, but a direct verification shows that this is impossible if $\sum_{l,j=1}^k |\alpha_{lj}| > 0$. Thus, $\sum_{l,j=1}^k |\alpha_{lj}| = 0$.

4.2. Main results

We introduce an additional notation. Let $\varphi(t) \in L^\infty_\sigma \cap L^2_\nu$. Then $\mathcal{G}_\varphi(E_\lambda)$ means the totality of operators from $\mathfrak{A}_0$ that are an original of the function $\varphi(t)$.

Theorem 4.7. *$A \in \mathfrak{A}_0$ if and only A satisfies Condition (4.2) and there is a function $\varphi(t) \in L^\infty_\sigma \cap L^2_\nu$ such that*

$$\left.\begin{aligned} &\text{a) } \varphi(t) \text{ and } A \text{ are related by Conditions (3.9);}\\ &\text{b) if } \psi(t) = \sum_{l,j=1}^{k} \alpha_{lj}\iota_{lj}(t) \in L^1_\sigma + L^2_\eta, \text{ then}\\ &\sum_{l,j=1}^{k} \alpha_{lj}(Ah_l, e_j) = \int_{-1}^{1} \varphi(t)\psi(t)d\sigma(t). \end{aligned}\right\} \tag{4.9}$$

Proof. The necessity of Conditions (4.2) and (3.9) for $A \in \mathfrak{A}_0$ was proved before, so we turn to Condition (4.9b). Let $\psi(t) = \sum_{l,j=1}^{k} \alpha_{lj}\iota_{lj}(t) \in L^1_\sigma + L^2_\eta$. By virtue of Proposition 4.1 the function $\psi(t)$ can be represented using vector-functions from $\mathcal{J}\text{-}\widetilde{L}^2_{\vec{\sigma}}(\mathfrak{E})$. With the notation from Proposition 4.1 and Theorem 3.1 set

$$y_{-1} = Wg_{-1}(t),\ y_0 = Wg_0(t),\ x_j = Wf_j(t),\ j = -1, 0, 1, \ldots, k. \tag{4.10}$$

By the definition of the subalgebra $\mathfrak{A}_0$ there is a sequence of continuous functions $\{\varphi_m(t)\}_1^\infty$ vanishing near zero, i.e., $\varphi_m(t) \equiv 0$ in some neighborhood of zero (generally speaking, this neighborhood depends on m), $m = 1, 2, \ldots$, such that

$$(Ah_l, e_j) = \lim_{m\to\infty} \int_{-1}^{1} \varphi_m(t)d(E_t h_l, e_j), \quad l, j = 1, 2, \ldots, k \tag{4.11}$$

and

$$\left.\begin{aligned} (Ay_{-1}, x_{-1}) &= \lim_{m\to\infty} \int_{-1}^{1} \varphi_m(t)d(E_t y_{-1}, x_{-1}),\\ (Ay_0, x_0) &= \lim_{m\to\infty} \int_{-1}^{1} \varphi_m(t)d(E_t y_0, x_0),\\ (Ah_j, x_j) &= \lim_{m\to\infty} \int_{-1}^{1} \varphi_m(t)d(E_t h_j, x_j), j = 1, 2, \ldots, k. \end{aligned}\right\} \tag{4.12}$$

On the other hand (see (4.10))

$$(Ay_{-1}, x_{-1}) = \int_{-1}^{1} \varphi(t)d(g_{-1}(t), f_{-1}(t))_{\mathfrak{E}}\, d\sigma(t) \tag{4.13}$$

and the rest of the expressions from (4.12) can be re-written in terms of a functional representation in a similar way. Moreover Formulae (4.5) and (4.11) imply

$$\sum_{l,j=1}^{k} \alpha_{lj}(Ah_l, e_j) = \lim_{m\to\infty} \int_{-1}^{1} \varphi_m(t)\psi(t)d\sigma(t). \tag{4.14}$$

Now the necessity of (4.9b) follows from Formulae (4.12), (4.13), (4.14) and the special choice of the vector collection $y_{-1}, y_0, x_{-1}, x_0, \ldots, x_k$.

Pass to the sufficiency of (4.9). Let A satisfies (4.9). Choose a maximal linear independent modulo $L^1_\sigma + L^2_\eta$ subset $\{\iota_{lj}(t)\}_{(l,j)\in M}$ of the set $\{\iota_{lj}(t)\}_1^k$. Suppose that $M \neq \emptyset$. Let $\mathfrak{B}$ be $L^1_\sigma + L^2_\eta$ and let an operator-valued function P_t be the multiplication operator by $\chi_{[-1;t)}(t)$ acting in $L^1_\sigma + L^2_\eta$. Take as vector sets $\mathfrak{C}_{\lambda,\mu}$

satisfying Conditions (2.3) the sets of continuous functions vanishing in the corresponding neighborhoods of zero. Then by Lemma 2.5 and Proposition 2.7 there is a sequence of continuous functions $\{\varphi_m(t)\}_1^\infty$ vanishing near zero and such that for every function $\psi(t) \in L^1_\sigma + L^2_\eta$ the equality

$$\lim_{m\to\infty} \int_{-1}^{1} \varphi_m(t)\psi(t)d\sigma(t) = \int_{-1}^{1} \varphi(t)\psi(t)d\sigma(t),$$

holds and for $(l,j) \in M$

$$\lim_{m\to\infty} \int_{-1}^{1} \varphi_m(t)\iota_{lj}(t)d\sigma(t) = (Ah_l, e_j). \tag{4.15}$$

If $f(t), g(t) \in J\text{-}\widetilde{L}^2_{\widetilde{\sigma}}(\mathfrak{E})$ and $j = 1, 2, \dots, k$, then evidently $[f(t), g(t)]_{\mathfrak{E}} \in L^1_\sigma$ and $[f(t), \widetilde{g}(t)]_{\mathfrak{E}} \in L^2_\eta$. The latter implies that $\mathrm{w}-\lim\limits_{m\to\infty} \int_{-1}^{1} \varphi_m(t)d\widetilde{E}_t = \widetilde{A}$ and $\mathrm{w}-\lim\limits_{m\to\infty} \int_{-1}^{1} \varphi_m(t)d\widetilde{E}_t^\uparrow = \widetilde{A}^\uparrow$. Taking into account (4.15) one can conclude that also $\mathrm{w}-\lim\limits_{m\to\infty} \int_{-1}^{1} \varphi_m(t)dE_t = A$. Finally note that if $M = \emptyset$, it only simplifies the construction of the sequence $\{\varphi_m(t)\}_1^\infty$ because in this case there are no Conditions (4.15). □

Corollary 4.8. *$\mathcal{G}_\varphi(E_\lambda) \neq \emptyset$ for every function $\varphi(t) \in L^1_\sigma + L^2_\eta$. If $B \in \mathcal{G}_\varphi(E_\lambda)$ is a fixed operator, then $\mathcal{G}_\varphi(E_\lambda) = \{B + S\}_{S\in\mathfrak{S}(E_\lambda)}$, in particular, $\mathcal{G}_0(E_\lambda) = \mathfrak{S}(E_\lambda)$.*

Corollary 4.9. *The set $\mathfrak{S}(E_\lambda)$ does not depend on a choice of canonical scalar product in $\mathfrak{H}$.*

Corollary 4.10. *The set $\mathfrak{A}_0$ coincides with the strong sequential closure of the set of all operators represented by (4.1).*

Proof. Let $A \in \mathfrak{A}_0$ and let $\varphi(t)$ be the portrait of A. For $\varphi(t)$ construct a sequence $\{\varphi_m(t)\}_{m=1}^\infty$ of continuous functions vanishing near zero such that

a) $\|\varphi_m(t)\|_{L^\infty_\sigma} \leq \|\varphi(t)\|_{L^\infty_\sigma}$;

b) $\varphi_m(t) \overset{\mu_\sigma}{\to} \varphi(t)$ for $m \to \infty$;

c) $\|\varphi_m(t) - \varphi(t)\|_{L^2_\nu} \to 0$ for $m \to \infty$.

Then $\mathrm{s}-\lim\limits_{m\to\infty} \int_{-1}^{1} \varphi_m(t)d\widetilde{E}_t = \widetilde{A}$, $\mathrm{s}-\lim\limits_{m\to\infty} \int_{-1}^{1} \varphi_m(t)d\widetilde{E}_t^\uparrow = \widetilde{A}^\uparrow$, where $\widetilde{A}$ and $\widetilde{A}^\uparrow$ are the same as in (3.9). The rest follows from Lemma 2.5 and Remark 2.6. □

Theorem 4.11. *For the given algebra $\mathfrak{A}$ there is a finite collection of J-s.a. operators $A_1, A_2, \dots, A_l \in \mathfrak{A}$, such that every operator $B \in \mathfrak{A}$ has a representation*

$$B = S + Q(A_1, A_2, \dots, A_l) + C, \tag{4.16}$$

where $S \in \mathfrak{A}$ is a nilpotent operator, $S|_{\widetilde{\mathfrak{H}}} = S^\#|_{\widetilde{\mathfrak{H}}} = 0$, $Q(t_1, t_2, \dots, t_l)$ is a polynomial of l variables, $C \in \mathfrak{A}_0$.

Proof. Consider the totality of portraits corresponding to operators from $\mathfrak{A}$. By Proposition 2.10 this totality contains only a finite number of functions linearly independent modulo $L^\infty_\sigma \cap L^2_\nu$ and by virtue of Remark 2.11 one can chose these functions real-valued. Let $\{\varphi_1(t), \varphi_2(t), \dots, \varphi_l(t)\}$ be a maximal system of real-valued portraits linearly independent modulo $L^\infty_\sigma \cap L^2_\nu$. Fix some J-s.a. operators $A_1, A_2, \dots$ and A_l from $\mathfrak{A}$ that are originals for functions $\varphi_1(t), \varphi_2(t), \dots$ and $\varphi_l(t)$ respectively. Let $B \in \mathfrak{A}$. Then for B there is a linear combination $\alpha_1 A_1 + \alpha_2 A_2 + \dots + \alpha_l A_l$ of these operators, such that the portrait $\varphi(t)$ of the operator $B - \alpha_1 A_1 - \alpha_2 A_2 - \dots - \alpha_l A_l$ belongs $L^\infty_\sigma \cap L^2_\nu$. By Theorem 4.7 there is an operator $C \in \mathcal{G}_\varphi(E_\lambda)$. Thus one need only to prove that the operator $S := B - \alpha_1 A_1 - \alpha_2 A_2 - \dots - \alpha_l A_l - C$ satisfies Conditions (4.16). By the construction $S|_{\widetilde{\mathfrak{H}}} = 0$ and $S\mathfrak{H} \subset \widetilde{\mathfrak{H}}^{[\perp]}$. On the other hand by the definition of e.s.f. of the algebra $\mathfrak{A}$ the subspace $\widetilde{\mathfrak{H}}^{[\perp]}$ is the root space or its part for S (this is true for every operator belonging to $\mathfrak{A}$). Since $\mathfrak{H}_1 \subset \widetilde{\mathfrak{H}}^{[\perp]}$ and $\mathfrak{H}_1 \subset \operatorname{Ker} S$, the rest is trivial. □

Remark 4.12. Recall (see [1] Part VI, Section 92), that a commutative W^*-algebra is monogenic, i.e., it contains an operator A, such that $\operatorname{Alg} A$ coincides with the algebra. For commutative WJ^*-algebras this is not true. In particular, if the subspace $\widetilde{\mathfrak{H}}^{[\perp]}$ has infinite dimension, then in general the nilpotent part of the algebra $\mathfrak{A}$ can have an infinite number of algebraically independent elements. An idea for the construction of an algebra with the desired property can be found in [7]. Next, Representation (4.16) contains the polynomial $Q(t_1, t_2, \dots, t_l)$. In the proof of Theorem 4.11 given above $Q(t_1, t_2, \dots, t_l)$ is a linear function. If one wants to minimize l in Representation (4.16) one needs to consider polynomials of general type. In fact, it can occur that one of the operators from $\{A_j\}_1^l$ has a representation of the type (4.16) (say, $A_l = S' + Q'(A_1, A_2, \dots, A_{l-1}) + C'$) and can be excluded from the right part of (4.16). Let us give an example where l cannot be reduced to one.

Example 4.13. We describe a main idea omitting some details. Here $\mathfrak{H}$ is a Pontryagin space with $\kappa = 3$, $\dim \mathfrak{H}_1 = 3$, $L^2_{\tilde{\sigma}}(\mathfrak{E})$ coincides with $L^2_t(\mathbb{C})$, unbounded elements take the form: $\widetilde{g}_1(t) = t^{-1}$, $\widetilde{g}_2(t) = t^{-4/5}$, $\widetilde{g}_3(t) = t^{-2}$. Then a maximal collection of functions-portraits for $\mathfrak{A}$ linearly independent modulo $L^\infty_\sigma \cap L^2_\nu$, is formed by three functions, say: $\varphi_1(t) \equiv 1$, $\varphi_2(t) = t^{6/5}$, $\varphi_3(t) = t$. Using the relations $\varphi_1(t) = -(\varphi_1(t) + \varphi_2(t))^2 + 2(\varphi_1(t) + \varphi_2(t)) \,|\, \mathrm{Mod}\ L^\infty_\sigma \cap L^2_\nu$, $\varphi_2(t) = (\varphi_1(t) + \varphi_2(t))^2 - (\varphi_1(t) + \varphi_2(t)) \,|\, \mathrm{Mod}\ L^\infty_\sigma \cap L^2_\nu)$ one can reduce the number l (in Representation (4.16)) of generating operators for the algebra in question to $l = 2$, but a further reduction of l is impossible because $(\varphi_2(t))^2 = (\varphi_3(t))^2 = \varphi_2(t)\varphi_3(t) = 0 \,|\, \mathrm{Mod}\ L^\infty_\sigma \cap L^2_\nu$.

Up to this point the integral $\int_{-1}^{1} \varphi(t) dE_\lambda$ was defined exclusively for a continuous function $\varphi(t)$ vanishing near zero. However it is clear that there is a more extended variety of functions such that the integral in question has some natural sense.

Proposition 4.14. *Let $\varphi(t)$ be a μ_σ-measurable function. Then the integral*

$$\int_{-1}^{1} \varphi(t) dE_\lambda \tag{4.17}$$

converges unconditionally in the weak topology as an improper integral with the singularity in zero if and only if the function $\varphi(t)$ satisfies the following conditions:

$$\left.\begin{aligned} &\text{a) } \varphi(t) \in L_\sigma^\infty \cap L_\nu^2; \\ &\text{b) } \int_{-1}^{1} |\varphi(t)| \sum_{l,j=1}^{k} |\iota_{lj}(t)| d\sigma(t) < \infty. \end{aligned}\right\} \tag{4.18}$$

Proof. Since $\int_{-1}^{1} \varphi(t) dE_\lambda|_{\mathfrak{H}_1} = 0$, the necessity of (4.18a) follows from Proposition 2.9, and the necessity of (4.18b) is a consequence of Formula (4.5) and the equivalence in the case of scalar functions between unconditional and absolute convergence of integrals. The sufficiency of Condition (4.18) for the weak unconditional convergence of the integral (4.17) can be justified as follows. For the component $P_2 \int_{-1}^{1} \varphi(t) dE_t|_{\mathfrak{H}_2}$ the problem of convergence is solved in a way similar to that for an orthogonal spectral function in a Hilbert space. Next, $|\varphi(t)| \widetilde{g}_j(t) \in L_{\widetilde{\sigma}}^2(\mathfrak{E})$ for every function $\varphi(t) \in L_\nu^2$, so (recall Notation (4.4)) the integrals $\displaystyle\int_{-1}^{1} \varphi(t) d(E_t h_j, x)$ and $\displaystyle\int_{-1}^{1} \varphi(t) d(E_t x, e_j)$, $j = 1, 2, \dots, k$, converge absolutely. Finally, integrals $\displaystyle\int_{-1}^{1} \varphi(t) d(E_t h_l, x_j)$, $l, j = 1, 2, \dots, k$ converge absolutely by (4.18b). □

Remark 4.15. Since weak convergence and strong convergence of the improper integral $P_2 \int_{-1}^{1} \varphi(t) dE_t|_{\mathfrak{H}_2}$ can only take place simultaneously, the weak convergence of the improper integral (4.17) implies its strong convergence.

Remark 4.16. Conditions (4.18) can be simplified by different ways, depending on properties of E_λ. In particular, if $k = 1$ and $\iota_{11}(t) \equiv 0$ (see Example 4.3), then Conditions (4.18) are reduced to Condition (4.18a). Note also that due to the inequality $\sum_{l,j=1}^{k} |\iota_{lj}(t)| < k^2 G(t)$ the condition

$$\varphi(t) \in L_\sigma^\infty \cap L_\nu^1 \tag{4.19}$$

is sufficient (but, generally speaking, not necessary) for the unconditional strong ($\equiv$weak) convergence of the improper integral (4.17). Moreover, if $E_\lambda \in \mathbf{H}$, then in some neighborhood of zero (see (4.8)) the inequality $G(t) < 1 + \sum_{l,j=1}^{k} |\iota_{lj}(t)|$ holds, therefore in this case Condition (4.19) is necessary and sufficient for the convergence in the indicated sense of the integral (4.17).

4.3. Monogenic algebras

Here a problem of a functional representation for $\mathfrak{A}$ generated by two operators A and I $(A = A^{\#})$ is considered. The operator A is assumed to be fixed. It is clear that $\mathfrak{A} = \operatorname{Alg} A$ so the e.s.f. of A is simultaneously an e.s.f. of $\mathfrak{A}$. Everywhere in this subsection E_λ means the e.s.f. of A. Condition (3.1) is assumed to be fulfilled.

Proposition 4.17. *If* $\dim \widetilde{\mathfrak{H}}^{[\perp]} = \infty$, *then there is a regular subspace* $\mathfrak{H}_4 \subset \operatorname{Ker} A$, *such that its co-dimension with respect to* $\widetilde{\mathfrak{H}}^{[\perp]}$ *is finite.*

This proposition is easy to prove. In practice it means that the case $\dim \widetilde{\mathfrak{H}}^{[\perp]} = \infty$ is negligible. Proposition 4.17 is mentioned here for case of reference.

Proposition 4.18. *Let* $B \in \mathfrak{A}$ *and* $B|_{\mathfrak{H}_1} = 0$. *Then*

$$B = S + B_0 \tag{4.20}$$

where $B_0 \in \mathfrak{A}_0$, S *is a nilpotent operator, and* $S|_{\widetilde{\mathfrak{H}}} = 0$.

Proposition 4.18 can be proved using a scheme similar to the one used in the proof of Theorem 4.11. Comparing (4.16) with (4.20), note that the summand $Q(A)$ in (4.20) is absent due to the condition $B|_{\mathfrak{H}_1} = 0$.

Start to construct a functional calculus for A.

Remark 4.19. If the integral (4.17) weakly converges in the unconditional sense, set

$$\overset{\diamond}{\varphi}(A) := \int_{-1}^{1} \varphi(t) dE_\lambda. \tag{4.21}$$

If $\varphi(\xi)$ is a function regular in some neighborhood of $\sigma(A)$, then, generally speaking, the operator $\varphi(A)$ standardly defined by the Riesz integral does not coincide with $\overset{\diamond}{\varphi}(A)$. If k_0 is the maximal length of Jordan chains of the operator $A|_{\mathfrak{H}_1}$ then $t^{k_0} \in L^\infty_\sigma \cap L^2_\nu$ and the integral $\int_{-1}^{1} \lambda^{2k_0} dE_\lambda$ unconditionally converges in the strong topology but it can occur (see Proposition 4.18) that $A^{2k_0} = Z_{2k_0} + \int_{-1}^{1} \lambda^{2k_0} dE_\lambda$, where the operator $Z_{2k_0} \neq 0$ has the same properties as the operator S from (4.20). But there exists a natural number m_0, such that

$$A^m = \int_{-1}^{1} \lambda^m dE_\lambda, \; m \geq m_0, \tag{4.22}$$

where the integral has the same sense as above. This implies the representation

$$\varphi(A) = \int_{-1}^{1} \varphi(\lambda) dE_\lambda, \tag{4.23}$$

valid for every function $\varphi(\xi)$ regular in some neighborhood of $\sigma(A)$ and having in zero a root with multiplicity no less that m_0. Below we use Formula (4.23) not only for the regular function $\varphi(t)$ but also accept it as definition in the case when $\varphi(t)$ is an arbitrary μ_σ-measurable function satisfying in some neighborhood of zero the condition

$$\varphi(t) = O(t^{m_0}). \tag{4.24}$$

Proposition 4.20. *Let $\varphi(t)$ be a fixed function satisfying Conditions (4.18). Then there is a positive constant c_φ, such that for every function $\theta(t) \in L^\infty_\sigma$ and the function $\psi(t) = \theta(t)\varphi(t)$ the inequality $\| \overset{\diamond}{\psi}(A)\| \leq c_\varphi \|\theta(t)\|_{L^\infty_\sigma}$ holds.*

Proof. The definition of $\overset{\diamond}{\psi}(A)$ implies

$$\begin{aligned}\| \overset{\diamond}{\psi}(A)h_j\|^2 &= \|\psi(t)\widetilde{g}_j(t)\|^2_{L^2_{\tilde{\sigma}}(\mathfrak{E})} + \sum_{l=1}^{k} \left| \int_{-1}^{1} \psi(t)\iota_{lj}(t)d\sigma(t)\right|^2 \\ &\leq \|\theta(t)\|^2_{L^\infty_\sigma} \times \left(\|\varphi(t)\widetilde{g}_j(t)\|^2_{L^2_{\tilde{\sigma}}(\mathfrak{E})} + \sum_{l=1}^{k} \left(\int_{-1}^{1} |\varphi(t)||\iota_{lj}(t)|d\sigma(t)\right)^2 \right);\\ \| \overset{\diamond}{\psi}(A)x\|^2 &= \|\psi(t)f(t)\|^2_{L^2_{\tilde{\sigma}}(\mathfrak{E})} + \sum_{j=1}^{k} \left| \int_{-1}^{1} [\psi(t)f(t), \widetilde{g}_j(t)]_{\mathfrak{E}} d\sigma(t)\right|^2 \\ &\leq \|\theta(t)\|^2_{L^\infty_\sigma} \cdot \|f(t)\|^2_{L^2_{\tilde{\sigma}}(\mathfrak{E})} \cdot \left(\|\varphi(t)\|^2_{L^\infty_\sigma} + \sum_{j=1}^{k} \int_{-1}^{1} \|\bar{\varphi}(t)\widetilde{g}_j(t)\|^2_{\mathfrak{E}} d\sigma(t) \right) \\ &= \|\theta(t)\|^2_{L^\infty_\sigma} \cdot \|x\|^2 \cdot \left(\|\varphi(t)\|^2_{L^\infty_\sigma} + \sum_{j=1}^{k} \int_{-1}^{1} \|\bar{\varphi}(t)\widetilde{g}_j(t)\|^2_{\mathfrak{E}} d\sigma(t) \right),\end{aligned}$$

where $x \in \mathfrak{H}_2$, $f(t) = W^{-1}x$. Since $\mathfrak{H}_0$ is a finite-dimensional subspace and $\overset{\diamond}{\psi}(A)|_{\mathfrak{H}_1} = 0$, the rest is trivial. ☐

Proposition 4.21. *Let $\psi(t)$ be a continuous function vanishing near zero. Then there is a sequence of polynomials $\{\mathcal{P}_j(t)\}_{j=1}^\infty$, such that $\lim\limits_{j\to\infty} \|\{\mathcal{P}_j(A) - \psi(A)\| = 0$.*

Proof. Let m_0 be as defined in (4.22). Take a sequence $\{Q_j(t)\}_{j=1}^\infty$ of polynomials uniformly converging on $[-1;1]$ to the continuous function $\psi(t) \cdot t^{-m_0}$ vanishing near zero. Set $\mathcal{P}_j(t) := Q(t) \cdot t^{m_0}$, $\psi_j := t^{m_0} \cdot (Q_j(t) - \psi(t) \cdot t^{-m_0})$, $j = 1, 2, \ldots$, $\varphi(t) = t^{m_0}$. Then by Proposition 4.20 $\|\psi(A) - \mathcal{P}_j(A)\| = \|\psi_j(A)\| \leq c_\varphi \max\limits_{t\in[-1;1]} \{\|Q_j(t) - \psi(t) \cdot t^{-m_0}\|\}$. ☐

Corollary 4.22. *The subalgebra $\mathfrak{A}_0$ coincides with the closure in the strong topology of the set of all operators $\{Q(A)\}$, where $Q(t)$ runs through the set of all polynomials satisfying (4.24).*

Theorem 4.23. *Let $B \in \mathfrak{A}$ $(= \operatorname{Alg} A)$. Then*

$$B = Q(A) + C, \tag{4.25}$$

where $Q(t)$ is a polynomial, $C \in \mathfrak{A}_0$. Conversely, if B is represented by (4.25), then $B \in \mathfrak{A}$.

Proof. Using the same reasoning as for the proof of Theorem 4.11 and Representation (4.25) one can assume that B is a nilpotent operator and

$$B|_{\widetilde{\mathfrak{H}}} = 0. \tag{4.26}$$

Since $\mathfrak{H}_1 \cap \operatorname{Ker} A \neq \{0\}$, then $B|_{\operatorname{Ker} A} = 0$ and thanks to Proposition (4.13) one can assume that $\dim \widetilde{\mathfrak{H}}^{[\perp]} = n < \infty$. Since $\mathfrak{H}_1 \subset \widetilde{\mathfrak{H}}$, it implies that $n \geq k$. For simplicity we re-denote the union of the systems $\{e_j\}_{j=1}^k$ and $\{h_j\}_{j=1}^k$, setting $y_1 = e_1, y_2 = e_2, \ldots, y_k = e_k, y_{k+1} = h_1, y_{k+2} = h_2, \ldots, y_{2k} = h_k$, and, if $n > k$, complete this system in an arbitrary way to a orthonormalized basis $\{y_j\}_{j=1}^{n+k}$ in the space $\widetilde{\mathfrak{H}}^{[\perp]} \oplus \mathfrak{H}_0$.

Next, fix the maximal number of linearly independent relations of the form

$$\psi_\upsilon(t) = \sum_{l,j=1}^{k} \alpha_{lj}^{(\upsilon)} \iota_{lj}(t) \in L^1_\sigma + L^2_\eta, \quad \upsilon = 1, 2, \ldots, \Upsilon. \tag{4.27}$$

It can occur that such relations are absent, i.e., the system $\{\iota_{lj}(t)\}_1^k$ is linearly independent modulo $L^1_\sigma + L^2_\eta$. Consider this case later, i.e., for now assume that $1 \leq \Upsilon \leq k^2$. Note also that the set of conditions

$$\sum_{l=k+1}^{2k} \sum_{j=1}^{k} \alpha_{lj}^{(\upsilon)} (Sy_l, y_j) = 0, \quad \upsilon = 1, 2, \ldots, \Upsilon, \tag{4.28}$$

is equivalent to Condition (4.6b).

By virtue of the condition $B \in \operatorname{Alg} A$ and Condition (4.26) there is a sequence of polynomials $\{\mathcal{R}_p(t)\}_1^\infty$, such that

$$\mathcal{R}_p(A)|_{\mathfrak{H}_1} = 0, \quad p = 1, 2, \ldots\ , \tag{4.29}$$

and

$$\left.\begin{aligned} &\text{a)}\ \lim_{p\to\infty} (\mathcal{R}_p(A)y_l, y_j) = (By_l, y_j), \quad l, j = 1, 2, \ldots, n+k; \\ &\text{b)}\ \lim_{p\to\infty} \int_{-1}^{1} \mathcal{R}_p(t)\psi_\upsilon(t) d\sigma(t) = 0, \quad \upsilon = 1, 2, \ldots, \Upsilon. \end{aligned}\right\} \tag{4.30}$$

Condition (4.29) means that the number $t_0 = 0$ is a root of all polynomials $\mathcal{R}_p(t)$ and its multiplicity for $\mathcal{R}_p(t)$ is no less than k_0, where k_0, as above, is the maximum length of Jordan chains for the operator $A|_{\mathfrak{H}_1}$. Thus, the operator $\mathcal{R}_p(A)$ from (4.29) and (4.30) represents a (finite) linear combination of operators from the set $\{A^{k_0+j}\}_{j=0}^\infty$. The latter operators have the representation (4.20)

$$A^m = S_m + (A^m)_0, \quad (A_m)_0 \in \mathfrak{A}_0, \quad m = k_o, k_0 + 1, \ldots\ . \tag{4.31}$$

The elements of Decomposition (4.31) are not, generally speaking, uniquely defined. This indefiniteness is not convenient in what follows, so we fix a decision making for picking S_m and $(A^m)_0$ in (4.31). First, choose a decomposition $A^{k_0} = S_{k_0} + (A^{k_0})_0$, fix it and, second, set

$$S_{k_0+l} = A^l S_{k_0}, \quad (A^{k_0+l})_0 = A^l (A^{k_0})_0. \tag{4.32}$$

By Corollary 4.10 $A^l(A^{k_0})_0 = \mathrm{s}-\lim\limits_{m\to\infty} A^l\varphi_m(A) = \mathrm{s}-\lim\limits_{m\to\infty}\psi_m(A)$, where $\{\varphi_m(t)\}_1^\infty$ is some sequence of continuous functions vanishing near zero, $\psi_m(t) = t^l\varphi_m(t)$, so $A^l(A^{k_0})_0 \in \mathfrak{A}_0$. Hence it is shown that (4.32) is a uniquely defined representation of the type (4.31).

Thus, every operator $\mathcal{R}_p(A)$, satisfying Condition (4.29) has the representation

$$\mathcal{R}_p(A) = Z_p + D_p, \tag{4.33}$$

where (see (4.32)) $Z_p \in \mathfrak{Z}:\ = \mathrm{Lin}\{S_{k_0+l}\}_{l=0}^\infty$, $D_p \in \mathcal{G}_{\mathcal{R}_p}(E_\lambda)$. Then $\mathfrak{Z}$ is a finite-dimensional linear space consisting of nilpotent operators $\{Z\}$. For every Z the subspace $\mathfrak{H}_2^{[\perp]} = \widetilde{\mathfrak{H}}^{[\perp]} \oplus \mathfrak{H}_0$ is Z-invariant and $Z|_{\mathfrak{H}_2} = 0$. Let $\dim\mathfrak{Z} = \mathcal{K}$. Since the operators from $\mathfrak{Z}$ commute, it implies that $\mathcal{K} < (k+n)^2$. Choose and fix $(k+n)^2 - \mathcal{K}$ independent conditions

$$\sum_{l,j=1}^{k+n} \beta_{lj}^{(m)} z_{lj} = 0, \quad m = 1,2,\dots,(k+n)^2 - \mathcal{K}, \tag{4.34}$$

where $z_{lj}:\ = (Zy_l, y_j)$, that hold if and only if $Z \in \mathfrak{Z}$. For the elements of Decomposition (4.33) set $r_{lj}^{(p)}:\ = (\mathcal{R}_p(A)y_l, y_j)$, $z_{lj}^{(p)}:\ = (Z_p y_l, y_j)$, $d_{lj}^{(p)}:\ = (D_p y_l, y_j)$, $l,j = 1,2,\dots,n+k$. These values satisfy the following relations (see (4.27), (4.28)):

$$\left.\begin{array}{l}
\text{a) } z_{lj}^{(p)} + d_{lj}^{(p)} = r_{lj}^{(p)}, \quad l,j = 1,2,\dots,k+n;\\
\text{b) } \displaystyle\sum_{l,j=1}^{k+n} \beta_{lj}^{(m)} z_{lj}^{(p)} = 0,\ m = 1,2,\dots,(k+n)^2 - \mathcal{K};\\
\text{c) } \displaystyle\sum_{l=k+1}^{2k}\sum_{j=1}^{k} \alpha_{lj}^{(v)} d_{lj}^{(p)} = \int_{-1}^{1} \mathcal{R}_p(t)\psi_v(t)d\sigma(t),\ v = 1,2,\dots,\Upsilon,\\
\text{d) } d_{lj}^{(p)} = 0 \text{ if } l \le k \text{ or } l > 2k \text{ or } j > k.
\end{array}\right\} \tag{4.35}$$

Set $b_{lj}:\ = (By_l, y_j)$, $l,j = 1,2,\dots,k+n$, and consider the system

$$\left.\begin{array}{l}
\text{a) } z_{lj} + d_{lj} = b_{lj}, \quad l,j = 1,2,\dots,k+n;\\
\text{b) } \displaystyle\sum_{l,j=1}^{k+n} \beta_{lj}^{(m)} z_{lj} = 0, \quad m = 1,2,\dots,(k+n)^2 - \mathcal{K};\\
\text{c) } \displaystyle\sum_{l=k+1}^{2k}\sum_{j=1}^{k} \alpha_{lj}^{(v)} d_{lj} = 0, \quad v = 1,2,\dots,\Upsilon;\\
\text{d) } d_{lj} = 0 \text{ for } l \le k \text{ or } l > 2k \text{ or } j > k.
\end{array}\right\} \tag{4.36}$$

Relations (4.36) is a system of linear equations with respect to the unknowns $\{z_{lj},\ d_{lj}\}_{l,j=1}^{k+n}$. It has the same principal matrix as System (4.35) and the right-hand side of System (4.36) is obtained from the right-hand side of System (4.35) via passage to the limit as $p \to \infty$. Since System (4.35) is solvable, System (4.36) is solvable by the well-known theorem of Kronecker-Capelli. Take some solution of (4.36). Then the operator Z given by the conditions

$$Z|_{\mathfrak{H}_2} = 0, \quad Z\mathfrak{H}_2^{[\perp]} \subset \mathfrak{H}_2^{[\perp]}, \quad (Zy_l, y_j) = z_{lj}, \quad l, j = 1, 2, \dots, k+n,$$

belongs to $\mathfrak{Z}$. Hence there is a polynomial $\mathcal{R}(t)$ satisfying Condition (4.29) and such that $\mathcal{R}(A) = Z + (\mathcal{R}(A))_0$, where $\mathcal{R}(A))_0 \in \mathcal{G}_{\mathcal{R}}(E_\lambda$. Moreover, the solvability of System (4.36) implies the existence of the operator $D \in \mathfrak{S}(E_\lambda)$, such that $(Dy_l, y_j) = d_{lj}$, $l, j = 1, 2, \dots, k+n$ and, finally, the representation $B = \mathcal{R}(A) - (\mathcal{R}(A))_0 + D$. Thus, the theorem has been proved under Hypothesis (4.27). If the system $\{\iota_{lj}(t)\}_1^k$ is linearly independent modulo $L^1_\sigma + L^2_\eta$, then the set $\mathfrak{S}(E_\lambda)$ is subjected only Condition (4.6a), i.e., the operator D satisfies Conditions $D|_{\mathfrak{H}_2} = 0$, $D\mathfrak{H}_2^{[\perp]} \subset \mathfrak{H}_2^{[\perp]}$ and Condition (4.36d), then $D \in \mathfrak{S}(E_\lambda)$. The latter means that if there are no Conditions (4.27), the proof follows the similar way, as above, with the exclusion of Equality (4.35c) and Equation (4.36c). □

Corollary 4.24. Alg A *is equal to the sequential closure of the set* $\{Q(A))$ *in the strong topology, where* $\{Q(t)\}$ *runs through the totality of (all) polynomials.*

Consider now a problem of representation of a monogenic uniformly closed algebra. Let C^0_σ be the space of functions $\{\phi(t)\}$ continuous on $\operatorname{Supp}(\mu_\sigma)$ and satisfying the condition $\phi(0) = 0$. The norm on C^0_σ is defined in the standard way, i.e., $\|\phi\|_{C^0_\sigma} = \max_{t \in \operatorname{Supp}(\mu_\sigma)} \{|\phi(t)|\}$. Then the space $C^0_\sigma \cap L^2_\nu$ is well defined as a Banach space with the norm

$$\|\phi\|_{C^0_\sigma \cap L^2_\nu} = \max\{\|\phi\|_{C^0_\sigma}, \|\phi\|_{L^2_\nu}\}.$$

Corollary 4.25. *An operator* B *belongs to the closure of the set* $\{Q(A)\}$ *(as above* $Q(t)$ *runs through the totality of polynomials) in the uniform topology if and only if there exists a function* $\varphi(t) \in C^0_\sigma \cap L^2_\nu$ *and a polynomial* $Q(t)$*, such that*

$$B = Q(A) + C, \qquad C \in \mathcal{G}_\varphi(E_\lambda). \tag{4.37}$$

Proof. Indeed, if B belongs to the uniform algebra, generated by A and I, then $B \in \operatorname{Alg} A$, therefore B has Representation (4.25). On the other hand it is clear that the portrait of B under the given conditions must belong to the set of functions continuous on $\operatorname{Supp}(\mu_\sigma)$. So, the necessity of Condition (4.37) has been established. In turn, the sufficiency of (4.37) follows from the corresponding results for self-adjoint operators, from the finite dimensionality of the spaces $\mathfrak{H}_0$ and $\mathfrak{H}_1$ and from Remark 2.6. □

5. WJ^*-algebras with an arbitrary number of peculiarities

Now we start to consider the problem of the functional description of the algebra $\mathfrak{A}$ assuming that some J-s.a. operators from $\mathfrak{A}$ have non-real spectra and that its e.s.f. E_λ does not satisfies Conditions (3.1b),c)). We assume (with no loss of generality) that $\operatorname{Supp}(E_\lambda) \subset (-1; 1)$. Then for the set of spectral peculiarities Λ

the relation $\Lambda \subset (-1;1)$ holds, $E_{-1} = E_{-1+0} = 0$, $E_\lambda = \text{const}$ for $\lambda \geq 1$, but, generally speaking, $E_1 \neq I$. If the latter holds, the operator $I - E_1$ is a J-orthoprojection onto a finite-dimensional $\mathfrak{A}$-invariant subspace, such that there exists at least one J-s.a. operator $A \in \mathfrak{A}$ with $\sigma(A|_{(I-E_1)\mathcal{H}}) \cap \mathbb{R} = \emptyset$.

For the construction of a functional representation for $\mathfrak{A}$ one needs to modify the description of some key function spaces, say $L^\infty_\sigma \cap L^1_\nu$, $L^1_\sigma + L^2_\eta$, etc. Let the subspace $\widetilde{\mathcal{H}}$ be the same as in (1.6d). The notation from (3.2) also remains valid. Consider (cf. Remark 3.3) the factor-space $\overset{*}{\mathcal{H}}: = \mathcal{H}/\mathcal{H}_1$. Then $\overset{*}{\mathcal{H}}$ is a Krein space and $\widetilde{E}_\lambda$ induces on this space the J-orth.sp.f. $\overset{*}{E}_\lambda$ without peculiarities. Assume (with no loss of generality) that $\overset{*}{E}_\lambda$ is both orthogonal and J-orthogonal. Then $\overset{*}{E}_\lambda$ is similar to the operator-valued function X_λ (the multiplication operator by the indicator of the set $(-\infty;t)$ denoted $\chi_{(-\infty;t)}(\tau)$) acting in some standard Krein space $\mathcal{J}\text{-}L^2_{\vec{\sigma}}(\mathcal{E})$. The corresponding operator of similarity is simultaneously symmetric and J-symmetric. Fix $\mathcal{J}\text{-}L^2_{\vec{\sigma}}(\mathcal{E})$ and the corresponding function $\sigma(t)$ giving the measure on the standard Krein space. Note that $\sigma(t)$ is continuous in all points from Λ.

Let $\{\alpha_j\}_{j=1}^{n-1} \subset (-1;1)$ be a set of numbers taken in ascending order. The set $(-1;1)\backslash\Lambda$ is decomposed as the union of disjoint open intervals. Choose $\{\alpha_j\}_{j=1}^{n-1} \subset (-1;1)$ such that every point from Λ belongs to one and only one of these intervals. Set $\alpha_0 = -1$ and $\alpha_n = 1$. Then the spectral function $E^{(j)}_\lambda: = E_\lambda|_{(E_{\alpha_{j+1}} - E_{\alpha_j})\mathcal{H}}$ has one and only one point of peculiarity, $j = 1, 2, \ldots n-1$. Denote $\mathcal{J}\text{-}L^{2,(j)}_{\vec{\sigma}}(\mathcal{E})$ the subspace of $\mathcal{J}\text{-}L^2_{\vec{\sigma}}(\mathcal{E})$ that contains all functions from $\mathcal{J}\text{-}L^2_{\vec{\sigma}}(\mathcal{E})$ vanishing outside the interval $[\alpha_j;\alpha_{j+1})$. There are two options for $E^{(j)}_\lambda$. The first is that its unique peculiarity is regular. Then set $\mathcal{J}\text{-}\widetilde{L}^{2,(j)}_{\vec{\sigma}}(\mathcal{E}): = \mathcal{J}\text{-}L^{2,(j)}_{\vec{\sigma}}(\mathcal{E})$ and fix a corresponding operator of similarity $W^{(j)}: \mathcal{J}\text{-}\widetilde{L}^{2,(j)}_{\vec{\sigma}}(\mathcal{E}) \mapsto (E_{\alpha_{j+1}} - E_{\alpha_j})\mathcal{H}$. The second is that the peculiarity is singular. Then construct a basic model space $\mathcal{J}\text{-}\widetilde{L}^{2,(j)}_{\vec{\sigma}}(\mathcal{E})$ for $E^{(j)}_\lambda$ expanding $\mathcal{J}\text{-}L^{2,(j)}_{\vec{\sigma}}(\mathcal{E})$ with the evident modifications related with the position of the peculiarity, that now does not coincide with zero. In this option fix some corresponding operator of similarity $W^{(j)}$ that here acts between $\mathcal{J}\text{-}\widetilde{L}^{2,(j)}_{\vec{\sigma}}(\mathcal{E})$ and $(E_{\alpha_{j+1}} - E_{\alpha_j})\widetilde{\mathcal{H}}$. Now set

$$\mathcal{J}\text{-}\widetilde{L}^2_{\vec{\sigma}}(\mathcal{E}): = \mathcal{J}\text{-}\widetilde{L}^{2,(0)}_{\vec{\sigma}}(\mathcal{E}) \oplus \cdots \oplus \mathcal{J}\text{-}\widetilde{L}^{2,(n)}_{\vec{\sigma}}(\mathcal{E}),\ W: = W^{(0)} \oplus \cdots \oplus W^{(n)}. \quad (5.1)$$

There are again two options. The first is that Λ has only regular peculiarities and the second one appears if there is a singular peculiarity in Λ. In the first case the space $\mathcal{J}\text{-}\widetilde{L}^2_{\vec{\sigma}}(\mathcal{E})$ coincides with $\mathcal{J}\text{-}L^2_{\vec{\sigma}}(\mathcal{E})$ and a functional representation for $\mathfrak{A}|_{\widetilde{\mathcal{H}}}$ is in fact the same as for a commutative W^*-algebra. This case is quite simple and will be considered later.

So assume that Λ contains at least one singular peculiarity and hence the subspace $\mathcal{H}_1$ is nontrivial. Then in $\mathcal{J}\text{-}\widetilde{L}^2_{\vec{\sigma}}(\mathcal{E})$ there is a system of functions $\{\widetilde{g}_j(t)\}_{j=1}^k$

that are linearly independent modulo $\mathcal{J}\text{-}L^2_{\widetilde{\sigma}}(\mathcal{E})$ and by the definition of $\mathcal{J}\text{-}\widetilde{L}^2_{\widetilde{\sigma}}(\mathcal{E})$ form an orthonormalized system. With no loss of generality assume that $\widetilde{g}_j(t) \equiv 0$ on $[\alpha_l; \alpha_{l+1})$ if this interval contains a regular singularity. Let us use formulas from (4.4) taking into account however that now W is defined by (5.1) and $W^{\uparrow} := JW$. Continuing to use the old notation define $G(t)$ by (4.8). Then thanks to the hypothesis assumed above $G(t) \equiv 1$ on $[\alpha_l; \alpha_{l+1})$ if this interval contains a regular singularity. Define $\nu(t)$ in the following way: set $\nu(t) := \int_{-1}^{t} G(t)d\sigma(t)$ for every t that is less than the first singular peculiarity, say λ_{j_0}; next, set $\nu(t) := \int_{\alpha_{j_0+1}}^{t} G(t)d\sigma(t)$ for every t greater than λ_{j_0} and less than the next singular peculiarity, say λ_{j_1}, etc. Finally, set

$$\eta(t) := \int_{-1}^{t} (1/G(\tau))d\sigma(\tau), \quad t \in [-1; 1].$$

Thus, the space $L^1_\sigma + L^2_\eta$ is well defined.

The space L^2_ν can be defined via the equality $L^\infty_\sigma \cap L^2_\nu = (L^1_\sigma + L^2_\eta)^*$ or via the following description. Let $f(t)$ and $g(t)$ be arbitrary continuous scalar functions defined on $[-1; 1]$ and vanishing near all singular peculiarities from Λ. Let $(f(t), g(t) := \int_{-1}^{1} f(t)\bar{g}(t)d\sigma(t)$. The set of all these functions equipped with the given scalar product forms a pre-Hilbert space. Its completion is by definition the space L^2_ν.

Pass to descriptions of some operator subalgebras of $\mathfrak{A}$. Let $\varphi(t)$ be a continuous scalar function vanishing near Λ. Set

$$B_\varphi = \int_{-1}^{1} \varphi(t)dE_\lambda, \tag{5.2}$$

where the improper integral has the obvious meaning.

Let $\mathfrak{S}(E_\lambda)$ be a collection of operators $\{S\}$ satisfying the conditions

a) $S = P_1 S P_0$;

b) if $\sum_{l,j=1}^{k} \alpha_{lj} \iota_{lj}(t) \in L^1_\sigma + L^2_\eta$, then $\sum_{l,j=1}^{k} \alpha_{lj}(Sh_l, e_j) = 0$.

These conditions coincide formally with ones given in (4.6) but really are a generalization of the last due to the arbitrariness in the location of Λ.

Up to this point it was assumed that Λ contains at least one singular peculiarity. Now we describe how to introduce function and operator sets needed below if there are no singular peculiarities in Λ. In particular, in this case $\nu(t) \equiv \eta(t) \equiv \sigma(t)$, so $L^\infty_\sigma \cap L^2_\nu = L^\infty_\sigma$ and $L^1_\sigma + L^2_\eta = L^1_\sigma$, moreover $\mathfrak{S}(E_\lambda) = \{0\}$. At the same time Formula (5.2) can be applied independently to the characterization in question of Λ.

Denote $\mathfrak{A}_\Lambda$ the weak closure of the operator set $\{B_\varphi\}$ generated by (5.2). Let $\mathcal{G}_\psi(E_\lambda)$ be the subset of the operators from $\mathfrak{A}_\Lambda$ which are originals for $\psi(t)$ with respect to E_λ.

Theorem 5.1. *An operator A belongs to $\mathfrak{A}_\Lambda$ if and only if the following conditions hold:*

$$\left.\begin{array}{l}\text{a) } AE_\lambda = E_\lambda A \text{ for every } \lambda \in [-1;1]\backslash\Lambda;\\ \text{b) } A|_{\widetilde{\mathfrak{H}}^{[\perp]}} = 0,\ A\mathfrak{H} \subset \widetilde{\mathfrak{H}};\\ \text{c) there is a function } \varphi(t) \in L^\infty_\sigma \cap L^2_\nu, \text{ such that } \varphi(t) \text{ is the}\\ \quad \text{portrait of } A \text{ with respect to } E_\lambda;\\ \text{d) if } \mathfrak{H}_1 \neq \{0\} \text{ and } \psi(t) = \sum_{l,j=1}^k \alpha_{lj}\iota_{lj}(t) \in L^1_\sigma + L^2_\eta, \text{ then}\\ \qquad \displaystyle\sum_{l,j=1}^k \alpha_{lj}(Ah_l, e_j) = \int_{-1}^1 \varphi(t)\psi(t)d\sigma(t).\end{array}\right\} \tag{5.3}$$

This theorem is, evidently, a generalization of Theorem 4.7 and is its corollary if one applies Theorem 4.7 separately for every singular peculiarity from Λ. If J-orth.sp.f. E_λ has no singular peculiarities, then $\mathfrak{H}_1 = \{0\}$ and Theorem 5.1 coincides in fact with the corresponding theorem for self-adjoint operators.

Corollary 5.2. *For every function $\varphi(t) \in L^\infty_\sigma \cap L^2_\nu$ the inequality $\mathcal{G}_\varphi(E_\lambda) \neq \emptyset$ holds and if $B \in \mathcal{G}_\varphi(E_\lambda)$ is a fixed operator, then $\mathcal{G}_\varphi(E_\lambda) = \{B + S\}_{S\in\mathfrak{S}(E_\lambda)}$; in particular, $\mathcal{G}_0(E_\lambda) = \mathfrak{S}(E_\lambda)$.*

Corollary 5.3. *$\mathfrak{S}(E_\lambda)$ coincides with the strong sequential closure of the set of operators given by (5.2).*

Theorem 5.4. *For a commutative WJ^*-algebra $\mathfrak{A} \in D^+_\kappa$ there is a finite collection of J-s.a. operators $A_1, A_2, \dots, A_l \in \mathfrak{A}$, such that every operator $B \in \mathfrak{A}$ has the representation*

$$B = S + Q(A_1, A_2, \dots, A_l) + C, \tag{5.4}$$

where $S \in \mathfrak{A}$ is a nilpotent operator, $S|_{\widetilde{\mathcal{H}}} = S^\#|_{\widetilde{\mathcal{H}}} = 0$, $Q(t_1, t_2, \dots, t_l)$ is a polynomial in l variables, and $C \in \mathfrak{A}_\Lambda$.

Proof. If $E(\mathbb{R}) = I$, one can take into account Theorem 5.1 and repeat the scheme used for the proof of Theorem 4.11. If $E(\mathbb{R}) \neq I$, then the first step step is the following. Find J-s.a. operators $\widehat{A}_1, \widehat{A}_2, \dots, \widehat{A}_l \in \widehat{\mathfrak{A}} = \mathfrak{A}|_{E(\mathbb{R})\mathcal{H}}$, such that every operator $\widehat{B} \in \widehat{\mathfrak{A}}$ has a representation of the type (5.4), say $\widehat{B} = \widehat{S} + \widehat{Q}(\widehat{A}_1, \widehat{A}_2, \dots, \widehat{A}_l) + \widehat{C}$. Next, consider the algebra $\breve{\mathfrak{A}} := \mathfrak{A}|_{\mathcal{H}_3}$ with $\mathcal{H}_3 := (I - E(\mathbb{R}))\mathcal{H}$. Since $\mathcal{H}_3$ is finite-dimensional, the latter algebra has a finite number of generators, so there exists a collection of J-s.a. operators $\breve{A}_{l+1}, \breve{A}_{l+2}, \dots, \breve{A}_n \in \breve{\mathfrak{A}}$, such that every operator $\breve{B} \in \breve{\mathfrak{A}}$ has a representation $\breve{B} = \breve{Q}(\breve{A}_{l+1}, \breve{A}_{l+2}, \dots, \breve{A}_n)$, where $\breve{Q}(t_{l+1}, t_{l+2}, \dots, t_n)$ is a polynomial. Let $x \in \mathcal{H}$. Put $A_j x := \widehat{A}_j E(\mathbb{R})x$ if $j = 1, 2, \dots, l$, and $A_j x := \breve{A}_j(I - E(\mathbb{R}))x$ whenever $j = l+1, l+2, \dots, n$. The rest is straightforward. □

Corollary 5.5. *Let $\mathfrak{A} \in D^+_\kappa$ be a commutative WJ^*-algebra and let $\mathcal{F}(\mathfrak{A})$ be the set of all portraits corresponding to operators in $\mathfrak{A}$. Then there is a finite* *collection of J-s.a. operators* $A_1, A_2, \dots, A_l \in \mathfrak{A}$ *with corresponding portraits* $f_{A_1}(\lambda), f_{A_2}(\lambda), \dots, f_{A_l}(\lambda)$*, such that* $f(\lambda) \in \mathcal{F}(\mathfrak{A})$ *if and only if*

$$f(\lambda) = Q(f_{A_1}(\lambda), f_{A_2}(\lambda), \dots, f_{A_l}(\lambda)) + \phi(\lambda),$$

where $Q(t_1, t_2, \dots, t_l)$ *is a polynomial in* l *variables, and* $\phi(\lambda) \in L^\infty_\sigma \cap L^2_\nu$*.*

Theorem 5.6. *Let* WJ^**-algebra* $\mathfrak{A} \in D^+_\kappa$ *be such that* $\mathfrak{A} = \operatorname{Alg} A$, $A = A^\#$*. Then every operator* $B \in \mathfrak{A}$ *has a representation*

$$B = Q(A) + C, \tag{5.5}$$

where $Q(t)$ *is a polynomial,* $C \in \mathfrak{A}_\Lambda$*. Conversely, if an operator* B *is as in* (5.5)*, then* $B \in \mathfrak{A}$*.*

Proof. With no loss of generality one can take the e.s.f. E_λ of A as an e.s.f of the algebra $\mathfrak{A}$. Next, let $\Lambda = \{\lambda_1, \lambda_2, \dots, \lambda_m\}$, and let open intervals $\Delta_1, \Delta_2, \dots, \Delta_m \subset (-1; 1)$ be disjoint and such that $\Delta_j \cap \Lambda = \{\lambda_j\}$, $j = 1, 2, \dots, m$. Then the algebra $\mathfrak{A}|_{E(\Delta_j)\mathcal{H}}$ coincides with $\operatorname{Alg}((A - \lambda_j I)|_{E(\Delta_j)\mathcal{H}})$. If λ_j is a singular peculiarity, then Theorem 4.23 is applicable, so

$$B|_{E(\Delta_j)\mathfrak{H}} = Q_j(A|_{E(\Delta_j)\mathfrak{H}}) + C_j, \quad C_j \in \mathfrak{A}_\Lambda|_{E(\Delta_j)\mathfrak{H}}. \tag{5.6}$$

If λ_j is a regular spectral peculiarity, then there exists the limit $E(\{\lambda_j\}) = \operatorname{s-}\lim\limits_{\epsilon \to +0} E((\lambda_j - \epsilon; \lambda_j + \epsilon))$ and the algebra $\mathfrak{A}|_{E(\{\lambda_j\})\mathfrak{H}}$ is finite-dimensional, so

$$B|_{E(\{\lambda_j\})\mathfrak{H}} = Q_j(A)|_{E(\{\lambda_j\})\mathfrak{H}}. \tag{5.7}$$

Finally, if $E(\mathbb{R}) \neq I$, then the algebra $\mathfrak{A}|_{(I - E(\mathbb{R}))\mathfrak{H}}$ is also non-trivial and finite-dimensional, so

$$B|_{(I - E(\mathbb{R}))\mathfrak{H}} = Q_{m+1}(A)|_{(I - E(\mathbb{R}))\mathfrak{H}}. \tag{5.8}$$

Now let $Q(t)$ be a polynomial "interpolating" the set of polynomials $\{Q_j(t)\}_1^{m+1}$ in the following sense. If λ_j is a singular peculiarity, then, first, the integral $\int_{\Delta_j} (Q(t) - Q_j(t)) dE_t$ unconditionally converges in the strong topology and, second, $Q(A|_{E(\Delta_j)\mathfrak{H}}) - Q_j(A|_{E(\Delta_j)\mathfrak{H}}) = \int_{\Delta_j} (Q(t) - Q_j(t)) dE_t|_{E(\Delta_j)\mathfrak{H}}$, where $Q_j(t)$ corresponds (5.6). If λ_j is a regular spectral peculiarity, then $Q(A|_{E(\{\lambda_j\})\mathfrak{H}}) = Q_j(A|_{E(\{\lambda_j\})\mathfrak{H}})$, where $Q_j(t)$ corresponds to (5.7). Finally, if $E(\mathbb{R}) \neq I$, then $Q(A)|_{(I - E(\mathbb{R}))\mathcal{H}} = Q_{m+1}(A)|_{(I - E(\mathbb{R}))\mathcal{H}}$. Note that if $E(\mathbb{R}) = I$, then the latter condition for $Q(t)$ as well as Representation (5.8) must be omitted. The rest is straightforward. □

Corollary 5.7. *Let* $A = A^\# \in D^+_\kappa$ *and let* $\mathcal{F}(A)$ *be the set of all portraits corresponding to operators in* $\operatorname{Alg} A$*. Then* $f(\lambda) \in \mathcal{F}(A)$ *if and only if*

$$f(\lambda) = Q(\lambda) + \phi(\lambda),$$

where $Q(\lambda)$ *is a polynomial and* $\phi(\lambda) \in L^\infty_\sigma \cap L^2_\nu$*.*

Up to this point we have assumed that $\mathfrak{A}$ is a WJ^*-algebra. Now we omit this preestablished condition and pass to conditions guaranteeing a weakly closed algebra to be a WJ^*-algebra.

Theorem 5.8. *Let an algebra $\mathfrak{A} \in D^+_\kappa$ be weakly closed, commutative and J-symmetric. Then $\mathfrak{A}$ is a WJ^*-algebra if and only if it is non-degenerate, i.e., for every vector $x \in \mathfrak{H}$ there is an operator $A \in \mathfrak{A}$, such that $Ax \neq 0$.*

Proof. One need only show that the identity belongs to a weakly closed commutative non-degenerate J-symmetric algebra $\mathfrak{A} \in D^+_\kappa$ because the remaining assertions of the theorem are trivial. Let, as above, a subspace $\mathcal{L}_+ \in \mathfrak{M}^+(\mathcal{H}) \cap h^+$ be $\mathfrak{Y}$-invariant and $\mathcal{L}_1 = \mathcal{L}_+ \cap \mathcal{L}_+^{[\perp]}$. Denote $\Omega_\lambda(A)$ the subspace of all eigenvectors and root vectors of an operator A that correspond to $\lambda \in \sigma_p(A)$. If $A = A^\#$, $\lambda \in \sigma(A|_{\mathcal{L}_1})$, $\lambda \neq 0$, then (recall that $\mathcal{L}_1$ is finite-dimensional) there is a polynomial $Q(t)$ with real coefficients such that $Q(0) = 0$ (0 here is a number!), $Q(A)x = x$ for every $x \in \Omega_\lambda(A)$ or (if $\lambda \notin \mathbb{R}$) $x \in \Omega_{\overline{\lambda}}(A)$ and $Q(A)y = 0$ for $y \in \Omega_\mu(A)$ if $\mu \in \sigma(A|_{\mathcal{L}_1})$, $\mu \neq \lambda, \overline{\lambda}$.

Thus ($\mathfrak{A}$ is non-degenerate and commutative!), there exists a J-s.a. operator A such that $A|_{\mathfrak{L}_1} = I|_{\mathfrak{L}_1}$. Fix this A. Let E_λ be the e.s.f. of A. Note that all spectral peculiarities of E_λ belongs to $\sigma(A|_{\mathcal{L}_1})$, and if $\lambda \notin \mathbb{R}$ and $\lambda \in \sigma_p(A)$, then $\lambda \in \sigma(A|_{\mathcal{L}_1})$ or $\overline{\lambda} \in \sigma(A|_{\mathcal{L}_1})$. But by the hypothesis $\sigma(A|_{\mathcal{L}_1}) = \{1\}$, therefore the spectrum of A is real and the unique spectral peculiarity (if any) of E_λ is $\lambda_0 = 1$. Moreover, the algebra $\mathfrak{A}|_{E_{1/2}\mathfrak{H}}$ has a pair of J-orthogonal uniformly definite invariant subspaces and therefore is a W^*-algebra with respect to a suitable canonical scalar product. So, the proof can be reduced to the testing of the condition $I - E_{1/2} \in \mathfrak{A}$. Let

$$\mathcal{N}(t) = 1 - (t-1)^{2n}; \quad \psi(t) = \begin{cases} -1/2, & t \leq -1/4; \\ 2t, & t \in (-1/4; 1/2); \\ 1, & t \geq 1/2; \end{cases}$$

$$\varphi(t) = \psi(t) - \mathcal{N}(t); \; \theta(t) = \begin{cases} \varphi(t)/t, & t \neq 0; \\ 2(1-n), & t = 0. \end{cases}$$

For a suitable choice of n (see Remark 4.19, Formula (4.23), Condition (4.24) and Proposition 4.20) there is a sequence of polynomials $\{Q_l(t)\}_1^\infty$, such that $\lim\limits_{l\to\infty} \|Q_l(A) - \theta(A)\| = 0$, where $\theta(A) = \int_{\mathbb{R}\setminus\{1\}} \theta(t)dE_t$, and the integral is treated in the same way as in (4.23). This implies that $\lim\limits_{l\to\infty} \|AQ_l(A) - A\theta(A)\| = 0$. At the same time $A\theta(A) = \varphi(A) = \int_{\mathbb{R}\setminus\{1\}} \varphi(t)dE_t$, so $\varphi(A) \in \mathfrak{A}$. Let $B := \varphi(A) + \mathcal{N}(A)$. Since $\mathcal{N}(0) = 0$, then $B \in \mathfrak{A}$. A direct verification shows that $BE_{1/2} = (-1/2)E_{-1/2} + \int_0^{1/2} 2t dE_t$, $B(I - E_{1/2}) = (I - E_{1/2})$. This implies that $\text{s}-\lim\limits_{l\to\infty} B^l = (I - E_{1/2})$. □

Note that the condition of commutativity of the algebra $\mathfrak{A}$ in the assertion of Theorem 5.8 is essential even if $\mathfrak{A}$ is defined on a finite-dimensional space.

Example 5.9. Assume that x_1, x_2, x_3, x_4 is a basis in a four-dimensional space. Make up an algebra $\mathfrak{A} = \{A\}$ spanning four operators:

A_1: $A_1x_1 = x_2$, $A_1x_2 = x_4$, $A_1x_3 = A_1x_4 = 0$; A_2: $A_2x_1 = A_2x_2 = A_2x_4 = 0$, $A_2x_3 = x_4$; A_3: $A_3x_1 = x_3$, $A_3x_2 = A_3x_3 = A_3x_4 = 0$; A_4: $A_4x_1 = x_1$, $A_4x_2 = x_2$, $A_4x_3 = 0$, $A_4x_4 = x_4$.

Give a canonical symmetry J by the following equalities: $Jx_1 = x_4$, $Jx_4 = x_1$, $Jx_2 = x_2$, $Jx_3 = x_3$.

A direct verification shows that $A_1^{\#} = A_1$, $A_2^{\#} = A_3$, $A_4^{\#} = A_4$, so the given algebra is J-symmetric. It is also clear that $\mathfrak{A}$ is non-degenerate. A calculation gives $A_1A_2 = A_1A_3 = A_2A_1 = A_2^2 = A_2A_4 = A_3A_1 = A_3A_2 = A_3^2 = A_4A_3 = 0$, $A_1A_4 = A_4A_1 = A_1$, $A_3A_4 = A_3$, $A_4A_2 = A_2$, $A_4^2 = A_4$. Finally, only two products lead to a new operator: $A_1^2 = A_2A_3 = A_5$, where A_5: $A_5x_1 = x_4$, $A_5x_2 = A_5x_3 = A_5x_4 = 0$. Since $A_1A_5 = A_5A_1 = A_2A_5 = A_5A_2 = A_3A_5 = A_5A_3 = A_5^2 = 0$, $A_5A_4 = A_4A_5 = A_5$, every operator $A \in \mathfrak{A}$ is a linear combination of the operators A_1, A_2, A_3, A_4 and A_5. Thus $I \notin A$.

6. Closing remarks

Spectral functions are a traditional source for creating a functional calculus for different classes of normal operators in Hilbert spaces, so it is natural to use the same technique in Krein spaces. The first theorem concerning the existence of spectral functions for π-s.a. operators was published by M.Krein and H. Langer in [23] (see also [26] for detailed proofs). The spectral functions introduced in Subsection 1.3 are a particular case of generalized spectral functions [12]. Our definition is motivated by the spectral measures arising in Operator Theory in Indefinite Metric Spaces but formally is independent (see [37]) of this Theory. A functional calculus based on the spectral resolution of so-called definitizable J-s.a. and J-unitary operators was given for the first time by H. Langer [25], [26]. Jonas [18] has results of a similar nature. The class D_κ^+ that is considered in the present paper as well as the class $K(H)$ (introduced by Azizov [3]) differ from the class of definitizable operators (see [7] for discussion) and represent another natural generalization of the class of self-adjoint operators in Pontryagin spaces, since (as follows from one of the theorems of Naimark [30]) every commutative family of π-s.a. (or π-unitary) operators belongs to the class D_κ^+. The conception of unbounded elements was introduced in [35] (see also [36]). A connection between the spaces of the type $L_\sigma^\infty \cap L_\nu^2$ and operator algebras was pointed out for the first time in [31] (for Pontryagin spaces of range 1) and [25]. Some part of Sections 4 and 5 represents a detailed account of results announced in [39]. Subsection 4.3 is a generalization of results announced in [38], where the case of a monogenic algebra in an arbitrary Pontryagin space was analyzed. The subject of this paper is very close to the theory of model representations for π-s.a. operators and algebras (a majority of works consider the case of range 1) [32], [29], [41], [28] (see also [17] for more references), [27]. The papers [19] and [20] deal with the case of a self-adjoint cyclic operator acting in arbitrary Pontryagin space.

Acknowledgment

The author would like to thank the Organizing Committee of the Vienna Colloquium for the opportunity to participate in this notable scientific event. He also is very grateful to Prof. Tom Berry for his assistance to making the paper more readable.

References

[1] N.I. Akhiezer, I.M. Glazman, *Theory of linear operators in Hilbert space.* Pitman: London, 1981.

[2] T.Ya. Azizov, I.S. Iokhvidov, *Linear operators in spaces with an indefinite metric.* "Matematicheskii analiz" (Itogi nauki i tehniki, VINITI) **17** (1979) Moscow, 113–205 (Russian).

[3] T.Ya. Azizov, I.S. Iokhvidov, *Foundation of the Theory of Linear Operators in Spaces with Indefinite Metric.* Nauka: Moscow, 1986 (Russian); *Linear Operators in Spaces with Indefinite Metric.* Wiley: New York, 1989 (English).

[4] T.Ya. Azizov, V.A. Strauss, *Spectral decomposition of self-adjoint and normal operators in Krein space.* Research Paper No 827 (2002) Department of Mathematics and Statistics, University of Calgary, Calgary, Canada.

[5] T.Ya. Azizov, V.A. Strauss, *Spectral decompositions for special classes of self-adjoint and normal operators on Krein spaces.* Spectral Theory and its Applications, Proceedings dedicated to the 70th birthday of Prof. I. Colojoară, Theta 2003, 45–67.

[6] T.Ya. Azizov, V.A. Strauss, *On a spectral decomposition of a commutative operator family in spaces with indefinite metric.* MFAT **11** (2005) No 1, 10–20.

[7] T.Ya. Azizov, L.I. Sukhocheva, V.A. Shtraus (=Strauss), *Operators in Krein space.* Matematicheskie Zametki **76** (2004) No 3, 324–334 (Russian); Math. Notes **76** (2004) No 3, 306–314 (English).

[8] J. Bergh, J. Löfström, *Interpolation spaces. An Introduction.* Springer Verlag, NY, 1976.

[9] J. Bognar, *Indefinite inner product spaces.* Springer-Verlag, NY, 1974.

[10] M.S. Birman, M.Z. Solomyak, *Spectral theory of self-ajoint operators in a Hilbert space.* Leningrad University, 1980 (Russian).

[11] O. Bratteli, D. W. Robinson, *Operator Algebras and Quantum Statistical Mechanics. Volume 1.* Springer-Verlag, NY, 1979.

[12] I. Colojoară, C. Foiaş, *Theory of generalized spectral operators.* Gordon and Breach, 1968.

[13] N. Dunford, J.T. Schwartz, *Linear Operators. General Theory.* John Wiley & Sons, 1958.

[14] N. Dunford, J.T. Schwartz, *Linear Operators. Part III. Spectral operators.* John Wiley & Sons, 1971.

[15] A. Gheondea, *Pseudo-regular spectral functions in Krein spaces.* J. Operator Theory **12** (1984), 349–358.

[16] I.S. Iokhvidov, M.G. Krein, H. Langer, *Introduction to the spectral theory of operators in spaces with an indefinite metric.* Akademie-Verlag, Berlin, 1982.

[17] R.S. Ismagilov, M.A. Naimark, *Representations of groups and algebras in spaces with indefinite metric.* "Matematicheskii analiz" (Itogi nauki i tehniki 1968, VINITI) (1969), Moscow, 73–105 (Russian).

[18] P. Jonas, *On the functional calculus and the spectral function for definitizable operators in Krein space.* Beitr. Anal. **16** (1981), 121–135.

[19] P. Jonas, H. Langer, *A model for π-selfadjont operators in π_1-spaces and a special linear pencil.* Integral Equations Opererator Theory **8** (1985) No 1, 13–35.

[20] P. Jonas, H. Langer, B. Textorius, *Models and unitary equivalence of ciclic selfadjoint operators in Pontrjagin spaces.* Operator Theory and complex analysis (Sapporo, 1991), 252–284, Oper. Theory Adv. Appl. **59** Birkhäuser, Basel, 1992.

[21] L.V. Kantorovich, G.P. Akilov, *Functional analysis, 2nd edition.* Nauka (Moscow) 1977.

[22] E. Kissin, V. Shulman, *Representations of Krein spaces and derivations of C^*-algebras.* Pitman Monographs and Surveys in Pure and Applied Mathematics **89** Addison-Vesely Longman, 1997.

[23] M.G. Krein, H. Langer, *On the spectral function of a self-adjoint operator in a space with indefinite metric.* Dokl. Akad. Nauk SSSR, **152** (1963), 39–42 (Russian).

[24] S.G. Krein, Yu.I. Petunin, Ye.M. Semyonov, *Interpolation of linear operators.* Nauka, Moscow, 1978 (Russian).

[25] H. Langer, *Spectraltheorie linearer Operatoren in J-räumen und enige Anwendungen auf die Shar $L(\lambda) = \lambda^2 I + \lambda B + C$.* Habilitationsschrift, Tech. Univer., Dresden, 1965.

[26] H. Langer, *Spectral functions of definitizable operators in Krein space.* Lect. Not. Math. **948** (1982), 1–46.

[27] H. Langer, B. Textorius, *L-resolvent matrices of symmetric linear relations with equal defect numbers; applications to canonical differential relations.* Integral Equations Operator Theory **5** (1982) No 2, 208–243.

[28] S.N. Litvinov, *Description of commutative symmetric algebras in the Pontryagin space Π_1.* DAN UzSSR (1987) No 1, 9–12 (Russian).

[29] A.I. Loginov, *Complete commutative symmetric algebras in Pontryagin spaces Π_1.* Mat. Sbornik **84** (1971) No 4, 575–582 (Russian).

[30] M.A. Naimark, *On commutative unitary operators in a space Π_κ.* Dokl. Akad. Nauk SSSR **149** (1963), 1261–1263 (Russian).

[31] M.A. Naimark, *On commutative operator algebras in a space Π_1.* Dokl. Akad. Nauk SSSR **156** (1964), 734–737 (Russian).

[32] M.A. Naimark, *Commutative algebras of operators in a space Π_1.* Rev. roum. math. pures et appl. **9** (1964) No 6, 499–529.(Russian)

[33] M.A. Naimark, *Normed Algebras.* Wolters-Nordhoff Publishing, Groningen, The Nedherlands, 1972.

[34] M. Reed, B. Simon *Methods of Modern Mathematical Physics. I: Functional Analysis.* 2nd Edition, Acad. Press, Inc., 1980.

[35] V.A. Strauss, *Some singularities of the spectral function of a π-selfadjoint operator.* Collection: Funktsionalnii analiz. Operator Theory, State Pedagogical Institute of Uliyanovsk, Ulyanovsk, USSR (1983), 135–146 (Russian)

[36] V.A. Strauss, *A model representation for a simplest π-selfadjoint operator.* Collection: Funktsionalnii analiz. Spectral theory, State Pedagogical Institute of Uliyanovsk, Ulyanovsk, USSR (1984), 123–133 (Russian).

[37] V.A. Strauss, *Integro-polynomial representation of regular functions of an operator whose spectral function has critical points.* Dokl. Akad. Nauk Ukrain. SSR. Ser. A **8** (1986), 26–29 (Russian).

[38] V.A. Strauss, *Functional representation of an algebra genereted by a selfadjoint operator in Pontryagin space.* Funktsionalnyi analiz i ego prilozheniya **20** (1986) No 1, 91–92 (Russian), English translation: Funct. Anal. Appl.

[39] V.A. Strauss, *The structure of a family of commuting J-self-adjoint operators* Ukrain. Mat.Zh., **41** (1989), No. 10, 1431–1433, 1441 (Russian).

[40] V.A. Strauss, *On models of functional type for a special class of commutative symmetric operator families in Krein spaces.* Submitted to Proceedings 4th Workshop Operator Theory in Krein Spaces and Applications, Berlin, 2004.

[41] V.S. Shulman, *Symmetric Banach algebras of operators in a space of type* Π_1. Mat. Sb. (N.S.) **89** (1972) No 2, 264–279 (Russian).

Vladimir Strauss
Department of Pure & Applied Mathematics
USB, Sartenejas-Baruta
Apartado 89.000 Caracas
1080-A, Venezuela
e-mail: str@usb.ve

Operator Theory:
Advances and Applications, Vol. 163, 337–350

On Normal Extensions of Unbounded Operators: IV. A Matrix Construction

Franciszek Hugon Szafraniec

to Heinz again, with pleasure

Abstract. A condition for an unbounded operator to have a normal extension, which is a matrix operator, is given. The circumstances under which this condition may become necessary are discussed as well and finally a question is posed. By the way some substantial facts concerning infinite operator matrices with unbounded entries are gathered.

Mathematics Subject Classification (2000). Primary 47B15, 47B12 .

Keywords. Normal operator, subnormal operator, matrix operator.

Bounded subnormals constitute a vital class in the theory of operators in Hilbert space. Furthermore, unbounded subnormal operators are present in quantum mechanic, see [5, Ch. III, Paragraph 10] and [12], which lays stress on their importance. Though the unbounded case bears the central ideas of subnormality, it causes a lot of hassle in implementing them. Some basic facts concerning the theory can be found in the trilogy [7], [8] and [9] as the most recent and far going results are in [10]. Unfortunately, all of them concern the case when an operator in question has invariant domain. The only results without this troublesome restriction are in [3] and [4].

In [1] Andô has answered to the need of providing a universal construction of a normal extension, independent of an operator one deals with. While Andô's approach has been intended as a way of shedding more light on normal extensions of *bounded* operators, ours, which bases upon that in [1], creates a chance to add one more criterion for subnormality of *unbounded* operators to the very few existing so far.

This work was supported by the KBN grant 2 P03A 037 024.

So as to achieve the goal a weighty piece of theory of operators coming from infinite *matrices* with *unbounded* entries has to be built up from scratch[1]. Description of the adjoint of those operators is especially underlined and it turns out column operators come into play in a natural way. This part may be of independent interest.

Let us make

Proclamation. All the Hilbert spaces are complex and the operators considered in them are densely defined.

Operator matrices versus matrix operators

1. Let $\mathcal{K} = \bigoplus_{n=0}^{\infty} \mathcal{H}_n$ be the orthogonal sum of Hilbert spaces $\mathcal{H}_n$ and let P_n stand for the orthogonal projection of $\mathcal{K}$ onto $\mathcal{H}_n$, $n \geq 0$. Denote by $\mathcal{K}_{\mathrm{fin}}$ the space of all vectors $f \in \mathcal{K}$ such that each $P_n f \in \mathcal{H}_n$ and $P_n f = 0$ for all but a finite number of indices n.

◇ We identify $g \in \mathcal{H}_n$ with $0 \oplus \cdots \oplus 0 \oplus g \oplus 0 \oplus \cdots \in \mathcal{K}$ if g is properly placed. This, in return, legitimizes us to write $P_n f \in \mathcal{H}_n$ for $f \in \mathcal{K}$. ◇

Remark 1. Let A be a closable operator[2] in $\mathcal{K}$ such that

$$P_n \mathcal{D}(A) \subset \mathcal{D}(A), \quad n = 0, 1, \ldots \tag{1}$$

Then for $f \in \mathcal{D}(A)$ and $g \in \mathcal{K}$ such that $P_n g \in \mathcal{D}(A^*)$ for all n

$$\begin{aligned}
\langle Af, g\rangle_{\mathcal{K}} &= \langle \bigoplus_{n=0}^{\infty} P_n Af, \bigoplus_{n=0}^{\infty} P_n g\rangle_{\mathcal{K}} = \sum_{n=0}^{\infty} \langle P_n Af, P_n g\rangle_{\mathcal{H}_n} = \sum_{n=0}^{\infty} \langle Af, P_n g\rangle_{\mathcal{K}} \\
&= \sum_{n=0}^{\infty} \langle f, A^* P_n g\rangle_{\mathcal{K}} = \sum_{n=0}^{\infty} \langle \bigoplus_{k=0}^{\infty} P_k f, \bigoplus_{k=0}^{\infty} P_k A^* P_n g\rangle_{\mathcal{K}} \\
&= \sum_{n=0}^{\infty}\sum_{k=0}^{\infty} \langle P_k f, P_k A^* P_n g\rangle_{\mathcal{H}_k} = \sum_{n=0}^{\infty}\sum_{k=0}^{\infty} \langle P_n A P_k f, P_n g\rangle_{\mathcal{H}_n} \\
&= \langle \bigoplus_{n=0}^{\infty}\sum_{k=0}^{\infty} P_n A P_k f, g\rangle_{\mathcal{K}}.
\end{aligned} \tag{2}$$

---

[1] A common association with *infinite* matrices is that of Jacobi ones; for some treatment in this matter we suggest to look into [11], [13] (scalar entries) and [2], [6] (bounded operator entries), the latter considers a bit more general matrices. Matrices with unbounded operator entries are dealt with rather occasionally, usually of finite dimension, and a thorough exposition of the topic would be difficult to point out in the literature.

[2] Notation: $\mathcal{D}(A)$, $\mathcal{R}(A)$, $\mathcal{N}(A)$, $\bar{A}$ and A^* stands for the domain, range, kernel, closure and adjoint of an operator A in a Hilbert space $\mathcal{H}$, resp. If $\mathcal{D} \subset \mathcal{D}(A)$ is dense in $\mathcal{H}$ then $A|_{\mathcal{D}}$ denotes the restriction of A to $\mathcal{D}$, that is an operator in $\mathcal{H}$ with domain $\mathcal{D}$.

The last equality is a heuristical step, anyway if this happens then

$$Af = \bigoplus_{n=0}^{\infty} \sum_{k=0}^{\infty} P_n A P_k f, \quad f \in \mathcal{D}(A),$$

as long as condition (1) holds. This shows the way the operator A determines a matrix $(P_i A P_j)_{i,j=0}^{\infty}$ of operators or, in other words, an *opcrator matrix.*

Formula (2) suggests the way back: from matrix to operator; however it requires to agree much in advance on the domain of the adjoint of the prospective operator; also the condition (1) would be difficult to meet when the operator is already closed. The case presented below tries to minimize this annoyance.

Suppose a matrix $\boldsymbol{A} = (A_{i,j})_{i,j=0}^{\infty}$ of closable operators $A_{i,j}$ from (a dense subspace of) $\mathcal{H}_j$ into $\mathcal{H}_i$ is given. Suppose, moreover, a sequence $\boldsymbol{D} = (\mathcal{D}_n)_{n=0}^{\infty}$ of linear spaces such that each D_n is dense in $\mathcal{H}_n$ and each $\mathcal{D}_k \subset \mathcal{D}(A_{n,k})$ for all n, is chosen. Apparently, the linear space

$$\mathcal{D}_{\mathrm{fin}} \stackrel{\mathsf{def}}{=} \mathcal{K}_{\mathrm{fin}} \cap \bigoplus_{n=0}^{\infty} \mathcal{D}_n$$

is dense in $\mathcal{K}$.

Under these formalities we define in two steps a relevant operator in $\mathcal{K}$:

① the linear space $\mathcal{D}_{\boldsymbol{A},\boldsymbol{D}}$ is composed of all these $f \in \mathcal{D}_{\mathrm{fin}}$ for which[3]

$$\sum_{n=0}^{\infty} \| \sum_{k} A_{n,k} P_k f \|_{\mathcal{H}_n}^2 < +\infty,$$

if $\mathcal{D}_{\boldsymbol{A},\boldsymbol{D}}$ is dense in $\mathcal{K}$, we say the couple $(\boldsymbol{A},\boldsymbol{D})$ *determines an operator;*

② if $(\boldsymbol{A},\boldsymbol{D})$ determines an operator, set $\mathcal{D}(M_{\boldsymbol{A},\boldsymbol{D}}) \stackrel{\mathsf{def}}{=} \mathcal{D}_{\boldsymbol{A},\boldsymbol{D}}$ and define

$$M_{\boldsymbol{A},\boldsymbol{D}} f \stackrel{\mathsf{def}}{=} \bigoplus_{n=0}^{\infty} \sum_{k} A_{n,k} P_k f, \quad f \in \mathcal{D}(M_{\boldsymbol{A},\boldsymbol{D}}).$$

Then the operator $M_{\boldsymbol{A},\boldsymbol{D}}$ is said to be *determined* by the couple $(\boldsymbol{A},\boldsymbol{D})$ and one may think of it as a *matrix operator.*

Remark 2. Notice, under our circumstances, always $\mathcal{D}_{\boldsymbol{A},\boldsymbol{D}} \subset \mathcal{D}_{\mathrm{fin}}$. So,

$$\mathcal{D}_{\mathrm{fin}} \subset \mathcal{D}_{\boldsymbol{A},\boldsymbol{D}} \quad \Longleftrightarrow \quad \mathcal{D}(M_{\boldsymbol{A},\boldsymbol{D}}) = \mathcal{D}_{\mathrm{fin}};$$

if this happens the couple $(\boldsymbol{A},\boldsymbol{D})$ certainly determines an operator. In particular, this is the case the matrix $\boldsymbol{A}$ has only finite number of diagonals different from 0.

When one defines the $^{\times}$ operation on matrices by $\boldsymbol{A}^{\times} \stackrel{\mathsf{def}}{=} (A_{i,j}^{\times})_{i,j=}^{\infty}$, $A_{i,j}^{\times} \stackrel{\mathsf{def}}{=} A_{j,i}^{*}$ for all i,j, then apparently $\boldsymbol{A}^{\times\times} = \boldsymbol{A}^{-}$ where $\boldsymbol{A}^{-} \stackrel{\mathsf{def}}{=} (\overline{A_{i,j}})_{i,j=0}^{\infty}$.

[3] Dropping limits under the summation sign brings to mind it to be finite.

2. Now a need may appear to know more about the adjoint of an operator determined by a matrix $\boldsymbol{A}$. In the sequel we are going to shed some additional light on this. For this suppose another sequence $\boldsymbol{D}^\times = (\mathcal{D}_n^\times)_{n=0}^\infty$ of linear spaces such that each $D_n^\times$ is dense in $\mathcal{H}_n$ and each $\mathcal{D}_k^\times \subset \mathcal{D}(A_{n,k}^\times)$ for all n is chosen.

Proposition 3. *If the couples* $(\boldsymbol{A}, \boldsymbol{D})$ *and* $(\boldsymbol{A}^\times, \boldsymbol{D}^\times)$ *determine operators then the operators* $M_{\boldsymbol{A},\boldsymbol{D}}$ *and* $M_{\boldsymbol{A}^\times,\boldsymbol{D}^\times}$ *satisfy*

$$M_{\boldsymbol{A},\boldsymbol{D}} \subset M^*_{\boldsymbol{A}^\times,\boldsymbol{D}^\times},$$
$$M_{\boldsymbol{A}^\times,\boldsymbol{D}^\times} \subset M^*_{\boldsymbol{A},\boldsymbol{D}};$$

hence, they are closable.

Proof. Take $f \in \mathcal{D}(M_{\boldsymbol{A},\boldsymbol{D}})$ and $g \in \mathcal{D}(M_{\boldsymbol{A}^\times,\boldsymbol{D}^\times})$. Then using twice ②

$$\begin{aligned}\langle M_{\boldsymbol{A},\boldsymbol{D}} f, g\rangle_{\mathcal{K}} &= \sum_{n=0}^\infty \langle \sum_k A_{n,k} P_k f, P_n g\rangle_{\mathcal{H}_n} = \sum_{n=0}^\infty \sum_k \langle A_{n,k} P_k f, P_n g\rangle_{\mathcal{H}_n} \\ &= \sum_k \sum_{n=0}^\infty \langle P_k f, A_{n,k}^* P_n g\rangle_{\mathcal{H}_k} = \sum_k \langle P_k f, \sum_{n=0}^\infty A_{n,k}^* P_n g\rangle_{\mathcal{H}_k} \\ &= \langle f, M_{\boldsymbol{A}^\times,\boldsymbol{D}^\times} g\rangle_{\mathcal{K}},\end{aligned}$$

with the fourth equality due to the fact that $(\boldsymbol{A}^\times, \boldsymbol{D}^\times)$ determines an operator. Therefore we get

$$\langle M_{\boldsymbol{A},\boldsymbol{D}} f, g\rangle = \langle f, M_{\boldsymbol{A}^\times,\boldsymbol{D}^\times} g\rangle, \quad f \in \mathcal{D}(M_{\boldsymbol{A},\boldsymbol{D}}),\ g \in \mathcal{D}(M_{\boldsymbol{A}^\times,\boldsymbol{D}^\times})$$

which completes the proof of both inclusions. □

As an immediate consequence of Proposition 3 we get the following.

Corollary 4. *If either*

(a) $\mathcal{D}(M_{\boldsymbol{A},\boldsymbol{D}})$ *is a core*[4] *of* $M^*_{\boldsymbol{A}^\times,\boldsymbol{D}^\times}$,

or

(b) $\mathcal{D}(M_{\boldsymbol{A}^\times,\boldsymbol{D}^\times})$ *is a core of* $M^*_{\boldsymbol{A},\boldsymbol{D}}$,

then

$$\overline{M_{\boldsymbol{A},\boldsymbol{D}}} = M^*_{\boldsymbol{A}^\times,\boldsymbol{D}^\times} \quad \text{and} \quad \overline{M_{\boldsymbol{A}^\times,\boldsymbol{D}^\times}} = M^*_{\boldsymbol{A},\boldsymbol{D}}.$$

Consequently, both (a) and (b) must necessarily hold.

3. The following simple observation will be useful in the sequel.

Fact 5. *For closable operators A and B*

1° *if* $\mathcal{D}(A) \subset \mathcal{D}(B)$ *and* $\|Af\| = \|Bf\|$ *for* $f \in \mathcal{D}(A)$, *then* $\mathcal{D}(\bar{A}) \subset \mathcal{D}(\bar{B})$ *and* $\|\bar{A}f\| = \|\bar{B}f\|$ *for* $f \in \mathcal{D}(\bar{A})$. *If, in addition,* $\mathcal{D}(A)$ *is a core of* B, *then* $\mathcal{D}(\bar{A}) = \mathcal{D}(\bar{B})$;

2° *if* $\mathcal{D}(\bar{A}) = \mathcal{D}(\bar{B})$ *and* $\|\bar{A}f\| = \|\bar{B}f\|$ *for* $f \in \mathcal{D}(\bar{A}) = \mathcal{D}(\bar{B})$, *then* $\mathcal{D}(A)$ *is a core of* B.

[4] $\mathcal{D}$ is a core of a closable operator A if $A \subset \overline{A|_{\mathcal{D}}}$.

Our efforts so far are surmounted by the result which follows. Let us remind first that an operator N is said to be *normal* if $\mathcal{D}(N) = \mathcal{D}(N^*)$ and $\|Nf\| = \|N^*f\|$ for $f \in \mathcal{D}(N)$; N is called *essentially normal* if its closure $\overline{N}$ is normal.

Theorem 6. *Suppose both* $(\boldsymbol{A}, \boldsymbol{D})$ *and* $(\boldsymbol{A}^\times, \boldsymbol{D})$ *determine operators. If* $\mathcal{D}(M_{\boldsymbol{A},\boldsymbol{D}})$ *is a core of* $M_{\boldsymbol{A}^\times,\boldsymbol{D}}$ *as well as that of* $M^*_{\boldsymbol{A},\boldsymbol{D}}$ *(or, equivalently, that of* $M^*_{\boldsymbol{A}^\times,\boldsymbol{D}}$*) and*

$$\|M_{\boldsymbol{A},\boldsymbol{D}}f\| = \|M_{\boldsymbol{A}^\times,\boldsymbol{D}}f\|, \quad f \in \mathcal{D}(M_{\boldsymbol{A},\boldsymbol{D}}), \tag{3}$$

then $M_{\boldsymbol{A},\boldsymbol{D}}$ *is essentially normal.*

Proof. As $\mathcal{D}(M_{\boldsymbol{A},\boldsymbol{D}})$ is a core of $M_{\boldsymbol{A}^\times,\boldsymbol{D}}$ and (3) holds, using Fact 5 we infer that $\mathcal{D}(\overline{M_{\boldsymbol{A},\boldsymbol{D}}}) = \mathcal{D}(\overline{M_{\boldsymbol{A}^\times,\boldsymbol{D}}})$ with (3) extended to this common domain. On the other hand, as $\mathcal{D}(M_{\boldsymbol{A},\boldsymbol{D}})$ is a core of $M^*_{\boldsymbol{A},\boldsymbol{D}}$ Corollary 4 leads directly to normality of $\overline{M_{\boldsymbol{A},\boldsymbol{D}}}$. □

4. So as to utilize the above theorem we work out some conditions for the most worrying part of its assumptions to be satisfied. First we expand our considerations a bit to make them easier to comply with. Let $B = (B_n)_n$ be a finite or infinite sequence of operators defined on a common dense subspace $\mathcal{D}$ of a Hilbert space $\mathcal{L}$, taking values in $\mathcal{H}_n$ respectively and such that

$$\sum_n \|B_n x\|^2_{\mathcal{H}_n} < +\infty, \quad x \in \mathcal{D}. \tag{4}$$

Then the operator C_B from $\mathcal{D}$ into $\mathcal{K} = \bigoplus_n \mathcal{H}_n$ defined by $C_B x \stackrel{\text{def}}{=} \bigoplus_n B_n x$ makes sense.

Proposition 7. *Let* C_B *be as above. Then*

1° *for every* n *and* $f \in \mathcal{K}$

$$P_n f \in \mathcal{D}(C_B^*) \iff P_n f \in \mathcal{D}(B_n^*);$$

2° *with* $\mathcal{D}_B \stackrel{\text{def}}{=} \{f \in \bigcap_n P_n^{-1}(\mathcal{D}(B_n^*))\colon \sum_n B_n^* P_n f$ *converges weakly*$\}$, *the equality*

$$\mathcal{D}_B = \{f \in \mathcal{D}(C_B^*)\colon \; P_n f \in \mathcal{D}(C_B^*),\, n = 0, 1, \ldots\}$$

holds and

$$C_B^* f = \sum_n B_n^* P_n f, \quad f \in \mathcal{D}_B.$$

Proof. Part 1° is obvious. For 2° take $x \in \mathcal{D}$, $f \in \bigcap_n P_n^{-1}(\mathcal{D}(B_n^*))$ and write

$$\langle C_B x, f\rangle_{\mathcal{K}} = \sum_n \langle B_n x, P_n f\rangle_{\mathcal{H}_n} = \sum_n \langle x, B_n^* P_n f\rangle_{\mathcal{L}}.$$

Because $\sum_{n=0}^\infty B_n^* P_n f$ converges weakly, the vector f is in $\mathcal{D}(C_B^*)$. If $f \in \mathcal{D}(C_B^*)$ is such that $f \in \bigcap_n P_n^{-1}\mathcal{D}(C_B^*)$ then, due to 1°,

$$\langle x, C_B^* f\rangle_{\mathcal{L}} = \langle C_B x, f\rangle_{\mathcal{K}} = \sum_n \langle B_n x, P_n f\rangle_{\mathcal{H}_n} = \sum_n \langle x, B_n^* f\rangle_{\mathcal{L}}$$

and f is in $\mathcal{D}_B$. The rest of the conclusion follows then. □

Corollary 8. *If all the B_n's are zero but a finite number and $f \in \mathcal{D}(C_B^*)$ implies $P_n f \in \mathcal{D}(C_B^*)$ for every n, then*

$$C_B^* f = \sum_n B_n^* P_n f, \quad f \in \mathcal{D}(C_B^*) = \bigcap_n P_n^{-1}(\mathcal{D}(B_n^*)).$$

Remark 9. If only two of the B_n's are different from zero, say B_i and B_j, then $f \in \mathcal{D}(C_B^*)$ and $P_i f \in \mathcal{D}(C_B^*)$ implies $P_j f \in \mathcal{D}(C_B^*)$; this happens for instance when one of the operators is bounded. It can be extended properly to any finite number of operators.

In order to adjust the above to the current situation make a shorthand notation

$$A_k \overset{\mathsf{def}}{=} (A_{i,k})_{i=0}^{\infty} \text{ with } \mathcal{D}(A_k) = \mathcal{D}_k, \text{ for } k = 0, 1, \dots,$$

which apparently applies to $\mathcal{L} = \mathcal{H}_k$. If $(\boldsymbol{A}, \boldsymbol{D})$ determines an operator, each C_{A_k} is well defined because (4) is automatically satisfied. We may think of the operator C_{A_k} as a *column* operator of the matrix $\boldsymbol{A}$; it maps $\mathcal{D}_k$ into $\mathcal{K}$. In the same way we define $C_{A_k^{\times}}$ replacing the matrix $\boldsymbol{A}$ by $\boldsymbol{A}^{\times}$.

Notice that, in the case of $\boldsymbol{D}_{\text{fin}} = \mathcal{D}(M_{\boldsymbol{A},\boldsymbol{D}})$, the defining formula ② takes the very convenient form in terms of column operators

$$M_{\boldsymbol{A},\boldsymbol{D}} = \Big(\sum_{i=0}^{\infty} C_{A_i} P_i\Big)|_{\mathcal{D}(M_{\boldsymbol{A},\boldsymbol{D}})} \tag{5}$$

with the series converging on $\mathcal{D}(M_{\boldsymbol{A},\boldsymbol{D}})$ due to the fact that the corresponding sum is effectively finite, that is it is so when applied to vectors in $\mathcal{D}(M_{\boldsymbol{A},\boldsymbol{D}})$. More precisely, for $f \in \mathcal{D}_{\text{fin}}$ each $P_i f \in \mathcal{D}_i = \mathcal{D}(A_{n,i})$ which means $P_i f$ can be inserted in ① yielding

$$\sum_{n=0}^{\infty} \|A_{n,i} P_i f\|_{\mathcal{H}_n}^2 < +\infty$$

and this in turn means $P_i f$ is in $\mathcal{D}(C_{A_i})$. Now starting from ① we have

$$\begin{aligned} +\infty > \sum_{n=0}^{\infty} \|\sum_k A_{n,k} P_k f\|_{\mathcal{H}_n}^2 &= \sum_{n=0}^{\infty} \sum_{k,l} \langle A_{n,k} P_k f, A_{n,l} P_l f\rangle_{\mathcal{H}_n} \\ &= \sum_{k,l} \sum_{n=0}^{\infty} \langle A_{n,k} P_k f, A_{n,l} P_l f\rangle_{\mathcal{H}_n} = \sum_{k,l} \langle \bigoplus_{n=0}^{\infty} A_{n,k} P_k f, \bigoplus_{n=0}^{\infty} A_{n,l} P_l f\rangle_{\mathcal{K}} \\ &= \|\sum_k C_k P_k f\|_{\mathcal{K}}^2 \end{aligned}$$

and this establishes (5).

Based on the above we take a further step towards depicting the operator $M_{\boldsymbol{A},\boldsymbol{D}}^*$, in which the column operators play the essential *rôle*. Let us remind that

$$\mathcal{D}\Big(\bigoplus_{n=0}^{\infty} C_{A_n}\Big) \overset{\mathsf{def}}{=} \{f \in \textstyle\bigcap_{k=0}^{\infty} \mathcal{D}(C_{A_k}^*) \colon \displaystyle\sum_{n=0}^{\infty} \|C_{A_n}^* f\|^2 < +\infty\}.$$

Theorem 10. *Suppose $\mathcal{D}_{\text{fin}} \subset \mathcal{D}(M_{\boldsymbol{A},\boldsymbol{D}})$. Then*

$$M^*_{\boldsymbol{A},\boldsymbol{D}} = \bigoplus_i C^*_{A_i}.$$

Proof. The formula (5) results in

$$M^*_{\boldsymbol{A},\boldsymbol{D}} = \Big((\sum_{i=0}^{\infty} C_{A_i} P_i)|_{\mathcal{D}(M_{\boldsymbol{A},\boldsymbol{D}})} \Big)^*.$$

Take $f \in \mathcal{D}(M_{\boldsymbol{A},\boldsymbol{D}}) = \mathcal{D}_{\text{fin}}$ and $g \in \mathcal{D}(\bigoplus_{i=0}^{\infty} C^*_{A_i})$. Then

$$\langle \sum_{i=0}^{\infty} C_{A_i} P_i f, g\rangle_{\mathcal{K}} = \sum_{i=0}^{\infty} \langle C_{A_i} P_i f, g\rangle_{\mathcal{K}} = \sum_{i=0}^{\infty} \langle P_i f, C^*_{A_i} g\rangle_{\mathcal{H}_i} = \langle f, \bigoplus_{i=0}^{\infty} C^*_{A_i} g\rangle_{\mathcal{K}}.$$

and, consequently, g is in $\mathcal{D}\big(\big((\sum_{i=0}^{\infty} C_{A_i} P_i)|_{\mathcal{D}(M_{\boldsymbol{A},\boldsymbol{D}})}\big)^*\big) = \mathcal{D}(M^*_{\boldsymbol{A},\boldsymbol{D}})$ as well as

$$\bigoplus_{i=0}^{\infty} C^*_{A_i} \subset \sum_{i=0}^{\infty} M^*_{\boldsymbol{A},\boldsymbol{D}}. \tag{6}$$

Notice that

$$\langle M^*_{\boldsymbol{A},\boldsymbol{D}} f, g\rangle_{\mathcal{K}} = \langle f, M_{\boldsymbol{A},\boldsymbol{D}}\, g\rangle_{\mathcal{K}} = \sum_k \langle f, C_{A_k} P_k g\rangle_{\mathcal{K}},$$
$$f \in \mathcal{D}(M^*_{\boldsymbol{A},\boldsymbol{D}}),\ g \in \mathcal{D}(M_{\boldsymbol{A},\boldsymbol{D}}).$$

If g is such that $P_k g = g$ for a temporarily fixed k, then, because all such g's fill up the whole of $\mathcal{D}_k$ (notice that the assumption $\mathcal{D}_{\text{fin}} \subset \mathcal{D}(M_{\boldsymbol{A},\boldsymbol{D}})$ has to be used here), we get immediately $f \in \mathcal{D}(C^*_{A_k})$ and

$$\langle M^*_{\boldsymbol{A},\boldsymbol{D}} f, g\rangle_{\mathcal{K}} = \langle C^*_{A_k} f, P_k g\rangle_{\mathcal{H}_k}. \tag{7}$$

Consequently,

$$\mathcal{D}(M^*_{\boldsymbol{A},\boldsymbol{D}}) \subset \bigcap_k \mathcal{D}(C^*_{A_k}).$$

As an arbitrary $g \in \mathcal{D}_{\text{fin}} = \mathcal{D}(M_{\boldsymbol{A},\boldsymbol{D}})$ is of the form $g = \bigoplus_k P_k g$ the equality (7) implies

$$\langle M^*_{\boldsymbol{A},\boldsymbol{D}} f, g\rangle_{\mathcal{K}} = \sum_k \langle C^*_{A_k} f, P_k g\rangle_{\mathcal{H}_k} = \langle \bigoplus_k C^*_{A_k} f, g\rangle_{\mathcal{K}}$$

the sum being finite. Therefore,

$$|\langle \bigoplus_k C^*_{A_k} f, g\rangle_{\mathcal{K}}| \leq \|M^*_{\boldsymbol{A},\boldsymbol{D}} f\|\, \|g\|, \quad f \in \mathcal{D}(M^*_{\boldsymbol{A},\boldsymbol{D}}),\ g \in \mathcal{D}(M_{\boldsymbol{A},\boldsymbol{D}}),$$

hence the sums $\bigoplus_k C^*_{A_k} f$ are uniformly bounded, which results in f belonging to $\mathcal{D}(\bigoplus_{k=0}^{\infty} C^*_{A_k})$. This gives us the opposite to the inclusion (6) and the conclusion follows. □

Corollary 11. *With assumption of* Corollary 8 *to be satisfied*

$$\mathcal{D}(M^*_{\boldsymbol{A},\boldsymbol{D}}) = \bigcap_{n=0}^{\infty} \bigcap_k P_n^{-1} \mathcal{D}(A^*_{n,k}).$$

The twin Theorem 10 is as follows and the proof can repeated verbatim.

Theorem 12. *Suppose* $\mathcal{D}_{\mathrm{fin}} \subset \mathcal{D}(M_{\boldsymbol{A}^\times,\boldsymbol{D}})$. *Then*

$$M^*_{\boldsymbol{A}^\times,\boldsymbol{D}} = \bigoplus\nolimits_k C^*_{A_k^\times}.$$

Using the graph norm of $M^*_{\boldsymbol{A},\boldsymbol{D}}$ leads straightforwardly to a characterization of $\mathcal{D}_{\mathrm{fin}}$ to be a core of the adjoint to $M_{\boldsymbol{A},\boldsymbol{D}}$, given in terms of column operators.

Corollary 13. *Suppose* $\mathcal{D}_{\mathrm{fin}} \subset \mathcal{D}(M_{\boldsymbol{A}^\times,\boldsymbol{D}}) \cap \mathcal{D}(M_{\boldsymbol{A},\boldsymbol{D}})$. *Then* $\mathcal{D}_{\mathrm{fin}}$ *is a core of* $M^*_{\boldsymbol{A},\boldsymbol{D}}$ *if and only if the following condition holds*

$$f \in \mathcal{D}\Big(\bigoplus\nolimits_k C^*_{A_k^\times}\Big),$$

$$\langle P_k f, g_k\rangle_{\mathcal{H}_k} + \sum_{n=0}^{\infty} \langle C^*_{A_n} f, A^*_{k,n} g_k\rangle_{\mathcal{H}_n} = 0, \quad g_k \in \mathcal{D}_k,\ k = 0,1,\dots$$

implies $f = 0$.

Andô's construction in unbounded circumstances; sufficiency

5. Having done all the preliminary work on matrix operators we state the working part of Andô's setting. Suppose we are given two sequences $(S_n)_{n=0}^\infty$ and $(D_n)_{n=0}^\infty$ of closable operators acting in a Hilbert space $\mathcal{H}$ as well as a sequence $\boldsymbol{D} = (\mathcal{D}_n)_{n=0}^\infty$ of dense subspaces of $\mathcal{H}$ such that

$$\mathcal{D}_n \subset \mathcal{D}(S_n) \cap \mathcal{D}(S_n^*) \cap \mathcal{D}(D_n) \cap \mathcal{D}(D_{n+1}^*) \text{ for } n = 0,1,\dots \tag{8}$$

It becomes clear that for the two diagonal matrix

$$\boldsymbol{N} \overset{\text{def}}{=} \begin{pmatrix} S_0 & D_1 & 0 & 0 & \\ 0 & S_1 & D_2 & 0 & \ddots \\ 0 & 0 & S_2 & D_3 & \ddots \\ & \ddots & \ddots & \ddots & \ddots \end{pmatrix}$$

the operators $M_{\boldsymbol{N},\boldsymbol{D}}$ and $M_{\boldsymbol{N}^\times,\boldsymbol{D}}$ are determined and they act in $\mathcal{K} = \mathcal{H} \oplus \mathcal{H} \oplus \cdots$ with domains $\mathcal{D}(M_{\boldsymbol{N},\boldsymbol{D}}) = \mathcal{D}(M_{\boldsymbol{N}^\times,\boldsymbol{D}}) = \mathcal{D}_{\mathrm{fin}}$.

Proposition 14. *Suppose* $D_0 \overset{\text{def}}{=} 0$. *Then the operator* $M_{\boldsymbol{N},\boldsymbol{D}}$ *satisfies* (3) *if and only if*

$$\left.\begin{aligned} \langle S_n f, D_{n+1} g\rangle_{\mathcal{H}} &= \langle D^*_{n+1} f, S^*_{n+1} g\rangle_{\mathcal{H}}, \quad f \in \mathcal{D}_n,\ g \in \mathcal{D}_{n+1}, \\ \|D^*_{n+1} f\|^2 + \|S^*_n f\|^2 &= \|D_n f\|^2 + \|S_n f\|^2, \ f \in \mathcal{D}_n, \end{aligned}\right\} \quad n = 0,1,\dots \tag{$*$}$$

Moreover, if under the above circumstances $\mathcal{D}_{\mathrm{fin}}$ *is a core of* $M^*_{\boldsymbol{N},\boldsymbol{D}}$, *then the operator* $M_{\boldsymbol{N},\boldsymbol{D}}$ *is essentially normal.*

Proof. The 'if' part of the first conclusion is direct. For the 'only if' part we suggest to plug into the formal normality condition $f = P_n f$ first and then $f = P_n f + P_{n+1} f$ and consider it step by step.

The additional conclusion follows from Theorem 6 if one notices that $\mathcal{D}_{\text{fin}}$ as equals $\mathcal{D}(M_{\boldsymbol{A}^\times,\boldsymbol{D}})$ is apparently a core of $M_{\boldsymbol{N}^\times,\boldsymbol{D}}$. □

Before we come to the main result of this paper recall an operator S in a Hilbert space $\mathcal{H}$ is called *subnormal* if there is another Hilbert space $\mathcal{K}$, which contains isometrically $\mathcal{H}$, and a normal operator N in it such that

$$Sf = Nf, \quad f \in \mathcal{D}(S).$$

Theorem 15. *Let S be a closable operator in a Hilbert space $\mathcal{H}$ and let two sequences $(S_n)_{n=0}^\infty$ and $(D_n)_{n=0}^\infty$ of closable operators acting in a Hilbert space $\mathcal{H}$ as well as a sequence $\boldsymbol{D} = (\mathcal{D}_n)_{n=0}^\infty$ of dense subspaces of $\mathcal{H}$ satisfying* (8) *be given. Suppose*

$$S_0 = S,\ D_0 \stackrel{\text{def}}{=} 0 \text{ and } \mathcal{D}_0 = \mathcal{D}(S)$$

as well as the condition $(*)$ *holds. Then S is subnormal with the operator $\overline{M_{\boldsymbol{N},\boldsymbol{D}}}$ being its normal extension in the space $\mathcal{H} \oplus \mathcal{H} \oplus \cdots$ if and only if the following condition is satisfied*[5]*:*

$$\begin{gathered} f \in \mathcal{D}\Big(\bigoplus_{k=0}^{\infty} C^*_{(D_k,S_k)}\Big), \\ \langle P_n f, g_n\rangle_{\mathcal{H}} + \langle C^*_{(D_n,S_n)} f, S_n^* g_n\rangle_{\mathcal{H}} + \langle C^*_{(D_{n+1},S_{n+1})} f, D_{n+1}^* g_n\rangle_{\mathcal{H}} = 0, \\ g_n \in \mathcal{D}_n,\ n = 0, 1, \ldots \\ \textit{implies } f = 0, \end{gathered}$$

with additional notation $g_{-1} \stackrel{\text{def}}{=} 0$.

Remark 16. Remark 9, if applicable, can provide some help in further identification of the domain of $C^*_{(D_n,S_n)}$.

Proof of Theorem 15. Use Proposition 14 and Corollary 13. □

Andô's construction in unbounded circumstances; necessity

6. Now we look upon the possibility of extorting conditions $(*)$ from the assumption that S has already a normal extension of any sort. In other words, we would like to know how much the limitations of Theorem 15 pose a challenge for subnormality of S.

It follows directly from the definition of subnormality that

$$\mathcal{D}(S) \subset \mathcal{D}(N) \cap \mathcal{H} = \mathcal{D}(N^*) \cap \mathcal{H} \subset \mathcal{D}(S^*), \tag{9}$$

[5] $C_{(D_k,S_k)}$ stands apparently for the (a bit overloaded in the present circumstances) hieroglyphics $C_{(0,\ldots,0,D_k,S_k,0,\ldots)}$ with positioning of D_k, S_k easy to guess; it should not mean for the formalists $B = (D_k, S_k)$ as referring to the definition of C_B, p. 341 – apology for this kind of perversity.

with inclusions to be proper sometimes and

$$\|S^*f\| \leq \|Sf\|, \quad f \in \mathcal{D}(S), \tag{10}$$

the latter means hyponormality of S. Also as normal operators are closed, subnormal ones must necessarily be closable.

The following result is a part of Theorem 37 in [10].

Theorem 17. *An operator S in $\mathcal{H}$ with invariant domain $\mathcal{D}(S)$ is subnormal if and only if there exists a form ψ on $\mathfrak{N}_+ \times \mathcal{D}(S)$ which is positive definite*[6] *and*

$$\psi(m, n, f, g) = \langle S^m f, S^n g \rangle, \quad m, n = 0, 1, \ldots, \; f, g \in \mathcal{D}(S).$$

As a consequence we get a way to transport subnormality from one space to another, which is an unbounded counterpart of Lemma 1 of [1].

Proposition 18. *Let S be an operator in $\mathcal{H}$ with invariant domain $\mathcal{D}(S)$ and let $V\colon \mathcal{H} \to \mathcal{H}_1$ be a bounded operator such that $V^*VS = S$. If S is subnormal in $\mathcal{H}$, then so is VSV^* in $\mathcal{H}_1$ provided it is densely defined.*

Proof. Define a form ψ_1 on $\mathfrak{N}_+ \times \mathcal{D}(VSV^*)$ by $\psi_1(m, n, x, y) \stackrel{\mathsf{def}}{=} \psi(m, n, V^*x, V^*y)$ and remembering that $(VSV^*)^k = VS^kV^*$ for $k = 1, 2, \ldots$ check that ψ_1 is positive definite and corresponds to the operator $S_1 \stackrel{\mathsf{def}}{=} VSV^*$ as in Theorem 17. □

Suppose S is a subnormal operator in $\mathcal{H}$. Let N be its normal extension in $\mathcal{K}$ (remember, $\mathcal{H} \subset \mathcal{K}$ isometrically). Let us point out some relations between S^* and N^*. The first of them is[7]

$$\langle N^*f - S^*f, g \rangle_{\mathcal{K}} = 0, \quad f \in \mathcal{D}(N^*) \cap \mathcal{H}, \; g \in \mathcal{H}. \tag{α}$$

Indeed,

$$\langle N^*f, g \rangle_{\mathcal{K}} = \langle f, Ng \rangle_{\mathcal{K}} = \langle f, Sg \rangle_{\mathcal{H}} = \langle S^*f, g \rangle_{\mathcal{H}} = \langle S^*f, g \rangle_{\mathcal{K}},$$

first for $g \in \mathcal{D}(S)$, then for all $g \in \mathcal{H}$; all this makes sense as (9) holds.

The second says

$$\|f + N^*g - S^*g\|_{\mathcal{K}}^2 = \|f\|_{\mathcal{H}}^2 + \|N^*g - S^*g\|_{\mathcal{K}}^2, \quad f \in \mathcal{H}, \; g \in \mathcal{D}(N^*) \cap \mathcal{H}. \tag{β}$$

This is so because, due to (α), $N^*g - S^*g \perp \mathcal{H}$.

The third states

$$\|N^*f - S^*f\|_{\mathcal{K}}^2 = \|Sf\|_{\mathcal{H}}^2 - \|S^*f\|_{\mathcal{H}}^2, \quad f \in \mathcal{D}(S). \tag{γ}$$

[6] For the relevant terminology and notation cf. [10].

[7] Sometimes we emphasize the underlying space by adding a subscript to the inner product.

For this take $f \in \mathcal{D}(S)$, then, employing normality of N and using (α), one gets [8]

$$\begin{aligned}\|N^*f - S^*f\|_{\mathcal{K}}^2 &= \|N^*f\|_{\mathcal{K}}^2 - 2\,\mathfrak{Re}\langle N^*f, S^*f\rangle_{\mathcal{K}} + \|S^*f\|_{\mathcal{H}}^2\\ &= \|Nf\|_{\mathcal{K}}^2 - 2\,\mathfrak{Re}\langle N^*f, S^*f\rangle_{\mathcal{K}} + \|S^*f\|_{\mathcal{H}}^2\\ &= \|Sf\|_{\mathcal{H}}^2 - 2\,\mathfrak{Re}\langle S^*f, S^*f\rangle_{\mathcal{K}} + \|S^*f\|_{\mathcal{H}}^2\\ &= \|Sf\|_{\mathcal{H}}^2 - \|S^*f\|_{\mathcal{H}}^2.\end{aligned}$$

Proposition 19. *If $(N^* - S^*)|_{\mathcal{D}(S)}$ is closable* [9]*, then there is a positive operator D in $\mathcal{H}$ having $\mathcal{D}(S)$ as its core and such that*

$$\|Df\|_{\mathcal{H}}^2 = \|N^*f - S^*f\|_{\mathcal{K}}^2 = \|Sf\|_{\mathcal{H}}^2 - \|S^*f\|_{\mathcal{H}}^2, \quad f \in \mathcal{D}(S). \tag{11}$$

Proof. Set $A \overset{\text{def}}{=} (N^* - S^*)|_{\mathcal{D}(S)}$ and consider it as an operator to $\mathcal{K}$. $A^*\bar{A}$ is a positive and selfadjoint operator in $\mathcal{H}$. Denote by D its positive square root. Then according to [13, Theorem 5.40]

$$\mathcal{D}(D) = \mathcal{D}(\bar{A}), \quad \|Df\|_{\mathcal{H}} = \|\bar{A}f\|_{\mathcal{K}},\ f \in \mathcal{D}(\bar{A}).$$

This implies in turn that $\mathcal{D}(S)$ is a core of D (use part 2° of Fact 5). □

Suppose now that $\mathcal{D}(S)$ is invariant for S, that is

$$S\mathcal{D}(S) \subset \mathcal{D}(S). \tag{12}$$

Then $\mathcal{D}(S) + N^*\mathcal{D}(S)$ is invariant for N. Indeed, because $\mathcal{D}(S) \subset \mathcal{D}(N) = \mathcal{D}(N^*)$,

$$N(f + N^*g) = Sf + NN^*g = Sf + N^*Ng = Sf + N^*Sg, \quad f, g \in \mathcal{D}(S).$$

Denote by $\mathcal{K}_1$ the closure in $\mathcal{K}$ of the linear space $\mathcal{D}(S) + N^*\mathcal{D}(S)$. Then the operator $N|_{\mathcal{D}(S)+N^*\mathcal{D}(S)}$ is subnormal in $\mathcal{K}_1$.

Due to (12), the second equality in (11) and (β),

$$V\colon \mathcal{D}(S) + (N^* - S^*)\mathcal{D}(S) \ni f + N^*g - S^*g \mapsto (f, Dg) \in \mathcal{H} \oplus \mathcal{H}$$

is a well defined isometry which we extend to the whole of $\mathcal{K}_1$. Assuming that $\mathcal{D}(S)$ is invariant for S^* as well, we get $\mathcal{D}(S) + N^*\mathcal{D}(S) = \mathcal{D}(S) + (N^* - S^*)\mathcal{D}(S)$ which make the operator $VN|_{\mathcal{D}(S)+N^*\mathcal{D}(S)}V^*$ densely defined in $\mathcal{H} \oplus \mathcal{H}$; according to Lemma 19 it is subnormal. Then the rest of Andô's arguments can be done with due caution.

The calculation which follows leads us to the matricial form of the operator $VN|_{\mathcal{D}(S)+N^*\mathcal{D}(S)}V^*$. Because V^* is a partial isometry with the initial space $\mathcal{R}(V) = V\mathcal{K}_1 = \overline{\{(f, Dg)\colon\ f, g \in \mathcal{D}(S)\}}$ we have

$$\mathcal{R}(V) = \overline{\mathcal{R}(D|_{\mathcal{D}(S)})} = \mathcal{H} \oplus \overline{\mathcal{R}(D)}$$

as $\mathcal{D}(S)$ is a core of D (Proposition 19). Therefore $\mathcal{R}(V)^{\perp} = 0 \oplus \mathcal{N}(D)$ and according to this for $f, h \in \mathcal{D}(S)$ we have the decomposition

$$(f, h) = (f, Dg) \oplus (0, h - Dg) \text{ with } D^2g = Dh. \tag{13}$$

[8] This gives an *ad hoc* argument for (10).

[9] Surprisingly we cannot prove it. If this is not because of our ignorance, it may be an interesting problem even in approximation theory in $\mathcal{L}^2$-spaces over noncompact sets.

Notice that $\mathcal{D}(S) \subset \mathcal{D}(D^2)$, This stems from formula (11) and the fact that $\mathcal{D}(S)$ is a core of D (Proposition 19 again). Now, using (13), for $f, h \in \mathcal{D}(S)$ we get

$$\begin{aligned} VN|_{\mathcal{D}(S)+N^*\mathcal{D}(S)}V^*(f,h) &= VN|_{\mathcal{D}(S)+N^*\mathcal{D}(S)}V^*(f,Dg) = VN(f+(N^*-S^*)g) \\ &= V(Sf+(S^*S-SS^*)g+(N^*-S^*)Sg) \\ &= (Sf+D^2g, DSg) = (Sf+Dh, DSD^{-1}h) \end{aligned}$$

because, as we have already assumed, $\mathcal{D}(S)$ is invariant for S^* as well (which is used for the third equality) and, what has to be assumed furthermore, that $\mathcal{D}(S)$ is <u>invariant</u> [10] for D (which helps to make the calculation for the last equality [11]). The latter also makes it sure the operator $VN|_{\mathcal{D}(S)+N^*\mathcal{D}(S)}V^*$ act in $\mathcal{H} \oplus \mathcal{H}$ as a matrix operator $M_{(\boldsymbol{S},(\mathcal{D}(S),\mathcal{D}(S))}$ determined by the matrix

$$\boldsymbol{S} \stackrel{\mathsf{def}}{=} \begin{pmatrix} S & D|_{\mathcal{D}(S)} \\ 0 & DSD^{-1}|_{\mathcal{D}(S)} \end{pmatrix}.$$

Notice that if $\mathcal{D}(S)$ is supposed to be invariant for S, S^* and D, as D is selfadjoint, $\mathcal{D}(S) \oplus \mathcal{D}(S)$ is invariant for the matrix operators $M_{(\boldsymbol{S},(\mathcal{D}(S),\mathcal{D}(S))}$ and $M_{(\boldsymbol{S}^\times,(\mathcal{D}(S),\mathcal{D}(S))}$ [12]. Nickname, for further references, the way we have passed from S to $\boldsymbol{S}$ *Andô's procedure.*

7. Now anticipating Andô's procedure to be a basic tool for the induction process we try to watch how it goes on. Suppose we have succeeded in constructing sequences $(S_n)_{n=0}^k$ and $(D_n)_{n=0}^{k-1}$ of closable operators, $S_0 = S$, $D_0 = 0$, such that conditions (8) and $(*)$ are satisfied up to $n = k-1$ with $\mathcal{D}_n = \mathcal{D}(S)$ and the matrix operator determined by the operator matrix

$$\boldsymbol{S}_k \stackrel{\mathsf{def}}{=} \begin{pmatrix} S_0 & D_1 & 0 & \cdots & \cdots & 0 \\ 0 & S_1 & D_2 & 0 & \ddots & 0 \\ 0 & 0 & S_2 & D_3 & \ddots & 0 \\ \vdots & \ddots & \ddots & \ddots & \ddots & 0 \\ 0 & \cdots & \cdots & 0 & S_{k-1} & D_k \\ 0 & \cdots & \cdots & \cdots & 0 & S_k \end{pmatrix} \tag{14}$$

is subnormal. If, in addition, we suppose all the D_n's are selfadjoint as well as $\mathcal{D}(S)$ is <u>invariant</u> for S_n, S_n^* and D_n, $n = 0, 1, \ldots, k-1$ then $\mathcal{D}(S) \oplus \cdots \oplus \mathcal{D}(S)$,

[10] Maybe this assumption is too rough.

[11] For a selfadjoint operator A in $\mathcal{H}$ its *partial inverse* A^{-1} here is understood as an operator

$$A^{-1}f \stackrel{\mathsf{def}}{=} \begin{cases} \left(A|_{\mathcal{D}(A)\cap\overline{\mathcal{R}(A)}}\right)^{-1}f & \text{if } f \in \mathcal{R}(A) \\ 0 & \text{if } f \in \mathcal{R}(A)^\perp = \mathcal{N}(A) \end{cases}, \quad f \in \mathcal{R}(A) \oplus \mathcal{N}(A),$$

which is in $\mathcal{H}$. Thus $A^{-1}A = P|_{\mathcal{D}(A)}$ and $AA^{-1} = P|_{\mathcal{R}(A)\oplus\mathcal{N}(A)}$, P being here a projection onto $\overline{\mathcal{R}(A)}$.

[12] In this case, because the matrix is finite, the definition of an operator determined by a matrix simplifies and allows us here and there to shorten notation using in particular $\boldsymbol{S}$ for $M_{(\boldsymbol{S},(\mathcal{D}(S),\mathcal{D}(S)))}$ and $\boldsymbol{S}^\times$ for $M_{(\boldsymbol{S}^\times,(\mathcal{D}(S),\mathcal{D}(S)))}$.

$k+1$-times, is invariant for $\boldsymbol{S}_k$ and $\boldsymbol{S}_k^*$. Moreover, cf. footnote 12,

$$\boldsymbol{S}_k^\times \boldsymbol{S}_k - \boldsymbol{S}_k \boldsymbol{S}_k^\times = \begin{pmatrix} 0 & \cdots & 0 & 0 \\ \vdots & \cdots & \vdots & \vdots \\ 0 & \cdots & 0 & 0 \\ 0 & \cdots & 0 & D_k^2 + S_k^* S_k - S_k S_k^* \end{pmatrix}. \tag{15}$$

After applying Andô's procedure to $\boldsymbol{S}_k$, we come to a subnormal operator

$$\tilde{\boldsymbol{S}}_{k+1} \stackrel{\mathsf{def}}{=} \begin{pmatrix} \boldsymbol{S}_k & \boldsymbol{D}_{k+1}|_{\mathcal{D}(\boldsymbol{S}_k)} \\ 0 & \boldsymbol{D}_{k+1} \boldsymbol{S}_k \boldsymbol{D}_{k+1}^{-1}|_{\mathcal{D}(\boldsymbol{S}_k)} \end{pmatrix} \tag{16}$$

in $\mathcal{H}^{k+1} \oplus \mathcal{H}^{k+1}$, where $\boldsymbol{D}_{k+1}$ determined as in Proposition 19 as a $(k+1)\times(k+1)$ matrix operator, due to (15), has the only non-zero entry in its downright corner; the latter is the operator D_{k+1} satisfying $\|D_{k+1}f\|^2 + \|S_k^* f\|^2 = \|D_k f\|^2 + \|S_k f\|^2$ for $f \in \mathcal{D}(S)$. Because of this, when one opens up the blocks in (16) to $(k+1)\times(k+1)$ (still operator) matrices, it turns out that there is a surplus of rows composed totaly of zero entries. Thus the operator

$$\boldsymbol{V}_{k+1} \colon \mathcal{H}^{k+1} \oplus \mathcal{H}^{k+1} \ni (f_0, \ldots, f_{k+1}, g_0, \ldots, g_{k+1}) \mapsto (f_0, \ldots, f_{k+1}, g_{k+1}) \in \mathcal{H}^{k+2},$$

crossing out needless coordinates makes [13] the operator [14]

$$\boldsymbol{S}_{k+1} \stackrel{\mathsf{def}}{=} \boldsymbol{V}_{k+1} \tilde{\boldsymbol{S}}_{k+1} \boldsymbol{V}_{k+1}^* = \begin{pmatrix} \boldsymbol{S}_k & C_{(0,\ldots,0,D_{k+1}|_{\mathcal{D}(S)})} \\ 0 & D_{k+1} S_k D_{k+1}^{-1}|_{\mathcal{D}(S)} \end{pmatrix}$$

subnormal in $\mathcal{H}^{k+2}$. The only missing thing here is the requirement $\mathcal{D}(S)$ to be invariant for D_{k+1}, which has to be kept going.

Summary. If S is subnormal operator in $\mathcal{H}$, then there are two sequences $(S_n)_{n=0}^\infty$ and $(D_n)_{n=0}^\infty$ having common domain $\mathcal{D}(S)$, $S_0 = S$, $D_0 = 0$, D_n's being positive, and such that $(*)$ holds provided all those 'ifs' concerning invariance of $\mathcal{D}(S)$ and the assumption of closability as in Proposition 19 on every stage can be released. If this happens then each of the operators (14) is a subnormal extension of S. This gives us a starting point to go backwards but for that we need the other assumptions of Theorem 15 to be satisfied but this is another story. Let us mention that for the famous creation operator, which is subnormal, all the S_n's are equal to D while all the D_n's are equal to the identity operator I. Another flicker of hope is in some room left by not requiring the operators D_n to be positive (in the sufficiency part). This a bit optimistic remark brings us on to the final question which follows.

Question. Though there is a lot of unbounded subnormal operators for which all the invariance assumptions are satisfied, a gap between the class of operators satisfying $(*)$ and all the subnormal ones under the present circumstances has been opened. So how wide is it and is it really the case? In other words, does this

[13] This is due to Proposition 18 as $\boldsymbol{V}_{k+1}^* \boldsymbol{V}_{k+1} \tilde{\boldsymbol{S}}_{k+1} = \tilde{\boldsymbol{S}}_{k+1}$.

[14] Notice at the upright corner the column operator appears again.

gap come from the fact that our way of realizing Andô's programme is not tuned finely enough or it is a result of intrinsic inflexibility of Andô's method itself?

Acknowlegement

Thanks are due to Jan Stochel for his criticism effecting the final version of this paper.

References

[1] T. Andô, Matrices of normal extensions of subnormal operators, *Acta Sci. Math. (Szeged)*, **24** (1963), 91–96.

[2] Yu.A. Berezanskii, *Eigenfunction expansions for selfadjoint operators*, Naukova Dumka, Kiev, **1965**.

[3] E. Bishop, Spectral theory of operators on a Banach space, *Trans. Amer. Math. Soc.*, **86** (1957), 414–445.

[4] C. Foiaş, Décomposition en opérateurs et vecteurs propres. I. Études de ces décompositions et leurs rapports avec les prolongements des opérateurs, *Rev. Roumaine Math. Pures Appl.*, **7** (1962), 241–282.

[5] A.S. Holevo, *Probabilistic and statistical aspects of quantum theory*, North-Holland, Amsterdam – New York – Oxford, **1982**.

[6] J. Janas and J. Stochel, Selfadjoint operators with finite rows, *Ann. Polon. Math.*, **66** (1997), 155–172.

[7] J. Stochel and F.H. Szafraniec, On normal extensions of unbounded operators. I, *J. Operator Theory*, **14** (1985), 31–55.

[8] ———, On normal extensions of unbounded operators. II, *Acta Sci. Math. (Szeged)*, **53** (1989), 153–177.

[9] ———, On normal extensions of unbounded operators. III. Spectral properties, *Publ. RIMS, Kyoto Univ.*, **25** (1989), 105–139.

[10] ———, The complex moment problem and subnormality: a polar decomposition approach, *J. Funct. Anal.*, **159** (1998), 432–491.

[11] M.H. Stone, *Linear transformations in Hilbert space*, Amer. Math. Soc., Providence, R.I, **1932**.

[12] F.H. Szafraniec, Subnormality in the quantum harmonic oscillator, *Commun. Math. Phys.*, **210** (2000), 323–334.

[13] J. Weidmann, *Linear operators in Hilbert spaces*, Springer-Verlag, New York, Heidelberg, Berlin, **1980**.

Franciszek Hugon Szafraniec
Instytut Matematyki
Uniwersytet Jagielloński
ul. Reymonta 4
PL-30059 Kraków
e-mail: `fhszafra@im.uj.edu.pl`

Operator Theory:
Advances and Applications, Vol. 163, 351–363

Directing Mappings in Kreĭn Spaces

Björn Textorius

To Heinz Langer on the occasion of his retirement.

Abstract. M.G. Kreĭn's method of directing mappings is extended to quasidensely defined symmetric operators with selfadjoint definitizable extensions in some Kreĭn space, by direct use of the spectral function of the extension, thus avoiding a reduction to the semidefinite situation.

Mathematics Subject Classification (2000). Primary 47B50, 46C20 ; Secondary 34B09.

Keywords. Kreĭn space, Pontrjagin space, definitizable operator, directing mapping.

1. Introduction

The method of directing functionals or directing mappings is often useful in order to prove spectral representations for operators and functions. Originally, this method was developed in [3] for a linear space $\mathcal{L}$ with a semidefinite inner product, a symmetric operator S in $\mathcal{L}$ with quasidense domain and a finite system of directing functionals; it is explained, e.g., in M.A. Naimark's book [11]. Generalizations to an infinite number of directing functionals or a directing mapping for a symmetric linear operator were given in [7]; for a symmetric linear relation in [9], [10], and analogous results for isometric operators have been formulated in [1].

It is the aim of this note to show that this method can also be generalized to definitizable operators in an indefinite inner product space $(\mathcal{L}, [\cdot,\cdot])$. First results in this direction were discussed by M.G. Kreĭn and H. Langer more than 30 years ago. There the main idea was to use a new semidefinite inner product $(f,g) := [p(S)f, g]$ on $\mathcal{D}(S^\pi)$, where π is the degree of the definitizing polynomial p of S, or $(f,g) := [q(S)f, q(S)g]$ on $\mathcal{D}(S^\rho), \rho = \deg q$, if the definitizing polynomial p is of the form $p = q\bar{q}$ and to apply the results concerning the case of a semidefinite inner product to the restrictions of S in these spaces. There arise, however, some technical difficulties and restrictions. It has to be shown that, e.g., the domain $\mathcal{D}(S^{\pi+1})$ is quasidense in $\mathcal{D}(S^{\pi})$ with respect to the inner product $(\cdot,\cdot)$. For differential operators the domains of these powers are often not so easy to deal with.

Several years ago Heinz Langer and the author began to study a way to avoid this difficulty. Instead of reducing the problem to the semidefinite situation, the idea is to assume the existence of a definitizable extension $\widetilde{S}$ of S in some Kreĭn space $\widetilde{\mathcal{K}}$ and make use of the spectral function $\widetilde{E}$ of $\widetilde{S}$ directly. It is then necessary to change the definition of a directing mapping slightly. While in the original definition the equation

$$(S-z)f = g$$

is considered for fixed g, f at single points z, we replace f by a holomorphic function $f(z)$ in some domain $\mathcal{D}_g \subset C$ and with values in $\mathcal{L}$, and suppose that the corresponding equation $(S-z)f(z) = g(z)$ has a holomorphic solution $f(z)$ if and only if $\Phi_z g(z) = 0\,(z \in \mathcal{D}_g)$. In the case of a semidefinite inner product, this assumption follows from from the original condition. It should be mentioned, however, that in applications where S is, e.g., an ordinary differential operator, this in some sense stronger condition dealing with these holomorphic functions can as easily be verified as the original one.

The results of this note can be applied to obtain representations of kernels and functions having a finite number of negative squares and to ordinary differential problems with an indefinite weight function.

I thank a referee for valuable remarks.

2. Definitizable operators

Let A be a definitizable selfadjoint operator in a Kreĭn space $\mathcal{K}$ with definitizing polynomial p. Recall that this means that the resolvent set $\rho(A)$ is not empty and that

$$[p(A)x,x] \geq 0 \ \text{ for all } \ x \in \mathcal{D}(A^{\pi}),$$

where π denotes the degree of the polynomial p. The non-real spectrum $\sigma_0(A)$ of the operator A consists of a finite number of eigenvalues of finite order, located symmetrically with respect to the real axis. Moreover, A has a spectral function E defined on the semiring $\mathcal{R}_A$ of all intervals of R with endpoints not in the set $\widetilde{c}(A)$ of critical points of A, and their complements in R. The spectral function can be extended in a natural way to not necessarily real sets $\widetilde{\Delta}$, for which $\widetilde{\Delta} \cap R \in \mathcal{R}_A$ by putting

$$E(\widetilde{\Delta}) := E(\widetilde{\Delta} \cap R) + \sum_{\lambda_j \in \widetilde{\Delta} \cap \sigma_0(A)} E_{\lambda_j},$$

where E_λ denotes the Riesz projection corresponding to the (non-real) eigenvalue λ of A. This extended semiring is denoted by $\widetilde{\mathcal{R}}_A$. Furthermore, as in [6] and [8], $c(A)$ denotes the set of finite critical points of A: $c(A) = \widetilde{c}(A) \cap R$, and we put $\widetilde{\sigma_0}(A) := c(A) \cup \sigma_0(A)$. A critical point is either regular or singular, see, e.g., [8].

Lemma 2.1. *Let $\alpha \in R$ be a real critical point, and let $\Delta \in \mathcal{R}_A$ be bounded and such that $\Delta \cap c(A) = \{\alpha\}$. If q is a nonnegative integer such that*

$$[(A-\alpha)^q f, f] \geq 0 \text{ for all } f \in E(\Delta)\mathcal{K},$$

then for all these elements f and $z \in \rho(A)$ we have

$$(A-z)^{-1}f = \int_\Delta \left[(\lambda-z)^{-1} + \sum_{k=0}^{q-1}(z-\alpha)^{-k-1}(\lambda-\alpha)^k\right] E(d\lambda)f$$
$$-\sum_{k=0}^{q-1}(z-\alpha)^{-k-1}A^k f - (z-\alpha)^{-q-1}Nf, \tag{2.1}$$

where N is a bounded selfadjoint operator in $\mathcal{K}$ with the properties $N^2 = 0, JN = 0$, and $NE(\Delta') = 0$ for all sets $\Delta' \in \widetilde{\mathcal{R}}_A$ such that $\alpha \notin \Delta'$, and the integral exists in the strong operator topology as an improper one with respect to the singularity of E at α. This singularity is, however, such that $\int_\Delta (\lambda-\alpha)^q E(d\lambda)$ converges.

Proof. Without loss of generality we assume that $\alpha = 0$, $\widetilde{\sigma}_0(A) = \{\alpha\}$ and that A is bounded: $\sigma(A) \subset \Delta$. The identity

$$\lambda^q z^{-q}(\lambda-z)^{-1} = (\lambda-z)^{-1} + z^{-1} + \lambda z^{-2} + \cdots + \lambda^{q-1}z^{-q}$$

implies

$$(A-z)^{-1} = z^{-q}A^q(A-z)^{-1} - z^{-1} - z^{-2}A - \cdots - z^{-q}A^{q-1}. \tag{2.2}$$

According to [6] and [8], the representation

$$z^{-q}A^q(A-z)^{-1}E(\Delta) = \int_\Delta z^{-q}\lambda^q(\lambda-z)^{-1}E(d\lambda) - z^{-q-1}N \tag{2.3}$$

holds. The relations (2.2), (2.3) evidently give (2.1). □

3. Preliminaries on integration

Let $\mathcal{H}, \mathcal{K}$ be two Kreĭn spaces, $\Psi : z \to \Psi_z$ a function defined and holomorphic on some set $\sigma \subset C$ and with values in $\mathcal{L}(\mathcal{H}, \mathcal{K})$, and let $A \in \mathcal{L}(\mathcal{K})$ be a selfadjoint operator in $\mathcal{K}$ such that $\sigma(A) \subset \sigma$. If $\mathcal{C}_A$ is a Jordan contour in σ surrounding $\sigma(A)$, symmetric with respect to the real axis, we define the bounded linear operator $\Psi_A \in \mathcal{L}(\mathcal{H}, \mathcal{K})$ by

$$\Psi_A := -\frac{1}{2\pi i}\int_{\mathcal{C}_A}(A-z)^{-1}\Psi_z dz. \tag{3.1}$$

If $\Theta_.$ is a function with the same properties as $\Psi_.$, then the following formula is easy to verify for arbitrary functions a, b, holomorphic on σ and with values in $\mathcal{H}$:

$$\left[-\frac{1}{2\pi i}\oint_{\mathcal{C}_A}(A-z)^{-1}\Psi_z a(z)dz, -\frac{1}{2\pi i}\oint_{\mathcal{C}_A}(A-z)^{-1}\Theta_z b(z)dz\right]$$
$$= -\frac{1}{2\pi i}\oint_{\mathcal{C}_A}[(A-z)^{-1}\Psi_z a(z), \Theta_{\bar{z}} b(\bar{z})]dz.$$

This formula does still hold if the operator is definitizable and $\sigma(A)$ is not necessarily surrounded by $\mathcal{C}_A$, but $\mathcal{C}_A$ intersects R nontangentially and only in points which do not belong to $c(A) \cup \sigma_p(A)$. In this case the integrals are to be understood as principal values in these points of R. In the following such integrals are denoted by $'$.

Now let $\mathcal{H}$ be a Hilbert space. By $\mathcal{V}$ we denote the following class of $\mathcal{L}(\mathcal{H})$-valued measures V:

1. V is defined on the semiring $\mathcal{R}_V$ of bounded intervals of R whose closures do not contain the points of a finite exceptional set $s(V) = \{\alpha_1, \ldots, \alpha_l\}$ (depending on V).
2. V is either nonnegative or nonpositive on each interval in $R \setminus \{\alpha_1, \ldots, \alpha_l\}$, which does not contain points from a given finite set (depending on V).
3. For each $\alpha \in s(V)$ there exists an integer $q \geq 0$ such that the integral $\int_{\alpha-\epsilon}^{\alpha+\epsilon} (\lambda - \alpha)^q V(d\lambda)$ exists if $\epsilon > 0$ is so small that $[\alpha - \epsilon, \alpha + \epsilon] \cap s(V) = \{\alpha\}$.

In the sequel we have to deal with integrals of holomorphic $\mathcal{H}$-valued functions on R with respect to the operator measure $V(d\lambda)$. For an interval Δ such that $\overline{\Delta} \cap s(V) = \emptyset$ these integrals are defined in [7]. Now assume that the interval Δ contains one exceptional point α in its interior Δ^i, that it has positive distance from all the other exceptional points of V, and that the integral $\int_{\alpha-\epsilon}^{\alpha+\epsilon} (\lambda-\alpha)^q V(d\lambda)$ exists. Let ϕ, ψ be holomorphic, $\mathcal{H}$-valued functions on $\overline{\Delta}$, and assume that Δ is so small that on $\overline{\Delta}$ the expansions

$$\phi(\lambda) = \sum_{\mu=0}^{\infty} \phi_\mu (\lambda - \alpha)^\mu, \quad \psi(\lambda) = \sum_{\nu=0}^{\infty} \psi_\nu (\lambda - \alpha)^\nu$$

hold. Then an improper integral with respect to the singularity of V at α is defined by means of the following regularization:

$$\begin{aligned}
&\int_\Delta (V(d\lambda)\phi(\lambda), \psi(\lambda))_{(\alpha;q)} \\
&:= \int_\Delta \left\{ (V(d\lambda)\phi(\lambda), \psi(\lambda)) - \sum_{l=0}^{q-1} (V(d\lambda)\phi(t), \psi(t))^{(l)}|_{t=\alpha} \frac{(\lambda-\alpha)^l}{l!} \right\} \\
&= \sum_{\mu,\nu\geq 0;\, \mu+\nu\geq q} \int_\Delta (V(d\lambda)\phi_\mu, \psi_\nu)\lambda^{\mu+\nu}. \qquad (3.2)
\end{aligned}$$

If, e.g., $\dim H = 1$ the second expression makes sense in the usual way. It then reads as

$$\int_\Delta V(d\lambda) \left(\phi(\lambda)\overline{\psi(\lambda)} - \sum_{l=0}^{q-1} (\phi(t)\overline{\psi(t)})^{(l)}|_{t=\alpha} \frac{(\lambda-\alpha)^l}{l!} \right).$$

For the integral defined above we need a kind of Stieltjes-Lifshits inversion formula.

Lemma 3.1. *Let $V, \alpha, q, \Delta, \phi, \psi$ be as before and assume that $\mathcal{C}_\Delta$ is a Jordan contour, symmetric with respect to R with Δ in its interior domain and all other exceptional points of V in its exterior domain. Assume that ϕ and ψ are holomorphic on $\mathcal{C}_\Delta$ and its interior and that V is continuous on the boundary of Δ. Then, for $z \notin \overline{\Delta}$*

$$-\frac{1}{2\pi i}\oint_{\mathcal{C}_\Delta}' \left\{\int_\Delta \left[(\lambda - z)^{-1} + \sum_{k=0}^{q-1}(z-\alpha)^{-k-1}(\lambda-\alpha)^k\right](V(d\lambda)\phi(z), \psi(\bar{z}))\right\} dz$$
$$= \int_\Delta (V(d\lambda)\phi(\lambda), \psi(\lambda))_{(\alpha;\, q)}. \tag{3.3}$$

Proof. Without loss of generality we assume $\alpha = 0$. The left-hand side of (3.3) then equals

$$-\frac{1}{2\pi i}\oint_{\mathcal{C}_\Delta}' \left\{\int_\Delta \lambda^q z^{-q}(\lambda - z)^{-1}(V(d\lambda)\phi(z), \psi(\bar{z}))\right\} dz$$
$$= -\frac{1}{2\pi i}\oint_{\mathcal{C}_\Delta}' \left\{\int_\Delta \lambda^q z^{-q}(\lambda - z)^{-1} \sum_{\mu,\nu=0}^{\infty} (V(d\lambda)\phi_\mu, \psi_\nu) z^{\mu+\nu}\right\} dz$$
$$= \int_\Delta \left(\sum_{\mu,\nu=0}^{\infty}\left(-\frac{1}{2\pi i}\oint_{\mathcal{C}_\Delta}' \lambda^q z^{-q}(\lambda - z)^{-1} z^{\mu+\nu}(V(d\lambda)\phi_\mu, \psi_\nu)\right)\right) dz$$
$$= \int_\Delta \left(\sum_{\mu,\nu\geq 0,\mu+\nu\geq q} \lambda^{\mu+\nu}(V(d\lambda)\phi_\mu, \psi_\nu\right),$$

which, according to (3.2), equals the right-hand side of (3.3).

In the last step we used the formula

$$-\frac{1}{2\pi i}\oint_{\mathcal{C}_\Delta} \lambda^q z^{-q}(\lambda - z)^{-1} z^{\mu+\nu} dz = \begin{cases} \lambda^{\mu+\nu} & \mu+\nu \geq q \\ 0 & \mu+\nu < q \end{cases},$$

which holds if λ is in the domain surrounded by $\mathcal{C}_\Delta$. □

4. Directing mappings

Let $\mathcal{L}$ be a linear space equipped with an inner product $[\cdot,\cdot]$, that is a possibly degenerate Hermitian sesquilinear form. We always assume that $\mathcal{L}$ is decomposable, that is there are subspaces $\mathcal{L}_+, \mathcal{L}_-, \mathcal{L}_0$ such that $\mathcal{L} = \mathcal{L}_+[\dot{+}]\mathcal{L}_-[\dot{+}]\mathcal{L}_0$, where $[f,f] \geq 0$ (≤ 0) if $0 \neq f \in \mathcal{L}_+$ ($\mathcal{L}_-$), and $\mathcal{L}_0 = \mathcal{L} \cap \mathcal{L}^{[\perp]}$ is the isotropic part of $\mathcal{L}$. Here $[\dot{+}]$ denotes a direct sum of linear manifolds, which are orthogonal with respect to $[\cdot,\cdot]$, and $[\perp]$ denotes the orthogonal complement with respect to $[\cdot,\cdot]$. Completing $\mathcal{L}_+$ ($\mathcal{L}_-$) with respect to the norms $\|f\| = [f,f]^{\frac{1}{2}}$ if $f \in \mathcal{L}_+$ and $\|f\| = (-[f,f])^{\frac{1}{2}}$ if $f \in \mathcal{L}_-$ and forming the orthogonal sum $\mathcal{K}$ of these completions, then $\mathcal{K}$ becomes in a natural way a Kreĭn space containing the factor space

$\widehat{\mathcal{L}} = \mathcal{L}/\mathcal{L}_0$ as a dense subspace. In the sequel, if $\mathcal{L}$ is a decomposable inner product space, we always fix one decomposition and everything is to be understood with respect to this decomposition. Recall that $\mathcal{L}$ is always decomposable in each of the following two cases:

(a) $\mathcal{L}$ is a Hilbert space with respect to some inner product $(\cdot,\cdot)$ and the indefinite inner product $[\cdot,\cdot]$ is continuous in this Hilbert space.
(b) The inner product $[\cdot,\cdot]$ has only a finite number κ, $0 \leq \kappa < \infty$ of negative squares.

Let S be a symmetric linear operator in $\mathcal{L}$ with a quasidense domain $\mathcal{D}(\mathcal{S}) \subset \mathcal{L}$. Here symmetric means that $[Sf,g] = [f,Sg]$ for all $f,g \in \mathcal{D}(\mathcal{S})$, and a set $\mathcal{D} \in \mathcal{L}$ is said to be quasidense in $\mathcal{L}$ if for each $f \in \mathcal{L}$ there exists a sequence $(f_n) \subset \mathcal{D}$ such that $\|\widehat{f}_n - \widehat{f}\| \to 0$ if $n \to \infty$. Then S generates a symmetric linear operator $\widehat{S}$ in $\mathcal{K}$ as follows:

$$\mathcal{D}(\widehat{S}) = \widehat{\mathcal{D}(S)} \text{ and } \widehat{S}\widehat{f} := \widehat{Sf}, f \in \mathcal{D}(S).$$

It is easy to see that this definition of $\widehat{S}$ is correct. If $f, f_1 \in \mathcal{D}(S)$ and $\widehat{f} = \widehat{f}_1$, then $[\widehat{Sf} - \widehat{Sf_1}, g] = [Sf - Sf_1, g] = [f - f_1, Sg] = 0$ for all $g \in \mathcal{D}(S)$. But the set of all $\widehat{g} \in \mathcal{K}$ where $g \in \mathcal{D}(S)$ is dense in $\mathcal{K}$, hence $\widehat{Sf} = \widehat{Sf_1}$.

In the following we shall always assume that $\hat{S}$ has a selfadjoint definitizable extension $\tilde{S}$ in some Kreĭn space $\tilde{\mathcal{K}} \supset \mathcal{K}$. It is well known (see, e.g., [7]) that in particular this holds if the inner product $[\cdot,\cdot]$ on $\mathcal{L}$ has a finite number of negative squares. It also holds if the Hermitian sesquilinear form $[Sf,g]$ on $\mathcal{D}(S)$ has a finite number of negative squares, and for at least one $\lambda_0 \in C$ the range $\mathcal{R}(\widehat{S} - \lambda_0)$ is closed in $\mathcal{K}$ (see, e.g., [2]).

A function $f : z \to f(z)$ with values in $\mathcal{L}$ is called holomorphic on $\mathcal{D}_f$ if $\widehat{f} : z \to \widehat{f}(z)$ with values in $\mathcal{K}$ is holomorphic.

Let again S be a symmetric linear operator in the decomposable inner product space $\mathcal{L}$, and let $\mathcal{H}$ be a Hilbert space. A mapping Φ from $\mathcal{L} \times C$ into $\mathcal{H}$

$$\Phi : (f; z) \to \Phi_z f,\ f \in \mathcal{L},\ z \in C,$$

is called a directing mapping for S if it has the following properties:

1. For each $z \in C$ the mapping $f \to \Phi_z f$, $f \in \mathcal{L}$ is linear.
2. For each $f \in \mathcal{L}$ the mapping $z \to \Phi_z f$, $z \in C$ is holomorphic on C.
3. If the function $z \to h(z)$ with values in $\mathcal{L}$ is holomorphic on $\mathcal{D}_h \subset C$, then $\Phi_z h(z) = 0$, $z \in \mathcal{D}_h$, if and only if there exists a holomorphic function $z \to g(z)$ on $\mathcal{D}_h$ with values in $\mathcal{D}(S)$ such that $(S - z)g(z) = h(z)$.
4. For each compact set $\Gamma \subset C$ there is a mapping $\Psi^{(\Gamma)} : (x; z) \to \Psi_z^{(\Gamma)} x$ from $\mathcal{H} \times \Gamma$ into $\mathcal{L}$ such that
   (a) for each $z \in \Gamma$ the mapping $z \to \Psi_z^{(\Gamma)} x$, $x \in \mathcal{H}$, is linear and for each $x \in \mathcal{H}$ the mapping $z \to \Psi_z^{(\Gamma)} x$ is holomorphic on the interior Γ^i of Γ;

(b) $\Phi_z \Psi_z^{(\Gamma)} x = x$, $x \subset \mathcal{H}$, $z \in \Gamma$;

(c) if $z \in \Gamma$ is fixed we have $\|\Psi_z^{(\Gamma)} x\| \leq C(z;\Gamma)\|x\|$ with some constant $C(z;\Gamma)$ depending on z and Γ.

Recall that in the original definition of a directing functional or a directing mapping (see [7]), concerning the case of a nonnegative semidefinite inner product $[\cdot,\cdot]$ on $\mathcal{L}$ the following condition (3_0) was imposed instead of (3):

3_0. For $z \in C$ and $h \in \mathcal{L}$ we have $\Phi_z h = 0$ if and only if there exists an element $g \in \mathcal{D}(S)$ such that $Sg - zg = h$.

Evidently, if we assume (3_0), then for a function $h(z)$ as in (3) there exists a function $g(\cdot)$ on $\mathcal{D}_h$ which solves $(S - z)g(z) = h(z)$ on $\mathcal{D}_h$. However, it is not clear whether $g(\cdot)$ is holomorphic on $\mathcal{D}_h$. The following lemma implies that this is the case if the inner product $[\cdot,\cdot]$ is nonnegative semidefinite on $\mathcal{L}$, i.e., in this case condition (3_0) implies condition (3).

Lemma 4.1. *Assume that the mapping Φ has the properties* (1), (2), (3_0). *Then the condition* (3) *is satisfied in each of the following two cases:*

(a) *The inner product $[\cdot,\cdot]$ on $\mathcal{L}$ is nonnegative semidefinite.*

(b) *$\widehat{S}$ has a definitizable extension $\widetilde{S}$ in some Kreĭn space $\widetilde{\mathcal{K}} \supset \mathcal{K}$ and for each function $z \to h(z)$ which is holomorphic on $\mathcal{D}_h \subset C$ and such that $\Phi_z h(z) = 0$, $z \in \mathcal{D}_h$, the function $z \to g(z)$ on $\mathcal{D}_h$ defined by the solutions of the equation $(S - z)g(z) = h(z)$ (which exist according to (3_0)) is locally bounded on $\mathcal{D}_h$.*

Proof. If the inner product $[\cdot,\cdot]$ on $\mathcal{L}$ is nonnegative, then the symmetric operator $\widehat{S}$ on the Hilbert space $\mathcal{K}$ has at least one selfadjoint extension $\widetilde{S}$ in some Hilbert space $\widetilde{\mathcal{K}} \supset \mathcal{K}$, which can also be regarded as a definitizable operator. In both cases we can consider a definitizable extension $\widetilde{S}$ of $\widehat{S}$ in some Kreĭn space $\widetilde{\mathcal{K}} \supset \mathcal{K}$. Let $z \to h(z)$ be as in condition (3), and let $z \to g(z)$ on $\mathcal{D}_h$ be a solution of the equation $(S - z)g(z) = h(z)$ (which exists according to (3_0)). Then $(\widehat{S} - z)\widehat{g(z)} = \widehat{h(z)}$, hence also

$$(\widetilde{S} - z)\widehat{g(z)} = \widehat{h(z)} \tag{4.1}$$

and if $z \notin \sigma_p(\widetilde{S})$, then $\widehat{g(z)} = (\widetilde{S} - z)^{-1}\widehat{h(z)}$.

On the set $\mathcal{D}_h \backslash \sigma(\widetilde{S})$ we consider the holomorphic function $\widetilde{g}$:

$$\widetilde{g}(z) := (\widetilde{S} - z)^{-1}\widehat{h(z)}. \tag{4.2}$$

Then we have $(\widetilde{S} - z)\widetilde{g}(z) = \widehat{h(z)}$, and from (4.1) $(\widetilde{S} - z)\widehat{g(z)} = \widehat{h(z)}$. It follows that $\widehat{g(z)}$ is holomorphic on $\mathcal{D}_h \backslash \sigma(\widetilde{S})$.

Let $\Gamma = (\alpha, \beta)$ be a real interval such that $\overline{\Gamma} \subset \mathcal{D}_h$. Assume that $\alpha, \beta \notin c(\widetilde{S}) \cup \sigma_p(S)$ and consider again the function $\widetilde{g}(z)$ given by (4.2) for $z \neq \bar{z}$, $z \notin \sigma_p(S)$. Then the function $z \to (\widetilde{S} - z)^{-1}\widehat{h(z)}$ is bounded if z approaches α or β nontangentially.

Indeed, we have, e.g.,

$$\begin{aligned}(\widetilde{S}-z)^{-1}\widehat{h(z)} \\ &= (\widetilde{S}-z)^{-1}(\widehat{h(z)}-\widehat{h(\alpha)}+((\widetilde{S}-z)^{-1}-(\widetilde{S}-\alpha)^{-1})\widehat{h(\alpha)}+(\widetilde{S}-\alpha)^{-1}\widehat{h(\alpha)} \\ &= (z-\alpha)(\widetilde{S}-z)^{-1}(\widehat{h(z)}-\widehat{h(\alpha)})(z-\alpha)^{-1}+(z-\alpha)(\widetilde{S}-z)^{-1}\widehat{g(\alpha)}+\widehat{g(\alpha)}\end{aligned}$$

and this remains bounded if $z\to\alpha$ nontangentially as $|z-\alpha|\|(\widetilde{S}-z)^{-1}\|$ has this property.

We now choose a closed Jordan contour $\mathcal{C}_\Gamma$ surrounding Γ, intersecting the real axis in α and β nontangentially and such that the non-real spectrum of $\widetilde{S}$ is outside of $\mathcal{C}_\Gamma$. Consider in the domain surrounded by $\mathcal{C}_\Gamma$ the function $\widehat{g_0}$

$$\widehat{g_0}(z) := \frac{1}{2\pi i}\oint'_{C_\Gamma}(\zeta-z)^{-1}(\widetilde{S}-\zeta)^{-1}\widehat{h(\zeta)}d\zeta.$$

It is holomorphic, and we have

$$(\widetilde{S}-z)\widehat{g_0}(z) = \frac{1}{2\pi i}\oint'_{C_\Gamma}(\zeta-z)^{-1}\widehat{h(\zeta)}d\zeta+\frac{1}{2\pi i}\oint'_{C_\Gamma}(\widetilde{S}-\zeta)^{-1}\widehat{h(\zeta)}d\zeta = \widehat{h(z)}+\widetilde{h_0} \quad (4.3)$$

with some element $\widetilde{h_0}\in\mathcal{K}$, defined by the last integral. Evidently, $\widetilde{h_0}\in\widetilde{E}(\Gamma)\widetilde{\mathcal{K}}$, where by $\widetilde{E}$ we denote the spectral function of $\widetilde{\mathcal{S}}$. It follows from (4.3) and (4.1) that for all z in the interior of C_Γ

$$(\widetilde{S}-z)(\widehat{g_0}(z)-\widehat{g(z)}) = \widetilde{h_0}. \quad (4.4)$$

Consider an interval $\Gamma_0\subset\Gamma$, $\Gamma_0\in\mathcal{R}_{\widetilde{S}}$ such that all the points of $\Gamma_0\cap\sigma(\widetilde{S})$ are of either positive or negative type (see [6], [8]). Then (4.4) implies

$$\left|\int_{\Gamma_0}\frac{[\widetilde{E}(d\lambda)\widetilde{h}_0,\widetilde{h}_0]}{|\lambda-z|^2}\right|<\infty$$

for all $z\in\Gamma_0$, hence $[\widetilde{E}(\Gamma_0)\widetilde{h}_0,\widetilde{h}_0]=0$ and $\widetilde{E}(\Gamma_0)\widetilde{h}_0=0$. From (4.4) follows $\widehat{g(z)}=\widehat{g_0}(z)$ on Γ_0, hence $g(z)$ is holomorphic there and the first statement of the lemma is proved.

In the second situation the above reasoning implies that $\widetilde{h}_0$ belongs to the linear span of the algebraic eigenspaces of $\widetilde{S}$ corresponding to eigenvalues in $c(\widetilde{S})\cap\Gamma$. Then, if $\widetilde{h}_0$ has a non-zero component in the algebraic eigenspace corresponding to $\lambda_0\in c(\widetilde{S})\cap\Gamma$, we find

$$(\widetilde{S}-z)^{-1}\widetilde{h}_0=\sum_{j=0}^{n_0}\frac{\widetilde{h}_0^{(j)}}{(\lambda_0-z)^{j+1}}+\cdots$$

for some $n_0\geq 0$, $\widetilde{h}_0^{(n_0)}\neq 0$, where $\ldots$ denotes terms which are holomorphic at λ_0.

But then (4.4) implies that $\widehat{g_0}(z)-\widehat{g(z)}$ is unbounded at λ_0, which is impossible, since $\widehat{g_0}(z)$ and $\widehat{g(z)}$ are bounded there. It follows that $\widetilde{h}_0=0$, so $\widehat{g_0}(z)=\widehat{g(z)}$.

The non-real points of $\sigma_0(\widetilde{S}) \cap \mathcal{D}_h$ can be treated in a similar way, and the lemma is proved. □

Let $\Gamma \in C$ be a closed disc around zero. By $\widehat{\Psi}_z^{(\Gamma)}$ we denote the mapping in $\mathcal{L}(\mathcal{H}, \mathcal{K})$ generated by $\Psi_z^{(\Gamma)}$, that is

$$\widehat{\Psi}_z^{(\Gamma)} x = \widehat{\Psi_z^{(\Gamma)} x},\ x \in \mathcal{H}.$$

Then $\widehat{\Psi}_\cdot^{(\Gamma)}$ is holomorphic on Γ^i and it can also be considered as a function with values in $\mathcal{L}(\mathcal{H}, \widetilde{\mathcal{K}})$. For $A \in \mathcal{L}(\widetilde{K})$ the operator $\widehat{\Psi}_A^{(\Gamma)}$ is defined according to (3.1).

Lemma 4.2. *Let the closed disc $\Gamma \subset C$ around zero be such that $\widetilde{\sigma_0}(S) \subset \Gamma^i$, and let Δ be a simply connected subset of Γ^i, symmetric with respect to R with smooth boundary $\partial\Delta$ which intersects the real axis nontangentially and is such that $(\partial\Delta) \cap (\sigma_p(\widetilde{S}) \cup c(\widetilde{S})) = \emptyset$. Denote by $\widetilde{S}_\Delta$ the restriction of $\widetilde{S}$ to $\widetilde{E}(\Delta)\mathcal{K}$. Then, for $f \in \mathcal{L}$*

$$\widetilde{E}(\Delta)\widehat{f} = -\frac{1}{2\pi i}\oint_{\partial\Delta} (\widetilde{S} - z)^{-1} \widehat{\Psi}_{\widetilde{S}_\Delta}^{(\Gamma)} \Phi_z f dz.$$

Proof. For any $g \in \widetilde{\mathcal{K}}$

$$\begin{aligned}
&-\frac{1}{2\pi i}\oint_{\partial\Delta}' \left[\widetilde{E}(\Delta)\widehat{f} + \frac{1}{2\pi i}\oint_{\partial\Delta} (\widetilde{S} - z)^{-1}\widehat{\Psi}_{\widetilde{S}_\Delta}^{(\Gamma)}\Phi_z f dz, (\widetilde{S} - \zeta)^{-1} g\right] d\bar{\zeta} \\
&= -\frac{1}{2\pi i}\oint_{\partial\Delta}' \left[\widehat{f} + \frac{1}{2\pi i}\oint_{\partial\Delta} (\widetilde{S}_\Delta - z)^{-1}\widehat{\Psi}_z^{(\Gamma)}\Phi_z f dz, (\widetilde{S}_\Delta - \zeta)^{-1} g\right] d\bar{\zeta} \\
&= \frac{1}{2\pi i}\oint_{\partial\Delta}' \left[\widehat{f} - \widehat{\Psi}_z^{(\Gamma)}\Phi_z f, (\widetilde{S}_\Delta - \bar{z})^{-1} g\right] dz.
\end{aligned}$$

Furthermore, $\Phi_z(f - \Psi_z^{(\Gamma)}\Phi_z f) = 0$, hence there exists a holomorphic solution $g(z) \in \mathcal{L}$ of the equation $(S - z)g(z) = f - \Psi_z^{(\Gamma)}\Phi_z f$, therefore $(\widehat{S} - z)\widehat{g(z)} = \widehat{f} - \widehat{\Psi}_z^{(\Gamma)}\Phi_z f$, the last integral becomes

$$\frac{1}{2\pi i}\oint_{\partial\Delta}' [\widehat{g(z)}, g] dz = 0,$$

and the lemma is proved. □

The proof of the following lemma is similar.

Lemma 4.3. *Let Γ, Γ' be discs around zero, both containing some set Δ as in Lemma 4.2. If Ψ, Ψ' are mappings with the properties (4), then*

$$\widehat{\Psi}_{\widetilde{S}_\Delta}^{(\Gamma)} = \widehat{(\Psi')}_{\widetilde{S}_\Delta}^{(\Gamma')}.$$

Proof. If $z \in \Gamma^i \cap (\Gamma')^i$, then for $x \in \mathcal{H}$ we have $\Phi_z(\Psi_z^{(\Gamma)}x - (\Psi')_z^{(\Gamma')}x) = 0$, hence $\Psi_z^{(\Gamma)}x - (\Psi')_z^{(\Gamma')}x = (S - z)g(z)$ with some function g holomorphic in $\overline{\Delta}$ and with values in $\mathcal{D}(S)$. For arbitrary $g \in \widetilde{\mathcal{K}}$ it follows

$$-\frac{1}{2\pi i}\oint_{\partial\Delta}' [\widehat{\Psi}_{\tilde{S}_\Delta}^{(\Gamma)}x - \widehat{(\Psi')}_{\tilde{S}_\Delta}^{(\Gamma')}x, (\widetilde{S} - z)^{-1}g]dz = -\frac{1}{2\pi i}\oint_{\delta\Delta}' [\widehat{g(z)}, g]dz = 0,$$

and the lemma is proved. □

5. The main theorem

Theorem 5.1. *Let $\mathcal{L}$ be a linear space with an inner product $[\cdot,\cdot]$ which is decomposable, let some decomposition be fixed, and let S be a symmetric linear operator in $\mathcal{L}$ with a quasidense domain $\mathcal{D}$, such that its corresponding operator $\widehat{S}$ in $\mathcal{K}$ has at least one definitizable extension in some Kreĭn space $\widetilde{\mathcal{K}} \supset \mathcal{K}$. Assume also that S has a directing mapping Φ with values in some Hilbert space $\mathcal{H}$.*

Then, for each definitizable selfadjoint extension $\widetilde{S}$ of $\widehat{S}$ there exist operators
$A_{i,\nu} = A_{i,\nu}^* \in \mathcal{L}(\mathcal{H})$, $\nu = 0, 1, \dots, q_i$, $i = 1, 2, \dots, l$, $A_{i,q_i} \geq 0$,
$B_{j,\nu} \in \mathcal{L}(\mathcal{H})$, $\nu = 0, 1 \dots, r_j$, $j = 1, 2 \dots, n$,
$C_{k,\nu} = C_{k,\nu}^* \in \mathcal{L}(\mathcal{H})$, $C_{k,p_k} \neq 0$; $\nu = 0, 1, \dots, p_k$; $p_k \geq 1$, $k = 0, 1, \dots, m$,
points $\beta_1, \dots, \beta_n \in C^+$,
an $\mathcal{L}(\mathcal{H})$-valued measure $V \in \mathcal{V}$ with $s(V) = \{\alpha_1, \dots, \alpha_l\}$,
points $\gamma_1, \dots, \gamma_m \in R \backslash s(V)$ where $V(\{\gamma_1, \dots, \gamma_m\}) = 0$,

such that for arbitrarily chosen intervals $\Delta_i \in \mathcal{R}_V$ around α_i which satisfy the condition

$$(\{\gamma_1, \dots, \gamma_m\} \cup s(V)) \cap \Delta_i = \{\alpha_i\},\ i = 1, 2, \dots, l,$$

the representation

$$\begin{aligned}
[f, g] = &\sum_{i=1}^{l}\left(\int_{\Delta_i}(V(d\lambda)\Phi_\lambda f, \Phi_\lambda g)_{(\alpha_i;q_i)} + \sum_{\nu=1}^{q_i}(A_{i,\nu}\Phi_t f, \Phi_t g)^{(\nu)}|_{t=\alpha_i}\right)\\
&+ \int_{R\backslash\cup_{i=1}^{l}\Delta_i}(V(d\lambda)\Phi_\lambda f, \Phi_\lambda g)\\
&+ \sum_{j=0}^{n}\sum_{\nu=0}^{r_j}\left[(B_{j,\nu}\Phi_t f, \Phi_t g)^{(\nu)})|_{t=\beta_j} + (B_{j,\nu}^*\Phi_t f, \Phi_t g)^{(\nu)}|_{t=\beta_j}\right]\\
&+ \sum_{k=1}^{m}\sum_{\nu=0}^{p_j}(C_{k,\nu}\Phi_t f, \Phi_t g)^{(\nu)}|_{t=\gamma_k} \qquad (5.1)
\end{aligned}$$

holds, if at least one of the elements f, g belongs to $\mathcal{D}(S)$, the other one being arbitrary in $\mathcal{L}$. If the degree of the definitizing polynomial of $\widetilde{S}$ is zero (that is $\widetilde{\mathcal{K}}$ is a Hilbert space) or even, then the representation holds for all $f, g \in \mathcal{L}$.

The representation can be chosen such that $s(V)$ is the set of singular critical points of $\widetilde{S}$, $\gamma_1,\dots,\gamma_m$ are the regular critical points of $\widetilde{S}$ which belong to $\sigma_p(\widetilde{S})$ and are such that $\widetilde{S}$ has a Jordan chain of length >1 in γ_j, and $\beta_1,\dots,\beta_n$ are the eigenvalues of $\tilde{S}$ in C^+.

If in particular the inner product $[\cdot,\cdot]$ has a finite number κ of negative squares and the numbers q_i, $i=1,2,\dots,l$ are chosen minimal: $q_i=2\kappa_i$, $\kappa_i\in\{0,\frac{1}{2},1,\dots\}$, then the representation can be chosen so that V with possible exception of $\kappa'\le\kappa$ points $\delta_1,\dots,\delta_{\kappa'}\in\operatorname{supp}V$ is nonnegative and

$$\sum_{\nu=1}^{\kappa'}\kappa_-(V(\{\delta_\nu\}))+\sum_{i=1}^{l}\kappa_i+\sum_{j=1}^{n}\operatorname{rank}B_j+\sum_{k=1}^{m}\operatorname{rank}C_k\le\kappa, \tag{5.2}$$

where

$$B_j=\begin{pmatrix} B_{j,r_j} & 0 & \dots & 0\\ B_{j,r_j-1} & B_{j,r_j} & \dots & 0\\ \vdots & \vdots & \vdots & \vdots\\ B_{j,0} & B_{j,1} & \dots & B_{j,r_j}\end{pmatrix}$$

and

$$C_k=\begin{pmatrix} C_{k,q_k} & 0 & \dots & 0\\ \vdots & \vdots & \vdots & \vdots\\ C_{k,0} & C_{k,1} & \dots & C_{k,p_k}\end{pmatrix}.$$

Remark 5.2. If $l=0, m=0$ etc. in the representation (5.1) it means that the corresponding sums are 0.

Proof. Let $\tilde{S}$ be a definitizable selfadjoint extension of $\hat{S}$ in some Kreĭn space $\tilde{\mathcal{K}}\supset\mathcal{K}$. If α is a finite critical point of S and Δ is a bounded interval around α such that $\Delta\cap c(\widetilde{S})=\{\alpha\}$, and the endpoints of Δ are not in $\sigma_p(\widetilde{S})$, then, for some nonnegative integer q, $[(\widetilde{S}-\alpha)^q\widetilde{f},\widetilde{f}]\ge 0$ for all $\widetilde{f}\in\widetilde{E}(\Delta)\mathcal{K}$ (see [8]). According to Lemma 2.1, for $f,g\in\mathcal{L}$ and with suitably chosen Γ and $\mathcal{C}_\Delta$, we then have

$$\begin{aligned}
[\widetilde{E}(\Delta)\hat{f},\hat{g}]&=\left[-\frac{1}{2\pi i}\oint_{\mathcal{C}_\Delta}(\tilde{S}_\Delta-z)^{-1}\widehat{\Psi}^{(\Gamma)}_{\tilde{S}_\Delta}\Phi_z f dz,-\frac{1}{2\pi i}\oint_{\mathcal{C}_\Delta}(\tilde{S}_\Delta-z)^{-1}\widehat{\Psi}^{(\Gamma)}_{\tilde{S}_\Delta}\Phi_z g dz\right]\\
&=-\frac{1}{2\pi i}\oint_{\mathcal{C}_\Delta}\left[\left(\tilde{S}_\Delta-z\right)^{-1}\widehat{\Psi}^{(\Gamma)}_{\tilde{S}_\Delta}\Phi_z f,\widehat{\Psi}^{(\Gamma)}_{\tilde{S}_\Delta}\Phi_{\bar{z}}g\right]dz \qquad (5.3)\\
&=-\frac{1}{2\pi i}\oint_{\mathcal{C}_\Delta}\left[\int_\Delta\left(\frac{1}{\lambda-z}+\frac{1}{z-\alpha}+\dots+\frac{(\lambda-\alpha)^{q-1}}{(z-\alpha)^q}\right)\tilde{E}(d\lambda)\widehat{\Psi}^{(\Gamma)}_{\tilde{S}_\Delta}\Phi_z f,\widehat{\Psi}^{(\Gamma)}_{\tilde{S}_\Delta}\Phi_{\bar{z}}g\right]dz\\
&+\frac{1}{2\pi i}\oint_{\mathcal{C}_\Delta}\left[\left(\frac{1}{z-\alpha}+\frac{\tilde{S}_\Delta}{(z-\alpha)^2}+\dots+\frac{\tilde{S}^{q-1}_\Delta}{(z-\alpha)^q}+\frac{N}{(z-\alpha)^{q+1}}\right)\widehat{\Psi}^{(\Gamma)}_{\tilde{S}_\Delta}\Phi_z f,\widehat{\Psi}^{(\Gamma)}_{\tilde{S}_\Delta}\Phi_{\bar{z}}g\right]dz.
\end{aligned}$$

If we put

$$V(d\lambda):=(\widehat{\Psi}^{(\Gamma)}_{\tilde{S}_\Delta})^*\tilde{E}(d\lambda)\widehat{\Psi}^{(\Gamma)}_{\tilde{S}_\Delta}, \tag{5.4}$$

then, according to (3.3), the first integral on the right-hand side equals

$$\int_{\Delta} (V(d\lambda)\Phi_\lambda f, \Phi_\lambda g)_{(\alpha;q)}.$$

With $\alpha = \alpha_i$, $\Delta = \Delta_i$, $q = q_i$, $i = 1, 2, \ldots, l$ this gives the first sum in (5.1) and with

$$A_{i,\nu} := (\widehat{\Psi}^{(\Gamma)}_{\tilde{S}_{\Delta_i}})^* \tilde{S}^{\nu}_{\Delta_i} \widehat{\Psi}^{(\Gamma)}_{\tilde{S}_{\Delta_i}}, \ \nu = 0, \ldots, q_i - 1; \ A_{i,q_i} := (\widehat{\Psi}^{(\Gamma)}_{\tilde{S}_{\Delta_i}})^* N \widehat{\Psi}^{(\Gamma)}_{\tilde{S}_{\Delta_i}},$$

$i = 1, 2, \ldots, l$, the second integral in (5.3) gives the second sum in (5.1).

The other terms in (5.1) follow in the same way from (5.3), if only $d\lambda$ has a positive distance from all the critical points.

The terms in (5.1) corresponding to the non-real points $\beta_1, \ldots \beta_n$ follow in the same way from the Laurent expansion of $(\tilde{S} - z)^{-1}$ around β_k, $k = 1, 2, \ldots, n$.

In the middle integral in (5.1) over $R \setminus \cup_{i=1}^{l} \Delta_i$ the measure $V(d\lambda)$ is again defined by (5.4), and this integral exists in the case that ∞ is a critical point of $\tilde{S}$ if at least one of the elements $\widehat{f}$ or $\widehat{g}$ belongs to $\mathcal{D}(\widehat{S}) \subset \mathcal{D}(\widetilde{S})$. If ∞ is not a critical point of $\tilde{S}$ the integral exists for all $f, g \in \mathcal{D}(\widetilde{S})$. The statement $V(\{\gamma_1, \ldots, \gamma_m\}) = 0$ follows by continuity.

For the statement in the case that the inner product $[\cdot,\cdot]$ has a finite number κ of negative squares we observe, e.g., (see [4], [5]) that the dimension of the algebraic eigenspace corresponding to an eigenvalue β_j in C^+, $j = 1, \ldots, n$, coincides with the dimension of the range of the matrix

$$\begin{pmatrix} E_{j,s_j} & 0 & \ldots & 0 \\ E_{j,s_j-1} & E_{j,s_j} & \ldots & 0 \\ \vdots & \vdots & \vdots & \vdots \\ E_{j,1} & E_{j,2} & E_{j,s_j-1} & E_{j,s_j} \end{pmatrix},$$

where $E_{j,\nu}$ are the coefficients of the Laurent expansion at $z = \beta_j$ of the resolvent:

$$(\tilde{S} - z)^{-1} = \frac{E_{j,s_j}}{(\beta_j - z)^{s_j}} + \cdots + \frac{E_{j,1}}{\beta_j - z} + \cdots,$$

and

$$B_{j,\nu} := (\widehat{\Psi}^{(\Gamma)}_{\tilde{S}_{\tilde{\Delta}_j}})^* E_{j,\nu} \widehat{\Psi}^{(\Gamma)}_{\tilde{S}_{\tilde{\Delta}_j}}, \ \beta_j \in \widetilde{\Delta}_j \subset C^+,$$

where $\sigma(\widetilde{S}) \cap \widetilde{\Delta}_j = \{\beta_j\}$.

On the other hand, this dimension equals the number of negative squares of $[\cdot,\cdot]$ on the eigenspace.

We also observe that for $i = 1, \ldots, l$, κ_i at most equals the number of negative squares of $[\cdot,\cdot]$ on $\widetilde{E}(\Delta_i)\widetilde{\Pi}_\kappa$ (see Definition 2.2 and Folgerung 3.11 in [6]). The other terms arise in the same way, thus the estimate (5.2) holds. □

References

[1] M.B. Bekker, *Isometric operators with directing mappings.* Current analysis and its applications **218**, 3–9, "Naukova Dumka", Kiev, 1989 (Russian).

[2] B. Ćurgus, H. Langer, *Kreĭn space approach to symmetric ordinary differential operators with an indefinite weight function.* J. Differential Equations **79** (1989), no. 1, 31–61.

[3] M.G. Kreĭn, *On hermitian operators with directing functionals.* Sb. Trud. Inst. Mat. Kiev **10** (1948), 83–106 (Ukrainian).

[4] M.G. Kreĭn, H. Langer, *Über die Q-Funktion eines π-hermiteschen Operators im Raume Π_κ.* Acta Sci. Math. (Szeged) **34** (1973), 191–230.

[5] M.G. Kreĭn, H. Langer, *Some propositions on analytic matrix functions related to the theory of operators in the space Π_κ.* Acta Sci. Math. (Szeged) **43** (1981), 181–205.

[6] H. Langer, *Zur Spektraltheorie linearer Operatoren in J-Räumen und einige Anwendungen auf die Schar $L(\lambda) = \lambda^2 I + \lambda B + C$.* Habilitationsschrift, Techn. Universität Dresden, 1965.

[7] H. Langer, *Über die Methode der richtenden Funktionale von M.G.Kreĭn.* Acta Math. Acad. Sci. Hungar. **21** (1970), 207–224.

[8] H. Langer, *Spectral functions of definitizable operators in Kreĭn spaces.* Functional analysis, Proc. Dubrovnik 1981, Lecture Notes in Math. **948**, pp 1–46, Springer-Verlag, Berlin, 1982.

[9] H. Langer, B. Textorius, *Spectral functions of a symmetric linear relation with a directing mapping.* Proc. Roy. Soc. Edinburgh **81A** (1978), 237–246.

[10] H. Langer, B. Textorius, *Spectral functions of a symmetric linear relation with a directing mapping, I.* Proc Roy. Soc. Edinburgh **97A** (1984), 165–176.

[11] M.A. Naimark, *Linear differential operators.* Ungar, New York 1968.

Björn Textorius
Linköping University
SE 58183 Linköping
Sweden
e-mail: bjtex@mai.liu.se

Operator Theory:
Advances and Applications, Vol. 163, 365–379

The Extension Problem for Positive Definite Functions. A Short Historical Survey

Zoltán Sasvári

Dedicated to Heinz Langer on the occasion of his retirement

Abstract. The aim of the present paper is to give a short historical survey on the extension problem for positive definite functions.

Mathematics Subject Classification (2000). Primary 43A35, 47A57, 34A55; Secondary 42A82, 46C20, 34B20.

Keywords. continuation, positive definite functions, functions with negative squares.

1. Introduction

Acknowledgments are usually placed at the end of a paper. In this case, however, I would like to use the first page to thank Heinz for drawing my attention to the extension problem, for posing many interesting questions and answering many of my questions. He and M.G. Krein studied many interpolation, moment and continuation problems.

In the present paper I try to give a short historical survey only in a special area of these continuation problems. I will consider only functions, not kernels. These functions will be complex-valued (i.e., not operator- or matrix-valued) and in most cases defined on the real line. However, I will also treat functions with negative squares, as a natural generalization of positive definite functions. Even in this special area, the survey is not complete, the choice of the topics was also influenced by the author's research interests.

2. The emergence of the concept of positive definite functions

The history of positive definite functions started in 1907 with Carathéodory's paper [10]. In this paper he studied functions of the form

$$1 + \sum_{k=1}^{\infty} (a_k + \mathrm{i}b_k) \cdot z^k \qquad (a_k, b_k \in \mathbb{R})$$

analytic in the unit disc and having positive real part. He characterized these functions by the property that for each $n = 1, 2, \ldots$ the point

$$(a_1, b_1, \ldots, a_n, b_n) \in \mathbb{R}^{2n}$$

should lie in the smallest convex set containing the points

$$2 \cdot (\cos\varphi, \sin\varphi, \cos 2\varphi, \sin 2\varphi, \ldots, \cos n\varphi, \sin n\varphi), \qquad 0 \le \varphi < 2\pi.$$

In 1911 Toeplitz [56] noticed that Carathéodory's condition is equivalent to the inequality

$$\sum_{k,l=1}^{n} d_{k-l} c_k \bar{c}_l \ge 0 \qquad (n = 1, 2, \ldots;\ c_k \in \mathbb{C}) \tag{2.1}$$

where $d_0 = 2,\ d_k = a_k - \mathrm{i}b_k,\ d_{-k} = \overline{d_k}$.

In the same year Herglotz [17] solved the so-called *trigonometric moment-problem*. He proved that condition (2.1) of Toeplitz holds if and only if there exists a non-negative measure μ on $[0, 2\pi)$ such that

$$d_n = \int_0^{2\pi} \mathrm{e}^{\mathrm{i}nt}\, \mathrm{d}\mu(t) \quad (n = 0, \pm 1, \pm 2, \ldots).$$

Note that if such a measure exists, then it is unique.

Inspired by the work of Carathéodory and Toeplitz, Mathias [41] introduced in 1923 the notion of a positive definite function on $\mathbb{R}$. He called a complex-valued function f on $\mathbb{R}$ *positive definite* if

$$f(-x) = \overline{f(x)} \quad (x \in \mathbb{R}) \tag{2.2}$$

and

$$\sum_{i,j=1}^{n} f(x_i - x_j) c_i \bar{c}_j \ge 0 \tag{2.3}$$

for every choice of $x_1, \ldots, x_n \in \mathbb{R}$ and $c_1, \ldots, c_n \in \mathbb{C}$. [1] He showed that a continuous function $f \in L^1(\mathbb{R})$ is positive definite if and only if

$$\int_{-\infty}^{\infty} f(t) \cdot \mathrm{e}^{\mathrm{i}tx}\, \mathrm{d}t \ge 0 \quad (x \in \mathbb{R}).$$

This condition can be used for example to show that the functions

$$x \mapsto \mathrm{e}^{-|x|}, \quad \mathrm{e}^{-x^2}, \quad \frac{1}{x^2+1} \quad \text{and} \quad \max(0, 1 - |x|)$$

[1] It is interesting to note that condition (2.2) remains part of the definition of a positive definite function until 1933, when F. Riesz [47] points out that it follows easily from (2.3).

are positive definite. Among the few examples, where the validity of (2.3) can easily be checked are the functions $x \mapsto \mathrm{e}^{\mathrm{i}xy}$ $(x, y \in \mathbb{R})$. Indeed,

$$\sum_{j,k=1}^{n} \mathrm{e}^{\mathrm{i}(x_j - x_k)y}\, c_j \overline{c_k} = \left| \sum_{j=1}^{n} \mathrm{e}^{\mathrm{i}x_j y} c_j \right|^2 \geq 0.$$

In 1932 Bochner [6] proved that these functions are the building stones of positive definite functions in the following sense: A continuous, complex-valued function f on $\mathbb{R}$ is positive definite if and only if there exists a non-negative finite measure μ on $\mathbb{R}$ such that

$$f(x) = \int_{-\infty}^{\infty} \mathrm{e}^{\mathrm{i}xt}\, \mathrm{d}\mu(t).$$

If such a measure exists, then it is unique. In 1933 Bochner [8] generalized this result to functions of several real variables.

Another important theorem of the same period is due to Stone [54, 55]: for each $t \in \mathbb{R}$ let U_t be a unitary operator in a Hilbert space H such that $U_{t+s} = U_t U_s$ and $t \longrightarrow (v, U_t w)$ is continuous for all $v, w \in H$. Then there exists a so-called *partition of unity* $E(\lambda)$ such that

$$U_t = \int_{-\infty}^{\infty} \mathrm{e}^{\mathrm{i}t\lambda}\, \mathrm{d}E(\lambda).$$

About the same time both F. Riesz [47] and Bochner [7] pointed out that Bochner's theorem can be used to prove Stone's result. Their proofs use the fact that for any $v \in H$ the function

$$t \mapsto (v, U_t v)$$

is positive definite:

$$\sum_{j,k=1}^{n} (U_{x_j - x_k} v, v) c_j \overline{c_k} = \left(\sum_{j=1}^{n} c_j U_{x_j} v, \sum_{k=1}^{n} c_k U_{x_k} v \right) \geq 0.$$

Another application of positive definite functions is in the theory of random processes. Let $Z = \{Z(x) : x \in \mathbb{R}\}$ be a random process, where the random variables $Z(x)$ are real-valued and are defined on a common probability space. The process Z is said to be *second-order stationary*, or simply *stationary*, if second moments exist and the expectation $E(Z(x))$ and the covariance $\mathrm{cov}(Z(x), Z(x+h))$ do not depend on x. We may then define the *covariance function*

$$C(h) = \mathrm{cov}(Z(x),\, Z(x+h)), \quad h \in \mathbb{R}.$$

Khintchin [23] used Bochner's theorem to show that a continuous real-valued function C is the covariance function of a continuous stationary random process if and only if there exists a non-negative measure μ on $\mathbb{R}$ such that

$$C(h) = \int_{-\infty}^{\infty} \cos ht\, \mathrm{d}\mu(t),$$

i.e., if C is a real-valued positive definite function.

In 1959 M.G. Krein [27] (see also [26, 19, 20, 21]) generalized the concept of positive definiteness in the following way:

A complex-valued Hermitian function f on $\mathbb{R}$ (that is, $f(-x) = \overline{f(x)}$ for all $x \in \mathbb{R}$) is said to have k *negative squares* if the Hermitian matrix

$$A = (f(x_i - x_j))_{i,j=1}^{n}$$

has at most k negative eigenvalues (counted with their multiplicities) for any choice of n and $x_1, \ldots, x_n \in \mathbb{R}$, and for some choice of n and $x_1, \ldots, x_n$ the matrix A has exactly k negative eigenvalues. We denote by $P_k(\mathbb{R})$ the set of all functions on $\mathbb{R}$ with k negative squares. Note that $P_0(\mathbb{R})$ is the set of positive definite functions.

We list a few examples:

$$x \mapsto x^k \mathrm{e}^x + (-x)^k \mathrm{e}^{-x} \in P_k(\mathbb{R})$$
$$x \mapsto (-1)^k x^{2k} \in P_k(\mathbb{R}),$$
$$x \mapsto (-1)^{k+1} x^{2k} \in P_{k+1}(\mathbb{R}).$$

More generally: if $a \in (2k-2, 2k]$ $(k > 0)$ then

$$x \mapsto (-1)^k |x|^a \in P_k(\mathbb{R}).$$

In [27] M.G. Krein proved that every continuous function $f \in P_k(\mathbb{R})$ is definitizable in the following sense: there exists a polynomial Q of degree k such that the inequality

$$\int_{-\infty}^{\infty} \int_{-\infty}^{\infty} f(x-y) Q\left(-\mathrm{i}\frac{\mathrm{d}}{\mathrm{d}y}\right) h(y)\, \overline{Q\left(-\mathrm{i}\frac{\mathrm{d}}{\mathrm{d}x}\right) h(x)}\, \mathrm{d}y\, \mathrm{d}x \geq 0$$

holds for every infinitely differentiable function h with compact support. He obtained the integral representation

$$f(x) = p(x) + \int_{-\infty}^{\infty} \frac{\mathrm{e}^{\mathrm{i}tx} - S(x,t)}{|Q_0(t)|^2}\, \mathrm{d}\mu(t)$$

where p is a Hermitian solution of the differential equation

$$\overline{Q}\left(-\mathrm{i}\frac{\mathrm{d}}{\mathrm{d}x}\right) Q\left(-\mathrm{i}\frac{\mathrm{d}}{\mathrm{d}x}\right) p(x) = 0 \qquad \left(\overline{Q}(t) = \overline{Q(\overline{t})}\right).$$

Q_0 is a polynomial that is obtained by deleting the non-real zeros of Q, S is a regularizing correction compensating for the real zeros of Q, and μ is a non-negative measure satisfying

$$\int_{-\infty}^{\infty} \frac{1}{(1+t^2)^m}\, \mathrm{d}\mu(t) < \infty$$

where m denotes the degree of Q_0. See the Notes to Section 6 in [50] for more historical remarks.

3. Results on the existence of extensions having certain properties

Now we turn to the extension problem that was first explicitly posed by M.G. Krein [24] in 1940, though, as we will see, Carathéodory has proved some results in the discrete case as early as 1911.

A complex-valued function f defined on the finite interval $(-2a, 2a)$, $a > 0$, is called *positive definite*, if the inequality

$$\sum_{j,k=1}^{n} f(x_j - x_k) c_j \overline{c_k} \geq 0$$

holds, whenever $c_j \in \mathbb{C}$ and $x_j \in (-a, a)$.

More generally, let V be a symmetric subset of a group G containing the identity of G. A complex-valued function f on V is called *positive definite* if the inequality

$$\sum_{i,j=1}^{n} f(x_i^{-1} x_j) c_i \overline{c_j} \geq 0$$

holds for all finite systems $c_1, \ldots, c_n$ of complex numbers and elements $x_1, \ldots, x_n$ of V with $x_i^{-1} x_j \in V$ $(i, j = 1, \ldots, n)$. We will denote by $P(V)$ the set of all positive definite functions on V.

Can every function $f \in P(V)$ be extended to a function in $P(G)$?

The case where G is the group $\mathbb{Z}$ of integers and V has the form $V = V_n = \{k \in \mathbb{Z} : |k| \leq n\}$ with some non-negative integer n has been investigated by Carathéodory [11, 10]. As we have seen in the previous section, he proved that the real part of an analytic function

$$f(z) = 1 + \sum_{k=1}^{\infty} (a_k + \mathrm{i} b_k) z^k \quad (|z| < 1)$$

is positive if and only if

$$(a_1, b_1, \ldots, a_n, b_n) \in Q_n$$

for all n where Q_n denotes the smallest convex set containing the points

$$2 \cdot (\cos\varphi, \sin\varphi, \ldots, \cos n\varphi, \sin n\varphi), \quad 0 \leq \varphi < 2\pi.$$

Carathéodory calls the point

$$(a_1, b_1, \ldots, a_n, b_n)$$

the *nth geometric representative* of f and proves:

An arbitrary point of Q_n is the nth geometric representative of at least one function f (as above). Moreover, for the points on the surface of Q_n this function f is uniquely determined.[2]

[2]He also proved that in this case f is a rational function with at most n zeros.

In [11] he shows, of course using a different terminology, that the points of Q_n can be identified with what we denoted by $P(V_n)$. Hence, we can reformulate the first statement as:

Theorem 3.1 (C. Carathéodory). *Every positive definite function on* $V_n = \{k \in \mathbb{Z} : |k| \leq n\}$ *can be extended to a positive definite function on* $\mathbb{Z}$.

This statement can be proved as follows. Suppose that f is positive definite on V_n and set

$$f(n+1) := z, \quad f(-n-1) := \overline{z}$$

where z is a complex number. Then f is positive definite on V_{n+1} if and only if

$$D(z) := \begin{vmatrix} f(0) & f(1) & \dots & f(n) & z \\ \overline{f(1)} & f(0) & \dots & f(n-1) & f(n) \\ & & \ddots & & \\ \overline{f(n)} & \overline{f(n-1)} & \dots & f(0) & f(1) \\ \overline{z} & \overline{f(n)} & \dots & \overline{f(1)} & f(0) \end{vmatrix} \geq 0.$$

It can be shown that the inequality above holds if and only if

$$|z - z_0| \leq r$$

with some $z_0 \in \mathbb{C}$ and $r \geq 0$. Thus, f can be extended to V_{n+1} in at least one way.

The continuous analogue of Carathéodory's result has been given by Krein.

Theorem 3.2 (M.G. Krein, 1940). *Any continuous positive definite function* f *on* $(-A, A)$ *can be extended to a positive definite function on* $\mathbb{R}$.

In [24] Krein gives the following sketch of the proof. Let L_A be the linear space of all functions φ of the form

$$\varphi(t) = \sum_k c_k \mathrm{e}^{\mathrm{i}x_k t} \quad \text{(finite sum)}$$

where $-A < x_k < A$ and $c_k \in \mathbb{C}$. On L_A we define a linear functional Φ by

$$\Phi(\varphi) = \sum_k c_k f(x_k).$$

Since f is positive definite this functional is non-negative (i.e., $\Phi(\varphi) \geq 0$ if $\varphi \geq 0$) and hence it can be extended to a non-negative linear functional on L_∞.

Setting

$$F(x) := \Phi(\mathrm{e}^{\mathrm{i}xt}), \quad -\infty < x < \infty$$

we have $f(x) = F(x)$ $(-A < x < A)$ and

$$\sum_{j,k} F(x_j - x_k) c_j \overline{c}_k = \Phi\left(\left| \sum_j c_j \mathrm{e}^{\mathrm{i}x_j t} \right|^2 \right) \geq 0.$$

In [46] D.A. Raikov gave another proof of Theorem 3.2.

Krein used the continuity of f to show that Φ is non-negative. It was A.P. Artjomenko [3] who pointed out that the continuity assumption can be dropped.

Theorem 3.3 (A.P. Artjomenko, 1941). *Any positive definite function on* $(-A, A)$ *can be extended to a positive definite function on* $\mathbb{R}$.

About one decade later Calderón and Pepinsky [9] showed that in higher dimensions the extension is not always possible.

Theorem 3.4 (A. Calderón, R. Pepinsky, 1952). *There exist positive definite functions defined on rectangles in* $\mathbb{Z}^n$ $(n \geq 2)$ *that cannot be extended to positive definite functions defined on all of* $\mathbb{Z}^n$.

The results of Calderón and Pepinsky were rediscovered and extended to $\mathbb{R}^n$ by Rudin [48].

Theorem 3.5 (W. Rudin, 1963). *There exist continuous positive definite functions defined on rectangles in* $\mathbb{R}^n$ $(n \geq 2)$ *that cannot be extended to positive functions defined on all of* $\mathbb{R}^n$.

The proofs of these nonexistence results use the fact that there exist positive polynomials of two real variables that cannot be represented as a finite sum of squares of polynomials with real coefficients. D. Hilbert showed in 1888 [18] that there exist non-negative homogeneous polynomials in three variables which are not the sum of squares of homogeneous polynomials. The first simple explicit example has been found by T.S. Motzkin [43]. A very simple example for $\mathbb{R}^2$ is the polynomial

$$F(s,t) = s^2t^2(s^2 + t^2 - 1) + 1$$

(see [4]).

Rudin [49] has also shown that positive definite extensions do exist in $\mathbb{R}^n$ if V is a ball and if the functions are isotropic.[3]

Theorem 3.6 (W. Rudin, 1970). *Continuous, isotropic positive definite functions defined on a ball*

$$\{x \in \mathbb{R}^n : \ \|x\| < r\}$$

can be extended to positive definite functions defined on all of $\mathbb{R}^n$.

In higher dimensions (e.g., analysis of geostatistical data) random processes are often assumed to be both stationary and isotropic (i.e., the covariance function is isotropic). Covariance models fitted to observed data frequently involve a so-called *nugget effect*, that is a discontinuity at the origin $0 \in \mathbb{R}^n$:

$$C_0(t) = \begin{cases} c & \text{if} \quad t = 0 \\ 0 & \text{if} \quad t \neq 0 \end{cases}$$

If C_1 is a continuous isotropic covariance function, then

$$C = C_0 + C_1$$

[3] A function f defined on a subset of $\mathbb{R}^n$ is called *isotropic* if $f(x) = \varphi(\|x\|)$ where φ is a real-valued function on $[0, \infty)$ and $\|\cdot\|$ denotes the Euclidean norm.

is an isotropic covariance function with a nugget effect. It was conjectured by I.J. Schoenberg [53] in 1938 that all isotropic covariance functions have this form if $n \geq 2$. A positive answer was given by T. Gneiting, Z. Sasvári [14]:

Theorem 3.7 (T. Gneiting, Z. Sasvári, 1999). *Any (Lebesgue) measurable, isotropic covariance function on a ball of $\mathbb{R}^n$ can be extended to a covariance function C on $\mathbb{R}^n$. If $n > 1$ then C has the form $C = C_0 + C_1$ with some continuous covariance function C_1.*

For $n = 1$, the second statement is not true: the indicator function of the rational numbers is positive definite and isotropic but nowhere continuous.

One of the basic problems of statistical estimation is to guess an unknown probability measure, given certain observations. In general there are infinitely many measures consistent with the data. The question, which of these is the best is an important one. The principle that nature favors the states of largest entropy was applied with success by physicists in statistical mechanics. For a probability measure μ with density p we define the *entropy* of μ by

$$K(\mu) := \frac{1}{\pi} \int_{-\infty}^{\infty} \frac{\log p(x)}{1+x^2}\, \mathrm{d}x - \log 4.$$

The next result was proved in [12]:

Theorem 3.8 (J. Chover, 1961). *Let f be a positive definite function on $(-A, A)$ satisfying $f(0) = 1$. Suppose that $f''(0)$ is finite and that there is at least one probability measure μ with density such that*

(i) $f(x) = \int_{-\infty}^{\infty} \mathrm{e}^{\mathrm{i}tx}\, \mathrm{d}\mu(t), \quad -A < x < A$

(ii) $K(\mu) > -\infty$.

Then there exists a unique measure μ satisfying (i) and (ii) for which $K(\mu)$ is maximal.

We refer to [38] for more information on the maximum entropy principle in connection with the trigonometric moment problem. This principle has also been intensively studied in connection with matrix completion problems. A partial matrix is an array in which some entries are exactly specified and the remaining ones are thought of as free variables. In a typical matrix completion problem, one asks for conditions under which the unspecified entries can be chosen so that the resulting ordinary matrix is of a desired type (e.g., positive definite like in the sketched proof of Theorem 3.1). There is a vast literature on matrix completion problems, we mention here the paper [16].

We close this section with results stating that certain local properties of positive definite functions imply global ones. The first one is due to A.P. Artjomenko and follows immediately from the inequality

$$|f(x) - f(y)|^2 \leq 2f(0) \cdot [f(0) - \operatorname{Re} f(y-x)] \quad (x, y \in \mathbb{R})$$

satisfied by an arbitrary positive definite function f.

Theorem 3.9. *If the real part of a function $f \in P(\mathbb{R})$ is continuous at 0 then f is uniformly continuous on $\mathbb{R}$.*

A positive definite function f is said to be *analytic* if there exists a complex-valued function θ of one complex variable, which is holomorphic in a circle around 0 and $f(t) = \theta(t)$ for all real t in this circle. The next result deals with analyticity and has been proved independently by P. Lévy [40] and D.A. Raikov [45].

Theorem 3.10. *If f is an analytic positive definite function then there exist $\alpha_f, \beta_f \in (0, \infty]$ such that f extends to a function which is holomorphic in the strip*

$$\{ z \in \mathbb{C} : \; -\alpha_f < \operatorname{Im} z < \beta_f \}$$

and such that α_f and β_f are maximal with this property.

We refer to Section 1.12 of [5] for more details. As a corollary we obtain that an analytic positive definite function on $(-A, A)$ has exactly one positive definite extension (which is analytic). The next theorem deals with differentiability. Unfortunately we cannot assign exact priorities to it.

Theorem 3.11. *If for some positive integer k the real part of f is $2k$-times differentiable at 0 then f is $2k$-times continuously differentiable on $\mathbb{R}$.*

The question, whether (Lebesgue) measurable positive definite functions have measurable positive definite extensions has been posed by M.G. Krein [25, 28]. An affirmative answer was given by the author [51].[4]

Theorem 3.12 (Z. Sasvári, 1986). *If f is measurable on $(-A, A)$ then f is measurable on $\mathbb{R}$.*

Combining this with Theorem 3.3 we see that any measurable positive definite function on $(-A, A)$ can be extended to a measurable positive definite function on $\mathbb{R}$.

4. Description of all extensions

In this section we deal with the following problem which was first investigated by M.G. Krein [24] in 1940 and which plays an important role in his method for solving inverse spectral problems [29, 30, 31].

Let f be a continuous positive definite function f on $[-2a, 2a]$. Describe all positive definite functions $\tilde{f}$ on $\mathbb{R}$ satisfying $\tilde{f}(t) = f(t), \; t \in [-2a, 2a]$.

We will call $\tilde{f}$ an *extension* or *continuation* of f.

---

[4] A personal remark: I learned Krein's problem from Heinz, who informed Krein in a letter about my solution. As a young mathematician I was very glad when Heinz showed me a part of Krein's answer in which he congratulated me.

There are two possibilities:

1. There exists exactly one extension. This is the case for example if f is analytic or $f(0) = 1$ and $|f(t_0)| = 1$ for some $t_0 \in [-2a, 2a]$, $t_0 \neq 0$.
2. There are infinitely many extensions (the so-called *indeterminate case*).

In [24] Krein gave several criteria for uniqueness of the extension by employing methods reminiscent of those used in the classical moment problems (see [1] or [52]). We mention also E. Akutowicz [2] and Devinatz [13] who gave criteria for uniqueness in terms of the self-adjointness of certain operators in Hilbert spaces. The classification of all possible extensions in the indeterminate case is of the same nature as that in the classical moment problems. An essential role is played by the description of all selfadjoint extensions of a given Hermitian operator with defect $(1, 1)$.

In the papers [32, 33, 34, 35, 36] and in the unpublished manuscript [37] M.G. Krein and H. Langer formulate and study indefinite analogues of interpolation, moment, and continuation problems. Some of their results were new even when restricted to the positive definite case. To formulate the indefinite continuation problem denote by $P_{k,a}$ the set of all continuous, Hermitian functions f on the interval $[-2a, 2a]$ having k *negative squares*, i.e., such that the maximum of the numbers of the negative eigenvalues, counted with multiplicities, of the matrices

$$(f(t_i - t_j))_{i,j}^{n}, \quad n \in \mathbb{N},\ t_1, \ldots, t_n \in (-a, a)$$

is equal to k. With this notation the indefinite continuation problem can be formulated as follows:

Given a function $f \in P_{k,a}$ find all functions
$\tilde{f} \in P_k(\mathbb{R})$ such that $\tilde{f}(t) = f(t)$, $t \in [-2a, 2a]$.

An even continuous function C defined on $[-2a, 2a]$ is said to have an accelerant H if (i) $\mathrm{H}(t) = -C''(t)$ exists for $t \neq 0$, (ii) $\mathrm{H}(t)$ is absolutely integrable over $[-2a, 2a]$, and (iii) $C'(0+) < 0$. With the accelerant H we associate the operator $\mathbf{H}$ in $L^2[0, 2a]$, defined by

$$(\mathbf{H}\,\varphi)(t) = \int_0^{2a} \mathrm{H}(t-s)\,\varphi(s)\,\mathrm{d}s \qquad (0 \leq t \leq 2a).$$

We are now able to formulate one of the numerous results of Krein and Langer [36]:

Theorem 4.1 (M.G. Krein, H. Langer, 1985). *Assume that $f \in P_{k,a}$ has an accelerant and that -1 does not belong to the spectrum of the integral operator $\mathbf{H}$. If f has more than one continuation then there exist four entire functions $w_{jk}(z) = w_{jk}(a, z)$, such that the equality*

$$\mathrm{i}\int_0^{\infty} \mathrm{e}^{\mathrm{i}zt}\,\overline{\tilde{f}(t)}\,\mathrm{d}t = \frac{w_{11}(z)T(z) + w_{12}(z)}{w_{21}(z)T(z) + w_{22}(z)}, \quad \operatorname{Im} z > \gamma$$

for some $\gamma \geq 0$ *establishes a bijective correspondence between all continuations* $\tilde{f} \in P_k(\mathbb{R})$ *and all functions* T *of the form* $T \equiv \infty$ *or*

$$T(z) = \alpha + \beta z + \int_{-\infty}^{\infty} \frac{tz+1}{t-z} \, \mathrm{d}\sigma(t), \; z \in \mathbb{C} \setminus \mathbb{R}$$

where $\alpha \in \mathbb{R}$, $\beta \geq 0$ *and* σ *is a non-negative finite measure.*[5]

The method of proof is based on the extension theory of symmetric operators in a Pontryagin space (see [34] and the references therein).

The matrix function

$$\mathcal{W}(\cdot; z) = \begin{pmatrix} w_{11}(\cdot; z) & w_{12}(\cdot; z) \\ w_{21}(\cdot; z) & w_{22}(\cdot; z) \end{pmatrix}$$

called the *resolvent matrix* of this continuation problem, is shown to satisfy the canonical system

$$\frac{\mathrm{d}\mathcal{W}(a; z)}{\mathrm{d}a} \mathcal{J} = z \mathcal{W}(a; z) \mathcal{H}(a) \tag{4.1}$$

and the initial condition

$$\mathcal{W}(0; z) = \begin{pmatrix} 1 & 0 \\ -z & 1 \end{pmatrix}. \tag{4.2}$$

Here $\mathcal{J} = \begin{pmatrix} 0 & -1 \\ 1 & 0 \end{pmatrix}$, and the *Hamiltonian* $\mathcal{H}(a)$ is a continuous 2×2-matrix function. The paper [39] contains a detailed study of the special case $f(t) = 1 - |t|$. This function is positive definite on $[-2a, 2a]$ if $a \leq 1$ and it has one negative square if $a > 1$. The resolvent matrix is given by

$$W(a, z) = \begin{pmatrix} \frac{\sin az - z \cos az}{(a-1)z} & \frac{(1-(a-1)z^2) \sin az - az \cos az}{z^2} \\ \frac{z \cos az}{a-1} & (a-1) z \sin az + \cos az \end{pmatrix}$$

and the Hamiltonian is

$$\mathcal{H}(a) = \begin{pmatrix} (a-1)^2 & 0 \\ 0 & \frac{1}{(a-1)^2} \end{pmatrix}.$$

The mentioned papers of Krein and Langer contain a wealth of interesting results and their connections to other parts of mathematics. To give a detailed survey would require a book rather than a paper.

M. Kaltenbäck and H. Woracek [22] gave a description of all extensions with k negative squares of a continuous function $f \in P_{k,a}$, without posing additional assumptions on f.

---

[5] T is a so-called Nevanlinna function: it is holomorphic in the upper half-plane and has a non-negative imaginary part there.

They proved that there are the following three cases:

1. f has infinitely many extensions with k negative squares.
2. f has a unique extension with k negative squares and no other extension with a finite number of negative squares.
3. f has a unique extension with k negative squares and there exists a positive integer $\Delta(f)$ such that the function f has no extensions with k' negative squares, where $k < k' < k + \Delta(f)$, and infinitely many extensions with $k' \geq k + \Delta(f)$ negative squares.

At last, we mention that the results in [36] have been successfully applied in [42] (see also [15]) to characterize a class of stationary processes that have polynomial covariance functions on an interval.

References

[1] N.I. Akhiezer, *The classical moment problem.* Edinburgh: Oliver and Boyd 1965.

[2] E.J. Akutowicz, *On extrapolating a positive definite function from a finite interval.* Math. Scand. **7** (1959), 157–169.

[3] A.P. Artjomenko, *Hermitian positive functions and positive functionals.* (Russian), Dissertation, Odessa State University (1941). Published in Teor. Funkciĭ, Funkcional. Anal. i Priložen. **41**, (1983) 1–16; **42** (1984), 1–21.

[4] C. Berg, J.P.R. Christensen, P. Ressel, *Harmonic Analysis on Semigroups. Theory of Positive Definite and Related Functions.* Berlin-Heidelberg-New York-Tokyo: Springer-Verlag 1984.

[5] T.M. Bisgaard, Z. Sasvári, *Characteristic Functions and Moment Sequences. Positive Definiteness in Probability.* Huntigton, NY: Nova Science Publishers 2000.

[6] S. Bochner, *Vorlesungen über Fouriersche Integrale.* Leipzig: Akademische Verlagsgesellschaft 1932.

[7] S. Bochner, *Spektralzerlegung linearer Scharen unitärer Operatoren.* Sitzungsber. Preuss. Akad. Wiss. phys.-math. (1933), 371–376.

[8] S. Bochner, *Monotone Funktionen, Stieltjessche Integrale und harmonische Analyse.* Math. Ann. **108** (1933), 378–410.

[9] A. Calderón, R. Pepinsky, *On the phases of Fourier coefficients for positive real periodic functions.* Computing Methods and the Phase Problem in X-Ray Crystal Analysis, published by The X-Ray Crystal Analysis Laboratory, Department of Physics, The Pennsylvania State College (1952), 339–348.

[10] C. Carathéodory, *Über den Variabilitätsbereich der Koeffizienten von Potenzreihen, die gegebene Werte nicht annehmen.* Math. Ann. **64** (1907), 95–115.

[11] C. Carathéodory, *Über den Variabilitätsbereich der Fourierschen Konstanten von positiven harmonischen Funktionen.* Rend. Circ. Mat. Palermo, **32** (1911), 193–217.

[12] J. Chover, *On normalized entropy and the extensions of a positive-definite function.* J. Math. Mech. **10** (1961), 927–945.

[13] A. Devinatz, *On the extensions of positive definite functions.* Acta Math. **102** (1959), 109–134.

[14] T. Gneiting, Z. Sasvári, *The Characterization Problem for Isotropic Covariance Functions* Math. Geology **31(1)** (1999), 105–111.

[15] T. Gneiting, Z. Sasvári, M. Schlather, *Analogies and correspondences between variograms and covariance functions.* Adv. Appl. Probab. **33** (2001), 617–630.

[16] I. Gohberg, M.A. Kaashoek, H.J. Woerdemann, *A Maximum Entropy Principle in the General Framework of the Band Method.* J. Funct. Anal. **95** (1991), 231–254.

[17] G. Herglotz, *Über Potenzreihen mit positivem, reellen Teil im Einheitskreis.* Leipziger Berichte, math.-phys. Kl. **63** (1911), 501–511.

[18] D. Hilbert, *Über die Darstellung definiter Formen als Summe von Formenquadraten.* Math. Ann. **32** (1988), 342–350.

[19] I.S. Iohvidov, *Unitary and selfadjoint operators in spaces with an indefinite metric.* (Russian) Dissertation, Odessa (1950).

[20] I.S. Iohvidov, *On the theory of indefinite Toeplitz forms.* (Russian) Dokl. Akad. Nauk SSSR **101(2)** (1955), 213–216.

[21] I.S. Iohvidov, M.G. Krein, *Spectral theory of operators in spaces with an indefinite metric II.* (Russian) Trudy Moskov. Mat. Obšč. **8** (1959), 413–496. English translation: Amer. Math. Soc. Translations **2(3)**, (1963), 283–373.

[22] M. Kaltenbäck, H. Woracek, *On extensions of hermitian functions with a finite number of negative squares.* J. Operator Theory **40** (1998), 147–183.

[23] A. Khintchin, *Korrelationstheorie der stationären stochastischen Prozesse.* Math. Ann. **109** (1934), 604–615.

[24] M.G. Krein, *Sur le problème du prolongement des fonctions hermitiennes positives et continues.* Dokl. Akad. Nauk SSSR **26** (1940), 17–22.

[25] M.G. Krein, *On the representation of functions by Fourier-Stieltjes integrals.* (Russian) Učenije Zapiski Kuibishevskogo Gosud. Pedag. i Učitelskogo Inst. **7** (1943), 123–148.

[26] M.G. Krein, *Screw lines in infinite-dimensional Lobachevski space and the Lorentz transformation.* (Russian) Usp. Mat. Nauk **3(3)** (1948), 158–160.

[27] M.G. Krein, *On the integral representation of a continuous Hermitian-indefinite function with a finite number of negative squares.* (Russian) Dokl. Akad. Nauk SSSR, **125(1)** (1959), 31–34.

[28] M.G. Krein, *On measurable Hermitian-positive functions.* (Russian) Mat. Zametki **23** (1978), 79–89. English translation: Math. Notes **23** (1978), 45–50.

[29] M.G. Krein, Solution of the inverse Sturm–Liouville problem. *Dokl. Akad. Nauk. SSSR* **76**:1 (1951), 21–24.

[30] M.G. Krein, On the transition function of the one-dimensional second order boundary value problem. *Dokl. Akad. Nauk. SSSR***88**:3 (1953), 405–408.

[31] M.G. Krein, On the determination of the potential of a particle by its S-function. *Dokl. Akad. Nauk. SSSR* **105**:3 (1955), 433–436.

[32] M.G. Krein, H. Langer, *On the indefinite power moment problem.* Soviet Math. Dokl. **17** (1976), 90–93.

[33] M.G. Krein, H. Langer, *Über einige Fortsetzungsprobleme, die eng mit der Theorie hermitescher Operatoren im Raume Π_κ zusammenhängen. I. Einige Funktionenklassen und ihre Darstellungen.* Math. Nachr. **77** (1977), 187–236.

[34] M.G. Krein, H. Langer, *Über einige Fortsetzungsprobleme, die eng mit der Theorie hermitescher Operatoren im Raume Π_k zusammenhängen. II. Verallgemeinerte Resolventen, u-Resolventen und ganze Operatoren.* J. Funct. Anal. **30** (1978), 390–447.

[35] M.G. Krein, H. Langer, *On some extension problems which are closely related with the theory of hermitian operators in a space π_k. III. Indefinite analogues of the Hamburger and Stieltjes problems. Part (I).* Beiträge zur Analysis **14** (1979), 25–40.

[36] M.G. Krein, H. Langer, *On some continuation problems which are closely related to the theory of operators in spaces Π_k. IV: Continuous analogues of orthogonal polynomials on the unit circle with respect to an indefinite weight and related coninuation problems for some classes of functions.* J. Operator Theory **13** (1985), 299–417.

[37] M.G. Krein, H. Langer, *Continuation of Hermitian positive definite functions and related questions.* (Manuscript)

[38] H.J. Landau, *Maximum Entropy and the Moment Problem*, Bull. Amer. Math. Soc. **16(1)**, (1987), 47–77.

[39] H. Langer, M. Langer, Z. Sasvári, *Continuations of Hermitian Indefinite Functions and Corresponding Canonical Systems: An Example.* Methods. of Funct. Anal. and Topology, **10(1)** 2004, 39–53.

[40] P. Lévy, *Théorie de l'addition des variables aléatoires.* Paris: Gauthier-Villars, 1937.

[41] M. Mathias, *Über positive Fourier-Integrale.* Math. Zeitschrift **16** (1923), 103–125.

[42] S. Mitra, T. Gneiting, Z. Sasvári, *Polynomial covariance functions on intervals.* Bernoulli **9(2)** 2003, 229–241.

[43] T.S. Motzkin, *The arithmetic-geometric inequality.* Published in: Inequalities (Ed. by O. Shisha). New York-London: Academic Press 1967.

[44] G. Pólya, *Remarks on characteristic functions.* Proc. First Berkeley Conf. on Math. Stat. and Prob. Berkeley: Univ. of Calif. Press (1949), 115–123.

[45] D.A. Raikov, *On the decomposition of Gauss and Poisson laws.* (Russian) Izv. Akad. Nauk. SSSR, ser. math. **2** (1937), 91–124.

[46] D.A. Raikov, *Sur les fonctions positivement définies.* Dokl. Akad. Nauk SSSR, **26** (1940), 860–865.

[47] F. Riesz, *Über Sätze von Stone und Bochner.* Acta Sci. Math. **6** (1932–1934), 184–198.

[48] W. Rudin, *The extension problem for positive-definite functions.* Illinois J. Math. **7** (1963), 532–539.

[49] W. Rudin, *An extension theorem for positive-definite functions.* Duke Math. J. **37** (1970), 49–53.

[50] Z. Sasvári, *Positive Definite and Definitizable Functions.* Berlin: Akademie Verlag 1994.

[51] Z. Sasvári, *The extension problem for measurable positive definite functions.* Math. Zeitschrift **191** (1986), 475–478.

[52] J. Shohat, J. Tamarkin, *The problem of moments. Math.* Surveys No. **1**, Providence, R. I.: Amer. Math. Soc. 1943.

[53] I.J. Schoenberg, *Metric spaces and competely monotone functions.* Ann. of Math. **39(4)** (1938), 811–841.

[54] M.H. Stone, *Linear transformations in Hilbert space. III. Operational methods and group theory.* Proc. Nat. Acad. Sci. U.S.A. **16** (1930), 172–175.

[55] M.H. Stone, *On one-parameter unitary groups in Hilbert space.* Ann. of Math. **33(2)** (1932) 643–648.

[56] O. Toeplitz, *Über die Fourier'sche Entwickelung positiver Funktionen.* Rend. Circ. Mat. Palermo **32** (1911), 191–192.

Zoltán Sasvári
Mommsenstr. 13
D-01062 Dresden
Germany
e-mail: sasvari@math.tu-dresden.de

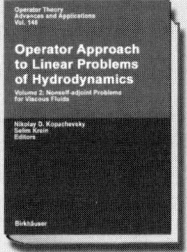